A COURSE OF MODERN ANALYSIS

This classic work has been a unique resource for thousands of mathematicians, scientists and engineers since its first appearance in 1902. Never out of print, its continuing value lies in its thorough and exhaustive treatment of special functions of mathematical physics and the analysis of differential equations from which they emerge. The book also is of historical value as it was the first book in English to introduce the then modern methods of complex analysis.

 This fifth edition preserves the style and content of the original, but it has been supplemented with more recent results and references where appropriate. All the formulas have been checked and many corrections made. A complete bibliographical search has been conducted to present the references in modern form for ease of use. A new foreword by Professor S. J. Patterson sketches the circumstances of the book's genesis and explains the reasons for its longevity. A welcome addition to any mathematician's bookshelf, this will allow a whole new generation to experience the beauty contained in this text.

E.T. WHITTAKER was Professor of Mathematics at the University of Edinburgh. He was awarded the Copley Medal in 1954, 'for his distinguished contributions to both pure and applied mathematics and to theoretical physics'.

G.N. WATSON was Professor of Pure Mathematics at the University of Birmingham. He is known, amongst other things, for the 1918 result now known as Watson's lemma and was awarded the De Morgan Medal in 1947.

VICTOR H. MOLL is Professor in the Department of Mathematics at Tulane University. He co-authored *Elliptic Curves* (Cambridge, 1997) and was awarded the Weiss Presidential Award in 2017 for his Graduate Teaching. He first received a copy of Whittaker and Watson during his own undergraduate studies at the Universidad Santa Maria in Chile.

(Left): Edmund Taylor Whittaker (1873–1956); (Right): George Neville Watson (1886–1965): Universal History Archive/Contributor/Getty Images.

A COURSE OF MODERN ANALYSIS

Fifth Edition

An introduction to the general theory of infinite
processes and of analytic functions with an account
of the principal transcendental functions

E.T. WHITTAKER AND G.N. WATSON

Fifth edition edited and prepared for publication by
Victor H. Moll
Tulane University, Louisiana

CAMBRIDGE
UNIVERSITY PRESS

CAMBRIDGE
UNIVERSITY PRESS

University Printing House, Cambridge CB2 8BS, United Kingdom

One Liberty Plaza, 20th Floor, New York, NY 10006, USA

477 Williamstown Road, Port Melbourne, VIC 3207, Australia

314–321, 3rd Floor, Plot 3, Splendor Forum, Jasola District Centre, New Delhi – 110025, India

103 Penang Road, #05–06/07, Visioncrest Commercial, Singapore 238467

Cambridge University Press is part of the University of Cambridge.

It furthers the University's mission by disseminating knowledge in the pursuit of
education, learning, and research at the highest international levels of excellence.

www.cambridge.org
Information on this title: www.cambridge.org/9781316518939
DOI: 10.1017/9781009004091

First edition 1902
Second edition 1915
Third edition 1920
Fourth edition 1927
Reprinted 1935, 1940, 1946, 1950, 1952, 1958, 1962, 1963
Reissued in the Cambridge Mathematical Library Series 1996
Sixth printing 2006
Fifth edition 2021

Printed in the United Kingdom by TJ Books Limited, Padstow Cornwall

A catalogue record for this publication is available from the British Library.

ISBN 978-1-316-51893-9 Hardback

Contents

Foreword

S.J. Patterson

There are few books which remain in print and in constant use for over a century; "Whittaker and Watson" belongs to this select group. In fact there were two books with the title "A Course in Modern Analysis", the first in 1902 by Edmund Whittaker alone, a textbook with a very specific agenda, and then the joint work, first published in 1915 as a second edition. It is an extension of the first edition but in such a fashion that it becomes a handbook for those working in analysis. As late as 1966 J.T. Whittaker, the son of E.T. Whittaker, wrote in his Biographical Memoir of Fellows of the Royal Society (i.e. obituary) of G.N. Watson that there were still those who preferred the first edition but added that for most readers the later edition was to be preferred. Indeed the joint work is superior in many different ways.

The first edition was written at a time when there was a movement for reform in mathematics at Cambridge. Edmund Whittaker's mentor Andrew Forsyth was one of the driving forces in this movement and had himself written a *Theory of Functions* (1893) which was, in its time, very influential but is now scarcely remembered. In the course of the nineteenth century the mathematics education had become centered around the Mathematical Tripos, an intensely competitive examination. Competitions and sports were salient features of Victorian Britain, a move away from the older system of patronage and towards a meritocracy. The reader familiar with Gilbert and Sullivan operettas will think of the Modern Major-General in *The Pirates of Penzance*. The Tripos had become not only a sport but a spectator sport, followed extensively in middle-class England[1]. The result of this system was that the colleges were in competition with one another and employed coaches to prepare the talented students for the Tripos. They developed the skills needed to answer difficult questions quickly and accurately – many Tripos questions can be found in the exercises in *Whittaker and Watson*. The Tripos system did not encourage the students to become mathematicians and separated them from the professors who were generally very well informed about the developments on the Continent. It was a very inward-looking, self-reproducing system. The system on the Continent, especially in the German universities, was quite different. The professors there sought contact with the students, either as note-takers for lectures or in seminar talks, and actively supported those by whom they were most impressed. The students vied with one another for the attention of the professor, a different and more fruitful form of competition. This

---

[1] Some idea of this may be gleaned from G.B. Shaw's play *Mrs Warren's Profession*, written in 1893 but held back by censorship until 1902. In this play Mrs Warren's daughter Vivie has distinguished herself in Cambridge – she tied with the third Wrangler, described as a "magnificent achievement" by a character who has no mathematical background. She herself could not be ranked as a Wrangler as she was female. She would have been a contemporary of Grace Chisholm, later Grace Chisholm Young, whose family background was by no means as colourful as that of the fictional Vivie Warren.

xvii

system allowed the likes of Weierstrass and Klein to build up groups of talented and highly motivated students. It had become evident to Andrew Forsyth and others that Cambridge was missing out on the developments abroad because of the concentration on the Tripos system[2].

It is interesting to read what Whittaker himself wrote about the situation at the end of the nineteenth century in Cambridge and so of the conditions under which *Whittaker and Watson* was written. We quote from his Royal Society Obituary Notice (1942) of Andrew Russell Forsyth:

> He had for some time past realized, as no one else did, the most serious deficiency of the Cambridge school, namely its ignorance of what had been and was being done on the continent of Europe. The college lecturers could not read German, and did not read French.
>
> $\vdots$
>
> The schools of Göttingen and Berlin to a great extent ignored each other (Berlin said that Göttingen proved nothing, and Göttingen retorted that Berlin had no ideas) and both of them ignored French work.
>
> But Cambridge had hitherto ignored them all: and the time was ripe for Forsyth's book. The younger men, even undergraduates, had heard in his lectures of the extraordinary riches and beauty of the domain beyond Tripos mathematics, and were eager to enter into it. From the day of its publication in 1893, the face of Cambridge was changed: the majority of the pure mathematicians who took their degrees in the next twenty years became function-theorists.

and further

> As head of the Cambridge school of mathematics he was conspicuously successful. British mathematicians were already indebted to him for the first introduction of the symbolic invariant-theory, the Weierstrassian elliptic functions, the Cauchy–Hermite applications of contour-integration, the Riemannian treatment of algebraic functions, the theory of entire functions, and the theory of automorphic functions: and the importation of novelties continued to occupy his attention. A great traveller and a good linguist, he loved to meet eminent foreigners and invite them to enjoy Trinity hospitality: and in this way his post-graduate students had opportunities of becoming known personally to such men as Felix Klein (who came frequently), Mittag-Leffler, Darboux and Poincaré. To the students themselves, he was devoted: young men fresh from the narrow examination routine of the Tripos were invited to his rooms and told of the latest research papers: and under his fostering care, many of the wranglers of the period 1894–1910 became original workers of distinction.

The two authors were very different people. Edmund Whittaker (1874–1956) went on from Cambridge in 1906 to become the Royal Astronomer in Ireland (then still a part of the

---

[2] For his arguments see A. Forsyth: Old Tripos Days at Cambridge, *Math. Gazette* **19** 162–179 (1935). For a dissenting opinion see K. Pearson: Old Tripos Days at Cambridge, as seen from another viewpoint, *Math. Gazette* **20** 27–36 (1936).

United Kingdom) and Director of Dunsink Observatory, thereby following in the footsteps of William Rowan Hamilton. In 1985, on the occasion of the bicentenary of Dunsink, the then Director, Patrick A. Wayman, singled out Whittaker as the greatest director aside from Hamilton and one who, despite his relatively short tenure of office, 1906–1912, had achieved most for the Observatory[3]. This appointment brought out his skills as an administrator. Following this he moved to Edinburgh where he exerted his influence to guide mathematics there into the new century. Some indication of the success is given by the fact that it was W.V.D. Hodge, a student of his, who, at the International Congress of Mathematicians in 1954, invited the International Mathematical Union to hold the next Congress in Edinburgh. Whittaker himself did not live to experience the event which reflected the status in which Edinburgh was held at the end of his life.

George Neville Watson (1886–1965) on the other hand was a retiring scholar who, after leaving Cambridge, at least in the flesh, spent four years (1914–1918) in London, and then became professor in Birmingham where he remained for the rest of his life[4], living a relatively withdrawn life devoted to his mathematical work and with stamp-collecting and the study of the history of railways as hobbies. His early work was very much in the direction of E.W. Barnes and A.G. Greenhill. After Ramanujan's death he took over from Hardy the analysis of many of Ramanujan's unpublished papers, especially those connected with the theory of modular forms and functions, and of complex multiplication. It is worth remarking that Greenhill, a student and ardent admirer of James Clerk Maxwell and primarily an applied mathematician, concerned himself with the computation of singular moduli, and it was probably he who aroused Ramanujan's interest in this topic. Watson's work in this area is, besides his books, that for which he is best remembered today.

Both authors wrote other books that are still used today. In Whittaker's case these are his *A Treatise on the Analytical Dynamics of Particles & Rigid Bodies*, reprinted in 1999, with a foreword by Sir William McCrea in the CUP series "Cambridge Mathematical Library", a source of much mathematics which is difficult to find elsewhere, and his *History of Theories of the Aether and Electricity* which, despite some unconventional views, is an invaluable source on the history of these parts of physics and the associated mathematics.

Watson, on the other hand, wrote his *A Treatise on the Theory of Bessel Functions*, published in 1922, which like *Whittaker and Watson* has not been out of print since its appearance. On coming across it for the first time as a student I was taken aback by such a thick book being devoted to what seemed to be a very circumscribed subject. One of the Fellows of my college, a physicist, replying to a fellow student who had made a similar observation, declared that it was a work of genius and he would have been proud to have written something like it. In the course of the years I have had recourse to it over and over again and would now concur with this opinion.

Watson's *Bessel Functions*, like *Whittaker and Watson*, despite being somewhat old-fashioned, has retained a freshness and relevance that has made both of them classics. Unlike many other books of this period the terminology, although not the style, is that of today. It is less a *Cours d'Analyse* and more of a *Handbuch der Funktionentheorie*. Perhaps my own experiences can illuminate this. My copy was given to me in 1967 by my mathematics teacher,

---

[3] *Irish Astronomical Journal* **17** 177–178 (1986).
[4] It is worth noting that from 1924 on E.W. Barnes was a disputative Bishop of Birmingham.

Mr Cecil Hawe, after I had been awarded a place to study mathematics in Cambridge. He had bought it 20 years earlier as a student. During my student years *the* textbook on second year analysis was J. Dieudonné's *Foundations of Modern Analysis*. People then were prone to be a bit supercilious at least about the "modern" in the title of *Whittaker and Watson*.[5] At that time it lay on my bookshelf unused. Five years later I was coming to terms with the theory of non-analytic automorphic forms, especially with Selberg's theory of Eisenstein series. At this point I discovered how useful a book it was, both for the treatment of Bessel functions and for that of the hypergeometric function. It also has a very useful chapter on Fredholm's theory of integral equations which Selberg had used. In the years since then several other chapters have proved useful, and ones I thought I knew became useful in novel ways. It became a constant companion. This was mainly in connection with doing mathematics but it also proved its worth in teaching – for example the chapter on Fourier series gives very useful results which can be obtained by relatively elementary methods and are suitable for undergraduate lectures. Dieudonné's book is tremendous for the university teacher; it gives the fundamentals of analysis in a concentrated form, something very useful when one has an overloaded syllabus and a limited number of hours to teach it in. On the other hand it is much less useful as a "*Handbuch*" for the working analyst, at least in my experience. Nor was it written for this purpose. *Whittaker and Watson* started, in the first edition, as such a book for teaching but in the second and later editions became that book which has remained on the bookshelves of generations of working mathematicians, be they formally mathematicians, natural scientists or engineers.

One aspect that probably contributed to the long popularity of *Whittaker and Watson* is the fact that it is not overloaded with many of the topics that are within range of the text. Thus, for example, the authors do not go into the arithmetic theory of the Riemann zeta-function beyond the Euler product over primes. Whereas they discuss the 24 solutions to the hypergeometric equation in terms of the hypergeometric series from Riemann's point of view they do not go into H.A. Schwarz' beautiful solution of Gauss' problem as to which of these functions is algebraic. Schwarz' theory is covered in Forsyth's *Function Theory*. The decision to leave this out must have been difficult for Whittaker for it is a topic close to his early research. Finally they touch on the theory of Hilbert spaces only very lightly, just enough for their purposes. On the other hand Fredholm's theory, well treated here, has often been pushed aside by the theory of Hilbert spaces in other texts and it is a topic about which an analyst should be aware.

So, gentle reader, you have in your hands a book which has been useful and instructive to those working in mathematics for well over a hundred years. The language is perhaps a little quaint but it is a pleasure to peruse. May you too profit from this new edition.

---

[5] B.L. v.d. Waerden's *Moderne Algebra* became simply *Algebra* from the 1955 edition on; with either name it remains a great text on algebra.

Preface to the Fifth Edition

In 1896 Edmund Whittaker was elected to a Fellowship at Trinity College, Cambridge. Amongst other duties, he was employed to teach students, many of whom would later become distinguished figures in science and mathematics. These included G.H. Hardy, Arthur Eddington, James Jeans, J.E. Littlewood and a certain G. Neville Watson. His course on mathematical analysis changed the way the subject was taught, and he decided to write a book. So was born *A Course of Modern Analysis*, which was first published in 1902. It introduced students to functions of a complex variable, to the 'methods and processes of higher mathematical analysis', much of which was then fairly modern, and above all to special functions associated with equations that were used to describe physical phenomena. It was one of the first books in English to describe material developed *on the continent*, mostly in France and Germany. Its breadth and depth of coverage were unparalleled at the time and it became an instant classic. A second edition was called for, but in 1906 Whittaker had left Cambridge, moving first to Dublin, and then in 1912 to Edinburgh. His various duties, and no doubt, the moves themselves, impeded work on the new edition, and Whittaker gratefully accepted the offer from Watson to help him. A greatly expanded second edition duly appeared in 1915. The third edition, published five years later, was also enlarged by the addition of chapters, but the fourth edition was not much more than a corrected reprint with added references. I do not know if a fifth edition was ever planned. Both authors remained active for many years (Watson wrote, amongst other publications, the definitive *Treatise on Bessel Functions*), but perhaps they had nothing more to say to warrant a new edition. Nevertheless, the book remained a classic, being continually in print and reissued in paperback, first in 1963, and again, in 1996, as a volume of the *Cambridge Mathematical Library*. It never lost its appeal and occupied a unique place in the heart and work of many mathematicians (in particular, me) as an indispensable reference.

The original editions were typeset using 'hot metal', and over the years successive reprintings led to the degrading of the original plates. Photographic printing methods slowed this decline, but David Tranah at Cambridge University Press had the idea to halt, indeed reverse, the degradation, by rekeying the book and at the same time updating it with new references and commentary. He spoke to me about this, and we agreed that if he arranged for the rekeying into LaTeX, I would do the updating. I did not need much persuading: it has been a labor of love. So much so that I have preserved the archaic spelling of the original, along with the Peano decimal system of numbering paragraphs, as described by Watson in the Preface to the fourth edition! This will make it straightforward for users of this fifth edition to refer to the previous one. I have however decided to create a complete reference list and to refer readers to that rather than to items in footnotes, items that were often hard to identify. Many of these items are now available in digital libraries and so for many people will be easier to access than they were in the authors' time.

I have made no substantial changes to the text: in particular, the original idea of adding commentaries on the text was abandoned. I have checked and rechecked the mathematics, and I have added some additional references. I have also written an introduction that describes what's in the book and how it may be used in contemporary teaching of analysis. I have also provided summaries of each chapter, and, within them, make mention of more recent work where appropriate.

As I said, preparing this edition has been a labor of love. I have also learned a lot of mathematics, evidence of the enduring quality and value of the original work. It has been a rewarding experience to edit *A Course of Modern Analysis*: I hope that it will be equally rewarding for readers.

Victor H. Moll
2020, New Orleans

Preface to the Fourth Edition

Advantage has been taken of the preparation of the fourth edition of this work to add a few additional references and to make a number of corrections of minor errors.

Our thanks are due to a number of our readers for pointing out errors and misprints, and in particular we are grateful to Mr E. T. Copson, Lecturer in Mathematics in the University of Edinburgh, for the trouble which he has taken in supplying us with a somewhat lengthy list.

E. T. W.
G. N. W.
June 18, 1927

The decimal system of paragraphing, introduced by Peano, is adopted in this work. The integral part of the decimal represents the number of the chapter and the fractional parts are arranged in each chapter in order of magnitude. Thus, e.g., on pp. 187, 188[6], §9.632 precedes §9.7 [because 9.632 < 9.7.]

G.N.W.
July 1920

---

[6] in the fourth edition

Preface to the Third Edition

Advantage has been taken of the preparation of the third edition of this work to add a chapter on Ellipsoidal Harmonics and Lamé's Equation and to rearrange the chapter on Trigonometric Series so that the parts which are used in Applied Mathematics come at the beginning of the chapter. A number of minor errors have been corrected and we have endeavoured to make the references more complete.

Our thanks are due to Miss Wrinch for reading the greater part of the proofs and to the staff of the University Press for much courtesy and consideration during the progress of the printing.

E. T. W.
G. N. W.
July, 1920

Preface to the Second Edition

When the first edition of my *Course of Modern Analysis* became exhausted, and the Syndics of the Press invited me to prepare a second edition, I determined to introduce many new features into the work. The pressure of other duties prevented me for some time from carrying out this plan, and it seemed as if the appearance of the new edition might be indefinitely postponed. At this juncture, my friend and former pupil, Mr G. N. Watson, offered to share the work of preparation; and, with his cooperation, it has now been completed.

The appearance of several treatises on the Theory of Convergence, such as Mr Hardy's *Course of Pure Mathematics* and, more particularly, Dr Bromwich's *Theory of Infinite Series*, led us to consider the desirability of omitting the first four chapters of this work; but we finally decided to retain all that was necessary for subsequent developments in order to make the book complete in itself. The concise account which will be found in these chapters is by no means exhaustive, although we believe it to be fairly complete. For the discussion of Infinite Series on their own merits, we may refer to the work of Dr Bromwich.

The new chapters of Riemann Integration, on Integral Equations, and on the Riemann Zeta-Function, are entirely due to Mr Watson: he has revised and improved the new chapters which I had myself drafted and he has enlarged or partly rewritten much of the matter which appeared in the original work. It is therefore fitting that our names should stand together on the title-page.

Grateful acknowledgement must be made to Mr W. H. A. Lawrence, B.A., and Mr C. E. Winn, B.A., Scholars of Trinity College, who with great kindness and care have read the proof-sheets, to Miss Wrinch, Scholar of Girton College, who assisted in preparing the index, and to Mr Littlewood, who read the early chapters in manuscript and made helpful criticisms. Thanks are due also to many readers of the first edition who supplied corrections to it; and to the staff of the University Press for much courtesy and consideration during the progress of the printing.

E.T. Whittaker
July 1915

Preface to the First Edition

The first half of this book contains an account of those methods and processes of higher mathematical analysis, which seem to be of greatest importance at the present time; as will be seen by a glance at the table of contents, it is chiefly concerned with the properties of infinite series and complex integrals and their applications to the analytical expression of functions. A discussion of infinite determinants and of asymptotic expansions has been included, as it seemed to be called for by the value of these theories in connexion with linear differential equations and astronomy.

In the second half of the book, the methods of the earlier part are applied in order to furnish the theory of the principal functions of analysis – the Gamma, Legendre, Bessel, Hypergeometric, and Elliptic Functions. An account has also been given of those solutions of the partial differential equations of mathematical physics which can be constructed by the help of these functions.

My grateful thanks are due to two members of Trinity College, Rev. E. M. Radford, M.A. (now of St John's School, Leatherhead), and Mr J. E. Wright, B.A., who with great kindness and care have read the proof-sheets; and to Professor Forsyth, for many helpful consultations during the progress of the work. My great indebtedness to Dr Hobson's memoirs on Legendre functions must be specially mentioned here; and I must thank the staff of the University Press for their excellent cooperation in the production of the volume.

E. T. WHITTAKER
Cambridge
1902 August 5

Introduction

The book is divided into two distinct parts. **Part I. The Processes of Analysis** discusses topics that have become standard in beginning courses. Of course the emphasis is in concrete examples and regrettably, this is different nowadays. Moreover the quality and level of the problems presented in this part is higher than what appears in more modern texts. During the second part of the last century, the tendency in introductory Analysis texts was to emphasize the topological aspects of the material. For obvious reasons, this is absent in the present text. There are 11 chapters in Part I.

For a student in an American university, the material presented here is roughly distributed along the following lines:

- Chapter 1 (Complex Numbers)
- Chapter 2 (The Theory of Convergence)
- Chapter 3 (Continuous Functions and Uniform Convergence)
- Chapter 4 (The Theory of Riemann Integration)

are covered in *Real Analysis* courses.

- Chapter 5 (The Fundamental Properties of Analytic Functions; Taylor's, Laurent's and Liouville's Theorems)
- Chapter 6 (The Theory of Residues, Applications to the Evaluations of Definite Integrals)
- Chapter 7 (The Expansion of Functions in Infinite Series)

are covered in *Complex Analysis*. These courses usually cover the more elementary aspects of

- Chapter 12 (The Gamma-Function)

appearing in Part II.

Most undergraduate programs also include basic parts of

- Chapter 9 (Fourier Series and Trigonometric Series)
- Chapter 10 (Linear Differential Equations)

and some of them will expose the student to the elementary parts of

- Chapter 8 (Asymptotic Expansions and Summable Series)
- Chapter 11 (Integral Equations)

The material covered in Part II is mostly absent from a generic graduate program. Students interested in Number Theory will be exposed to some parts of the contents in

- Chapter 12 (The Gamma-Function)
- Chapter 13 (The Zeta-Function of Riemann)
- Chapter 14 (The Hypergeometric Function)

and a glimpse of

- Chapter 17 (Bessel Functions)
- Chapter 20 (Elliptic Functions. General Theorems and the Weierstrassian Functions)
- Chapter 21 (The Theta-Functions)
- Chapter 22 (The Jacobian Elliptic Functions).

Students interested in Applied Mathematics will be exposed to

- Chapter 15 (Legendre Functions)
- Chapter 16 (The Confluent Hypergeometric Function)
- Chapter 18 (The Equations of Mathematical Physics)

and some parts of

- Chapter 19 (Mathieu Functions)
- Chapter 23 (Ellipsoidal Harmonics and Lamé's Equation)

It is perfectly possible to complete a graduate education without touching upon the topics in Part II. For instance, in the most commonly used textbooks for Analysis, such as Royden [565] and Wheeden and Zygmund [666] there is no mention of special functions. On the complex variables side, in Ahlfors [13] and Greene–Krantz [260] one finds some discussion on the Gamma function, but not much more.

This is not a new phenomenon. Fleix Klein [377] in 1928 (quoted in [91, p. 209]) writes *'When I was a student, Abelian functions were, as an effect of the Jacobian tradition, considered the uncontested summit of mathematics, and each of us was ambitious to make progress in this field. And now? The younger generation hardly knows Abelian functions.*

During the last two decades, the trend towards the abstraction is being complemented by a group of researchers who emphasize concrete examples as developed by Whittaker and Watson. Among the factors influencing this return to the classics one should include[7] the appearance of symbolic languages and algorithms producing automatic proofs of identities. The work initiated by Wilf and Zeilberger, described in [518], shows that many identities have *automatic proofs*. A second influential factor is the monumental work by B. Berndt, G. Andrews and collaborators to provide context and proofs of all results appearing in S. Ramanujan's work. This has produced a collection of books, starting with [60] and currently at [25]. The third example in this list is the work developed by J. M. Borwein and his collaborators in the propagation of *Experimental Mathematics*. In the volumes [88, 89] the authors present their ideas on how to transform mathematics into a subject, similar in flavor to other experimental sciences. The point of view expressed in the three examples mentioned above has attracted a new generation of researchers to get involved in this point of view type of mathematics. This is just one direction in which Whittaker and Watson has been a profound influence in modern authors.

---

[7] This list is clearly a subjective one.

The remainder of this chapter outlines the content of the book and a comparison with modern practices.

The first part is named **The Processes of Analysis**. It consists of 11 chapters. A brief description of each chapter is provided next.

Chapter 1: Complex Numbers. The authors begin with an informal description of positive integers and move on to rational numbers. Stating that *from the logical standpoint it is improper to introduce geometrical intuition to supply deficiencies in arithmetical arguments*, they adopt Dedekind's point of view on the construction of real numbers as classes of rational numbers, later called *Dedekind's cuts*. An example is given to show that there is no rational number whose square is 2. The arithmetic of real numbers is defined in terms of these cuts. Complex numbers are then introduced with a short description of *Argand diagrams*. The current treatment offers two alternatives: some authors present the real number from a collection of axioms (as an ordered infinite field) and other approach them from Cauchy's theory of sequences: *a real number is an equivalence class of Cauchy sequences of rational numbers*. The reader will find the first point of view in [304] and the second one is presented in [599].

Chapter 2. The Theory of Convergence. This chapter introduces the notion of convergence of sequences of real or complex numbers starting with the definition of $\lim_{n \to \infty} x_n = L$ currently given in introductory texts. The authors then consider monotone sequences of real numbers and show that, for bounded sequences, there is a natural Dedekind cut (that is, a real number) associated to them. A presentation of Bolzano's theorem *a bounded sequence of real numbers contains a limit point* and Cauchy's formulation of the completeness of real numbers; that is, the existence of the limit of a sequence in terms of elements being arbitrarily close, is discussed. These ideas are then illustrated in the analysis of convergence of series. The discussion begins with *Dirichlet's test for convergence: Assume a_n is a sequence of complex numbers and f_n is a sequence of positive real numbers. If the partial sums $\sum_{n=1}^{p} a_n$ are uniformly bounded and f_n is decreasing and converges to 0, then $\sum_{n=1}^{\infty} a_n f_n$ converges.* This is used to give examples of convergence of Fourier series (discussed in detail in Chapter 9). The convergence of the geometric series $\sum_{n=1}^{\infty} x^n$ and the series $\sum_{n=1}^{\infty} \frac{1}{n^s}$, for real s, are presented in detail. This last series defines the *Riemann zeta function* $\zeta(s)$, discussed in Chapter 13. The elementary ratio test states that $\sum_{n=1}^{\infty} a_n$ converges if $\lim_{n \to \infty} |a_{n+1}/a_n| < 1$ and diverges if the limit is strictly above 1. A discussion of the case when the limit is 1 is presented and illustrated with the convergence analysis of the *hypergeometric series* (presented in detail in Chapter 14). The chapter contains some standard material on the convergence of power series as well as some topics not usually found in modern textbooks: discussion on double series, convergence of infinite products and infinite determinants. The final exercise[8] in this chapter presents the evaluation of an infinite determinant considered by Hill in his analysis of the Schrödinger

---

[8] In this book, Examples are often what are normally known as Exercises and are numbered by section, i.e., 'Example a.b.c'. At the end of most chapters are Miscellaneous Examples, all of which are Exercises, and which are numbered by chapter: thus 'Example a.b'. This is how to distinguish them.

equation with periodic potential (this is now called the *Hill equation*). The reader will find in [451] and [536] information about this equation.

Chapter 3. Continuous Functions and Uniform Convergence. This chapter also discusses functions $f(x, y)$ of two real variables as well as functions of one complex variable $g(z)$. The notion of *uniform convergence* of a series is discussed in the context of the limiting function of a series of functions. This is normally covered in every introductory course in Analysis. The classical M-test of Weierstrass is presented. The reader will also find a test for uniform convergence, due to Hardy, and its application to the convergence of Fourier series. The chapter also contains a discussion of the series

$$g(z) = \sum_{m,n} \frac{1}{(z + 2m\omega_1 + 2n\omega_2)^\alpha}$$

which will be used to analyze the Weierstrass $\wp$-function: one of the fundamental *elliptic functions* (discussed in Chapter 20). The chapter contains a discussion on the fact that a continuous function defined of a compact set (in the modern terminology) attains its maximum/minimum value. This is nowadays a standard result in elementary analysis courses.

Chapter 4. The Theory of Riemann Integration. The authors present the notion of the Riemann integral on a finite interval $[a, b]$, as it is currently done: as limiting values of upper and lower sums. The fact that a continuous function is integrable is presented. The case with finite number of discontinuities is given as an exercise. Basic results, such as integration by parts, differentiation with respect to the limits of integration, differentiation with respect to a parameter, the mean value theorem for integrals and the representation of a double integral as iterated integral are presented. This material has become standard. The chapter also contains a discussion on integrals defined on an infinite interval. There is a variety of tests to determine convergence and criteria that can be used to evaluate the integrals. Two examples of integral representations of the *beta integral* (discussed in Chapter 12) are presented. A basic introduction to *complex integration* is given at the end of the chapter; the reader is referred to Watson [650] for more details. This material is included in basic textbooks in Complex Analysis (for instance, see [13, 26, 155, 260, 455, 552]).

Chapter 5. The Fundamental Properties of Analytic Functions; Taylor's, Laurent's and Liouville's Theorems. This chapter presents the basic properties of analytic functions that have become standard in elementary books in complex analysis. These include the Cauchy–Riemann equations and Cauchy's theorem on the vanishing of the integral of an analytic function taken over a closed contour. This is used to provide an integral representation as

$$f(z) = \frac{1}{2\pi i} \int_\Gamma \frac{f(\xi)}{z - \xi} \, d\xi$$

where Γ is a closed contour containing ξ in its interior. This is then used to establish classical results on analytic functions such as bounds on the derivatives and Taylor theorem. There is also a small discussion on the process of analytic continuation and many-valued functions. This chapter contains also basic properties on functions having poles as isolated singularities: Laurent's theorem on expansions and Liouville's theorem on the fact that every entire function that is bounded must be constant (a result that plays an important role in the presentation of elliptic functions in Chapter 20). The Bessel function J_n, defined by its integral representation

$$J_n(x) = \frac{1}{2\pi} \int_0^{2\pi} \cos(n\theta - x\sin\theta)\, d\theta$$

makes its appearance in an exercise. This function is discussed in detail in Chapter 17. The chapter also contains a proof of the following fact: *any function that is analytic, including at* ∞, *except for a number of non-essential singularities, must be a rational function.* This has become a standard result. It represents the most elementary example of characterizing *functions of rational character on a Riemann surface.* This is the case of $\mathbb{P}^1$, the Riemann sphere. The next example corresponds to the torus $\mathbb{C}/\mathbb{L}$, where $\mathbb{L}$ is a lattice. This is the class of elliptic functions described in Chapters 20, 21 and 22. The reader is referred to [461, 553, 600, 665] for more details.

Chapter 6. The Theory of Residues: Application to the Evaluation of Definite Integrals. This chapter presents application of Cauchy's integral representation of functions analytic except for a certain number of poles. Most of the material discussed here has become standard. One of the central concepts is that of the *residue* of a function at a pole $z = z_k$, defined as the coefficient of $(z - z_k)^{-1}$ in the expansion of f near $z = z_k$. As a first sign of the importance of these residues is the statement that the integral of $f(z)$ over the boundary of a domain Ω is given by the sum of the residues of f inside Ω, the so-called *argument principle* which gives the difference between zeros and poles of a function as a contour integral. This chapter also presents methods based on residues to evaluate a variety of definite integrals including rational functions of $\cos\theta$, $\sin\theta$ over $[0, 2\pi]$, integrals over the whole real line via deformation of a semicircle, integrals involving some of the kernels such as $1/(e^{2\pi z} - 1)$ (coming from the Fermi–Dirac distribution in Statistical Mechanics) and $1/(1 - 2a\cos x + a^2)$ related to Legendre polynomials (discussed in Chapter 15). An important function makes its appearance as Exercise 17:

$$\psi(t) = \sum_{n=-\infty}^{\infty} e^{-n^2\pi t},$$

introduced by Poisson in 1823. The exercise outlines a proof of the transformation rule

$$\psi(t) = t^{-1/2}\psi(1/t)$$

known as *Poisson summation formula.* It plays a fundamental role in many problems in Number Theory, including the proof of the *prime number theorem.* This states that, for $x > 0$, the number of primes up to x, denoted by $\pi(x)$, has the asymptotic behavior $\pi(x) \sim x/\log x$ as $x \to \infty$. The reader will find in [492] how to use contour integration and the function $\psi(t)$ to provide a proof of the asymptotic behavior of $\psi(t)$. This function reappears in Chapter 21 in the study of *theta functions.*

Chapter 7. The Expansion of Functions in Infinite Series. This chapter begins with a result of Darboux on the expansion of an analytic function defined on a region Ω. For points a, x, with the segment from a to x contained in Ω, one has the expansion

$$\phi^{(n)}(0)[f(z) - f(0)] = \sum_{k=1}^{n}(-1)^{k-1}(z-a)^k\left[\phi^{(n-k)}(1)f^{(m)}(z) - \phi^{n-k}(0)f^{(k)}(a)\right]$$

$$+ (-1)^n(z-a)^{n+1}\int_0^1 \phi(t)f^{(n+1)}(a+t(z-a))\,dt,$$

for any polynomial ϕ. The formula is then applied to the *Bernoulli polynomials* currently defined by the generating function

$$\frac{te^{zt}}{e^t - 1} = \sum_{n=0}^{\infty}\frac{B_n(z)}{n!}t^n.$$

(The text employs the notation $\phi_n(t)$ without giving the value for $n = 0$.) Darboux's theorem then becomes the classical *Euler–MacLaurin summation formula*

$$\sum_{j=0}^{n} f(j) = \int_0^n f(x)\,dx + \frac{f(n) + f(0)}{2} + \sum_{k=1}^{\lfloor p/2 \rfloor}\frac{B_{2k}}{(2k)!}\left[f^{(2k-1)}(n) - f^{(2k-1)}(0)\right]$$

$$+ (-1)^{p-1}\int_0^n f^{(p)}(x)\frac{B_p(x - \lfloor x \rfloor)}{p!}\,dx.$$

The quantity $x - \lfloor x \rfloor$ is the *fractional part* of x, denoted by $\{x\}$. This formula is used to estimate partial sums of series of values of an analytic function in terms of the corresponding integrals. The important example of the *Riemann zeta function* $\zeta(s)$ is presented in Chapter 13.

The chapter contains a couple of examples of expansions of one function in terms of another one. The first one, due to Bürmann, starts with an analytic function $\phi(z)$ defined on a region and $\phi(a) = b$ with $\phi'(a) \neq 0$. Define $\psi(z) = (z - a)/(\phi(z) - a)$, then one obtains the expansion

$$f(z) = f(a) + \sum_{k=1}^{n-1}\frac{[\phi(z) - b]^k}{k!}\left(\frac{d}{da}\right)^{k-1}\left[f'(a)\psi^k(a)\right] + R_n$$

where the error term has the integral representation

$$R_n = \frac{1}{2\pi i}\int_a^z\int_\gamma\left[\frac{\phi(z) - b}{\phi(t) - b}\right]^{n-1}\frac{f'(t)\phi'(z)}{\phi(t) - \phi(z)}\,dt\,dz,$$

where γ is a contour in the t-plane, enclosing a and t and such that, for any μ interior to γ, the equation $\phi(t) = \phi(\mu)$ has a unique solution $t = \mu$. The discussion also contains results of Teixeira on conditions for the convergence of the series for $f(z)$ obtained by letting $n \to \infty$. This type of result also contains an expansion of Lagrange for solutions of the equation $\mu = a + t\phi(\mu)$, for analytic function ϕ satisfying $|t\phi(z)| < |z - a|$. The theorem states that any analytic function f of the solution μ can be expanded as

$$f(\mu) = f(a) + \sum_{n=1}^{\infty}\frac{t^n}{n!}\left(\frac{d}{da}\right)^{n-1}\left[f'(a)\phi^n(a)\right].$$

This expansion has interesting applications in Combinatorics; see [681] for details. The last type of series expansion described here corresponds to the classical partial fraction expansions of a rational function and its extensions to trigonometric functions.

The results of this chapter are then used to prove representations of an entire function f in the form

$$f(z) = f(0)e^{G(z)} \prod_{n=1}^{\infty} \left\{ \left(1 - \frac{z}{a_n}\right) e^{g_n(z)} \right\}^{m_n}$$

where a_n is a zero of f of multiplicity m_n and $G(z)$ is an entire function. The function $g_n(z)$ is a polynomial, introduced by Weierstrass, which makes the product converge. An application to $1/\Gamma(z)$ is discussed in Chapter 12.

Chapter 8. Asymptotic Expansions and Summable Series. This chapter presents an introduction to the basic concepts behind asymptotic expansion. The initial example considers $f(x) = \int_x^{\infty} t^{-1} e^{x-t}\, dt$. A direct integration by parts shows that the sum $S_n(x) = \sum_{k=0}^{\infty} \frac{(-1)^k k!}{x^{k+1}}$ satisfies, for fixed x, the inequality $|f(x) - S_n(x)| \leq n!/x^{n+1}$. Therefore, for $x \geq 2n$, one obtains $|f(x) - S_n(x)| < 1/n^2 2^{n+1}$. It follows that the integral f can be evaluated with great accuracy for large values of x by computing the partial sum of the divergent series $S_n(x)$. This type of behavior is written as $f(x) \sim \sum_{n=0}^{\infty} A_n x^{-n}$ and the series is called *the asymptotic expansion* of f.

The chapter covers the basic properties of asymptotic series: such expansions can be multiplied and integrated *but not* differentiated. Examples of asymptotic expansions of special functions appear in later chapters: for the Gamma function in Chapter 12 and for the Bessel function in Chapter 17.

The final part of the chapter deals with *summation methods*, concentrating on methods assigning a value to a function given by a power series outside its circle of convergence D. The first example, due to Borel, starts with the identity

$$\sum_{n=0}^{\infty} a_n z^n = \int_0^{\infty} e^{-t} \phi(tz)\, dt \text{ where } \phi(u) = \sum_{n=0}^{\infty} \frac{a_n}{n!} u^n \text{ valid for } z \in D.$$

The series $\sum_{n=0}^{\infty} a_n z^n$ is said to be *Borel summable* if the integral on the right converges for z outside D. For such z, the Borel sum of the series is assigned to be the value of the integral. The discussion continues with *Cesàro summability*, a notion to be discussed in the context of Fourier series in Chapter 9. Extensions by Riesz and Hardy are mentioned. More details on asymptotic expansions can be found in [468, 508].

Chapter 9. Fourier Series and Trigonometric Series. The authors discuss *trigonometrical series* defined as series of the form

$$\frac{1}{2} a_0 + \sum_{n=1}^{\infty} (a_n \cos nx + b_n \sin nx)$$

for two sequences of real numbers $\{a_n\}$ and $\{b_n\}$. Such series are named *Fourier series* if there is a function f, with finite integral over $(-\pi, \pi)$, such that the coefficients are given by

$$a_n = \frac{1}{\pi} \int_{-\pi}^{\pi} f(t) \cos nt\, dt \quad \text{and} \quad b_n = \frac{1}{\pi} \int_{-\pi}^{\pi} f(t) \sin nt\, dt.$$

The chapter contains a variety of results dealing with conditions under which the Fourier series associated to a function f converges to f. These include Dirichlet's theorem stating that, under some technical conditions, the Fourier series converges to $\frac{1}{2}[f(x+0) + f(x-0)]$. This is followed by Fejer's theorem that the Fourier series is Césaro summable at all points where the limits $f(x\pm0)$ exist. The proofs are based on the analysis of the so-called Dirichlet–Féjer kernel. Examples are provided where there is not a single analytic expression for the Fourier series. The notion of orthogonality of the sequence of trigonometric functions makes an implicit appearance in all the proofs. The so-called Riemann–Lebesgue theorem, on the behavior of Fourier coefficients, is established. This result states that if $\psi(\theta)$ is integrable on the interval (a, b), then $\lim\limits_{n\to\infty} \int_a^b \psi(\theta)\sin(\lambda\theta)\, d\theta = 0$. The chapter contains results on the function f which imply pointwise convergence of the Fourier series. The results of Dini and Jordan, with conditions on the expressions $f(x \pm 2\theta) - f(x \pm 0)$ near $\theta = 0$, are presented. The reader will find more information about convergence of Fourier series in [368] and in the treatise [690]. The results of Kolmogorov [381, 382] on an integrable function with a Fourier series diverging everywhere, as well as the theorem of Carleson [118] on the almost-everywhere convergence of the Fourier series of a continuous function, are some of the high points of this difficult subject.

The chapter also includes a discussion on the uniqueness of the representation of a Fourier series for a function f and also of the *Gibbs phenomenon* on the behavior of a Fourier series in a neighborhood of a point of discontinuity of f.

Chapter 10. Linear Differential Equations. This chapter discusses properties of solutions of second order linear differential equations

$$\frac{d^2u}{dz^2} + p(z)\frac{du}{dz} + q(z)u = 0,$$

where p, q are analytic functions of z except for a finite number of points. The discussion is local; that is, in a neighborhood of a point $c \in \mathbb{C}$. The points c are classified as *ordinary*, where the functions p, q are assumed to be analytic at c and otherwise *singular*.

The question of existence and uniqueness of solutions of the equation is discussed. The equation is transformed first into the form $\dfrac{d^2v}{dz^2} + J(z)v = 0$, by an elementary change of variables. Existence of solutions is obtained from an integral equation equivalent to the original problem. An iteration process is used to produce a sequence of analytic functions $\{v_n\}$. Then it is shown that, in a neighborhood of an ordinary point, this sequence converges uniformly to a solution of the equation. Uniqueness of the solution comes also from this process.

The solutions near an ordinary point are presented in the case of an *ordinary singular point*. These are points $c \in \mathbb{C}$ where p or q have a pole, but $(z-c)p(z)$ and $(z-c)^2q(z)$ are analytic functions in a deleted neighborhood of $z = c$. The so-called *method of Frobenius* is then used to seek formal series solutions in the form

$$u(z) = (z-c)^\alpha \left[1 + \sum_{n=1}^{\infty} a_n(z-c)^n \right].$$

The so-called *indicial equation* $\alpha^2 + (p_0 - 1)\alpha + q_0 = 0$ and its roots α_1, α_2, control the properties of these formal power series. The numbers p_0, q_0 are the leading terms of $(z-c)p(z)$ and $(z-c)^2q(z)$, respectively. It is shown that if α_1, α_2 do not differ by an integer, there are two formal solutions and these series actually converge and thus represent actual solutions. Otherwise one of the formal series is an actual solution and there is a procedure to obtain a second solution containing a logarithmic term. The reader will find in [151] all the details.

It is a remarkable fact that the behavior of the singularities determines the equation itself. For example, the most general differential equation of second order which has every point except a_1, a_2, a_3, a_4 and ∞ as ordinary points and these five points as regular points, must be of the form

$$
\frac{d^2u}{dz^2} + \left\{ \sum_{r=1}^{4} \frac{1 - \alpha_r - \beta_r}{z - a_r} \right\} \frac{du}{dz}
$$
$$
+ \left\{ \sum_{r=1}^{4} \frac{\alpha_r \beta_r}{(z - a_r)^2} + \frac{Az^2 + Bz + C}{(z - a_1)(z - a_2)(z - a_3)(z - a_4)} \right\} u = 0,
$$

for some constants α_r, β_r, A, B, C. F. Klein [376] describes how all the classical equations of Mathematical Physics appear in this class. Six classes, carrying the names of their discoverers (Lamé, Mathieu, Legendre, Bessel, Weber–Hermite and Stokes) are discussed in later chapters.

The chapter finally discusses the so-called Riemann P-function. This is a mechanism used to write a solution of an equation with three singular points and the corresponding roots of the indicial equation. Some examples of formal rules on P, which allow to transform a solution with expansion at one singularity to another are presented. The chapter concludes showing that a second order equation with three regular singular points may be converted to the *hypergeometric equation*. This is the subject of Chapter 14.

The modern theory of this program, to classify differential equations by their singularities, is its extension to nonlinear equations. A singularity of an ordinary differential equation is called *movable* if its location depends on the initial condition. An equation is called a *Painlevé equation* if its only movable singularities are poles. Poincaré and Fuchs proved that any first-order equation with this property may be transformed into the Ricatti equation or it may be solved in terms of the Weierstrass elliptic function. Painlevé considered the case of second order, transformed them into the form $u'' = R(u, u', z)$, where R is a rational function. Then he put them into 50 *canonical forms* and showed that all but six may be solved in terms of previously known functions. The six remaining cases gave rise to the *six Painlevé functions* $P_{\mathrm{I}}, \ldots, P_{\mathrm{VI}}$. See [261, 310, 336] for details. It is a remarkable fact that these functions, created for an analytic study, have recently appeared in a large variety of problems. See [37] and [562] for their appearance in combinatorial questions, [76, 636] for their relations to classical functions, [640] for connections to orthogonal polynomials, [632] for their appearance in Statistical Physics. The reader will find in [212] detailed information about their asymptotic behavior.

Chapter 11. Integral Equations. Given a function f, continuous on an interval $[a, b]$ and a *kernel* $K(x, y)$, say continuous on both variables or in the region $a \leq y \leq x \leq b$ and vanishing for $y > x$, the equation

$$\phi(x) = f(x) + \lambda \int_a^b K(x, y)\phi(y)\, dy$$

for the unknown function ϕ, is called the *Fredholm integral equation of the second kind*. The solution presented in this chapter is based on the construction of functions $D(x, y, \lambda)$ and $D(\lambda)$, both entire in λ, as a series in which the nth-term consists of determinants of order $n \times n$ based on the function $K(x, y)$. The solution is then expressed as

$$\phi(x) = f(x) + \frac{1}{D(\lambda)} \int_a^b D(x, \xi, \lambda) f(\xi)\, d\xi.$$

In particular, in the homogeneous case $f \equiv 0$, there is a unique solution $\phi \equiv 0$ for those values of λ with $D(\lambda) \neq 0$. A process to obtain a solution for those values of λ with $D(\lambda) = 0$ is also described.

Volterra introduced the concept of *reciprocal functions* for a pair of functions $K(x, y)$ and $k(x, y; \lambda)$ satisfying the relation

$$K(x, y) + k(x, y; \lambda) = \lambda \int_a^b k(x, \xi; \lambda) K(\xi, y) d\xi.$$

Then the solution to the Fredholm equation is given by

$$f(x) = \phi(x) + \lambda \int_a^b k(x, \xi; \lambda)\phi(\xi) d\xi.$$

The last part of the chapter discusses the equation

$$\Phi(x) = f(x) + \lambda \int_a^b K(x, \xi)\Phi(\xi)\, d\xi$$

and the solution is expressed as a series in terms of a sequence of orthonormal functions and the sequence $\{\lambda_n\}$ of eigenvalues of the kernel $K(x, y)$. In detail, if $f(x) = \sum b_n \phi_n(x)$, then the solution Φ is given by $\Phi(x) = \sum \frac{b_n \lambda_n}{\lambda - \lambda_n} \phi_n(x)$.

The Fredholm equation is written formally as $\Phi = f + K\Phi$ and this gives $\Phi = f + Kf + K^2\Phi$. Iteration of this process gives the so-called *Neumann series* $\Phi = \sum_{n=0}^{\infty} K^n f$, expressing the unknown Φ in terms of iterations of the functional defined by the kernel K.

The study of Fredholm integral equations is one of the beginnings of modern Functional Analysis. The reader will find more details in P. Lax [415]. The ideas of Fredholm have many applications: the reader will find in H. P. McKean [460] a down-to-earth explanation of Fredholm's work and applications to *integrable systems* (such as the Korteweg–de Vries equation $u_t = u_{xxx} + 6uu_x$ and some special solutions called *solitons*), to the calculations of some integrals involving Brownian paths (such as P. Lévy's formula for the area generated by a two-dimensional Brownian path) and finally to explain the appearance of the so-called *sine kernel* in the limiting distribution of eigenvalues of random unitary matrices. This subject has some mysterious connections to the *Riemann hypothesis* as described by B. Conrey [154].

The second part of the book is called **The Transcendental Functions** and it consists of 12 chapters. A brief description of them is provided next.

Chapter 12. The Gamma-Function. This function, introduced by Euler, represents an extension of factorials $n!$ from positive integers to complex values of n. The presentation begins with the infinite product

$$P(z) = ze^{\gamma z} \prod_{n=1}^{\infty} \left(1 + \frac{z}{n}\right) e^{-z/n}$$

where $\gamma = \lim_{n \to \infty} \left(1 + \frac{1}{2} + \cdots + \frac{1}{n} - \log n\right)$ is nowadays called the *Euler–Mascheroni constant*. The product is an entire function of $z \in \mathbb{C}$ and the Gamma function is defined by $\Gamma(z) = 1/P(z)$. Therefore $\Gamma(z)$ is an analytic function except for simple poles at $z = 0, -1, 2, \ldots$. The constant γ is identified as $-\Gamma'(1)$. The fact that Γ is a transcendental function is reflected by the fact, mentioned in this chapter, that Γ does not satisfy a differential equation with coefficients being rational functions of z. The chapter contains proofs of a couple of representations by Euler

$$\Gamma(z) = \frac{1}{z} \prod_{n=1}^{\infty} \left(1 + \frac{1}{z}\right)^z \left(1 + \frac{z}{n}\right)^{-1}$$

$$= \lim_{n \to \infty} \frac{(n-1)!}{z(z+1)\cdots(z+n-1)} n^z.$$

The functional equation $\Gamma(z+1) = z\Gamma(z)$ follows directly from here. Using the value $\Gamma(1) = 1$, this leads to $\Gamma(n) = (n-1)!$ for $n \in \mathbb{N}$, showing that Γ interpolates factorials.

The chapter also presents proofs of the *reflection formula*

$$\Gamma(z)\Gamma(1-z) = \frac{\pi}{\sin \pi z}$$

leading to the special value $\Gamma(\frac{1}{2}) = \sqrt{\pi}$. There is also a discussion of the *multiplication formula* due to Gauss

$$\Gamma(nz) = (2\pi)^{-(n-1)/2} n^{-1/2+nz} \prod_{k=0}^{n-1} \Gamma\left(z + \frac{k}{n}\right)$$

and the special *duplication formula* of Legendre

$$\Gamma(2z) = \frac{1}{\sqrt{\pi}} 2^{2z-1} \Gamma(z)\Gamma(z + \tfrac{1}{2}).$$

This may be used to derive the relation $\Gamma(\frac{1}{3})\Gamma(\frac{2}{3}) = \frac{2\pi}{\sqrt{3}}$. Arithmetical properties of these values are difficult to establish. The reader is referred to [92] and [166] for an elementary presentation of the Gamma function, and to [106] for an introduction to issues of transcendence.

There are several integral representations of the Gamma function established in this chapter. Most of them appear in the collection of integrals by Gradshetyn and Ryzhik [258]. The first one, due to Euler, is

$$\Gamma(z) = \int_0^{\infty} t^{z-1} e^{-t}\, dt,$$

valid for $\mathrm{Re}\, z > 0$. This may be transformed to the logarithmic scale

$$\Gamma(z) = \int_0^1 \left(\log \frac{1}{x}\right)^{z-1} dx.$$

There is also a presentation of *Hankel's contour integral*

$$\Gamma(z) = -\frac{1}{2i \sin \pi z} \int_C (-t)^{z-1} e^{-t}\, dt, \quad z \notin \mathbb{Z}$$

where C is a thin contour enclosing the positive real axis.

The chapter also contains a discussion of two functions related to Γ: its logarithm $\log \Gamma(z)$ and the *digamma function*, $\psi(z) = \Gamma'(z)/\Gamma(z) = (\log \Gamma(z))'$. Integral representations for $\psi(z)$ include

$$\psi(z) = \int_0^\infty \left(\frac{e^{-t}}{t} - \frac{e^{-zt}}{1 - e^{-t}} \right) dt$$

$$= \int_0^\infty \left(e^{-x} - \frac{1}{(1+x)^z} \right) \frac{dx}{x};$$

the first one is due to Gauss and the second one to Dirichlet. The chapter also contains a multi-dimensional integral due to Dirichlet that can be reduced to a single variable problem:

$$\int_{\mathbb{R}_+^n} f(t_1 + \cdots + t_n) t_1^{a_1-1} \cdots t_n^{a_n-1}\, dt_1 \cdots dt_n$$

$$= \frac{\Gamma(a_1) \cdots \Gamma(a_n)}{\Gamma(a_1 + \cdots + a_n)} \int_0^1 f(\tau) \tau^{a_1 + \cdots + a_n - 1}\, d\tau.$$

Other multi-dimensional integrals appear in the modern literature. For a description of a remarkable example due to Selberg, the reader is referred to [214].

The properties of $\log \Gamma(z)$ presented in this chapter include a proof of the identity

$$\frac{d^2}{dz^2} \log \Gamma(z + 1) = \sum_{k=1}^\infty \frac{1}{(z + k)^2},$$

showing that $\Gamma(z+1)$ is log-convex. This property, the functional equation and the value $\Gamma(1) = 1$ characterize the Gamma function. The reader will also find two integral representations due to Binet

$$\log \Gamma(z) = \left(z - \frac{1}{2} \right) \log z - z + \frac{1}{2} \log 2\pi + \int_0^\infty \left(\frac{1}{2} - \frac{1}{t} + \frac{1}{e^t - 1} \right) \frac{e^{-tz}}{t}\, dt$$

and

$$\log \Gamma(z) = \left(z - \frac{1}{2} \right) \log z - z + \frac{1}{2} \log 2\pi + 2 \int_0^\infty \frac{\tan^{-1}\left(\frac{t}{z}\right)}{e^{2\pi t} - 1}\, dt.$$

Integrals involving $\log \Gamma(z)$ present interesting challenges. The value

$$\int_0^1 \log \Gamma(z)\, dz = \log \sqrt{2\pi}$$

due to Euler, may be obtained from the reflection formula for $\Gamma(z)$. The generalization of the previous evaluation to

$$L_n = \int_0^1 (\log \Gamma(z))^n\, dz$$

is discussed next. The value of L_2 is presented in [196] as an expression involving the

Riemann zeta function and its derivatives. The values of L_3 and L_4 were obtained in [38] and they involve more advanced objects: *multiple zeta values*. At the present time an evaluation of L_n, for $n \geq 5$, is an open question.

The chapter also contains a discussion of the asymptotic behavior of $\log \Gamma(z)$, as a generalization of Stirling's formula for factorials and also a proof of the expression for the Fourier series of $\log \Gamma(z)$ due to Kummer. The *Barnes G-function*, an important generalization of $\Gamma(z)$, appears in the exercises at the end of this chapter. A detailed presentation of these and other topics may be found in [20].

Chapter 13. The Zeta-Function of Riemann. For $s = \sigma + it \in \mathbb{C}$, the function

$$\zeta(s) = \sum_{n=1}^{\infty} \frac{1}{n^s}$$

is the *Riemann zeta function*. This had been considered by Euler for $s \in \mathbb{R}$. For $\delta > 0$, the series defines an analytic function of s on the half-plane $\sigma = \operatorname{Re} s \geq 1 + \delta$. The function admits the integral representation

$$\zeta(s) = \frac{1}{\Gamma(s)} \int_0^{\infty} \frac{x^{s-1} e^{-x}}{1 - e^{-x}} \, dx.$$

Euler produced the infinite product

$$\zeta(s) = \prod_p (1 - p^{-s})^{-1}$$

where the product extends over all prime numbers. This formula shows that $\zeta(s)$ has no zeros in the open half-plane $\operatorname{Re} s > 1$. The auxiliary function

$$\xi(s) = \frac{1}{2} \pi^{-s/2} s(s - 1) \Gamma\left(\frac{s}{2}\right) \zeta(s)$$

is analytic and satisfies the identity $\xi(s) = \xi(1 - s)$. This function now shows that $\zeta(s)$ has no zeros for $\operatorname{Re} s < 0$, aside for the so-called *trivial zeros* at $s = -2, -4, -6, \ldots$ coming from the poles of $\Gamma(s/2)$. Thus all the non-trivial zeros lie on the strip $0 \leq \operatorname{Re} s \leq 1$. The *Riemann hypothesis* states that all the roots of $\zeta(s) = 0$ are on the *critical line* $\operatorname{Re} s = \frac{1}{2}$. At the end of §13.3 the authors state that:

It was conjectured by Riemann, but it has not yet been proved, that all the zeros of $\zeta(s)$ in this strip lie on the line $\sigma = \frac{1}{2}$; while it has quite recently been proved by Hardy [279] that an infinity of zeros of $\zeta(s)$ actually lie on $\sigma = \frac{1}{2}$. It is highly probable that Riemann's conjecture is correct, and the proof of it would have far-reaching consequences in the theory of Prime Numbers.

The reader will find in [93, 153, 458] more information about the Riemann hypothesis. In a remarkable new connection, it seems that the distribution of the zeros of $\zeta(s)$ is related to the eigenvalues of random matrices [367, 369].

This chapter establishes the identity

$$\zeta(2n) = \frac{(-1)^{n-1} B_{2n} (2\pi)^{2n}}{2 \, (2n)!}$$

where $n \in \mathbb{N}$ and B_{2n} is the Bernoulli number. This is a generalization of the so-called *Basel problem* $\zeta(2) = \pi^2/6$. The solution of this problem won the young Euler instant fame. It follows that $\zeta(2n)$ is a rational multiple of π^{2n}, therefore this is a transcendental number. The arithmetic properties of the odd zeta values are more difficult to obtain. Apéry proved in 1979 that $\zeta(3)$ is not a rational number; see [27, 72, 689]. It is still unknown whether $\zeta(5)$ is irrational, but Zudilin [688] proved that one of the numbers $\zeta(5)$, $\zeta(7)$, $\zeta(9)$, $\zeta(11)$ is irrational. It is conjectured that *all odd zeta values* are irrational.

The literature contains a large variety of extensions of the Riemann zeta function. The chapter contains information about some of them: the *Hurwitz zeta function*

$$\zeta(s,a) = \sum_{n=0}^{\infty} \frac{1}{(n+a)^s}, \quad \text{with } 0 < a \le 1$$

with integral representation

$$\zeta(s,a) = \frac{1}{\Gamma(s)} \int_0^{\infty} \frac{x^{s-1} e^{-ax}}{1 - e^{-x}} \, dx.$$

The chapter establishes the values of $\zeta(-m, a)$ in terms of derivatives of the Bernoulli polynomials and presents a proof of Lerch's theorem

$$\frac{d}{ds} \zeta(s,a) \bigg|_{s=0} = \log\left(\frac{\Gamma(a)}{\sqrt{2\pi}}\right).$$

The chapter mentions two further generalizations: one introduced by Lerch (see [414] for details)

$$\phi(x, a; s) = \sum_{n=0}^{\infty} \frac{e^{2\pi i n x}}{(n+a)^s},$$

and another one by Barnes [43, 44, 45, 46]

$$\zeta_N(s, w \mid a_1, \dots, a_N) = \sum_{n_1, \dots, n_N} \frac{1}{(w + n_1 a_1 + \cdots + n_N a_N)^s}.$$

The reader will find in [566] more recent information on this function.

Chapter 14. The Hypergeometric Function. This function is defined by the series

$$F(a, b; c, z) = \sum_{n=0}^{\infty} \frac{(a)_n (b)_n}{(c)_n n!} z^n,$$

provided c is not a negative integer. Here $(u)_n = \Gamma(u + n)/\Gamma(u)$ is the Pochhammer symbol. The series converges for $|z| < 1$ and on the unit circle $|z| = 1$ if $\text{Re}\,(c - a - b) > 0$. Many elementary functions can be expressed in hypergeometric form, for instance

$$F(1, 1; 1; z) = \frac{1}{1-z} \quad \text{and} \quad e^z = \lim_{b \to \infty} F\left(1, b; 1; \frac{z}{b}\right).$$

The chapter begins with Gauss' evaluation $F(a, b; c; 1)$ in the form

$$F(a, b; c; 1) = \frac{\Gamma(c)\Gamma(c - a - b)}{\Gamma(c - a)\Gamma(c - b)}.$$

The function F satisfies the differential equation

$$z(1-z)\frac{d^2u}{dz^2} + [c - (a+b+1)z]\frac{du}{dz} - abu = 0.$$

This equation has 0, 1, ∞ as regular singular points and every other point is ordinary. The generalization to singular points at a, b, c with exponents given by $\{\alpha, \alpha'\}$, $\{\beta, \beta'\}$, $\{\gamma, \gamma'\}$, respectively, is the *Riemann differential equation*

$$\frac{d^2w}{dz^2} + \left[\frac{1-\alpha-\alpha'}{z-a} + \frac{1-\beta-\beta'}{z-b} + \frac{1-\gamma-\gamma'}{z-c}\right]\frac{dw}{dz}$$

$$+ \left[\frac{\alpha\alpha'(a-b)(a-c)}{z-a} + \frac{\beta\beta'(b-c)(b-a)}{z-b} + \frac{\gamma\gamma'(c-a)(c-b)}{z-c}\right]\frac{w}{(z-a)(z-b)(z-c)} = 0.$$

It is shown that

$$\left(\frac{z-a}{z-b}\right)^\alpha \left(\frac{z-c}{z-b}\right)^\gamma F\left(\alpha+\beta+\gamma, \alpha+\beta'+\gamma; 1+\alpha-\alpha'; \frac{(z-a)(c-b)}{(z-b)(c-a)}\right)$$

solves the Riemann differential equation. Using the invariance of this equation with respect to some permutations of the parameters (for example, the exchange of α and α') produces from $F(a,b;c;z)$ Kummer's 24 new solutions of Riemann's equation, for example

$$(1-z)^{-a}F\left(a, c-b; c; \frac{z}{z-1}\right) \quad \text{and} \quad (1-z)^{-b}F\left(c-a, b; c; \frac{z}{z-1}\right).$$

Since the solutions of a second-order differential equation form a two dimensional vector space, this type of transformation can be used to generate identities among hypergeometric series. The reader will find in [20, 339, 534, 641] more details on these ideas. The corresponding equation with *four regular singular points* at 0, 1, ∞, a is called the *Heun equation*

$$\frac{d^2u}{dx^2} + \left[\frac{\gamma}{x} + \frac{\delta}{x-1} + \frac{\varepsilon}{x-a}\right]\frac{du}{dx} + \left[\frac{\alpha\beta x - q}{x(x-1)(x-a)}\right]u = 0.$$

The corresponding process on the symmetries of the equation now gives 192 solutions. These are described in [452]. The reader will find in [190] an example of the appearance of Heun's equation in integrable systems.

The chapter also contains a presentation of *Barnes' integral representation*

$$F(a,b;c;z) = \frac{1}{2\pi i}\int_{-i\infty}^{i\infty} \frac{\Gamma(a+s)\Gamma(b+s)\Gamma(-s)}{\Gamma(c+s)}(-z)^s \, ds$$

and its use in producing an analytic continuation of the hypergeometric series. Finally, the identities of Clausen

$$\left[F(a,b;a+b+\tfrac{1}{2};x)\right]^2 = {}_3F_2(2a, a+b, 2b; a+b+\tfrac{1}{2}, 2a+2b; x)$$

where ${}_3F_2$ is the analog of the hypergeometric series, now with three Pochhammer symbols on top and two in the bottom of the summand and Kummer's *quadratic transformation*

$$F(2a, 2b; a+b+\tfrac{1}{2}; x) = F(a, b; a+b+\tfrac{1}{2}; 4x(1-x))$$

appear as exercises in this chapter. The reader will find in [20] a detailed analysis of these topics.

Chapter 15. Legendre Functions. This chapter discusses *Legendre polynomials* $P_n(z)$ and some of their extensions. These days, the usual starting point for these functions is defining them as orthogonal polynomials on the interval $(-1, 1)$; that is,

$$\int_{-1}^{1} P_n(z)P_m(z)\, dz = 0 \quad \text{if } n \neq m,$$

plus some normalization in the case $n = m$. The starting point in this chapter is the generating function

$$(1 - 2zh + h^2)^{-1/2} = \sum_{n=0}^{\infty} P_n(z)h^n.$$

It is established from here that

$$P_n(z) = \sum_{r=0}^{\lfloor \frac{n}{2} \rfloor} (-1)^r \frac{(2n - 2r)!}{2^n r!(n - r)!(n - 2r)!} z^{n-2r}$$

showing that $P_n(z)$ is a polynomial of degree n with leading coefficient $2^{-n}\binom{2n}{n}$.

The properties of these polynomials established in this chapter include

Rodriguez formula

$$P_n(z) = \frac{1}{2^n\, n!} \left(\frac{d}{dz} \right)^n (z^2 - 1)^n.$$

Legendre's differential equation The polynomials $P_n(z)$ are solutions of the differential equation

$$(1 - z^2)\frac{d^2u}{dz^2} - 2z\frac{du}{dz} + n(n + 1)u = 0.$$

In the new scale $x = z^2$, this equation takes its hypergeometric form

$$x(1 - x)\frac{d^2y}{dx^2} + \frac{1}{2}(1 - 3x)\frac{du}{dx} + \frac{1}{4}n(n + 1)u = 0.$$

The (more convenient) hypergeometric form $P_n(z) = {}_2F_1(n + 1, -n; 1; \frac{1}{2}(1 - z))$ is also established.

Recurrences The chapter presents proofs of the recurrences

$$(n + 1)P_{n+1}(z) - (2n + 1)zP_n(z)zP_n(z) - nzP_n(z) = 0$$

and

$$P'_{n+1}(z) - zP'_n(z) - (n + 1)P_n(z) = 0.$$

Integral representations A variety of integral representations for the Legendre polynomials are presented:

- *Schläfli*:

$$P_n(z) = \frac{1}{2\pi i} \oint_C \frac{(t^2 - 1)^n}{2^n (t - z)^{n+1}} \, dt$$

where C is a contour enclosing z. This is then used to prove the *orthogonality relation*

$$\int_{-1}^{1} P_n(z) P_m(z) \, dz = \begin{cases} 0 & \text{if } n \neq m \\ 2/(2n + 1) & \text{if } n = m. \end{cases}$$

- *Laplace*:

$$P_n(z) = \frac{1}{\pi} \int_0^{\pi} \left[z + (z^2 - 1)^{1/2} \cos \theta \right]^n \, d\theta$$

- *Mehler–Dirichlet*:

$$P_n(\cos \theta) = \frac{1}{\pi} \int_{-\theta}^{\theta} \frac{e^{(n+\frac{1}{2})i\varphi}}{(2 \cos \varphi - 2 \cos \theta)^{1/2}} \, d\varphi.$$

The formula of Schläfli given above is then used to extend the definition of $P_n(z)$ for $n \notin \mathbb{N}$. In order to obtain a single-valued function, the authors introduce a cut from -1 to $-\infty$ in the domain of integration.

Since the differential equation for the Legendre polynomials is of second order, it has a second solution independent of $P_n(z)$. This is called the *Legendre function of degree n of the second type*. It is denoted by $Q_n(z)$. The chapter discusses integral representations and other properties similar to those described for $P_n(z)$. For example, one has the hypergeometric expression

$$Q_n(z) = \frac{\sqrt{\pi} \Gamma(n + 1)}{2^{n+1} \Gamma \left(n + \frac{3}{2} \right)} \frac{1}{z^{n+1}} F\left(\frac{n + 1}{2}, \frac{n}{2} + 1; n + \frac{3}{2}; z^{-2} \right).$$

One obtains from here

$$Q_0(z) = \frac{1}{2} \log \frac{z + 1}{z - 1}$$

$$Q_1(z) = \frac{1}{2} z \log \frac{z + 1}{z - 1} - 1.$$

In general $Q_n(z) = A_n(z) + B_n(z) \log \frac{z+1}{z-1}$ for polynomials A_n, B_n.

The chapter also includes further generalizations of the Legendre functions introduced by Ferrer and Hobson. These are called *associated Legendre functions*. Some of their properties are presented. There is also a discussion of the *addition theorem* for Legendre polynomial, as well as a short section on the Gegenbauer function. The reader will find in the *Digital Library of Mathematical Functions* developed at NIST [443] more information about these functions.

Chapter 16. The Confluent Hypergeometric Function. This chapter discusses the second-order differential equation with singularities at $\{0, \infty, c\}$ and corresponding exponents $\{\{\frac{1}{2} + m, \frac{1}{2} - m\}, \{-c, 0\}, \{c - k, k\}\}$ in the limiting situation $c \to \infty$. This is the case of *confluent singularities* (the limiting equation now has only two singularities: 0 and ∞, with 0 remaining regular and ∞ becomes an irregular singularity). After a change of variables to eliminate the first derivative term, the limiting equation becomes

$$\frac{d^2W}{dz^2} + \left(-\frac{1}{4} + \frac{k}{z} + \frac{\frac{1}{4} - m^2}{z^2}\right) W = 0.$$

This is called *Whittaker equation*.

The authors introduce the functions

$$M_{k,m}(z) = z^{1/2+m} e^{-z/2} \left(1 + \frac{\frac{1}{2} + m - k}{1!\,(2m+1)} z + \frac{(\frac{1}{2} + m - k)(\frac{3}{2} + m - k)}{2!\,(2m+1)(2m+2)} z^2 + \cdots\right)$$

and show that, when $2m \notin \mathbb{N}$, the functions $M_{k,m}(z)$ and $M_{k,-m}(z)$ form a fundamental set of solutions.

It turns out that it is more convenient to work with the functions $W_{k,m}(z)$ defined by the integral representation

$$W_{k,m}(z) = \frac{z^k e^{-k/2}}{\Gamma(\frac{1}{2} - k + m)} \int_0^\infty t^{-k-1/2+m} \left(1 + \frac{t}{z}\right)^{k-1/2+m} e^{-t}\, dt.$$

The reader is referred to [59, Chapter 6] for a readable description of the basic properties of these functions, called *Whittaker functions* in the literature.

The chapter also presents a selection of special functions that can be expressed in terms of $W_{k,m}(z)$. This includes the *incomplete gamma function*

$$\gamma(a,x) = \int_0^x t^{a-1} e^{-t}\, dt$$

that can be expressed as

$$\gamma(a,x) = \Gamma(a) - x^{(a-1)/2} e^{-x/2} W_{\frac{1}{2}(a-1),\frac{1}{2}a}(x),$$

as well as the *logarithmic integral function*, defined by

$$\mathrm{li}(z) = \int_0^x \frac{dt}{\log t} = -(-\log z)^{-1/2} z^{1/2} W_{-\frac{1}{2},0}(-\log z).$$

This function appears in the description of the asymptotic behavior of the function

$$\pi(x) = \text{number of primes } p \le x.$$

The *prime number theorem* may be written as $\pi(x) \sim \mathrm{li}(x)$ as $x \to \infty$. See [191] for details. The final example is the function

$$D_n(z) = 2^{n/2+1/4} z^{-1/2} W_{\frac{n}{2}+\frac{1}{4},-\frac{1}{4}}\left(\frac{z^2}{2}\right),$$

related in a simple manner to the *Hermite polynomials*, defined by

$$H_n(z) = (-1)^n e^{z^2/2} \left(\frac{d}{dz}\right)^n e^{-z^2/2}.$$

See [20] for information on this class of orthogonal polynomials.

Chapter 17. Bessel Functions. This chapter discusses the *Bessel functions* defined, for $n \in \mathbb{Z}$, by the expansion

$$\exp\left(\frac{z}{2}\left(t - \frac{1}{t}\right)\right) = \sum_{n=-\infty}^{\infty} J_n(z)t^n.$$

Some elementary properties of $J_n(z)$ are derived directly from this definition, such as $J_{-n}(z) = (-1)^n J_n(z)$, the series

$$J_n(z) = \sum_{r=0}^{\infty} \frac{(-1)^r}{r!\,(n+r)!}\left(\frac{z}{2}\right)^{n+2r},$$

and the *addition theorem*

$$J_n(y+z) = \sum_{m=-\infty}^{\infty} J_m(y)J_{n-m}(z).$$

The Cauchy integral formula is then used to produce the representation

$$J_n(z) = \frac{1}{2\pi i}\left(\frac{z}{2}\right)^n \oint_C t^{-n-1} e^{t-z^2/4t}\, dt,$$

where C is a closed contour enclosing the origin. From here it is possible to extend the definition of $J_n(z)$ to values $n \notin \mathbb{Z}$ and produce the series representation

$$J_n(z) = \sum_{r=0}^{\infty} \frac{(-1)^r z^{n+2r}}{2^{n+2r} r!\,\Gamma(n+r+1)}.$$

This function is called the *Bessel function of the first kind of order n*. The integral representation of $J_n(z)$ is then used to show that $y(z) = J_n(z)$ is a solution of the differential equation

$$\frac{d^2 y}{dz^2} + \frac{1}{z}\frac{dy}{dz} + \left(1 - \frac{n^2}{z^2}\right)y = 0,$$

called the *Bessel equation*. In the case $n \notin \mathbb{Z}$, the functions $J_n(z)$ and $J_{-n}(z)$ form a basis for the space of solutions. In the case $n \in \mathbb{Z}$ a second solution, independent of $J_n(z)$, is given by

$$Y_n(z) = \lim_{\varepsilon \to 0} 2\pi e^{\pi i(n+\varepsilon)}\left(\frac{J_{n+\varepsilon}(z)\cos(\pi(n+\varepsilon)) - J_{-(n+\varepsilon)}(z)}{\sin(2\pi(n+\varepsilon))}\right).$$

The functions $Y_n(z)$ are called the *Bessel functions of the second kind*.

This chapter also contains some information on some variations of the Bessel function such as

$$I_n(z) = i^{-n} J_n(iz) \quad \text{and} \quad K_n(z) = \frac{\pi}{2}\left[I_{-n}(z) - I_n(z)\right]\cot(\pi n).$$

Among the results presented here one finds

Recurrences such as

$$J_{n-1}(z) + J_{n+1}(z) = \frac{2n}{z} J_n(z),$$

$$J_n'(z) = \frac{n}{z} J_n(z) - J_{n+1}(z)$$

and

$$z^{-n-1} J_{n+1}(z) = -\frac{1}{z}\frac{d}{dz}\left[z^{-n} J_n(z)\right]$$

which produces relations of Bessel functions of consecutive indices.

Zeros of Bessel functions it is shown that between any two non-zero consecutive roots of $J_n(z) = 0$ there is a unique root of $J_{n+1}(z) = 0$.

Integral representations such as

$$J_n(z) = \frac{1}{\pi}\int_0^\pi \cos(n\theta - z\sin\theta)\,d\theta - \frac{\sin\pi n}{\pi}\int_0^\infty e^{-n\theta - z\sinh\theta}\,d\theta,$$

where, for $n \in \mathbb{Z}$, the second term vanishes.

Hankel representation in the form

$$J_n(z) = \frac{\Gamma(\frac{1}{2} - n)}{2\pi i \sqrt{\pi}}\left(\frac{z}{2}\right)^n \int_C (t^2 - 1)^{n-1/2}\cos(zt)\,dt$$

where C is a semi-infinite contour on the real line.

Evaluation of definite integrals such as one due to Mahler

$$K_0(x) = \int_0^\infty \frac{t}{1 + t^2} J_0(tx)\,dt$$

and an example due to Sonine giving an expression for

$$\int_0^\infty x^{1-m} J_m(ax) J_m(bx) J_m(cx)\,dx.$$

A large selection of integrals involving Bessel functions may be found in [105], [258] and [544].

Series expansion The chapter also contains information about expansions of a function $f(z)$ in a series of the form

$$f(z) = \sum_{n=0}^\infty a_n J_n(z) \quad\text{or}\quad f(z) = \sum_{n=0}^\infty a_n J_0(nz).$$

The reader will find in [20] and [59] more information on these functions at the level discussed in this chapter. Much more appears in the volume [653].

There are many problems whose solutions involve the Bessel functions. As a current problem of interest, consider the symmetric group $\mathfrak{S}_N$ of permutations π of N symbols. An increasing sequence of length k is a collection of indices $1 \leq i_1 < \cdots < i_k \leq N$ such that $\pi(i_1) < \pi(i_2) < \cdots < \pi(i_k)$. Define on $\mathfrak{S}_N$ a uniform probability distribution; that is, $\mathbb{P}(\pi) = 1/N!$ for each permutation π. Then the maximal length of an increasing subsequence of π is a random variable, denoted by $\ell_N(\pi)$, and its distribution is of interest. This is the *Ulam problem*. Introduce the centered and scaled function

$$\chi_N(\pi) = \frac{\ell_N(\pi) - 2\sqrt{N}}{N^{1/6}}.$$

Baik–Deift–Johannson [36] proved that $\lim_{N\to\infty} \mathbb{P}(\chi_N(\pi) \le x) = F(x)$, where $F(x)$, the so-called *Tracy–Widom distribution*, is given by

$$F(x) = \exp\left(-\int_x^\infty (y-x)u^2(y)\,dy\right).$$

Here $u(x)$ is the solution of the Painlevé P_{II} equation $u''(x) = 2u^3(x) + xu(x)$, with asymptotic behavior $u(x) \sim \mathrm{Ai}(x)$ as $x \to \infty$. The *Airy function* $\mathrm{Ai}(x)$ is defined by $\mathrm{Ai}(x) = \sqrt{x}K_{1/3}\left(\frac{2}{3}x^{3/2}\right)/\pi\sqrt{3}$. The reader will find in [37] an introduction to this fascinating problem.

Chapter 18. The Equations of Mathematical Physics. This chapter contains a brief description of methods of solutions for the basic equations encountered in Mathematical Physics. The results are given for *Laplace's equation*

$$\Delta V = \frac{\partial^2 V}{\partial x^2} + \frac{\partial^2 V}{\partial y^2} + \frac{\partial^2 V}{\partial z^2}$$

on a domain $\Omega \subset \mathbb{R}^3$. The chapter has a presentation on the physical problems modeled by this equation.

The results include the integral representation of the solution

$$V(x,y,z) = \int_{-\pi}^{\pi} f(z + ix\cos u + iy\sin u, u)\,du$$

as the 3-dimensional analog of the form $V(x,y) = f(x+iy) + g(x-iy)$ valid in the 2-dimensional case as well as an expression for $V(x,y,z)$ as a series with terms of the form

$$\int_{-\pi}^{\pi} (z + ix\cos u + iy\sin u)^n \begin{pmatrix} \cos mu \\ \sin mu \end{pmatrix} du.$$

This series is then converted into one of the form

$$V = \sum_{n=0}^{\infty} r^n \left\{ A_n P_n(\cos\theta) + \sum_{m=1}^{\infty} \left(A_n^{(m)}\cos m\phi + B_n^{(m)}\sin m\phi\right) P_n^m(\cos\theta)\right\}$$

where P_n^m is Ferrer's version of the associated Legendre function.

The chapter also contains similar results for Laplace's equation on a sphere. For this type of domain, the authors obtain the formula

$$V(r,\theta,\phi) = \frac{a(a^2-r^2)}{4\pi}\int_{-\pi}^{\pi}\int_0^{\pi}\frac{f(\theta',\phi')\sin\theta'\,d\theta'\,d\phi'}{[r^2 - 2ar\{\cos\theta\cos\theta' + \sin\theta\sin\theta'\cos(\phi-\phi')\} + a^2]^{3/2}},$$

and refer to Thompson and Tait [628] for further discussions on the *theory of Green's functions*. A similar analysis for an equation on a cylinder also appears in this chapter. In that case the Legendre functions are replaced by Bessel functions. Some of the material discussed in this chapter has become standard in basic textbooks in Mathematical Physics; see for instance [476].

Chapter 19. Mathieu Functions. This chapter discusses the *wave equation* $V_{tt} = c^2\Delta V$ and assuming a special form $V(x,y,t) = u(x,y)\cos(pt+\varepsilon)$ of the unknown V in a special system of coordinates (ξ, η) (introduced by Lamé) yields an equation for $u(x, y)$. Using the classical method of separation of variables $(u * x, y) = F(\xi)G(\eta)$ produces the equation

$$\frac{d^2 y}{dz^2} + (a + 16q \cos(2z))\, y = 0.$$

This is called *Mathieu's equation*. The value of a is determined by the periodicity condition $G(\eta + 2\pi) = G(\eta)$ and q is determined by a vanishing condition at the boundary. This type of equation is now called *Hill's equation*, considered by Hill [306] in a study on lunar motion. Details about this equation appear in [451] and connections to integrable systems appear in [459, 462, 463].

The authors show that $G(\eta)$ satisfies the integral equation

$$G(\eta) = \lambda \int_{-\pi}^{\pi} e^{k \cos \eta \cos \theta} G(\theta)\, d\theta$$

and this λ must be a characteristic value as described in Chapter 11.

A sequence of functions, named *Mathieu functions*, are introduced from the study of Mathieu's equation. In the case $q = 0$, the solutions are $\{1, \cos nz, \sin nz\}_{n \in \mathbb{N}}$, and via Fourier series the authors introduce functions $\{ce_0(z, q), ce_n(z, q), se_n(z, q)\}_{n \in \mathbb{N}}$, reducing to the previous set as $q \to 0$. Some expressions for the first coefficients in the Fourier series of these functions are produced (it looks complicated to obtain exact expressions for them).

The authors present basic aspects of *Floquet theory* (more details appear in [451]). One looks for solution of Mathieu's equation in the form $y(z) = e^{\mu z} \phi(z)$, with ϕ periodic. The values of μ producing such solutions are obtained in terms of a determinant (called the *Hill determinant*). The modern theory yields these values in terms of a discriminant attached to the equation. The chapter also discusses results of Lindemann, transforming Mathieu's equation into the form

$$4\xi(1 - \xi)u'' + 2(1 - 2\xi)u' + (a - 16q + 32q\xi)u = 0.$$

This equation is *not* of hypergeometric type: the points 0, 1 are regular, but ∞ is an irregular singular point. Finally, the chapter includes some description of the asymptotic behavior of Mathieu functions. More details appear in [34] and [509].

Chapter 20. Elliptic Functions. General Theorems and the Weierstrassian Functions. Consider two complex numbers ω_1, ω_2 with non-real ratio. An *elliptic function* is a doubly-periodic functions: $f(z + 2\omega_1) = f(z + 2\omega_2) = f(z)$ where its singularities are at worst poles. The chapter discusses basic properties of the class $\mathcal{E}$ of elliptic functions. It is simple to verify that $\mathcal{E}$ is closed under differentiation and that the values of $f \in \mathcal{E}$ are determined by its values on the parallelogram with vertices 0, $2\omega_1$, $2\omega_1 + 2\omega_2$, $2\omega_2$. (Observe the factor of 2 in the periods.) This is called a *fundamental cell* and is denoted by $\mathbb{L}$. One may always assume that there are no poles of the function on the boundary of the cell. The first type of results deal with basic properties of an elliptic function:

(1) the number of poles is always finite; the same is true for the number of solutions of $f(z) = c$. This is independent of $c \in \mathbb{C}$ and is called the *degree* of the function f.

(2) any elliptic function without poles must be constant.

This result is used throughout the chapter to establish a large number of identities. The fundamental example

$$\wp(z) = \frac{1}{z^2} + \sum \left(\frac{1}{(z-\omega)^2} - \frac{1}{\omega^2} \right)$$

where the sum runs over all non-zero $\omega = 2n\omega_1 + 2m\omega_2$, was introduced by Weierstrass. It is an elliptic function of order 2. It has a double pole at $z = 0$. It is an even function, so its zeros in the fundamental cell are of the form $\pm z_0 \mod \mathbb{L}$. A remarkably recent formula for z_0 is given by Eichler and Zagier [192]. The $\wp$ function satisfies a differential equation

$$\left(\frac{d\wp(z)}{dz} \right)^2 = 4\wp(z)^3 - g_2\wp(z) - g_3,$$

where g_2, g_3 are the so-called *invariants* of the lattice $\mathbb{L}$. This function is then used to parametrize the algebraic curve $y^2 = 4x^3 + ax + b$, for $a, b \in \mathbb{C}$ with $a^3 + 27b^2 \neq 0$. The subject is also connected to differential equation by showing that if $y = \wp(z)$, then the inverse $z = \wp^{-1}(y)$ (given by an elliptic integral) can be written as the quotient of two solutions of

$$\frac{d^2v}{dy^2} + \left(\frac{3}{16} \sum_{r=1}^{3} (y - e_r)^{-2} - \frac{3}{8} y \prod_{r=1}^{3} (y - e_r)^{-1} \right) v = 0.$$

Here e_r are the roots of the cubic polynomial appearing in the differential equation for $\wp(z)$.

The *addition theorem*

$$\wp(z + y) = \frac{1}{4} \left[\frac{\wp'(z) - \wp'(y)}{\wp(z) - \wp(y)} \right]^2 - \wp(z) - \wp(y)$$

is established by a variety of methods. One presented by Abel deals with the intersection of the cubic curve $y^2 = 4x^3 + ax + b$ and a line and it is the basis for an *addition on the elliptic curve*, as the modern language states. Take two points $\mathfrak{a}$, $\mathfrak{b}$ on the curve and compute the line joining them. This line intersects the cubic at three points: the third is declared $-\mathfrak{a} \oplus \mathfrak{b}$. The points on the curve now form an abelian group: this is expected since the cubic may be identified with a torus $\mathbb{C}/\mathbb{L}$. The remarkable fact is that the addition of points preserves points with rational coordinates, so this set is also an abelian group. A theorem of Mordell and Weil states that this group is finitely generated. More information about the arithmetic of elliptic curves may be found in [331, 461, 592, 593]. The chapter also contains some information about two additional functions: the Weierstrass ζ-function, defined by $\zeta'(z) = -\wp(z)$ with $\lim_{z \to 0} \zeta(z) - 1/z = 0$ and the Weierstrass σ-function, defined by $(\log \sigma(z))' = \zeta(z)$ with $\lim_{z \to 0} \sigma(z)/z = 1$. These are the elliptic analogs of the cotangent and sine functions. The chapter contains some identities for them, for instance one due to Stickelberger: if $x + y + z = 0$, then

$$[\zeta(x) + \zeta(y) + \zeta(z)]^2 + \zeta'(x) + \zeta'(y) + \zeta'(z) = 0,$$

as well as the identity

$$\wp(z) - \wp(y) = -\frac{\sigma(z + y)\,\sigma(z - y)}{\sigma^2(z)\sigma^2(y)},$$

just to cite two of many. Among the many important results established in this chapter, we select three:

(1) any elliptic function f can be written in the form $R_1(\wp) + R_2(\wp)\wp'(z)$, with R_1, R_2 rational functions;
(2) every elliptic function f satisfies an *algebraic differential equation*;
(3) any curve of genus 1 can be parametrized by elliptic functions.

The chapter contains a brief discussion on *the uniformization of curves of higher genus*. This problem is discussed in detail in [7, 12, 477].

Chapter 21. The Theta-Functions. The study of the function

$$\vartheta(z, q) = \sum_{n=-\infty}^{\infty} (-1)^n q^{n^2} e^{2niz}$$

with $q = \exp(\pi i \tau)$ and $\operatorname{Im} \tau > 0$ was initiated by Euler and perfected by Jacobi in [349]. This is an example of a *theta function*. It is a non-constant analytic function of $z \in \mathbb{C}$, so it cannot be elliptic, but it has a simple transformation rule under $z \mapsto z + \tau$. This chapter considers ϑ, relabelled as ϑ_1 as well as three other companion functions ϑ_2, ϑ_3 and ϑ_4. These functions have a single zero in the fundamental cell $\mathbb{L}$ and since they transform in a predictable manner under the elements of $\mathbb{L}$, it is easy to produce elliptic functions from them. This leads to a remarkable series of identities such as

$$\vartheta_3(z, q) = \vartheta_3(2z, q^4) + \vartheta_2(2z, q^4)$$

and

$$\vartheta_2^4(0, q) + \vartheta_4^4(0, q) = \vartheta_3^4(0, q)$$

that represents a parametrization of the Fermat projective curve $x^4 + y^4 = z^4$. The chapter also discusses the *addition theorem*

$$\vartheta_3(z + y)\vartheta_3(z - y)\vartheta_3^2(0) = \vartheta_3^2(y)\vartheta_3^2(z) + \vartheta_1^2(y)\vartheta_1^2(z)$$

(where the second variable q has been omitted) as well as an identity of Jacobi

$$\vartheta_1'(0) = \vartheta_2(0)\vartheta_3(0)\vartheta_4(0).$$

This corresponds to the *triple product identity*, written as

$$\prod_{n=1}^{\infty} (1 - q^{2n})(1 + q^{2n-1}p^2)(1 + q^{2m-1}p^{-2}) = \sum_{n=-\infty}^{\infty} q^{n^2} p^{2n},$$

using the representation of theta functions as infinite products. The literature contains a variety of proofs of this fundamental identity; see Andrews [19] for a relatively simple one, Lewis [433] and Wright [683] for enumerative proofs and [311] for more general information on the so-called *q-series*. The chapter also shows that a quotient of theta functions ξ satisfies the differential equation

$$\left(\frac{d\xi}{d\tau}\right)^2 = \left(\vartheta_2^2(0) - \xi^2 \vartheta_3^2(0)\right) \left(\vartheta_3^2(0) - \xi^2 \vartheta_2^2(0)\right).$$

This is Jacobi's version of the differential equation satisfied by the Weierstrass $\wp$-function. The properties of solutions of this equation form the subject of the next chapter.

The reader will find in Baker [39, 40] a large amount of information on these functions from the point of view of the 19th century, Mumford [478, 479, 480] for a more modern point of view and [208, 209] for their connections to Riemann surfaces. Theta functions appeared scattered in the magnificent collection by Berndt [60, 61, 62, 63, 64] and Andrews–Berndt [21, 22, 23, 24, 25] on the formulas stated by Ramanujan.

Chapter 22. The Jacobian Elliptic Functions. Each elliptic function f has a *degree* attached to it. This is defined as the number of solutions to $f(z) = c$ in a fundamental cell. Constants have degree 0 and there are no functions of degree 1. A function of degree 2 either has a double pole (say at the origin) or two simple poles. The first case corresponds to the Weierstrass $\wp$ function described in Chapter 20. The second case is discussed in this chapter. The starting point is to show that any such function $y = y(u)$ may be written as a quotient of theta functions. From here the authors show that y must satisfy the equation

$$\left(\frac{dy}{du}\right)^2 = \left(1 - y^2\right)\left(1 - k^2 y^2\right)$$

where $k \in \mathbb{C}$ is the *modulus*. An expression for k as a ratio of null-values of theta functions is provided. Then $y = y(u)$ is seen to come from the inversion of the relation

$$u = \int_0^y \left(1 - t^2\right)^{-1/2} \left(1 - k^2 t^2\right)^{-1/2} dt$$

and, following Jacobi, the function y is called the *sinus amplitudinis* and is denoted by $y = \operatorname{sn}(u, k)$. This function becomes the trigonometrical $y = \sin u$ when $k \to 0$. Two companion functions $\operatorname{cn}(u, k)$ and $\operatorname{dn}(u, k)$ are also introduced. These functions satisfy a system of nonlinear differential equations

$$\dot{X} = YZ, \quad \dot{Y} = -ZX, \quad \dot{Z} = -k^2 XY,$$

and they are shown to parametrize the curve $\xi^2 = (1 - \eta^2)(1 - k^2 \eta^2)$.

The chapter also contains an addition theorem for these functions, such as

$$\operatorname{sn}(u + v) = \frac{\operatorname{sn} u \operatorname{cn} v \operatorname{dn} v + \operatorname{sn} v \operatorname{cn} u \, dnu}{1 - k^2 \operatorname{sn}^2 u \operatorname{sn}^2 v}$$

and other similar expressions.

The complete elliptic integral of the first kind $K(k)$ (and the complementary one $K'(k)$) appears here from $\operatorname{sn}(K(k), k) = 1$. The authors establish an expression for $K(k)$ in terms of theta values, prove Legendre's identity

$$\frac{d}{dk}\left(k(k')^2 \frac{dK}{dk}\right) = kK,$$

and present a discussion of the periods of the (Jacobian elliptic) functions sn, cn, dn in terms of elliptic integrals. The reader will find details of these properties in [90, 461]. Other results appearing here include product representations of Jacobi functions, the *Landen transformation* and several definite integrals involving these functions. There is also a discussion on the so-called *singular values*: these are special values of the modulus k such that the ratio

$K'(k)/K(k)$ has the form $(a + b\sqrt{n})/(c + d\sqrt{n})$ with a, b, c, d and $n \in \mathbb{Z}$. These values of k satisfy polynomial equations with integer coefficients. The authors state that the study of these equation *lies beyond the scope of this book*. The reader will find information about these equations in [90].

Chapter 23. Ellipsoidal Harmonics and Lamé's Equation. This chapter presents the basic theory of *ellipsoidal harmonics*. It begins with the expression

$$\Theta_p = \frac{x^2}{a^2 + \theta_p} + \frac{y^2}{b^2 + \theta_p} + \frac{z^2}{c^2 + \theta_p} - 1$$

where $a > b > c$ are the semi-axis of the ellipsoid $\Theta_p = 0$. A function of the form $\Pi_m(\Theta) = \Theta_1 \cdots \Theta_m$ is called an *ellipsoidal harmonic of the first species*. The chapter describes harmonic functions (that is, one satisfying $\Delta u = 0$) of this form. It turns out that every such function (with n even) has the form

$$\prod_{p=1}^{n/2} \left(\frac{x^2}{a^2 + \theta_p} + \frac{y^2}{b^2 + \theta_p} + \frac{z^2}{c^2 + \theta_p} - 1 \right)$$

where $\theta_1, \ldots, \theta_{n/2}$ are zeros of a polynomial $\Lambda(\theta)$ of degree $n/2$. This polynomial solves the *Lamé equation*

$$4\sqrt{(a^2 + \theta)(b^2 + \theta)(c^2 + \theta)} \frac{d}{d\theta} \left[\sqrt{(a^2 + \theta)(b^2 + \theta)(c^2 + \theta)} \frac{d\Lambda}{d\theta} \right] = [n(n + 1)\theta + C] \Lambda(\theta).$$

The value C is constant and it is shown that there are $\frac{1}{2}n + 1$ possible choices. There are three other types of ellipsoidal harmonics with a similar theory behind them.

The chapter contains many versions of Lamé's equation: the *algebraic form*

$$\frac{d^2\Lambda}{d\lambda^2} + \frac{1}{2} \left(\frac{1}{a^2 + \lambda} + \frac{1}{b^2 + \lambda} + \frac{1}{c^2 + \lambda} \right) \frac{d\Lambda}{d\lambda} = \frac{[n(n + 1)\lambda + C]\Lambda}{4(a^2 + \lambda)(b^2 + \lambda)(c^2 + \lambda)}$$

as well as the *Weierstrass elliptic form*

$$\frac{d^2\Lambda}{du^2} = [n(n + 1)\wp(u) + B]\Lambda$$

and finally the *Jacobi elliptic form*

$$\frac{d^2\Lambda}{d\alpha^2} = \left[n(n + 1)k^2\mathrm{sn}^2\alpha + A \right]\Lambda.$$

These equations are used to introduce *Lamé functions*. These are used to show that there are $2n + 1$ ellipsoidal harmonics that form a fundamental system of the harmonic functions of degree n.

The chapter contains a brief comment on work by Heun [300, 301] mentioning the study of an equation with *four* singular points. The reader will find in Ronveaux [563] more information about this equation.

Part I

The Process of Analysis

1

Complex Numbers

1.1 Rational numbers

The idea of a set of numbers is derived in the first instance from the consideration of the set of *positive integral numbers*, or *positive integers*; that is to say, the numbers $1, 2, 3, 4, \ldots$. (Strictly speaking, a more appropriate epithet would be, not *positive*, but *signless*.) Positive integers have many properties, which will be found in treatises on the Theory of Integral Numbers; but at a very early stage in the development of Mathematics it was found that the operations of Subtraction and Division could only be performed among them subject to inconvenient restrictions; and consequently, in elementary Arithmetic, classes of numbers are constructed such that the operations of subtraction and division can always be performed among them.

To obtain a class of numbers among which the operation of subtraction can be performed without restraint we construct the class of *integers*, which consists of the class of positive integers (in the strict sense) $(+1, +2, +3, \ldots)$ and of the class of negative integers $(-1, -2, -3, \ldots)$ and the number 0.

To obtain a class of numbers among which the operations both of subtraction and of division can be performed freely, with the exception of division by the rational number 0, we construct the class of *rational numbers*. Symbols which denote members of this class are $\frac{1}{2}, 3, 0, -\frac{15}{7}$. We have thus introduced three classes of numbers, (i) the *signless integers*, (ii) the *integers*, (iii) the *rational numbers*.

It is not part of the scheme of this work to discuss the construction of the class of integers or the logical foundations of the theory of rational numbers. Such a discussion, defining a rational number as an ordered number-pair of integers in a similar manner to that in which a complex number is defined in §1.3 as an ordered number-pair of real numbers, will be found in Hobson [315, §1-12].

The extension of the idea of number, which has just been described, was not effected without some opposition from the more conservative mathematicians. In the latter half of the eighteenth century, Maseres (1731–1824) and Frend (1757–1841) published works on Algebra, Trigonometry, etc., in which the use of negative numbers was disallowed, although Descartes had used them unrestrictedly more than a hundred years before.

A rational number x may be represented to the eye in the following manner: If, on a straight line, we take an origin O and a fixed segment OP_1 (P_1 being on the right of O), we can measure from O a length OP_x such that the ratio OP_x/OP_1 is equal to x; the point P_x is taken on the right or left of O according as the number x is positive or negative. We may regard either the *point* P_x or the *displacement* OP_x (which will be written $\overline{OP_x}$) as representing the number x.

All the rational numbers can thus be represented by points on the line, but the converse is not true. For if we measure off on the line a length OQ equal to the diagonal of a square of which OP_1 is one side, it can be proved that Q does not correspond to any rational number.

Points on the line which do not represent rational numbers may be said to represent irrational numbers; thus the point Q is said to represent the irrational number $\sqrt{2} = 1.414213 \cdots$. But while such an explanation of the existence of irrational numbers satisfied the mathematicians of the eighteenth century and may still be sufficient for those whose interest lies in the applications of mathematics rather than in the logical upbuilding of the theory, yet from the logical standpoint it is improper to introduce geometrical intuitions to supply deficiencies in arithmetical arguments; and it was shewn by Dedekind [169] in 1858 that the theory of irrational numbers can be established on a purely arithmetical basis without any appeal to geometry.

1.2 Dedekind's theory of irrational numbers

The geometrical property of points on a line which suggested the starting point of the arithmetical theory of irrationals was that, if all points of a line are separated into two classes such that every point of the first class is on the right of every point of the second class, there exists one and only one point at which the line is thus severed. The theory, though elaborated in 1858, was not published before the appearance of Dedekind's tract [169]. Other theories are due to Weierstrass (see [642]) and Cantor [116].

Following up this idea, Dedekind considered rules by which a separation or *section* of *all* rational numbers into two classes can be made. This procedure formed the basis of the treatment of irrational numbers by the Greek mathematicians in the sixth and fifth centuries B.C. The advance made by Dedekind consisted in observing that a purely *arithmetical* theory could be built up on it.

These classes, which will be called the *L*-class and the *R*-class, or the left class and the right class, being such that they possess the following properties:

 (i) At least one member of each class exists.
(ii) Every member of the *L*-class is less than every member of the *R*-class.

It is obvious that such a section is made by any rational number x; and x is either the greatest number of the *L*-class or the least number of the *R*-class. But sections can be made in which no rational number x plays this part. Thus, since there is no rational number[1] whose square is 2, it is easy to see that we may form a section in which the *R*-class consists of the positive rational numbers whose squares exceed 2, and the *L*-class consists of all other rational numbers.

Then this section is such that the *R*-class has no least member and the *L*-class has no greatest member; for, if x be any positive rational fraction, and $y = \frac{x(x^2+6)}{3x^2+2}$, then $y - x = \frac{2x(2-x^2)}{3x^2+2}$ and $y^2 - 2 = \frac{(x^2-2)^3}{(3x^2+2)^2}$, so x^2, y^2 and 2 are in order of magnitude; and therefore given any member

---

[1] For if p/q be such a number, this fraction being in its lowest terms, it may be seen that $(2q - p)/(p - q)$ is another such number, and $0 < p - q < q$, so that p/q is not in its lowest terms. The contradiction implies that such a rational number does not exist.

x of the L-class, we can always find a greater member of the L-class, or given any member x' of the R-class, we can always find a smaller member of the R-class, such numbers being, for instance, y and y', where y' is the same function of x' as y of x.

If a section is made in which the R-class has a least member A_2, or if the L-class has a greatest member A_1, the section determines a *rational-real number*; which it is convenient to denote by the *same* symbol A_2 or A_1. This causes no confusion in practice.

If a section is made, such that the R-class has no least member and the L-class has no greatest member, *the section determines an irrational-real number*.

Note B. A. W. Russell [567] defines the class of real numbers as *actually being* the class of all L-classes; the class of real numbers whose L-classes have a greatest member corresponds to the class of rational numbers, and though the rational-real number x which corresponds to a rational number x is conceptually distinct from it, no confusion arises from denoting both by the same symbol.

If x, y are real numbers (defined by sections) we say that x is greater than y if the L-class defining x contains at least two members of the R-class defining y. If the classes had only one member in common, that member might be the greatest member of the L-class of x and the least member of the R-class of y.

Let $\alpha, \beta, \ldots$ be real numbers and let $A_1, B_1, \ldots$ be any members of the corresponding L-classes while $A_2, B_2, \ldots$ are any members of the corresponding R-classes. The classes of which $A_1, A_2, \ldots$ are respectively members will be denoted by the symbols $(A_1), (A_2), \ldots$.

Then the *sum* (written $\alpha + \beta$) of two real numbers α and β is defined as the real number (rational or irrational) which is determined by the L-class $(A_1 + B_1)$ and the R-class $(A_2 + B_2)$.

It is, of course, necessary to prove that these classes determine a section of the rational numbers. It is evident that $A_1 + B_1 < A_2 + B_2$ and that at least one member of each of the classes $(A_1 + B_1)$, $(A_2 + B_2)$ exists. It remains to prove that there is, at most, *one* rational number which is greater than every $A_1 + B_1$ and less than every $A_2 + B_2$; suppose, if possible, that there are two, x and y, $(y > x)$. Let α_1 be a member of (A_1) and let α_2 be a member of (A_2); and let N be the integer next greater than $(\alpha_2 - \alpha_1)/\{\frac{1}{2}(y - x)\}$. Take the last of the numbers $\alpha_1 + \frac{m}{N}(\alpha_2 - \alpha_1)$, (where $m = 0, 1, \ldots, N$), which belongs to (A_1) and the first of them which belongs to (A_2); let these two numbers be c_1, c_2. Then

$$c_2 - c_1 = \frac{1}{N}(\alpha_2 - \alpha_1) < \frac{1}{2}(y - x).$$

Choose d_1, d_2 in a similar manner from the classes defining β; then

$$c_2 + d_2 - c_1 - d_1 < y - x.$$

But $c_2 + d_2 \geq y$, $c_1 + d_1 \leq x$, and therefore $c_2 + d_2 - c_1 - d_1 \geq y - x$; we have therefore arrived at a contradiction by supposing that two rational numbers x, y exist belonging neither to $(A_1 + B_2)$ nor to $(A_2 + B_2)$.

If every rational number belongs either to the class $(A_1 + B_1)$ or to the class $(A_2 + B_2)$, then the classes $(A_1 + B_1), (A_2 + B_2)$ define an irrational number. If one rational number x exists belonging to neither class, then the L-class formed by x and $(A_1 + B_1)$ and the R-class $(A_2 + B_2)$ define the rational number-real x. In either case, the number defined is called the sum $\alpha + \beta$.

The difference $\alpha - \beta$ of two real numbers is defined by the L-class $(A_1 - B_2)$ and the R-class $(A_2 - B_1)$.

The product of two positive real numbers α, β is defined by the R-class $(A_2 B_2)$ and the L-class of all other rational numbers.

The reader will see without difficulty how to define the product of negative real numbers and the quotient of two real numbers; and further, it may be shewn that real numbers may be combined in accordance with the associative, distributive and commutative laws.

The aggregate of rational-real and irrational-real numbers is called the aggregate of real numbers; for brevity, rational-real numbers and irrational-real numbers are called rational and irrational numbers respectively.

1.3 Complex numbers

We have seen that a real number may be visualised as a displacement along a definite straight line. If, however, P and Q are any two points in a plane, the displacement $\overline{PQ}$ needs two real numbers for its specification; for instance, the differences of the coordinates of P and Q referred to fixed rectangular axes. If the coordinates of P be (ξ, η) and those of $Q(\xi + x, \eta + y)$, the displacement $\overline{PQ}$ may be described by the symbol $[x, y]$. We are thus led to consider the association of real numbers in ordered pairs. The order of the two terms distinguishes the ordered number-pair $[x, y]$ from the ordered number-pair $[y, x]$. The natural definition of the sum of two displacements $[x, y]$, $[x', y']$ is the displacement which is the result of the successive applications of the two displacements; it is therefore convenient to define the sum of two number-pairs by the equation

$$[x, y] + [x', y'] = [x + x', y + y'].$$

The product of a number-pair and a real number x' is then naturally defined by the equation

$$x' \times [x, y] = [x'x, x'y].$$

We are at liberty to define the *product* of two number-pairs in any convenient manner; but the only definition, which does not give rise to results that are merely trivial, is that symbolised by the equation

$$[x, y] \times [x', y'] = [xx' - yy', xy' + x'y].$$

It is then evident that

$$[x, 0] \times [x', y'] = [xx', xy'] = x \times [x', y']$$

and

$$[0, y] \times [x', y'] = [-yy', x'y] = y \times [-y', x'].$$

The geometrical interpretation of these results is that the effect of multiplying by the displacement $[x, 0]$ is the same as that of multiplying by the real number x; but the effect of multiplying a displacement by $[0, y]$ is to multiply it by a real number y and turn it through a right angle.

It is convenient to denote the number-pair $[x, y]$ by the compound symbol $x + iy$; and a number-pair is now conveniently called (after Gauss) a *complex number*; in the fundamental operations of Arithmetic, the complex number $x + i0$ may be replaced by the real number x

and, defining i to mean $[0, 1]$, we have $i^2 = [0, 1] \times [0, 1] = [-1, 0]$; and so i^2 may be replaced by -1.

The reader will easily convince himself that the definitions of addition and multiplication of number-pairs have been so framed that we may perform the ordinary operations of algebra with complex numbers in exactly the same way as with real numbers, treating the symbol i as a number and replacing the product ii by -1 wherever it occurs.

Thus he will verify that, if a, b, c are complex numbers, we have

$$a + b = b + a,$$
$$ab = ba,$$
$$(a + b) + c = a + (b + c),$$
$$(ab)c = a(bc),$$
$$a(b + c) = ab + ac,$$

and if ab is zero, then either a or b is zero.

It is found that algebraical operations, direct or inverse, when applied to complex numbers, do not suggest numbers of any fresh type; the complex number will therefore for our purposes be taken as the most general type of number.

The introduction of the complex number has led to many important developments in mathematics. Functions which, when real variables only are considered, appear as essentially distinct, are seen to be connected when complex variables are introduced: thus the circular functions are found to be expressible in terms of exponential functions of a complex argument, by the equations

$$\cos x = \frac{1}{2} (e^{ix} + e^{-ix}), \quad \sin x = \frac{1}{2i} (e^{ix} - e^{-ix}).$$

Again, many of the most important theorems of modern analysis are not true if the numbers concerned are restricted to be real; thus, the theorem that every algebraic equation of degree n has n roots is true in general only when regarded as a theorem concerning complex numbers.

Hamilton's quaternions furnish an example of a still further extension of the idea of number. A quaternion

$$w + xi + yj + zk$$

is formed from four real numbers w, x, y, z, and four number-units $1, i, j, k$, in the same way that the ordinary complex number $x + iy$ might be regarded as being formed from two real numbers x, y, and two number-units $1, i$. Quaternions however do not obey the commutative law of multiplication.

1.4 The modulus of a complex number

Let $x + iy$ be a complex number, x and y being real numbers. Then the positive square root of $x^2 + y^2$ is called the *modulus* of $(x + iy)$, and is written

$$|x + iy|.$$

Let us consider the complex number which is the sum of two given complex numbers, $x + iy$ and $u + iv$. We have

$$(x + iy) + (u + iv) = (x + u) + i(y + v).$$

The modulus of the sum of the two numbers is therefore

$$\{(x+u)^2+(y+v)^2\}^{1/2} = \{(x^2+y^2)+(u^2+v^2)+2(xu+yv)\}^{1/2}.$$

But

$$\begin{aligned}
\{|\,x+iy\,|+|\,u+iv\,|\}^2 &= \{(x^2+y^2)^{1/2}+(u^2+v^2)^{1/2}\}^2 \\
&= (x^2+y^2)+(u^2+v^2)+2(x^2+y^2)^{1/2}(u^2+v^2)^{1/2} \\
&= (x^2+y^2)+(u^2+v^2)+2\,\{(xu+yv)^2+(xv-yu)^2\}^{1/2},
\end{aligned}$$

and this latter expression is greater than (or at least equal to)

$$(x^2+y^2)+(u^2+v^2)+2(xu+yv).$$

We have therefore

$$|x+iy|+|u+iv| \geq |(x+iy)+(u+iv)|,$$

i.e. *the modulus of the sum of two complex numbers cannot be greater than the sum of their moduli*; and it follows by induction that the modulus of the sum of any number of complex numbers cannot be greater than the sum of their moduli.

Let us consider next the complex number which is the product of two given complex numbers, $x+iy$ and $u+iv$, we have

$$(x+iy)(u+iv) = (xu-yv)+i(xv+yu),$$

and so

$$\begin{aligned}
|(x+iy)(u+iv)| &= \{(xu-yv)^2+(xv+yu)^2\}^{1/2} \\
&= \{(x^2+y^2)(u^2+v^2)\}^{1/2} \\
&= |x+iy||u+iv|.
\end{aligned}$$

The modulus of the product of two complex numbers (and hence, by induction, of any number of complex numbers) *is therefore equal to the product of their moduli.*

1.5 The Argand diagram

We have seen that complex numbers may be represented in a geometrical diagram by taking rectangular axes Ox, Oy in a plane. Then a point P whose coordinates referred to these axes are x, y may be regarded as representing the complex number $x+iy$. In this way, to every point of the plane there corresponds some one complex number; and, conversely, to every possible complex number there corresponds one, and only one, point of the plane. The complex number $x+iy$ may be denoted by a single letter z. It is convenient to call x and y the *real* and *imaginary* parts of z respectively. We frequently write $x = \operatorname{Re} z$, $y = \operatorname{Im} z$. The point P is then called the *representative point* of the number z; we shall also speak of the number z as being the *affix* of the point P.

If we denote $(x^2+y^2)^{1/2}$ by r and choose θ so that $r\cos\theta = x$, $r\sin\theta = y$, then r and θ are clearly the radius vector and vectorial angle of the point P, referred to the origin O and axis Ox.

The representation of complex numbers thus afforded is often called the *Argand diagram*.

It was published by J. R. Argand [33]; it had however previously been used by Gauss [235] in his Helmstedt dissertation in 1799, who had discovered it in Oct. 1797 [375]; and Caspar Wessel had discussed it in a memoir presented to the Danish Academy in 1797 and published by that Society in 1798–9 [664]. The phrase *complex number* first occurs in [237, p. 102].

By the definition already given, it is evident that r is the modulus of z. The angle θ is called the *argument* or *phase*, of z. We write $\theta = \arg z$.

From geometrical considerations, it appears that (although the modulus of a complex number is unique) the argument is not unique (see the *Appendix*, §A.521) if θ be a value of the argument, the other values of the argument of a complex number are comprised in the expression $2n\pi + \theta$ where n is any integer, not zero. The *principal value of the argument of a complex number* value of $\arg z$ is that which satisfies the inequality $-\pi < \arg z \leq \pi$.

If P_1 and P_2 are the representative points corresponding to values z_1 and z_2 respectively of z, then the point which represents the value $z_1 + z_2$ is clearly the terminus of a line drawn from P_1, equal and parallel to that which joins the origin to P_2.

To find the point which represents the complex number $z_1 z_2$, where z_1 and z_2 are two given complex numbers, we notice that if

$$z_1 = r_1(\cos \theta_1 + i \sin \theta_1),$$
$$z_2 = r_2(\cos \theta_2 + i \sin \theta_2)$$

then, by multiplication,

$$z_1 z_2 = r_1 r_2 \{\cos(\theta_1 + \theta_2) + i \sin(\theta_1 + \theta_2)\}.$$

The point which represents the number $z_1 z_2$ has therefore a radius vector measured by the product of the radii vectors of P_1 and P_2 and a vectorial angle equal to the sum of the vectorial angles of P_1 and P_2.

1.6 Miscellaneous examples

Example 1.1 Shew that the representative points of the complex numbers $1 + 4i$, $2 + 7i$, $3 + 10i$, are collinear.

Example 1.2 Shew that a parabola can be drawn to pass through the representative points of the complex numbers

$$2 + i, \ 4 + 4i, \ 6 + 9i, \ 8 + 16i, \ 10 + 25i.$$

Example 1.3 (Math. Trip. 1895). Determine the nth roots of unity by aid of the Argand diagram; and shew that the number of primitive roots (roots the powers of each of which give all the roots) is the number of integers (including unity) less than n and prime to it.

Prove that if $\theta_1, \theta_2, \theta_3, \ldots$ be the arguments of the primitive roots, $\sum \cos p\theta = 0$ when p is a positive integer less than $\dfrac{n}{abc \cdots k}$, where $a, b, c, \ldots, k$ are the different constituent primes of n; and that, when

$$p = \frac{n}{abc \cdots k}, \quad \text{then} \quad \sum \cos p\theta = \frac{(-1)^{\mu} n}{abc \cdots k},$$

where μ is the number of constituent primes.

2

The Theory of Convergence

2.1 The definition of the limit of a sequence

Let $z_1, z_2, z_3, \ldots$ be an unending sequence of numbers, real or complex. Then, if a number ℓ exists such that, corresponding to every positive[1] number ε, no matter how small, a number n_0 can be found, such that $|z_n - \ell| < \varepsilon$ for all values of n greater than n_0, *the sequence (z_n) is said to tend to the limit ℓ as n tends to infinity*. A definition equivalent to this was first given by John Wallis in 1655 [645, p. 382].

Symbolic forms of the statement, 'the limit of the sequence (z_n), as n tends to infinity, is ℓ' are:

$$\lim_{n \to \infty} z_n = \ell, \quad \lim z_n = \ell, \quad z_n \to \ell \quad as \quad n \to \infty.$$

The arrow notation is due to Leathem (see [416]).

If the sequence be such that, given an arbitrary number N (no matter how large), we can find n_0 such that $|z_n| > N$ for all values of n greater than n_0, we say that '$|z_n|$ tends to infinity as n tends to infinity', and we write $|z_n| \to \infty$. In the corresponding case when $-x_n > N$ when $n > n_0$ we say that $x_n \to -\infty$. If a sequence of real numbers does not tend to a limit or to ∞ or to $-\infty$, the sequence is said to *oscillate*.

2.11 Definition of the phrase 'of the order of'

If (ζ_n) and (z_n) are two sequences such that a number n_0 exists such that $|(\zeta_n/z_n)| < K$ whenever $n > n_0$, where K is *independent of n*, we say that ζ_n is 'of the order of' z_n, and we write $\zeta_n = O(z_n)$; thus $\frac{15n+19}{1+n^3} = O\left(\frac{1}{n^2}\right)$. This notation is due to Bachmann [35, p. 401] and Landau [405, p. 61].

Note If $\lim \frac{\zeta_n}{z_n} = 0$, we write $\zeta_n = o(z_n)$.

2.2 The limit of an increasing sequence

Let (x_n) be a sequence of real numbers such that $x_{n+1} \geq x_n$ for all values of n; then *the sequence tends to a limit or else tends to infinity* (and so it does not oscillate).

Let x be any rational-real number; then either:

(i) $x_n \geq x$ for all values of n greater than some number n_0 depending on the value of x; or
(ii) $x_n < x$ for every value of n.

---

[1] The number zero is excluded from the class of positive numbers.

If (ii) is not the case for *any* value of x (no matter how large), then $x_n \to \infty$.

But if values of x exist for which (ii) holds, we can divide the rational numbers into two classes, the L-class consisting of those rational numbers x for which (i) holds and the R-class of those rational numbers x for which (ii) holds. This section defines a real number α, rational or irrational.

And if ε be an arbitrary positive number, $\alpha - \frac{1}{2}\varepsilon$ belongs to the L-class which defines α, and so we can find n_1 such that $x_n \geq \alpha - \frac{1}{2}\varepsilon$ whenever $n > n_1$; and $\alpha + \frac{1}{2}\varepsilon$ is a member of the R-class and so $x_n < \alpha + \frac{1}{2}\varepsilon$. Therefore, whenever $n > n_1$, $|\alpha - x_n| < \varepsilon$. Therefore $x_n \to \alpha$.

Corollary 2.2.1 *A decreasing sequence tends to a limit or to* $-\infty$.

Example 2.2.1 If $\lim z_m = \ell$, $\lim z'_m = \ell'$, then $\lim(z_m + z'_m) = \ell + \ell'$.

For, given ε, we can find n and n' such that

(i) when $m > n$, $|z_m - \ell| < \frac{1}{2}\varepsilon$;
(ii) when $m > n'$, $|z'_m - \ell'| < \frac{1}{2}\varepsilon$.

Let n_1 be the greater of n and n'; then, when $m > n_1$,

$$\left|(z_m + z'_m) - (\ell + \ell')\right| \leq \left|(z_m - \ell)\right| + \left|(z'_m - \ell')\right|,$$
$$< \varepsilon;$$

and this is the condition that $\lim(z_m + z'_m) = \ell + \ell'$.

Example 2.2.2 Prove similarly that $\lim(z_m - z'_m) = \ell - \ell'$, $\lim(z_m z'_m) = \ell\ell'$, and, if $\ell' \neq 0$, $\lim(z_m / z'_m) = \ell/\ell'$.

Example 2.2.3 If $0 < x < 1$, $x^n \to 0$. For if $x = (1 + \alpha)^{-1}$, $\alpha > 0$, and

$$0 < x^n = \frac{1}{(1 + a)^n} < \frac{1}{1 + na},$$

by the binomial theorem for a positive integral index. And it is obvious that, given a positive number ε, we can choose n_0 such that $(1 + na)^{-1} < \varepsilon$ when $n > n_0$; and so $x^n \to 0$.

2.21 Limit-points and the Bolzano–Weierstrass theorem

This theorem, frequently ascribed to Weierstrass, was proved by Bolzano [81]. It seems to have been known to Cauchy.

Let (x_n) be a sequence of real numbers. If any number G exists such that, for every positive value of ε, no matter how small, an unlimited number of terms of the sequence can be found such that

$$G - \varepsilon < x_n < G + \varepsilon,$$

then G is called a *limit-point*, or *cluster-point* of the sequence.

Theorem 2.2.2 (Bolzano) *If* $\lambda \leq x_n \leq \rho$, *where* λ, ρ *are independent of* n, *then the sequence* (x_n) *has at least one limit-point.*

To prove the theorem, choose a section in which (i) the *R*-class consists of all the rational numbers which are such that, if *A* be any one of them, there are only a limited number of terms x_n satisfying $x_n > A$; and (ii) the *L*-class is such that there are an unlimited number of terms x_n such that $x_n \geq \alpha$ for all members α of the *L*-class.

This section defines a real number *G*; and, if ε be an arbitrary positive number, $G - \frac{1}{2}\varepsilon$ and $G + \frac{1}{2}\varepsilon$ are members of the *L*- and *R*-classes respectively, and so there are an unlimited number of terms of the sequence satisfying

$$G - \varepsilon < G - \frac{1}{2}\varepsilon \leq x_n \leq G + \frac{1}{2}\varepsilon < G + \varepsilon,$$

and so *G* satisfies the condition that it should be a limit-point.

2.211 Definition of 'the greatest and the least of the limits'

The number *G* obtained in §2.21 is called 'the greatest of the limits of the sequence (x_n).' The sequence (x_n) cannot have a limit-point greater than *G*; for if *G'* were such a limit-point, and $\varepsilon = \frac{1}{2}(G' - G)$, $G' - \varepsilon$ is a member of the *R*-class defining *G*, so that there are only a limited number of terms of the sequence which satisfy $x_n > G' - \varepsilon$. This condition is inconsistent with *G'* being a limit-point. We write

$$G = \overline{\lim_{n \to \infty}} x_n.$$

The 'least of the limits' *L*, of the sequence (written $\varliminf_{n \to \infty} x_n$) is defined to be

$$-\overline{\lim_{n \to \infty}} (-x_n).$$

2.22 Cauchy's theorem on the necessary and sufficient condition for the existence of a limit [120, p. 125]

We shall now shew that the necessary and sufficient condition for the existence of a limiting value of a sequence of numbers $z_1, z_2, z_3, \ldots$ is that, *corresponding to any given positive number ε, however small, it shall be possible to find a number n such that*

$$\left| z_{n+p} - z_n \right| < \varepsilon$$

for all positive integral values of p. This result is one of the most important and fundamental theorems of analysis. It is sometimes called the *Principle of Convergence*.

First, we have to shew that this condition is *necessary*, i.e. that it is satisfied whenever a limit exists. Suppose then that a limit ℓ exists; then (§2.1) corresponding to any positive number ε, however small, an integer *n* can be chosen such that

$$\left| z_n - \ell \right| < \tfrac{1}{2}\varepsilon, \qquad \left| z_{n+p} - \ell \right| < \tfrac{1}{2}\varepsilon,$$

for all positive values of *p*; therefore

$$\left| z_{n+p} - z_n \right| = \left| (z_{n+p} - \ell) - (z_n - \ell) \right|$$
$$\leq \left| z_{n+p} - \ell \right| + \left| z_n - \ell \right| < \varepsilon,$$

which shews the *necessity* of the condition

$$\left|z_{n+p} - z_n\right| < \varepsilon,$$

and thus establishes the first half of the theorem.

Second, we have to prove that this condition is *sufficient*, i.e. that if it is satisfied, then a limit exists. This proof is given in Stolz–Gmeiner [613, p. 144].

(I) Suppose that the sequence of *real* numbers (x_n) satisfies Cauchy's condition; that is to say that, corresponding to any positive number ε, an integer n can be chosen such that

$$\left|x_{n+p} - x_n\right| < \varepsilon$$

for all positive integral values of p.

Let the value of n, corresponding to the value 1 of ε, be m. Let λ_1, ρ_1 be the least and greatest of $x_1, x_2, \ldots, x_m$; then

$$\lambda_1 - 1 < x_n < \rho_1 + 1,$$

for all values of n; write $\lambda_1 - 1 = \lambda$, $\rho_1 + 1 = \rho$.

Then, for all values of n, $\lambda < x_n < \rho$. *Therefore by Theorem 2.2.2, the sequence (x_n) has at least one limit-point G.*

Further, there cannot be more than one limit-point; for if there were two, G and H $(H < G)$, take $\varepsilon < \frac{1}{4}(G - H)$. Then, by hypothesis, a number n exists such that $\left|x_{n+p} - x_n\right| < \varepsilon$ for every positive value of p. But since G and H are limit-points, positive numbers q and r exist such that

$$\left|G - x_{n+q}\right| < \varepsilon, \quad \left|H - x_{n+r}\right| < \varepsilon.$$

Then $\left|G - x_{n+q}\right| + \left|x_{n+q} - x_n\right| + \left|x_n - x_{n+r}\right| + \left|x_{n+r} - H\right| < 4\varepsilon$. But, by §1.4, the sum on the left is greater than or equal to $\left|G - H\right|$. Therefore $G - H < 4\varepsilon$, which is contrary to hypothesis; so there is only one limit-point. Hence there are only a finite number of terms of the sequence outside the interval $(G - \delta, G + \delta)$, where δ is an arbitrary positive number; for, if there were an unlimited number of such terms, these would have a limit-point which would be a limit-point of the given sequence and which would not coincide with G; *and therefore G is the limit of (x_n).*

(II) Now let the sequence (z_n) of real or complex numbers satisfy Cauchy's condition; and let $z_n = x_n + iy_n$, where x_n and y_n are real; then for all values of n and p

$$\left|x_{n+p} - x_n\right| \le \left|z_{n+p} - z_n\right|, \quad \left|y_{n+p} - y_n\right| \le \left|z_{n+p} - z_n\right|.$$

Therefore the sequences of real numbers (x_n) and (y_n) satisfy Cauchy's condition; and so, by (I), the limits of (x_n) and (y_n) exist. Therefore, by Example 2.2.1, the limit of (z_n) exists. The result is therefore established.

2.3 Convergence of an infinite series

Let $u_1, u_2, u_3, \ldots, u_n, \ldots$ be a sequence of numbers, real or complex. Let the sum

$$u_1 + u_2 + \cdots + u_n$$

be denoted by S_n.

Then, if S_n tends to a limit S as n tends to infinity, the infinite series

$$u_1 + u_2 + u_3 + u_4 + \cdots$$

is said to *be convergent*, or to *converge to the sum S*. In other cases, the infinite series is said to be *divergent*. When the series converges, the expression $S - S_n$, which is the sum of the series

$$u_{n+1} + u_{n+2} + u_{n+3} + \cdots,$$

is called the *remainder after n terms*, and is frequently denoted by the symbol R_n. The sum $u_{n+1} + u_{n+2} + \cdots + u_{n+p}$ will be denoted by $S_{n,p}$.

It follows at once, by combining the above definition with the results of the last paragraph, that the necessary and sufficient condition for the convergence of an infinite series is that, given an arbitrary positive number ε, we can find n such that $\left| S_{n,p} \right| < \varepsilon$ for every positive value of p.

Since $u_{n+1} = S_{n,1}$, it follows as a particular case that $\lim u_{n+1} = 0$, in other words, the nth term of a convergent series must tend to zero as n tends to infinity. But this last condition, though necessary, is not sufficient in itself to ensure the convergence of the series, as appears from a study of the series

$$\frac{1}{1} + \frac{1}{2} + \frac{1}{3} + \frac{1}{4} + \frac{1}{5} + \cdots.$$

In this series, $S_{n,n} = \dfrac{1}{n + 1} + \dfrac{1}{n + 2} + \dfrac{1}{n + 3} + \cdots + \dfrac{1}{2n}$. The expression on the right is diminished by writing $(2n)^{-1}$ in place of each term, and so $S_{n,n} > \frac{1}{2}$. Therefore

$$S_{2^{n+1}} = 1 + S_{1,1} + S_{2,2} + S_{4,4} + S_{8,8} + S_{16,16} + \cdots + S_{2^n,2^n}$$

$$> \frac{1}{2}(n + 3) \to \infty;$$

so the series is divergent; this result was noticed by Leibniz in 1673.

There are two general classes of problems which we are called upon to investigate in connexion with the convergence of series:

1. We may arrive at a series by some formal process, e.g. that of solving a linear differential equation by a series, and then to justify the process it will usually have to be proved that the series thus formally obtained is convergent. Simple conditions for establishing convergence in such circumstances are obtained in §§2.31–2.61.

2. Given an expression S, it may be possible to obtain a development $S = \sum\limits_{m=1}^{n} u_m + R_n$, valid for all values of n; and, from the definition of a limit, it follows that, if we can prove that $R_n \to 0$, then the series $\sum\limits_{m=1}^{\infty} u_m$ converges and its sum is S. An example of this problem occurs in §5.4.

Infinite series were used by Lord Brouncker in [103, pp. 645–649], and the term convergent was introduced by James Gregory, Professor of Mathematics at Edinburgh, in the same year; the term divergent was introduced by N. Bernoulli in 1713. Infinite series were used systematically by Newton [494, pp. 206–247], and he investigated the convergence of hypergeometric series (§14.1) in 1704. (See also the convergence of products in §2.7.) But the great mathematicians of the eighteenth century used infinite series freely without, for the most part, examining their convergence. Thus Euler gave the sum of the series

$$\cdots + \frac{1}{z^3} + \frac{1}{z^2} + \frac{1}{z} + 1 + z + z^2 + z^3 + \cdots \tag{a}$$

as zero, on the ground that

$$z + z^2 + z^3 + \cdots = \frac{z}{1 - z} \tag{b}$$

and

$$1 + \frac{1}{z} + \frac{1}{z^2} + \cdots = \frac{z}{z - 1} \tag{c}$$

The error of course arises from the fact that the series (b) converges only when $|z| < 1$, and the series (c) converges only when $|z| > 1$, so the series (a) never converges.

For the history of researches on convergence, see Pringsheim and Molk [543] and Reiff [551].

2.301 Abel's inequality.

This appears in [1, pp. 311–339]. A particular case of Corollary 2.3.3 also appears in that memoir.

Theorem 2.3.1 *Let $f_n \geq f_{n+1} > 0$ for all integer values of n. Then*

$$\left| \sum_{n=1}^{m} a_n f_n \right| \leq A f_1,$$

where A is the greatest of the sums

$$|a_1|, \; |a_1 + a_2|, \; |a_1 + a_2 + a_3|, \ldots, |a_1 + a_2 + \cdots + a_m|.$$

For, writing $a_1 + a_2 + \cdots + a_n = s_n$, we have

$$\sum_{n=1}^{m} a_n f_n = s_1 f_1 + (s_2 - s_1) f_2 + (s_3 - s_2) f_3 + \cdots + (s_m - s_{m-1}) f_m$$

$$= s_1(f_1 - f_2) + s_2(f_2 - f_3) + \cdots + s_{m-1}(f_{m-1} - f_m) + s_m f_m.$$

Since $f_1 - f_2, f_2 - f_3, \ldots$ are not negative, we have, when $n = 2, 3, \ldots, m$, $|s_{n-1}| (f_{n-1} - f_n) \leq A(f_{n-1} - f_n)$ also $|s_m| f_m \leq A f_m$, and so, summing and using §1.4, we get

$$\left| \sum_{n=1}^{m} a_n f_n \right| \leq A f_1.$$

Corollary 2.3.2 (Hardy)   *If $a_1, a_2, \ldots, w_1, w_2, \ldots$ are any numbers, real or complex,*

$$\left| \sum_{n=1}^{m} a_n w_n \right| \leq A \left\{ \sum_{n=1}^{m-1} |w_{n+1} - w_n| + |w_m| \right\},$$

where A is the greatest of the sums $\left| \sum_{n=1}^{p} a_n \right|$, $(p = 1, 2, \ldots, m)$.

2.31 Dirichlet's test for convergence

This appears in [177, pp. 253–255]. Before the publication of the Second edition of Jordan's *Cours d'Analyse* [361], Dirichlet's test and Abel's test were frequently jointly described as the Dirichlet–Abel test, see e.g. Pringsheim [537, p. 423].

Let $\left| \sum_{n=1}^{p} a_n \right| < K$, *where* K is independent *of* p. *Then, if* $f_n \geq f_{n+1} > 0$ *and* $\lim f_n = 0$, $\sum_{n=1}^{\infty} a_n f_n$ *converges.*

In these circumstances, we say $f_n \to 0$ *steadily*.

Proof For, since $\lim f_n = 0$, given an arbitrary positive number ε, we can find m such that $f_{m+1} < \varepsilon/2K$. Then

$$\left| \sum_{n=m+1}^{m+q} a_n \right| \leq \left| \sum_{n=1}^{m+q} a_n \right| + \left| \sum_{n=1}^{m} a_n \right| < 2K,$$

for all positive values of q; so that, by Abel's inequality, we have, for all positive values of p,

$$\left| \sum_{n=m+1}^{m+p} a_n f_n \right| \leq A f_{m+1},$$

where $A < 2K$.

Therefore $\left| \sum_{n=m+1}^{m+p} a_n f_n \right| < 2K f_{m+1} < \varepsilon$; and so, by §2.3, $\sum_{n=1}^{\infty} a_n f_n$ converges.

Corollary 2.3.3 *Abel's test for convergence. If* $\sum_{n=1}^{\infty} a_n$ *converges and the sequence* (u_n) *is monotonic (i.e.* $u_n \geq u_{n+1}$ *always or else* $u_n \leq u_{n+1}$ *always) and* $|u_n| < \kappa$, *where* κ *is independent of* n, *then* $\sum_{n=1}^{\infty} a_n u_n$ *converges.*

For, by §2.2, u_n tends to a limit u; let $| u - u_n | = f_n$. Then $f_n \to 0$ steadily; and therefore $\sum_{n=1}^{\infty} a_n f_n$ converges. But, if (u_n) is an increasing sequence, $f_n = u - u_n$, and so $\sum_{n=1}^{\infty} (u - u_n) a_n$ converges; therefore since $\sum_{n=1}^{\infty} u a_n$ converges, $\sum_{n=1}^{\infty} u_n a_n$ converges. If (u_n) is a decreasing sequence $f_n = u_n - u$, and a similar proof holds.

Corollary 2.3.4 *Taking* $a_n = (-1)^{n-1}$ *in Dirichlet's test, it follows that, if* $f_n \geq f_{n+1}$ *and* $\lim f_n = 0, f_1 - f_2 + f_3 - f_4 + \cdots$ *converges.*

Example 2.3.1 Shew that if $0 < \theta < 2\pi, \left| \sum_{n=1}^{p} \sin n\theta \right| < \operatorname{cosec} \tfrac{1}{2}\theta$; and deduce that, if $f_n \to 0$ steadily, $\sum_{n=1}^{\infty} f_n \sin n\theta$ converges for all real values of θ, and that $\sum_{n=1}^{\infty} f_n \cos n\theta$ converges if θ is not an even multiple of π.

Example 2.3.2 Shew that, if $f_n \to 0$ steadily, $\sum_{n=1}^{\infty} (-1)^n f_n \cos n\theta$ converges if θ is real and not an odd multiple of π and $\sum_{n=1}^{\infty} (-1)^n f_n \sin n\theta$ converges for all real values of θ. *Hint.* Write $\pi + \theta$ for θ in Example 2.3.1.

2.32 Absolute and conditional convergence

In order that a series $\sum\limits_{n=1}^{\infty} u_n$ of real or complex terms may converge, it is *sufficient* (but not necessary) that the series of moduli $\sum\limits_{n=1}^{\infty} |u_n|$ should converge. For, if $\sigma_{n,p} = |u_{n+1}| + |u_{n+2}| + \cdots + |u_{n+p}|$ and if $\sum\limits_{n=1}^{\infty} |u_n|$ converges, we can find n, corresponding to a given number ε, such that $\sigma_{n,p} < \varepsilon$ for all values of p. But $|S_{n,p}| \leq \sigma_{n,p} < \varepsilon$, and so $\sum\limits_{n=1}^{\infty} u_n$ converges.

The condition is not necessary; for, writing $f_n = 1/n$ in Corollary 2.3.4, we see that $\frac{1}{1} - \frac{1}{2} + \frac{1}{3} - \frac{1}{4} + \cdots$ converges, though (§2.3) the series of moduli $\frac{1}{1} + \frac{1}{2} + \frac{1}{3} + \frac{1}{4} + \cdots$ is known to diverge.

In this case, therefore, the divergence of the series of moduli does not entail the divergence of the series itself.

Series which are such that the series formed by the moduli of their terms are convergent, possess special properties of great importance, and are called *absolutely convergent* series. Series which though convergent are not absolutely convergent (i.e. the series themselves converge, but the series of moduli diverge) are said to be *conditionally convergent*.

2.33 The geometric series, and the series $\sum_{n=1}^{\infty} \frac{1}{n^s}$

The convergence of a particular series is in most cases investigated, not by the direct consideration of the sum $S_{n,p}$, but (as will appear from the following articles) by a comparison of the given series with some other series which is known to be convergent or divergent. We shall now investigate the convergence of two of the series which are most frequently used as standards for comparison.

(I) The geometric series. The geometric series is defined to be the series

$$1 + z + z^2 + z^3 + z^4 + \cdots .$$

Consider the series of moduli

$$1 + |z| + |z|^2 + |z|^3 + \cdots ; \tag{2.1}$$

for this series

$$S_{n,p} = |z|^{n+1} + |z|^{n+2} + \cdots + |z|^{n+p}$$
$$= |z|^{n+1} \frac{1 - |z|^p}{1 - |z|} .$$

Hence, if $|z| < 1$, then $S_{n,p} < \frac{|z|^{n+1}}{1-|z|}$ for all values of p, and, by Example 2.2.3, given any positive number ε, we can find n such that

$$|z|^{n+1} \{1 - |z|\}^{-1} < \varepsilon.$$

Thus, given ε, we can find n such that, for all values of p, $S_{n,p} < \varepsilon$. Hence, by §2.22, the series (2.1) is convergent so long as $|z| < 1$, and therefore *the geometric series is absolutely convergent if $|z| < 1$*.

When $|z| \geq 1$, the terms of the geometric series do not tend to zero as n tends to infinity, and the series is therefore divergent.

(II) The series $\frac{1}{1^s} + \frac{1}{2^s} + \frac{1}{3^s} + \frac{1}{4^s} + \frac{1}{5^s} + \cdots$. Consider now the series $S_n = \sum\limits_{m=1}^{n} \frac{1}{m^s}$, where s is greater than 1. We have

$$\frac{1}{2^s} + \frac{1}{3^s} < \frac{2}{2^s} = \frac{1}{2^{s-1}},$$

$$\frac{1}{4^s} + \frac{1}{5^s} + \frac{1}{6^s} + \frac{1}{7^s} < \frac{4}{4^s} = \frac{1}{4^{s-1}},$$

and so on. Thus the sum of $2^p - 1$ terms of the series is less than

$$\frac{1}{1^{s-1}} + \frac{1}{2^{s-1}} + \frac{1}{4^{s-1}} + \frac{1}{8^{s-1}} + \cdots + \frac{1}{2^{(p-1)(s-1)}} < \frac{1}{1 - 2^{1-s}},$$

and so the sum of *any* number of terms is less than $(1 - 2^{1-s})^{-1}$. Therefore the increasing sequence $\sum\limits_{m=1}^{n} m^{-s}$ cannot tend to infinity; *therefore, by §2.2, the series* $\sum\limits_{n=1}^{\infty} \frac{1}{n^s}$ *is convergent if* $s > 1$; and since its terms are all real and positive, they are equal to their own moduli, and so the series of moduli of the terms is convergent; that is, *the convergence is absolute*.

If $s = 1$, the series becomes

$$\frac{1}{1} + \frac{1}{2} + \frac{1}{3} + \frac{1}{4} + \cdots,$$

which we have already shewn to be divergent; and when $s < 1$, it is *a fortiori* divergent, since the effect of diminishing s is to increase the terms of the series. *The series* $\sum\limits_{n=1}^{\infty} \frac{1}{n^s}$ *is therefore divergent if* $s \leq 1$.

2.34 The comparison theorem

We shall now shew that *a series* $u_1 + u_2 + u_3 + \cdots$ *is absolutely convergent, provided that* $|u_n|$ *is always less than* $C|v_n|$, *where C is some number independent of n, and* v_n *is the nth term of another series which is known to be absolutely convergent.*

For, under these conditions, we have

$$|u_{n+1}| + |u_{n+2}| + \cdots + |u_{n+p}| < C\left\{|v_{n+1}| + |v_{n+2}| + \cdots + |v_{n+p}|\right\},$$

where n and p are any integers. But since the series $\sum v_n$ is absolutely convergent, the series $\sum |v_n|$ is convergent, and so, given ε, we can find n such that

$$|v_{n+1}| + |v_{n+2}| + \cdots + |v_{n+p}| < \varepsilon/C,$$

for all values of p. It follows therefore that we can find n such that

$$|u_{n+1}| + |u_{n+2}| + \cdots + |u_{n+p}| < \varepsilon,$$

for all values of p, i.e. the series $\sum |u_n|$ is convergent. The series $\sum u_n$ is therefore absolutely convergent.

Corollary 2.3.5 *A series is absolutely convergent if the ratio of its nth term to the nth term of a series which is known to be absolutely convergent is less than some number independent of n.*

Example 2.3.3 Shew that the series

$$\cos z + \frac{1}{2^2}\cos 2z + \frac{1}{3^2}\cos 3z + \frac{1}{4^2}\cos 4z + \cdots$$

is absolutely convergent for all real values of z.

When z is real, we have $|\cos nz| \le 1$, and therefore $\left|\frac{\cos nz}{n^2}\right| \le \frac{1}{n^2}$. The moduli of the terms of the given series are therefore less than, or at most equal to, the corresponding terms of the series

$$\frac{1}{1} + \frac{1}{2^2} + \frac{1}{3^2} + \frac{1}{4^2} + \cdots,$$

which by §2.33 is absolutely convergent. The given series is therefore absolutely convergent.

Example 2.3.4 Shew that the series

$$\frac{1}{1^2(z - z_1)} + \frac{1}{2^2(z - z_2)} + \frac{1}{3^2(z - z_3)} + \frac{1}{4^2(z - z_4)} + \cdots,$$

where $z_n = e^{in}$, $(n = 1, 2, 3, \ldots)$ is convergent for all values of z that are not on the circle $|z| = 1$.

The geometric representation of complex numbers is helpful in discussing a question of this kind. Let values of the complex number z be represented on a plane; then the numbers $z_1, z_2, z_3, \ldots$ will give a sequence of points which lie on the circumference of the circle whose centre is the origin and whose radius is unity; and it can be shewn that every point on the circle is a limit-point (§2.21) of the points z_n.

For these special values z_n of z, the given series does not exist, since the denominator of the nth term vanishes when $z = z_n$. For simplicity we do not discuss the series for any point z situated on the circumference of the circle of radius unity.

Suppose now that $|z| \ne 1$. Then for all values of n, $|z - z_n| \ge |\{1 - |z|\}| > c^{-1}$, for some value of c; so the moduli of the terms of the given series are less than the corresponding terms of the series

$$\frac{c}{1^2} + \frac{c}{2^2} + \frac{c}{3^2} + \frac{c}{4^2} + \cdots,$$

which is known to be absolutely convergent. The given series is therefore absolutely convergent for all values of z, except those which are on the circle $|z| = 1$.

It is interesting to notice that the area in the z-plane over which the series converges is divided into two parts, between which there is no intercommunication, by the circle $|z| = 1$.

Example 2.3.5 Shew that the series

$$2\sin\frac{z}{3} + 4\sin\frac{z}{9} + 8\sin\frac{z}{27} + \cdots + 2^n\sin\frac{z}{3^n} + \cdots$$

converges absolutely for all values of z.

Since $\lim_{n\to\infty} 3^n \sin(z/3^n) = z$ (this is evident from results proved in the *Appendix*), we can find a number k, *independent of n* (but depending on z), such that $|3^n \sin(z/3^n)| < k$; and therefore

$$\left| 2^n \sin \frac{z}{3^n} \right| < k \left(\frac{2}{3} \right)^n.$$

Since $\sum\limits_{n=1}^{\infty} k \left(\frac{2}{3} \right)^n$ converges, the given series converges absolutely.

2.35 Cauchy's test for absolute convergence

This appears in [120, p. 132–135].

If $\overline{\lim\limits_{n\to\infty}} |u_n|^{1/n} < 1$, *then* $\sum\limits_{n=1}^{\infty} u_n$ *converges absolutely.*

For we can find m such that, when $n \geq m$, $|u_n|^{1/n} \leq \rho < 1$, where ρ is independent of n. Then, when $n > m$, $|u_n| < \rho^n$; and since $\sum\limits_{n=m+1}^{\infty} \rho^n$ converges, it follows from §2.34 that $\sum\limits_{n=m+1}^{\infty} u_n$ (and therefore $\sum\limits_{n=1}^{\infty} u_n$) converges absolutely.

Note If $\overline{\lim} |u_n|^{1/n} > 1$, u_n does not tend to zero, and, by §2.3, $\sum\limits_{n=1}^{\infty} u_n$ does not converge.

2.36 D'Alembert's ratio test for absolute convergence

This appears in [159, pp. 171–182]. We shall now shew that *a series*

$$u_1 + u_2 + u_3 + u_4 + \cdots$$

is absolutely convergent, provided that for all values of n greater than some fixed value r, the ratio $\left| \frac{u_{n+1}}{u_n} \right|$ *is less than ρ, where ρ is a positive number* independent *of n and less than unity.*

For the terms of the series

$$| u_{r+1} | + | u_{r+2} | + | u_{r+3} | + \cdots$$

are respectively less than the corresponding terms of the series

$$| u_{r+1} | (1 + \rho + \rho^2 + \rho^3 + \cdots),$$

which is absolutely convergent when $\rho < 1$; therefore $\sum\limits_{n=r+1}^{\infty} u_n$ (and hence the given series) is absolutely convergent.

A particular case of this theorem is that if $\lim\limits_{n\to\infty} |u_{n+1}/u_n| = \ell < 1$, the series is absolutely convergent.

For, by the definition of a limit, we can find r such that

$$\left| \left| \frac{u_{n+1}}{u_n} \right| - \ell \right| < \frac{1}{2}(1 - \ell), \quad \text{when} \quad n > r,$$

and then

$$\left| \frac{u_{n+1}}{u_n} \right| < \frac{1}{2}(1 + \ell) < 1, \quad \text{when} \quad n > r.$$

Note If $\lim |u_{n+1}/u_n| > 1$, then u_n does not tend to zero, and, by §2.3, $\sum\limits_{n=1}^{\infty} u_n$ does not converge.

Example 2.3.6 If $|c| < 1$, shew that the series

$$\sum_{n=1}^{\infty} c^{n^2} e^{nz}$$

converges absolutely for all values of z. *Hint.* For $u_{n+1}/u_n = c^{(n+1)^2-n^2} e^z = c^{2n+1} e^z \to 0$, as $n \to \infty$, if $|c| < 1$.

Example 2.3.7 Shew that the series

$$z + \frac{(a-b)}{2!}z^2 + \frac{(a-b)(a-2b)}{3!}z^3 + \frac{(a-b)(a-2b)(a-3b)}{4!}z^4 + \cdots$$

converges absolutely if $|z| < |b|^{-1}$. *Hint.* For $\frac{u_{n+1}}{u_n} = \frac{a-nb}{n+1}z \to -bz$, as $n \to \infty$; so the condition for absolute convergence is $|bz| < 1$, i.e. $|z| < |b|^{-1}$.

Example 2.3.8 Shew that the series $\sum\limits_{n=1}^{\infty} \frac{nz^{n-1}}{z^n-(1+n^{-1})^n}$ converges absolutely if $|z| < 1$. *Hint.* For, when $|z| < 1$, $|z^n - (1+n^{-1})^n| \geq (1+n^{-1})^n - |z^n| \geq 1 + 1 + \frac{n-1}{2n} + \cdots - 1 > 1$, so the moduli of the terms of the series are less than the corresponding terms of the series $\sum\limits_{n=1}^{\infty} n|z^{n-1}|$; but this latter series is absolutely convergent, and so the given series converges absolutely.

2.37 A general theorem on series for which $\lim\limits_{n\to\infty} \left|\dfrac{u_{n+1}}{u_n}\right| = 1$

It is obvious that if, for all values of n greater than some fixed value r, $|u_{n+1}|$ is greater than $|u_n|$, then the terms of the series do not tend to zero as $n \to \infty$, and the series is therefore divergent. On the other hand, if $\left|\frac{u_{n+1}}{u_n}\right|$ is less than some number which is itself less than unity and independent of n (when $n > r$), we have shewn in §2.36 that the series is absolutely convergent. The critical case is that in which, as n increases, $\left|\frac{u_{n+1}}{u_n}\right|$ tends to the value unity. In this case a further investigation is necessary.

We shall now shew that a *series $u_1 + u_2 + u_3 + \cdots$, in which* $\lim\limits_{n\to\infty} \left|\dfrac{u_{n+1}}{u_n}\right| = 1$ *will be absolutely convergent if a positive number c exists such that*

$$\overline{\lim_{n\to\infty}}\, n\left\{\left|\frac{u_{n+1}}{u_n}\right| - 1\right\} = -1 - c.$$

This is the second (D'Alembert's theorem given in §2.36 being the first) of a hierarchy of theorems due to De Morgan. See Chrystal [146, p. xxvi] for an historical account of these theorems.

For, compare the series $\sum |u_n|$ with the convergent series $\sum v_n$, where

$$v_n = An^{-1-\frac{1}{2}c}$$

and A is a constant; we have

$$\frac{v_{n+1}}{v_n} = \left(\frac{n}{n+1}\right)^{1+\frac{1}{2}c} = \left(1 + \frac{1}{n}\right)^{-(1+\frac{1}{2}c)} = 1 - \frac{1+\frac{1}{2}c}{n} + O\left(\frac{1}{n^2}\right).$$

As $n \to \infty$, $n\left\{\frac{v_{n+1}}{v_n} - 1\right\} \to -1 - \frac{1}{2}c$, and hence we can find m such that, when $n > m$,

$$\left|\frac{u_{n+1}}{u_n}\right| \le \frac{v_{n+1}}{v_n}.$$

By a suitable choice of the constant A, we can therefore secure that for all values of n we shall have

$$|u_n| < v_n.$$

As $\sum v_n$ is convergent, $\sum |u_n|$ is also convergent, and so $\sum u_n$ is absolutely convergent.

Corollary 2.3.6 *If* $\left|\frac{u_{n+1}}{u_n}\right| = 1 + \frac{A_1}{n} + O\left(\frac{1}{n^2}\right)$, *where* A_1 *is independent of n, then the series is absolutely convergent if* $A_1 < -1$.

Example 2.3.9 Investigate the convergence of $\sum_{n=1}^{\infty} n^r \exp\left(-k \sum_{m=1}^{n} \frac{1}{m}\right)$, when $r > k$ and when $r < k$.

2.38 Convergence of the hypergeometric series

The theorems which have been given may be illustrated by a discussion of the convergence of the *hypergeometric series*

$$1 + \frac{a \cdot b}{1 \cdot c}z + \frac{a(a+1)b(b+1)}{1 \cdot 2 \cdot c(c+1)}z^2 + \frac{a(a+1)(a+2)b(b+1)(b+2)}{1 \cdot 2 \cdot 3 \cdot c(c+1)(c+2)}z^3 + \cdots,$$

which is generally denoted (see Chapter 14) by $F(a, b; c; z)$.

If c is a negative integer, all the terms after the $(1-c)$th have zero denominators; and if either a or b is a negative integer the series will terminate at the $(1-a)$th or $(1-b)$th term as the case may be. We shall suppose these cases set aside, so that a, b, and c are assumed not to be negative integers. In this series

$$\left|\frac{u_{n+1}}{u_n}\right| = \left|\frac{(a+n-1)(b+n-1)}{n(c+n-1)}z\right| \to |z|,$$

as $n \to \infty$. We see therefore, by §2.36, that *the series is absolutely convergent when* $|z| < 1$, *and divergent when* $|z| > 1$.

When $|z| = 1$, we have[2]

$$\left|\frac{u_{n+1}}{u_n}\right| = \left|1 + \frac{a-1}{n}\right|\left|1 + \frac{b-1}{n}\right|\left|1 - \frac{c-1}{n} + O\left(\frac{1}{n^2}\right)\right|$$

$$= \left|1 + \frac{a+b-c-1}{n} + O\left(\frac{1}{n^2}\right)\right|.$$

---

[2] The symbol $O(1/n^2)$ does not denote the same function of n throughout. See §2.11.

Let a, b, c be complex numbers, and let them be given in terms of their real and imaginary parts by the equations

$$a = a' + ia'', \quad b = b' + ib'', \quad c = c' + ic''.$$

Then we have

$$\left| \frac{u_{n+1}}{u_n} \right| = \left| 1 + \frac{a' + b' - c' - 1 + i(a'' + b'' - c'')}{n} + O\left(\frac{1}{n^2}\right) \right|$$

$$= \left\{ \left(1 + \frac{a' + b' - c' - 1}{n} \right)^2 + \left(\frac{a'' + b'' - c''}{n} \right)^2 + O\left(\frac{1}{n^2}\right) \right\}^{1/2}$$

$$= 1 + \frac{a' + b' - c' - 1}{n} + O\left(\frac{1}{n^2}\right).$$

By Corollary 2.3.6, a condition for absolute convergence is

$$a' + b' - c' < 0.$$

Hence *when $|z| = 1$, a sufficient condition for the absolute convergence of the hypergeometric series is that the real part of $a + b - c$ shall be negative.* The condition is also necessary. See Bromwich [102, pp. 202–204].

2.4 Effect of changing the order of the terms in a series

In an ordinary sum the order of the terms is of no importance, for it can be varied without affecting the result of the addition. In an infinite series, however, this is no longer the case[3], as will appear from the following example.

Let

$$T = 1 + \frac{1}{3} - \frac{1}{2} + \frac{1}{5} + \frac{1}{7} - \frac{1}{4} + \frac{1}{9} + \frac{1}{11} - \frac{1}{6} + \cdots \text{ and } S = 1 - \frac{1}{2} + \frac{1}{3} - \frac{1}{4} + \frac{1}{5} - \frac{1}{6} + \cdots,$$

and let T_n and S_n denote the sums of their first n terms. These infinite series are formed of the same terms, but the order of the terms is different, and so T_n and S_n are quite distinct functions of n.

Let $\sigma_n = \frac{1}{1} + \frac{1}{2} + \cdots + \frac{1}{n}$, so that $S_{2n} = \sigma_{2n} - \sigma_n$. Then

$$T_{3n} = \frac{1}{1} + \frac{1}{3} + \cdots + \frac{1}{4n - 1} - \frac{1}{2} - \frac{1}{4} - \cdots - \frac{1}{2n}$$

$$= \sigma_{4n} - \tfrac{1}{2}\sigma_{2n} - \tfrac{1}{2}\sigma_n$$

$$= (\sigma_{4n} - \sigma_{2n}) + \tfrac{1}{2}(\sigma_{2n} - \sigma_n)$$

$$= S_{4n} + \tfrac{1}{2}S_{2n}.$$

Making $n \to \infty$, we see that $T = S + \tfrac{1}{2}S$; and so the derangement of the terms of S has altered its sum.

---

[3] We say that the series $\sum_{n=1}^{\infty} v_n$ consists of the terms of $\sum_{n=1}^{\infty} u_n$ in a different order if a law is given by which corresponding to each positive integer p we can find one (and only one) integer q and *vice versa*, and v_q is taken equal to u_p. The result of this section was noticed by Dirichlet [173, p. 48]. See also Cauchy [125, p. 57].

Example 2.4.1 (Manning)   If in the series

$$1 - \frac{1}{2} + \frac{1}{3} - \frac{1}{4} + \cdots$$

the order of the terms be altered, so that the ratio of the number of positive terms to the number of negative terms in the first n terms is ultimately a^2, shew that the sum of the series will become $\log(2a)$.

2.41 The fundamental property of absolutely convergent series

We shall shew that the sum of an absolutely convergent series is *not* affected by changing the order in which the terms occur.

Let $S = u_1 + u_2 + u_3 + \cdots$ be an absolutely convergent series, and let S' be a series formed by the same terms in a different order.

Let ε be an arbitrary positive number, and let n be chosen so that

$$|u_{n+1}| + |u_{n+2}| + \cdots + |u_{n+p}| < \frac{1}{2}\varepsilon$$

for all values of p.

Suppose that in order to obtain the first n terms of S we have to take m terms of S'; then if $k > m$,

$$S_k' - S_n + \text{ terms of } S \text{ with suffices greater than } n,$$

so that

$$S_k' - S = S_n - S + \text{ terms of } S \text{ with suffices greater than } n.$$

Now the modulus of the sum of any number of terms of S with suffices greater than n does not exceed the sum of their moduli, and therefore is less than $\frac{1}{2}\varepsilon$. Therefore $\left|S_k' - S\right| < |S_n - S| + \frac{1}{2}\varepsilon$. But

$$|S_n - S| \le \lim_{p \to \infty} \left\{|u_{n+1}| + |u_{n+2}| + \cdots + |u_{n+p}|\right\} \le \frac{1}{2}\varepsilon.$$

Therefore given ε we can find m such that $\left| S_k' - S \right| < \varepsilon$ when $k > m$; therefore $S_m' \to S$, which is the required result.

If a series of real terms converges, but not absolutely, and if S_p be the sum of the first p positive terms, and if σ_n be the sum of the first n negative terms, then $S_p \to \infty, \sigma_n \to -\infty$; and $\lim(S_p + \sigma_n)$ does not exist unless we are given some relation between p and n. It has, in fact, been shewn by Riemann [558, p. 221], that it is possible, by choosing a suitable relation, to make $\lim(S_p + \sigma_n)$ equal to *any* given real number.

2.5 Double series

A complete theory of double series, on which this account is based, is given by Pringsheim [541]. See further memoirs by that writer [542] and by London [442], and also Bromwich [102], which, in addition to an account of Pringsheim's theory, contains many developments of the subject. Other important theorems are given by Bromwich [101].

Let $u_{m,n}$ be a number determinate for all positive integral values of m and n; consider the array

$$
\begin{array}{cccc}
u_{1,1} & u_{1,2} & u_{1,3} & \cdots \\
u_{2,1} & u_{2,2} & u_{2,3} & \cdots \\
u_{3,1} & u_{3,2} & u_{3,3} & \cdots \\
\vdots & \vdots & \vdots & \vdots
\end{array}
$$

Let the sum of the terms inside the rectangle, formed by the first m rows of the first n columns of this array of terms, be denoted by $S_{m,n}$.

If a number S exists such that, given any arbitrary positive number ε, it is possible to find integers m and n such that $\left| S_{\mu,\nu} - S \right| < \varepsilon$ whenever both $\mu > m$ and $\nu > n$, we say[4] that the *double series of which the general element is $u_{\mu,\nu}$ converges to the sum S*, and we write

$$
\lim_{\mu \to \infty, \nu \to \infty} S_{\mu,\nu} = S.
$$

If the double series, of which the general element is $|u_{\mu,\nu}|$, is convergent, we say that the given double series is *absolutely convergent*.

Since $u_{\mu,\nu} = (S_{\mu,\nu} - S_{\mu,\nu-1}) - (S_{\mu-1,\nu} - S_{\mu-1,\nu-1})$, it is easily seen that, if the double series is convergent, then

$$
\lim_{\mu \to \infty, \nu \to \infty} u_{\mu,\nu} = 0.
$$

Stolz' necessary and sufficient condition for convergence. This condition, stated by Stolz [612], appears to have been first proved by Pringsheim. A condition for convergence which is obviously necessary (see §2.22) is that, given ε, we can find m and n such that $\left| S_{\mu+\rho,\nu+\sigma} - S_{\mu,\nu} \right| < \varepsilon$ whenever $\mu > m$ and $\nu > n$ and ρ, σ may take *any* of the values $0, 1, 2, \ldots$. The condition is also sufficient; for, suppose it satisfied; then, when $\mu > m + n$, $\left| S_{\mu+\rho,\mu+\rho} - S_{\mu,\mu} \right| < \varepsilon$.

Therefore, by §2.22, $S_{\mu,\mu}$ has a limit S; and then making ρ and σ tend to infinity in such a way that $\mu + \rho = \nu + \sigma$, we see that $\left| S - S_{\mu,\nu} \right| \leq \varepsilon$ whenever $\mu > m$ and $\nu > n$; that is to say, the double series converges.

Corollary 2.5.1 *An absolutely convergent double series is convergent. For if the double series converges absolutely and if $t_{m,n}$ be the sum of m rows of n columns of the series of moduli, then, given ε, we can find μ such that, when $\rho > m > \mu$ and $q > n > \mu$, $t_{p,q} - t_{m,n} < \varepsilon$. But $\left| S_{p,q} - S_{m,n} \right| \leq t_{p,q} - t_{m,n}$ and so $\left| S_{p,q} - S_{m,n} \right| < \varepsilon$ when $p > m > \mu$, $q > n > \mu$; and this is the condition that the double series should converge.*

2.51 Methods of summing a double series

These methods are due to Cauchy. Let us suppose that $\displaystyle\sum_{\nu=1}^{\infty} u_{\mu,\nu}$ converges to the sum S_μ.

Then $\displaystyle\sum_{\mu=1}^{\infty} S_\mu$ is called the *sum by rows* of the double series; that is to say, the sum by rows is

$\displaystyle\sum_{\mu=1}^{\infty} \left(\sum_{\nu=1}^{\infty} u_{\mu,\nu} \right)$. Similarly, the *sum by columns* is defined as $\displaystyle\sum_{\nu=1}^{\infty} \left(\sum_{\mu=1}^{\infty} u_{\mu,\nu} \right)$. That these two sums

---

[4] This definition is practically due to Cauchy [120, p. 540].

are not necessarily the same is shewn by the example $S_{\mu,\nu} = \dfrac{\mu - \nu}{\mu + \nu}$, in which the sum by rows is -1, the sum by columns is $+1$; and S does not exist.

Theorem 2.5.2 (Pringsheim's theorem)   *[541, p. 117]. If S exists and the sums by rows and columns exist, then each of these sums is equal to S.*

For since S exists, then we can find m such that $\left| S_{\mu,\nu} - S \right| < \varepsilon$, if $\mu > m, \nu > m$. And therefore, since $\lim_{\nu \to \infty} S_{\mu,\nu}$ exists, $\left| \left(\lim_{\nu \to \infty} S_{\mu,\nu} \right) - S \right| \leq \varepsilon$; that is to say, $\left| \sum_{p=1}^{\mu} S_p - S \right| \leq \varepsilon$ when $\mu > m$, and so (§2.22) the sum by rows converges to S. In like manner the sum by columns converges to S.

2.52 Absolutely convergent double series

We can prove the analogue of §2.41 for double series, namely that *if the terms of an absolutely convergent double series are taken in* any *order as a simple series, their sum tends to the same limit, provided that every term occurs in the summation.*

Let $\sigma_{\mu,\nu}$ be the sum of the rectangle of μ rows and ν columns of the double series whose general element is $\left| u_{\mu,\nu} \right|$; and let the sum of this double series be σ. Then given ε we can find m and n such that $\sigma - \sigma_{\mu,\nu} < \varepsilon$ whenever both $\mu > m$ and $\nu > n$.

Now suppose that it is necessary to take N terms of the deranged series (in the order in which the terms are taken) in order to include all the terms of $S_{M+1,M+1}$, and let the sum of these terms be t_N.

Then $t_N - S_{M+1,M+1}$ consists of a sum of terms of the type $u_{p,q}$ in which $p > m$, $q > n$ whenever $M > m$ and $M > n$; and therefore

$$\left| t_N - S_{M+1,M+1} \right| \leq \sigma - \sigma_{M+1,M+1} < \tfrac{1}{2}\varepsilon.$$

Also, $S - S_{M+1,M+1}$ consists of terms $u_{p,q}$ in which $p > m, q > n$; therefore $\left| S - S_{M+1,M+1} \right| \leq \sigma - \sigma_{M+1,M+1} < \tfrac{1}{2}\varepsilon$; therefore $\left| S - t_N \right| < \varepsilon$; and, corresponding to any given number ε, we can find N; and therefore $t_N \to S$.

Example 2.5.1   Prove that in an absolutely convergent double series, $\sum_{n=1}^{\infty} u_{m,n}$ exists, and thence that the sums by rows and columns respectively converge to S. *Hint.* Let the sum of μ rows of ν columns of the series of moduli be $t_{\mu,\nu}$, and let t be the sum of the series of moduli. Then $\sum_{\nu=1}^{\infty} \left| u_{\mu,\nu} \right| < t$, and so $\sum_{\nu=1}^{\infty} u_{\mu,\nu}$ converges; let its sum be b_{μ}; then

$$\left| b_1 \right| + \left| b_2 \right| + \cdots + \left| b_{\mu} \right| \leq \lim_{\nu \to \infty} t_{\mu,\nu} \leq t,$$

and so $\sum_{\mu=1}^{\infty} b_{\mu}$ converges absolutely. Therefore the sum by rows of the double series exists, and similarly the sum by columns exists; and the required result then follows from Pringsheim's theorem.

Example 2.5.2 Shew from first principles that if the terms of an absolutely convergent double series be arranged in the order

$$u_{1,1} + (u_{2,1} + u_{1,2}) + (u_{3,1} + u_{2,2} + u_{1,3}) + (u_{4,1} + \cdots + u_{1,4}) + \cdots,$$

this series converges to S.

2.53 Cauchy's theorem on the multiplication of absolutely convergent series

This appears in [120, Note VII]. We shall now shew that *if two series*

$$S = u_1 + u_2 + u_3 + \cdots \quad and \quad T = v_1 + v_2 + v_3 + \cdots$$

are absolutely convergent, then the series

$$P = u_1 v_1 + u_2 v_1 + u_1 v_2 + \cdots,$$

formed by the products of their terms, written in any order, is absolutely convergent, and has for sum ST.

Let

$$S_n = u_1 + u_2 + \cdots + u_n,$$
$$T_n = v_1 + v_2 + \cdots + v_n.$$

Then $ST = \lim S_n \lim T_n = \lim(S_n T_n)$ by Example 2.2.2. Now

$$
\begin{array}{ccccccccc}
S_n T_n = & & u_1 v_1 & + & u_2 v_1 & + & \cdots & + & u_n v_1 \\
 & + & u_1 v_2 & + & u_2 v_2 & + & \cdots & + & u_n v_2 \\
 & + & \vdots & & \vdots & & & & \vdots \\
 & + & u_1 v_n & + & u_2 v_n & + & \cdots & + & u_n v_n.
\end{array}
$$

But this double series is absolutely convergent; for if these terms are replaced by their moduli, the result is $\sigma_n \tau_n$, where

$$\sigma_n = |u_1| + |u_2| + \cdots + |u_n|,$$

$$\tau_n = |v_1| + |v_2| + \cdots + |v_n|,$$

and $\sigma_n \tau_n$ is known to have a limit. Therefore, by §2.52, if the elements of the double series, of which the general term is $u_m v_n$, be taken in any order, their sum converges to ST.

Example 2.5.3 Shew that the series obtained by multiplying the two series

$$1 + \frac{z}{2} + \frac{z^2}{2^2} + \frac{z^3}{2^3} + \frac{z^4}{2^4} + \cdots \quad and \quad 1 + \frac{1}{z} + \frac{1}{z^2} + \frac{1}{z^3} + \cdots,$$

and rearranging according to powers of z, converges so long as the representative point of z lies in the ring-shaped region bounded by the circles $|z| = 1$ and $|z| = 2$.

2.6 Power series

The results of this section are due to Cauchy [120, Ch. IX]. A series of the type

$$a_0 + a_1 z + a_2 z^2 + a_3 z^3 + \cdots,$$

in which the coefficients $a_0, a_1, a_2, a_3, \ldots$ are independent of z, is called a *series proceeding according to ascending powers of z*, or briefly a *power series*.

We shall now shew that *if a power series converges for any value z_0 of z, it will be absolutely convergent for all values of z whose representative points are within a circle which passes through z_0 and has its centre at the origin.*

For, if z be such a point, we have $|z| < |z_0|$. Now, since $\sum_{n=0}^{\infty} a_n z_0^n$ converges, $a_n z_0^n$ must tend to zero as $n \to \infty$, and so we can find M (independent of n) such that $|a_n z_0^n| < M$. Thus

$$|a_n z^n| < M \left| \frac{z}{z_0} \right|^n.$$

Therefore every term in the series $\sum_{n=0}^{\infty} a_n z^n$ is less than the corresponding term in the convergent geometric series $\sum_{n=0}^{\infty} M |z/z_0|^n$; the series is therefore convergent; and so the power series is *absolutely* convergent, as the series of moduli of its terms is a convergent series; the result stated is therefore established.

Let $\varliminf |a_n|^{-1/n} = r$; then, from §2.35, $\sum_{n=0}^{\infty} a_n z^n$ converges absolutely when $|z| < r$; if $|z| > r$, $a_n z^n$ does not tend to zero and so $\sum_{n=0}^{\infty} a_n z^n$ diverges (§2.3). The circle $|z| = r$, which includes all the values of z for which the power series

$$a_0 + a_1 z + a_2 z^2 + a_3 z^3 + \cdots$$

converges, is called the *circle of convergence* of the series. The radius of the circle is called the *radius of convergence*.

In practice there is usually a simpler way of finding r, derived from d'Alembert's test (§2.36); r is $\lim(a_n/a_{n+1})$ if this limit exists.

A power series may converge for all values of the variable, as happens, for instance, in the case of the series[5]

$$z - \frac{z^3}{3!} + \frac{z^5}{5!} - \cdots,$$

which represents the function $\sin z$; in this case the series converges over the whole z-plane.

On the other hand, the radius of convergence of a power series may be zero; thus in the case of the series

$$1 + 1!z + 2!z^2 + 3!z^3 + 4!z^4 + \cdots$$

we have $|u_{n+1}/u_n| = n|z|$, which, for all values of n after some fixed value, is greater than

---

[5] The series for e^z, $\sin z$, $\cos z$ and the fundamental properties of these functions and of $\log z$ will be assumed throughout. A brief account of the theory of the functions is given in the *Appendix*.

unity when z has any value different from zero. The series converges therefore only at the point $z = 0$, and the radius of its circle of convergence vanishes.

A power series may or may not converge for points which are actually *on* the periphery of the circle; thus the series

$$1 + \frac{z}{1^s} + \frac{z^2}{2^s} + \frac{z^3}{3^s} + \frac{z^4}{4^s} + \cdots,$$

whose radius of convergence is unity, converges or diverges at the point $z = 1$ according as s is greater or not greater than unity, as was seen in §2.33.

Corollary 2.6.1 *If (a_n) be a sequence of positive terms such that $\lim(a_{n+1}/a_n)$ exists, this limit is equal to $\varliminf a_n^{1/n}$.*

2.61 Convergence of series derived from a power series

Let $a_0 + a_1 z + a_2 z^2 + a_3 z^3 + a_4 z^4 + \cdots$ be a power series, and consider the series

$$a_1 + 2a_2 z + 3a_3 z^2 + 4a_4 z^3 + \cdots,$$

which is obtained by differentiating the power series term by term. We shall now shew that *the derived series has the same circle of convergence as the original series.*

For let z be a point within the circle of convergence of the power series; and choose a positive number r_1, intermediate in value between $|z|$ and r the radius of convergence. Then, since the series $\sum_{n=0}^{\infty} a_n r_1^n$ converges absolutely, its terms must tend to zero as $n \to \infty$; and it must therefore be possible to find a positive number M, independent of n, such that $|a_n| < M r_1^{-n}$ for all values of n.

Then the terms of the series $\sum_{n=1}^{\infty} n|a_n| \, |z|^{n-1}$ are less than the corresponding terms of the series

$$\frac{M}{r_1} \sum_{n=1}^{\infty} \frac{n|z|^{n-1}}{r_1^{n-1}}.$$

But this series converges, by §2.36, since $|z| < r_1$. Therefore, by §2.34, the series

$$\sum_{n=1}^{\infty} n|a_n||z|^{n-1}$$

converges; that is, the series $\sum_{n=1}^{\infty} na_n z^{n-1}$ converges absolutely for all points z situated within the circle of convergence of the original series $\sum_{n=0}^{\infty} a_n z^n$. When $|z| > r$, $a_n z^n$ does not tend to zero, and *a fortiori* $na_n z^n$ does not tend to zero; and so the two series have the same circle of convergence.

Corollary 2.6.2 *The series $\sum_{n=0}^{\infty} \frac{a_n}{n+1} z^{n+1}$, obtained by integrating the original power series term by term, has the same circle of convergence as $\sum_{n=0}^{\infty} a_n z^n$.*

2.7 Infinite products

We next consider a class of limits, known as *infinite products*. Let $1 + a_1, 1 + a_2, 1 + a_3, \ldots$ be a sequence such that none of its members vanish. If, as $n \to \infty$, the product

$$(1 + a_1)(1 + a_2)(1 + a_3) \cdots (1 + a_n)$$

(which we denote by $\prod_n$) tends to a definite limit other than zero, this limit is called the value of the infinite product

$$\prod = (1 + a_1)(1 + a_2)(1 + a_3) \cdots,$$

and the product is said to be *convergent*. (The convergence of the product in which $a_{n-1} = -1/n^2$ was investigated by Wallis as early as 1655.) It is almost obvious that a *necessary* condition for convergence is that $\lim a_n = 0$, since $\lim \prod_{n-1} = \lim \prod_n \neq 0$. The limit of the product is written $\prod_{n=1}^{\infty}(1 + a_n)$.

Now

$$\prod_{n=1}^{m}(1 + a_n) = \exp\left(\sum_{n=1}^{m} \log(1 + a_n)\right), \tag{2.2}$$

and (see Appendix §A.2), $\exp\left(\lim_{m \to \infty} u_m\right) = \lim_{m \to \infty} (\exp u_m)$ if the former limit exists; hence a sufficient condition that the product should converge is that $\sum_{n=1}^{\infty} \log(1 + a_n)$ should converge when the logarithms have their principal values. If this series of logarithms converges absolutely, the convergence of the product is said to be *absolute*.

The condition for absolute convergence is given by the following theorem: *in order that the infinite product*

$$(1 + a_1)(1 + a_2)(1 + a_3) \cdots$$

may be absolutely convergent, it is necessary and sufficient that the series

$$a_1 + a_2 + a_3 + \cdots$$

should be absolutely convergent.

For, by definition, $\prod$ is absolutely convergent or not according as the series

$$\log(1 + a_1) + \log(1 + a_2) + \log(1 + a_3) + \cdots$$

is absolutely convergent or not.

Now, since $\lim a_n = 0$, we can find m such that, when $n > m$, $|a_n| < \frac{1}{2}$; and then

$$\left|a_n^{-1} \log(1 + a_n) - 1\right| = \left|-\frac{a_n}{2} + \frac{a_n^{2}}{3} - \frac{a_n^{3}}{4} + \cdots\right|$$

$$< \frac{1}{2^2} + \frac{1}{2^3} + \cdots = \frac{1}{2}.$$

And thence, when $n > m$,

$$\frac{1}{2} \leq \left|\frac{\log(1 + a_n)}{a_n}\right| \leq \frac{3}{2};$$

therefore, by the comparison theorem, the absolute convergence of $\sum \log(1 + a_n)$ entails that of $\sum a_n$ and conversely, provided that $a_n \neq -1$ for any value of n. This establishes the result.

Note   A discussion of the convergence of infinite products, in which the results are obtained without making use of the logarithmic function, is given by Pringsheim [539], and also by Bromwich [102, Ch. VI].

If, in a product, a finite number of factors vanish, and if, when these are suppressed, the resulting product converges, the original product is said to *converge* to zero. But such a product as $\prod_{n=2}^{\infty}(1 - n^{-1})$ is said to *diverge* to zero.

Corollary 2.7.1   *Since, if $S_n \to \ell$, $\exp(S_n) \to \exp \ell$, it follows from §2.41 that the factors of an absolutely convergent product can be deranged without affecting the value of the product.*

Example 2.7.1   Shew that if $\prod_{n=1}^{\infty}(1 + a_n)$ converges, so does $\sum_{n=1}^{\infty} \log(1 + a_n)$, if the logarithms have their principal values.

Example 2.7.2   Shew that the infinite product

$$\frac{\sin z}{z} \cdot \frac{\sin \frac{1}{2}z}{\frac{1}{2}z} \cdot \frac{\sin \frac{1}{3}z}{\frac{1}{3}z} \cdot \frac{\sin \frac{1}{4}z}{\frac{1}{4}z} \cdots$$

is absolutely convergent for all values of z. *Hint.* For $\left(\sin \frac{z}{n}\right) / \left(\frac{z}{n}\right)$ can be written in the form $1 - \frac{\lambda_n}{n^2}$, where $|\lambda_n| < k$ and k is independent of n; and the series $\sum_{n=1}^{\infty} \frac{\lambda_n}{n^2}$ is absolutely convergent, as is seen on comparing it with $\sum_{n=1}^{\infty} \frac{1}{n^2}$. The infinite product is therefore absolutely convergent.

2.71 Some examples of infinite products

Consider the infinite product

$$\left(1 - \frac{z^2}{\pi^2}\right)\left(1 - \frac{z^2}{2^2\pi^2}\right)\left(1 - \frac{z^2}{3^2\pi^2}\right) \cdots,$$

which, as will be proved later (§7.5), represents the function $\dfrac{\sin z}{z}$.

In order to find whether it is absolutely convergent, we must consider the series $\sum_{n=1}^{\infty} \frac{z^2}{n^2\pi^2}$, or $\dfrac{z^2}{\pi^2} \sum_{n=1}^{\infty} \dfrac{1}{n^2}$; this series is absolutely convergent, and so the product is absolutely convergent for all values of z.

Now let the product be written in the form

$$\left(1 - \frac{z}{\pi}\right)\left(1 + \frac{z}{\pi}\right)\left(1 - \frac{z}{2\pi}\right)\left(1 + \frac{z}{2\pi}\right) \cdots.$$

The absolute convergence of this product depends on that of the series

$$-\frac{z}{\pi} + \frac{z}{\pi} - \frac{z}{2\pi} + \frac{z}{2\pi} - \cdots .$$

But this series is only conditionally convergent, since its series of moduli

$$\frac{|z|}{\pi} + \frac{|z|}{\pi} + \frac{|z|}{2\pi} + \frac{|z|}{2\pi} + \cdots$$

is divergent. In this form therefore the infinite product is not absolutely convergent, and so, if the order of the factors $\left(1 \pm \frac{z}{n\pi}\right)$ is deranged, there is a risk of altering the value of the product.

Lastly, let the same product be written in the form

$$\left\{\left(1 - \frac{z}{\pi}\right) e^{\frac{z}{\pi}}\right\} \left\{\left(1 + \frac{z}{\pi}\right) e^{-\frac{z}{\pi}}\right\} \left\{\left(1 - \frac{z}{2\pi}\right) e^{\frac{z}{2\pi}}\right\} \left\{\left(1 + \frac{z}{2\pi}\right) e^{-\frac{z}{2\pi}}\right\} \cdots ,$$

in which each of the expressions

$$\left(1 \pm \frac{z}{m\pi}\right) e^{\mp \frac{z}{m\pi}}$$

is counted as a single factor of the infinite product. The absolute convergence of this product depends on that of the series of which the $(2m - 1)$th and $(2m)$th terms are

$$\left(1 \mp \frac{z}{m\pi}\right) e^{\pm \frac{z}{m\pi}} - 1.$$

But it is easy to verify that

$$\left(1 \mp \frac{z}{m\pi}\right) e^{\pm \frac{z}{m\pi}} = 1 + O\left(\frac{1}{m^2}\right),$$

and so the absolute convergence of the series in question follows by comparison with the series

$$1 + 1 + \frac{1}{2^2} + \frac{1}{2^2} + \frac{1}{3^2} + \frac{1}{3^2} + \frac{1}{4^2} + \frac{1}{4^2} + \cdots .$$

The infinite product in this last form is therefore again absolutely convergent, the adjunction of the factors $e^{\pm \frac{z}{n\pi}}$ having changed the convergence from conditional to absolute. This result is a particular case of the first part of the factor theorem of Weierstrass (§7.6).

Example 2.7.3 Prove that $\prod\limits_{n=1}^{\infty} \left\{\left(1 - \frac{z}{c+n}\right) e^{\frac{z}{n}}\right\}$ is absolutely convergent for all values of z, if c is a constant other than a negative integer. *Hint.* For the infinite product converges absolutely with the series

$$\sum_{n=1}^{\infty} \left\{\left(1 - \frac{z}{c+n}\right) e^{\frac{z}{n}} - 1\right\}.$$

Now the general term of this series is

$$\left(1 - \frac{z}{c+n}\right) \left\{1 + \frac{z}{n} + \frac{z^2}{2n^2} + O\left(\frac{1}{n^3}\right)\right\} - 1 = \frac{zc - \frac{1}{2}z^2}{n^2} + O\left(\frac{1}{n^3}\right) = O\left(\frac{1}{n^2}\right).$$

But $\sum\limits_{n=1}^{\infty} \frac{1}{n^2}$ converges, and so, by §2.34, $\sum\limits_{n=1}^{\infty} \left\{\left(1 - \frac{z}{c+n}\right) e^{\frac{z}{n}} - 1\right\}$ converges absolutely, and therefore the product converges absolutely.

Example 2.7.4 Shew that $\prod_{n=2}^{\infty} \left\{ 1 - \left(1 - \frac{1}{n}\right)^{-n} z^{-n} \right\}$ converges for all points z situated outside a circle whose centre is the origin and radius unity. *Hint*. For the infinite product is absolutely convergent provided that the series

$$\sum_{n=2}^{\infty} \left(1 - \frac{1}{n}\right)^{-n} z^{-n}$$

is absolutely convergent. But $\lim\limits_{n \to \infty} \left(1 - \frac{1}{n}\right)^{-n} = e$, so the limit of the ratio of the $(n+1)$th term of the series to the nth term is $1/z$; there is therefore absolute convergence when $|1/z| < 1$, i.e. when $|z| > 1$.

Example 2.7.5 Shew that

$$\frac{1 \cdot 2 \cdot 3 \cdots (m-1)}{(z+1)(z+2) \cdots (z+m-1)} m^z$$

tends to a finite limit as $m \to \infty$, unless z is a negative integer. *Hint*. For the expression can be written as a product of which the nth factor is

$$\frac{n}{z+n} \left(\frac{n+1}{n}\right)^s = \left(1 + \frac{1}{n}\right)^s \left(1 + \frac{z}{n}\right)^{-1} = \left\{ 1 + \frac{z(z-1)}{2n^2} + O\left(\frac{1}{n^3}\right) \right\}.$$

This product is therefore absolutely convergent, provided the series

$$\sum_{n=1}^{\infty} \left\{ \frac{z(z-1)}{2n^2} + O\left(\frac{1}{n^3}\right) \right\}$$

is absolutely convergent; and a comparison with the convergent series $\sum_{n=1}^{\infty} \frac{1}{n^2}$ shews that this is the case. When z is a negative integer the expression does not exist because one of the factors in the denominator vanishes.

Example 2.7.6 Prove that

$$z \left(1 - \frac{z}{\pi}\right) \left(1 - \frac{z}{2\pi}\right) \left(1 + \frac{z}{\pi}\right) \left(1 - \frac{z}{3\pi}\right) \left(1 - \frac{z}{4\pi}\right) \left(1 + \frac{z}{2\pi}\right) \cdots = e^{-z \log 2/\pi} \sin z.$$

For the given product

$$\lim_{k \to \infty} z \left(1 - \frac{z}{\pi}\right) \left(1 - \frac{z}{2\pi}\right) \left(1 + \frac{z}{\pi}\right) \cdots \left(1 - \frac{z}{(2k-1)\pi}\right) \left(1 - \frac{z}{2k\pi}\right) \left(1 + \frac{z}{k\pi}\right)$$

$$= \lim_{k \to \infty} \left[\begin{array}{c} e^{\frac{z}{\pi}\left(-1 - \frac{1}{2} + 1 - \frac{1}{3} - \frac{1}{4} + \frac{1}{2} - \cdots - \frac{1}{2k-1} - \frac{1}{2k} + \frac{1}{k}\right)} \\ \times z \left(1 - \frac{z}{\pi}\right) e^{\frac{z}{\pi}} \cdot \left(1 - \frac{z}{2\pi}\right) e^{\frac{z}{2\pi}} \cdots \left(1 - \frac{z}{2k\pi}\right) e^{\frac{z}{2k\pi}} \cdot \left(1 + \frac{z}{k\pi}\right) e^{-\frac{z}{k\pi}} \end{array} \right]$$

$$= \lim_{k \to \infty} e^{-\frac{z}{\pi}\left(1 - \frac{1}{2} + \frac{1}{3} - \cdots + \frac{1}{2k-1} - \frac{1}{2k}\right)}$$

$$\times z \left(1 - \frac{z}{\pi}\right) e^{\frac{z}{\pi}} \left(1 + \frac{z}{\pi}\right) e^{-\frac{z}{\pi}} \left(1 - \frac{z}{2\pi}\right) e^{\frac{z}{2\pi}} \left(1 + \frac{z}{2\pi}\right) e^{-\frac{z}{2\pi}} \cdots,$$

since the product whose factors are $\left(1 - \frac{z}{r\pi}\right) e^{r\pi}$ is *absolutely* convergent, and so the order of its factors can be altered. Since $\log 2 = 1 - \frac{1}{2} + \frac{1}{3} - \frac{1}{4} + \frac{1}{5} - \cdots$, this shews that the given product is equal to $e^{-z \log 2/\pi} \sin z$.

2.8 Infinite determinants

Infinite series and infinite products are not by any means the only known cases of limiting processes which can lead to intelligible results. The researches of G. W. Hill in the Lunar Theory, reprinted in [306], brought into notice the possibilities of *infinite determinants*. Infinite determinants had previously occurred in the researches of Fürstenau [230] on the algebraic equation of the nth degree. Special types of infinite determinants (known as *continuants*) occur in the theory of infinite continued fractions; see Sylvester [617, p. 504] and [618, p. 249].

The actual investigation of the convergence is due not to Hill but to Poincaré [528]. We shall follow the exposition given by H. von Koch [643, p. 217].

Let A_{ik} be defined for all integer values (positive and negative) of i, k, and denote by

$$D_m = [A_{ik}]_{i,k=-m,\ldots,+m}$$

the determinant formed of the numbers A_{ik} $(i, k = -m, \ldots, +m)$; then if, as $m \to \infty$, the expression D_m tends to a determinate limit D, we shall say that the infinite determinant

$$[A_{ik}]_{i,k=-\infty,\ldots,+\infty}$$

is *convergent* and has the value D. If the limit D does not exist, the determinant in question will be said to be *divergent*.

The elements A_{ii} (where i takes all values) are said to form the *principal diagonal* of the determinant D; the elements A_{ik} (where i is fixed and k takes all values) are said to form the *row* i; and the elements A_{ik} (where k is fixed and i takes all values) are said to form the *column* k. Any element A_{ik} is called a *diagonal* or a *non-diagonal* element, according as $i = k$ or $i \neq k$. The element $A_{0,0}$ is called the *origin* of the determinant.

2.81 Convergence of an infinite determinant

We shall now shew that *an infinite determinant converges, provided the product of the diagonal elements converges absolutely, and the sum of the non-diagonal elements converges absolutely.*

For let the diagonal elements of an infinite determinant D be denoted by $1 + a_{ii}$, and let the non-diagonal elements be denoted by a_{ik}, $(i \neq k)$, so that the determinant is

$$\begin{vmatrix} \vdots & \vdots & \vdots & \vdots & \vdots \\ \cdots & 1 + a_{-1-1} & a_{-10} & a_{-11} & \cdots \\ \cdots & a_{0-1} & 1 + a_{00} & a_{01} & \cdots \\ \cdots & a_{1-1} & a_{10} & 1 + a_{11} & \cdots \\ \vdots & \vdots & \vdots & \vdots & \vdots \end{vmatrix}$$

Then, since the series $\sum\limits_{t,k=-\infty}^{\infty} |a_{ik}|$ is convergent, the product

$$\overline{P} = \prod_{i=-\infty}^{\infty} \left(1 + \sum_{k=-\infty}^{\infty} |a_{ik}| \right)$$

is convergent.

Now form the products

$$P_m = \prod_{i=-m}^{m} \left(1 + \sum_{k=-m}^{m} a_{ik}\right), \quad \overline{P}_m = \prod_{i=-m}^{m} \left(1 + \sum_{k=-m}^{m} |a_{ik}|\right);$$

then if, in the expansion of P_m, certain terms are replaced by zero and certain other terms have their signs changed, we shall obtain D_m; thus, to each term in the expansion of D_m there corresponds, in the expansion of $\overline{P}_m$, a term of equal or greater modulus. Now $D_{m+p} - D_m$ represents the sum of those terms in the determinant D_{m+p} which vanish when the numbers a_{ik} $\{i, k = \pm(m+1) \cdots \pm(m+p)\}$ are replaced by zero; and to each of these terms there corresponds a term of equal or greater modulus in $\overline{p}_{m+p} - \overline{p}_m$.

Hence $\left|D_{m+p} - D_m\right| \le \overline{P}_{m+p} - \overline{P}_m$. Therefore, since P_m tends to a limit as $m \to \infty$, so also D_m tends to a limit. This establishes the proposition.

2.82 The rearrangement theorem for convergent infinite determinants

We shall now shew that *a determinant, of the convergent form already considered, remains convergent when the elements of any row are replaced by any set of elements whose moduli are all less than some fixed positive number.*

Replace, for example, the elements

$$\ldots, A_{0,-m}, \ldots, A_0, \ldots, A_{0,m}, \ldots$$

of the row through the origin by the elements

$$\ldots, \mu_{-m}, \ldots, \mu_0, \ldots, \mu_m, \ldots$$

which satisfy the inequality

$$|\mu_r| < \mu,$$

where μ is a positive number; and let the new values of D_m and D be denoted by $D_{m'}$, and D'. Moreover, denote by $\overline{P}_{m'}$ and $\overline{P}'$ the products obtained by suppressing in $\overline{P}_m$ and $\overline{P}$ the factor corresponding to the index zero; we see that no terms of $D_{m'}$ can have a greater modulus than the corresponding term in the expansion of $\mu\overline{P}_{m'}$; and consequently, reasoning as in the last article, we have

$$\left|D'_{m+p} - D_{m'}\right| < \mu\overline{P}'_{m+p} - \mu\overline{P}_m,$$

which is sufficient to establish the result stated.

Example 2.8.1 (von Koch)   Shew that the necessary and sufficient condition for the absolute convergence of the infinite determinant

$$\lim_{m\to\infty} \begin{vmatrix} 1 & \alpha_1 & 0 & 0 & \cdots & 0 \\ \beta_1 & 1 & \alpha_2 & 0 & \cdots & 0 \\ 0 & \beta_2 & 1 & \alpha_3 & \cdots & 0 \\ \vdots & \vdots & \vdots & \vdots & \vdots & \vdots \\ 0 & \cdots & 0 & \beta_m & & 1 \end{vmatrix}$$

is that the series $\alpha_1\beta_1 + \alpha_2\beta_2 + \alpha_3\beta_3 + \cdots$ shall be absolutely convergent.

2.9 Miscellaneous examples

Example 2.1   Evaluate $\lim_{n\to\infty} e^{-na}n^b$, $\lim_{n\to\infty} n^{-a}\log n$ when $a>0, b>0$.

Example 2.2 (Trinity, 1904)   Investigate the convergence of

$$\sum_{n=1}^{\infty}\left\{1-n\log\frac{2n+1}{2n-1}\right\}.$$

Example 2.3 (Peterhouse, 1906)   Investigate the convergence of

$$\sum_{n=1}^{\infty}\left\{\frac{1\cdot3\cdots2n+1}{2\cdot4\cdots2n}\cdot\frac{4n+3}{2n+2}\right\}^2.$$

Example 2.4   Find the range of values of z for which the series

$$2\sin^2 z-4\sin^4 z+8\sin^6 z-\cdots+(-1)^{n+1}2^n\sin^{2n}z+\cdots$$

is convergent.

Example 2.5 (Simon)   Shew that the series

$$\frac{1}{z}-\frac{1}{z+1}+\frac{1}{z+2}-\frac{1}{z+3}+\cdots$$

is conditionally convergent, except for certain exceptional values of z; but that the series

$$\frac{1}{z}+\frac{1}{z+1}+\cdots+\frac{1}{z+p-1}-\frac{1}{z+p}-\frac{1}{z+p+1}-\cdots$$

$$-\frac{1}{z+2p+q-1}+\frac{1}{z+2p+q}+\cdots,$$

in which $(p+q)$-negative terms always follow p positive terms, is divergent.

Example 2.6 (Trinity, 1908)   Shew that

$$1-\frac{1}{2}-\frac{1}{4}+\frac{1}{3}-\frac{1}{6}-\frac{1}{8}+\frac{1}{5}-\cdots=\frac{1}{2}\log 2.$$

Example 2.7 (Cesàro)   Shew that the series

$$\frac{1}{1^{\alpha}}+\frac{1}{2^{\beta}}+\frac{1}{3^{\alpha}}+\frac{1}{4^{\beta}}+\cdots\quad(1<\alpha<\beta)$$

is convergent, although $u_{2n+1}/u_{2n}\to\infty$.

Example 2.8 (Cesàro)   Shew that the series $\alpha+\beta^2+\alpha^3+\beta^4+\cdots$ (with $0<\alpha<\beta<1$) is convergent although $u_{2n}/u_{2n-1}\to\infty$.

Example 2.9   Shew that the series

$$\sum_{n=1}^{\infty}\frac{nz^{n-1}\left\{(1+n^{-1})^n-1\right\}}{(z^n-1)\left\{z^n-(1+n^{-1})^n\right\}}$$

converges absolutely for all values of z, except the values

$$z=\left(1+\frac{a}{m}\right)e^{2k\pi i/m}\quad(a=0,1;\ k=0,1,\ldots,m-1;\ m=1,2,3,\ldots).$$

Example 2.10 (de la Vallée Poussin [638])   Shew that, when $\delta > 1$,

$$\sum_{n=1}^{\infty} \frac{1}{n^\delta} = \frac{1}{\delta - 1} + \sum_{n=1}^{\infty} \left[\frac{1}{n^\delta} + \frac{1}{\delta - 1} \left\{ \frac{1}{(n+1)^{\delta-1}} - \frac{1}{n^{\delta-1}} \right\} \right],$$

and shew that the series on the right converges when $0 < \delta < 1$.

Example 2.11   In the series whose general term is $u_n = q^{n-\nu} x^{\frac{1}{2}\nu(\nu+1)}$, $(0 < q < 1 < x)$ where ν denotes the number of digits in the expression of n in the ordinary decimal scale of notation, shew that

$$\lim_{n \to \infty} u_n^{1/n} = q,$$

and that the series is convergent, although $\lim_{n \to \infty} u_{n+1}/u_n = \infty$.

Example 2.12 (Cesàro)   Shew that the series

$$q_1 + q_1^2 + q_2^3 + q_1^4 + q_2^5 + q_3^6 + q_1^7 + \cdots,$$

where $q_n = q^{1+4/n}$, $(0 < q < 1)$ is convergent, although the ratio of the $(n+1)$th term to the nth is greater than unity when n is not a triangular number.

Example 2.13   Shew that the series

$$\sum_{n=0}^{\infty} \frac{e^{2n\pi ix}}{(w+n)^s},$$

where w is real, and where $(w+n)^s$ is understood to mean $e^{s \log(w+n)}$, the logarithm being taken in its arithmetic sense, is convergent for all values of s, when Im x is positive, and is convergent for values of s whose real part is positive, when x is real and not an integer.

Example 2.14   If $u_n > 0$, shew that if $\sum u_n$ converges, then $\varliminf_{n \to \infty} (nu_n) = 0$, and that, if in addition $u_n \geq u_{n+1}$, then $\lim_{n \to \infty} nu_n = 0$.

Example 2.15 (Trinity, 1904)   If

$$a_{m,n} = \begin{cases} \frac{m-n}{2^{m+n}} \frac{(m+n-1)!}{m!\, n!} & m, n > 0, \\ 2^{-m} & n = 0,\ m \neq 0, \\ -2^{-n} & m = 0,\ n \neq 0, \\ 0 & n = m = 0. \end{cases}$$

shew that

$$\sum_{m=0}^{\infty} \left(\sum_{n=0}^{\infty} a_{m,n} \right) = -1, \quad \sum_{n=0}^{\infty} \left(\sum_{m=0}^{\infty} a_{m,n} \right) = 1.$$

Example 2.16 (Jacobi)   By converting the series

$$1 + \frac{8q}{1-q} + \frac{16q^2}{1+q^2} + \frac{24q^3}{1-q^3} + \cdots,$$

(in which $|q| < 1$), into a double series, shew that it is equal to

$$1 + \frac{8q}{(1-q)^2} + \frac{8q^2}{(1+q^2)^2} + \frac{8q^3}{(1-q^3)^2} + \cdots .$$

Example 2.17 (Math. Trip. 1904)   Assuming that $\sin z = z \prod\limits_{r=1}^{\infty} \left(1 - \frac{z^2}{r^2\pi^2}\right)$, shew that if $m \to \infty$ and $n \to \infty$ in such a way that $\lim m/n = k$, where k is finite, then

$$\lim \prod\limits_{r=-n}^{m}{}' \left(1 + \frac{z}{r\pi}\right) = k^{z/\pi}\frac{\sin z}{z},$$

the prime indicating that the factor for which $r = 0$ is omitted.

Example 2.18 (Math. Trip. 1906)   If $u_0 = u_1 = u_2 = 0$, and if, when $n > 1$,

$$u_{2n-1} = -\frac{1}{\sqrt{n}}, \qquad u_{2n} = \frac{1}{\sqrt{n}} + \frac{1}{n} + \frac{1}{n\sqrt{n}},$$

then $\prod\limits_{n=0}^{\infty}(1 + u_n)$ converges, though $\sum\limits_{n=0}^{\infty} u_n$ and $\sum\limits_{n=0}^{\infty} u_n^2$ are divergent.

Example 2.19   Prove that

$$\prod\limits_{n=1}^{\infty}\left\{\left(1 - \frac{z}{n}\right)^{nk} \exp\left(\sum\limits_{m=1}^{k+1} \frac{n^{k-m}z^m}{m}\right)\right\},$$

where k is any positive integer, converges absolutely for all values of z.

Example 2.20 (Cauchy)   If $\sum\limits_{n=1}^{\infty} a_n$ be a conditionally convergent series of real terms, then $\prod\limits_{n=1}^{\infty}(1 + a_n)$ converges (but not absolutely) or diverges to zero according as $\sum\limits_{n=1}^{\infty} a_n^2$ converges or diverges.

Example 2.21 (Hill; see §19.42)   Let $\sum\limits_{n=1}^{\infty} \theta_n$ be an absolutely convergent series. Shew that the infinite determinant

$$\Delta(c) = \begin{vmatrix} \vdots & \vdots & \vdots & \vdots & \vdots & \vdots & \vdots \\ \cdots & \frac{(c-4)^2-\theta_0}{4^2-\theta_0} & \frac{-\theta_1}{4^2-\theta_0} & \frac{-\theta_2}{4^2-\theta_0} & \frac{-\theta_3}{4^2-\theta_0} & \frac{-\theta_3}{4^2-\theta_0} & \cdots \\ \cdots & \frac{-\theta_1}{2^2-\theta_0} & \frac{(c-2)^2-\theta_0}{2^2-\theta_0} & \frac{-\theta_1}{2^2-\theta_0} & \frac{-\theta_2}{2^2-\theta_0} & \frac{-\theta_3}{2^2-\theta_0} & \cdots \\ \cdots & \frac{-\theta_2}{0^2-\theta_0} & \frac{-\theta_1}{0^2-\theta_0} & \frac{c^2-\theta_0}{0^2-\theta_0} & \frac{-\theta_1}{0^2-\theta_0} & \frac{-\theta_2}{0^2-\theta_0} & \cdots \\ \cdots & \frac{-\theta_3}{2^2-\theta_0} & \frac{-\theta_2}{2^2-\theta_0} & \frac{-\theta_1}{2^2-\theta_0} & \frac{(c+2)^2-\theta_0}{2^2-\theta_0} & \frac{-\theta_1}{2^2-\theta_0} & \cdots \\ \cdots & \frac{-\theta_4}{4^2-\theta_0} & \frac{-\theta_3}{4^2-\theta_0} & \frac{-\theta_2}{4^2-\theta_0} & \frac{-\theta_1}{4^2-\theta_0} & \frac{(c+4)^2-\theta_0}{4^2-\theta_0} & \\ & \vdots & \vdots & \vdots & \vdots & \vdots & \vdots \end{vmatrix}$$

converges; and shew that the equation $\Delta(c) = 0$ is equivalent to the equation

$$\sin^2(\pi c/2) = \Delta(0)\sin^2\left(\pi\sqrt{\theta_0}/2\right). \tag{2.3}$$

3

Continuous Functions and Uniform Convergence

3.1 The dependence of one complex number on another

The problems with which Analysis is mainly occupied relate to the *dependence* of one complex number on another. If z and ζ are two complex numbers, so connected that, if z is given any one of a certain set of values, corresponding values of ζ can be determined, e.g. if ζ is the square of z, or if $\zeta = 1$ when z is real and $\zeta = 0$ for all other values of z, then ζ is said to be a function of z.

This dependence must not be confused with the most important case of it, which will be explained later under the title of *analytic functionality*.

If ζ is a real function of a real variable z, then the relation between ζ and z, which may be written $\zeta = f(z)$, can be visualised by a curve in a plane, namely the locus of a point whose coordinates referred to rectangular axes in the plane are (z, ζ). No such simple and convenient geometrical method can be found for visualising an equation $\zeta = f(z)$, considered as defining the dependence of one complex number $\zeta = \xi + i\eta$ on another complex number $z = x + iy$. A representation strictly analogous to the one already given for real variables would require four-dimensional space, since the number of variables ξ, η, x, y is now four.

One suggestion (made by Lie and Weierstrass) is to use a doubly-manifold system of lines in the quadruply-manifold totality of lines in three-dimensional space. Another suggestion is to represent ξ and η separately by means of surfaces $\xi = \xi(x, y)$, $\eta = \eta(x, y)$. A third suggestion, due to Heffter [284], is to write $\zeta = re^{i\theta}$, then draw the surface $r = r(x, y)$, which may be called the *modular-surface of the function*, and on it to express the values of θ by surface-markings. It might be possible to modify this suggestion in various ways by representing θ by curves drawn on the surface $r = r(x, y)$.

3.2 Continuity of functions of real variables

The reader will have a general idea (derived from the graphical representation of functions of a real variable) as to what is meant by continuity.

We now have to give a precise definition which shall embody this vague idea. Let $f(x)$ be a function of x defined when $a \leq x \leq b$. Let x_1 be such that $a \leq x_1 \leq b$. If there exists a number ℓ such that, corresponding to an arbitrary positive number ε, we can find a positive number η such that $|f(x) - \ell| < \varepsilon$, whenever $|x - x_1| < \eta$, $x \neq x_1$, and $a \leq x \leq b$, then ℓ is called the limit of $f(x)$ as $x \to x_1$.

It may happen that we can find a number ℓ_+ (even when ℓ does not exist) such that

$|f(x) - \ell_+| < \varepsilon$ when $x_1 < x < x_1 + \eta$. We call ℓ_+ the limit of $f(x)$ when x approaches x_1 from the right and denote it by $f(x_1 + 0)$; in a similar manner we define $f(x_1 - 0)$ if it exists.

If $f(x_1 + 0), f(x_1), f(x_1 - 0)$ all exist and are equal, we say that $f(x)$ is *continuous* at x_1; so that if $f(x)$ is continuous at x_1, then, given ε, we can find η such that $|f(x) - f(x_1)| < \varepsilon$, whenever $|x - x_1| < \eta$, and $a \le x \le b$.

If ℓ_+ and ℓ_- exist but are unequal, $f(x)$ is said to have an *ordinary discontinuity*[1] at x_1; and if $\ell_+ = \ell_- \ne f(x_1)$, $f(x)$ is said to have a *removable discontinuity* at x_1.

If $f(x)$ is a complex function of a real variable, and if $f(x) = g(x) + ih(x)$ where $g(x)$ and $h(x)$ are real, the continuity of $f(x)$ at x_1 implies the continuity of $g(x)$ and of $h(x)$. For when $|f(x) - f(x_1)| < \varepsilon$, then $|g(x) - g(x_1)| < \varepsilon$ and $|h(x) - h(x_1)| < \varepsilon$; and the result stated is obvious.

Example 3.2.1 From Examples 2.2.1 and 2.2.2 deduce that if $f(x)$ and $\phi(x)$ are continuous at x_1, so are $f(x) \pm \phi(x)$, $f(x) \times \phi(x)$ and, if $\phi(x_1) \ne 0$, $f(x)/\phi(x)$.

The popular idea of continuity, so far as it relates to a real variable $f(x)$ depending on another real variable x, is somewhat different from that just considered, and may perhaps best be expressed by the statement "The function $f(x)$ is said to depend continuously on x if, as x passes through the set of all values intermediate between any two adjacent values x_1 and x_2, $f(x)$ passes through the set of all values intermediate between the corresponding values $f(x_1)$ and $f(x_2)$."

The question thus arises, how far this popular definition is equivalent to the precise definition given above.

Cauchy shewed that if a real function $f(x)$, of a real variable x, satisfies the precise definition, then it also satisfies what we have called the popular definition; this result will be proved in §3.63. But the converse is not true, as was shewn by Darboux. This fact may be illustrated by the following example due to Mansion [454].

Between $x = -1$ and $x = +1$ (except at $x = 0$), let $f(x) = \sin\frac{\pi}{2x}$; and let $f(0) = 0$. It can then be proved that $f(x)$ depends continuously on x near $x = 0$, in the sense of the popular definition, but is not continuous at $x = 0$ in the sense of the precise definition.

Example 3.2.2 If $f(x)$ be defined and be an increasing function in the range (a, b), the limits $f(x \pm 0)$ exist at all points in the interior of the range. *Hint.* If $f(x)$ be an increasing function, a section of rational numbers can be found such that, if a, A be any members of its L-class and its R-class, $a < f(x + h)$ for every positive value of h and $A \ge f(x + h)$ for some positive value of h. The number defined by this section is $f(x + 0)$.

3.21 Simple curves. Continua

Let x and y be two real functions of a real variable t which are continuous for every value of t such that $a \le t \le b$. We denote the dependence of x and y on t by writing

$$x = x(t), \quad y = y(t) \qquad (a \le t \le b)$$

The functions $x(t), y(t)$ are supposed to be such that they do not assume the same pair of

---

[1] If a function is said to have ordinary discontinuities at certain points of an interval it is implied that it is continuous at all other points of the interval.

values for any two different values of t in the range $a < t < b$. Then the set of points with coordinates (x, y) corresponding to these values of t is called a *simple curve*. If

$$x(a) = x(b), \quad y(a) = y(b),$$

the simple curve is said to be *closed*.

Example 3.2.3 The circle $x^2 + y^2 = 1$ is a simple closed curve; for we may write[2]

$$x = \cos t, \quad y = \sin t, \quad (0 \leq t \leq 2\pi).$$

A *two-dimensional continuum* is a set of points in a plane possessing the following two properties:

(i) If (x, y) be the Cartesian coordinates of any point of it, a *positive* number δ (depending on x and y) can be found such that every point whose distance from (x, y) is less than δ belongs to the set.
(ii) Any two points of the set can be joined by a simple curve consisting entirely of points of the set.

Example 3.2.4 The points for which $x^2 + y^2 < 1$ form a continuum. For if P be any point inside the unit circle such that $OP = r < 1$, we may take $\delta = 1 - r$; and any two points inside the circle may be joined by a straight line lying wholly inside the circle.

The following two theorems will be assumed in this work; simple cases of them appear obvious from geometrical intuitions and, generally, theorems of a similar nature will be taken for granted, as formal proofs are usually extremely long and difficult. Formal proofs will be found in Watson [650].

(I) A simple closed curve divides the plane into two continua (the *interior* and the *exterior*).
(II) If P be a point on the curve and Q be a point not on the curve, the angle between QP and Ox increases by $\pm 2\pi$ or by zero, as P describes the curve, according as Q is an interior point or an exterior point. If the increase is $+2\pi$, P is said to describe the curve *counter-clockwise*.

A continuum formed by the interior of a simple curve is sometimes called *an open two-dimensional region*, or briefly an *open region*, and the curve is called its *boundary*; such a continuum with its boundary is then called *a closed two-dimensional region*, or briefly a *closed region* or *domain*. A simple curve is sometimes called *a closed one-dimensional region*; a simple curve *with its end-points omitted* is then called an *open one-dimensional region*.

3.22 Continuous functions of complex variables

Let $f(z)$ be a function of z defined at all points of a closed region (one- or two-dimensional) in the Argand diagram, and let z_1 be a point of the region.

---

[2] For a proof that the sine and cosine are continuous functions, see the *Appendix*, §A.41.

Then $f(z)$ is said to be continuous at z_1, if given any positive number ε, we can find a corresponding positive number η such that

$$|f(z) - f(z_1)| < \varepsilon,$$

whenever $|z - z_1| < \eta$ and z is a point of the region.

3.3 Series of variable terms. Uniformity of convergence

Consider the series

$$x^2 + \frac{x^2}{1 + x^2} + \frac{x^2}{(1 + x^2)^2} + \cdots + \frac{x^2}{(1 + x^2)^n} + \cdots .$$

This series converges absolutely (§2.33) for all real values of x. If $S_n(x)$ be the sum of n terms, then

$$S_n(x) = 1 + x^2 - \frac{1}{(1 + x^2)^{n-1}};$$

and so $\lim_{n \to \infty} S_n(x) = 1 + x^2$; $(x \neq 0)$, but $S_n(0) = 0$, and therefore $\lim_{n \to \infty} S_n(0) = 0$.

Consequently, although the series is an absolutely convergent series of continuous functions of x, the sum is a discontinuous function of x. We naturally enquire the reason of this rather remarkable phenomenon, which was investigated in 1841–1848 by Stokes [608], Seidel [590] and Weierstrass [662, pp. 67, 75], who shewed that it cannot occur except in connexion with another phenomenon, that of *non-uniform* convergence, which will now be explained.

Let the functions $u_1(z), u_2(z), \ldots$ be defined at all points of a closed region of the Argand diagram. Let

$$S_n(z) = u_1(z) + u_2(z) + \cdots + u_n(z).$$

The condition that the series $\sum_{n=1}^{\infty} u_n(z)$ should *converge* for any particular value of z is that, given ε, a number n should exist such that

$$|S_{n+p}(z) - S_n(z)| < \varepsilon$$

for *all* positive values of p, the value of n of course depending on ε.

Let n have the smallest integer value for which the condition is satisfied. This integer will *in general* depend on the particular value of z which has been selected for consideration. We denote this dependence by writing $n(z)$ in place of n. Now it *may happen* that we can find a number N, *independent of z*, such that $n(z) < N$ for all values of z in the region under consideration. If this number N exists, the series is said to *converge uniformly* throughout the region. If no such number N exists, the convergence is said to be non-uniform. The reader who is unacquainted with the concept of uniformity of convergence will find it made much clearer by consulting Bromwich [102], where an illuminating account of Osgood's graphical investigation is given.

Uniformity of convergence is thus a property depending on a whole *set* of values of z, whereas previously we have considered the convergence of a series for various particular values of z, the convergence for each value being considered *without reference to the other values.*

We define the phrase 'uniformity of convergence *near* a point z' to mean that there is a definite positive number δ such that the series converges uniformly in the domain common to the circle $|z - z_1| \le \delta$ and the region in which the series converges.

3.31 On the condition for uniformity of convergence

This section shews that it is indifferent whether uniformity of convergence is defined by means of the partial remainder $R_{n,p}(z)$ or by $R_n(z)$. Writers differ in the definition taken as fundamental.

If $R_{n,p}(z) = u_{n+1}(z) + u_{n+2}(z) + \cdots + u_{n+p}(z)$, we have seen that the necessary and sufficient condition that $\sum_{n=1}^{\infty} u_n(z)$ should converge uniformly in a region is that, given any positive number ε, it should be possible to choose N *independent of z* (but depending on ε) such that

$$|R_{N,p}(z)| < \varepsilon$$

for all positive integral values of p.

If the condition is satisfied, by §2.22, $S_n(z)$ tends to a limit, $S(z)$, say for each value of z under consideration; and then, since ε is *independent of p*,

$$|\lim_{p \to \infty} R_{N,p}(z)| \le \varepsilon,$$

and therefore, when $n > N$,

$$S(z) - S_n(z) = \lim_{p \to \infty} R_{N,p}(z) - R_{N,n-N}(z),$$

and so $|S(z) - S_n(z)| < 2\varepsilon$.

Thus (writing $\varepsilon/2$ for ε) a *necessary* condition for uniformity of convergence is that $|S(z) - S_n(z)| < \varepsilon$, whenever $n > N$ and N is *independent of z*; the condition is also *sufficient*; for if it is satisfied it follows as in §2.22 (I) that $|R_{N,p}(z)| < 2\varepsilon$, which, by definition, is the condition for uniformity.

Example 3.3.1   Shew that, if x be real, the sum of the series

$$\frac{x}{1(x + 1)} + \frac{x}{(x + 1)(2x + 1)} + \cdots + \frac{x}{\{(n - 1)x + 1\}\{nx + 1\}} + \cdots$$

is discontinuous at $x = 0$ and the series is non-uniformly convergent near $x = 0$.

Solution   The sum of the first n terms is easily seen to be $1 - \frac{1}{nx+1}$; so when $x = 0$ the sum is 0; when $x \ne 0$, the sum is 1. The value of $R_n(x) = S(x) - S_n(x)$ is $\frac{1}{nx+1}$ if $x \ne 0$; so when x is small, say $x =$ one-hundred-millionth, the remainder after a million terms is $\frac{1}{\frac{1}{100}+1}$ or $1 - \frac{1}{101}$, so the first million terms of the series do not contribute one per cent of the sum. And in general, to make $\frac{1}{nx+1} < \varepsilon$, it is necessary to take $n > \frac{1}{x}\left(\frac{1}{\varepsilon} - 1\right)$. Corresponding to a given ε, no number N exists, independent of x, such that $n < N$ for all values of x in any interval including $x = 0$; for by taking x sufficiently small we can make n greater than any number N which is independent of x. There is therefore non-uniform convergence near $x = 0$.

Example 3.3.2   Discuss the series

$$\sum_{n=1}^{\infty} \frac{x\{n(n+1)x^2 - 1\}}{\{1 + n^2 x^2\}\{1 + (n+1)^2 x^2\}},$$

in which x is real.

The nth term can be written $\frac{nx}{1+n^2 x^2} - \frac{(n+1)x}{1+(n+1)^2 x^2}$, so $S(x) = \frac{x}{1+x^2}$, and

$$R_n(x) = \frac{(n+1)x}{1 + (n+1)^2 x^2}.$$

Note   In this example the sum of the series is not discontinuous at $x = 0$.

But (taking $\varepsilon < \frac{1}{2}$, and $x \neq 0$), $|R_n(x)| < \varepsilon$ if $\varepsilon^{-1}(n+1)|x| < 1 + (n+1)^2 x^2$; i.e. if

$$n + 1 > \frac{1}{2}\{\varepsilon^{-1} + \sqrt{\varepsilon^{-2} - 4}\}|x|^{-1} \quad, \text{or} \quad n + 1 < \frac{1}{2}\{\varepsilon^{-1} - \sqrt{\varepsilon^{-2} - 4}\}|x|^{-1}.$$

Now it is not the case that the second inequality is satisfied for all values of n *greater* than a certain value and for all values of x; and the first inequality gives a value of $n(x)$ which tends to infinity as $x \to 0$; so that, corresponding to any interval containing the point $x = 0$, there is no number N *independent of* x. The series, therefore, is non-uniformly convergent near $x = 0$.

The reader will observe that $n(x)$ is discontinuous at $x = 0$; for $n(x) \to \infty$ as $x \to 0$, but $n(0) = 0$.

3.32 Connexion of discontinuity with non-uniform convergence

We shall now shew that *if a series of continuous functions of z is uniformly convergent for all values of z in a given closed domain, the sum is a continuous function of z at all points of the domain.*

For let the series be $f(z) = u_1(z) + u_2(z) + \cdots + u_n(z) + \cdots = S_n(z) + R_n(z)$, where $R_n(z)$ is the remainder after n terms. Since the series is uniformly convergent, given any positive number ε, we can find a corresponding integer n *independent of* z, such that $|R_n(z)| < \frac{1}{3}\varepsilon$ for *all* values of z within the domain. Now n and ε being thus fixed, we can, on account of the continuity of $S_n(z)$, find a positive number η such that

$$|S_n(z) - S_n(z')| < \tfrac{1}{3}\varepsilon,$$

whenever $|z - z'| < \eta$. We have then

$$|f(z) - f(z')| = |S_n(z) - S_n(z') + R_n(z) - R_n(z')|$$
$$< |S_n(z) - S_n(z')| + |R_n(z)| + |R_n(z')|$$
$$< \varepsilon,$$

which is the condition for continuity at z.

Example 3.3.3   Shew that near $x = 0$ the series $u_1(x) + u_2(x) + u_3(x) + \cdots$, where

$$u_1(x) = x, \qquad u_n(x) = x^{\frac{1}{2n-1}} - x^{\frac{1}{2n-3}},$$

and real values of x are concerned, is discontinuous and non-uniformly convergent.

In this example it is convenient to take a slightly different form of the test; we shall shew that, given an arbitrarily small number ε, it is possible to choose values of x, as small as we please, depending on n in such a way that $|R_n(x)|$ is *not* less than ε for any value of n, no matter how large. The reader will easily see that the existence of such values of x is inconsistent with the condition for uniformity of convergence.

The value of $S_n(x)$ is $x^{1/2n-1}$; as n tends to infinity, $S_n(x)$ tends to $+1$, 0, or -1, according as x is positive, zero, or negative. The series is therefore absolutely convergent for all values of x, and has a discontinuity at $x = 0$.

In this series $R_n(x) = 1 - x^{1/(2n-1)}$, $(x > 0)$; however great n may be, by taking $x = e^{-(2n-1)}$ (this value of x satisfies the condition $|x| < \delta$ whenever $2n - 1 > \log \delta^{-1}$), we can cause this remainder to take the value $1 - e^{-1}$, which is not arbitrarily small. The series is therefore non-uniformly convergent near $x = 0$.

Example 3.3.4   Shew that near $z = 0$ the series

$$\sum_{n=1}^{\infty} \frac{-2z(1 + z)^{n-1}}{\{1 + (1 + z)^{n-1}\}\{1 + (1 + z)^n\}}$$

is non-uniformly convergent and its sum is discontinuous.

The nth term can be written

$$\frac{1 - (1 + z)^n}{1 + (1 + z)^n} - \frac{1 - (1 + z)^{n-1}}{1 + (1 + z)^{n-1}},$$

so the sum of the first n terms is $\dfrac{1 - (1 + z)^n}{1 + (1 + z)^n}$. Thus, considering real values of z greater than -1, it is seen that the sum to infinity is $+1$, 0, or -1, according as z is negative, zero, or positive. There is thus a discontinuity at $z = 0$. This discontinuity is explained by the fact that the series is non-uniformly convergent near $z = 0$; for the remainder after n terms in the series when z is positive is $\dfrac{-2}{1 + (1 + z)^n}$, and, however great n may be, by taking $z = \dfrac{1}{n}$, this can be made numerically greater than $\dfrac{2}{1 + e}$, which is not arbitrarily small. The series is therefore non-uniformly convergent near $z = 0$.

3.33 *The distinction between absolute and uniform convergence*

The *uniform* convergence of a series in a domain does not necessitate its *absolute* convergence at any points of the domain, nor conversely. Thus the series $\sum_{n=1}^{\infty} \frac{z^2}{(1+z^2)^n}$ converges *absolutely*, but (near $z = 0$) not *uniformly*; while in the case of the series $\sum_{n=1}^{\infty} \frac{(-1)^{n-1}}{z^2+n}$, the series of moduli is $\sum_{n=1}^{\infty} \frac{1}{|z^2+n|}$, which is divergent, so the series is only *conditionally convergent*; but for all real values of z, the terms of the series are alternately positive and negative and numerically decreasing, so the sum of the series lies between the sum of its first n terms and of its first $(n + 1)$ terms, and so the remainder after n terms is numerically less than the nth term. Thus we only need take a finite number (independent of z) of terms in order to ensure that for all real values of z the remainder is less than any assigned number ε, and so the series is *uniformly* convergent.

Absolutely convergent series behave like series with a finite number of terms in that we can multiply them together and transpose their terms. *Uniformly convergent* series behave like series with a finite number of terms in that they are continuous if each term in the series is continuous and (as we shall see) the series can then be integrated term by term.

3.34 A condition, due to Weierstrass, for uniform convergence

This appears in [661, p. 70]. The test given by this condition is usually described (e.g. by Osgood, [512]) as the *M*-test for uniform convergence.

A sufficient, though not *necessary*, condition for the uniform convergence of a series may be enunciated as follows:

If, for all values of z within a domain, the moduli of the terms of a series $S = u_1(z) + u_2(z) + u_3(z) + \cdots$ are respectively less than the corresponding terms in a convergent series of positive terms $T = M_1 + M_2 + M_3 + \cdots$, where M_n is *independent of* z, then the series S is uniformly convergent in this region. This follows from the fact that, the series T being convergent, it is always possible to choose n so that the remainder after the first n terms of T, and therefore the modulus of the remainder after the first n terms of S, is less than an assigned positive number ε; and since the value of n thus found is independent of z, it follows (§3.31) that the series S is uniformly convergent; by §2.34, the series S also converges absolutely.

Example 3.3.5 The series

$$\cos z + \frac{1}{2^2} \cos^2 z + \frac{1}{3^2} \cos^3 z + \cdots$$

is uniformly convergent for all real values of z, because the moduli of its terms are not greater than the corresponding terms of the convergent series

$$1 + \frac{1}{2^2} + \frac{1}{3^2} + \cdots,$$

whose terms are positive constants.

3.341 Uniformity of convergence of infinite products

The definition is, effectively, that given by Osgood [513, p. 462]. The condition here given for uniformity of convergence is also established in that work.

A convergent product $\prod_{n=1}^{\infty} (1 + u_n(z))$ is said to converge uniformly in a domain of values of z if, given ε, we can find m *independent of* z such that

$$\left| \prod_{n=1}^{m+p} (1 + u_n(z)) - \prod_{n=1}^{m} (1 + u_n(z)) \right| < \varepsilon$$

for all positive integral values of p.

The only condition for uniformity of convergence which will be used in this work is that the product converges uniformly if $|u_n(z)| < M_n$ where M_n is independent of z and $\sum_{n=1}^{\infty} M_n$ converges.

To prove the validity of the condition we observe that $\prod_{n=1}^{\infty}(1 + M_n)$ converges (§2.7), and so we can choose m such that

$$\prod_{n=1}^{m+p}(1 + M_n) - \prod_{n=1}^{m}(1 + M_n) < \varepsilon;$$

and then we have

$$\left|\prod_{n=1}^{m+p}(1 + u_n(z)) - \prod_{n=1}^{m}(1 + u_n(z))\right| = \left|\prod_{n=1}^{m}(1 + u_n(z))\left[\prod_{n=m+1}^{m+p}(1 + u_n(z)) - 1\right]\right|$$

$$\leq \prod_{n=1}^{m}(1 + M_n)\left[\prod_{n=m+1}^{m+p}(1 + M_n) - 1\right]$$

$$< \varepsilon,$$

and the choice of m is independent of z.

3.35 *Hardy's tests for uniform convergence*

These results, which are generalizations of Abel's theorem (§3.71, below), though well known, do not appear to have been published before 1907 [276]. From their resemblance to the tests of Dirichlet and Abel for convergence, Bromwich proposes to call them Dirichlet's and Abel's tests respectively.

The reader will see, from §2.31, that if, in a given domain, $\left|\sum_{n=1}^{p} a_n(z)\right| \leq k$ where $a_n(z)$ is real and k is finite and independent of p and z, and if $f_n(z) \geq f_{n+1}(z)$ and $f_n(z) \to 0$ *uniformly* as $n \to \infty$, then $\sum_{n=1}^{\infty} a_n(z)f_n(z)$ converges uniformly.

Also that if $k \geq u_n(z) \geq u_{n+1}(z) \geq 0$, where k is *independent of z* and $\sum_{n=1}^{\infty} a_n(z)$ converges uniformly, then $\sum_{n=1}^{\infty} a_n(z)u_n(z)$ converges uniformly. *Hint.* To prove the latter, observe that m can be found such that

$$a_{m+1}(z), a_{m+1}(z) + a_{m+2}(z), \dots, a_{m+1}(z) + a_{m+2}(z) + \cdots + a_{m+p}(z)$$

are numerically less than ε/k; and therefore (§2.301)

$$\left|\sum_{n=m+1}^{m+p} a_n(z)u_n(z)\right| < \varepsilon u_{m+1}(z)/k < \varepsilon,$$

and the choice of ε and m is *independent of z*.

Example 3.3.6   Shew that, if $\delta > 0$, the series

$$\sum_{n=1}^{\infty} \frac{\cos n\theta}{n}, \quad \sum_{n=1}^{\infty} \frac{\sin n\theta}{n}$$

converge uniformly in the range $\delta \leq \theta \leq 2\pi - \delta$. Obtain the corresponding result for the series

$$\sum_{n=1}^{\infty} \frac{(-1)^n \cos n\theta}{n}, \quad \sum_{n=1}^{\infty} \frac{(-1)^n \sin n\theta}{n},$$

by writing $\theta + \pi$ for θ.

Example 3.3.7 (Hardy)   If, when $a \leq x \leq b$, $|\omega_n(x)| < k_1$ and $\sum_{n=1}^{\infty} |\omega_{n+1}(x) - \omega_n(x)| < k_2$, where k_1, k_2 are independent of n and x, and if $\sum_{n=1}^{\infty} a_n$ is a convergent series independent of x, then $\sum_{n=1}^{\infty} a_n \omega_n(x)$ converges uniformly when $a \leq x \leq b$. *Hint.* For we can choose m, independent of x, such that $\left| \sum_{n=m+1}^{m+p} a_n \right| < \varepsilon$, and then, by Corollary 2.3.2, we have $\left| \sum_{n=m+1}^{m+p} a_n \omega_n(x) \right| < (k_1 + k_2)\varepsilon$.

3.4 Discussion of a particular double series

Let ω_1 and ω_2 be any constants whose ratio is not purely real; and let α be positive. The series $\sum_{m,n} \dfrac{1}{(z + 2m\omega_1 + 2n\omega_2)^\alpha}$, in which the summation extends over all positive and negative integral and zero values of m and n, is of great importance in the theory of Elliptic Functions. At each of the points $z = -2m\omega_1 - 2n\omega_2$ the series does not exist. It can be shewn that the series converges absolutely for all other values of z if $\alpha > 2$, and the convergence is uniform for those values of z such that $|z + 2m\omega_1 + 2n\omega_2| \geq \delta$ for all integral values of m and n, where δ is an arbitrary positive number.

Let $\sum'$ denote a summation for all integral values of m and n, the term for which $m = n = 0$ being omitted.

Now, if m and n are not both zero, and if $|z + 2m\omega_1 + 2n\omega_2| \geq \delta > 0$ for all integral values of m and n, then we can find a positive number C, depending on δ but not on z, such that

$$\left| \frac{1}{(z + 2m\omega_1 + 2n\omega_2)^\alpha} \right| < C \left| \frac{1}{(2m\omega_1 + 2n\omega_2)^\alpha} \right|.$$

Consequently, by §3.34, the given series is absolutely and uniformly convergent in the domain considered if $\sum' \dfrac{1}{|m\omega_1 + n\omega_2|^\alpha}$ converges. (The reader will easily define uniformity of convergence of double series (see §3.5).) To discuss the convergence of the latter series, let

$$\omega_1 = \alpha_1 + i\beta_1, \quad \omega_2 = \alpha_2 + i\beta_2,$$

where $\alpha_1, \alpha_2, \beta_1, \beta_2$, are real. Since ω_2/ω_1 is not real, $\alpha_1\beta_2 - \alpha_2\beta_1 \neq 0$. Then the series is

$$\sum' \frac{1}{\{(\alpha_1 m + \alpha_2 n)^2 + (\beta_1 m + \beta_2 n)^2\}^{\alpha/2}}.$$

This converges (Corollary 2.5.1) if the series

$$S = \sum{}' \frac{1}{(m^2 + n^2)^{\alpha/2}}$$

converges; for the quotient of corresponding terms is

$$\left\{ \frac{(\alpha_1 + \alpha_2\mu)^2 + (\beta_1 + \beta_2\mu)^2}{1 + \mu^2} \right\}^{\alpha/2},$$

where $\mu = n/m$. This expression, *qua* function of a continuous real variable μ, can be proved to have a positive minimum[3] (not zero) since $\alpha_1\beta_2 - \alpha_2\beta_1 \neq 0$; and so the quotient is always greater than a *positive* number K (independent of μ). We have therefore only to study the convergence of the series S. Let

$$S_{p,q} = \sum_{m=-p}^{p} \sum_{n=-q}^{q}{}' \frac{1}{(m^2 + n^2)^{\alpha/2}} \leq 4 \sum_{m=0}^{\infty} \sum_{m=0}^{\infty}{}' \frac{1}{(m^2 + n^2)^{\alpha/2}}.$$

Separating $S_{p,q}$ into the terms for which $m = n$, $m > n$, and $m < n$, respectively, we have

$$\frac{1}{4}S_{p,q} = \sum_{m=1}^{p} \frac{1}{(2m^2)^{\alpha/2}} + \sum_{m=1}^{p}\sum_{n=0}^{m-1} \frac{1}{(m^2 + n^2)^{\alpha/2}} + \sum_{n=1}^{q}\sum_{m=0}^{n-1} \frac{1}{(m^2 + n^2)^{\alpha/2}}.$$

But

$$\sum_{n=0}^{m-1} \frac{1}{(m^2 + n^2)^{\alpha/2}} < \frac{m}{(m^2)^{\alpha/2}} = \frac{1}{m^{\alpha-1}}; \tag{3.1}$$

therefore

$$\frac{1}{4}S \leq \sum_{m=1}^{\infty} \frac{1}{2^{\alpha/2}m^\alpha} + \sum_{m=1}^{\infty} \frac{1}{m^{\alpha-1}} + \sum_{n=1}^{\infty} \frac{1}{n^{\alpha-1}}. \tag{3.2}$$

But these last series are known to be convergent if $\alpha - 1 > 1$. So the series S is convergent if $\alpha > 2$. The original series is therefore absolutely and uniformly convergent, when $\alpha > 2$, for the specified range of values of z.

Example 3.4.1 (Eisenstein [193])   Prove that the series

$$\sum \frac{1}{(m_1^2 + m_2^2 + \cdots + m_r^2)^\mu},$$

in which the summation extends over all positive and negative integral values and zero values of $m_1, m_2, \ldots, m_r$, except the set of simultaneous zero values, is absolutely convergent if $\mu > \frac{1}{2}r$.

---

[3] The reader will find no difficulty in verifying this statement; the minimum value in question is given by

$$K^{2/\alpha} = \tfrac{1}{2}[\alpha_1^2 + \alpha_2^2 + \beta_1^2 + \beta_2^2 - \{(\alpha_1 - \beta_2)^3 + (\alpha_2 + \beta_1)^2\}^{1/2} \{(\alpha_1 + \beta_2)^2 + (\alpha_2 - \beta_1)^2\}^{1/2}].$$

3.5 The concept of uniformity

There are processes other than that of summing a series in which the idea of uniformity is of importance.

Let ε be an arbitrary positive number; and let $f(z, \zeta)$ be a function of two variables z and ζ, which for each point z of a closed region, satisfies the inequality $|f(z, \zeta)| < \varepsilon$ when ζ is given any one of a certain set of values which will be denoted by (ζ_z); the particular set of values of course depends on the particular value of z under consideration. If a set $(\zeta)_0$ can be found such that every member of the set $(\zeta)_0$ is a member of *all* the sets (ζ_z), the function $f(z, \zeta)$ is said to satisfy the inequality *uniformly* for all points z of the region. And if a function $\phi(z)$ possesses some property, for every positive value of ε, in virtue of the inequality $|f(z, \zeta)| < \varepsilon$, $\phi(z)$ is then said to *possess the property uniformly*.

In addition to the uniformity of convergence of series and products, we shall have to consider uniformity of convergence of integrals and also uniformity of continuity; thus a series is uniformly convergent when $|R_n(z)| < \varepsilon$, $\zeta(= n)$ assuming integer values independent of z only.

Further, a function $f(z)$ is continuous in a closed region if, given ε, we can find a positive number η_s such that $|f(z + \zeta_s) - f(z)| < \varepsilon$ whenever $0 < |\zeta_s| < \eta_s$ and $z + \zeta$ is a point of the region.

The function will be *uniformly* continuous if we can find a positive number η *independent of z*, such that $\eta < \eta_s$ and $|f(z + \zeta) - f(z)| < \varepsilon$ whenever $0 < |\zeta| < \eta$ and $z + \zeta$ is a point of the region (in this case the set $(\zeta)_0$ is the set of points whose moduli are less than η).

We shall find later (§3.61) that continuity involves uniformity of continuity; this is in marked contradistinction to the fact that convergence does not involve uniformity of convergence.

3.6 The modified Heine–Borel theorem

The following theorem is of great importance in connexion with properties of uniformity; we give a proof for a one-dimensional closed region. (A formal proof of the theorem for a two-dimensional region will be found in Watson [650].)

Given (i) a straight line CD and (ii) a law by which, corresponding to each point P of CD, we can determine a closed interval $I(P)$ of CD, P being an interior point of $I(P)$ (except when P is at C or D, when it is an end point). Examples of such laws associating intervals with points will be found in §3.61 and §5.13.

Then the line CD can be divided into a finite number of closed intervals $J_1, J_2, \ldots, J_k$, such that each interval J_r contains at least one point (not an end point) P_r, such that no point of J_r lies outside the interval $I(P_r)$ associated (by means of the given law) with that point P_r. This statement of the Heine–Borel theorem (which is sometimes called the Borel–Lebesgue theorem) is due to Baker [41]. Hobson [316] points out that the theorem is practically given in Goursat [254]; the ordinary form of the Heine–Borel theorem will be found in the treatise cited.

A closed interval of the nature just described will be called a *suitable* interval, and will be said to satisfy condition (*A*).

If CD satisfies condition (*A*), what is required is proved. If not, bisect CD; if either or both of the intervals into which CD is divided is not suitable, bisect it or them. A suitable interval is not to be bisected; for one of the parts into which it is divided might not be suitable.

This process of bisecting intervals which are not suitable either will terminate or it will not. If it does terminate, the theorem is proved, for CD will have been divided into suitable intervals.

Suppose that the process does not terminate; and let an interval, which *can* be divided into suitable intervals by the process of bisection just described, be said to satisfy condition (B). Then, by hypothesis, CD does not satisfy condition (B); therefore at least one of the bisected portions of CD does not satisfy condition (B). Take that one which does not (if neither satisfies condition (B) take the left-hand one); bisect it and select that bisected part which does not satisfy condition (B). This process of bisection and selection gives an unending sequence of intervals $s_0, s_1, s_2, \ldots$ such that:

  (i) The length of s_n is $2^{-n}CD$.
 (ii) No point of s_{n+1} is outside s_n.
(iii) The interval s_n does not satisfy condition (A).

Let the distances of the end points of s_n from C be x_n, y_n; then $x_n \leq x_{n+1} < y_{n+1} \leq y_n$. Therefore, by §2.2, x_n and y_n have limits; and, by the condition (i) above, these limits are the same, say ξ; let Q be the point whose distance from C is ξ. But, by hypothesis, there is a number δ_Q such that every point of CD, whose distance from Q is less than δ_Q, is a point of the associated interval $I(Q)$. Choose n so large that $2^{-n}CD < \delta_Q$; then Q is an internal point or end point of s_n and the distance of every point of s_n from Q is less than δ_Q. And therefore the interval s_n satisfies condition (A), which is contrary to condition (iii) above. The hypothesis that the process of bisecting intervals does not terminate therefore involves a contradiction; therefore the process does terminate and the theorem is proved.

In the two-dimensional form of the theorem[4], the interval CD is replaced by a closed two-dimensional region, the interval $I(P)$ by a circle, or the portion of the circle which lies inside the region, with centre P, and the interval J_r by a square with sides parallel to the axes.

3.61 Uniformity of continuity

From the theorem just proved, it follows without difficulty that if a function $f(x)$ of a real variable x is continuous when $a \leq x \leq b$, then $f(x)$ is *uniformly* continuous throughout the range $a \leq x \leq b$. This result is due to Heine [288].

For let ε be an arbitrary positive number; then, in virtue of the continuity of $f(x)$, corresponding to any value of x, we can find a positive number δ_x, depending on x, such that $|f(x') - f(x)| < \varepsilon/4$ for all values of x' such that $|x' - x| < \delta_x$.

Then by §3.6 we can divide the range (a, b) into a *finite* number of closed intervals with the property that in each interval there is a number x_1 such that $|f(x') - f(x_1)| < \frac{1}{4}\varepsilon$, whenever x' lies in the interval in which x_1 lies.

Let δ_0 be the length of the smallest of these intervals; and let ξ, ξ' be *any* two numbers

---

[4] The reader will see that a proof may be constructed on similar lines by drawing a square circumscribing the region and carrying out a process of dividing squares into four equal squares.

in the closed range (a, b) such that $|\xi - \xi'| < \delta_0$. Then ξ, ξ' lie in the same or in adjacent intervals; if they lie in adjacent intervals let ξ_0 be the common end point. Then we can find numbers x_1, x_2, one in each interval, such that

$$|f(\xi) - f(x_1)| < \tfrac{1}{4}\varepsilon, \quad |f(\xi_0) - f(x_1)| < \tfrac{1}{4}\varepsilon,$$

$$|f(\xi') - f(x_2)| < \tfrac{1}{4}\varepsilon, \quad |f(\xi_0) - f(x_2)| < \tfrac{1}{4}\varepsilon,$$

so that

$$|f(\xi) - f(\xi')| = |\{f(\xi) - f(x_1)\} - \{f(\xi_0) - f(x_1)\}$$
$$- \{f(\xi') - f(x_2)\} + \{f(\xi_0) - f(x_2)\}| < \varepsilon.$$

If ξ, ξ' lie in the same interval, we can prove similarly that $|f(\xi) - f(\xi')| < \varepsilon/2$. In either case we have shewn that, for *any* number ξ in the range, we have

$$|f(\xi) - f(\xi + \zeta)| < \varepsilon$$

whenever $\xi + \zeta$ is in the range and $-\delta_0 < \zeta < \delta_0$ where δ_0 is *independent of* ξ. The *uniformity* of the continuity is therefore established.

Corollary 3.6.1 *From the two-dimensional form of the theorem of §3.6 we can prove that a function of a complex variable, continuous at all points of a closed region of the Argand diagram, is uniformly continuous throughout that region.*

Corollary 3.6.2 *A function $f(x)$ which is continuous throughout the range $a \leq x \leq b$ is bounded in the range; that is to say we can find a number κ independent of x such that $|f(x)| < \kappa$ for all points x in the range.*

Let n be the number of parts into which the range is divided. Let $a, \xi_1, \xi_2, \ldots, \xi_{n-1}, b$ be their end points; then if x be any point of the rth interval we can find numbers $x_1, x_2, \ldots, x_n$ such that

$$|f(a) - f(x_1)| < \tfrac{1}{4}\varepsilon, \quad |f(x_1) - f(\xi_1)| < \tfrac{1}{4}\varepsilon, \quad |f(\xi_1) - f(x_2)| < \tfrac{1}{4}\varepsilon,$$
$$|f(x_2) - f(\xi_2)| < \tfrac{1}{4}\varepsilon, \ldots, |f(x_{r-1}) - f(x)| < \tfrac{1}{4}\varepsilon.$$

Therefore $|f(a) - f(x)| < \tfrac{1}{2}r\varepsilon$, and so $|f(x)| < |f(a)| + \tfrac{1}{2}n\varepsilon$, which is the required result, since the right-hand side is independent of x. The corresponding theorem for functions of complex variables is left to the reader.

3.62 A real function, of a real variable, continuous in a closed interval, attains its upper bound

Let $f(x)$ be a real continuous function of x when $a \leq x \leq b$. Form a section in which the R-class consists of those numbers r such that $r > f(x)$ for all values of x in the range (a, b), and the L-class of all other numbers. This section defines a number α such that $f(x) \leq \alpha$, but, if δ be *any* positive number, values of x in the range exist such that $f(x) > \alpha - \delta$. Then α is called the *upper bound* of $f(x)$; and the theorem states that a number x' in the range can be found such that $f(x') = \alpha$.

For, no matter how small δ may be, we can find values of x for which $|f(x) - \alpha|^{-1} > \delta^{-1}$;

therefore $|f(x) - \alpha|^{-1}$ is not bounded in the range; therefore (Corollary 3.6.2) it is not continuous at some point or points of the range; but since $|f(x) - \alpha|$ is continuous at all points of the range, its reciprocal is continuous at all points of the range (Example 3.2.1) except those points at which $f(x) = \alpha$; therefore $f(x) = \alpha$ at some point of the range; the theorem is therefore proved.

Corollary 3.6.3   *The lower bound of a continuous function may be defined in a similar manner; and a continuous function attains its lower bound.*

Corollary 3.6.4   *If $f(z)$ be a function of a complex variable continuous in a closed region, $|f(z)|$ attains its upper bound.*

3.63 A real function, of a real variable, continuous in a closed interval, attains all values between its upper and lower bounds

Let M, m be the upper and lower bounds of $f(x)$; then we can find numbers $\bar{x}, \underline{x}$, by §3.62, such that $f(\bar{x}) = M$, $f(\underline{x}) = m$; let μ be any number such that $m < \mu < M$. Given any positive number ε, we can (by §3.61) divide the range $(\bar{x}, \underline{x})$ into a *finite* number, r, of closed intervals such that

$$|f(x_{1,r}) - f(x_{2,r})| < \varepsilon,$$

where $x_{1,r}, x_{2,r}$ are any points of the rth interval; take $x_{1,r}, x_{2,r}$ to be the end points of the interval; then there is at least one of the intervals for which $f(x_{1,r}) - \mu$ and $f(x_{2,r}) - \mu$ have opposite signs; and since $\left|\{f(x_{1,r}) - \mu\} - \{f(x_{2,r}) - \mu\}\right| < \varepsilon$, it follows that $|f(x_{1,r}) - \mu| < \varepsilon$.

Since we can find a number $x_{1,r}$ to satisfy this inequality for all values of ε, no matter how small, the lower bound of the function $|f(x) - \mu|$ is zero; since this is a continuous function of x, it follows from Corollary 3.6.3 that $f(x) - \mu$ vanishes for some value of x.

3.64 The fluctuation of a function of a real variable

The terminology of this section is partly that of Hobson [316] and partly that of Young [687]. Let $f(x)$ be a real bounded function, defined when $a \le x \le b$. Let $a \le x_1 \le x_2 \le \cdots \le x_n \le b$.

Then $|f(a) - f(x_1)| + |f(x_1) - f(x_2)| + \cdots + |f(x_n) - f(b)|$ is called the *fluctuation* of $f(x)$ in the range (a, b) for the set of subdivisions $x_1, x_2, \ldots, x_n$. If the fluctuation have an upper bound F_a^b, independent of n, for all choices of $x_1, x_2, \ldots, x_n$, then $f(x)$ is said to have *limited* total *fluctuation* in the range (a, b). F_a^b is called the *total fluctuation* in the range.

Example 3.6.1   If $f(x)$ be monotonic; that is, $(f(x) - f(x'))/(x - x')$ is one-signed or zero for all pairs of different values of x and x', in the range (a, b), its total fluctuation in the range is $|f(a) - f(b)|$.

Example 3.6.2   A function with limited total fluctuation can be expressed as the difference of two positive increasing monotonic functions. *Hint.* These functions may be taken to be $\frac{1}{2}\{F_a^z + f(x)\}, \frac{1}{2}\{F_a^z - f(x)\}$.

Example 3.6.3   If $f(x)$ have limited total fluctuation in the range (a, b), then the limits $f(x \pm 0)$ exist at all points in the interior of the range. [See Example 3.2.2].

Example 3.6.4 If $f(x), g(x)$ have limited total fluctuation in the range (a, b) so has $f(x)g(x)$.
Hint. For

$$|f(x')g(x') - f(x)g(x) \leq |f(x')| \cdot |g(x') - g(x)| + |g(x)| \cdot |f(x') - f(x)|, \tag{3.3}$$

and so the total fluctuation of $f(x)g(x)$ cannot exceed $g \cdot F_a^b + f \cdot G_a^b$, where f, g are the upper bounds of $|f(x)|, |g(x)|$.

3.7 Uniformity of convergence of power series

Let the power series $a_0 + a_1 z + \cdots + a_n z^n + \cdots$ converge absolutely when $z = z_0$. Then, if $|z| \leq |z_0|, |a_n z^n| \leq |a_n z_0^n|$. But since $\sum_{n=0}^{\infty} |a_n z_0^n|$ converges, it follows, by §3.34, that $\sum_{n=0}^{\infty} a_n z^n$ converges uniformly with regard to the variable z when $|z| \leq |z_0|$. Hence, by §3.32, a power series is a continuous function of the variable throughout the closed region formed by the interior and boundary of any circle concentric with the circle of convergence and of smaller radius (§2.6).

3.71 Abel's theorem

Abel's proof [1] employs directly the arguments by which the theorems of §3.32 and §3.35 are proved. In the case when $\sum |a_n|$ converges, the theorem is obvious from §3.7 on continuity up to the circle of convergence.

Let $\sum_{n=0}^{\infty} a_n z^n$ be a power series, whose radius of convergence is unity, and let it be such that $\sum_{n=0}^{\infty} a_n$ converges; and let $0 \leq x \leq 1$; then Abel's theorem asserts that

$$\lim_{x \to 1} \left(\sum_{n=0}^{\infty} a_n x^n \right) = \sum_{n=0}^{\infty} a_n. \tag{3.4}$$

For, with the notation of §3.35, the function x^n satisfies the conditions laid on $u_n(x)$, when $0 \leq x \leq 1$; consequently $f(x) = \sum_{n=0}^{\infty} a_n x^n$ converges *uniformly* throughout the range $0 \leq x \leq 1$; it is therefore, by §3.32, a continuous function of x throughout the range, and so $\lim_{x \to 1^-} f(x) = f(1)$, which is the theorem stated.

3.72 Abel's theorem on multiplication of convergent series

This is a modification of the theorem of §2.53 for absolutely convergent series. This is Abel's original proof [1, Theorem VI]. In some textbooks a more elaborate proof, by the use of Cesàro's sums (§8.43), is given.

Let $c_n = a_0 b_n + a_1 b_{n-1} + \cdots + a_n b_0$. *Then the convergence of* $\sum_{n=0}^{\infty} a_n, \sum_{n=0}^{\infty} b_n$, *and* $\sum_{n=0}^{\infty} c_n$ *is a sufficient condition that*

$$\left(\sum_{n=0}^{\infty} a_n \right) \left(\sum_{n=0}^{\infty} b_n \right) = \sum_{n=0}^{\infty} c_n.$$

For, let

$$A(x) = \sum_{n=0}^{\infty} a_n x^n, \quad B(x) = \sum_{n=0}^{\infty} b_n x^n, \quad C(x) = \sum_{n=0}^{\infty} c_n x^n.$$

Then the series for $A(x)$, $B(x)$, $C(x)$ are absolutely convergent when $|x| < 1$ (§2.6); and consequently, by §2.53, $A(x)B(x) = C(x)$ when $0 < x < 1$; therefore, by Example 2.2.2,

$$\lim_{x \to 1^-} A(x) \cdot \lim_{x \to 1^-} B(x) = \lim_{x \to 1^-} C(x)$$

provided that these three limits exist; but, by §3.71, these three limits are $\sum_{n=0}^{\infty} a_n$, $\sum_{n=0}^{\infty} b_n$, $\sum_{n=0}^{\infty} c_n$; and the theorem is proved.

3.73 Power series which vanish identically

If a convergent power series vanishes for all values of z such that $|z| \leq r_1$, where $r_1 > 0$, then all the coefficients in the power series vanish.

For, if not, let a_m be the first coefficient which does not vanish. Then $a_m + a_{m+1}z + a_{m+2}z^2 + \cdots$ vanishes for all values of z (zero excepted) and converges absolutely when $|z| \leq r < r_1$; hence, if $s = a_{m+1} + a_{m+2}z + \cdots$, we have

$$|s| \leq \sum_{n=1}^{\infty} |a_{m+n}| r^{n-1},$$

and so we can find[5] *a positive number $\delta \leq r$ such that*, whenever $|z| \leq \delta$,

$$\left| a_{m+1}z + a_{m+2}z^2 + \cdots \right| \leq \tfrac{1}{2} |a_m|;$$

and then $|a_m + s| \geq |a_m| - |s| > \tfrac{1}{2} |a_m|$, and so $|a_m + s| \neq 0$ when $|z| < \delta$. We have therefore arrived at a contradiction by supposing that some coefficient does not vanish. Therefore all the coefficients vanish.

Corollary 3.7.1 *We may equate corresponding coefficients in two power series whose sums are equal throughout the region $|z| < \delta$, where $\delta > 0$.*

Corollary 3.7.2 *We may also equate coefficients in two power series which are proved equal only when z is real.*

3.8 Miscellaneous examples

Example 3.1   Shew that the series

$$\sum_{n=1}^{\infty} \frac{z^{n-1}}{(1 - z^n)(1 - z^{n+1})}$$

is equal to $\frac{1}{(1-z)^2}$ when $|z| < 1$ and is equal to $\frac{1}{z(1-z)^2}$ when $|z| > 1$. Is this fact connected with the theory of uniform convergence?

---

[5] It is sufficient to take δ to be the smaller of the numbers r and $\tfrac{1}{2} |a_m| \div \sum_{n=1}^{\infty} |a_{m+n}| r^{n-1}$.

Example 3.2 Shew that the series

$$2\sin\frac{1}{3z} + 4\sin\frac{1}{9z} + \cdots + 2^n\sin\frac{1}{3^n z} + \cdots$$

converges absolutely for all values of z ($z = 0$ excepted), but does not converge uniformly near $z = 0$.

Example 3.3 (Math. Trip. 1907) If $u_n(x) = -2(n-1)^2 x e^{-(n-1)^2 x^2} + 2n^2 x e^{-n^2 x^2}$, shew that $\sum_{n=1}^{\infty} u_n(x)$ does not converge uniformly near $x = 0$.

Example 3.4 Shew that the series $\frac{1}{\sqrt{1}} - \frac{1}{\sqrt{2}} + \frac{1}{\sqrt{3}} - \cdots$ is convergent, but that its square formed by Abel multiplication,

$$\frac{1}{1} - \frac{2}{\sqrt{2}} + \left(\frac{2}{\sqrt{3}} + \frac{1}{2}\right) - \left(\frac{2}{\sqrt{4}} + \frac{2}{\sqrt{6}}\right) - \cdots,$$

is divergent.

Example 3.5 (Cauchy, Cajori) If the convergent series $s = \frac{1}{1^r} - \frac{1}{2^r} + \frac{1}{3^r} - \frac{1}{4^r} + \cdots$ (with $r > 0$) be multiplied by itself the terms of the product being arranged as in Abel's result, shew that the resulting series diverges if $r \leq \frac{1}{2}$ but converges to the sum s^2 if $r > \frac{1}{2}$.

Example 3.6 (Cajori) If the two conditionally convergent series

$$\sum_{n=1}^{\infty} \frac{(-1)^{n+1}}{n^r} \quad \text{and} \quad \sum_{n=1}^{\infty} \frac{(-1)^{n+1}}{n^s},$$

where r and s lie between 0 and 1, be multiplied together, and the product arranged as in Abel's result, shew that the necessary and sufficient condition for the convergence of the resulting series is $r + s > 1$.

Example 3.7 (Cajori) Shew that if the series $1 - \frac{1}{3} + \frac{1}{5} - \frac{1}{7} + \cdots$ be multiplied by itself any number of times, the terms of the product being arranged as in Abel's result, the resulting series converges.

Example 3.8 Shew that the qth power of the series

$$\alpha_1 \sin\theta + \alpha_2 \sin 2\theta + \cdots + \alpha_n \sin n\theta + \cdots$$

is convergent whenever $q(1 - r) < 1$, r being the greatest number satisfying the relation $\alpha_n \leq n^{-r}$ for all values of n.

Example 3.9 (Math. Trip. 1896) Shew that if θ is not equal to 0 or a multiple of 2π, and if $u_0, u_1, u_2, \ldots$ be a sequence such that $u_n \to 0$ steadily, then the series $\sum u_n \cos(n\theta + a)$ is convergent. Shew also that, if the limit of u_n is not zero, but u_n is still monotonic, the sum of the series is oscillatory if θ/π is rational, but that, if θ/π is irrational, the sum may have any value between certain bounds whose difference is $\alpha \operatorname{cosec}(\theta/2)$, where $\alpha = \lim_{n\to\infty} u_n$.

4

The Theory of Riemann Integration

4.1 The concept of integration

The reader is doubtless familiar with the idea of integration as the operation inverse to that of differentiation; and he is equally well aware that the integral (in this sense) of a given elementary function is not always expressible in terms of elementary functions. In order therefore to give a definition of the integral of a function which shall be always available, even though it is not practicable to obtain a function of which the given function is the differential coefficient, we have recourse to the result that the integral of $f(x)$, defined as the (elementary) function whose differential coefficient is $f(x)$, between the limits a and b is the area bounded by the curve $y = f(x)$, the axis of x and the ordinates $x = a$, $x = b$. We proceed to frame a formal definition of integration with this idea as the starting-point.

4.11 Upper and lower integrals

The following procedure for establishing existence theorems concerning integrals is based on that given by Goursat [255, I, Ch. IV]. The concepts of upper and lower integrals are due to Darboux, [160, p. 64].

Let $f(x)$ be a bounded function of x in the range (a, b). Divide the interval at the points $x_1, x_2, \ldots, x_{n-1}$, $(a \leq x_1 \leq x_2 \leq \cdots \leq x_{n-1} \leq b)$. Let U, L be the bounds of $f(x)$ in the range (a, b), and let U_r, L_r, be the bounds of $f(x)$ in the range (x_{r-1}, x_r), where $x_0 = a$, $x_n = b$.

The reader will find a figure of great assistance in following the argument of this section. S_n and s_n represent the sums of the areas of a number of rectangles which are respectively greater and less than the area bounded by $y = f(x)$, $x = a$, $x = b$ and $y = 0$, if this area be assumed to exist.

Consider the sums

$$S_n = U_1(x_1 - a) + U_2(x_2 - x_1) + \cdots + U_n(b - x_{n-1}),$$
$$s_n = L_1(x_1 - a) + L_2(x_2 - x_1) + \cdots + L_n(b - x_{n-1}).$$

Then $U(b - a) \geq S_n \geq s_n \geq L(b - a)$.

For a given n, S_n and s_n are bounded functions of $x_1, x_2, \ldots, x_{n-1}$. Let their lower and upper bounds[1] respectively be $\underline{S}_n, \bar{s}_n$, so that $\underline{S}_n, \bar{s}_n$ depend only on n and on the form of $f(x)$, and not on the particular way of dividing the interval into n parts.

---

[1] The bounds of a function of n variables are defined in just the same manner as the bounds of a function of a single variable (§3.62).

58

Let the lower and upper bounds of these functions of n be S, s. Then $S_n \geq S$, $s_n \leq s$. We proceed to shew that s is *at most* equal to S; i.e. $S \geq s$.

Let the intervals (a, x_1), (x_1, x_2), ... be divided into smaller intervals by new points of subdivision, and let

$$a, y_1, y_2, \ldots, y_{k-1}, \ y_k(= x_1), \ y_{k+1}, \ldots, y_{l-1}, \ y_l(= x_2), \ y_{l+1}, \ldots, y_{m-1}, b$$

be the end points of the smaller intervals; let U_r', L_r' be the bounds of $f(x)$ in the interval (y_{r-1}, y_r). Let

$$T_m = \sum_{r=1}^{m}(y_r - y_{r-1})U_r', \quad t_m = \sum_{r=1}^{m}(y_r - y_{r-1})L_r'.$$

Since U_1', U_2', ..., U_k' do not exceed U_1, it follows without difficulty that

$$S_n \geq T_m \geq t_m \geq s_n.$$

Now consider the subdivision of (a, b) into intervals by the points $x_1, x_2, \ldots, x_{n-1}$, and also the subdivision by a different set of points $x_1', x_2', \ldots, x_{n'-1}'$. Let $S_{n'}'$, $s_{n'}'$ be the sums for the second kind of subdivision which correspond to the sums S_n, s_n for the first kind of subdivision. Take *all* the points $x_1, \ldots, x_{n-1}$; $x_1', \ldots, x_{n'-1}'$ as the points $y_1, y_2, \ldots, y_m$. Then

$$S_n \geq T_m \geq t_m \geq s_n, \quad \text{and} \quad S_{n'}' \geq T_m \geq t_m \geq s_{n'}'.$$

Hence every expression of the type S_n *exceeds* (or at least equals) every expression of the type $s_{n'}'$; and therefore S cannot be less than s. For if $S < s$ and $s - S = 2\eta$ we could find an S_n and an $s_{n'}'$ such that $S_n - S < \eta$, $s - s_{n'}' < \eta$ and so $s_{n'}' > S_n$, which is impossible.

The bound S is called the *upper* integral of $f(x)$, and is written $U \int_a^b f(x)dx$; the bound s is called the *lower* integral, and written $L \int_a^b f(x)dx$. If $S = s$, their common value is called the *integral* of $f(x)$ taken between the limits[2] of integration a and b. The integral is written

$$\int_a^b f(x)dx.$$

We define $\int_b^a f(x)dx$, when $a < b$, to mean $-\int_a^b f(x)dx$.

Example 4.1.1   Prove that $\int_a^b \{f(x) + \phi(x)\}dx = \int_a^b f(x)dx + \int_a^b \phi(x)dx$.

Example 4.1.2   By means of Example 4.1.1, define the integral of a continuous complex function of a real variable.

4.12 Riemann's condition of integrability

Riemann [558, p. 239] bases his definition of an integral on the limit of the sum occurring in §4.13; but it is then difficult to prove the uniqueness of the limit. A more general definition

---

[2]   'Extreme' values would be a more appropriate term but 'limits' has the sanction of custom. 'Termini' has been suggested by Lamb [399, p. 207].

of integration (which is of very great importance in the modern theory of Functions of Real Variables) has been given by Lebesgue [417]. See also [418].

A function is said to be 'integrable in the sense of Riemann' if (with the notation of §4.11) S_n and s_n have a common limit (called the *Riemann integral* of the function) when the number of intervals (x_{r-1}, x_r) tends to infinity in such a way that the length of the longest of them tends to zero.

The necessary and sufficient condition that a bounded function should be integrable is that $S_n - s_n$ should tend to zero when the number of intervals (x_{r-1}, x_r) tends to infinity in such a way that the length of the longest tends to zero.

The condition is obviously necessary, for if S_n and s_n have a common limit $S_n - s_n \to 0$ as $n \to \infty$. And it is sufficient; for, since $S_n \geq S \geq s \geq s_n$, it follows that if $\lim(S_n - s_n) = 0$, then

$$\lim S_n = \lim s_n = S = s.$$

Remark 4.1.1   A continuous function $f(x)$ is *integrable*. For, given ε, we can find δ such that $|f(x') - f(x'')| < \varepsilon/(b-a)$ whenever $|x' - x''| < \delta$. Take all the intervals $(x_{\delta-1}, x_\delta)$ less than δ, and then $U_\delta - L_\delta < \varepsilon/(b-a)$ and so $S_n - s_n < \varepsilon$; therefore $S_n - s_n \to 0$ under the circumstances specified in the condition of integrability.

Corollary 4.1.2   *If S_n and s_n have the same limit S for one mode of subdivision of (a, b) into intervals of the specified kind, the limits of S_n and of s_n for any other such mode of subdivision are both S.*

Example 4.1.3   The product of two integrable functions is an integrable function.

Example 4.1.4   A function which is continuous except at a finite number of ordinary discontinuities is integrable. *Hint.* If $f(x)$ have an ordinary discontinuity at c, enclose c in an interval of length δ_1; given ε, we can find δ so that $|f(x') - f(x)| < \varepsilon$ when $|x' - x| < \delta$ and x, x' are not in this interval. Then $S_n - s_n \leq \varepsilon(b - a - \delta_1) + k\delta_1$, where k is the greatest value of $|f(x') - f(x)|$, when x, x' lie in the interval. When $\delta_1 \to 0$, $k \to |f(c+0) - f(c-0)|$, and hence $\lim_{n\to\infty}(S_n - s_n) = 0$.

Example 4.1.5   A function with limited total fluctuation and a finite number of ordinary discontinuities is integrable. (See Example 3.6.2.)

4.13 A general theorem on integration

Let $f(x)$ be integrable, and let ε be any positive number. Then it is possible to choose δ so that

$$\left| \sum_{p=1}^{n} (x_p - x_{p-1}) f(x'_{p-1}) - \int_a^b f(x)dx \right| < \varepsilon,$$

provided that $x_p - x_{p-1} \leq \delta$, and $x_{p-1} \leq x'_{p-1} \leq x_p$.

To prove the theorem we observe that, given ε, we can choose the length of the longest interval, δ, so small that $S_n - s_n < \varepsilon$. Also

$$S_n \geq \sum_{p=1}^{n} (x_p - x_{p-1}) f(x'_{p-1}) \geq s_n,$$

$$S_n \geq \int_a^b f(x)dx \geq s_n.$$

Therefore

$$\left| \sum_{p=1}^{n} (x_p - x_{p-1}) f(x'_{p-1}) - \int_a^b f(x)dx \right| \leq S_n - s_n < \varepsilon.$$

As an example (see Netto [484]) of the evaluation of a definite integral directly from the theorem of this section consider $\int_0^X \dfrac{dx}{(1-x^2)^{1/2}}$, where $X < 1$. Take $\delta = \dfrac{1}{p}$ arcsin X and let $x_s = \sin s\delta$, $(0 < s\delta < \frac{1}{2}\pi)$, so that

$$x_{\delta+1} - x_\delta = 2\sin\left(\tfrac{\delta}{2}\right)\cos\left(\delta + \tfrac{1}{2}\right) < \delta;$$

also let $x_\delta' = \sin(\delta + \tfrac{1}{2})\delta$. Then

$$\sum_{\delta=1}^{p} \frac{x_\delta - x_{\delta-1}}{(1 - x'^2_{\delta-1})^{1/2}} = \sum_{\delta=1}^{p} \frac{\sin s\delta - \sin(\delta-1)\delta}{\cos(\delta - \tfrac{1}{2})\delta}$$

$$= 2p \sin \tfrac{1}{2}\delta$$

$$= \text{arc}\sin X \cdot \left\{ \frac{\sin \tfrac{1}{2}\delta}{\tfrac{1}{2}\delta} \right\}.$$

By taking p sufficiently large we can make

$$\left| \int_0^X \frac{dx}{(1-x^2)^{1/2}} - \sum_{\delta=1}^{p} \frac{x_s - x_{s-1}}{(1 - x'^2_{s-1})^{1/2}} \right|$$

arbitrarily small. We can also make arcsin $X \cdot \left\{ \frac{\sin(\delta/2)}{\delta/2} - 1 \right\}$ arbitrarily small. That is, given an arbitrary number ε, we can make

$$\left| \int_0^X \frac{dx}{(1-x^2)^{1/2}} - \text{arcsin } X \right| < \varepsilon$$

by taking p sufficiently large. But the expression now under consideration *does not depend on p*; and therefore it must be zero; for if not we could take ε to be less than it, and we should have a contradiction. That is to say

$$\int_0^X \frac{dx}{(1-x^2)^{1/2}} = \text{arcsin } X. \tag{4.1}$$

Example 4.1.6 Shew that

$$\lim_{n\to\infty} \frac{1}{n}\left(1 + \cos\frac{x}{n} + \cos\frac{2x}{n} + \cdots + \cos\frac{(n-1)x}{n}\right) = \frac{\sin x}{x}.$$

Example 4.1.7   If $f(x)$ has ordinary discontinuities at the points $a_1, a_2, \ldots, a_k$, then

$$\int_a^b f(x)dx = \lim \left\{ \int_a^{a_1-\delta_1} + \int_{a_1+\varepsilon_1}^{a_2-\delta_2} + \cdots + \int_{a_k+\varepsilon_k}^b f(x)\,dx \right\},$$

where the limit is taken by making $\delta_1, \delta_2, \ldots, \delta_k, \varepsilon_1, \varepsilon_2, \ldots, \varepsilon_k$ tend to $+0$, independently.

Example 4.1.8   If $f(x)$ is integrable when $a_1 \leq x \leq b_1$ and if, when $a_1 \leq a < b < b_1$, we write

$$\int_a^b f(x)dx = \phi(a, b),$$

and if $f(b + 0)$ exists, then

$$\lim_{\delta \to +0} \frac{\phi(a, b + \delta) - \phi(a, b)}{\delta} = f(b + 0).$$

Deduce that, if $f(x)$ is continuous at a and b,

$$\frac{d}{da} \int_a^b f(x)dx = -f(a), \quad \frac{d}{db} \int_a^b f(x)dx = f(b).$$

Example 4.1.9   Prove by differentiation that, if $\phi(x)$ is a continuous function of x and $\dfrac{dx}{dt}$ a continuous function of t, then

$$\int_{x_0}^{x_1} \phi(x)dx = \int_{t_0}^{t_1} \phi(x)\frac{dx}{dt}dt.$$

Example 4.1.10   If $f'(x)$ and $\phi'(x)$ are continuous when $a \leq x \leq b$, shew from Example 4.1.8 that

$$\int_a^b f'(x)\phi(x)dx + \int_a^b \phi'(x)f(x)dx = f(b)\phi(b) - f(a)\phi(a).$$

Example 4.1.11   If $f(x)$ is integrable in the range (a, c) and $a \leq b \leq c$, shew that $\displaystyle\int_a^b f(x)dx$ is a continuous function of b.

4.14 Mean-value theorems

The two following general theorems are frequently useful.

(I) Let U and L be the upper and lower bounds of the integrable function $f(x)$ in the range (a, b). Then from the definition of an integral it is obvious that

$$\int_a^b (U - f(x))\,dx, \quad \int_a^b (f(x) - L)\,dx$$

are not negative; and so

$$U(b - a) \geq \int_a^b f(x)dx \geq L(b - a).$$

This is known as the *First Mean-Value Theorem*.

If $f(x)$ is *continuous* we can find a number ξ such that $a \le \xi \le b$ and such that $f(\xi)$ has any given value lying between U and L (§3.63). Therefore we can find ξ such that

$$\int_a^b f(x)dx = (b-a)f(\xi).$$

If $F(x)$ has a continuous differential coefficient $F'(x)$ in the range (a,b), we have, on writing $F'(x)$ for $f(x)$,

$$F(b) - F(a) = (b-a)F'(\xi)$$

for some value of ξ such that $a \le \xi \le b$.

Example 4.1.12 If $f(x)$ is continuous and $\phi(x) \ge 0$, shew that ξ can be found such that

$$\int_a^b f(x)\phi(x)\,dx = f(\xi)\int_a^b \phi(x)\,dx.$$

(II) Let $f(x)$ and $\phi(x)$ be integrable in the range (a,b) and let $\phi(x)$ be a *positive decreasing function of x*. Then *Bonnet's form of the Second Mean-Value Theorem* [83] is that a number ξ exists such that $a \le \xi \le b$, and

$$\int_a^b f(x)\phi(x)\,dx = \phi(a)\int_a^\xi f(x)\,dx.$$

The proof given is a modified form of an investigation due to Hölder [326].

For, with the notation of §4.1 and §4.13, consider the sum

$$S = \sum_{s=1}^p (x_s - x_{s-1})f(x_{s-1})\phi(x_{s-1}).$$

Writing $(x_s - x_{s-1})f(x_{s-1}) = a_{s-1}$, $\phi(x_{s-1}) = \phi_{s-1}$, $a_0 + a_1 + \cdots + a_s = b_s$, we have

$$S = \sum_{s=1}^{p-1} b_{s-1}(\phi_{s-1} - \phi_s) + b_{p-1}\phi_{p-1}.$$

Each term in the summation is increased by writing $\bar{b}$ for b_{s-1} and decreased by writing $\underline{b}$ for b_{s-1}, if $\bar{b}$, and $\underline{b}$ be the greatest and least of $b_0, b_1, \ldots, b_{p-1}$; and so $\underline{b}\phi_0 \le S \le \bar{b}\phi_0$. Therefore S lies between the greatest and least of the sums $\phi(x_0)\sum_{s=1}^m (x_s - x_{s-1})f(x_{s-1})$ where $m = 1, 2, 3, \ldots, p$. But, given ε, we can find δ such that, when $x_s - x_{s-1} < \delta$,

$$\left| \sum_{s=1}^p (x_s - x_{s-1})f(x_{s-1})\phi(x_{s-1}) - \int_{x_0}^{x_p} f(x)\phi(x)\,dx \right| < \varepsilon,$$

$$\left| \phi(x_0)\sum_{s=1}^m (x_s - x_{s-1})f(x_{s-1}) - \phi(x_0)\int_{x_0}^{x_m} f(x)\,dx \right| < \varepsilon,$$

and so, writing a, b for x_0, x_p, we find that $\int_a^b f(x)\phi(x)\,dx$ lies between the upper and lower bounds of $\phi(a)\int_a^{\xi_1} f(x)\,dx \pm 2\varepsilon$, where ξ_1 may take all values between a and b. (By

Example 4.1.11, since $f(x)$ is bounded, $\int_a^{\xi_1} f(x)\,dx$ is a continuous function of ξ_1.) Let U and L be the upper and lower bounds of $\phi(a)\int_a^{\xi_1} f(x)\,dx$. Then $U + 2\varepsilon \geq \int_a^b f(x)\phi(x)dx \geq L - 2\varepsilon$ for *all* positive values of ε; therefore

$$U \geq \int_a^b f(x)\phi(x)\,dx \geq L.$$

Since $\phi(a)\int_a^{\xi_1} f(x)\,dx$ *qua* function of ξ_1 takes all values between its upper and lower bounds, there is some value ξ, say, of ξ_1 for which it is equal to $\int_a^b f(x)\phi(x)\,dx$. This proves the Second Mean-Value Theorem.

Example 4.1.13 (Du Bois Reymond)　By writing $|\phi(x) - \phi(b)|$ in place of $\phi(x)$ in Bonnet's form of the mean-value theorem, shew that if $\phi(x)$ is a monotonic function, then a number ξ exists such that $a \leq \xi \leq b$ and

$$\int_a^b f(x)\phi(x)dx = \phi(a)\int_a^{\xi} f(x)dx + \phi(b)\int_{\xi}^b f(x)dx.$$

4.2 Differentiation of integrals containing a parameter

The equation

$$\frac{d}{d\alpha}\int_a^b f(x,\alpha)dx = \int_a^b \frac{\partial f}{\partial \alpha}\,dx \tag{4.2}$$

is true *if $f(x,\alpha)$ possesses a Riemann integral with respect to x and f_α, which equals $\dfrac{\partial f}{\partial \alpha}$, is a continuous function of the variables x and α.* This formula was given by Leibniz, without specifying the restrictions laid on $f(x,a)$.

Note　$\phi(x,y)$ is defined to be a continuous function of *both* variables if, given ε, we can find δ such that $|\phi(x',y') - \phi(x,y)| < \varepsilon$ whenever $\{(x'-x)^2 + (y'-y)^2\}^{1/2} < \delta$. It can be shewn by §3.6 that if $\phi(x,y)$ is a continuous function of both variables at all points of a closed region in a Cartesian diagram, it is *uniformly* continuous throughout the region (the proof is almost identical with that of §3.61). It should be noticed that, if $\phi(x,y)$ is a continuous function of *each* variable, it is *not* necessarily a continuous function of both; as an example take

$$\phi(x,y) = \frac{(x+y)^2}{x^2+y^2}, \quad \phi(0,0) = 1;$$

this is a continuous function of x and of y at $(0,0)$, but not of both x and y.

For

$$\frac{d}{d\alpha}\int_a^b f(x,\alpha)\,dx = \lim_{h\to 0}\int_a^b \frac{f(x,\alpha+h) - f(x,\alpha)}{h}\,dx \tag{4.3}$$

if this limit exists. But, by the first mean-value theorem, since f_α is a continuous function of

α, the second integrand is $f_a(x, \alpha + \theta h)$, where $0 \le \theta \le 1$. But, for any given ε, a number δ *independent* of x exists (since the continuity of f_α is uniform with respect to the variable x) such that

$$|f_a(x, \alpha') - f_a(x, \alpha)| < \varepsilon/(b - a),$$

whenever $|\alpha' - \alpha| < \delta$. It is obvious that it would have been sufficient to assume that f_α had a Riemann integral and was a continuous function of a (the continuity being uniform with respect to x), instead of assuming that f_a was a continuous function of both variables. This is actually done by Hobson [315, p. 599].

Taking $|h| < \delta$ we see that $|\theta h| < \delta$, and so *whenever* $|h| < \delta$,

$$\left| \int_a^b \frac{f(x, \alpha + h) - f(x, \alpha)}{h} \, dx - \int_a^b f_\alpha(x, \alpha) \, dx \right|$$

$$\le \int_a^b |f_\alpha(x, \alpha + \theta h) - f_\alpha(x, \alpha)| \, dx < \varepsilon.$$

Therefore by the definition of a limit of a function (§3.2),

$$\lim_{h \to 0} \int_a^b \frac{f(x, \alpha + h) - f(x, \alpha)}{h} \, dx$$

exists and is equal to $\int_a^b f_\alpha \, dx$.

Example 4.2.1   If a, b be not constants but functions of α with continuous differential coefficients, shew that

$$\frac{d}{d\alpha} \int_a^b f(x, \alpha) dx = f(b, \alpha) \frac{db}{d\alpha} - f(a, \alpha) \frac{da}{d\alpha} + \int_a^b \frac{\partial f}{\partial \alpha} dx.$$

Example 4.2.2   If $f(x, \alpha)$ is a continuous function of both variables, $\int_a^b f(x, \alpha) dx$ is a continuous functions of α.

4.3 Double integrals and repeated integrals

Let $f(x, y)$ be a function which is continuous with regard to both of the variables x and y, when $a \le x \le b, \alpha \le y \le \beta$. By Example 4.2.2 it is clear that

$$\int_a^b \left\{ \int_\alpha^\beta f(x, y) \, dy \right\} dx, \quad \int_\alpha^\beta \left\{ \int_a^b f(x, y) \, dx \right\} dy$$

both exist. These are called *repeated integrals*.

Also, as in §3.62, $f(x, y)$, being a continuous function of both variables, attains its upper and lower bounds.

Consider the range of values of x and y to be the points inside and on a rectangle in a Cartesian diagram; divide it into $n\nu$ rectangles by lines parallel to the axes. Let $U_{m,\mu}$, $L_{m,\mu}$

be the upper and lower bounds of $f(x, y)$ in one of the smaller rectangles whose area is, say, $A_{m,\mu}$; and let

$$\sum_{m=1}^{n} \sum_{\mu=1}^{\nu} U_{m,\mu} A_{m,\mu} = S_{n,\nu}, \quad \sum_{m=1}^{n} \sum_{\mu=1}^{\nu} L_{m,\mu} A_{m,\mu} = s_{n,\nu}.$$

Then $S_{n,\nu} > s_{n,\nu}$, and, as in §4.11, we can find numbers $\underline{S}_{n,\nu}$, $\bar{s}_{n,\nu}$, which are the lower and upper bounds of $S_{n,\nu}$, $s_{n,\nu}$ respectively, the values of $\underline{S}_{n,\nu}$, $\bar{s}_{n,\nu}$ depending only on the number of the rectangles and not on their shapes; and $\underline{S}_{n,\nu} \geq \bar{s}_{n,\nu}$. We then find the lower and upper bounds (S and s) respectively of $\underline{S}_{n,\nu}$, $\bar{s}_{n,\nu}$ *qua* functions of n and ν; and $S_{n,\nu} \geq S \geq s \geq s_{n,\nu}$, as in §4.11.

Also, from the uniformity of the continuity of $f(x, y)$, given ε, we can find δ such that $U_{m,\mu} - L_{m,\mu} < \varepsilon$, (for all values of m and μ) whenever the sides of all the small rectangles are less than the number δ which depends only on the form of the function $f(x, y)$ and on ε.

And then $S_{n,\nu} - s_{n,\nu} < \varepsilon(b - a)(\beta - \alpha)$, and so $S - s < \varepsilon(b - a)(\beta - \alpha)$. But S and s are *independent* of ε, and so $S = s$.

The common value of S and s is called the *double integral* of $f(x, y)$ and is written

$$\int_a^b \int_\alpha^\beta f(x, y)\, dy\, dx.$$

It is easy to shew that the repeated integrals and the double integral are all equal when $f(x, y)$ is a continuous function of both variables. For let Y_m, Λ_m be the upper and lower bounds of

$$\int_\alpha^\beta f(x, y)\, dy$$

as x varies between x_{m-1} and x_m.

Then

$$\sum_{m=1}^{n} Y_m(x_m - x_{m-1}) \geq \int_a^b \left\{ \int_\alpha^\beta f(x, y)\, dy \right\} dx \geq \sum_{m=1}^{n} \Lambda_m(x_m - x_{m-1}).$$

But the upper bound of $f(x, y)$ in the rectangle $A_{m,\mu}$ is not less than the upper bound of $f(x, y)$ on that portion of the line $x = \xi$ which lies in the rectangle, therefore

$$\sum_{\mu=1}^{\nu} U_{m,\mu}(y_\mu - y_{\mu-1}) \geq Y_m \geq \Lambda_m \geq \sum_{\mu=1}^{\nu} L_{m,\mu}(y_\mu - y_{\mu-1}).$$

Multiplying these last inequalities by $x_m - x_{m-1}$, using the preceding inequalities and summing, we get

$$\sum_{m=1}^{n} \sum_{\mu=1}^{\nu} U_{m,\mu} A_{m,\mu} \geq \int_a^b \left\{ \int_\alpha^\beta f(x, y)\, dy \right\} dx \geq \sum_{m=1}^{n} \sum_{\mu=1}^{\nu} L_{m,\mu} A_{m,\mu};$$

and so, proceeding to the limit,

$$S \geq \int_a^b \left\{ \int_\alpha^\beta f(x, y)\, dy \right\} dx \geq s.$$

But $S = \int_a^b \int_a^\beta f(x, y) \, dx \, dy$, and so one of the repeated integrals is equal to the double integral. Similarly the other repeated integral is equal to the double integral.

Corollary 4.3.1 *If $f(x, y)$ be a continuous function of both variables,*

$$\int_0^1 dx \left\{ \int_0^{1-x} f(x, y) dy \right\} = \int_0^1 dy \left\{ \int_0^{1-v} f(x, y) dx \right\}.$$

4.4 Infinite integrals

If $\lim_{b \to \infty} \left(\int_a^b f(x) dx \right)$ exists, we denote it by $\int_a^\infty f(x) dx$; and the limit in question is called an *infinite integral*. This phrase, due to Hardy [274, p. 16], suggests the analogy between an infinite integral and an infinite series.

Example 4.4.1 1. $\int_a^\infty \frac{dx}{x^2} = \lim_{b \to \infty} \left(\frac{1}{a} - \frac{1}{b} \right) = \frac{1}{a}.$

2. $\int_0^\infty \frac{x \, dx}{(x^2 + a^2)^2} = \lim_{b \to \infty} \left(-\frac{1}{2(b^2 + a^2)} + \frac{1}{2a^2} \right) = \frac{1}{2a^2}.$

3. (Euler). By integrating by parts, shew that $\int_0^\infty t^n e^{-t} dt = n!.$

Similarly we define $\int_{-\infty}^b f(x) \, dx$ to mean $\lim_{a \to -\infty} \int_a^b f(x) \, dx$; if this limit exists; and $\int_{-\infty}^\infty f(x) \, dx$ is defined as $\int_{-\infty}^a f(x) \, dx + \int_a^\infty f(x) \, dx$. In this last definition the choice of a is a matter of indifference.

4.41 Infinite integrals of continuous functions. Conditions for convergence

A necessary and sufficient condition for the convergence of $\int_a^\infty f(x) dx$ is that, corresponding to any positive number ε, a positive number X should exist such that $\left| \int_{x'}^{x''} f(x) dx \right| < \varepsilon$ whenever $x'' \geq x' \geq X$.

The condition is obviously necessary; to prove that it is sufficient, suppose it is satisfied; then, if $n \geq X - a$ and n be a positive integer and $S_n = \int_a^{a+n} f(x) dx$, we have $|S_{n+p} - S_n| < \varepsilon$. Hence, by §2.22, S_n tends to a limit, S; and then, if $\xi > a + n$,

$$\left| S - \int_a^\xi f(x) dx \right| \leq \left| S - \int_a^{a+n} f(x) dx \right| + \left| \int_{a+n}^\xi f(x) dx \right|$$
$$< 2\varepsilon;$$

and so $\lim_{\xi \to \infty} \int_a^\xi f(x) \, dx = S$; so that the condition is sufficient.

4.42 Uniformity of convergence of an infinite integral

The integral $\int_a^\infty f(x,\alpha)\,dx$ is said to converge uniformly with regard to α in a given domain of values of α if, corresponding to an arbitrary positive number ε, there exists a number X *independent* of α such that

$$\left| \int_{x'}^\infty f(x,\alpha)\,dx \right| < \varepsilon$$

for all values of α in the domain and all values of $x' \geq X$.

The reader will see without difficulty on comparing §2.22 and §3.31 with §4.41 that a necessary and sufficient condition that $\int_a^\infty f(x,\alpha)\,dx$ should converge uniformly in a given domain is that, corresponding to any positive number ε, there exists a number X independent of α such that

$$\left| \int_{x'}^{x''} f(x,\alpha)\,dx \right| < \varepsilon$$

for all values of α in the domain whenever $x'' \geq x' \geq X$.

4.43 Tests for the convergence of an infinite integral

There are conditions for the convergence of an infinite integral analogous to those given in Chapter 2 for the convergence of an infinite series. The following tests are of special importance.

(I) *Absolutely convergent integrals.* It may be shewn that $\int_a^\infty f(x)\,dx$ certainly converges if $\int_a^\infty |f(x)|\,dx$ does so; and the former integral is then said to be absolutely convergent. The proof is similar to that of §2.32.

Example 4.4.2 *The comparison test. If $|f(x)| \leq g(x)$ and $\int_a^\infty g(x)\,dx$ converges, then $\int_a^\infty f(x)\,dx$ converges absolutely.*

Note It was observed by Dirichlet [175] (with the example $f(x) = \sin x^2$) that it is *not necessary* for the convergence of $\int_a^\infty f(x)\,dx$ that $f(x) \to 0$ as $x \to \infty$: the reader may see this by considering the function

$$f(x) = \begin{cases} 0 & (n \leq x \leq n+1-(n+1)^{-2}), \\ (n+1)^4(n+1-x)\left[x-(n+1)+(n+1)^{-2}\right] & (n+1-(n+1)^{-2} \leq x \leq n+1, \end{cases}$$

where n takes all integral values.

For $\int_0^\xi f(x)\,dx$ increases with ξ and $\int_n^{n+1} f(x)\,dx = \frac{1}{6}(n+1)^{-2}$; whence it follows without difficulty that $\int_a^\infty f(x)\,dx$ converges. But when $x = n+1-\frac{1}{2}(n+1)^{-2}$, $f(x) = \frac{1}{4}$; and so $f(x)$ does *not* tend to zero.

(II) *The Maclaurin–Cauchy test.*[3] If $f(x) > 0$ and $f(x) \to 0$ steadily, $\int_1^\infty f(x)\,dx$ and $\sum_{n=1}^\infty f(n)$ converge or diverge together.

For $f(m) \geq \int_m^{m+1} f(x)\,dx \geq f(m+1)$, and so

$$\sum_{m=1}^n f(m) \geq \int_1^{n+1} f(x)\,dx \geq \sum_{m=2}^{n+1} f(m).$$

The first inequality shews that, if the series converges, the increasing sequence $\int_1^{n+1} f(x)\,dx$ converges (§2.2) when $n \to \infty$ through integral values, and hence it follows without difficulty that $\int_1^{x'} f(x)\,dx$ converges when $x' \to \infty$; also if the integral diverges, so does the series. The second shews that if the series diverges so does the integral, and if the integral converges so does the series (§2.2).

(III) *Bertrand's test* [67, p. 38–39]. If $f(x) = O(x^{\lambda-1})$, $\int_n^\infty f(x)\,dx$ converges when $\lambda < 0$; and if $f(x) = O(x^{-1}\{\log x\}^{\lambda-1})$, $\int_a^\infty f(x)\,dx$ converges when $\lambda < 0$.

These results are particular cases of the comparison test given in (I).

(IV) *Chartier's test* [143] *for integrals involving periodic functions.*[4]

If $f(x) \to 0$ steadily as $x \to \infty$ and if $\left| \int_a^x \phi(x)\,dx \right|$ is bounded as $x \to \infty$, then $\int_a^\infty f(x)\phi(x)\,dx$ is convergent.

For if the upper bound of $\left| \int_a^x \phi(x)\,dx \right|$ be A, we can choose X such that $f(x) < \varepsilon/2A$ when $x \geq X$; and then by the second mean-value theorem, when $x'' \geq x' \geq X$, we have

$$\left| \int_{x'}^{x''} f(x)\,\phi(x)\,dx \right| = \left| f(x') \int_{x'}^\xi \phi(x)\,dx \right| = f(x') \left| \int_a^\xi \phi(x)\,dx - \int_a^{x'} \phi(x)\,dx \right|$$
$$\leq 2A f(x') < \varepsilon,$$

which is the condition for convergence.

Example 4.4.3 $\int_0^\infty \dfrac{\sin x}{x}\,dx$ converges.

Example 4.4.4 $\int_0^\infty \dfrac{\sin(x^3 - ax)}{x}\,dx$ converges.

---

[3] Maclaurin [449, vol. I, p. 289–290] makes a verbal statement practically equivalent to this result. Cauchy's result is given in [129, v. 7, p. 269].

[4] It is remarkable that this test for *conditionally* convergent integrals should have been given some years before formal definitions of absolutely convergent integrals.

4.431 Tests for uniformity of convergence of an infinite integral.

The results of this section and of §4.44 are due to de la Vallée Poussin [637].

(I) *De la Vallée Poussin's test.* This name is due to Osgood. The reader will easily see by using the reasoning of §3.34 that $\int_a^\infty f(x, \alpha)\, dx$ converges uniformly with regard to α in a domain of values of α if $|f(x, \alpha)| < \mu(x)$, where $\mu(x)$ is independent of α and $\int_a^\infty \mu(x)\, dx$ converges.

For, choosing X so that $\int_{x'}^{x''} \mu(x)\, dx < \varepsilon$ when $x'' \geq x' \geq X$, we have $\left| \int_{x'}^{x''} f(x, \alpha)\, dx \right| < \varepsilon$, and the choice of X is independent of α.

Example 4.4.5 $\int_0^\infty x^{a-1} e^{-x}\, dx$ converges uniformly in any interval (A, B) such that $1 \leq A \leq B$.

(II) *The method of change of variable.* This may be illustrated by an example. Consider $\int_0^x \frac{\sin ax}{x}\, dx$ where a is real. We have

$$\int_{x'}^{x''} \frac{\sin ax}{x}\, dx = \int_{ax'}^{ax''} \frac{\sin y}{y}\, dy.$$

Since $\int_0^\infty \frac{\sin y}{y}\, dy$ converges we can find Y such that $\left| \int_{y'}^{y''} \frac{\sin y}{y}\, dy \right| < \varepsilon$ when $y'' \geq y' \geq Y$.

So $\left| \int_{x'}^{x''} \frac{\sin ax}{x}\, dx \right| < \varepsilon$ whenever $|ax'| \geq Y$; if $|a| \geq \delta > 0$, we therefore get

$$\left| \int_{x'}^{x''} \frac{\sin ax}{x}\, dx \right| < \varepsilon$$

when $x'' \geq x' \geq X = Y/\delta$; and this choice of X is independent of a. So the convergence is uniform when $a \geq \delta > 0$ and when $a \leq -\delta < 0$.

Example 4.4.6 (de la Vallée Poussin)   Prove that $\int_1^\infty \left\{ \int_0^a \sin(b^2 x^3)\, db \right\} dx$ is uniformly convergent in any range of real values of a.

Write $b^2 x^3 = z$, and observe that $\left| \int_0^{a^2 x^3} z^{-1/2} \sin z\, dz \right|$ does not exceed a constant independent of a and x since $\int_0^\infty z^{-1/2} \sin z\, dz$ converges.

(III) *The method of integration by parts.* If $\int f(x, a)\, dx = \phi(x, a) + \int \chi(x, a)\, dx$ and if $\phi(x, a) \to 0$ uniformly as $x \to \infty$ and $\int_a^\infty \chi(x, a)\, dx$ converges uniformly with regard to a, then obviously $\int_a^\infty f(x, a)\, dx$ converges uniformly with regard to a.

(IV) *The method of decomposition.*

Example 4.4.7

$$\int_0^\infty \frac{\cos x \sin ax}{x}\, dx = \frac{1}{2}\int_0^\infty \frac{\sin(a+1)x}{x}\, dx + \frac{1}{2}\int_0^\infty \frac{\sin(a-1)x}{x}\, dx;$$

both of the latter integrals converge uniformly in any closed domain of real values of a from which the points $a = \pm 1$ are excluded.

4.44 Theorems concerning uniformly convergent infinite integrals

(I) *Let* $\int_a^\infty f(x,\alpha)\, dx$ *converge uniformly when* a *lies in a domain S. Then, if* $f(x,\alpha)$ *is a continuous function of both variables when* $x \geq a$ *and* a *lies in S,* $\int_a^\infty f(x,a)\, dx$ *is a continuous function of* α. This result is due to Stokes. His statement is that the integral is a continuous function of a if it does not 'converge infinitely slowly'.

For, given ε, we can find X *independent of* α, such that $\left|\int_\xi^\infty f(x,a)\, dx\right| < \varepsilon$ whenever $\xi \geq X$. Also we can find δ *independent of* x *and* a, such that

$$|f(x,\alpha) - f(x,\alpha')| < \varepsilon/(X-a)$$

whenever $|\alpha - \alpha'| < \delta$. That is to say, given ε, we can find δ independent of a, such that

$$\left|\int_a^\infty f(x,\alpha')\, dx - \int_a^\infty f(x,\alpha)\, dx\right| \leq \left|\int_a^X \{f(x,\alpha) - f(x,\alpha')\}\, dx\right|$$
$$+ \left|\int_X^\infty f(x,\alpha')\, dx\right| + \left|\int_X^\infty f(x,\alpha)\, dx\right|$$
$$< 3\varepsilon,$$

whenever $|\alpha' - \alpha| < \delta$; and this is the condition for continuity.

(II) *If* $f(x,\alpha)$ *satisfies the same conditions as in* (I), *and if* α *lies in S when* $A \leq \alpha \leq B$, *then*

$$\int_A^B \left\{\int_a^\infty f(x,\alpha)\, dx\right\} d\alpha = \int_a^\infty \left\{\int_A^B f(x,\alpha)\, d\alpha\right\} dx.$$

For, by §4.3,

$$\int_A^B \left\{\int_a^\xi f(x,\alpha)\, dx\right\} d\alpha = \int_a^\xi \left\{\int_A^B f(x,\alpha)\, d\alpha\right\} dx.$$

Therefore

$$\left|\int_A^B \left\{\int_a^\infty f(x,\alpha)\, dx\right\} d\alpha - \int_a^\xi \left\{\int_A^B f(x,\alpha)\, d\alpha\right\} dx\right|$$
$$= \left|\int_A^B \left\{\int_\infty^\xi f(x,\alpha)\, dx\right\} d\alpha\right| < \int_A^B \varepsilon\, d\alpha < \varepsilon(B-A),$$

for all sufficiently large values of ξ.

But, from §2.1 and §4.41, this is the condition that

$$\lim_{\xi \to \infty} \int_a^\xi \left\{ \int_A^B f(x,\alpha)\,d\alpha \right\} dx$$

should exist, and be equal to

$$\int_A^B \left\{ \int_a^\infty f(x,\alpha)\,dx \right\} d\alpha.$$

Corollary 4.4.1 *The equation* $\dfrac{d}{da} \displaystyle\int_a^\infty \phi(x,a)\,dx = \int_a^\infty \dfrac{\partial \phi}{\partial a}\,dx$ *is true if the integral on the right converges uniformly and the integrand is a continuous function of both variables, when $x \ge a$ and a lies in a domain S, and if the integral on the left is convergent.*

Let A be a point of S, and let $\dfrac{\partial \phi}{\partial a} = f(x,a)$, so that, by Example 4.1.8,

$$\int_A^a f(x,a)\,da = \phi(x,a) - \phi(x,A).$$

Then $\displaystyle\int_a^x \left\{ \int_A^a f(x,a)\,da \right\} dx$ *converges, that is* $\displaystyle\int_a^\infty \{\phi(x,a) - \phi(x,A)\}\,dx$ *converges,*

and therefore, since $\displaystyle\int_a^\infty \phi(x,a)\,dx$ *converges, so does* $\displaystyle\int_a^\infty \phi(x,A)\,dx.$

Then

$$\frac{d}{da} \left[\int_a^\infty \phi(x,a)dx \right] = \frac{d}{da} \left[\int_a^\infty \{\phi(x,a) - \phi(x,A)\}dx \right]$$

$$= \frac{d}{da} \left[\int_a^\infty \left\{ \int_A^a f(x,a)da \right\} dx \right]$$

$$= \frac{d}{da} \int_A^a \left\{ \int_a^\infty f(x,a)dx \right\} da$$

$$= \int_a^\infty f(x,a)\,dx = \int_a^\infty \frac{\partial \phi}{\partial a}\,dx,$$

which is the required result; the change of the order of the integrations has been justified above, and the differentiation of $\int_A^a$ with regard to a is justified by §4.44 (I) and Example 4.1.8.

4.5 Improper integrals. Principal values

If $|f(x)| \to \infty$ as $x \to a + 0$, then $\displaystyle\lim_{\delta \to +0} \int_{a+\delta}^b f(x)\,dx$ may exist, and is written simply $\displaystyle\int_a^b f(x)dx$; this limit is called an *improper integral*. If $|f(x)| \to \infty$ as $x \to c$, where $a < c < b$, then

$$\lim_{\delta \to +0} \int_a^{c-\delta} f(x)dx + \lim_{\delta' \to +0} \int_{c+\delta'}^b f(x)dx$$

may exist; this is also written $\int_a^b f(x)dx$, and is also called an improper integral; it might however happen that neither of these limits exists when $\delta, \delta' \to 0$ independently, but

$$\lim_{\delta \to +0} \left\{ \int_a^{c-\delta} f(x)dx + \int_{c+\delta}^b f(x)dx \right\}$$

exists; this is called *Cauchy's principal value* of $\int_a^b f(x)dx$ and is written for brevity

$$P \int_a^b f(x)dx.$$

Results similar to those of §4.4–§4.44 may be obtained for improper integrals. But all that is required in practice is (i) the idea of absolute convergence, (ii) the analogue of Bertrand's test for convergence, (iii) the analogue of de la Vallée Poussin's test for uniformity of convergence. The construction of these is left to the reader, as is also the consideration of integrals in which the integrand has an infinite limit at more than one point of the range of integration. For a detailed discussion of improper integrals, the reader is referred either to Hobson [315] or to Pierpont [519]. The connexion between infinite integrals and improper integrals is exhibited by Bromwich [102, §164].

Example 4.5.1   1. $\int_0^\pi x^{-1/2} \cos x \, dx$ is an improper integral.

2. $\int_0^1 x^{\lambda-1}(1-x)^{\mu-1} \, dx$ is an improper integral if $0 < \lambda < 1, 0 < \mu < 1$. It does not converge for negative values of λ and μ.

3. $P \int_0^2 \dfrac{x^{\alpha-1}}{1-x} \, dx$ is the principal value of an improper integral when $0 < \alpha < 1$.

4.51 The inversion of the order of integration of a certain repeated integral

General conditions for the legitimacy of inverting the order of integration when the integrand is not continuous are difficult to obtain. The following is a good example of the difficulties to be overcome in inverting the order of integration in a repeated improper integral.

Let $f(x, y)$ be a continuous function of both variables, and let $0 < \lambda, \mu, \nu \leq 1$; then

$$\int_0^1 dx \left\{ \int_0^{1-x} x^{\lambda-1} y^{\mu-1} (1 - x - y)^{\nu-1} f(x,y) \, dy \right\}$$
$$= \int_0^1 dy \left\{ \int_0^{1-y} x^{\lambda-1} y^{\mu-1} (1 - x - y)^{\nu-1} f(x,y) \, dx \right\}.$$

This integral, which was first employed by Dirichlet, is of importance in the theory of integral equations; the investigation which we shall give is due to W. A. Hurwitz [329, p. 183].

Let $x^{\lambda-1} y^{\mu-1} (1 - x - y)^{\nu-1} f(x, y) = \phi(x, y)$; and let M be the upper bound of $|f(x, y)|$. Let δ be any positive number less than $1/\delta$. Draw the triangle whose sides are $x = \delta$, $y = \delta$, $x + y = 1 - \delta$; at all points on and inside this triangle $\phi(x, y)$ is continuous, and hence, by

Corollary 4.3.1

$$\int_\delta^{1-2\delta} dx \left\{ \int_\delta^{1-x-\delta} \phi(x,y)dy \right\} = \int_\delta^{1-2\delta} dy \left\{ \int_\delta^{1-y-\delta} \phi(x,y)dx \right\}.$$

Now

$$\int_\delta^{1-2\delta} dx \left\{ \int_0^{1-x} \phi(x,y)\,dy \right\} = \int_\delta^{1-2\delta} dx \left\{ \int_\delta^{1-x-\delta} \phi(x,y)\,dy \right\}$$

$$+ \int_\delta^{1-2\delta} I_1\, dx + \int_\delta^{1-2\delta} I_2\, dx,$$

where $I_1 = \int_0^\delta \phi(x,y)\,dy$, and $I_2 = \int_{1-x-\delta}^{1-x} \phi(x,y)\,dy$. But

$$|I_1| \le \int_0^\delta M x^{\lambda-1} y^{\mu-1}(1-x-y)^{\nu-1} dy \le M x^{\lambda-1}(1-x-\delta)^{\nu-1} \int_0^\delta y^{\mu-1}dy,$$

since $(1-x-y)^{\nu-1} \le (1-x-\delta)^{\nu-1}$. Therefore, writing $x = (1-\delta)x_1$, and since $\int_0^1 x_1^{\lambda-1}(1-x_1)^{\nu-1}dx_1 = B(\lambda,\nu)$ exists if $\lambda > 0$, $\nu > 0$ (see (2) of Example 4.5.1), we have

$$\left| \int_\delta^{1-2\delta} I_1\, dx \right| \le M \delta^\mu \mu^{-1} \int_0^{1-\delta} x^{\lambda-1}(1-x-\delta)^{\nu-1}\, dx$$

$$\le M \delta^\mu \mu^{-1}(1-\delta)^{\lambda+\nu-1} \int_0^1 x_1^{\lambda-1}(1-x_1)^{\nu-1}\, dx_1$$

$$< M \delta^\mu \mu^{-1}(1-\delta)^{\lambda+\nu-1} B(\lambda,\nu) \to 0 \text{ as } \delta \to 0.$$

The reader will prove similarly that $I_2 \to 0$ as $\delta \to 0$.
Hence[5]

$$\int_0^1 dx \left\{ \int_0^{1-x} \phi(x,y)\,dy \right\} = \lim_{\delta \to 0} \int_\delta^{1-2\delta} dx \left\{ \int_0^{1-x} \phi(x,y)\,dy \right\}$$

$$= \lim_{\delta \to 0} \int_\delta^{1-2\delta} dx \left\{ \int_\delta^{1-x-\delta} \phi(x,y)\,dy \right\}$$

$$= \lim_{\delta \to 0} \int_\delta^{1-2\delta} dy \left\{ \int_\delta^{1-y-\delta} \phi(x,y)\,dx \right\},$$

by what has been already proved; but, by a precisely similar piece of work, the last integral is

$$\int_0^1 dy \left\{ \int_0^{1-\nu} \phi(x,y)\,dx \right\}.$$

[5] The repeated integral exists, and is, in fact, absolutely convergent; for

$$\int_0^{1-x} |x^{\lambda-1}y^{\mu-1}(1-x-y)^{\nu-1} f(x,y)\,dy| < M x^{\lambda-1}(1-x)^{\mu+\nu-1} \int_0^1 \delta^{\mu-1}(1-\delta)^{\nu-1}\,ds,$$

writing $y = (1-x)\delta$; and $\int_0^1 M x^{\lambda-1}(1-x)^{\mu+\nu-1}\,dx \cdot \int_0^1 \delta^{\mu-1}(1-\delta)^{\nu-1}\,d\delta$ exists. And since the integral exists, its value which is $\lim_{\delta,\varepsilon \to 0} \int_\delta^{1-\varepsilon}$ may be written $\lim_{\delta \to 0} \int_\delta^{1-2\delta}$.

We have consequently proved the theorem in question.

Corollary 4.5.1 *Writing $\xi = a + (b - a)x$, $\eta = b - (b - a)y$, we see that, if $\phi(\xi, \eta)$ is continuous,*

$$\int_a^b d\xi \left\{ \int_\xi^b (\xi - a)^{\lambda-1} (b - \eta)^{\mu-1} (\eta - \xi)^{\nu-1} \phi(\xi, \eta) \, d\eta \right\}$$
$$= \int_a^b d\eta \left\{ \int_\alpha^\eta (\xi - a)^{\lambda-1} (b - \eta)^{\mu-1} (\eta - \xi)^{\nu-1} \phi(\xi, \eta) \, d\xi \right\}.$$

This is called *Dirichlet's formula*.

Note What are now called infinite and improper integrals were defined by Cauchy [124], though the idea of infinite integrals seems to date from Maclaurin [449]. The test for convergence was employed by Chartier [143]. Stokes (1847) distinguished between 'essentially' (absolutely) and non-essentially convergent integrals though he did not give a formal definition. Such a definition was given by Dirichlet [179] in 1854 and 1858 (see [179, p. 39]). In the early part of the nineteenth century improper integrals received more attention than infinite integrals, probably because it was not fully realised that an infinite integral is really the *limit* of an integral.

4.6 Complex integration

A treatment of complex integration based on a different set of ideas and not making so many assumptions concerning the curve AB will be found in Watson [650].

Integration with regard to a real variable x may be regarded as integration along a particular path (namely part of the real axis) in the Argand diagram. Let $f(z) (= P + iQ)$, be a function of a complex variable z, which is continuous along a simple curve AB in the Argand diagram.

Let the equations of the curve be $x = x(t)$, $y = y(t)$ $(a \leq t \leq b)$. Let $x(a) + iy(a) = z_0$, $x(b) + iy(b) = Z$.

Then if $x(t)$, $y(t)$ have continuous differential coefficients[6] (see Example 4.1.9) we *define* $\int_{z_0}^{Z} f(z) \, dz$ taken along the simple curve AB to mean

$$\int_a^b (P + iQ) \left(\frac{dx}{dt} + i \frac{dy}{dt} \right) dt.$$

The 'length' of the curve AB will be defined as $\int_a^b \sqrt{\left(\frac{dx}{dt} \right)^2 + \left(\frac{dy}{dt} \right)^2} \, dt$. It obviously exists if $\frac{dx}{dt}, \frac{dy}{dt}$ are continuous; we have thus reduced the discussion of a complex integral to the discussion of four real integrals, viz.

$$\int_a^b P \frac{dx}{dt} dt, \qquad \int_a^b P \frac{dy}{dt} dt, \qquad \int_a^b Q \frac{dx}{dt} dt, \qquad \int_a^b Q \frac{dy}{dt} dt.$$

---

[6] This assumption will be made throughout the subsequent work.

By Example 4.1.9, this definition is consistent with the definition of an integral when AB happens to be part of the real axis.

Example 4.6.1 $\displaystyle\int_{z_0}^{Z} f(z)\,dz = -\int_{Z}^{z_0} f(z)\,dz$, the paths of integration being the same (but in opposite directions) in each integral.

$$\int_{z_0}^{Z} dz = Z - z_0, \qquad \int_{z_0}^{Z} z\,dz = \int_{a}^{b}\left\{ x\,\frac{dx}{dt} - y\,\frac{dy}{dt} + i\left(x\,\frac{dy}{dt} + y\,\frac{dx}{dt}\right)\right\} dt$$

$$= \left[\frac{1}{2}x^2 - \frac{1}{2}y^2 + ixy\right]_{t=a}^{t=b} = \frac{1}{2}(Z^2 - z_0^2).$$

4.61 The fundamental theorem of complex integration

From §4.13, the reader will easily deduce the following theorem:

Let a sequence of points be taken on a simple curve $z_0 Z$; and let the first n of them, rearranged in order of magnitude of their parameters, be called

$$z_1^{(n)}, z_2^{(n)}, \ldots, z_n^{(n)} \qquad (z_0^{(n)} = z_0, \quad z_{n+1}^{(n)} = Z);$$

let their parameters be $t_1^{(n)}, t_2^{(n)}, \ldots, t_n^{(n)}$, and let the sequence be such that, given any number δ, we can find N such that, when $n > N$, $t_{r+1}^{(n)} - t_r^{(n)} < \delta$, for $r = 0, 1, 2, \ldots, n$; let $\zeta_r^{(n)}$ be any point whose parameter lies between $t_r^{(n)}$ and $t_{r+1}^{(n)}$; then we can make

$$\left| \sum_{r=0}^{n}(Z_{r+1^{(n)}} - Z_{r^{(n)}})f(\zeta_r^{(n)}) - \int_{z_0}^{Z} f(z)\,dz \right|$$

arbitrarily small by taking n sufficiently large.

4.62 An upper limit to the value of a complex integral

Let M be the upper bound of the continuous function $|f(z)|$. Then

$$\left| \int_{z_0}^{Z} f(z)\,dz \right| \le \int_{a}^{b} |f(z)|\left|\left(\frac{dx}{dt} + i\frac{dy}{dt}\right)\right| dt$$

$$\le \int_{a}^{b} M\left\{\left(\frac{dx}{dt}\right)^2 + \left(\frac{dy}{dt}\right)^2\right\}^{1/2} dt$$

$$\le ML,$$

where L is the 'length' of the curve $z_0 Z$. That is to say, $\left| \displaystyle\int_{z_0}^{Z} f(z)\,dz \right|$ cannot exceed ML.

4.7 Integration of infinite series

We shall now shew that if $S(z) = u_1(z) + u_2(z) + \cdots$ is a uniformly convergent series of continuous functions of z, for values of z contained within some region, then the series

$$\int_C u_1(z)dz + \int_C u_2(z)dz + \cdots,$$

(where all the integrals are taken along some path C in the region) is convergent, and has for sum $\int_C S(z)\,dz$.

For, writing

$$S(z) = u_1(z) + u_2(z) + \cdots + u_n(z) + R_n(z),$$

we have

$$\int_C S(z)dz = \int_C u_1(z)dz + \cdots + \int_C u_n(z)\,dz + \int_C R_n(z)dz.$$

Now since the series is uniformly convergent, to every positive number ε there corresponds a number r *independent* of z, such that when $n > r$ we have $|R_n(z)| < \varepsilon$, for all values of z in the region considered. Therefore if L be the length of the path of integration, we have (§4.62)

$$\left| \int_C R_n(z)\,dz \right| < \varepsilon L.$$

Therefore the modulus of the difference between $\int_C S(z)\,dz$ and $\sum_{m=1}^{n} \int_C u_m(z)\,dz$ can be made less than any positive number, by giving n any sufficiently large value. This proves both that the series $\sum_{m=1}^{\infty} \int_C u_m(z)\,dz$ is convergent, and that its sum is $\int_C S(z)\,dz$.

Corollary 4.7.1 *As in Corollary 4.4.1, it may be shewn that*[7]

$$\frac{d}{dz} \sum_{n=0}^{\infty} u_n(z) = \sum_{n=0}^{\infty} \frac{d}{dz} u_n(z)$$

if the series on the right converges uniformly and the series on the left is convergent.

Example 4.7.1 Consider the series

$$\sum_{n=1}^{\infty} \frac{2x\{n(n + 1)\sin^2 x^2 - 1\}\cos x^2}{\{1 + n^2 \sin^2 x^2\}\{1 + (n + 1)^2 \sin^2 x^2\}},$$

in which x is real.

The nth term is

$$\frac{2xn\cos x^2}{1 + n^2 \sin^2 x^2} - \frac{2x(n + 1)\cos x^2}{1 + (n + 1)^2 \sin^2 x^2},$$

---

[7] $\frac{df(z)}{z}$ means $\lim_{h \to 0} \frac{f(z+h)-f(z)}{h}$ where $h \to 0$ along a definite simple curve; this definition is modified slightly in §5.12 in the case when $f(z)$ is an *analytic* function.

and the sum of n terms is therefore

$$\frac{2x\cos x^2}{1 + \sin^2 x^2} - \frac{2x(n+1)\cos x^2}{1 + (n+1)^2 \sin^2 x^2}.$$

Hence the series is absolutely convergent for all real values of x except $\pm\sqrt{m\pi}$ where $m = 1, 2, \ldots$; but

$$R_n(x) = \frac{2x\,(n+1)\cos x^2}{1 + (n+1)^2 \sin^2 x^2},$$

and if n be any integer, by taking $x = (n+1)^{-1}$ this has the limit 2 as $n \to \infty$. The series is therefore non-uniformly convergent near $x = 0$.

Now the sum to infinity of the series is $\dfrac{2x\cos x^2}{1 + \sin^2 x^2}$, and so the integral from 0 to x of the sum of the series is $\arctan\left(\sin x^2\right)$. On the other hand, the sum of the integrals from 0 to x of the first n terms of the series is

$$\arctan\left(\sin x^2\right) - \arctan\left((n+1)\sin x^2\right),$$

and as $n \to \infty$ this tends to $\arctan\left(\sin x^2\right) - \frac{\pi}{2}$. Therefore the integral of the sum of the series differs from the sum of the integrals of the terms by $\frac{\pi}{2}$.

Example 4.7.2 Discuss, in a similar manner, the series

$$\sum_{n=1}^{\infty} \frac{2e^n x\{1 - n(e-1) + e^{n+1}x^2\}}{n(n+1)(1 + e^n x^2)(1 + e^{n+1}x^2)}$$

for real values of x.

Example 4.7.3 Discuss the series

$$u_1 + u_2 + u_3 + \cdots,$$

where

$$u_1 = ze^{-z^2}, \qquad u_n = nze^{-nz^2} - (n-1)ze^{-(n-1)z^2},$$

for real values of z. *Hint.* The sum of the first n terms is nze^{-nz^2}, so the sum to infinity is 0 for all real values of z. Since the terms u_n are real and ultimately all of the same sign, the convergence is absolute.

In the series

$$\int_0^z u_1\,dz + \int_0^z u_2\,dz + \int_0^z u_3\,dz + \cdots,$$

the sum of n terms is $\frac{1}{2}(1 - e^{-nz^2})$, and this tends to the limit $\frac{1}{2}$ as n tends to infinity; this is not equal to the integral from 0 to z of the sum of the series $\sum u_n$.

The explanation of this discrepancy is to be found in the non-uniformity of the convergence near $z = 0$, for the remainder after n terms in the series $u_1 + u_2 + \cdots$ is $-nze^{-nz^2}$; and by taking $z = n^{-1}$ we can make this equal to $e^{-1/n}$, which is not arbitrarily small; the series is therefore non-uniformly convergent near $z = 0$.

Example 4.7.4 (Trinity, 1903)   Compare the values of

$$\int_0^z \left\{ \sum_{n=1}^{\infty} u_n \right\} dz \quad \text{and} \quad \sum_{n=1}^{\infty} \int_0^z u_n \, dz,$$

where

$$u_n = \frac{2n^2 z}{(1 + n^2 z^2) \log(n + 1)} - \frac{2(n + 1)^2 z}{\{1 + (n + 1)^2 z^2\} \log(n + 2)}.$$

4.8 Miscellaneous examples

Example 4.1 (Dirichlet, Du Bois Reymond)   Shew that the integrals

$$\int_0^{\infty} \sin(x^2) \, dx, \qquad \int_0^{\infty} \cos(x^2) \, dx, \qquad \int_0^{\infty} x \, \exp(-x^6 \sin^2 x) \, dx$$

converge.

Example 4.2 (Stokes)   If a be real, the integral

$$\int_0^{\infty} \frac{\cos(ax)}{1 + x^2} \, dx$$

is a continuous function of a.

Example 4.3 (de la Vallée Poussin)   Discuss the uniformity of the convergence of $\int_0^{\infty} x \sin(x^3 - ax) \, dx$. *Hint.* Use

$$3 \int x \sin(x^3 - ax) dx = -\left(\frac{1}{x} + \frac{a}{3x^3}\right) \cos(x^3 - ax) - \int \left(\frac{1}{x^2} + \frac{a}{x^4}\right) \cos(x^3 - ax) \, dx$$

$$+ \frac{1}{3}a^2 \int \frac{\sin(x^3 - ax)}{x^3} \, dx.$$

Example 4.4 (Stokes)   Shew that $\int_0^{\infty} \exp[-e^{ia}(x^3 - nx)] \, dx$ converges uniformly in the range $\left(-\frac{1}{2}\pi, \frac{1}{2}\pi\right)$ of values of a.

Example 4.5 (Hardy [275])   Discuss the convergence of $\int_0^{\infty} \frac{x^{\mu} dx}{1 + x^{\nu}|\sin x|^p}$ when μ, ν, p are positive.

Example 4.6 (Math. Trip. 1914)   Examine the convergence of the integrals

$$\int_0^{\infty} \left(\frac{1}{x} - \frac{1}{2}e^{-x} + \frac{1}{1 - e^x}\right) \frac{dx}{x}, \qquad \int_0^{\infty} \frac{\sin(x + x^2)}{x^n} \, dx.$$

Example 4.7   Shew that $\int_{\pi}^{\infty} \frac{dx}{x^2(\sin x)^{2/3}}$ exists.

Example 4.8 (Math. Trip. 1908)   Shew that $\displaystyle\int_a^\infty x^{-n} e^{\sin x} \sin 2x \, dx$ converges if $a > 0$, $n > 0$.

Example 4.9 (Lerch [430])   If a series $g(z) = \sum\limits_{v=0}^{\infty} (C_v - C_{v+1}) \sin(2v + 1)\pi z$, (in which $C_0 = 0$) converges uniformly in an interval, shew that $g(z)\dfrac{\pi}{\sin \pi z}$ is the derivative of the series $f(z) = \sum\limits_{v=1}^{\infty} \dfrac{C_v}{v} \sin 2v\pi z$.

Example 4.10 (Math. Trip. 1904)   Shew that

$$\int^\infty \int^\infty \cdots \int^\infty \frac{dx_1 \, dx_2 \cdots dx_n}{(x_1^2 + x_2^2 + \cdots + x_n^2)^\alpha}$$

$$\text{and} \quad \int^\infty \int^\infty \cdots \int^\infty \frac{dx_1 \, dx_2 \cdots dx_n}{x_1^\alpha + x_2^\beta + \cdots + x_n^\lambda}$$

converge when $\alpha > \frac{1}{2} n$ and $\alpha^{-1} + \beta^{-1} + \cdots + \lambda^{-1} < 1$ respectively.

Example 4.11 (Bôcher)   If $f(x, y)$ be a continuous function of both x and y in the ranges $(a \le x \le b)$, $(a \le y \le b)$ except that it has ordinary discontinuities at points on a finite number of curves, with continuously turning tangents, each of which meets any line parallel to the coordinate axes only a finite number of times, then $\displaystyle\int_a^b f(x, y) \, dx$ is a continuous function of y. *Hint.* Consider

$$\int_a^{a_1 - \delta_1} + \int_{a_1 + \varepsilon_1}^{a_2 - \delta_2} + \cdots + \int_{a_n + \varepsilon_n}^b \{f(x, y + h) - f(x, y)\} \, dx,$$

where the numbers $\delta_1, \delta_2, \ldots, \varepsilon_1, \varepsilon_2, \ldots$ are so chosen as to exclude the discontinuities of $f(x, y + h)$ from the range of integration; $a_1, a_2, \ldots$ being the discontinuities of $f(x, y)$.

5

The Fundamental Properties of Analytic Functions; Taylor's, Laurent's and Liouville's Theorems

5.1 Property of the elementary functions

The reader will be already familiar with the term *elementary function*, as used (in textbooks on Algebra, Trigonometry, and the Differential Calculus) to denote certain analytical expressions[1] depending on a variable z, the symbols involved therein being those of elementary algebra together with exponentials, logarithms and the trigonometrical functions; examples of such expressions are

$$z^2, \quad e^z, \quad \log z, \quad \arcsin z^{3/2}.$$

Such combinations of the elementary functions of analysis have in common a remarkable property, which will now be investigated.

Take as an example the function e^z. Write $e^z = f(z)$. Then, if z be a fixed point and if z' be any other point, we have

$$\frac{f(z') - f(z)}{z' - z} = \frac{e^{z'} - e^z}{z' - z} = e^z \cdot \frac{e^{(z'-z)} - 1}{z' - z}$$
$$= e^z \left\{ 1 + \frac{z' - z}{2!} + \frac{(z' - z)^2}{3!} + \cdots + \right\};$$

and since the last series in brackets is uniformly convergent for all values of z', it follows (§3.7) that, as $z' \to z$, the quotient

$$\frac{f(z') - f(z)}{z' - z}$$

tends to the limit e^z, uniformly for all values of $\arg(z' - z)$. This shews that *the limit of*

$$\frac{f(z') - f(z)}{z' - z}$$

is in this case independent of the path by which the point z' tends towards coincidence with z.
It will be found that this property is shared by many of the well-known elementary functions; namely, that if $f(z)$ be one of these functions and h be any complex number, the limiting value of

$$\frac{1}{h} \{ f(z + h) - f(z) \}$$

---

[1] The reader will observe that this is not the sense in which the term function is defined (§3.1) in this work. Thus e.g. $x - iy$ and $|z|$ are *functions* of z (= $x + iy$) in the sense of §3.1, but are not elementary functions of the type under consideration.

exists and is independent of the mode in which h tends to zero.

The reader will, however, easily prove that, if $f(z) = x - iy$, where $z = x + iy$, then $\lim \dfrac{f(z+h) - f(z)}{h}$ is *not* independent of the mode in which $h \to 0$.

5.11 Occasional failure of the property

For each of the elementary functions, however, there will be certain points z at which this property will cease to hold good. Thus it does not hold for the function $1/(z-a)$ at the point $z = a$, since

$$\lim_{h\to 0} \frac{1}{h} \left\{ \frac{1}{z-a+h} - \frac{1}{z-a} \right\}$$

does not exist when $z = a$. Similarly it does not hold for the functions $\log z$ and $z^{1/2}$ at the point $z = 0$. These exceptional points are called *singular points* or *singularities* of the function $f(z)$ under consideration; at other points $f(z)$ is said to be *analytic*. The property does not hold good at *any* point for the function $|z|$.

5.12 Cauchy's definition of an analytic function of a complex variable

(See the memoir [121]). The property considered in §5.11 will be taken as the basis of the definition of an *analytic function*, which may be stated as follows.

Let a two-dimensional region in the z-plane be given; and let u be a function of z defined uniquely at all points of the region. Let z, $z + \delta z$ be values of the variable z at two points, and u, $u + \delta u$ the corresponding values of u. Then, if, at any point z within the area, $\dfrac{\delta u}{\delta z}$ tends to a limit when $\delta x \to 0$, $\delta y \to 0$, independently (where $\delta z = \delta x + i\delta y$), u is said to be a function of z, which is *monogenic* or *analytic* at the point. The words 'regular' and 'holomorphic' are sometimes used. A distinction has been made by Borel [86, p. 137–138], [87] between 'monogenic' and 'analytic' functions in the case of functions with an infinite number of singularities. See §5.51. If the function is analytic and *one-valued* at all points of the region, we say that the function is *analytic throughout the region*. See the footnote after Corollary 5.2.2.

We shall frequently use the word 'function' alone to denote an analytic function, as the functions studied in this work will be almost exclusively analytic functions. In the foregoing definition, the function u has been defined only within a certain region in the z-plane. As will be seen subsequently, however, the function u can generally be defined for other values of z not included in this region; and (as in the case of the elementary functions already discussed) may have *singularities*, for which the fundamental property no longer holds, at certain points outside the limits of the region. We shall now state the definition of analytic functionality in a more arithmetical form.

Let $f(z)$ be analytic at z, and let ε be an arbitrary positive number; then we can find numbers ℓ and δ (with δ depending on ε) such that

$$\left| \frac{f(z') - f(z)}{z' - z} - \ell \right| < \varepsilon$$

whenever $|z' - z| < \delta$.

If $f(z)$ is analytic at all points z of a region, ℓ obviously depends on z; we consequently write $\ell = f'(z)$. Hence $f(z') = f(z) + (z' - z)f'(z) + v(z' - z)$, where v is a function of z and z' such that $|v| < \varepsilon$ when $|z' - z| < \delta$.

Example 5.1.1   Find the points at which the following functions are not analytic:

1. z^2;
2. $\operatorname{cosec} z$ $(z = n\pi, n$ any integer);
3. $\dfrac{z - 1}{z^2 - 5z + 6}$ $(z = 2, 3)$;
4. $e^{1/z}$ $(z = 0)$;
5. $\{(z - 1)z\}^{1/3}$ $(z = 0, 1)$.

Example 5.1.2 (Riemann)   If $z = x + iy$, $f(z) = u + iv$, where u, v, x, y are real and f is an analytic function, shew that,

$$\frac{\partial u}{\partial x} = \frac{\partial v}{\partial y}, \quad \frac{\partial u}{\partial y} = -\frac{\partial v}{\partial x}.$$

5.13 An application of the modified Heine–Borel theorem

Let $f(z)$ be analytic at all points of a continuum; and on any point z of the boundary of the continuum let numbers $f_1(z), \delta$ (δ depending on z) exist such that

$$|f(z') - f(z) - (z' - z) f_1(z)| < \varepsilon |z' - z|$$

whenever $|z' - z| < \delta$ and z' is a point of the continuum or its boundary. *Hint.* We write $f_1(z)$ instead of $f'(z)$ as the differential coefficient might not exist when z' approaches z from outside the boundary so that $f_1(z)$ is not necessarily a unique derivative.

The above inequality is obviously satisfied for all points z of the continuum as well as boundary points.

Applying the two-dimensional form of the theorem of §3.6, we see that the region formed by the continuum and its boundary can be divided into a *finite* number of parts (squares with sides parallel to the axes and their interiors, or portions of such squares) such that *inside* or on the boundary of any part there is one point z_1 such that the inequality

$$|f(z') - f(z_1) - (z' - z_1)f_1(z_1)| < \varepsilon |z' - z_1|$$

is satisfied by all points z' inside or on the boundary of that part.

5.2 Cauchy's theorem on the integral of a function round a contour

The results here are due to Cauchy [121]. The proof here given is that due to Goursat [254].

A simple closed curve C in the plane of the variable z is often called a *contour*; if A, B, D be points taken in order in the counter-clockwise sense along the arc of the contour, and if

$f(z)$ be a one-valued continuous[2] function of z (not necessarily analytic) at all points on the arc, then the integral

$$\int_{ABDA} f(z)\, dz \quad \text{or} \quad \int_C f(z)\, dz$$

taken round the contour, starting from the point A and returning to A again, is called *the integral of $f(z)$ taken along the contour.* Clearly the value of the integral taken along the contour is unaltered if some point in the contour other than A is taken as the starting-point.

We shall now prove a result due to Cauchy, which may be stated as follows. *If $f(z)$ is a function of z, analytic at all points on and inside a contour C, then*

$$\int_C f(z)\, dz = 0.$$

Note   It is not necessary that $f(z)$ should be analytic on C (it is sufficient that it be continuous on and inside C), but if $f(z)$ is not analytic on C, the theorem is much harder to prove. This proof merely assumes that $f'(z)$ *exists* at all points on and inside C. Earlier proofs made more extended assumptions; thus Cauchy's proof assumed the *continuity* of $f'(z)$. Riemann's proof made an equivalent assumption. Goursat's first proof assumed that $f(z)$ was *uniformly* differentiable throughout C.

For divide up the interior of C by lines parallel to the real and imaginary axes in the manner of §5.13; then the interior of C is divided into a number of regions whose boundaries are squares $C_1, C_2, \ldots, C_M$ and other regions whose boundaries $D_1, D_2, \ldots, D_N$ are portions of sides of squares and parts of C; consider

$$\sum_{n=1}^{M} \int_{C_n} f(z)\, dz + \sum_{n=1}^{N} \int_{D_n} f(z) dz,$$

each of the paths of integration being taken counter-clockwise; in the complete sum each side of each square appears twice as a path of integration, and the integrals along it are taken in opposite directions and consequently cancel (see Example 4.6.1); the only parts of the sum which survive are the integrals of $f(z)$ taken along a number of arcs which together make up C, each arc being taken in the same sense as in $\int_C f(z)\, dz$; these integrals therefore just make up $\int_C f(z)\, dz$.

Now consider $\int_{C_n} f(z)\, dz$. With the notation of §5.12,

$$\int_{C_n} f(z)\, dz = \int_{C_n} \{f(z_1) + (z - z_1)f'(z_1) + (z - z_1)v\}\, dz$$

$$= \{f(z_1) - z_1 f'(z_1)\} \int_{C_n} dz + f'(z_1) \int_{C_n} z\, dz + \int_{C_n} (z - z_1)\, v\, dz.$$

But

$$\int_{C_n} dz = [z]_{C_n} = 0, \qquad \int_{C_n} z\, dz = \left[\tfrac{1}{2}z^2\right]_{C_n} = 0,$$

---

[2] It is sufficient for $f(z)$ to be continuous when variations of z *along the arc only* are considered.

by Example 4.6.1, since the end points of C_n coincide. Now let l_n be the side of C_n and A_n the area of C_n. Then, using §4.62,

$$\left| \int_{C_n} f(z)\, dz \right| = \left| \int_{C_n} (z - z_1) v\, dz \right| \leq \int_{C_n} |(z - z_1) v\, dz|$$

$$< \varepsilon l_n \sqrt{2} \int_{C_n} |dz| = \varepsilon l_n \sqrt{2} \cdot 4 l_n = 4 \varepsilon A_n \sqrt{2}.$$

In like manner

$$\left| \int_{D_n} f(z)\, dz \right| \leq \int_{D_n} |(z - z_1) v\, dz|$$

$$\leq 4\varepsilon (A_n' + l_n' \lambda_n) \sqrt{2},$$

where A_n' is the area of the complete square of which D_n is part, l_n' is the side of this square and λ_n is the length of the part of C which lies inside this square. Hence, if λ be the whole length of C, while l is the side of a square which encloses all the squares C_n and D_n,

$$\left| \int_C f(z)dz \right| \leq \sum_{n=1}^{M} \left| \int_{C_n} f(z)\, dz \right| + \sum_{n=1}^{N} \left| \int_{D_n} f(z)\, dz \right|$$

$$< 4\varepsilon\sqrt{2} \left\{ \sum_{n=1}^{M} A_n + \sum_{n=1}^{N} A_n' + l \sum_{n=1}^{N} \lambda_n \right\}$$

$$< 4\varepsilon\sqrt{2}(l^2 + l\lambda).$$

Now ε is arbitrarily small, and l, λ and $\int_C f(z)\, dz$ are *independent of* ε. It therefore follows from this inequality that the only value which $\int_C f(z)\, dz$ can have is zero; and this is Cauchy's result.

Corollary 5.2.1 *If there are two paths $z_0 AZ$ and $z_0 BZ$ from z_0 to Z, and if $f(z)$ is a function of z analytic at all points on these curves and throughout the domain enclosed by these two paths, then $\int_{z_0}^{Z} f(z)\, dz$ has the same value whether the path of integration is $z_0 AZ$ or $z_0 BZ$.* This follows from the fact that $z_0 AZBz_0$ is a contour, and so the integral taken round it (which is the difference of the integrals along $z_0 AZ$ and $z_0 BZ$) is zero. Thus, if $f(z)$ be an analytic function of z, the value of $\int_{AB} f(z)\, dz$ is to a certain extent independent of the choice of the arc AB, and depends only on the terminal points A and B. It must be borne in mind that this is only the case when $f(z)$ is an analytic function in the sense of §5.12.

Corollary 5.2.2 *Suppose that two simple closed curves C_0 and C_1 are given, such that C_0 completely encloses C_1, as e.g. would be the case if C_0 and C_1 were confocal ellipses.*
    Suppose moreover that $f(z)$ is a function which is analytic[3] *at all points on C_0 and C_1 and throughout the ring-shaped region contained between C_0 and C_1. Then by drawing a network*

---

[3] The phrase *analytic throughout a region*, implies one-valuedness (§5.12); that is to say that after z has described a closed path surrounding C_0, $f(z)$ has returned to its initial value. A function such as $\log z$ considered in the region $1 \leq |z| \leq 2$ will be said to be *analytic at all points of the region*.

of intersecting lines in this ring-shaped space, we can shew, exactly as in the theorem just proved, that the integral

$$\int f(z)\,dz$$

is zero, where the integration is taken round the whole boundary of the ring-shaped space; this boundary consisting of two curves C_0 and C_1, the one described in the counter-clockwise direction and the other described in the clockwise direction.

Corollary 5.2.3 *In general, if any connected region be given in the z-plane, bounded by any number of simple closed curves $C_0, C_1, C_2, \ldots$, and if $f(z)$ be any function of z which is analytic and one-valued everywhere in this region, then $\int f(z)\,dz$ is zero, where the integral is taken round the whole boundary of the region; this boundary consisting of the curves $C_0, C_1, \ldots$, each described in such a sense that the region is kept either always on the right or always on the left of a person walking in the sense in question round the boundary.*

An extension of Cauchy's theorem $\int f(z)\,dz = 0$, to curves lying on a cone whose vertex is at the origin, has been made by Ravut [549], Morera [474] and Osgood [511] have shewn that the property $\int f(z)\,dz = 0$ may be taken as the property defining an analytic function, the other properties being deducible from it. (See Chapter 5, Example 5.16).

Example 5.2.1 A ring-shaped region is bounded by the two circles $|z| = 1$ and $|z| = 2$ in the z-plane. Verify that the value of $\int \dfrac{dz}{z}$, where the integral is taken round the boundary of this region, is zero. *Solution.* For the boundary consists of the circumference $|z| = 1$, described in the clockwise direction, together with the circumference $|z| = 2$, described in the counter-clockwise direction. Thus, if for points on the first circumference we write $z = e^{i\theta}$, and for points on the second circumference we write $z = 2e^{i\phi}$, then θ and ϕ are real, and the integral becomes

$$\int_0^{-2\pi} \frac{i \cdot e^{i\theta}\,d\theta}{e^{i\theta}} + \int_0^{2\pi} \frac{i \cdot 2e^{i\phi}\,d\phi}{2e^{i\phi}} = -2\pi i + 2\pi i = 0.$$

5.21 *The value of an analytic function at a point, expressed as an integral taken round a contour enclosing the point*

Let C be a contour within and on which $f(z)$ is an analytic function of z. Then, if a be any point within the contour,

$$\frac{f(z)}{z - a}$$

is a function of z, which is analytic at all points within the contour C except the point $z = a$. Now, given ε, we can find δ such that

$$|f(z) - f(a) - (z - a)f'(a)| \leq \varepsilon\,|z - a|$$

whenever $|z - a| < \delta$; with the point a as centre describe a circle γ of radius $r < \delta$, r being so small that γ lies wholly inside C. Then in the space between γ and C the function $f(z)/(z-a)$ is analytic, and so, by Corollary 5.2.2 we have

$$\int_C \frac{f(z)\,dz}{z-a} = \int_\gamma \frac{f(z)\,dz}{z-a},$$

where $\int_C$ and $\int_\gamma$ denote integrals taken counter-clockwise along the curves C and γ respectively. But, since $|z - a| < \delta$ on γ, we have

$$\int_\gamma \frac{f(z)\,dz}{z-a} = \int_\gamma \frac{f(a) + (z-a)f'(a) + v(z-a)}{z-a}\,dz,$$

where $|v| < \varepsilon$; and so

$$\int_C \frac{f(z)\,dz}{z-a} = f(a) \int_\gamma \frac{dz}{z-a} + f'(a) \int_\gamma dz + \int_\gamma v\,dz.$$

Now, if z be on γ, we may write $z - a = re^{i\theta}$, where r is the radius of the circle γ, and consequently

$$\int_\gamma \frac{dz}{z-a} = \int_0^{2\pi} \frac{ire^{i\theta}\,d\theta}{re^{i\theta}} = i \int_0^{2\pi} d\theta = 2\pi i,$$

and

$$\int_\gamma dz = \int_0^{2\pi} ire^{i\theta}\,d\theta = 0;$$

also, by §4.62,

$$\left| \int_\gamma v\,dz \right| \le \varepsilon \cdot 2\pi r.$$

Thus

$$\left| \int_C \frac{f(z)\,dz}{z-a} - 2\pi i f(a) \right| = \left| \int_\gamma v\,dz \right| \le 2\pi r \varepsilon.$$

But the left-hand side is independent of ε, and so it must be zero, since ε is arbitrary; that is to say

$$f(a) = \frac{1}{2\pi i} \int_C \frac{f(z)\,dz}{z-a}.$$

This remarkable result expresses the value of a function $f(z)$, (which is *analytic* on and inside C) at any point a *within* a contour C, in terms of an integral which depends only on the value of $f(z)$ at points *on* the contour itself.

Corollary 5.2.4 *If $f(z)$ is an analytic one-valued function of z in a ring-shaped region bounded by two curves C and C', and a is a point in the region, then*

$$f(a) = \frac{1}{2\pi i} \int_C \frac{f(z)}{z-a}\,dz - \frac{1}{2\pi i} \int_{C'} \frac{f(z)}{z-a}\,dz,$$

where C is the outer of the curves and the integrals are taken counter-clockwise.

5.22 *The derivatives of an analytic function* $f(z)$

The function $f'(z)$, which is the limit of

$$\frac{f(z+h)-f(z)}{h}$$

as h tends to zero, is called the *derivate* or *derivative* of $f(z)$. We shall now shew that $f'(z)$ is itself an analytic function of z, and consequently itself possesses a derivative.

For if C be a contour surrounding the point a, and situated entirely within the region in which $f(z)$ is analytic, we have

$$f'(a) = \lim_{h\to 0} \frac{f(a+h)-f(a)}{h}$$

$$= \lim_{h\to 0} \frac{1}{2\pi i h}\left\{\int_C \frac{f(z)\,dz}{z-a-h} - \int_C \frac{f(z)\,dz}{z-a}\right\}$$

$$= \lim_{h\to 0} \frac{1}{2\pi i}\int_C \frac{f(z)\,dz}{(z-a)(z-a-h)}$$

$$= \frac{1}{2\pi i}\int_C \frac{f(z)\,dz}{(z-a)^2} + \lim_{h\to 0}\frac{h}{2\pi i}\int_C \frac{f(z)\,dz}{(z-a)^2(z-a-h)}.$$

Now, on C, $f(z)$ is continuous and therefore bounded, and so is $(z-a)^{-2}$; while we can take $|h|$ less than the lower bound of $\frac{1}{2}|z-a|$. Therefore $\left|\dfrac{f(z)}{(z-a)^2(z-a-h)}\right|$ is bounded; let its upper bound be K. Then, if l be the length of C,

$$\left|\lim_{h\to 0}\frac{h}{2\pi i}\int_c \frac{f(z)\,dz}{(z-a)^2(z-a-h)}\right| \le \lim_{h\to 0}|h|(2\pi)^{-1}Kl = 0,$$

and consequently

$$f'(a) = \frac{1}{2\pi i}\int_C \frac{f(z)\,dz}{(z-a)^2}, \tag{5.1}$$

a formula which expresses the value of the derivative of a function at a point as an integral taken along a contour enclosing the point.

From this formula we have, if the points a and $a+h$ are inside C,

$$\frac{f'(a+h)-f'(a)}{h} = \frac{1}{2\pi i}\int_C \frac{f(z)}{h}\left\{\frac{1}{(z-a-h)^2} - \frac{1}{(z-a)^2}\right\}dz$$

$$= \frac{1}{2\pi i}\int_C f(z)\frac{2(z-a-\frac{1}{2}h)}{(z-a-h)^2(z-a)^2}\,dz$$

$$= \frac{2}{2\pi i}\int_C \frac{f(z)\,dz}{(z-a)^3} + hA_h,$$

and it is easily seen that A_h is a bounded function of z when $|h| < \frac{1}{2}|z-a|$. Therefore, as h tends to zero, $h^{-1}\{f'(a+h)-f'(a)\}$ tends to a limit, namely

$$\frac{2}{2\pi i}\int_C \frac{f(z)\,dz}{(z-a)^3}.$$

Since $f'(a)$ has a unique differential coefficient, it is an analytic function of a; its derivative, which is represented by the expression just given, is denoted by $f''(a)$, and is called the *second derivative* of $f(a)$. Similarly it can be shewn that $f''(a)$ is an analytic function of a, possessing a derivative equal to

$$\frac{2 \cdot 3}{2\pi i} \int_C \frac{f(z)\, dz}{(z-a)^4};$$

this is denoted by $f'''(a)$, and is called the *third derivative* of $f(a)$. And in general an nth derivative $f^{(n)}(a)$ of $f(a)$ exists, expressible by the integral

$$\frac{n!}{2\pi i} \int_C \frac{f(z)dz}{(z-a)^{n+1}},$$

and having itself a derivative of the form

$$\frac{(n+1)!}{2\pi i} \int_C \frac{f(z)dz}{(z-a)^{n+2}};$$

the reader will see that this can be proved by induction without difficulty.

A function which possesses a first derivative with respect to the *complex* variable z at all points of a closed two-dimensional region in the z-plane therefore possesses derivatives of all orders at all points *inside* the region.

5.23 Cauchy's inequality for $f^{(n)}(a)$

Let $f(z)$ be analytic on and inside a circle C with centre a and radius r. Let M be the upper bound of $f(z)$ on the circle. Then, by §4.62,

$$\left| f^{(n)}(a) \right| \le \frac{n!}{2\pi} \int_C \frac{M}{r^{n+1}} |dz|$$

$$\le \frac{M n!}{r^n}.$$

Example 5.2.2 (Trinity, 1910)   If $f(z)$ is analytic, $z = x + iy$ and

$$\nabla^2 = \frac{\partial^2}{\partial x^2} + \frac{\partial^2}{\partial y^2},$$

shew that $\nabla^2 \log |f(z)| = 0$; and $\nabla^2 |f(z)| > 0$ unless $f(z) = 0$ or $f'(z) = 0$.

5.3 Analytic functions represented by uniformly convergent series

Let $\sum_{n=0}^{\infty} f_n(z)$ be a series such that: (i) it converges uniformly along a contour C; (ii) $f_n(z)$ is analytic throughout C and its interior. Then $\sum_{n=0}^{\infty} f_n(z)$ converges, and the sum of the series is an analytic function throughout C and its interior.

For let a be any point inside C; on C, let $\sum_{n=0}^{\infty} f_n(z) = \Phi(z)$. Then

$$\frac{1}{2\pi i} \int_C \frac{\Phi(z)}{z-a} dz = \frac{1}{2\pi i} \int_C \left\{ \sum_{n=0}^{\infty} f_n(z) \right\} \frac{dz}{z-a}$$

$$= \sum_{n=0}^{\infty} \left\{ \frac{1}{2\pi i} \int_C \frac{f_n(z)}{z-a} dz \right\},$$

by §4.7. Since $|z - a|^{-1}$ is bounded when a is fixed and z is on C, the uniformity of the convergence of $\sum_{n=0}^{\infty} f_n(z)/(z - a)$ follows from that of $\sum_{n=0}^{\infty} f_n(z)$. But this last series, by §5.21, is $\sum_{n=0}^{\infty} f_n(a)$; the series under consideration therefore converges at all points inside C; let its sum inside C (as well as on C) be called $\Phi(z)$. Then the function is analytic if it has a unique differential coefficient at all points inside C. But if a and $a + h$ are inside C,

$$\frac{\Phi(a + h) - \Phi(a)}{h} = \frac{1}{2\pi i} \int_C \frac{\Phi(z)\, dz}{(z - a)(z - a - h)},$$

and hence, as in §5.22, $\lim_{h \to 0} \left[\{\Phi(a + h) - \Phi(a)\}\, h^{-1} \right]$ exists and is equal to

$$\frac{1}{2\pi i} \int_C \frac{\Phi(z)\, dz}{(z - a)^2};$$

and therefore $\Phi(z)$ is analytic inside C. Further, by transforming the last integral in the same way as we transformed the first one, we see that $\Phi'(a) = \sum_{n=0}^{\infty} f_n'(a)$, so that $\sum_{n=0}^{\infty} f_n(a)$ may be *differentiated term by term*.

If a series of analytic functions converges only at points of a curve which is *not* closed nothing can be inferred as to the convergence of the derived series. This might have been anticipated as the main theorem of this section deals with uniformity of convergence over a *two-dimensional* region.

Thus $\sum_{n=1}^{\infty} (-1)^n \dfrac{\cos nx}{n^2}$ converges uniformly for real values of x (§3.34). But the derived series $\sum_{n=1}^{\infty} (-1)^{n-1} \dfrac{\sin nx}{n}$ converges non-uniformly near $x = (2m + 1)\pi$, (m any integer); and the derived series of this, viz. $\sum_{n=1}^{\infty} (-1)^{n-1} \cos nx$, does not converge at all.

Corollary 5.3.1 *By §3.7, the sum of a power series is analytic inside its circle of convergence.*

5.31 Analytic functions represented by integrals

Let $f(t, z)$ satisfy the following conditions when t lies on a certain path of integration (a, b) and z is any point of a region S:

1. f and $\dfrac{\partial f}{\partial z}$ are continuous functions of t.
2. f is an analytic function of z.
3. The continuity of $\dfrac{\partial f}{\partial z}$ *qua* function of z is uniform with respect to the variable t.

Then $\displaystyle \int_a^b f(t, z)\, dt$ is an analytic function of z. For, by §4.2, it has the unique derivative

$$\int_a^b \frac{\partial f(t, z)}{\partial z}\, dt.$$

5.32 Analytic functions represented by infinite integrals

From Corollary 4.4.1, it follows that $\int_a^\infty f(t, z)\, dt$ is an analytic function of z at all points of a region S if

(i) the integral converges,

(ii) $f(t, z)$ is an analytic function of z when t is on the path of integration and z is on S,

(iii) $\dfrac{\partial f(t, z)}{\partial z}$ is a continuous function of both variables,

(iv) $\int_a^\infty \dfrac{\partial f(t, z)}{\partial z}\, dt$ converges uniformly throughout S.

For if these conditions are satisfied $\int_a^\infty f(t, z)\, dt$ has the unique derivative

$$\int_a^\infty \frac{\partial f(t, z)}{\partial z}\, dt.$$

A case of very great importance is afforded by the integral $\int_0^\infty e^{-tz} f(t)\, dt$, where $f(t)$ is continuous and $|f(t)| < K e^{rt}$ where K, r are independent of t; it is obvious from the conditions stated that the integral is an analytic function of z when $R(z) \geq r_1 > r$. Condition (iv) is satisfied, by §4.431 (I), since $\int_0^\infty t e^{(r-r_1)t}\, dt$ converges.

5.4 Taylor's theorem

Consider a function $f(z)$, which is analytic in the neighborhood of a point $z = a$. Let C be a circle with a as centre in the z-plane, which does not have any singular point of the function $f(z)$ on or inside it; so that $f(z)$ is analytic at all points on and inside C. Let $z = a + h$ be any point inside the circle C. Then, by §5.21, we have

$$
\begin{aligned}
f(a + h) &= \frac{1}{2\pi i} \int_C \frac{f(z)\, dz}{z - a - h} \\
&= \frac{1}{2\pi i} \int_C f(z) \left\{ \frac{1}{z - a} + \frac{h}{(z - a)^2} + \cdots + \frac{h^n}{(z - a)^{n+1}} + \frac{h^{n+1}}{(z - a)^{n+1}(z - a - h)} \right\} dz \\
&= f(a) + h f'(a) + \frac{h^2}{2!} f''(a) + \cdots + \frac{h^n}{n!} f^{(n)}(a) + \frac{1}{2\pi i} \int_C \frac{f(z)\, dz}{(z - a)^{n+1}(z - a - h)} h^{n+1}.
\end{aligned}
$$

But when z is on C, the modulus of $f(z)/(z-a-h)$ is continuous, and so, by Corollary 3.6.2, will not exceed some finite number M.

Therefore, by §4.62,

$$\left| \frac{1}{2\pi i} \int_C \frac{f(z)\, dz \cdot h^{n+1}}{(z - a)^{n+1}(z - a - h)} \right| \leq \frac{M \cdot 2\pi R}{2\pi} \left(\frac{|h|}{R} \right)^{n+1},$$

where R is the radius of the circle C, so that $2\pi R$ is the length of the path of integration in the last integral, and $R = |z - a|$ for points z on the circumference of C.

The right-hand side of the last inequality tends to zero as $n \to \infty$. We have therefore

$$f(a + h) = f(a) + hf'(a) + \frac{h^2}{2!}f''(a) + \cdots + \frac{h^n}{n!}f^{(n)}(a) + \cdots ,$$

which we can write

$$f(z) = f(a) + (z - a)f'(a) + \frac{(z - a)^2}{2!}f''(a) + \cdots + \frac{(z - a)^n}{n!}f^{(n)}(a) + \cdots .$$

This result is known as *Taylor's theorem*; and the proof given is due to Cauchy. The formal expansion was first published by Dr. Brook Taylor [621].

It follows that *the radius of convergence of a power series is always at least so large as only just to exclude from the interior of the circle of convergence the nearest singularity of the function represented by the series*. And by Corollary 5.3.1 it follows that the radius of convergence is *not larger than the number just specified*. Hence the radius of convergence is just such as to exclude from the interior of the circle that singularity of the function which is nearest to a.

At this stage we may introduce some terms which will be frequently used.

If $f(a) = 0$, the function $f(z)$ is said to have a *zero* at the point $z = a$. If at such a point $f'(a)$ is different from zero, the zero of $f(a)$ is said to be *simple*; if, however, $f'(a), f''(a), \ldots, f^{(n-1)}(a)$ are all zero, so that the Taylor's expansion of $f(z)$ at $z = a$ begins with a term in $(z - a)^n$, then the function $f(z)$ is said to have a *zero of the nth order* at the point $z = a$.

Example 5.4.1 Find the function $f(z)$, which is analytic throughout the circle C and its interior, whose centre is at the origin and whose radius is unity, and has the value

$$\frac{a - \cos\theta}{a^2 - 2a\cos\theta + 1} + i\frac{\sin\theta}{a^2 - 2a\cos\theta + 1}$$

(where $a > 1$ and θ is the vectorial angle) at points on the circumference of C.
We have

$$
\begin{aligned}
f^{(n)}(0) &= \frac{n!}{2\pi i}\int_C \frac{f(z)\,dz}{z^{n+1}} \\
&= \frac{n!}{2\pi i}\int_0^{2\pi} e^{-ni\theta} \cdot i\,d\theta \cdot \frac{a - \cos\theta + i\sin\theta}{a^2 - 2a\cos\theta + 1}, \quad \text{(putting } z = e^{i\theta}) \\
&= \frac{n!}{2\pi}\int_0^{2\pi} \frac{e^{-ni\theta}\,d\theta}{a - e^{i\theta}} = \frac{n!}{2\pi i}\int_C \frac{dz}{z^{n+1}(a - z)} = \left[\frac{d^n}{dz^n}\frac{1}{a - z}\right]_{z=0} \\
&= \frac{n!}{a^{n+1}}.
\end{aligned}
$$

Therefore by Maclaurin's theorem[4],

$$f(z) = \sum_{n=0}^{\infty} \frac{z^n}{a^{n+1}},$$

---

[4] The result $f(z) = f(0) + zf'(0) + \frac{z^2}{2}f''(0) + \cdots$, obtained by putting $a = 0$ in Taylor's theorem, is usually called *Maclaurin's theorem*; it was discovered by Stirling in 1717 and published by Maclaurin [449].

or $f(z) = (a - z)^{-1}$ for all points within the circle.

This example raises the interesting question, *will it still be convenient to define $f(z)$ as $(a - z)^{-1}$ at points outside the circle?* This will be discussed in §5.51.

Example 5.4.2 Prove that the arithmetic mean of all values of $z^{-n} \sum\limits_{v=0}^{\infty} a_v z^v$ for points z on the circumference of the circle $|z| = 1$, is a_n; if $\sum a_v z^v$ is analytic throughout the circle and its interior. *Solution.* Let $\sum\limits_{v=0}^{\infty} a_v z^v = f(z)$, so that $a_v = \dfrac{f^{(v)}(0)}{v!}$. Then, writing $z = e^{i\theta}$, and calling C the circle $|z| = 1$,

$$\frac{1}{2\pi} \int_0^{2\pi} \frac{f(z)\, d\theta}{z^n} = \frac{1}{2\pi i} \int_C \frac{f(z)\, dz}{z^{n+1}} = \frac{f^{(n)}(0)}{n!} = a_n.$$

Example 5.4.3 Let $f(z) = z^r$; then $f(z + h)$ is an analytic function of h when $|h| < |z|$ for all values of r; and so $(z + h)^r = z^r + rz^{r-1}h + \dfrac{r(r - 1)}{2} z^{r-2}h^2 + \cdots$, this series converging when $|h| < |z|$. This is the binomial theorem.

Example 5.4.4 Prove that if h is a positive constant, and $(1 - 2zh + h^2)^{-1/2}$ is expanded in the form

$$1 + hP_1(z) + h^2 P_2(z) + h^3 P_3(z) + \cdots \tag{5.2}$$

(where $P_n(z)$ is easily seen to be a polynomial of degree n in z), then this series converges so long as z is in the interior of an ellipse whose foci are the points $z = 1$ and $z = -1$, and whose semi-major axis is $\frac{1}{2}(h + h^{-1})$.

Let the series be first regarded as a function of h. It is a power series in h, and therefore converges so long as the point h lies within a circle in the h-plane. The centre of this circle is the point $h = 0$, and its circumference will be such as to pass through that singularity of $(1 - 2zh + h^2)^{-1/2}$ which is nearest to $h = 0$. But

$$1 - 2zh + h^2 = \{h - z + (z^2 - 1)^{1/2}\}\{h - z - (z^2 - 1)^{1/2}\}, \tag{5.3}$$

so the singularities of $(1 - 2zh + h^2)^{-1/2}$ are the points $h = z - (z^2 - 1)^{1/2}$ and $h = z + (z^2 - 1)^{1/2}$. These singularities are branch points (see §5.7).

Thus the series (5.2) converges so long as $|h|$ is less than both

$$|z - (z^2 - 1)^{1/2}| \quad \text{and} \quad |z + (z^2 - 1)^{1/2}|.$$

Draw an ellipse in the z-plane passing through the point z and having its foci at ± 1. Let a be its semi-major axis, and θ the eccentric angle of z on it. Then $z = a \cos\theta + i(a^2 - 1)^{1/2} \sin\theta$, which gives $z \pm (z^2 - 1)^{1/2} = \{a \pm (a^2 - 1)^{1/2}\}(\cos\theta \pm i \sin\theta)$, so $|z \pm (z^2 - 1)^{1/2}| = a \pm (a^2 - 1)^{1/2}$. Thus the series (5.2) converges so long as h is less than the smaller of the numbers $a + (a^2 - 1)^{1/2}$ and $a - (a^2 - 1)^{1/2}$, i.e. so long as h is less than $a - (a^2 - 1)^{1/2}$. But $h = a - (a^2 - 1)^{1/2}$ when $a = \frac{1}{2}(h + h^{-1})$. Therefore the series (5.2) converges so long as z is within an ellipse whose foci are 1 and -1, and whose semi-major axis is $\frac{1}{2}(h + h^{-1})$.

5.41 Forms of the remainder in Taylor's series

Let $f(x)$ be a *real* function of a *real* variable; and let it have continuous differential coefficients of the first n orders when $a \leq x \leq a + h$. If $0 \leq t \leq 1$, we have

$$\frac{d}{dt}\left\{\sum_{m=1}^{n-1} \frac{h^m}{m!}(1-t)^m f^{(m)}(a+th)\right\} = \frac{h^n(1-t)^{n-1}}{(n-1)!} f^{(n)}(a+th) - hf'(a+th).$$

Integrating this between the limits 0 and 1, we have

$$f(a+h) = f(a) + \sum_{m=1}^{n-1} \frac{h^m}{m!} f^{(m)}(a) + \int_0^1 \frac{h^n(1-t)^{n-1}}{(n-1)!} f^{(n)}(a+th)\, dt.$$

Let

$$R_n = \frac{h^n}{(n-1)!} \int_0^1 (1-t)^{n-1} f^{(n)}(a+th)\, dt;$$

and let p be a positive integer such that $p \leq n$. Then

$$R_n = \frac{h^n}{(n-1)!} \int_0^1 (1-t)^{p-1} \cdot (1-t)^{n-p} f^{(n)}(a+th)\, dt.$$

Let U, L be the upper and lower bounds of $(1-t)^{n-p} f^{(n)}(a+th)$. Then

$$\int_0^1 L(1-t)^{p-1} dt < \int_0^1 (1-t)^{p-1} \cdot (1-t)^{n-p} f^{(n)}(a+th)\, dt < \int_0^1 U(1-t)^{p-1} dt.$$

Since $(1-t)^{n-p} f^{(n)}(a+th)$ is a continuous function it passes through all values between U and L, and hence we can find θ such that $0 \leq \theta \leq 1$, and

$$\int_0^1 (1-t)^{n-1} f^{(n)}(a+th)\, dt = p^{-1}(1-\theta)^{n-p} f^{(n)}(a+\theta h).$$

Therefore $R_n = \dfrac{h^n}{(n-1)!p}(1-\theta)^{n-p} f^{(n)}(a+\theta h)$. Writing $p = n$, we get $R_n = \dfrac{h^n}{n!} f^{(n)}(a+\theta h)$, which is *Lagrange's form for the remainder*; and writing $p = 1$, we get

$$R_n = \frac{h^n}{(n-1)!}(1-\theta)^{n-1} f^{(n)}(a+\theta h),$$

which is *Cauchy's form for the remainder*.

Taking $n = 1$ in this result, we get

$$f(a+h) - f(a) = hf'(a+\theta h)$$

if $f'(x)$ is continuous when $a \leq x \leq a+h$; this result is usually known as the *First Mean-Value Theorem* (see also §4.14).

Darboux [162, p. 291] gave a form for the remainder in Taylor's series, which is applicable to complex variables and resembles the above form given by Lagrange for the case of real variables.

5.5 The process of continuation

Near every point P, z_0, in the neighbourhood of which a function $f(z)$ is analytic, we have seen that an expansion exists for the function as a series of ascending positive integral powers of $(z - z_0)$, the coefficients in which involve the successive derivatives of the function at z_0.

Now let A be the singularity of $f(z)$ which is nearest to P. Then the circle within which this expansion is valid has P for centre and PA for radius.

Suppose that we are merely given the values of a function at all points of the circumference of a circle slightly smaller than the circle of convergence and concentric with it together with the condition that the function is to be analytic throughout the interior of the larger circle. Then the preceding theorems enable us to find its value at all points *within* the smaller circle and to determine the coefficients in the Taylor series proceeding in powers of $z - z_0$. The question arises, Is it possible to define the function at points *outside* the circle in such a way that the function is analytic throughout a larger domain than the interior of the circle?

In other words, *given a power series which converges and represents a function only at points within a circle, to define by means of it the values of the function at points outside the circle.*

For this purpose choose any point P_1 within the circle, not on the line PA. We know the value of the function and all its derivatives at P_1, from the series, and so we can form the Taylor series (for the same function) with P_1 as origin, which will define a function analytic throughout some circle of centre P_1. Now this circle will extend as far as the singularity (of the function defined by the new series) which is nearest to P_1, which may or may not be A; but in either case, this new circle will usually[5] lie partly outside the old circle of convergence, and *for points in the region which is included in the new circle but not in the old circle, the new series may be used to define the values of the function, although the old series failed to do so.*

Similarly we can take any other point P_2, in the region for which the values of the function are now known, and form the Taylor series with P_2 as origin, which will in general enable us to define the function at other points, at which its values were not previously known; and so on.

This process is called *continuation*. By means of it, starting from a representation of a function by any one power series we can find any number of other power series, which between them define the value of the function at all points of a domain, any point of which can be reached from P without passing through a singularity of the function; and the aggregate[6] of all the power series thus obtained constitutes the analytical expression of the function.

Note It is important to know whether continuation by two different paths PBQ, $PB'Q$ will give the same final power series; it will be seen that this is the case, if the function have no singularity inside the closed curve $PBQB'P$, in the following way: Let P_1 be any point on PBQ, inside the circle C with centre P; obtain the continuation of the function with P_1 as origin, and let it converge inside a circle C_1; let P_1' be any point inside both circles and also inside the curve $PBQB'P$; let S, S_1, S_1' be the power series with P, P_1, P_1' as origins; then

---

[5] The word 'usually' must be taken as referring to the cases which are likely to come under the reader's notice while studying the less advanced parts of the subject.

[6] Such an aggregate of power series has been obtained for various functions by M. J. M. Hill [307], by purely algebraical processes.

(since each is equal to S), $S_1 \equiv S_1'$ over a certain domain which will contain P_1, if P_1' be taken sufficiently near P_1; and hence S_1 will be the continuation of S_1'; for if T_1 were the continuation of S_1', we would have $T_1 \equiv S_1$ over a domain containing P_1, and so (§3.73) corresponding coefficients in S_1 and T_1 are the same. By carrying out such a process a sufficient number of times, we deform the path PBQ into the path $PB'Q$ if no singular point is inside $PBQB'P$. The reader will convince himself by drawing a figure that the process can be carried out in a finite number of steps.

Example 5.5.1   The series

$$\frac{1}{a} + \frac{z}{a^2} + \frac{z^2}{a^3} + \frac{z^3}{a^4} + \cdots$$

represents the function

$$f(z) = \frac{1}{a - z}$$

only for points z within the circle $|z| = |a|$. But any number of other power series exist, of the type

$$\frac{1}{a - b} + \frac{z - b}{(a - b)^2} + \frac{(z - b)^2}{(a - b)^3} + \frac{(z - b)^3}{(a - b)^4} + \cdots ;$$

if b/a is not real and positive these converge at points inside a circle which is partly inside and partly outside $|z| = |a|$; these series represent this same function at points outside this circle.

5.501 On functions to which the continuation-process cannot be applied

It is not always possible to carry out the process of continuation. Take as an example the function $f(z)$ defined by the power series

$$f(z) = 1 + z^2 + z^4 + z^8 + z^{16} + \cdots + z^{2^n} + \cdots ,$$

which clearly converges in the interior of a circle whose radius is unity and whose centre is at the origin.

Now it is obvious that, as $z \to 1^-$, $f(z) \to +\infty$; the point $+1$ is therefore a singularity of $f(z)$. But $f(z) = z^2 + f(z^2)$, and if $z^2 \to 1^-$, $f(z^2) \to \infty$ and so $f(z) \to \infty$, and hence the points for which $z^2 = 1$ are singularities of $f(z)$; the point $z = -1$ is therefore also a singularity of $f(z)$. Similarly since

$$f(z) = z^2 + z^4 + f(z^4),$$

we see that if z is such that $z^4 = 1$, then z is a singularity of $f(z)$; and, in general, any root of any of the equations

$$z^2 = 1, \quad z^4 = 1, \quad z^8 = 1, \quad z^{16} = 1, \ldots,$$

is a singularity of $f(z)$. But these points all lie on the circle $|z| = 1$; and in any arc of this circle, however small, there are an unlimited number of them. The attempt to carry out the process of continuation will therefore be frustrated by the existence of this unbroken front of singularities, beyond which it is impossible to pass.

In such a case the function $f(z)$ *cannot be continued at all* to points z situated outside the circle $|z| = 1$; such a function is called a *lacunary function*, and the circle is said to be a *limiting circle* for the function.

5.51 The identity of two functions

The two series

$$1 + z + z^2 + z^3 + \cdots$$

and $-1 + (z - 2) - (z - 2)^2 + (z - 2)^3 - (z - 2)^4 + \cdots$ do not both converge for any value of z, and are distinct expansions. Nevertheless, we generally say that they represent *the same function*, on the strength of the fact that they can both be represented by the same rational expression $1/(1 - z)$.

This raises the question of the *identity* of two functions. When can two *different* expansions be said to represent the *same* function?

We might define a function (after Weierstrass), by means of the last article, as consisting of one power series together with all the other power series which can be derived from it by the process of continuation. Two different analytical expressions will then define the same function, if they represent power series derivable from each other by continuation.

Since if a function is analytic (in the sense of Cauchy §5.12) at and near a point it can be expanded into a Taylor's series, and since a convergent power series has a unique differential coefficient (§5.3), it follows that the definition of Weierstrass is really equivalent to that of Cauchy.

It is important to observe that *the limit of a combination of analytic functions can represent different analytic functions in different parts of the plane*. This can be seen by considering the series

$$\frac{1}{2}\left(z + \frac{1}{z}\right) + \sum_{n=1}^{\infty} \left(z - \frac{1}{z}\right)\left(\frac{1}{1 + z^n} - \frac{1}{1 + z^{n-1}}\right).$$

The sum of the first $n + 1$ terms of this series is

$$\frac{1}{z} + \left(z - \frac{1}{z}\right) \cdot \frac{1}{1 + z^n}.$$

The series therefore converges for all values of z (zero excepted) not on the circle $|z| = 1$. But, as $n \to \infty$, $|z^n| \to 0$ or $|z^n| \to \infty$ according as $|z|$ is less or greater than unity; hence we see that the sum to infinity of the series is z when $|z| < 1$, and $1/z$ when $|z| > 1$. *This series therefore represents one function at points in the interior of the circle $|z| = 1$, and an entirely different function at points outside the same circle.* The reader will see from §5.3 that this result is connected with the non-uniformity of the convergence of the series near $|z| = 1$.

Note It has been shewn by Borel [86] that if a region C is taken and a set of points S such that points of the set S are arbitrarily near every point of C, it may be possible to define a function which has a unique differential coefficient (i.e. is monogenic) at all points of C which do not belong to S; but the function is not analytic in C in the sense of Weierstrass. The functions are not monogenic strictly in the sense of §5.1 because, in the example quoted, in working out $\{f(z+h) - f(z)\}/h$, it must be supposed that $\mathrm{Re}(z+h)$ and $\mathrm{Im}(z+h)$ are not both rational fractions. Such a function is

$$f(z) = \sum_{n=1}^{\infty} \sum_{p=0}^{n} \sum_{q=0}^{n} \frac{\exp(-\exp n^4)}{z - (p + qi)/n}.$$

5.6 Laurent's theorem

A very important theorem was published in 1843 by Laurent [413]. The theorem is contained in a paper which was written by Weierstrass [662, p. 51–66], but apparently not published before 1894. It relates to expansions of functions to which Taylor's theorem cannot be applied.

Let C and C' be two concentric circles of centre a, of which C' is the inner; and let $f(z)$ be a function which is analytic at all points on C and C' and throughout the annulus between C and C'. Let $a + h$ be any point in this ring-shaped space. Then we have (Corollary 5.2.4)

$$f(a + h) = \frac{1}{2\pi i} \int_C \frac{f(z)}{z - a - h} dz - \frac{1}{2\pi i} \int_{C'} \frac{f(z)}{z - a - h} dz,$$

where the integrals are supposed taken in the positive or counter-clockwise direction round the circles. This can be written as

$$f(a + h) = \frac{1}{2\pi i} \int_C f(z) \left\{ \frac{1}{z - a} + \frac{h}{(z - a)^2} + \cdots + \frac{h^n}{(z - a)^{n+1}} + \right.$$
$$\left. \frac{h^{n+1}}{(z - a)^{n+1}(z - a - h)} \right\} dz +$$
$$\frac{1}{2\pi i} \int_{C'} f(z) \left\{ \frac{1}{h} + \frac{z - a}{h^2} + \cdots + \frac{(z - a)^n}{h^{n+1}} - \frac{(z - a)^{n+1}}{h^{n+1}(z - a - h)} \right\} dz.$$

We find, as in the proof of Taylor's theorem, that

$$\int_C \frac{f(z)\, dz}{(z - a)^{n+1}(z - a - h)} h^{n+1} \quad \text{and} \quad \int_{C'} \frac{f(z)(z - a)^{n+1}}{(z - a - h)h^{n+1}} \, dz$$

tend to zero as $n \to \infty$; and thus we have

$$f(a + h) = a_0 + a_1 h + a_2 h^2 + \cdots + \frac{b_1}{h} + \frac{b_2}{h^2} + \cdots,$$

where $a_n = \dfrac{1}{2\pi i} \displaystyle\int_C \frac{f(z)\, dz}{(z - a)^{n+1}}$ and $b_n = \dfrac{1}{2\pi i} \displaystyle\int_{C'} (z - a)^{n-1} f(z)\, dz$. We cannot write $a_n = f^{(n)}(a)/n!$ as in Taylor's theorem since $f(z)$ is not necessarily analytic inside C'.

This result is *Laurent's theorem*; changing the notation, it can be expressed in the following form: *If $f(z)$ be analytic on the concentric circles C and C' of centre a, and throughout the annulus between them, then at any point z of the annulus $f(z)$ can be expanded in the form*

$$f(z) = a_0 + a_1(z - a) + a_2(z - a)^2 + \cdots + \frac{b_1}{(z - a)} + \frac{b_2}{(z - a)^2} + \cdots,$$

where

$$a_n = \frac{1}{2\pi i} \int_C \frac{f(t)\, dt}{(t - a)^{n+1}} \quad \text{and} \quad b_n = \frac{1}{2\pi i} \int_{C'} (t - a)^{n-1} f(t)\, dt.$$

An important case of Laurent's theorem arises when there is only one singularity within the inner circle C', namely at the centre a. In this case the circle C' can be taken as small as we please, and so Laurent's expansion is valid for all points in the interior of the circle C, except the centre a.

Example 5.6.1 Prove that

$$e^{\frac{x}{2}(z-1/z)} = J_0(x) + zJ_1(x) + z^2 J_2(x) + \cdots + z^n J_n(x) + \cdots$$
$$-\frac{1}{z}J_1(x) + \frac{1}{z^2}J_2(x) - \cdots + \frac{(-)^n}{z^n}J_n(x) + \cdots,$$

where $J_n(x) = \dfrac{1}{2\pi}\displaystyle\int_0^{2\pi} \cos(n\theta - x\sin\theta)\,d\theta$.

For the function of z under consideration is analytic in any domain which does not include the point $z = 0$; and so by Laurent's theorem,

$$e^{\frac{x}{2}(z-1/z)} = a_0 + a_1 z + a_2 z^2 + \cdots + \frac{b_1}{z} + \frac{b_2}{z^2} + \cdots,$$

where

$$a_n = \frac{1}{2\pi i}\int_C e^{\frac{x}{2}\left(z-\frac{1}{z}\right)}\frac{dz}{z^{n+1}}$$

and

$$b_n = \frac{1}{2\pi i}\int_{C'} e^{\frac{x}{2}\left(z-\frac{1}{z}\right)} z^{n-1}\,dz,$$

and where C and C' are any circles with the origin as centre. Taking C to be the circle of radius unity, and writing $z = e^{i\theta}$, we have

$$a_n = \frac{1}{2\pi i}\int_0^{2\pi} e^{ix\sin\theta}\cdot e^{-ni\theta}\,id\theta = \frac{1}{2\pi}\int_0^{2\pi}\cos(n\theta - x\sin\theta)\,d\theta,$$

since $\displaystyle\int_0^{2\pi}\sin(n\theta - x\sin\theta)\,d\theta$ vanishes, as may be seen by writing $2\pi - \phi$ for θ. Thus $a_n = J_n(x)$, and $b_n = (-1)^n a_n$, since the function expanded is unaltered if $-z^{-1}$ be written for z, so that $b_n = (-1)^n J_n(x)$, and the proof is complete.

Example 5.6.2 Shew that, in the annulus defined by $|a| < |z| < |b|$, the function

$$\left\{\frac{bz}{(z-a)(b-z)}\right\}^{1/2}$$

can be expanded in the form

$$S_0 + \sum_{n=1}^{\infty} S_n\left(\frac{a^n}{z^n} + \frac{z^n}{b^n}\right),$$

where

$$S_n = \sum_{l=0}^{\infty} \frac{1\cdot3\cdots(2l-1)\cdot1\cdot3\cdots(2l+2n-1)}{2^{2l+n}\cdot l!\,(l+n)!}\left(\frac{a}{b}\right)^l.$$

The function is one-valued and analytic in the annulus (see §5.7), for the branch-points $0, a$

neutralise each other, and so, by Laurent's theorem, if C denotes the circle $|z| = r$, where $|a| < r < |b|$, the coefficient of z^n in the required expansion is

$$\frac{1}{2\pi i} \int_C \frac{dz}{z^{n+1}} \left\{ \frac{bz}{(z-a)(b-z)} \right\}^{\frac{1}{2}}.$$

Putting $z = re^{i\theta}$, this becomes

$$\frac{1}{2\pi} \int_0^{2\pi} e^{-ni\theta} r^{-n} \, d\theta \left(1 - \frac{r}{b} e^{i\theta} \right)^{-1/2} \left(1 - \frac{a}{r} e^{-i\theta} \right)^{-1/2},$$

or

$$\frac{1}{2\pi} \int_0^{2\pi} e^{-in\theta} r^{-n} \, d\theta \sum_{k=0}^\infty \frac{1 \cdot 3 \cdots (2k-1)}{2^k \cdot k!} \frac{r^k e^{ik\theta}}{b^k} \sum_{l=0}^\infty \frac{1 \cdot 3 \cdots (2l-1)}{2^l \cdot l!} \frac{a^l e^{-il\theta}}{r^l},$$

the series being absolutely convergent and uniformly convergent with regard to θ.

The only terms which give integrals different from zero are those for which $k = l + n$. So the coefficient of z^n is

$$\frac{1}{2\pi} \int_0^{2\pi} d\theta \sum_{l=0}^\infty \frac{1 \cdot 3 \cdots (2l-1)}{2^l \cdot l!} \frac{1 \cdot 3 \cdots (2l+2n-1)}{2^{l+n}(l+n)!} \frac{a^l}{b^{l+n}} = \frac{S_n}{b^n}.$$

Similarly it can be shewn that the coefficient of z^{-n} is $S_n a^n$.

Example 5.6.3   Shew that

$$e^{nz+v/z} = a_0 + a_1 z + a_2 z^2 + \cdots + \frac{b_1}{z} + \frac{b_2}{z^2} + \cdots,$$

where

$$a_n = \frac{1}{2\pi} \int_0^{2\pi} e^{(u+v)\cos\theta} \cos\{(u-v)\sin\theta - n\theta\} \, d\theta,$$

and

$$b_n = \frac{1}{2\pi} \int_0^{2\pi} e^{(u+v)\cos\theta} \cos\{(v-u)\sin\theta - n\theta\} \, d\theta.$$

5.61   The nature of the singularities of one-valued functions

Consider first a function $f(z)$ which is analytic throughout a closed region S, except at a single point a inside the region.

Let it be possible to define a function $\phi(z)$ such that

(i) $\phi(z)$ is analytic throughout S,

(ii) when $z \neq a$, $f(z) = \phi(z) + \dfrac{B_1}{z-a} + \dfrac{B_2}{(z-a)^2} + \cdots + \dfrac{B_n}{(z-a)^n}$.

Then $f(z)$ is said to have a *pole of order n at a*; and the terms $\dfrac{B_1}{z-a} + \dfrac{B_2}{(z-a)^2} + \cdots + \dfrac{B_n}{(z-a)^n}$ are called the *principal part* of $f(z)$ near a. By the definition of a singularity (§5.12) a pole is a singularity. If $n = 1$, the singularity is called a *simple* pole.

Any singularity of a one-valued function other than a pole is called an *essential singularity*.

If the essential singularity, a, is isolated (i.e. if a region, of which a is an interior point, can be found containing no singularities other than a), then a Laurent expansion can be found, in ascending and descending powers of $(z - a)$ valid when $\Delta > |z - a| > \delta$, where Δ depends on the other singularities of the function, and δ is arbitrarily small. Hence the 'principal part' of a function near an isolated essential singularity consists of an infinite series.

It should be noted that a pole is, by definition, an isolated singularity, so that all singularities which are not isolated (e.g. the limiting point of a sequence of poles) are essential singularities.

Note   There does not exist, in general, an expansion of a function valid near a non-isolated singularity in the way that Laurent's expansion is valid near an isolated singularity.

Corollary 5.6.1   *If $f(z)$ has a pole of order n at a, and*

$$\psi(z) = (z - a)^n f(z) \, (z \neq a), \quad \psi(a) = \lim_{z \to a} (z - a)^n f(z),$$

then $\psi(z)$ is analytic at a.

Example 5.6.4   A function is not bounded near an isolated essential singularity. *Hint.* Prove that if the function were bounded near $z = a$, the coefficients of negative powers of $z - a$ would all vanish.

Example 5.6.5   Find the singularities of the function

$$\frac{e^{c/(z-a)}}{e^{z/a} - 1}. \tag{5.4}$$

At $z = 0$, the numerator is analytic, and the denominator has a simple zero. Hence the function has a simple pole at $z = 0$. Similarly there is a simple pole at each of the points $2\pi nia \, (n = \pm 1, \pm 2, \pm 3, \ldots)$; the denominator is analytic and does not vanish for other values of z. At $z = a$, the numerator has an isolated singularity, so Laurent's theorem is applicable, and the coefficients in the Laurent expansion may be obtained from the quotient

$$\frac{1 + \frac{c}{z-a} + \frac{c^2}{2!(z-a)^2} + \cdots}{\exp\left(1 + \frac{z-a}{a} + \cdots\right) - 1}, \tag{5.5}$$

which gives an expansion involving all positive and negative powers of $(z - a)$. So there is an essential singularity at $z = a$.

Example 5.6.6 (Math. Trip. 1899)   Shew that the function defined by the series

$$\sum_{n=1}^{\infty} \frac{nz^{n-1}\{(1 + n^{-1})^n - 1\}}{(z^n - 1)\{z^n - (1 + n^{-1})^n\}}$$

has simple poles at the points $z = (1 + n^{-1})\, e^{2ki\pi/n}$, $(k = 0, 1, 2, \ldots, n - 1; n = 1, 2, 3, \ldots)$.

5.62 The 'point at infinity'

The behaviour of a function $f(z)$ as $|z| \to \infty$ can be treated in a similar way to its behaviour as z tends to a finite limit.

If we write $z = 1/z'$, so that large values of z are represented by small values of z' in the z'-plane, there is a one-one correspondence between z and z', provided that neither is zero;

and to make the correspondence complete it is sometimes convenient to say that when z' is the origin, z is the 'point at infinity'. But the reader must be careful to observe that this is *not* a definite point, and any proposition about it is really a proposition concerning the point $z' = 0$.

Let $f(z) = \phi(z')$. Then $\phi(z')$ is not defined *at* $z' = 0$, but its behaviour near $z' = 0$ is determined by its Taylor (or Laurent) expansion in powers of z'; and we define $\phi(0)$ as $\lim_{z' \to 0} \phi(z')$ if that limit exists. For instance the function $\phi(z')$ may have a zero of order m at the point $z' = 0$; in this case the Taylor expansion of $\phi(z')$ will be of the form

$$A z'^{m} + B z'^{m+1} + C z'^{m+2} + \cdots ,$$

and so the expansion of $f(z)$ valid for sufficiently large values of $|z|$ will be of the form

$$f(z) = \frac{A}{z^m} + \frac{B}{z^{m+1}} + \frac{C}{z^{m+2}} + \cdots .$$

In this case, $f(z)$ is said to have a *zero of order m at 'infinity'*.

Again, the function $\phi(z')$ may have a pole of order m at the point $z' = 0$; in this case

$$\phi(z') = \frac{A}{z'^m} + \frac{B}{z'^{m-1}} + \frac{C}{z'^{m-2}} + \cdots + \frac{L}{z'} + M + Nz' + Pz'^{2} + \cdots ;$$

and so, for sufficiently large values of $|z|$, $f(z)$ can be expanded in the form

$$f(z) = Az^m + Bz^{m-1} + Cz^{m-2} + \cdots + Lz + M + \frac{N}{z} + \frac{P}{z^2} + \cdots .$$

In this case, $f(z)$ is said to have a *pole of order m at 'infinity'*.

Similarly $f(z)$ is said to have an *essential singularity* at infinity, if $\phi(z')$ has an essential singularity at the point $z' = 0$. Thus the function e^z has an essential singularity at infinity, since the function $e^{1/z'}$ or

$$1 + \frac{1}{z'} + \frac{1}{2! \, z'^{2}} + \frac{1}{3! \, z'^{3}} + \cdots$$

has an essential singularity at $z' = 0$.

Example 5.6.7　Discuss the function represented by the series

$$\sum_{n=0}^{\infty} \frac{1}{n!} \frac{1}{1 + a^{2n} z^2}, \qquad (a > 1).$$

Hint. The function represented by this series has singularities at $z = a^{-n}$ and $z = -ia^{-n}$, $(n = 1, 2, 3, \ldots)$, since at each of these points the denominator of one of the terms in the series is zero. These singularities are on the imaginary axis, and have $z = 0$ as a limiting point; so no Taylor or Laurent expansion can be formed for the function valid throughout any region of which the origin is an interior point.

For values of z, other than these singularities, the series converges absolutely, since the limit of the ratio of the $(n + 1)$th term to the nth is $\lim_{n \to \infty} (n + 1)^{-1} a^{-2} = 0$. The function is an even function of z (i.e. is unchanged if the sign of z be changed), tends to zero as $|z| \to \infty$, and is analytic on and outside a circle C of radius greater than unity and centre at the origin. So, for points outside this circle, it can be expanded in the form

$$\frac{b_2}{z^2} + \frac{b_4}{z^4} + \frac{b_6}{z^6} + \cdots,$$

where, by Laurent's theorem,

$$b_{2k} = \frac{1}{2\pi i} \int_C z^{2k-1} \sum_{n=0}^{\infty} \frac{1}{n!} \frac{a^{-2n}}{a^{-2n} + z^2} \, dz.$$

Now

$$\sum_{n=0}^{\infty} \frac{a^{-2n} z^{2k-1}}{n!(a^{-2n} + z^2)} = \sum_{n=0}^{\infty} \sum_{m=0}^{\infty} \frac{z^{2k-3} a^{-2n}}{n!} (-1)^m a^{-2nm} z^{-2m}.$$

This double series converges absolutely when $|z| > 1$, and if it be rearranged in powers of z it converges uniformly.

Since the coefficient of z^{-1} is $\sum_{n=0}^{\infty} \frac{(-1)^{k-1} a^{-2kn}}{n!}$ and the only term which furnishes a non-zero integral is the term in z^{-1}, we have

$$b_{2k} = \frac{1}{2\pi i} \int_C \sum_{n=0}^{\infty} \frac{(-1)^{k-1} a^{-2kn}}{n!} \frac{dz}{z}$$

$$= \sum_{n=0}^{\infty} \frac{(-1)^{k-1}}{n! \, a^{2kn}}$$

$$= (-1)^{k-1} e^{1/a^{2k}}.$$

Therefore, when $|z| > 1$, the function can be expanded in the form

$$\frac{e^{1/a^2}}{z^2} - \frac{e^{1/a^4}}{z^4} + \frac{e^{1/a^6}}{z^6} - \cdots.$$

The function has a zero of the second order at infinity, since the expansion begins with a term in z^{-2}.

5.63 Liouvillle's theorem

This theorem, which is really due to Cauchy [127], was given this name by Borchardt [84], who heard it in Liouville's lectures in 1847.

Let $f(z)$ be analytic for all values of z and let $|f(z)| < K$ for all values of z, where K is a constant (so that $|f(z)|$ is bounded as $|z| \to \infty$). Then $f(z)$ is a constant.

Let z, z' be any two points and let C be a contour such that z, z' are inside it. Then, by §5.21,

$$f(z') - f(z) = \frac{1}{2\pi i} \int_C \left\{ \frac{1}{\zeta - z'} - \frac{1}{\zeta - z} \right\} f(\zeta) \, d\zeta;$$

take C to be a circle whose centre is z and whose radius is $\rho \geq 2|z' - z|$; on C write

$\zeta = z + \rho e^{i\theta}$; since $|\zeta - z'| \geq \frac{1}{2}\rho$ when ζ is on C it follows from §4.62 that

$$|f(z') - f(z)| = \left| \frac{1}{2\pi} \int_C \frac{z' - z}{(\zeta - z')(\zeta - z)} f(\zeta)\, d\zeta \right|$$

$$< \frac{1}{2\pi} \int_0^{2\pi} \frac{|z' - z| \cdot K}{\frac{1}{2}\rho}\, d\theta$$

$$= 2|z' - z|K\rho^{-1}.$$

Make $\rho \to \infty$, keeping z and z' fixed; then it is obvious that $f(z') - f(z) = 0$; that is to say, $f(z)$ is constant.

As will be seen in the next article, and again frequently in the latter half of this volume (Chapters 20, 21 and 22), Liouville's theorem furnishes short and convenient proofs of some of the most important results in Analysis.

5.64 Functions with no essential singularities

We shall now shew that *the only one-valued functions which have no singularities, except poles, at any point* (*including* ∞) *are rational functions*.

For let $f(z)$ be such a function; let its singularities in the finite part of the plane be at the points $c_1, c_2, \ldots, c_k$: and let the principal part (§5.61) of its expansion at the pole c_r be

$$\frac{a_{r,1}}{z - c_r} + \frac{a_{r,2}}{(z - c_r)^2} + \cdots + \frac{a_{r,n_r}}{(z - c_r)^{n_r}}.$$

Let the principal part of its expansion at the pole at infinity be

$$a_1 z + a_2 z^2 + \cdots + a_n z^n ;$$

if there is not a pole at infinity, then all the coefficients in this expansion will be zero.

Now the function

$$f(z) - \sum_{r=1}^{k} \left\{ \frac{a_{r,1}}{z - c_r} + \frac{a_{r,2}}{(z - c_r)^2} + \cdots + \frac{a_{r,n_r}}{(z - c_r)^{n_r}} \right\} - a_1 z - a_2 z^2 - \cdots - a_n z^n$$

has clearly no singularities at the points $c_1, c_2, \ldots, c_k$ or at infinity; it is therefore analytic everywhere and is bounded as $|z| \to \infty$, and so, by Liouville's theorem, is a constant; that is,

$$f(z) = C + a_1 z + a_2 z^2 + \cdots + a_n z^n + \sum_{r=1}^{k} \left\{ \frac{a_{r,1}}{z - c_r} + \frac{a_{r,2}}{(z - c_r)^2} + \cdots + \frac{a_{r,n_r}}{(z - c_r)^{n_r}} \right\},$$

where C is constant; $f(z)$ is therefore a rational function, and the theorem is established.

It is evident from Liouville's theorem (combined with Corollary 3.6.2 that a function which is analytic everywhere (including ∞) is merely a constant. Functions which are analytic everywhere *except* at ∞ are of considerable importance; they are known as *integral functions*. Examples of such functions are e^z, $\sin z$, e^{e^z}. From §5.4 it is apparent that there is no finite radius of convergence of a Taylor's series which represents an integral function; and from the result of this section it is evident that all integral functions (except mere polynomials) have essential singularities at ∞.

5.7 Many-valued functions

In all the previous work, the functions under consideration have had a unique value (or limit) corresponding to each value (other than singularities) of z. But functions may be defined which have more than one value for each value of z; thus if $z = r(\cos\theta + i\sin\theta)$, the function $z^{1/2}$ has the two values

$$r^{1/2}\left(\cos\tfrac{1}{2}\theta + i\sin\tfrac{1}{2}\theta\right), \quad r^{1/2}\left\{\cos\tfrac{1}{2}(\theta + 2\pi) + i\sin\tfrac{1}{2}(\theta + 2\pi)\right\};$$

and the function arctan x (x real) has an unlimited number of values, viz. Arctan $x + n\pi$, where $-\frac{\pi}{2} < $ Arctan $x < \frac{\pi}{2}$ and n is any integer; further examples of many-valued functions are $\log z, z^{-5/3}, \sin(z^{1/2})$.

Either of the two functions which $z^{1/2}$ represents is, however, analytic except at $z = 0$, and we can apply to them the theorems of this chapter; and the two functions are called '*branches* of the many-valued function $z^{1/2}$'. There will be certain points in general at which two or more branches coincide or at which one branch has an infinite limit; these points are called 'branch-points'. Thus $z^{1/2}$ has a branch-point at 0; and, if we consider the change in $z^{1/2}$ as z describes a circle counter-clockwise round 0, we see that θ increases by 2π, r remains unchanged, and *either branch of the function passes over into the other branch*. This will be found to be a general characteristic of branch-points. It is not the purpose of this book to give a full discussion of the properties of many-valued functions, as we shall always have to consider particular branches of functions in regions not containing branch-points, so that there will be comparatively little difficulty in seeing whether or not Cauchy's theorem may be applied.

Note Thus we cannot apply Cauchy's theorem to such a function as $z^{3/2}$ when the path of integration is a circle surrounding the origin; but it is permissible to apply it to one of the branches of $z^{3/2}$ when the path of integration is like that shewn in §6.24, for throughout the contour and its interior the function has a single definite value.

Example 5.7.1 (Math. Trip. 1899)   Prove that if the different values of a^z, corresponding to a given value of z, are represented on an Argand diagram, the representative points will be the vertices of an equiangular polygon inscribed in an equiangular spiral, the angle of the spiral being independent of a.

The idea of the different *branches* of a function helps us to understand such a paradox as the following. Consider the function $y = x^x$, for which

$$\frac{dy}{dx} = x^x(1 + \log x). \tag{5.6}$$

When x is negative and real, $\dfrac{dy}{dx}$ is not real. But if x is negative and of the form $p/(2q+1)$ (where p and q are positive or negative integers), y is real. If therefore we draw the real curve $y = x^x$, we have for negative values of x a set of conjugate points, one point corresponding to each rational value of x with an odd denominator; and then we might think of proceeding to form the tangent as the limit of the chord, just as if the curve were continuous; and thus $\dfrac{dy}{dx}$, when derived from the inclination of the tangent to the axis of x, would appear to be real. The question thus arises, Why does the ordinary process of differentiation give a non-real value for $\frac{dy}{dx}$? The explanation is, that these conjugate points do not all arise from the same *branch* of the function $y = x^x$. We have in fact $y = e^{x \log x + 2k\pi i x}$, where k is any integer. To each value of k corresponds one branch of the function y. Now in order to get a real value of y when x is negative, we have to choose a suitable value for k : and *this value of k varies as we go from one conjugate point to an adjacent one*. So the conjugate points do not represent values of y arising from the same branch of the function $y = x^x$, and consequently we cannot expect the value of $\frac{dy}{dx}$ when evaluated for a definite branch to be given by the tangent of the inclination to the axis of x of the line joining two arbitrarily close members of the series of conjugate points.

5.8 Miscellaneous examples

Example 5.1   Obtain the expansion

$$f(z) =$$

$$f(a) + 2\left\{ \frac{z-a}{2} f'\left(\frac{z+a}{2}\right) + \frac{(z-a)^3}{2^3 \cdot 3} f'''\left(\frac{z+a}{2}\right) + \frac{(z-a)^5}{2^5 \cdot 5!} f^{(5)}\left(\frac{z+a}{2}\right) + \cdots \right\},$$

and determine the circumstances and range of its validity.

Example 5.2 (Corey [156])   Obtain, under suitable circumstances, the expansion

$$f(z) = f(a) + \frac{z-a}{m}\left[f'\left(a + \frac{z-a}{2m}\right) + f'\left\{a + \frac{3(z-a)}{2m}\right\} + \cdots \right.$$

$$\left. + f'\left\{a + \frac{(2m-1)(z-a)}{2m}\right\}\right] + \frac{2}{3!}\left(\frac{z-a}{2m}\right)^3$$

$$\left[f'''\left(a + \frac{z-a}{2m}\right) + f'''\left\{a + \frac{3(z-a)}{2m}\right\} + \cdots \right.$$

$$\left. + f'''\left\{a + \frac{(2m-1)(z-a)}{2m}\right\}\right] + \frac{2}{5!}\left(\frac{z-a}{2m}\right)^5\left[f^{(5)}\left(a + \frac{z-a}{2m}\right)\right.$$

$$\left. + f^{(5)}\left\{a + \frac{3(z-a)}{2m}\right\} + \cdots + f^{(5)}\left\{a + \frac{(2m-1)(z-a)}{2m}\right\}\right] + \cdots .$$

Example 5.3 (Weierstrass [660])   Shew that for the series

$$\sum_{n=0}^{\infty} \frac{1}{z^n + z^{-n}},$$

the region of convergence consists of two distinct areas, namely outside and inside a circle of radius unity, and that in each of these the series represents one function and represents it completely.

Example 5.4 (Lerch [425])   Shew that the function

$$\sum_{n=0}^{\infty} z^{n!}$$

tends to infinity as $z \to \exp(2\pi ip/m!)$ along the radius through the point; where m is any integer and p takes the values $0, 1, 2, \ldots, (m! - 1)$. Deduce that the function cannot be continued beyond the unit circle.

Example 5.5 (Jacobi [348] and Scheibner [574])   Shew that, if $z^2 - 1$ is not a positive real number, then

$$
\left(1 - z^2\right)^{-1/2} = 1 + \frac{1}{2}z^2 + \frac{1 \cdot 3}{2 \cdot 4}z^4 + \cdots + \frac{1 \cdot 3 \cdots (2n - 1)}{2 \cdot 4 \cdots 2n}z^{2n}
$$
$$
+ \frac{3 \cdot 5 \cdots (2n - 1)}{2 \cdot 4 \cdots (2n)}(1 - z^2)^{-1/2} \int_0^z t^{2n+1}(1 - t^2)^{-1/2}\, dt.
$$

Example 5.6 (Jacobi [348] and Scheibner [574])   Shew that, if $z - 1$ is not a positive real number, then

$$
(1 - z)^{-m} = 1 + \frac{m}{1}z + \frac{m(m + 1)}{2!}z^2 + \cdots + \frac{m(m + 1) \cdots (m + n - 1)}{n!}z^n
$$
$$
+ \frac{m(m + 1) \cdots (m + n)}{n!}(1 - z)^{-m} \int_0^z t^n(1 - t)^{m-1}\, dt.
$$

Example 5.7 (Jacobi [348] and Scheibner [574])   Shew that, if z and $1 - z$ are not negative real numbers, then

$$
\left(1 - z^2\right)^{-1/2} \int_0^z t^m(1 - t^2)^{-1/2}\, dt
$$
$$
= \frac{z^{m+1}}{m + 1}\left\{1 + \frac{m + 2}{m + 3}z^2 + \cdots + \frac{(m + 2) \cdots (m + 2n - 2)}{(m + 3) \cdots (m + 2n - 1)}z^{2n-2}\right\}
$$
$$
+ \left(1 - z^2\right)^{-1/2} \frac{(m + 2)(m + 4) \cdots (m + 2n)}{(m + 1)(m + 3) \cdots (m + 2n - 1)} \int_0^z t^{m+2n}(1 - t^2)^{-1/2}\, dt.
$$

Example 5.8 (Scheibner [574])   If, in the expansion of $(a_0 + a_1 z + a_2 z^2)^m$ by the multinomial theorem, the remainder after n terms be denoted by $R_n(z)$, so that

$$
\left(a_0 + a_1 z + a_2 z^2\right)^m = A_0 + A_1 z + A_2 z^2 + \cdots + A_{n-1}z^{n-1} + R_n(z),
$$

shew that

$$
R_n(z) = (a + a_1 z + a_2 z^2)^m \int_0^z \frac{na A_n t^{n-1} + (2m - n + 1)a_2 A_{n-1}t^n}{(a + a_1 t + a_2 t^2)^{m+1}}\, dt.
$$

Example 5.9 (Scheibner [574])   If

$$
(a_0 + a_1 z + a_2 z^2)^{-m-1} \int_0^z (a_0 + a_1 t + a_2 t^2)^m\, dt
$$

be expanded in ascending powers of z in the form $A_1 z + A_2 z^2 + \cdots$, shew that the remainder after $n - 1$ terms is

$$
(a_0 + a_1 z + a_2 z^2)^{-m-1} \int_0^z (a_0 + a_1 t + a_2 t^2)^m \{na_0 A_n - (2m + n + 1)a_2 A_{n-1}t\}t^{n-1}\, dt.
$$

The results of Examples 5.5, 5.6 and 5.7 are special cases of formulae contained in Jacobi's dissertation (Berlin, 1825) published in [354, vol. 3, pp. 1-44]. Jacobi's formulae were generalized by Scheibner [574].

Example 5.10 (Pincherle [523])   Shew that the series

$$\sum_{n=0}^{\infty} \{1 + \lambda_n(z)e^s\} \frac{d^n \phi(z)}{dz^n},$$

where

$$\lambda_n(z) = -1 + z - \frac{z^2}{2!} + \frac{z^3}{3!} - \cdots + (-1)^n \frac{z^n}{n!},$$

and where $\phi(z)$ is analytic near $z = 0$, is convergent near the point $z = 0$; and shew that if the sum of the series be denoted by $f(z)$, then $f(z)$ satisfies the differential equation

$$f'(z) = f(z) - \phi(z).$$

Example 5.11 (Gutzmer [264])   Shew that the arithmetic mean of the squares of the moduli of all the values of the series $\sum_{n-0}^{\infty} a_n z^n$ on a circle $|z| = r$, situated within its circle of convergence, is equal to the sum of the squares of the moduli of the separate terms.

Example 5.12 (Lerch [431])   Shew that the series

$$\sum_{m=1}^{\infty} e^{-2(am)^{1/2}} z^{m-1}$$

converges when $|z| < 1$; and that, when $a > 0$, the function which it represents can also be represented when $|z| < 1$ by the integral

$$\left(\frac{a}{\pi}\right)^{1/2} \int_0^\infty \frac{e^{-a/x}}{e^x - z} \frac{dx}{x^{3/2}},$$

and that it has no singularities except at the point $z = 1$.

Example 5.13 (Weierstrass [660])   Shew that the series

$$\frac{2}{\pi}(z + z^{-1}) + \frac{2}{\pi}\sum\left\{\frac{z}{(1 - 2v - 2v'zi)(2v + 2v'zi)^2} + \frac{z^{-1}}{(1 - 2v - 2v'z^{-1}i)(2v + 2v'z^{-1}i)^2}\right\},$$

in which the summation extends over all integral values of v, v', except the combination $(v = 0, v' = 0)$, converges absolutely for all values of z except purely imaginary values; and that its sum is $+1$ or -1, according as the real part of z is positive or negative.

Example 5.14   Shew that $\sin(u(z + 1/z))$ can be expanded in a series of the type

$$a_0 + a_1 z + a_2 z^2 + \cdots + \frac{b_1}{z} + \frac{b_2}{z^2} + \cdots,$$

in which the coefficients, both of z^n and of z^{-n}, are

$$\frac{1}{2\pi} \int_0^{2\pi} \sin(2u \cos \theta) \cos n\theta \, d\theta.$$

Example 5.15 If $f(z) = \sum_{n=1}^{\infty} \dfrac{z^2}{n^2 z^2 + a^2}$, shew that $f(z)$ is finite and continuous for all real values of z, but cannot be expanded as a Maclaurin's series in ascending powers of z; and explain this apparent anomaly. For other cases of failure of Maclaurin's theorem, see a posthumous memoir by Cellérier [140]; Lerch [427]; Pringsheim [540]; and Du Bois Reymond [189].

Example 5.16 If $f(z)$ be a *continuous* one-valued function of z throughout a two-dimensional region, and if

$$\int_C f(z)\, dz = 0$$

for all closed contours C lying inside the region, then $f(z)$ is an analytic function of z throughout the interior of the region. *Hint.* Let a be any point of the region and let

$$F(z) = \int_a^z f(z)\, dz.$$

It follows from the data that $F(z)$ has the unique derivative $f(z)$. Hence $F(z)$ is analytic (§5.1) and so (§5.22) its derivative $f(z)$ is also analytic. This important converse of Cauchy's theorem is due to Morera [474].

6

The Theory of Residues; Application to the Evaluation of Definite Integrals

6.1 Residues

If the function $f(z)$ has a pole of order m at $z = a$, then, by the definition of a pole, an equation of the form

$$f(z) = \frac{a_{-m}}{(z-a)^m} + \frac{a_{-m+1}}{(z-a)^{m-1}} + \cdots + \frac{a_{-1}}{z-a} + \phi(z),$$

where $\phi(z)$ is analytic near and at a, is true near a.

The coefficient a_{-1} in this expansion is called the *residue* of the function $f(z)$ relative to the pole a.

Consider now the value of the integral $\int_\alpha f(z)dz$, where the path of integration is a circle α, whose centre is the point a and whose radius ρ is so small that $\phi(z)$ is analytic inside and on the circle. The existence of such a circle is implied in the definition of a pole as an isolated singularity.

We have

$$\int_\alpha f(z)\,dz = \sum_{r-1}^m a_{-r} \int_\alpha \frac{dz}{(z-a)^r} + \int_\alpha \phi(z)\,dz.$$

Now $\int_\alpha \phi(z)\,dz = 0$ by §5.2; and (putting $z - a = \rho e^{i\theta}$) we have, if $r \neq 1$,

$$\int_\alpha \frac{dz}{(z-a)^r} = \int_0^{2\pi} \frac{\rho e^{i\theta} i\,d\theta}{\rho^r e^{ri\theta}} = \rho^{-r+1} \int_0^{2\pi} e^{(1-r)i\theta} i\,d\theta = \rho^{-r+1} \left[\frac{e^{(1-r)i\theta}}{1-r} \right]_0^{2\pi} = 0.$$

But when $r = 1$, we have

$$\int_\alpha \frac{dz}{z-a} = \int_0^{2\pi} i\,d\theta = 2\pi i.$$

Hence finally

$$\int_\alpha f(z)\,dz = 2\pi i a_{-1}.$$

Now let C be any contour, containing in the region interior to it a number of poles $a, b, c, \ldots$ of a function $f(z)$, with residues $a_{-1}, b_{-1}, c_{-1}, \ldots$ respectively: and suppose that the function $f(z)$ is analytic throughout C and its interior, except at these poles. Surround the points $a, b, c, \ldots$ by circles $\alpha, \beta, \gamma, \ldots$ so small that their respective centres are the only singularities

110

inside or on each circle; then the function $f(z)$ is analytic in the closed region bounded by $C, \alpha, \beta, \gamma, \ldots$.

Hence, by Corollary 5.2.3

$$\int_C f(z) \, dz = \int_\alpha f(z) \, dz + \int_\beta f(z) \, dz + \cdots$$

$$= 2\pi i a_{-1} + 2\pi i b_{-1} + \cdots .$$

Thus we have the *theorem of residues*, namely that *if $f(z)$ be analytic throughout a contour C and its interior except at a number of poles inside the contour, then*

$$\int_C f(z) \, dz = 2\pi i \sum R,$$

where $\sum R$ denotes the sum of the residues of the function $f(z)$ at those of its poles which are situated within the contour C. This is an extension of the theorem of §5.21 (giving (5.21)).

Note If a is a *simple* pole of $f(z)$ the residue of $f(z)$ at that pole is $\lim\limits_{z \to a} (z - a) f(z)$.

6.2 The evaluation of definite integrals

We shall now apply the result of §6.1 to evaluating various classes of definite integrals; the methods to be employed in any particular case may usually be seen from the following typical examples.

6.21 *The evaluation of the integrals of certain periodic functions taken between the limits 0 and 2π*

An integral of the type

$$\int_0^{2\pi} R(\cos \theta, \sin \theta) \, d\theta$$

where the integrand is a rational function of $\cos \theta$ and $\sin \theta$, finite on the range of integration, can be evaluated by writing $e^{i\theta} = z$; since

$$\cos \theta = \frac{1}{2}(z + z^{-1}), \quad \sin \theta = \frac{1}{2i}(z - z^{-1}),$$

the integral takes the form $\int_C S(z) \, dz$, where $S(z)$ is a rational function of z finite on the path of integration C, the circle of radius unity whose centre is the origin.

Therefore, by §6.1, the integral is equal to $2\pi i$ times the sum of the residues of $S(z)$ at those of its poles which are inside that circle.

Example 6.2.1 If $0 < p < 1$,

$$\int_0^{2\pi} \frac{d\theta}{1 - 2p \cos \theta + p^2} = \int_C \frac{dz}{i(1 - pz)(z - p)}.$$

The only pole of the integrand inside the circle is a simple pole at p; and the residue there is

$$\lim_{z \to p} \frac{z - p}{i(1 - pz)(z - p)} = \frac{1}{i(1 - p^2)}.$$

Hence

$$\int_0^{2\pi} \frac{d\theta}{1 - 2p \cos \theta + p^2} = \frac{2\pi}{1 - p^2}.$$

Example 6.2.2   If $0 < p < 1$,

$$\int_0^{2\pi} \frac{\cos^2 3\theta}{1 - 2p \cos 2\theta + p^2} \, d\theta = \int_c \frac{1}{iz} \left(\frac{1}{2} z^3 + \frac{1}{2} z^{-3} \right)^2 \frac{dz}{(1 - pz^2)(1 - pz^{-2})}$$

$$= 2\pi \sum R,$$

where $\sum R$ denotes the sum of the residues of $\dfrac{(z^6 + 1)^2}{4z^5(1 - pz^2)(z^2 - p)}$ at its poles inside C; these poles are $0, -\sqrt{p}, \sqrt{p}$; and the residues at them are

$$-\frac{1 + p^2 + p^4}{4p^3}, \quad \frac{(p^3 + 1)^2}{8p^3(1 - p^2)}, \quad \frac{(p^3 + 1)^2}{8p^3(1 - p^2)};$$

and hence the integral is equal to

$$\frac{\pi(1 - p + p^2)}{1 - p}.$$

Example 6.2.3   If n be a positive integer,

$$\int_0^{2\pi} e^{\cos \theta} \cos(n\theta - \sin \theta) \, d\theta = \frac{2\pi}{n!}, \qquad \int_0^{2\pi} e^{\cos \theta} \sin(n\theta - \sin \theta) \, d\theta = 0.$$

Example 6.2.4   If $a > b > 0$,

$$\int_0^{2\pi} \frac{d\theta}{(a + b\cos \theta)^2} = \frac{2\pi a}{(a^2 - b^2)^{3/2}}, \qquad \int_0^{2\pi} \frac{d\theta}{(a + b\cos^2 \theta)^2} = \frac{\pi(2a + b)}{a^{3/2}(a + b)^{3/2}}.$$

6.22 The evaluation of certain types of integrals taken between the limits $-\infty$ and $+\infty$

We shall now evaluate $\displaystyle\int_{-\infty}^{\infty} Q(x) \, dx$, where $Q(z)$ is a function such that:

(i) it is analytic when the imaginary part of z is positive or zero (except at a finite number of poles);

(ii) it has no poles on the real axis;

(iii) as $|z| \to \infty$, $z \, Q(z) \to 0$ uniformly for all values of arg z such that $0 \le \arg z \le \pi$;

(iv) provided that when x is real, $xQ(x) \to 0$, as $x \to \pm\infty$, in such a way[1] that $\displaystyle\int_0^{\infty} Q(x) \, dx$

and $\displaystyle\int_{-\infty}^0 Q(x) \, dx$ both converge.

---

[1] The condition $xQ(x) \to 0$ is not in itself sufficient to secure the convergence of $\displaystyle\int_0^{\infty} Q(x) \, dx$; consider $Q(x) = (x \log x)^{-1}$.

Given ε, we can choose ρ_0 (independent of $\arg z$) such that $|z\,Q(z)| < \varepsilon/\pi$ whenever $|z| > \rho_0$ and $0 \le \arg z \le \pi$. Consider $\int_C Q(z)\,dz$ taken round a contour C consisting of the part of the real axis joining the points $\pm\rho$ (where $\rho > \rho_0$) and a semicircle Γ, of radius ρ, having its centre at the origin, above the real axis. Then, by §6.1, $\int_C Q(z)\,dz = 2\pi i \sum R$, where $\sum R$ denotes the sum of the residues of $Q(z)$ at its poles above the real axis ($Q(z)$ has no poles above the real axis outside the contour). Therefore

$$\left| \int_{-\rho}^{\rho} Q(z)\,dz - 2\pi i \sum R \right| = \left| \int_{\Gamma} Q(z)\,dz \right|.$$

In the last integral write $z = \rho e^{i\theta}$, and then

$$\left| \int_{\Gamma} Q(z)\,dz \right| = \left| \int_0^{\pi} Q(\rho e^{i\theta})\rho e^{i\theta} i\,d\theta \right| < \int_0^{\pi} (\varepsilon/\pi)d\theta = \varepsilon,$$

by §4.62. Hence

$$\lim_{\rho \to \infty} \int_{-\rho}^{\rho} Q(z)\,dz = 2\pi i \sum R. \tag{6.1}$$

But the meaning of $\int_{-\infty}^{\infty} Q(x)\,dx$ is $\lim_{\rho,\sigma \to \infty} \int_{-\rho}^{\sigma} Q(x)\,dx$; and since $\lim_{\sigma \to \infty} \int_0^{\sigma} Q(x)\,dx$ and $\lim_{\rho \to \infty} \int_{-\rho}^{0} Q(x)\,dx$ both exist, this double limit is the same as $\lim_{\rho \to \infty} \int_{-\rho}^{\rho} Q(x)\,dx$. Hence we have proved that

$$\int_{-\infty}^{\infty} Q(x)\,dx = 2\pi i \sum R.$$

This theorem is particularly useful in the special case when $Q(x)$ is a rational function.

Note   Even if condition (iv) is not satisfied, we still have

$$\int_0^{\infty} \{Q(x) + Q(-x)\}\,dx = \lim_{\rho \to \infty} \int_{-\rho}^{\rho} Q(x)\,dx = 2\pi i \sum R. \tag{6.2}$$

Example 6.2.5   The only pole of $(z^2 + 1)^{-3}$ in the upper half plane is a pole at $z = i$ with residue there $-\frac{3}{16}i$. Therefore

$$\int_{-\infty}^{\infty} \frac{dx}{(x^2 + 1)^3} = \frac{3\pi}{8}.$$

Example 6.2.6   If $a > 0$, $b > 0$, shew that

$$\int_{-\infty}^{\infty} \frac{x^4 dx}{(a + bx^2)^4} = \frac{\pi}{16a^{3/2}b^{5/2}}.$$

Example 6.2.7   By integrating $\int e^{-\lambda z^2}\,dz$ around a parallelogram whose corners are $-R$, R, $R + ai$, $-R + ai$ and making $R \to \infty$, shew that, if $\lambda > 0$, then

$$\int_{-\infty}^{\infty} c^{-\lambda x^2} \cos(2\lambda ax)\,dx = e^{-\lambda a^2} \int_{-\infty}^{\infty} e^{-\lambda x^2}\,dx = 2\lambda^{-\frac{1}{2}}e^{-\lambda a^2} \int_0^{\infty} e^{-x^2}\,dx.$$

6.221 Certain infinite integrals involving sines and cosines

If $Q(z)$ satisfies the conditions (i), (ii) and (iii) of §6.22, and $m > 0$, then $Q(z)e^{miz}$ also satisfies those conditions. Hence $\int_0^\infty (Q(x)e^{mix} + Q(-x)e^{-mix})\,dx$ is equal to $2\pi i \sum R'$, where $\sum R'$ means the sum of the residues of $Q(z)e^{miz}$ at its poles in the upper half plane; and so

(i) If $Q(x)$ is an even function, i.e. if $Q(-x) = Q(x)$,

$$\int_0^\infty Q(x)\cos(mx)\,dx = \pi i \sum R'.$$

(ii) If $Q(x)$ is an odd function,

$$\int_0^\infty Q(x)\sin(mx)\,dx = \pi \sum R'.$$

6.222 Jordan's lemma

(Jordan [362, p. 285–286]). The results of §6.221 are true if $Q(z)$ be subject to the less stringent condition $Q(z) \to 0$ uniformly when $0 \le \arg z \le \pi$ as $|z| \to \infty$ in place of the condition $z\,Q(z) \to 0$ uniformly. To prove this we require a theorem known as Jordan's lemma, viz.

If $Q(z) \to 0$ uniformly with regard to $\arg z$ as $|z| \to \infty$ when $0 \le \arg z \le \pi$, and if $Q(z)$ is analytic when both $|z| > c$ (a constant) and $0 \le \arg z \le \pi$, then

$$\lim_{\rho \to \infty} \left(\int_\Gamma e^{miz} Q(z)dz \right) = 0,$$

where Γ is a semicircle of radius ρ above the real axis with centre at the origin.

Given ε, choose ρ_0 so that $|Q(z)| < \varepsilon/\pi$ when $|z| > \rho_0$ and $0 \le \arg z \le \pi$; then, if $\rho > \rho_0$,

$$\left| \int_\Gamma e^{miz} Q(z)\,dz \right| = \left| \int_0^\pi e^{mi(\rho\cos\theta + i\rho\sin\theta)} Q(\rho e^{i\theta})\rho e^{i\theta} i\,d\theta \right|.$$

But $\left| e^{mi\rho\cos\theta} \right| = 1$, and so

$$\left| \int_\Gamma e^{miz} Q(z)\,dz \right| < \int_0^\pi \frac{\varepsilon}{\pi} \rho e^{-m\rho\sin\theta}\,d\theta$$

$$= \frac{2\varepsilon}{\pi} \int_0^{\pi/2} \rho e^{-m\rho\sin\theta}\,d\theta.$$

Now $\sin\theta \ge 2\theta/\pi$, when[2] $0 \le \theta \le \pi/2$, and so

$$\left| \int_\Gamma e^{imz} Q(z)\,dz \right| < \frac{2\varepsilon}{\pi} \int_0^{\pi/2} \rho e^{-2m\rho\theta/\pi}\,d\theta$$

$$= \frac{2\varepsilon}{\pi} \cdot \frac{\pi}{2m} \left[-e^{-2m\rho\theta/\pi} \right]_0^{\pi/2}$$

$$< \frac{\varepsilon}{m}.$$

[2] This inequality appears obvious when we draw the graphs $y = \sin x$, $y = 2x/\pi$; it may be proved by shewing that $(\sin\theta)/\theta$ decreases as θ increases from 0 to $\pi/2$.

Hence

$$\lim_{\rho \to \infty} \int_{\Gamma} e^{miz} Q(z)\, dz = 0.$$

This result is Jordan's lemma.

Now

$$\int_0^\rho \{e^{mix} Q(x) + e^{-mix} Q(-x)\}\, dx = 2\pi i \sum R' - \int_{\Gamma} e^{miz} Q(z)\, dz,$$

and, making $\rho \to \infty$, we see at once that

$$\int_0^\infty \{e^{mix} Q(x) + e^{-mix} Q(-x)\}\, dx = 2\pi i \sum R',$$

which is the result corresponding to the result of §6.221.

Example 6.2.8   Shew that, if $a > 0$, then

$$\int_0^\infty \frac{\cos x}{x^2 + a^2}\, dx = \frac{\pi}{2a} e^{-a}.$$

Example 6.2.9   Shew that, if $a \geq 0$, $b \geq 0$, then

$$\int_0^\infty \frac{\cos 2ax - \cos 2bx}{x^2}\, dx = \pi(b - a).$$

Hint. Take a contour consisting of a large semicircle of radius ρ, a small semicircle of radius δ, both having their centres at the origin, and the parts of the real axis joining their ends; then make $\rho \to \infty$, $\delta \to 0$.

Example 6.2.10   Shew that, if $b > 0$, $m \geq 0$, then

$$\int_0^\infty \frac{3x^2 - a^2}{(x^2 + b^2)^2} \cos mx\, dx = \frac{\pi e^{-mb}}{4b^3} \{3b^2 - a^2 - mb(3b^2 + a^2)\}.$$

Example 6.2.11   Shew that, if $k > 0$, $a > 0$, then

$$\int_0^\infty \frac{x \sin ax}{x^2 + k^2}\, dx = \frac{1}{2} \pi e^{-ka}.$$

Example 6.2.12   Shew that, if $m \geq 0$, $a > 0$, then

$$\int_0^\infty \frac{\sin mx}{x(x^2 + a^2)^2}\, dx = \frac{\pi}{2a^4} - \frac{\pi e^{-ma}}{4a^3}\left(m + \frac{2}{a}\right).$$

(Take the contour of Example 6.2.9).

Example 6.2.13   Shew that, if the real part of z be positive, then

$$\int_0^\infty (e^{-t} - e^{-tz}) \frac{dt}{t} = \log z.$$

Solution. We have

$$\int_0^\infty (e^{-t} - e^{-tz}) \frac{dt}{t} = \lim_{\delta \to 0,\, \rho \to \infty} \left\{ \int_\delta^\rho \frac{e^{-t}}{t} dt - \int_\delta^\rho \frac{e^{-tz}}{t} dt \right\}$$

$$= \lim_{\delta \to 0,\, \rho \to \infty} \left\{ \int_\delta^\rho \frac{e^{-t}}{t} dt - \int_{\delta z}^{\rho z} \frac{e^{-u}}{u} du \right\}$$

$$= \lim_{\delta \to 0,\, \rho \to \infty} \left\{ \int_\delta^{\delta z} \frac{e^{-t}}{t} dt - \int_\rho^{\rho z} \frac{e^{-t}}{t} dt \right\},$$

since $t^{-1} e^{-t}$ is analytic inside the quadrilateral whose corners are δ, δz, ρz, ρ. Now $\int_\rho^{\rho z} t^{-1} e^{-t} \, dt \to 0$ as $\rho \to \infty$ when Re $z > 0$; and

$$\int_\delta^{\delta z} t^{-1} e^{-t} \, dt = \log z - \int_\delta^{\delta z} t^{-1} (1 - e^{-t}) \, dt \to \log z,$$

since $t^{-1}(1 - e^{-t}) \to 1$ as $t \to 0$.

6.23 *Principal values of integrals*

It was assumed in §6.22, §6.221, and §6.222 that the function $Q(x)$ had no poles on the real axis; if the function has a finite number of *simple* poles on the real axis, we can obtain theorems corresponding to those already obtained, except that the integrals are all principal values (§4.5) and $\sum R$ has to be replaced by $\sum R + \frac{1}{2} \sum R_0$, where $\sum R_0$ means the sum of the residues at the poles on the real axis. To obtain this result we saw that, instead of the former contour, we had to take as contour a circle of radius ρ and the portions of the real axis joining the points

$$-\rho,\ a - \delta_1;\ a + \delta_1,\ b - \delta_2;\ b + \delta_2,\ c - \delta_3,\ \ldots$$

and small semicircles above the real axis of radii $\delta_1, \delta_2, \ldots$ with centres $a, b, c, \ldots$, where $a, b, c, \ldots$ are the poles of $Q(z)$ on the real axis; and then we have to make $\delta_1, \delta_2, \ldots \to 0$; call these semicircles $\gamma_1, \gamma_2, \ldots$. Then instead of the equation

$$\int_{-\rho}^\rho Q(z) \, dz + \int_\Gamma Q(z) \, dz = 2\pi i \sum R,$$

we get

$$P \int_{-\rho}^\rho Q(z) \, dz + \sum_n \lim_{\delta_n \to 0} \int_{\gamma_n} Q(z) \, dz + \int_\Gamma Q(z) \, dz = 2\pi i \sum R.$$

Let a' be the residue of $Q(z)$ at a; then writing $z = a + \delta_1 e^{i\theta}$ on γ_1, we get

$$\int_{\gamma_1} Q(z) \, dz = \int_\pi^0 Q(a + \delta_1 e^{i\theta}) \delta_1 e^{i\theta} i \, d\theta.$$

But $Q(a + \delta_1 e^{i\theta}) \delta_1 e^{i\theta} \to a'$ uniformly as $\delta_1 \to 0$; and therefore

$$\lim_{\delta_1 \to 0} \int_{\gamma_1} Q(z) \, dz = -\pi i a';$$

we thus obtain

$$P \int_{-\rho}^{\rho} Q(z)dz + \int_{\Gamma} Q(z)dz = 2\pi i \sum R + \pi i \sum R_0,$$

and hence, using the arguments of §6.22, we get

$$P \int_{-\infty}^{\infty} Q(x)\, dx = 2\pi i \left(\sum R + \tfrac{1}{2} \sum R_0 \right).$$

The reader will see at once that the theorems of §6.221 and §6.222 have precisely similar generalisations. The process employed above of inserting arcs of small circles so as to diminish the area of the contour is called *indenting* the contour.

6.24 Evaluation of integrals of the form $\int_0^\infty x^{a-1} Q(x)\, dx$

Let $Q(x)$ be a rational function of x such that it has no poles on the positive part of the real axis and $x^a Q(x) \to 0$ both when $x \to 0$ and when $x \to \infty$.

Consider $\int (-z)^{a-l} Q(z)\, dz$ taken round the contour C shewn in the figure,

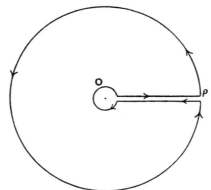

consisting of the arcs of circles of radii ρ, δ and the straight lines joining their end points; $(-z)^{a-1}$ is to be interpreted as

$$\exp\{(a-1)\log(-z)\}$$

and

$$\log(-z) = \log|z| + i\arg(-z),$$

where $-\pi \le \arg(-z) \le \pi$; with these conventions the integrand is one-valued and analytic on and within the contour save at the poles of $Q(z)$.

Hence, if $\sum r$ denote the sum of the residues of $(-z)^{a-1} Q(z)$ at all its poles,

$$\int_C (-z)^{a-1} Q(z)\, dz = 2\pi i \sum r.$$

On the small circle write $-z = \delta e^{i\theta}$, and the integral along it becomes

$$-\int_\pi^{-\pi} (-z)^a\, Q(z) i\, d\theta,$$

which tends to zero as $\delta \to 0$. On the large semicircle write $-z = \rho e^{i\theta}$, and the integral along it becomes

$$-\int_{-\pi}^{\pi} (-z)^a \, Q(z) i \, d\theta,$$

which tends to zero as $\rho \to \infty$. On one of the lines we write $-z = xe^{\pi i}$, on the other $-z = xe^{-\pi i}$ and $(-z)^{a-1}$ becomes $x^{a-1} e^{\pm(a-1)\pi i}$. Hence

$$\lim_{\delta \to 0, \, \rho \to \infty} \int_{\delta}^{\rho} \{ x^{a-1} e^{-(a-1)\pi i} Q(x) - x^{a-1} e^{a-1} xi \, Q(x) \} \, dx = 2\pi i \sum r;$$

and therefore

$$\int_0^{\infty} x^{a-1} Q(x) \, dx = \pi \operatorname{cosec}(a\pi) \sum r.$$

Corollary 6.2.1 *If $Q(x)$ have a number of simple poles on the positive part of the real axis, it may be shewn by indenting the contour that*

$$P \int_0^{\infty} x^{a-1} Q(x) \, dx = \pi \operatorname{cosec}(a\pi) \sum r - \pi \cot(a\pi) \sum r',$$

where $\sum r'$ is the sum of the residues of $z^{a-1} Q(z)$ at these poles.

Example 6.2.14 If $0 < a < 1$,

$$\int_0^{\infty} \frac{x^{a-1}}{1+x} \, dx = \pi \operatorname{cosec}(\pi a), \qquad P \int_0^{\infty} \frac{x^{a-1}}{1-x} \, dx = \pi \cot(\pi a).$$

Example 6.2.15 (Minding)   If $0 < z < 1$ and $-\pi < a < \pi$,

$$\int_0^{\infty} \frac{t^{z-1} \, dt}{t + e^{ia}} = \frac{\pi e^{i(z-1)a}}{\sin \pi z}.$$

Example 6.2.16   Shew that, if $-1 < z < 3$, then

$$\int_0^{\infty} \frac{x^z \, dx}{(1+x^2)^2} = \frac{\pi(1-z)}{4 \cos \frac{1}{2} \pi z}.$$

Example 6.2.17 (Euler)   Shew that, if $-1 < p < 1$ and $-\pi < \lambda < \pi$, then

$$\int_0^{\infty} \frac{x^{-p} \, dx}{1 + 2x \cos \lambda + x^2} = \frac{\pi}{\sin p\pi} \frac{\sin p\lambda}{\sin \lambda}.$$

6.3 Cauchy's integral

We shall next discuss a class of contour-integrals which are sometimes found useful in analytical investigations.

Let C be a contour in the z-plane, and let $f(z)$ be a function analytic inside and on C. Let $\phi(z)$ be another function which is analytic inside and on C except at a finite number of poles; let the zeros of $\phi(z)$ in the interior[3] of C be $a_1, a_2, \ldots$, and let their degrees of multiplicity be $r_1, r_2, \ldots$; and let its poles in the interior of C be $b_1, b_2, \ldots$, and let their degrees of multiplicity be $s_1, s_2, \ldots$.

---

[3] $\phi(z)$ must not have any zeros or poles on C.

Then, by the fundamental theorem of residues, $\dfrac{1}{2\pi i} \displaystyle\int_C f(z)\dfrac{\phi'(z)}{\phi(z)}\, dz$ is equal to the sum

of the residues of $\dfrac{f(z)\phi'(z)}{\phi(z)}$ at its poles inside C. Now $\dfrac{f(z)\phi'(z)}{\phi(z)}$ can have singularities

only at the poles and zeros of $\phi(z)$. Near one of the zeros, say a_1, we have $\phi(z) = A(z -$
$a_1)^{r_1} + B(z - a_1)^{r_1+1} + \cdots$. Therefore $\phi'(z) = Ar_1(z - a_1)^{r_1-1} + B(r_1 + 1)(z - a_1)^{r_1} + \cdots$,

and $f(z) = f(a_1) + (z - a_1)f'(a_1) + \cdots$. Therefore $\dfrac{f(z)\phi'(z)}{\phi(z)} - \dfrac{r_1 f(a_1)}{z - a_1}$ is analytic at a_1.

Thus the residue of $\dfrac{f(z)\phi'(z)}{\phi(z)}$, at the point $z = a_1$, is $r_1 f(a_1)$. Similarly the residue at $z = b_1$

is $-s_1 f(b_1)$; for near $z = b_1$, we have $\phi(z) = C(z - b_1)^{-s_1} + D(z - b_1)^{-s_1+1} + \cdots$, and

$f(z) = f(b_1) + (z - b_1)f'(b_1) + \cdots$, so $\dfrac{f(z)\phi'(z)}{\phi(z)} + \dfrac{s_1 f(b_1)}{z - b_1}$ is analytic at b_1. Hence

$$\frac{1}{2\pi i}\int_C f(z)\frac{\phi'(z)}{\phi(z)}\, dz = \sum r_1 f(a_1) - \sum s_1 f(b_1),$$

the summations being extended over all the zeros and poles of $\phi(z)$.

6.31 The number of roots of an equation contained within a contour

The result of the preceding paragraph can be at once applied to find how many roots of an
equation $\phi(z) = 0$ lie within a contour C. For, on putting $f(z) = 1$ in the preceding result,
we obtain the result that

$$\frac{1}{2\pi i}\int_C \frac{\phi'(z)}{\phi(z)}\, dz$$

is equal to the excess of the number of zeros over the number of poles of $\phi(z)$ contained in
the interior of C, each pole and zero being reckoned according to its degree of multiplicity.

Example 6.3.1 Shew that a polynomial $\phi(z)$ of degree m has m roots. *Hint.* Let $\phi(z) =$
$a_0 z^m + a_1 z^{m-1} + \cdots + a_m, (a_0 \neq 0)$. Then

$$\frac{\phi'(z)}{\phi(z)} = \frac{ma_0 z^{m-1} + \cdots + a_{m-1}}{a_0 z^m + \cdots + a_m}. \tag{6.3}$$

Consequently, for large values of $|z|$,

$$\frac{\phi'(z)}{\phi(z)} = \frac{m}{z} + O\left(\frac{1}{z^2}\right).$$

Thus, if C be a circle of radius ρ whose centre is at the origin, we have

$$\frac{1}{2\pi i}\int_C \frac{\phi'(z)}{\phi(z)}\, dz = \frac{m}{2\pi i}\int_C \frac{dz}{z} + \frac{1}{2\pi i}\int_C O\left(\frac{1}{z^2}\right) dz = m + \frac{1}{2\pi i}\int_C O\left(\frac{1}{z^2}\right) dz.$$

But, as in §6.22,

$$\int_C O\left(\frac{1}{z^2}\right) dz \to 0 \quad \text{as} \quad \rho \to 0;$$

and hence as $\phi(z)$ has no poles in the interior of C, the total number of zeros of $\phi(z)$ is

$$\lim_{\rho \to \infty} \frac{1}{2\pi i} \int_C \frac{\phi'(z)}{\phi(z)} \, dz = m.$$

Example 6.3.2   If at all points of a contour C the inequality

$$|a_k z^k| > |a_0 + a_1 z + \cdots + a_{k-1} z^{k-1} + a_{k+1} z^{k+1} + \cdots + a_m z^m|$$

is satisfied, then the contour contains k roots of the equation

$$a_m z^m + a_{m-1} z^{m-1} + \cdots + a_1 z + a_0 = 0.$$

For write $f(z) = a_m z^m + a_{m-1} z^{m-1} + \cdots + a_1 z + a_0$. Then

$$f(z) = a_k z^k \left(1 + \frac{a_m z^m + \cdots + a_{k+1} z^{k+1} + a_{k-1} z^{k-1} + \cdots + a_0}{a_k z^k} \right)$$

$$= a_k z^k (1 + U),$$

where $|U| \le a < 1$ on the contour, a being independent[4] of z. Therefore the number of roots of $f(z)$ contained in C is

$$\frac{1}{2\pi i} \int_C \frac{f'(z)}{f(z)} \, dz = \frac{1}{2\pi i} \int_C \left(\frac{k}{z} + \frac{1}{1+U} \frac{dU}{dz} \right) dz.$$

But $\int_C \frac{dz}{z} = 2\pi i$; and, since $|U| < 1$, we can expand $(1+U)^{-1}$ in the uniformly convergent series $1 - U + U^2 - U^3 + \cdots$, so

$$\int_C \frac{1}{1+U} \frac{dU}{dz} \, dz = \left[U - \tfrac{1}{2} U^2 + \tfrac{1}{3} U^3 - \cdots \right]_C = 0.$$

Therefore the number of roots contained in C is equal to k.

Example 6.3.3 (Clare, 1900)   Find how many roots of the equation

$$z^6 + 6z + 10 = 0$$

lie in each quadrant of the Argand diagram.

6.4 Connexion between the zeros of a function and the zeros of its derivative

MacDonald [444] has shewn that if $f(z)$ *is a function of z analytic throughout the interior of a single closed contour C, defined by the equation $|f(z)| = M$, where M is a constant, then the number of zeros of $f(z)$ in this region exceeds the number of zeros of the derived function $f'(z)$ in the same region by unity.*

On C let $f(z) = Me^{i\theta}$; then at points on C

$$f'(z) = Me^{i\theta} i \frac{d\theta}{dz}, \qquad f''(z) = Me^{i\theta} \left\{ i \frac{d^2\theta}{dz^2} - \left(\frac{d\theta}{dz} \right)^2 \right\}.$$

---

[4] $|U|$ is a continuous function of z on C, and so attains its upper bound (§3.62). Hence its upper bound a must be less than 1.

Hence, by §6.31, the excess of the number of zeros of $f(z)$ over the number of zeros of $f'(z)$ inside[5] C is

$$\frac{1}{2\pi i}\int_C \frac{f'(z)}{f(z)}dz - \frac{1}{2\pi i}\int_C \frac{f''(z)}{f'(z)}dz = -\frac{1}{2\pi i}\int_C \left(\frac{d\theta}{dz}\right)^{-1}\frac{d^2\theta}{dz^2}dz.$$

Let s be the arc of C measured from a fixed point and let ψ be the angle the tangent to C makes with $0x$; then

$$-\frac{1}{2\pi i}\int_C \left(\frac{d\theta}{dz}\right)^{-1}\frac{d^2\theta}{dz^2}dz = -\frac{1}{2\pi i}\left[\log\frac{d\theta}{dz}\right]_C$$

$$= -\frac{1}{2\pi i}\left[\log\frac{d\theta}{ds} - \log\frac{dz}{ds}\right]_C.$$

Now $\log\dfrac{d\theta}{ds}$ is purely real and its initial value is the same as its final value; and $\log\dfrac{dz}{ds} = i\psi$; hence the excess of the number of zeros of $f(z)$ over the number of zeros of $f'(z)$ is the change in $\psi/2\pi$ in describing the curve C; and it is obvious (for a formal proof, see [651]) that if C is any ordinary curve, ψ increases by 2π as the point of contact of the tangent describes the curve C; this gives the required result.

Example 6.4.1   Deduce from Macdonald's result the theorem that a polynomial of degree n has n zeros.

Example 6.4.2   Prove that, if a polynomial $f(z)$ has real coefficients and if its zeros are all real and different, then between two consecutive zeros of $f(z)$ there is one zero and one only of $f'(z)$. Pólya has pointed out that this result is not necessarily true for functions other than polynomials, as may be seen by considering the function $(z^2 - 4)\exp(z^2/3)$.

6.5 Miscellaneous examples

Example 6.1 (Trinity, 1898)   A function $\phi(z)$ is zero when $z = 0$, and is real when z is real, and is analytic when $|z| \le 1$; if $f(x, y)$ is the coefficient of i in $\phi(x + iy)$, prove that if $-1 < x < 1$, then

$$\int_0^{2\pi} \frac{x\sin\theta}{1 - 2x\cos\theta + x^2}f(\cos\theta, \sin\theta)\,d\theta = \pi\phi(x).$$

Example 6.2 (Legendre)   By integrating $\dfrac{e^{\pm aiz}}{e^{2\pi z} - 1}$ round a contour formed by the rectangle whose corners are $0, R, R+i, i$ (the rectangle being indented at 0 and i) and making $R \to \infty$, shew that

$$\int_0^\infty \frac{\sin ax}{e^{2\pi x} - 1}dx = \frac{1}{4}\frac{e^a + 1}{e^a - 1} - \frac{1}{2a}.$$

---

[5] $f'(z)$ does not vanish on C unless C has a node or other singular point; for, if $f = \phi + i\psi$, where ϕ and ψ are real, since $i\dfrac{\partial f}{\partial x} = \dfrac{\partial f}{\partial y}$, it follows that if $f'(z) = 0$ at any point, then $\dfrac{\partial\phi}{\partial x}, \dfrac{\partial\phi}{\partial y}, \dfrac{\partial\psi}{\partial x}, \dfrac{\partial\psi}{\partial y}$ all vanish; and these are sufficient conditions for a singular point on $\phi^2 + \psi^2 = M^2$.

Example 6.3   By integrating $\log(-z)\,Q(z)$ round the contour of §6.24, where $Q(z)$ is a rational function such that $zQ(z) \to 0$ as $|z| \to 0$ and as $|z| \to \infty$, shew that if $Q(z)$ has no poles on the positive part of the real axis, $\displaystyle\int_0^\infty Q(x)\,dx$ is equal to minus the sum of the residues of $\log(-z)\,Q(z)$ at the poles of $Q(z)$; where the imaginary part of $\log(-z)$ lies between $\pm\pi$.

Example 6.4   Shew that, if $a > 0$, $b > 0$,

$$\int_0^\infty e^{a\cos bx}\sin(a\sin bx)\frac{dx}{x} = \frac{1}{2}\pi(e^a - 1).$$

Example 6.5 (Cauchy)   Shew that

$$\int_0^{\pi/2} \frac{a\sin 2x}{1 - 2a\cos 2x + a^2}x\,dx = \begin{cases} \frac{\pi}{4}\log(1 + a), & (-1 < a < 1) \\ \frac{\pi}{4}\log(1 + a^{-1}), & (a^2 > 1) \end{cases}$$

Example 6.6 (Störmer [614])   Shew that

$$\int_0^\infty \frac{\sin\phi_1 x}{x}\frac{\sin\phi_2 x}{x}\cdots\frac{\sin\phi_n x}{x}\cos a_1 x\cdots\cos a_m x\frac{\sin ax}{x}\,dx = \frac{\pi}{2}\phi_1\phi_2\cdots\phi_n,$$

if $\phi_1, \phi_2, \ldots, \phi_n, a_1, a_2, \ldots, a_m$ are real and a be positive and

$$a > |\phi_1| + |\phi_2| + \cdots + |\phi_n| + |a_1| + \cdots + |a_m|.$$

Example 6.7 (Amigues [17])   If a point z describes a circle C of centre a, and if $f(z)$ be analytic throughout C and its interior except at a number of poles inside C, then the point $u = f(z)$ will describe a closed curve γ in the u-plane. Shew that if to each element of γ be attributed a mass proportional to the corresponding element of C, the centre of gravity of γ is the point r, where r is the sum of the residues of $\dfrac{f(z)}{z - a}$ at its poles in the interior of C.

Example 6.8   Shew that

$$\int_{-\infty}^\infty \frac{dx}{(x^2 + b^2)(x^2 + a^2)^2} = \frac{\pi(2a + b)}{2a^3 b(a + b)^2}.$$

Example 6.9   Shew that

$$\int_0^\infty \frac{dx}{(a + bx^2)^n} = \frac{\pi}{2^n b^{1/2}}\frac{1\cdot 3\cdots(2n - 3)}{1\cdot 2\cdots(n - 1)}\frac{1}{a^{n-1/2}}.$$

Example 6.10 (Laurent [412])   If $F_n(z) = \displaystyle\prod_{m=1}^{n-1}\prod_{p=1}^{n-1}(1 - z^{mp})$, shew that the series

$$f(z) = -\sum_{n=2}^\infty \frac{F_n(zn^{-1})}{(z^n n^{-n} - 1)n^{n-1}}$$

is an analytic function when z is not a root of any of the equations $z^n = n^n$; and that the sum of the residues of $f(z)$ contained in the ring-shaped space included between two circles whose centres are at the origin, one having a small radius and the other having a radius between n and $n + 1$, is equal to the number of prime numbers less than $n + 1$.

Example 6.11 (Grace [257]) If A and B represent on the Argand diagram two given roots (real or imaginary) of the equation $f(z) = 0$ of degree n, with real or imaginary coefficients, shew that there is at least one root of the equation $f'(z) = 0$ within a circle whose centre is the middle point of AB and whose radius is $\frac{1}{2}AB \cot(\pi/n)$.

Example 6.12 (Kronecker [386]) Shew that, if $0 < v < 1$,

$$\frac{e^{2\pi ivx}}{1 - e^{2\pi ix}} = \frac{1}{2\pi i} \lim_{n\to\infty} \sum_{k=-n}^{n} \frac{e^{2\pi ikv}}{k - x}.$$

Hint. Consider $\displaystyle\int \frac{e^{(2v-1)\pi iz}}{\sin \pi z} \frac{dz}{z - x}$ round a circle of radius $n + \frac{1}{2}$; and make $n \to \infty$.

Example 6.13 Shew that, if $m > 0$, then

$$\int_0^\infty \frac{\sin^n mt}{t^n} dt = \frac{\pi m^{n-1}}{2^n (n-1)!} \left\{ n^{n-1} - \frac{n}{1}(n-2)^{n-1} + \frac{n(n-1)}{2!}(n-4)^{n-1} \right.$$
$$\left. - \frac{n(n-1)(n-2)}{3!}(n-6)^{n-1} + \cdots \right\}.$$

Discuss the discontinuity of the integral at $m = 0$.

Example 6.14 (Wolstenholme) If $A + B + C + \cdots = 0$ and $a, b, c, \ldots$ are positive, shew that

$$\int_0^\infty \frac{A\cos ax + B\cos bx + \cdots + K\cos kx}{x} dx$$
$$= -A\log a - B\log b - \cdots - K\log k.$$

Example 6.15 By considering $\displaystyle\int \frac{e^{x(k+ti)}}{k + ti} dt$ taken around a rectangle indented at the origin, shew that, if $k > 0$,

$$i \lim_{\rho\to\infty} \int_{-\rho}^{\rho} \frac{e^{x(k+ti)}}{k + ti} dt = \pi i + \lim_{\rho\to\infty} P \int_{-\rho}^{\rho} \frac{e^{xti}}{t} dt,$$

and thence deduce, by using the contour of Example 6.2.9, or its reflexion in the real axis (according as $x \geq 0$ and $x < 0$), that

$$\lim_{\rho\to\infty} \frac{1}{\pi} \int_{-\rho}^{\rho} \frac{e^{x(k+ti)}}{k + ti} dt = 2, 1 \text{ or } 0,$$

according as $x > 0$, $x = 0$ or $x < 0$. This integral is known as *Cauchy's discontinuous factor*.

Example 6.16 Shew that, if $0 < a < 2, b > 0, r > 0$, then

$$\int_0^\infty x^{a-1} \sin\left(\frac{\pi a}{2} - bx\right) \frac{r dx}{x^2 + r^2} = \frac{\pi}{2} r^{a-1} e^{-br}.$$

Example 6.17 (Poisson [531]; Jacobi [352]) Let $t > 0$ and let $\displaystyle\sum_{n=-\infty}^{\infty} e^{-n^2\pi t} = \psi(t)$. By considering $\displaystyle\int \frac{e^{-s^2\pi t}}{e^{2\pi i s} - 1}\, dz$ around a rectangle whose corners are $\pm(N + \frac{1}{2}) \pm i$, where N is an integer, and making $N \to \infty$, shew that

$$\psi(t) = \int_{-\infty-i}^{\infty-i} \frac{e^{-s^2\pi t}}{e^{2\pi i z} - 1}\, dz - \int_{-\infty+i}^{\infty+i} \frac{e^{-s^2\pi t}}{e^{2\pi i z} - 1}\, dz.$$

By expanding these integrands in powers of $e^{-2\pi i s}, e^{2\pi i s}$ respectively and integrating term-by-term, deduce from Example 6.2.7 that

$$\psi(t) = \frac{1}{(\pi t)^{\frac{1}{2}}} \psi(1/t) \int_{-\infty}^{\infty} e^{-x^2}\, dx.$$

Hence, by putting $t = 1$ shew that

$$\psi(t) = t^{-\frac{1}{2}} \psi(1/t).$$

Example 6.18 (Poisson [532], Jacobi [346] and Landsberg [407])   Shew that, if $t > 0$,

$$\sum_{n=-\infty}^{\infty} e^{-n^2 \pi t - 2n\pi a t} = t^{-1/2} e^{\pi a^2 t} \left(1 + 2 \sum_{n=1}^{\infty} e^{-n^2 \pi / t} \cos 2n\pi a\right).$$

See also §21.51.

7

The Expansion of Functions in Infinite Series

7.1 A formula due to Darboux

Let $f(z)$ be analytic at all points of the straight line joining a to z, and let $\phi(t)$ be any polynomial of degree n in t. Then if $0 \le t \le 1$, we have by differentiation

$$\frac{d}{dt} \sum_{m=1}^{n} (-1)^m (z-a)^m \, \phi^{(n-m)}(t) f^{(m)} (a + t (z-a))$$

$$= -(z-a)\, \phi^{(n)}(t) f'(a + t(z-a)) + (-1)^n (z-a)^{n+1} \, \phi(t) f^{(n+1)}(a + t(z-a)).$$

Noting that $\phi^{(n)}(t)$ is constant $= \phi^{(n)}(0)$, and integrating between the limits 0 and 1 of t, we get

$$\phi^{(n)}(0)\{f(z) - f(a)\}$$

$$= \sum_{m=1}^{n} (-1)^{m-1}(z-a)^m \{\phi^{(n-m)}(1) f^{(m)}(z) - \phi^{(n-m)}(0) f^m(a)\}$$

$$+ (-1)^n (z-a)^{n+1} \int_0^1 \phi(t) f^{(n+1)}(a + t(z-a))\, dt,$$

which is the formula in question. It appears in Darboux [162].

Taylor's series may be obtained as a special case of this by writing $\phi(t) = (t-1)^n$ and making $n \to \infty$.

Example 7.1.1 By substituting $2n$ for n in the formula of Darboux, and taking $\phi(t) = t^n (t-1)^n$, obtain the expansion (supposed convergent)

$$f(z) - f(a) = \sum_{n=1}^{\infty} \frac{(-1)^{n-1}(z-a)^n}{2^n n!} \{f^{(n)}(z) + (-1)^{n-1} f^{(n)}(a)\},$$

and find the expression for the remainder after n terms in this series.

7.2 The Bernoullian numbers and the Bernoullian polynomials

The function $\frac{1}{2}z \cot \frac{1}{2}z$ is analytic when $|z| < 2\pi$ and, since it is an even function of z, it can be expanded into a Maclaurin series, thus

$$\frac{z}{2} \cot \frac{z}{2} = 1 - B_1 \frac{z^2}{2!} - B_2 \frac{z^4}{4!} - B_3 \frac{z^6}{6!} - \cdots ;$$

then B_n is called the nth *Bernoullian number*. These numbers were introduced by Jakob Bernoulli [66]. It is found that

$$B_1 = \frac{1}{6}, B_2 = \frac{1}{30}, B_3 = \frac{1}{42}, B_4 = \frac{1}{30}, B_5 = \frac{5}{66}, \ldots$$

The first sixty-two Bernoullian numbers were computed by Adams [11]; the first nine significant figures of the first 250 Bernoullian numbers were subsequently published by Glaisher [246].

These numbers can be expressed as definite integrals as follows:

We have, by Chapter 6, Example 6.2,

$$\int_0^\infty \frac{\sin px\, dx}{e^{\pi x} - 1} = -\frac{1}{2p} + \frac{i}{2} \cot ip$$

$$= -\frac{1}{2p} + \frac{1}{2p}\left[1 + B_1\frac{(2p)^2}{2!} - B_2\frac{(2p)^4}{4!} + \cdots\right].$$

Since $\displaystyle\int_0^\infty \frac{x^n \sin\left(px + \frac{1}{2}n\pi\right)}{e^{\pi x} - 1}\, dx$ converges uniformly (by de la Vallée Poussin's test) near $p = 0$ we may, by Corollary 4.4.1, differentiate both sides of this equation any number of times and then put $p = 0$; doing so and writing $2t$ for x, we obtain

$$B_n = 4^n \int_0^\infty \frac{t^{2n-1}\,dt}{e^{2\pi t} - 1}.$$

A proof of this result, depending on contour integration, is given by Carda [117].

Example 7.2.1   Shew that

$$B_n = \frac{2n}{\pi^{2n}(2^{2n} - 1)} \int_0^\infty \frac{x^{2n-1}\,dx}{\sinh x} > 0.$$

Now consider the function $t\dfrac{e^{zt} - 1}{e^t - 1}$, which may be expanded into a Maclaurin series in powers of t valid when $|t| < 2\pi$.

The *Bernoullian polynomial of order* n is defined to be the coefficient of $\frac{t^n}{n!}$ in this expansion. It is denoted by $\phi_n(z)$, so that

$$t\frac{e^{zt} - 1}{e^t - 1} = \sum_{n=1}^\infty \frac{\phi_n(z)t^n}{n!}.$$

The name was given by Raabe [547]. For a full discussion of their properties, see Nörlund [507].

This polynomial possesses several important properties. Writing $z + 1$ for z in the preceding equation and subtracting, we find that

$$te^{zt} = \sum_{n=1}^\infty \{\phi_n(z+1) - \phi_n(z)\}\frac{t^n}{n!}.$$

On equating coefficients of t^n on both sides of this equation we obtain

$$nz^{n-1} = \phi_n(z+1) - \phi_n(z),$$

which is a difference-equation satisfied by the function $\phi_n(z)$.

An explicit expression for the Bernoullian polynomials can be obtained as follows. We have

$$e^{zt} - 1 = zt + \frac{z^2 t^2}{2!} + \frac{z^3 t^3}{2!} + \cdots,$$

and

$$\frac{t}{e^t - 1} = \frac{t}{2i} \cot \frac{t}{2i} - \frac{t}{2} = 1 - \frac{t}{2} + \frac{B_1 t^2}{2!} - \frac{B_2 t^4}{4!} + \cdots. \tag{7.1}$$

Hence

$$\sum_{n=1}^{\infty} \frac{\phi_n(z) t^n}{n!} = \left\{ zt + \frac{z^2 t^2}{2!} + \frac{z^3 t^3}{3!} + \cdots \right\} \left\{ 1 - \frac{t}{2} + \frac{B_1 t^2}{2!} - \frac{B_2 t^4}{4!} + \cdots \right\}.$$

From this, by equating coefficients of t^n (§3.73), we have

$$\phi_n(z) = z^n - \frac{1}{2} n z^{n-1} + \binom{n}{2} B_1 z^{n-2} - \binom{n}{4} B_2 z^{n-4} + \binom{n}{6} B_3 z^{n-6} - \cdots,$$

the last term being that in z or z^2 and $\binom{n}{2}, \binom{n}{4}, \ldots$ being the binomial coefficients; this is the Maclaurin series for the nth Bernoullian polynomial.

When z is an integer, it may be seen from the difference-equation that

$$\frac{\phi_n(z)}{n} = 1^{n-1} + 2^{n-1} + \cdots + (z-1)^{n-1}.$$

The Maclaurin series for the expression on the right was given by Bernoulli.

Example 7.2.2 Shew that, when $n > 1$, $\phi_n(z) = (-1)^n \phi_n(1 - z)$.

7.21 The Euler–Maclaurin expansion

In the formula of Darboux (§7.1) write $\phi_n(t)$ for $\phi(t)$, where $\phi_n(t)$ is the nth Bernoullian polynomial. Differentiating the equation

$$\phi_n(t+1) - \phi_n(t) = nt^{n-1}$$

$n - k$ times, we have

$$\phi_n^{(n-k)}(t+1) - \phi_n^{(n-k)}(t) = n(n-1) \cdots k t^{k-1}.$$

Putting $t = 0$ in this, we have $\phi_n^{(n-k)}(1) = \phi_n^{(n-k)}(0)$. Now, from the Maclaurin series for $\phi_n(z)$, we have if $k > 0$

$$\phi_n^{(n-2k-1)}(0) = 0, \quad \phi_n^{(n-2k)}(0) = \frac{n!}{(2k)!} (-1)^{k-1} B_k,$$

$$\phi_n^{(n-1)}(0) = -\frac{1}{2} n!, \quad \phi_n^{(n)}(0) = n!.$$

Substituting these values of $\phi_n^{(n-k)}(1)$ and $\phi_n^{(n-k)}(0)$ in Darboux's result, we obtain the

Euler–Maclaurin sum formula[1],

$$(z-a)f'(a) = f(z) - f(a) - \frac{z-a}{2}\{f'(z) - f'(a)\}$$

$$+ \sum_{m=1}^{n-1} \frac{(-1)^{m-1} B_m (z-a)^{2m}}{(2m)!} \{f^{(2m)}(z) - f^{(2m)}(a)\}$$

$$- \frac{(z-a)^{2n+1}}{(2n)!} \int_0^1 \phi_{2n}(t) f^{(2n+1)}\{a + (z-a)t\}\, dt.$$

In certain cases the last term tends to zero as $n \to \infty$, and we can thus obtain an infinite series for $f(z) - f(a)$. If we write ω for $z - a$ and $F(x)$ for $f'(x)$, the last formula becomes

$$\int_a^{a+\omega} F(x)\, dx = \frac{1}{2}\omega\{F(a) + F(a + \omega)\}$$

$$+ \sum_{m=1}^{n-1} \frac{(-1)^m B_m \omega^{2m}}{(2m)!} \{F^{(2m-1)}(a + \omega) - F^{(2m-1)}(a)\}$$

$$+ \frac{\omega^{2n+1}}{(2n)!} \int_0^1 \phi_{2n}(t) F^{(2n)}(a + \omega t)\, dt.$$

Writing $a + \omega, a + 2\omega, \ldots, a + (r-1)\omega$ for a in this result and adding up, we get

$$\int_a^{a+r\omega} F(x)\, dx = \omega \left\{ \frac{1}{2}F(a) + F(a + \omega) + F(a + 2\omega) + \cdots + \frac{1}{2}F(a + r\omega) \right\}$$

$$+ \sum_{m=1}^{n-1} \frac{(-1)^m B_m \omega^{2m}}{(2m)!} \{F^{(2m-1)}(a + r\omega) - F^{(2m-1)}(a)\} + R_n,$$

where

$$R_n = \frac{\omega^{2n+1}}{(2n)!} \int_0^1 \phi_{2n}(t) \left\{ \sum_{m=0}^{r-1} F^{(2n)}(a + m\omega + \omega t) \right\} dt.$$

This last formula is of the utmost importance in connexion with the numerical evaluation of definite integrals. It is valid if $F(x)$ is analytic at all points of the straight line joining a to $a + r\omega$.

Example 7.2.3 If $f(z)$ be an odd function of z, shew that

$$zf'(z) = f(z) + \sum_{m=2}^{n} (-1)^m \frac{B_{m-1}(2z)^{2m-2}}{(2m-2)!} f^{(2m-2)}(z)$$

$$- \frac{2^{2n} 2^{2n+1}}{(2n)!} \int_0^1 \phi_{2n}(t) f^{(2n+1)}(-z + 2zt)\, dt.$$

---

[1] A history of the formula is given by Barnes [47]. It was discovered by Euler (1732), but was not published at the time. Euler communicated it (June 9, 1736) to Stirling who replied (April 16, 1738) that it included his own theorem (see §12.33) as a particular case, and also that the more general theorem had been discovered by Maclaurin; and Euler, in a lengthy reply, waived his claims to priority. The theorem was published by Euler [204] and by Maclaurin [449]. For information concerning the correspondence between Euler and Stirling, we are indebted to Mr. C. Tweedie.

Example 7.2.4 (Math. Trip. 1904)   Shew, by integrating by parts, that the remainder after n terms of the expansion of $\frac{z}{2} \cot \frac{z}{2}$ may be written in the form

$$\frac{(-1)^{n+1} z^{2n+1}}{(2n)! \sin z} \int_0^1 \phi_{2n}(t) \cos(zt)\, dt.$$

7.3 Bürmann's theorem

We shall next consider several theorems which have for their object *the expansion of one function in powers of another function*. This appears in [109]; see also Dixon [182].

Let $\phi(z)$ be a function of z which is analytic in a closed region S of which a is an interior point; and let

$$\phi(a) = b.$$

Suppose also that $\phi'(a) \neq 0$. Then Taylor's theorem furnishes the expansion

$$\phi(z) - b = \phi'(a)(z-a) + \frac{\phi''(a)}{2!}(z-a)^2 + \cdots,$$

and if it is legitimate to revert this series we obtain

$$z - a = \frac{1}{\phi'(a)}\{\phi(z) - b\} - \frac{1}{2}\frac{\phi''(a)}{\{\phi'(a)\}^3}\{\phi(z) - b\}^2 + \cdots,$$

which expresses z as an analytic function of the variable $\{\phi(z) - b\}$, for sufficiently small values of $|z - a|$. If then $f(z)$ be analytic near $z = a$, it follows that $f(z)$ is an analytic function of $\{\phi(z) - b\}$ when $|z - a|$ is sufficiently small, and so there will be an expansion of the form

$$f(z) = f(a) + a_1\{\phi(z) - b\} + \frac{a_2}{2!}\{\phi(z) - b\}^2 + \frac{a_3}{3!}\{\phi(z) - b\}^3 + \cdots.$$

The actual coefficients in the expansion are given by the following theorem, which is generally known as *Bürmann's theorem*.

Let $\psi(z)$ be a function of z defined by the equation

$$\psi(z) = \frac{z-a}{\phi(z) - b};$$

then an analytic function $f(z)$ can, in a certain domain of values of z, be expanded in the form

$$f(z) = f(a) + \sum_{m=1}^{n-1} \frac{\{\phi(z) - b\}^m}{m!} \frac{d^{m-1}}{da^{m-1}}[f'(a)\{\psi(a)\}^m] + R_n,$$

where

$$R_n = \frac{1}{2\pi i} \int_a^z \int_\gamma \left[\frac{\phi(z) - b}{\phi(t) - b}\right]^{n-1} \frac{f'(t)\phi'(z)dt\, dz}{\phi(t) - \phi(z)},$$

and γ is a contour in the t-plane, enclosing the points a and z and such that, if ζ be any point inside it, the equation $\phi(t) = \phi(\zeta)$ has no roots on or inside the contour except[2] a simple root $t = \zeta$.

---

[2] It is assumed that such a contour can be chosen if $|z - a|$ be sufficiently small; see §7.31.

To prove this, we have

$$f(z) - f(a) = \int_a^z f'(\zeta)\, d\zeta = \frac{1}{2\pi i} \int_a^z \int_\gamma \frac{f'(t)\, \phi'(\zeta)\, dt\, d\zeta}{\phi(t) - \phi(\zeta)}$$

$$= \frac{1}{2\pi i} \int_a^z \int_\gamma \frac{f'(t)\, \phi'(\zeta)\, dt\, d\zeta}{\phi(t) - b}$$

$$\times \left[\sum_{m=0}^{n-2} \left\{ \frac{\phi(\zeta) - b}{\phi(t) - b} \right\}^m + \frac{\{\phi(\zeta) - b\}^{n-1}}{\{\phi(t) - b\}^{n-2}\{\phi(t) - \phi(\zeta)\}} \right].$$

But, by §4.3,

$$\frac{1}{2\pi i} \int_a^z \int_\gamma \left[\frac{\phi(\zeta) - b}{\phi(t) - b} \right]^m \frac{f'(t)\, \phi'(\zeta)\, dt\, d\zeta}{\phi(t) - b} = \frac{\{\phi(z) - b\}^{m+1}}{2\pi i (m+1)} \int_\gamma \frac{f'(t)\, dt}{\{\phi(t) - b\}^{m+1}}$$

$$= \frac{\{\phi(z) - b\}^{m+1}}{2\pi i (m+1)} \int_\gamma \frac{f'(t)\{\psi(t)\}^{m+1}\, dt}{(t-a)^{m+1}} = \frac{\{\phi(z) - b\}^{m+1}}{(m+1)!} \frac{d^m}{da^m}[f'(a)\{\psi(a)\}^{m+1}].$$

Therefore, writing $m - 1$ for m,

$$f(z) = f(a) + \sum_{m=1}^{n-1} \frac{\{\phi(z) - b\}^m}{m!} \frac{d^{m-1}}{da^{m-1}}[f'(a)\{\psi(a)\}^m]$$

$$+ \frac{1}{2\pi i} \int_a^z \int_\gamma \left[\frac{\phi(\zeta) - b}{\phi(t) - b} \right]^{n-1} \frac{f'(t)\, \phi'(\zeta)\, dt\, d\zeta}{\phi(t) - \phi(\zeta)}.$$

If the last integral tends to zero as $n \to \infty$, we may write the right-hand side of this equation as an infinite series.

Example 7.3.1 Prove that

$$z = a + \sum_{n=1}^{\infty} \frac{(-1)^{n-1} C_n\, (z - a)^n\, e^{n(z^2 - a^2)}}{n!},$$

where

$$C_n = (2na)^{n-1} - \frac{n(n-1)(n-2)}{1!} (2na)^{n-3} + \frac{n^2(n-1)(n-2)(n-3)(n-4)}{2!} (2na)^{n-5} - \cdots .$$

To obtain this expansion, write

$$f(z) = z, \qquad \phi(z) - b = (z - a)e^{x^2 - a^2}, \qquad \psi(z) = e^{a^2 - z^2},$$

in the above expression of Bürmann's theorem; we thus have

$$z = a + \sum_{n=1}^{\infty} \frac{1}{n!} (z - a)^n e^{n(z^2 - a^2)} \left\{ \frac{d^{n-1}}{dz^{n-1}} e^{n(a^2 - z^2)} \right\}_{z=a}.$$

But, putting $z = a + t$,

$$\left\{\frac{d^{n-1}}{dz^{n-1}} e^{n(a^2-z^2)}\right\}_{z=a} = \left\{\frac{d^{n-1}}{dt^{n-1}} e^{-n(2at+t^2)}\right\}_{t=0}$$

$$= (n-1)! \times \text{ the coefficient of } t^{n-1} \quad \text{in the expansion of } e^{-nt(2a+t)}$$

$$= (n-1)! \times \text{ the coefficient of } t^{n-1} \quad \text{in } \sum_{r=0}^{\infty} \frac{(-1)^r n^r t^r (2a+t)^r}{r!}$$

$$= (n-1)! \times \sum_{r=0}^{n-1} \frac{(-1)^r n^r (2a)^{2r-n+1}}{(n-1-r)!(2r-n+1)!}.$$

The highest value of r which gives a term in the summation is $r = n - 1$. Arranging therefore the summation in descending indices r, beginning with $r = n - 1$, we have

$$\left\{\frac{d^{n-1}}{dz^{n-1}} e^{n(a^2-z^2)}\right\}_{z=a} = (-1)^{n-1}\left\{(2na)^{n-1} - \frac{n(n-1)(n-2)}{1!}(2na)^{n-3} + \cdots\right\}$$

$$= (-1)^{n-1} C_n,$$

which gives the required result.

Example 7.3.2 Obtain the expansion

$$z^2 = \sin^2 z + \frac{2}{3} \cdot \frac{\sin^4 z}{2} + \frac{2 \cdot 4}{3 \cdot 5} \cdot \frac{\sin^6 z}{3} + \cdots .$$

Example 7.3.3 Let a line p be drawn through the origin in the z-plane, perpendicular to the line which joins the origin to any point a. If z be any point on the z-plane which is on the same side of the line p as the point a is, shew that

$$\log z = \log a + 2 \sum_{m=1}^{\infty} \frac{1}{2m+1} \left(\frac{z-a}{z+a}\right)^{2m+1}.$$

7.31 Teixeira's extended form of Bürmann's theorem

In the last section we have not investigated closely the conditions of convergence of Bürmann's series, for the reason that a much more general form of the theorem will next be stated; this generalisation bears the same relation to the theorem just given that Laurent's theorem bears to Taylor's theorem: viz., in the last paragraph we were concerned only with the expansion of a function in *positive* powers of another function, whereas we shall now discuss the expansion of a function in *positive and negative* powers of the second function.

The general statement of the theorem is due to Teixeira [622], whose exposition we shall follow in this section. See also Bateman [56].

Suppose

(i) that $f(z)$ is a function of z analytic in a ring-shaped region A, bounded by an outer curve C and an inner curve c;

(ii) that $\theta(z)$ is a function analytic on and inside C, and has only one zero a within this contour, the zero being a simple one;

(iii) that x is a given point within A;

(iv) that for all points z of C we have $|\theta(x)| < |\theta(z)|$, and for all points z of c we have $|\theta(x)| > |\theta(z)|$.

The equation $\theta(z) - \theta(x) = 0$ has, in this case, a single root $z = x$ in the interior of C, as is seen from the equation[3]

$$\frac{1}{2\pi i} \int_C \frac{\theta'(z)\, dz}{\theta(z) - \theta(x)} = \frac{1}{2\pi i} \left[\int_C \frac{\theta'(z)}{\theta(z)}\, dz + \theta(x) \int_C \frac{\theta'(z)}{\{\theta(z)\}^2}\, dz + \cdots \right]$$

$$= \frac{1}{2\pi i} \int_C \frac{\theta'(z)\, dz}{\theta(z)},$$

of which the left-hand and right-hand members represent respectively the number of roots of the equation considered (§6.31) and the number of the roots of the equation $\theta(z) = 0$ contained within C.

Cauchy's theorem therefore gives

$$f(x) = \frac{1}{2\pi i} \left[\int_C \frac{f(z)\, \theta'(z)\, dz}{\theta(z) - \theta(x)} - \int_c \frac{f(z)\, \theta'(z)\, dz}{\theta(z) - \theta(x)} \right].$$

The integrals in this formula can be expanded, as in Laurent's theorem, in powers of $\theta(x)$, by the formulae

$$\int_C \frac{f(z)\, \theta'(z)\, dz}{\theta(z) - \theta(x)} = \sum_{n=0}^{\infty} \{\theta(x)\}^n \int_C \frac{f(z)\, \theta'(z)\, dz}{\{\theta(z)\}^{n+1}},$$

$$\int_c \frac{f(z)\, \theta'(z)\, dz}{\theta(z) - \theta(x)} = -\sum_{n=1}^{\infty} \frac{1}{\{\theta(x)\}^n} \int_c f(z)\{\theta(z)\}^{n-1} \theta'(z)\, dz.$$

We thus have the formula

$$f(x) = \sum_{n=0}^{\infty} A_n \{\theta(x)\}^n + \sum_{n=1}^{\infty} \frac{B_n}{\{\theta(x)\}^n},$$

where

$$A_n = \frac{1}{2\pi i} \int_C \frac{f(z)\, \theta'(z)\, dz}{\{\theta(z)\}^{n+1}}, \quad B_n = \frac{1}{2\pi i} \int_c f(z)\{\theta(z)\}^{n-1} \theta'(z)\, dz.$$

Integrating by parts, we get, if $n \neq 0$,

$$A_n = \frac{1}{2\pi i n} \int_C \frac{f'(z)}{\{\theta(z)\}^n}\, dz, \quad B_n = -\frac{1}{2\pi i n} \int_c \{\theta(z)\}^n f'(z)\, dz.$$

This gives a development of $f(x)$ in positive and negative powers of $\theta(x)$, valid for all points x within the ring-shaped space A.

If the zeros and poles of $f(z)$ and $\theta(z)$ inside C are known, A_n and B_n can be evaluated by §5.22 or by §6.1.

---

[3] The expansion is justified by §4.7, since $\sum_{n=1}^{\infty} \left\{ \frac{\theta(x)}{\theta(z)} \right\}^n$ converges uniformly when z is on C.

Example 7.3.4   Shew that, if $|x| < 1$, then

$$x = \frac{1}{2}\left(\frac{2x}{1+x^2}\right) + \frac{1}{2\cdot 4}\left(\frac{2x}{1+x^2}\right)^3 + \frac{1\cdot 3}{2\cdot 4\cdot 6}\left(\frac{2x}{1+x^2}\right)^5 + \cdots .$$

Shew that, when $|x| > 1$, the second member represents x^{-1}.

Example 7.3.5 (Teixeira)   If $S_{2n}^{(m)}$ denote the sum of all combinations of the numbers 2^2, 4^2, $6^2, \ldots, (2n-2)^2$, taken m at a time, shew that

$$\frac{1}{z} = \frac{1}{\sin z} + \sum_{n=0}^{\infty} \frac{(-1)^{n+1}}{(2n+2)!}\left\{\frac{1}{2n+3} - \frac{S_{2(n+1)}^{(1)}}{2n+1} + \cdots + \frac{(-1)^n S_{2(n+1)}^{(n)}}{3}\right\}(\sin z)^{2n+1},$$

the expansion being valid for all values of z represented by points within the oval whose equation is $|\sin z| = 1$ and which contains the point $z = 0$.

7.32 Lagrange's theorem

Suppose now that the function $f(z)$ of §7.31 is analytic at all points in the interior of C, and let $\theta(x) = (x-a)\theta_1(x)$. Then $\theta_1(x)$ is analytic and not zero on or inside C and the contour c can be dispensed with; therefore the formulae which give A_n and B_n now become, by §5.22 and §6.1,

$$A_n = \frac{1}{2\pi i n}\int_C \frac{f'(z)\,dz}{(z-a)^n\{\theta_1(z)\}^n} = \frac{1}{n!}\frac{d^{n-1}}{da^{n-1}}\left\{\frac{f'(a)}{\theta_1^n(a)}\right\} \qquad (n \geq 1),$$

$$A_0 = \frac{1}{2\pi i}\int_C \frac{f(z)\,\theta'(z)}{\theta_1(z)}\frac{dz}{z-a} = f(a),$$

$$B_n = 0.$$

The theorem of the last section accordingly takes the following form, if we write $\theta_1(z) = 1/\phi(z)$:

Let $f(z)$ and $\phi(z)$ be functions of z analytic on and inside a contour C surrounding a point a, and let t be such that the inequality

$$|t\phi(z)| < |z-a|$$

is satisfied at all points z on the perimeter of C; then the equation

$$\zeta = a + t\phi(\zeta),$$

regarded as an equation in ζ, has one root in the interior of C; and further any function of ζ analytic on and inside C can be expanded as a power series in t by the formula

$$f(\zeta) = f(a) + \sum_{n=1}^{\infty} \frac{t^n}{n!}\frac{d^{n-1}}{da^{n-1}}[f'(a)\{\phi(a)\}^n].$$

This result was published by Lagrange [395] in 1770.

Example 7.3.6   Within the contour surrounding b defined by the inequality $|z(z - \alpha)| > |\alpha|$, where $|\alpha| < \frac{1}{2}|b|$, the equation

$$z - \alpha - \frac{b}{z} = 0$$

has one root ζ, the expansion of which is given by Lagrange's theorem in the form

$$\zeta = b + \sum_{n=1}^{\infty} \frac{(-1)^{n-1}(2n-2)!}{n!(n-1)!b^{2n-1}} \alpha^n.$$

Now, from the elementary theory of quadratic equations, we know that the equation

$$z - \alpha - \frac{b}{z} = 0$$

has two roots, namely $\frac{\alpha}{2}\left\{1 + \sqrt{1 + 4b/\alpha^2}\right\}$ and $\frac{\alpha}{2}\left\{1 - \sqrt{1 + 4b/\alpha^2}\right\}$; and our expansion *represents the former of these only* (the latter is outside the given contour) – an example of the need for care in the discussion of these series.

Example 7.3.7   If y be that one of the roots of the equation

$$y = 1 + zy^2$$

which tends to 1 when $z \to 0$, shew that

$$y^n = 1 + nz + \frac{n(n+3)}{2!}z^2 + \frac{n(n+4)(n+5)}{3!}z^3$$
$$+ \frac{n(n+5)(n+6)(n+7)}{4!}z^4 + \frac{n(n+6)(n+7)(n+8)(n+9)}{5!}z^5 + \cdots$$

so long as $|z| < \frac{1}{4}$.

Example 7.3.8 (McClintock)   If x be that one of the roots of the equation

$$x = 1 + yx^a$$

which tends to 1 when $y \to 0$, shew that

$$\log x = y + \frac{2a-1}{2}y^2 + \frac{(3a-1)(3a-2)}{2.3}y^3 + \cdots,$$

the expansion being valid so long as $|y| < |(a-1)^{a-1}a^{-a}|$.

7.4 The expansion of a class of functions in rational fractions

This appears in Mittag-Leffler [470], see also [471]. Consider a function $f(z)$, whose only singularities in the finite part of the plane are simple poles $a_1, a_2, a_3, \ldots$, where $|a_1| \leq |a_2| \leq |a_3| \leq \cdots$. Let $b_1, b_2, b_3, \ldots$ be the residues at these poles, and let it be possible to choose a sequence of circles C_m (the radius of C_m being R_m) with centre at O, not passing through any poles, such that $|f(z)|$ is bounded on C_m. (The function cosec z may be cited as an example of the class of functions considered, and we take $R_m = (m + \frac{1}{2})\pi$.) Suppose further that $R_m \to \infty$ as $m \to \infty$ and that the upper bound (which is a function of m) of $|f(z)|$ on C_m is itself bounded as[4] $m \to \infty$; so that, for all points on the circle C_m, $|f(z)| < M$, where M is independent of m.

Then, if x be not a pole of $f(z)$, since the only poles of the integrand are the poles of $f(z)$ and the point $z = x$, we have, by §6.1,

$$\frac{1}{2\pi i}\int_{C_m}\frac{f(z)}{z-x}dz = f(x) + \sum_r \frac{b_r}{a_r - x},$$

where the summation extends over all poles in the interior of C_m. But

$$\frac{1}{2\pi i}\int_{C_m}\frac{f(z)\,dz}{z-x} = \frac{1}{2\pi i}\int_{C_m}\frac{f(z)\,dz}{z} + \frac{x}{2\pi i}\int_{C_m}\frac{f(z)\,dz}{z(z-x)}$$

$$= f(0) + \sum_r \frac{b_r}{a_r} + \frac{x}{2\pi i}\int_{C_m}\frac{f(z)\,dz}{z(z-x)},$$

if we suppose the function $f(z)$ to be analytic at the origin.

Now as $m \to \infty$, $\displaystyle\int_{C_m}\frac{f(z)\,dz}{z(z-x)}$ is $O(R_m^{-1})$, and so tends to zero as m tends to infinity. Therefore, making $m \to \infty$, we have

$$0 = f(x) - f(0) + \sum_{n=1}^{\infty} b_n\left(\frac{1}{a_n - x} - \frac{1}{a_n}\right) - \lim_{m\to\infty}\frac{x}{2\pi i}\int_{C_m}\frac{f(z)\,dz}{z(z-x)},$$

i.e. $f(x) = f(0) + \sum_{n=1}^{\infty} b_n\left\{\dfrac{1}{x - a_n} + \dfrac{1}{a_n}\right\}$, which is an expansion of $f(x)$ in rational fractions of x; and the summation extends over *all* the poles of $f(x)$.

If $|a_n| < |a_{n+1}|$ this series converges uniformly throughout the region given by $|x| \le a$, where a is any constant (except near the points a_n). For if R_m be the radius of the circle which encloses the points $|a_1|, \ldots, |a_n|$, the modulus of the remainder of the terms of the series after the first n is

$$\left|\frac{x}{2\pi i}\int_{C_m}\frac{f(z)\,dz}{z(z-x)}\right| < \frac{Ma}{R_m - a},$$

by §4.62; and, given ε, we can choose n *independent* of x such that $\dfrac{Ma}{R_m - a} < \varepsilon$.

The convergence is obviously still uniform even if $|a_n| \le |a_{n+1}|$ provided the terms of the series are grouped so as to combine the terms corresponding to poles of equal moduli.

If, instead of the condition $|f(z)| < M$, we have the condition $|z^{-p} f(z)| < M$, where M is independent of m when z is on C_m, and p is a positive integer, then we should have to expand $\displaystyle\int_C \frac{f(z)\,dz}{z-x}$ by writing

$$\frac{1}{z-x} = \frac{1}{z} + \frac{x}{z^2} + \cdots + \frac{x^{p+1}}{z^{p+1}(z-x)},$$

and should obtain a similar but somewhat more complicated expansion.

---

[4] Of course R_m need not (and frequently must not) tend to infinity continuously; e.g. in the example taken $R_m = (m + \frac{1}{2})\pi$, where m assumes only integer values.

Example 7.4.1 Prove that

$$\operatorname{cosec} z = \frac{1}{z} + \sum (-1)^n \left(\frac{1}{z - n\pi} + \frac{1}{n\pi} \right),$$

the summation extending to all positive and negative values of n.

To obtain this result, let $\operatorname{cosec} z - 1/z = f(z)$. The singularities of this function are at the points $z = n\pi$, where n is any positive or negative integer. The residue of $f(z)$ at the singularity $n\pi$ is therefore $(-1)^n$, and the reader will easily see that $|f(z)|$ is bounded on the circle $|z| = (n + \frac{1}{2})\pi$ as $n \to \infty$.

Applying now the general theorem

$$f(z) = f(0) + \sum c_n \left(\frac{1}{z - a_n} + \frac{1}{a_n} \right),$$

where c_n is the residue at the singularity a_n, we have

$$f(z) = f(0) + \sum (-1)^n \left(\frac{1}{z - n\pi} + \frac{1}{n\pi} \right).$$

But $f(0) = \lim_{z \to 0} \dfrac{z - \sin z}{z \sin z} = 0$. Therefore

$$\operatorname{cosec} z = \frac{1}{z} + \sum (-1)^n \left(\frac{1}{z - n\pi} + \frac{1}{n\pi} \right), \tag{7.2}$$

which is the required result.

Example 7.4.2 If $0 < a < 1$, shew that

$$\frac{e^{az}}{e^z - 1} = \frac{1}{z} + \sum_{n=1}^{\infty} \frac{2z \cos 2na\pi - 4n\pi \sin 2na\pi}{z^2 + 4n^2\pi^2}.$$

Example 7.4.3 Prove that

$$\frac{1}{2\pi x^2 (\cosh x - \cos x)} = \frac{1}{2\pi x^4} - \frac{1}{e^\pi - e^{-\pi}} \frac{1}{\pi^4 + \frac{1}{4}x^4} + \frac{2}{e^{2\pi} - e^{-2\pi}} \frac{1}{(2\pi)^4 + \frac{1}{4}x^4}$$
$$- \frac{3}{e^{3\pi} - e^{-3\pi}} \frac{1}{(3\pi)^4 + \frac{1}{4}x^4} + \cdots.$$

The general term of the series on the right is

$$\frac{(-r)^r r}{(e^{r\pi} - e^{-r\pi}) \left\{ (r\pi)^4 + \frac{1}{4}x^4 \right\}},$$

which is the residue at each of the four singularities $r, -r, ri, -ri$ of the function

$$\frac{\pi z}{(\pi^4 z^4 + \frac{1}{4}x^4)(e^{\pi z} - e^{-\pi z}) \sin \pi z}.$$

The singularities of this latter function which are not of the type $r, -r, ri, -ri$ are at the five points $0, (\pm 1 \pm i)x/2\pi$. At $z = 0$ the residue is $2/\pi x^4$; at each of the four points $z = (\pm 1 \pm i)x/2\pi$, the residue is $\{2\pi x^2(\cos x - \cosh x)\}^{-1}$. Therefore

$$4\sum_{r=1}^{\infty} \frac{(-1)^r r}{e^{r\pi} - e^{-r\pi}} \frac{1}{(r\pi)^4 + \frac{1}{4}x^4} + \frac{2}{\pi x^4} - \frac{2}{\pi x^2(\cosh x - \cos x)}$$

$$= \frac{1}{2\pi i} \lim_{n\to\infty} \int_C \frac{\pi z \, dz}{(\pi^4 z^4 + \frac{1}{4}x^4)(e^{\pi z} - e^{-\pi z})\sin \pi z},$$

where C is the circle whose radius is $n + \frac{1}{2}$ (here n an integer), and whose centre is the origin. But, at points on C, this integrand is $O(|z|^{-3})$; the limit of the integral round C is therefore zero.

From the last equation the required result is now obvious.

Example 7.4.4 Prove that

$$\sec x = 4\pi \left(\frac{1}{\pi^2 - 4x^2} - \frac{3}{9\pi^2 - 4x^2} + \frac{5}{25\pi^2 - 4x^2} - \cdots \right). \tag{7.3}$$

Example 7.4.5 Prove that

$$\operatorname{cosech} x = \frac{1}{x} - 2x \left(\frac{1}{\pi^2 + x^2} - \frac{1}{4\pi^2 + x^2} + \frac{1}{9\pi^2 + x^2} - \cdots \right). \tag{7.4}$$

Example 7.4.6 Prove that

$$\operatorname{sech} x = 4\pi \left(\frac{1}{\pi^2 + 4x^2} - \frac{3}{9\pi^2 + 4x^2} + \frac{1}{25\pi^2 + 4x^2} - \cdots \right). \tag{7.5}$$

Example 7.4.7 Prove that

$$\coth x = \frac{1}{x} + 2x \left(\frac{1}{\pi^2 + x^2} + \frac{1}{4\pi^2 + x^2} + \frac{1}{9\pi^2 + x^2} + \cdots \right). \tag{7.6}$$

Example 7.4.8 (Math. Trip. 1899)   Prove that

$$\sum_{m=-\infty}^{\infty} \sum_{n=-\infty}^{\infty} \frac{1}{(m^2 + a^2)(n^2 + b^2)} = \frac{\pi^2}{ab} \coth \pi a \coth \pi b. \tag{7.7}$$

7.5 The expansion of a class of functions as infinite products

The theorem of the last article can be applied to the expansion of a certain class of functions as infinite products.

For let $f(z)$ be a function which has simple zeros at the points[5] $a_1, a_2, a_3, \ldots$, where $\lim_{n\to\infty} |a_n|$ is infinite; and let $f(z)$ be analytic for all values of z. Then $f'(z)$ is analytic for all values of z (§5.22), and so $f'(z)/f(z)$ can have singularities only at the points $a_1, a_2, a_3, \ldots$

Consequently, by Taylor's theorem,

$$f(z) = (z - a_r) f'(a_r) + \frac{(z - a_r)^2}{2} f''(a_r) + \cdots$$

[5] These being the only zeros of $f(z)$; and $a_n \neq 0$.

and

$$f'(z) = f'(a_r) + (z - a_r)f''(a_r) + \cdots .$$

It follows immediately that at each of the points a_r, the function $f'(z)/f(z)$ has a simple pole, with residue $+1$.

If then we can find a sequence of circles C_m of the nature described in §7.4, such that $f'(z)/f(z)$ is bounded on C_m as $m \to \infty$, it follows, from the expansion given in §7.4, that

$$\frac{f'(z)}{f(z)} = \frac{f'(0)}{f(0)} + \sum_{n=1}^{\infty} \left(\frac{1}{z - a_n} + \frac{1}{a_n} \right).$$

Since this series converges uniformly when the terms are suitably grouped (§7.4), we may integrate term-by-term (§4.7). Doing so, and taking the exponential of each side, we get

$$f(z) = ce^{f'(0)z/f(0)} \prod_{n=1}^{\infty} \left\{ \left(1 - \frac{z}{a_n} \right) e^{z/a_n} \right\},$$

where c is independent of z. Putting $z = 0$, we see that $f(0) = c$, and thus the general result becomes

$$f(z) = f(0)e^{f'(0)z/f(0)} \prod_{n=1}^{\infty} \left\{ \left(1 - \frac{z}{a_n} \right) e^{z/a_n} \right\}.$$

This furnishes the expansion, in the form of an infinite product, of any function $f(z)$ which fulfils the conditions stated.

Example 7.5.1 Consider the function $f(z) = \dfrac{\sin z}{z}$, which has simple zeros at the points $r\pi$, where r is any positive or negative integer. In this case we have $f(0) = 1, f'(0) = 0$, and so the theorem gives immediately

$$\frac{\sin z}{z} = \prod_{n=1}^{\infty} \left\{ \left(1 - \frac{z}{n\pi} \right) e^{\frac{z}{n\pi}} \right\} \left\{ \left(1 + \frac{z}{n\pi} \right) e^{-\frac{z}{n\pi}} \right\};$$

for it is easily seen that the condition concerning the behaviour of $\frac{f'(z)}{f(z)}$ as $|z| \to \infty$ is fulfilled.

Example 7.5.2 (Trinity, 1899) Prove that

$$\left\{ 1 + \left(\frac{k}{x} \right)^2 \right\} \left\{ 1 + \left(\frac{k}{2\pi - x} \right)^2 \right\} \left\{ 1 + \left(\frac{k}{2\pi + x} \right)^2 \right\} \left\{ 1 + \left(\frac{k}{4\pi - x} \right)^2 \right\}$$

$$\times \left\{ 1 + \left(\frac{k}{4\pi + x} \right)^2 \right\} \cdots = \frac{\cosh k - \cos x}{1 - \cos x}.$$

7.6 The factor theorem of Weierstrass

This appears in [663, pp. 77–124]. The theorem of §7.5 is very similar to a more general theorem in which the character of the function $f(z)$, as $|z| \to \infty$, is not so narrowly restricted.

Let $f(z)$ be a function of z with no essential singularities (except at 'the point infinity'); and let the zeros and poles of $f(z)$ be at $a_1, a_2, a_3, \ldots$, where $0 < |a_1| \le |a_2| \le |a_3| \le \cdots$.

Let the zero at a_n be of (integer) order m_n. (We here regard a pole as being a zero of negative order.)

If the number of zeros and poles is unlimited, it is necessary that $|a_n| \to \infty$, as $n \to \infty$; for, if not, the points a_n would have a limit point[6], which would be an essential singularity of $f(z)$.

We proceed to shew first of all that it is possible to find polynomials $g_n(z)$ such that

$$\prod_{n=1}^{\infty} \left[\left\{ \left(1 - \frac{z}{a_n} \right) e^{g_n(z)} \right\}^{m_n} \right]$$

converges for all finite values of z, provided that z is not at one of the points a_n for which m_n is negative.

Let K be any constant, and let $|z| < K$; then, since $|a_n| \to \infty$, we can find N such that, when $n > N, |a_n| > 2K$. The first N factors of the product do not affect its convergence; consider any value of n greater than N, and let

$$g_n(z) = \frac{z}{a_n} + \frac{1}{2} \left(\frac{z}{a_n} \right)^2 + \cdots + \frac{1}{k_n - 1} \left(\frac{z}{a_n} \right)^{k_n - 1} .$$

Then

$$\left| -\sum_{m=1}^{\infty} \frac{1}{m} \left(\frac{z}{a_n} \right)^m + g_n(z) \right| = \left| \sum_{m=k_n}^{\infty} \frac{1}{m} \left(\frac{z}{a_n} \right)^m \right|$$

$$< \left| \frac{z}{a_n} \right|^{k_n} \sum_{m=0}^{\infty} \left| \frac{z}{a_n} \right|^m$$

$$< 2 \left| (K a_n^{-1})^{k_n} \right| ,$$

since $|z a_n^{-1}| < \frac{1}{2}$. Hence

$$\left\{ \left(1 - \frac{z}{a_n} \right) e^{g_n(z)} \right\}^{m_n} = e^{u_n(z)}, \tag{7.8}$$

where $|u_n(z)| \le 2 \left| m_n (K a_n^{-1})^{k_n} \right|$.

Now m_n and a_n are given, but k_n is at our disposal; since $K a_n^{-1} < \frac{1}{2}$, we choose k_n to be the smallest number such that $2 \left| m_n (K a_n^{-1})^{k_n} \right| < b_n$, where $\sum_{n=1}^{\infty} b_n$ is any convergent series of positive terms (e.g. we might take $b_n = 2^{-n}$). Hence

$$\prod_{n=N+1}^{\infty} \left[\left\{ \left(1 - \frac{z}{a_n} \right) e^{g_n(z)} \right\}^{m_n} \right] = \prod_{n=N+1}^{\infty} e^{u_n(z)}, \tag{7.9}$$

where $|u_n(z)| < b_n$; and therefore, since b_n is independent of z, the product converges absolutely and uniformly when $|z| < K$, except near the points a_n. Now let

$$F(z) = \prod_{n=1}^{\infty} \left[\left\{ \left(1 - \frac{z}{a_n} \right) e^{g_n(z)} \right\}^{m_n} \right] . \tag{7.10}$$

---

[6] From the two-dimensional analogue of §2.21.

Then, if $\dfrac{f(z)}{F(z)} = G_1(z)$, it follows that $G_1(z)$ is an integral function (§5.64) of z and has no

zeros. It follows that $\dfrac{1}{G_1(z)} \dfrac{d}{dz} G_1(z)$ is analytic for all finite values of z; and so, by Taylor's

theorem, this function can be expressed as a series $\sum\limits_{n=1}^{\infty} nb_n z^{n-1}$ converging everywhere;

integrating, it follows that

$$G_1(z) = ce^{G(z)},$$

where $G(z) = \sum\limits_{n=1}^{\infty} b_n z^n$ and c is a constant; this series converges everywhere, and so $G(z)$ is
an integral function.

Therefore, finally,

$$f(z) = f(0)e^{G(z)} \prod_{n=1}^{\infty} \left[\left\{ \left(1 - \frac{z}{a_n} \right) e^{g_n(z)} \right\}^{m_n} \right],$$

where $G(z)$ is some integral function such that $G(0) = 0$.

Note   The presence of the arbitrary element $G(z)$ which occurs in this formula for $f(z)$ is
due to the lack of conditions as to the behaviour of $f(z)$ as $|z| \to \infty$.

Corollary 7.6.1   *If $m_n = 1$, it is sufficient to take $k_n = n$, by §2.36.*

7.7 The expansion of a class of periodic functions in a series of cotangents

Let $f(z)$ be a periodic function of z, analytic except at a certain number of simple poles; for
convenience, let π be the period of $f(z)$ so that $f(z) = f(z + \pi)$.

Let $z = x + iy$ and let $f(z) \to \ell$ uniformly with respect to x as $y \to +\infty$, when $0 \le x \le \pi$;
similarly let $f(z) \to \ell'$ uniformly as $y \to -\infty$. Let the poles of $f(z)$ in the strip $0 < x \le \pi$
be at $a_1, a_2, \ldots, a_n$; and let the residues at them be $c_1, c_2, \ldots, c_n$. Further, let $ABCD$ be a
rectangle whose corners are[7] $-i\rho, \pi - i\rho, \pi + i\rho'$ and $i\rho'$ in order.

Consider $\dfrac{1}{2\pi i} \displaystyle\int f(t) \cot(t - z)\, dt$ taken round this rectangle; the residue of the integrand
at a_r is $c_r \cot(a_r - z)$, and the residue at z is $f(z)$. Also the integrals along DA and CB cancel
on account of the periodicity of the integrand; and as $\rho \to \infty$, the integrand on AB tends
uniformly to $i\ell'$, while as $\rho' \to \infty$ the integrand on CD tends uniformly to $-i\ell$; therefore

$$\frac{1}{2}(\ell + \ell') = f(z) + \sum_{r=1}^{n} c_r \cot(a_r - z).$$

That is to say, we have the expansion

$$f(z) = \frac{1}{2}(\ell + \ell') + \sum_{r=1}^{n} c_r \cot(z - a_r).$$

---

[7] If any of the poles are on $x = \pi$, shift the rectangle slightly to the right; ρ, ρ' are to be taken so large that
$a_1, a_2, \ldots, a_n$ are inside the rectangle.

Example 7.7.1 Prove that

$$\cot(x - a_1)\cot(x - a_2)\cdots\cot(x - a_n) =$$

$$\sum_{r=1}^{n}\cot(a_r - a_1)\cdots * \cdots\cot(a_r - a_n) \times \cot(x - a_r) + (-1)^{n/2},$$

$$\text{or} \quad \sum_{r=1}^{n}\cot(a_r - a_1)\cdots * \cdots\cot(a_r - a_n)\cot(x - a_r),$$

according as n is even or odd; the $*$ means that the factor $\cot(a_r - a_r)$ is omitted.

Example 7.7.2 Prove that

$$\frac{\sin(x - b_1)\sin(x - b_2)\cdots\sin(x - b_n)}{\sin(x - a_1)\sin(x - a_2)\cdots\sin(x - a_n)} = \frac{\sin(a_1 - b_1)\cdots\sin(a_1 - b_n)}{\sin(a_1 - a_2)\cdots\sin(a_1 - a_n)}\cot(x - a_1)$$

$$+ \frac{\sin(a_2 - b_1)\cdots\sin(a_2 - b_n)}{\sin(a_2 - a_1)\cdots\sin(a_2 - a_n)}\cot(x - a_2)$$

$$+ \qquad \vdots$$

$$+ \cos(a_1 + a_2 + \cdots + a_n - b_1 - b_2 - \cdots - b_n).$$

7.8 Borel's theorem

This appears in [85, p. 94] and the memoirs there cited. Let $f(z) = \sum_{n=0}^{\infty} a_n z^n$ be analytic when $|z| \le r$, so that, by §5.23, $|a_n r^n| < M$, where M is independent of n. Hence, if $\phi(z) = \sum_{n=0}^{\infty} \frac{a_n z^n}{n!}$, then $\phi(z)$ is an integral function, and

$$|\phi(z)| < \sum_{n=0}^{\infty} \frac{M |z^n|}{r^n \cdot n!} = M e^{|z|/r}, \tag{7.11}$$

and similarly $|\phi^{(n)}(z)| < M e^{|z|/r}/r^n$.

Now consider $f_1(z) = \int_0^{\infty} e^{-t}\phi(zt)\,dt$; this integral is an analytic function of z when $|z| < r$, by §5.32. Also, if we integrate by parts,

$$f_1(z) = \left[-e^{-t}\phi(zt)\right]_0^{\infty} + z\int_0^{\infty} e^{-t}\phi'(zt)\,dt$$

$$= \sum_{m=0}^{n} z^m \left[-e^{-t}\phi^{(m)}(zt)\right]_0^{\infty} + z^{n+1}\int_0^{\infty} e^{-t}\phi^{(n+1)}(zt)\,dt.$$

But $\lim_{t \to 0} e^{-t}\phi^{(m)}(zt) = a_m$; and, when $|z| < r$, $\lim_{t \to \infty} e^{-t}\phi^{(m)}(zt) = 0$. Therefore

$$f_1(z) = \sum_{m=0}^{n} a_m z^m + R_n, \text{ where}$$

$$|R_n| < |z^{n+1}| \int_0^\infty e^{-t} \times M e^{|zt|/r} r^{-n-1} dt$$

$$= |zr^{-1}|^{n+1} M\{1 - |z| r^{-1}\}^{-1} \to 0, \quad \text{as} \quad n \to \infty.$$

Consequently, when $|z| < r$,

$$f_1(z) = \sum_{m=0}^{\infty} a_m z^m = f(z);$$

and so

$$f(z) = \int_0^\infty e^{-t} \phi(zt) \, dt,$$

where $\phi(z) = \sum_{n=0}^{\infty} \frac{a_n z^n}{n!}$; $\phi(z)$ is called *Borel's function* associated with $\sum_{n=0}^{\infty} a_n z^n$.

If $S = \sum_{n=0}^{\infty} a_n$ and $\phi(z) = \sum_{n=0}^{\infty} \frac{a_n z^n}{n!}$ and if we can establish the relation $S = \int_0^\infty e^{-t} \phi(t) \, dt$,

the series S is said (§8.41) to be '*summable* (*B*)'; so that the theorem just proved shews that a Taylor's series representing an analytic function is summable (B).

7.81 *Borel's integral and analytic continuation*

We next obtain Borel's result that his integral represents an analytic function in a more extended region than the interior of the circle $|z| = r$.

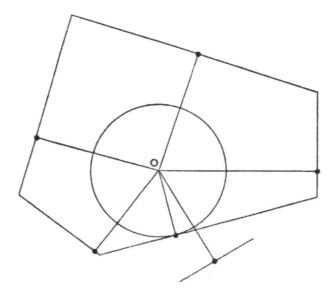

This extended region is obtained as follows: take the singularities $a, b, c, \dots$ of $f(z)$ and through each of them draw a line perpendicular to the line joining that singularity to the origin. The lines so drawn will divide the plane into regions of which one is a polygon with the origin inside it.

Then Borel's integral represents an analytic function (which, by §5.5 and §7.8, is obviously that defined by $f(z)$ and its continuations) *throughout the interior of this polygon*. The reader will observe that this is the first actual formula obtained for the analytic continuation of a function, except the trivial one of Example 5.5.1.

For, take any point P with affix ζ inside the polygon; then the circle on OP as diameter has no singularity on or inside it[8]; and consequently we can draw a slightly larger concentric circle (the difference of the radii of the circles being, say, δ) C with no singularity on or inside it. Then, by §5.4,

$$a_n = \frac{1}{2\pi i} \int_C \frac{f(z)}{z^{n+1}} \, dz,$$

and so

$$\phi(\zeta t) = \frac{1}{2\pi i} \sum_{n=0}^{\infty} \frac{\zeta^n t^n}{n!} \int_C \frac{f(z)}{z^{n+1}} dz;$$

but

$$\sum_{n=0}^{\infty} \frac{\zeta^n t^n}{n!} \frac{f(z)}{z^{n+1}}$$

converges uniformly (§3.34) on C since $f(z)$ is bounded and $|z| \geq \delta > 0$, where δ is independent of z; therefore, by §4.7,

$$\phi(\zeta t) = \frac{1}{2\pi i} \int_C z^{-1} f(z) \exp(\zeta t z^{-1}) \, dz,$$

and so, when t is real, $|\phi(\zeta t)| < F(\zeta)e^{\lambda t}$, where $F(\zeta)$ is bounded in any closed region lying wholly *inside* the polygon and is independent of t; and λ is the greatest value of the real part of ζ/z on C.

If we draw the circle traced out by the point z/ζ, we see that the real part of ζ/z is greatest when z is at the extremity of the diameter through ζ, and so the value of λ is $|\zeta| \cdot (|\zeta| + \delta)^{-1} < 1$. We can get a similar inequality for $\phi'(\zeta t)$ and hence, by §5.32, $\int_0^{\infty} e^{-t} \phi(\zeta t) \, dt$ is analytic at ζ and is obviously a one-valued function of ζ. This is the result stated above.

7.82 *Expansions in series of inverse factorials*

A mode of development of functions, which, after being used by Nicole [498] and Stirling (see [635], and [634]) in the eighteenth century, was systematically investigated by Schlömilch [583] in 1863, is that of expansion in a series of inverse factorials. More recent investigations are due to Kluyver [379], Nielsen [499, 501, 502] and Pincherle [524, 525]. Properties of functions defined by series of inverse factorials have been studied in an important memoir by Nörlund [506].

---

[8] The reader will see this from the figure; for if there were such a singularity the corresponding side of the polygon would pass between O and P; i.e. P would be outside the polygon.

To obtain such an expansion of a function analytic when $|z| > r$, we let the function be

$$f(z) = \sum_{n=0}^{\infty} a_n z^{-n}, \text{ and use the formula } f(z) = \int_0^{\infty} z e^{-tz} \phi(t)\, dt, \text{ where } \phi(t) = \sum_{n=0}^{\infty} a_n t^n / n!; \text{ this}$$

result may be obtained in the same way as that of §7.8. Modify this by writing $e^{-t} = 1 - \xi$,

$\phi(t) = F(\xi)$; then $f(z) = \int_0^1 z(1 - \xi)^{z-1} F(\xi)\, d\xi$. Now if $t = u + iv$ and if t be confined to the strip $-\pi < v < \pi$, t is a one-valued function of ξ and $F(\xi)$ is an analytic function of ξ; and ξ is restricted so that $-\pi < \arg(1 - \xi) < \pi$. Also the interior of the circle $|\xi| = 1$ corresponds to the interior of the curve traced out by the point $t = -\log(2\cos\frac{1}{2}\theta) + \frac{1}{2}i\theta$, (writing $\xi = \exp\{i(\theta + \pi)\}$); and inside this curve $|t| - R(t) \le [R(t)^2 + \pi^2]^{1/2} - R(t) \to 0$, as $R(t) \to \infty$.

It follows that, when $|\xi| \le 1$, $|F(\xi)| < M e^{r|t|} < M_1 |e^{rt}|$, where M_1 is independent of t; and so $F(\xi) < M_1(1-\xi)^{-r}$. Now suppose that $0 \le \xi < 1$; then, by §5.23, $|F^{(n)}(\xi)| < M_2 n! \rho^{-n}$, where M_2 is the upper bound of $|F(z)|$ on a circle with centre ξ and radius $\rho < 1 - \xi$.

Taking $\rho = n(1 - \xi)/(n + 1)$ and observing that $(1 + n^{-1})^n < e$ we find that[9]

$$\left| F^{(n)}(\xi) \right| < M_1 \left[1 - \left\{ \xi + \frac{n}{n+1}\xi \right\} \right]^{-r} n! \left\{ \frac{n(1-\xi)}{n+1} \right\}^{-n}$$

$$< M_1 e (n+1)^r n! (1-\xi)^{-r-n}.$$

Remembering that, by §4.5, $\int_0^1$ means $\lim\limits_{\varepsilon \to 0^+} \int_0^{1-\varepsilon}$, we have, by repeated integrations by parts,

$$f(z) = \lim_{\varepsilon \to 0^+} \left[-(1 - \xi)^z F(\xi) \right]_0^{1-\varepsilon} + \int_0^{1-\varepsilon} (1 - \xi)^z F'(\xi)\, d\xi$$

$$= \lim_{\varepsilon \to 0^+} \left[-(1 - \xi)^z F(\xi) \right]_0^{1-\varepsilon} + \frac{1}{z+1} \left[-(1 - \xi)^{z+1} F'(\xi) \right]_0^{1-\varepsilon}$$

$$+ \frac{1}{z+1} \int_0^{1-\varepsilon} (1 - \xi)^{z+1} F''(\xi)\, d\xi$$

$$= \quad \vdots$$

$$= b_0 + \frac{b_1}{z+1} + \frac{b_2}{(z+1)(z+2)} + \cdots + \frac{b_n}{(z+1)(z+2)\cdots(z+n)} + R_n,$$

where

$$b_n = \lim_{\varepsilon \to 0} \left[-(1 - \xi)^{z+n} F^{(n)}(\xi) \right]_0^{1-\varepsilon} = F^{(n)}(0),$$

---

[9] $(1 + x^{-1})^x$ increases with x; for $\frac{1}{1-y} > e^y$, when $y < 1$, and so $\log\left(\frac{1}{1-y}\right) > y$. That is to say, putting $y^{-1} = 1 + x$, $\frac{d}{dx} x \log(1 + x^{-1}) = \log(1 + x^{-1}) - \frac{1}{1+x} > 0$.

if the real part of $z + n - r - n > 0$, i.e. if $\operatorname{Re} z > r$; further

$$|R_n| \leq \frac{1}{|(z+1)(z+2)\cdots(z+n)|} \lim_{\varepsilon \to 0} \int_0^{1-\varepsilon} \left|(1-\xi)^{z+n} F^{(n+1)}(\xi)\right| d\xi$$

$$< \frac{M_1 e (n+2)^r n!}{|(z+1)(z+2)\cdots(z+n)| \cdot R(z-r)}$$

$$< \frac{M_1 e (n+2)^r n!}{(r+1+\delta)(r+2+\delta)\cdots(r+n+\delta) \cdot \delta},$$

where $\delta = R(z - r)$.

Now $\prod_{m=1}^{n} \left\{ \left(1 + \dfrac{r+\delta}{m}\right) e^{-\frac{r+\delta}{m}} \right\}$ tends to a limit (§2.71) as $n \to \infty$, and so $|R_{n'}| \to 0$ if

$(n+2)^r e^{-(r+\delta)\sum_1^n 1/m}$ tends to zero; but

$$\sum_{m=1}^{n} \frac{1}{m} > \int_1^{n+1} \frac{dx}{x} = \log(n+1),$$

by §4.43, and $(n+2)^r (n+1)^{-r-\delta} \to 0$ when $\delta > 0$; therefore $R_n \to 0$ as $n \to \infty$, and so, when $R(z) > r$, we have the convergent expansion

$$f(z) = b_0 + \frac{b_1}{z+1} + \frac{b_2}{(z+1)(z+2)} + \cdots + \frac{b_n}{(z+1)(z+2)\cdots(z+n)} + \cdots.$$

Example 7.8.1 Obtain the same expansion by using the results

$$\frac{1}{(z+1)(z+2)\cdots(z+n+1)} = \frac{1}{n!} \int_0^1 u^n (1-u)^z \, du, \tag{7.12}$$

$$\int_C \frac{f(t)\,dt}{z-t} = \int_C dt \int_0^1 f(t)(1-u)^{z-t-1} \, du.$$

Example 7.8.2 (Schlömilch)   Obtain the expansion

$$\log\left(1 + \frac{1}{z}\right) = \frac{1}{z} - \frac{a_1}{z(z+1)} - \frac{a_2}{z(z+1)(z+2)} - \cdots, \tag{7.13}$$

where $a_n = \displaystyle\int_0^1 t(1-t)(2-t)\cdots(n-1-t)\,dt$, and discuss the region in which it converges.

7.9 Miscellaneous examples

Example 7.1 (Levi-Città [432])   If $y - x - \phi(y) = 0$, where ϕ is a given function of its argument, obtain the expansion

$$f(y) = f(x) + \sum_{m=1}^{\infty} \frac{1}{m!} \{\phi(x)\}^m \left(\frac{1}{1-\phi'(x)} \frac{d}{dx}\right)^m f(x),$$

where f denotes any analytic function of its argument, and discuss the range of its validity.

Example 7.2 Obtain (from the formula of Darboux or otherwise) the expansion

$$f(z) - f(a) = \sum_{n=1}^{\infty} \frac{(-1)^{n-1}(z-a)^n}{n!(1-r)^n}\{f^{(n)}(z) - r^n f^{(n)}(a)\};$$

find the remainder after n terms, and discuss the convergence of the series.

Example 7.3 Shew that

$$f(x+h) - f(x) = \sum_{m=1}^{n}(-1)^{m-1}\frac{1\cdot3\cdot5\cdots(2m-1)}{(m!)^2}\frac{h^m}{2^m}\{f^{(m)}(x+h) - (-1)^m f^{(m)}(x)\}$$

$$+(-1)^n h^{n+1}\int_0^1 \gamma_n(t)f^{(n+1)}(x+ht)\,dt,$$

where

$$\gamma_n(x) = \frac{x^{n+\frac12}(1-x)^{n+\frac12}}{(n!)^2}\frac{d^n}{dx^n}\{x^{-\frac12}(1-x)^{-\frac12}\} = \frac{1}{\pi n!}\int_0^1 (x-z)^n z^{-\frac12}(1-z)^{-\frac12}\,dz,$$

and shew that $\gamma_n(x)$ is the coefficient of $n!t^n$ in the expansion of $\{(1-tx)(1+t-tx)\}^{-1/2}$ in ascending powers of t.

Example 7.4 By taking

$$\phi(x+1) = \frac{1}{n!}\left[\frac{d^n}{du^n}\left\{\frac{(1-r)e^{xu}}{1-re^{-u}}\right\}\right]_{u=0}$$

in the formula of Darboux, shew that

$$f(x+h) - f(x) = -\sum_{m=1}^{n} a_m \frac{h^m}{m!}\left\{f^{(m)}(x+h) - \frac{1}{r}f^{(m)}(x)\right\}$$

$$+(-1)^n h^{n+1}\int_0^1 \phi(t)f^{(n+1)}(x+ht)\,dt,$$

where

$$\frac{1-r}{1-re^{-u}} = 1 - a_1\frac{u}{1!} + a_2\frac{u^2}{2!} - a_3\frac{u^3}{3!} + \cdots.$$

Example 7.5 Shew that

$$f(z) - f(a) = \sum_{m=1}^{n}(-1)^{m-1}\frac{2B_m(2^{2n}-1)(z-a)^{2m-1}}{(2m)!}\{f^{(2m-1)}(a) + f^{(2m-1)}(z)\}$$

$$+\frac{(z-a)^{2n+1}}{(2n)!}\int_0^1 \psi_{2n}(t)f^{(2n+1)}\{a+t(z-a)\}\,dt,$$

where

$$\psi_n(t) = \frac{2}{n+1}\left[\frac{d^{n+1}}{du^{n+1}}\left(\frac{ue^{tu}}{e^u+1}\right)\right]_{u=0}.$$

Example 7.6 (Trinity, 1899)   Prove that

$$f(z_2) - f(z_1) = C_1(z_2 - z_1)f'(z_2) + C_2(z_2 - z_1)^2 f''(z_1) - C_3(z_2 - z_1)^3 f'''(z_2)$$
$$- C_4(z_2 - z_1)^4 f^{(iv)}(z_1) + \cdots + (-1)^n (z_2 - z_1)^{n+1}$$
$$\times \int_0^1 \left\{ \frac{d^n}{du^n}(e^{tu} \operatorname{sech} u) \right\}_{u=0} f^{(n+1)}(z_1 + tz_2 - tz_1)\, dt;$$

in the series plus signs and minus signs occur in pairs, and the last term before the integral is that involving $(z_2 - z_1)^n$; also C_n is the coefficient of z^n in the expansion of $\cot\left(\frac{\pi}{4} - \frac{z}{2}\right)$ in ascending powers of z.

Example 7.7   If x_1 and x_2 are integers, and $\phi(z)$ is a function which is analytic and bounded for all values of z such that $x_1 \le R(z) \le x_2$, shew (by integrating

$$\int \frac{\phi(z)\, dz}{e^{\pm 2\pi i z} - 1}$$

round indented rectangles whose corners are x_1, x_2, $x_2 \pm \infty i$, $x_1 \pm \infty i$) that

$$\tfrac{1}{2}\phi(x_1) + \phi(x_1 + 1) + \phi(x_1 + 2) + \cdots + \phi(x_2 - 1) + \tfrac{1}{2}\phi(x_2)$$
$$= \int_{x_1}^{x_3} \phi(z)\, dz + \frac{1}{i} \int_0^\infty \frac{\phi(x_2 + iy) - \phi(x_1 + iy) - \phi(x_2 - iy) + \phi(x_1 - iy)}{e^{2\pi y} - 1}\, dy.$$

Hence, by applying the theorem

$$4n \int_0^\infty \frac{y^{2n-1}}{e^{2\pi y} - 1}\, dy = B_n,$$

where $B_1, B_2, \ldots$ are Bernoulli's numbers, shew that

$$\phi(1) + \phi(2) + \cdots + \phi(n)$$
$$= C + \frac{1}{2}\phi(n) + \int^n \phi(z)\, dz + \sum_{r=1}^\infty \frac{(-1)^{r-1} B_r}{2r!}\phi^{(2r-1)}(n),$$

(where C is a constant not involving n), provided that the last series converges. (This important formula is due to Plana [526]; a proof by means of contour integration was published by Kronecker [386]. For a detailed history, see Lindelöf [436]. Some applications of the formula are given in Chapter 12.)

Example 7.8   Obtain the expansion

$$u = \frac{x}{2} + \sum_{n=2}^\infty (-1)^{n-1} \frac{1 \cdot 3 \cdots (2n - 3)}{n!} \frac{x^n}{2^n}$$

for one root of the equation $x = 2u + u^2$, and shew that it converges so long as $|x| < 1$.

Example 7.9 (Teixeira)   If $S_{2n+1}^{(m)}$ denote the sum of all combinations of the numbers

$$1^2,\ 3^2,\ 5^2, \ldots, (2n - 1)^2,$$

taken m at a time, shew that

$$\frac{\cos z}{z} = \frac{1}{\sin z} + \sum_{n=0}^{\infty} \frac{(-1)^{n+1}}{(2n+2)\,!}$$

$$\times \left\{ \frac{2^{2(n+1)}}{2n+3} - S_{2(n+1)}^{(1)} \frac{2^{2n}}{2n+1} + \cdots + (-1)^n S_{2(n+1)}^{(n)} \frac{2^2}{3} \right\} \sin^{2n+1} z.$$

Example 7.10 (Teixeira)   If the function $f(z)$ is analytic in the interior of that one of the ovals whose equation is $|\sin z| = C$ (where $C \le 1$), which includes the origin, shew that $f(z)$ can, for all points z within this oval, be expanded in the form

$$f(z) = f(0) + \sum_{n=1}^{\infty} \frac{f^{(2n)}(0) + S_{2n}^{(1)} f^{(2n-2)}(0) + \cdots + S_{2n}^{(n-1)} f''(0)}{2n\,!} \sin^{2n} z$$

$$+ \sum_{n=0}^{\infty} \frac{f^{(2n+1)}(0) + S_{2n+1}^{(1)} f^{(2n-1)}(0) + \cdots + S_{2n+1}^{(n)} f'(0)}{(2n+1)\,!} \sin^{2n+1} z,$$

where $S_{2n}^{(m)}$ is the sum of all combinations of the numbers

$$2^2,\ 4^2,\ 6^2,\ldots,(2n-2)^2,$$

taken m at a time, and $S_{2n+1}^{(m)}$ denotes the sum of all combinations of the numbers

$$1^2,\ 3^2,\ 5^2,\ldots,(2n-1)^2,$$

taken m at a time.

Example 7.11 (Kapteyn [366])   Shew that the two series

$$2z + \frac{2z^3}{3^2} + \frac{2z^5}{5^2} + \cdots,$$

and

$$\frac{2z}{1-z^2} - \frac{2}{1\cdot 3^2} \left(\frac{2z}{1-z^2} \right)^3 + \frac{2\cdot 4}{3\cdot 5^2} \left(\frac{2z}{1-z^2} \right)^6 - \cdots,$$

represent the same function in a certain region of the z-plane, and can be transformed into each other by Bürmann's theorem.

Example 7.12   If a function $f(z)$ is periodic, of period 2π, and is analytic at all points in the infinite strip of the plane included between the two branches of the curve $|\sin z| = C$ (where $C > 1$), shew that at all points in the strip it can be expanded in an infinite series of the form

$$f(z) = A_0 + A_1 \sin z + \cdots + A_n \sin^n z + \cdots +$$
$$+ \cos z\, (B_1 + B_2 \sin z + \cdots + B_n \sin^{n-1} z + \cdots);$$

and find the coefficients A_n and B_n.

Example 7.13   If ϕ and f are connected by the equation

$$\phi(x) + \lambda f(x) = 0,$$

of which one root is a, shew that

$$F(x) = F - \frac{\lambda}{1}\frac{1}{\phi'}fF' + \frac{\lambda^2}{1!\,2!}\frac{1}{\phi'^3}\begin{vmatrix}\phi' & f^2F' \\ \phi'' & (f^2F')'\end{vmatrix}$$

$$- \frac{\lambda^3}{1!\,2!\,3!}\frac{1}{\phi'^6}\begin{vmatrix}\phi' & (\phi^2)' & (f^3F') \\ \phi'' & (\phi^2)'' & (f^3F')' \\ \phi''' & (\phi^2)''' & (f^3F')''\end{vmatrix} + \cdots,$$

the general term being

$$(-1)^m \frac{\lambda^m}{1!\,2!\cdots m!\,(\phi')^{\frac{1}{2}m(m+1)}}$$

multiplied by a determinant in which the elements of the first row are

$$\phi', (\phi^2)', (\phi^3)', \ldots, (\phi^{m-1})', (f^m F')$$

and each row is the differential coefficient of the preceding one with respect to a; and $F, f, F', \ldots$ denote $F(a), f(a), F'(a), \ldots$. (See Wronski [684]. For proofs of the theorem see Cayley [135], Transon [633] and Ch. Lagrange [391].)

Example 7.14 (Ježek)   If the function $W(a, b, x)$ be defined by the series

$$W(a, b, x) = x + \frac{a - b}{2!}x^2 + \frac{(a - b)(a - 2b)}{3!}x^3 + \cdots,$$

which converges so long as $|x| < 1/|b|$, shew that

$$\frac{d}{dx}W(a, b, x) = 1 + (a - b)W(a - b, b, x);$$

and shew that if $y = W(a, b, x)$, then $x = W(b, a, y)$. Examples of this function are

$$W(1, 0, x) = e^x - 1,$$
$$W(0, 1, x) = \log(1 + x),$$
$$W(a, 1, x) = \frac{(1 + x)^a - 1}{a}.$$

Example 7.15 (Mangeot [453])   Prove that

$$\frac{1}{\sum\limits_{n=0}^{\infty} a_n x^n} = \frac{1}{a_0} + \sum_{n=1}^{\infty}\frac{(-1)^n x^n}{n!\,a_0^{n+1}}G_n,$$

where

$$G_n = \begin{vmatrix} 2a_1 & a_0 & 0 & 0 & \cdots & 0 \\ 4a_2 & 3a_1 & 2a_0 & 0 & \cdots & 0 \\ 6a_3 & 5a_2 & 4a_1 & 3a_0 & \cdots & 0 \\ \vdots & \vdots & \vdots & \vdots & \vdots & \vdots \\ (2n-2)a_{n-1} & \cdots & \cdots & \cdots & \cdots & (n-1)a_0 \\ na_n & (n-1)a_{n-1} & \cdots & \cdots & \cdots & a_1 \end{vmatrix},$$

and obtain a similar expression for $\left\{\sum\limits_{n=0}^{\infty} a_n x^n\right\}^{1/2}$.

Example 7.16 (Gambioli [232])   Shew that

$$\frac{1}{\sum\limits_{r=0}^{\infty} a_r x^r} = -\sum_{r=0}^{\infty} \frac{1}{r+1} \frac{\partial S_{r+1}}{\partial a_1} x^r,$$

where S_r is the sum of the rth powers of the reciprocals of the roots of the equation $\sum\limits_{r=0}^{n} a_r x^r = 0$.

Example 7.17 (Guichard)   If $f_n(z)$ denote the nth derivate of $f(z)$, and if $f_{-n}(z)$ denote that one of the nth integrals of $f(z)$ which has an n-tuple zero at $z = 0$, shew that if the series

$$\sum_{n=-\infty}^{\infty} f_n(z) g_{-n}(x)$$

is convergent it represents a function of $z + x$; and if the domain of convergence includes the origin in the x-plane, the series is equal to

$$\sum_{n=0}^{\infty} f_{-n}(z + x) g_n(0).$$

Obtain Taylor's series from this result, by putting $g(z) = 1$.

Example 7.18 (Math. Trip. 1895)   Shew that, if x be not an integer,

$$\sum_{m=-\nu}^{\nu} \sum_{n=-\nu}^{\nu} \frac{2x + m + n}{(x + m)^2 (x + n)^2} \to 0$$

as $\nu \to \infty$, provided that all terms for which $m = n$ are omitted from the summation.

Example 7.19 (Math. Trip. 1896)   Sum the series

$$\sum_{n=-q}^{p} \left(\frac{1}{(-1)^n x - a - n} + \frac{1}{n} \right),$$

where the value $n = 0$ is omitted, and p, q are positive integers to be increased without limit.

Example 7.20 (Trinity, 1898)   If $F(x) = \exp\left(\int_0^x x\pi \cot(x\pi) \, dx \right)$, shew that

$$F(x) = e^x \frac{\prod\limits_{n=1}^{\infty} \left\{ \left(1 - \frac{x}{n}\right)^n e^{x + \frac{x^2}{2n}} \right\}}{\prod\limits_{n=1}^{\infty} \left\{ \left(1 + \frac{x}{n}\right)^n e^{-x + \frac{x^2}{2n}} \right\}},$$

and that the function thus defined satisfies the relations

$$F(-x) = \frac{1}{F(x)}, \quad F(x)F(1 - x) = 2 \sin x\pi.$$

Further, if

$$\psi(z) = z + \frac{z^2}{2^2} + \frac{z^3}{3^2} + \cdots = -\int_0^z \log(1 - t) \frac{dt}{t},$$

shew that

$$F(x) = \exp\left(\frac{1}{2}\pi i x^2 - \frac{1}{2\pi i}\psi(1 - e^{-2\pi i x})\right)$$

when $|1 - e^{-2\pi i x}| < 1$.

Example 7.21 (Mildner) Shew that

$$\left[1 + \left(\frac{k}{x}\right)^n\right]\left[1 + \left(\frac{k}{2\pi - x}\right)^n\right]\left[1 + \left(\frac{k}{2\pi + x}\right)^n\right]$$

$$\left[1 + \left(\frac{k}{4\pi - x}\right)^n\right]\left[1 + \left(\frac{k}{4\pi + x}\right)^n\right]\cdots$$

$$= \frac{\prod_{g=1}^{\leq n/2}\left\{1 - 2e^{-\alpha_g}\cos(x + \beta_g) + e^{-2\alpha_g}\right\}^{1/2}\left\{1 - 2e^{-\alpha_g}\cos(x - \beta_g) + e^{-2\alpha_g}\right\}^{1/2}}{2^{n/2}(1 - \cos x)^{n/2}e^{-k\cos\pi/n}},$$

where $\alpha_g = k\sin\frac{2g-1}{n}\pi$, $\beta_g = k\cos\frac{2g-1}{n}\pi$, and $0 < x < 2\pi$.

Example 7.22 (Lerch [428]) If $|x| < 1$ and a is not a positive integer, shew that

$$\sum_{n=1}^{\infty}\frac{x^n}{n - \alpha} = \frac{2\pi i x^\alpha}{1 - e^{2\alpha\pi i}} + \frac{x}{1 - e^{2\alpha\pi i}}\int_C \frac{t^{\alpha-1} - x^{\alpha-1}}{t - x}dt,$$

where C is a contour in the t-plane enclosing the points $0, x$.

Example 7.23 If $\phi_1(z), \phi_2(z), \ldots$ are any polynomials in z, and if $F(z)$ be any integrable function, and if $\psi_1(z), \psi_2(z), \ldots$ be polynomials defined by the equations

$$\int_a^b F(x)\frac{\phi_1(z) - \phi_1(x)}{z - x}dx = \psi_1(z),$$

$$\int_a^b F(x)\phi_1(x)\frac{\phi_2(z) - \phi_2(x)}{z - x}dx = \psi_2(z),$$

$$\int_a^b F(x)\,\phi_1(x)\phi_2(x)\cdots\phi_{m-1}(x)\frac{\phi_m(z) - \phi_m(x)}{z - x}dx = \psi_m(z),$$

shew that

$$\int_a^b \frac{F(x)\,dx}{z - x} = \frac{\psi_1(z)}{\phi_1(z)} + \frac{\psi_2(z)}{\phi_1(z)\,\phi_2(z)} + \frac{\psi_3(z)}{\phi_1(z)\,\phi_2(z)\,\phi_3(z)} + \cdots$$

$$+ \frac{\psi_m(z)}{\phi_1(z)\phi_2(z)\cdots\phi_m(z)}$$

$$+ \frac{1}{\phi_1(z)\phi_2(z)\cdots\phi_m(z)}\int_a^b F(x)\phi_1(x)\phi_2(x)\cdots\phi_m(x)\frac{dx}{z - x}.$$

Example 7.24 (Pincherle [521]) A system of functions $p_0(z), p_1(z), p_2(z), \ldots$ is defined by the equations

$$p_0(z) = 1, \quad p_{n+1}(z) = (z^2 + a_n z + b_n)p_n(z),$$

where a_n and b_n are given functions of n, which tend respectively to the limits 0 and -1 as $n \to \infty$.

Shew that the region of convergence of a series of the form $\sum e_n p_n(z)$, where $e_1, e_2, \ldots$ are independent of z, is a Cassini's oval with the foci $+1, -1$.

Shew that every function $f(z)$, which is analytic on and inside the oval, can, for points inside the oval, be expanded in a series

$$f(z) = \sum (c_n + z c'_n) p_n(z),$$

where

$$c_n = \frac{1}{2\pi i} \int (a_n + z) q_n(z) f(z) \, dz, \quad c'_n = \frac{1}{2\pi i} \int q_n(z) f(z) \, dz,$$

the integrals being taken round the boundary of the region, and the functions $q_n(z)$ being defined by the equations

$$q_0(z) = \frac{1}{z^2 + a_0 z + b_0}, \quad q_{n+1}(z) = \frac{1}{z^2 + a_{n+1} z + b_{n+1}} q_n(z).$$

Example 7.25   Let C be a contour enclosing the point a, and let $\phi(z)$ and $f(z)$ be analytic when z is on or inside C. Let $|t|$ be so small that $|t\phi(z)| < |z - a|$ when z is on the periphery of C. By expanding

$$\frac{1}{2\pi i} \int_C f(z) \frac{1 - t\phi'(z)}{z - a - t\phi(z)} \, dz$$

in ascending powers of t, shew that it is equal to

$$f(a) + \sum_{n=1}^{\infty} \frac{t^n}{n!} \frac{d^{n-1}}{da^{n-1}} \left[f'(a) \{\phi(a)\}^n \right].$$

Hence, by using §6.3, §6.31, obtain Lagrange's theorem.

8

Asymptotic Expansions and Summable Series

8.1 Simple example of an asymptotic expansion

Consider the function $f(x) = \int_x^\infty t^{-1} e^{x-t}\, dt$, where x is real and positive, and the path of integration is the real axis. By repeated integrations by parts, we obtain

$$f(x) = \frac{1}{x} - \frac{1}{x^2} + \frac{2!}{x^3} - \cdots + \frac{(-1)^{n-1}(n-1)!}{x^n} + (-1)^n n! \int_x^\infty \frac{e^{x-t}\, dt}{t^{n+1}}.$$

In connexion with the function $f(x)$, we therefore consider the expression

$$u_{n-1} = \frac{(-1)^{n-1}(n-1)!}{x^n},$$

and we shall write

$$\sum_{m=0}^{n} u_m = \frac{1}{x} - \frac{1}{x^2} + \frac{2!}{x^3} - \cdots + \frac{(-1)^n n!}{x^{n+1}} = S_n(x).$$

Then we have $|u_m/u_{m-1}| = mx^{-1} \to \infty$ as $m \to \infty$. *The series $\sum u_m$ is therefore divergent for all values of x.* In spite of this, however, the series can be used for the calculation of $f(x)$; this can be seen in the following way.

Take any fixed value for the number n, and calculate the value of S_n. We have

$$f(x) - S_n(x) = (-1)^{n+1}(n+1)! \int_x^\infty \frac{e^{x-t}\, dt}{t^{n+2}},$$

and therefore, since $e^{x-t} \leq 1$,

$$|f(x) - S_n(x)| = (n+1)! \int_x^\infty \frac{e^{x-t}\, dt}{t^{n+2}} < (n+1)! \int_x^\infty \frac{dt}{t^{n+2}} = \frac{n!}{x^{n+1}}.$$

For values of x which are sufficiently large, the right-hand member of this equation is very small. Thus, if we take $x \geq 2n$, we have

$$|f(x) - S_n(x)| < \frac{1}{2^{n+1} n^2},$$

which for large values of n is very small. It follows therefore that *the value of the function $f(x)$ can be calculated with great accuracy for large values of x, by taking the sum of a suitable number of terms of the series $\sum u_m$.*

Taking even fairly small values of x and n

$$S_5(10) = 0.09152, \quad \text{and} \quad 0 < f(10) - S_5(10) < 0.00012.$$

153

The series is on this account said to be an *asymptotic expansion* of the function $f(x)$. The precise definition of an asymptotic expansion will now be given.

8.2 Definition of an asymptotic expansion

A divergent series

$$A_0 + \frac{A_1}{z} + \frac{A_2}{z^2} + \cdots + \frac{A_n}{z^n} + \cdots,$$

in which the sum of the first $(n + 1)$ terms is $S_n(z)$, is said to be an *asymptotic expansion* of a function $f(z)$ for a given range of values of arg z, if the expression $R_n(z) = z^n [f(z) - S_n(z)]$ satisfies the condition

$$\lim_{|z| \to \infty} R_n(z) = 0 \quad (n \text{ fixed}),$$

even though

$$\lim_{n \to \infty} |R_n(z)| = \infty \quad (z \text{ fixed}).$$

When this is the case, we can make

$$| z^n (f(z) - S_n(z)) | < \varepsilon,$$

where ε is arbitrarily small, by taking $|z|$ sufficiently large.

We denote the fact that the series is the asymptotic expansion of $f(z)$ by writing

$$f(z) \sim \sum_{n=0}^{\infty} A_n z^{-n}.$$

The definition which has just been given is due to Poincaré [529]. Special asymptotic expansions had, however, been discovered and used in the eighteenth century by Stirling, MacLaurin and Euler. Asymptotic expansions are of great importance in the theory of Linear Differential Equations, and in Dynamical Astronomy; some applications will be given in subsequent chapters of the present work.

The example discussed in §8.1 clearly satisfies the definition just given: for, when x is positive, $|x^n (f(x) - S_n(x))| < n!x^{-1} \to 0$ as $x \to \infty$. For the sake of simplicity, in this chapter we shall for the most part consider asymptotic expansions only in connexion with real positive values of the argument. The theory for complex values of the argument may be discussed by an extension of the analysis.

8.21 Another example of an asymptotic expansion

As a second example, consider the function $f(x)$, represented by the series

$$f(x) = \sum_{k=1}^{\infty} \frac{c^k}{x + k},$$

where $x > 0$ and $0 < c < 1$.

The ratio of the kth term of this series to the $(k - 1)$th is less than c, and consequently the series converges for all positive values of x. We shall confine our attention to positive values of x. We have, when $x > k$,

$$\frac{1}{x+k} = \frac{1}{x} - \frac{k}{x^2} + \frac{k^2}{x^3} - \frac{k^3}{x^4} + \frac{k^4}{x^5} - \cdots .$$

If, therefore, it were allowable[1] to expand each fraction $\dfrac{1}{x+k}$ in this way, and to rearrange the series for $f(x)$ in descending powers of x, we should obtain the formal series

$$\frac{A_1}{x} + \frac{A_2}{x^2} + \cdots + \frac{A_n}{x^n} + \cdots ,$$

where $A_n = (-1)^{n-1} \sum\limits_{k=1}^{\infty} k^{n-1} c^k$. But this procedure is not legitimate, and in fact $\sum\limits_{n=1}^{\infty} A_n x^{-n}$ diverges. We can, however, shew that it is an asymptotic expansion of $f(x)$.

For let $S_n(x) = \dfrac{A_1}{x} + \dfrac{A_2}{x^2} + \cdots + \dfrac{A_n}{x^{n+1}}$. Then

$$S_n(x) = \sum_{k=1}^{\infty} \left(\frac{c^k}{x} - \frac{kc^k}{x^2} + \frac{k^2 c^k}{x^3} + \cdots + \frac{(-)^n k^n c^k}{x^{n+1}} \right) \tag{8.1}$$

$$= \sum_{k=1}^{\infty} \left\{ 1 - \left(-\frac{k}{x} \right)^{n+1} \right\} \frac{c^k}{x+k}$$

so that $\left| f(x) - S_n(x) \right| = \left| \sum\limits_{k=1}^{\infty} \left(-\frac{k}{x} \right)^{n+1} \frac{c^k}{x+k} \right| < x^{-n-2} \sum\limits_{k=1}^{\infty} k^n c^k$.

Now $\sum\limits_{k=1}^{\infty} k^n c^k$ converges for any given value of n and is equal to C_n, say, and hence $\left| f(x) - S_n(x) \right| < C_n x^{-n-2}$. Consequently $f(x) \sim \sum\limits_{n=1}^{\infty} A_n x^{-n}$.

Example 8.2.1   If $f(x) = \displaystyle\int_x^{\infty} e^{x^2 - t^2} dt$, where x is positive and the path of integration is the real axis, prove that

$$f(x) \sim \frac{1}{2x} - \frac{1}{2^2 x^3} + \frac{1 \cdot 3}{2^3 x^5} - \frac{1 \cdot 3 \cdot 5}{2^4 x^7} + \cdots .$$

In fact, it was shewn by Stokes [610] in 1857 that

$$\int_0^x e^{x^2 - t^2} dt \sim \pm \frac{1}{2} e^{x^2} \sqrt{\pi} - \left(\frac{1}{2x} - \frac{1}{2^2 x^3} + \frac{1 \cdot 3}{2^3 x^5} - \frac{1 \cdot 3 \cdot 5}{2^4 x^7} + \cdots \right);$$

the upper or lower sign is to be taken according as

$$-\frac{1}{2}\pi < \arg x < \frac{1}{2}\pi \quad \text{or} \quad \frac{1}{2}\pi < \arg x < \frac{3}{2}\pi.$$

---

[1] It is not allowable, since $k > x$ for all terms of the series after some definite term.

8.3 Multiplication of asymptotic expansions

We shall now shew that two asymptotic expansions, valid for a common range of values of arg z, can be multiplied together in the same way as ordinary series, the result being a new asymptotic expansion.

For let $f(z) \sim \sum\limits_{m=0}^{\infty} A_m z^{-m}$, $\phi(z) \sim \sum\limits_{m=0}^{\infty} B_m z^{-m}$, and let $S_n(z)$ and $T_n(z)$ be the sums of their first $(n+1)$ terms; so that, n being fixed,

$$f(z) - S_n(z) = o(z^{-n}), \quad \phi(z) - T_n(z) = o(z^{-n}).$$

Then, if $C_m = A_0 B_m + A_1 B_{m-1} + \cdots + A_m B_0$, it is obvious that[2]

$$S_n(z) T_n(z) = \sum_{m=0}^{n} C_m z^{-m} + o(z^{-n}).$$

But

$$f(z)\phi(z) = (S_n(z) + o(z^{-n})) (T_n(z) + o(z^{-n}))$$
$$= S_n(z) T_n(z) + o(z^{-n})$$
$$= \sum_{m=0}^{n} C_m z^{-m} + o(z^{-n}).$$

This result being true for *any* fixed value of n, we see that

$$f(z)\phi(z) \sim \sum_{m=0}^{\infty} C_m z^{-m}.$$

8.31 *Integration of asymptotic expansions*

We shall now shew that it is permissible to integrate an asymptotic expansion term by term, the resulting series being the asymptotic expansion of the integral of the function represented by the original series.

For let $f(x) \sim \sum\limits_{m=2}^{\infty} A_m x^{-m}$, and let $S_n(x) = \sum\limits_{m=2}^{n} A_m x^{-m}$. Then, given any positive number ε, we can find x_0 such that

$$|f(x) - S_n(x)| < \varepsilon |x|^{-n} \text{ when } x > x_0,$$

and therefore

$$\left| \int_x^{\infty} f(x)dx - \int_x^{\infty} S_n(x)dx \right| \le \int_x^{\infty} |f(x) - S_n(x)|dx$$
$$< \frac{\varepsilon}{(n-1)x^{n-1}}.$$

---

[2] See §2.11; we use $o(z^{-n})$ to denote *any* function $\psi(z)$ such that $z^n \psi(z) \to 0$ as $|z| \to x$.

But $\displaystyle\int_x^\infty S_n(x)dx = \frac{A_2}{x} + \frac{A_3}{2x^2} + \cdots + \frac{A_n}{(n-1)x^{n-1}}$, and therefore

$$\int_x^\infty f(x)dx \sim \sum_{m=2}^\infty \frac{A_m}{(m-1)x^{m-1}}. \tag{8.2}$$

On the other hand, it is not in general permissible to differentiate an asymptotic expansion; this may be seen by considering $e^{-x}\sin(e^x)$. For a theorem concerning differentiation of asymptotic expansions representing analytic functions, see Ritt [560].

8.32 Uniqueness of an asymptotic expansion

A question naturally suggests itself, as to whether a given series can be the asymptotic expansion of several distinct functions. The answer to this is in the affirmative. To shew this, we first observe that there are functions $L(x)$ which are represented asymptotically by a series all of whose terms are zero, i.e. functions such that $\lim_{x\to\infty} x^n L(x) = 0$ for every fixed value of n. The function e^{-x} is such a function when x is positive. The asymptotic expansion[3] of a function $J(x)$ is therefore also the asymptotic expansion of $J(x) + L(x)$.

On the other hand, a function cannot be represented by more than one distinct asymptotic expansion over the whole of a given range of values of z; for, if

$$f(z) \sim \sum_{m=0}^\infty A_m z^{-m}, \quad f(z) \sim \sum_{m=0}^\infty B_m z^{-m},$$

then

$$\lim_{z\to\infty} z^n \left(A_0 + \frac{A_1}{z} + \cdots + \frac{A_n}{z^n} - B_0 - \frac{B_1}{z} - \cdots - \frac{B_n}{z^n} \right) = 0,$$

which can only be if $A_0 = B_0$, $A_1 = B_1, \ldots$.

Important examples of asymptotic expansions will be discussed later, in connexion with the Gamma-function (Chapter 12) and Bessel functions (Chapter 17).

8.4 Methods of summing series

We have seen that it is possible to obtain a development of the form $f(x) = \sum_{m=0}^n A_m x^{-m} + R_n(x)$, where $R_n(x) \to \infty$ as $n \to \infty$, and the series $\sum_{m=0}^\infty A_m x^{-m}$ does not converge. We now consider what meaning, if any, can be attached to the *sum* of a non-convergent series. That is to say, given the numbers $a_0, a_1, a_2, \ldots$, we wish to formulate definite rules by which we can obtain from them a number S such that $S = \sum_{n=0}^\infty a_n$ if $\sum_{n=0}^\infty a_n$ converges, and such that S exists when this series does not converge.

---

[3] It has been shewn that when the coefficients in the expansion satisfy certain inequalities, there is only one *analytic* function with that asymptotic expansion. See Watson [648].

8.41 Borel's method of summation [85, p. 97–115]

We have seen (§7.81) that $\sum_{n=0}^{\infty} a_n z^n = \int_0^{\infty} e^{-t}\phi(tz)\,dt$, where $\phi(tz) = \sum_{n=0}^{\infty} \frac{a_n t^n z^n}{n!}$, the equation certainly being true inside the circle of convergence of $\sum_{n=0}^{\infty} a_n z^n$. If the integral exists at points z outside this circle, we define the *Borel sum* of $\sum_{n=0}^{\infty} a_n z^n$ to mean the integral.

Thus, whenever $\operatorname{Re} z < 1$, the 'Borel sum' of the series $\sum_{n=0}^{\infty} z^n$ is

$$\int_0^{\infty} e^{-t} e^{tz}\,dt = (1-z)^{-1}.$$

If the Borel sum exists we say that the series is *summable (B)*.

8.42 Euler's method of summation [85, 201]

A method, practically due to Euler, is suggested by the theorem of §3.71; the *sum* of $\sum_{n=0}^{\infty} a_n$ may be defined as $\lim_{x \to 1^-} \sum_{n=0}^{\infty} a_n x^n$, when this limit exists.

Thus the *sum* of the series $1 - 1 + 1 - 1 + \cdots$ would be

$$\lim_{x \to 1^-} (1 - x + x^2 - \cdots) = \lim_{x \to 1^-} (1+x)^{-1} = \frac{1}{2}.$$

8.43 Cesàro's method of summation [141]

Let $s_n = a_1 + a_2 + \cdots + a_n$; then *if* $S = \lim_{n \to \infty} \frac{1}{n}(s_1 + s_2 + \cdots + s_n)$ *exists*, we say that $\sum_{n=1}^{\infty} a_n$ is 'summable (C1)', and that its sum (C1) is S. It is necessary to establish the *condition of consistency*, namely that $S = \sum_{n=1}^{\infty} a_n$ when this series is convergent. (See the end of §8.4.)

To obtain the required result, let $\sum_{m=1}^{\infty} a_m = s$, $\sum_{m=1}^{n} s_m = nS_n$; then we have to prove that $S_n \to s$. Given ε, we can choose n such that $\left| \sum_{m=n+1}^{n+p} a_m \right| < \varepsilon$ for all values of p, and so $|s - s_n| \le \varepsilon$. Then, if $v > n$, we have

$$S_v = a_1 + a_2 \left(1 - \frac{1}{v}\right) + \cdots + a_n \left(1 - \frac{n-1}{v}\right)$$

$$+ a_{n+1}\left(1 - \frac{n}{v}\right) + \cdots + a_v \left(1 - \frac{v-1}{v}\right).$$

Since $1, 1 - v^{-1}, 1 - 2v^{-1}, \ldots$ is a positive decreasing sequence, it follows from Abel's inequality (§2.301) that

$$\left| a_{n+1}\left(1 - \frac{n}{v}\right) + a_{n+2}\left(1 - \frac{n+1}{v}\right) + \cdots + a_v\left(1 - \frac{v-1}{v}\right) \right| < \left(1 - \frac{n}{v}\right)\varepsilon.$$

Therefore

$$\left| s_\nu - \left\{ a_1 + a_2 \left(1 - \frac{1}{\nu} \right) + \cdots + a_n \left(1 - \frac{n-1}{\nu} \right) \right\} \right| < \left(1 - \frac{n}{\nu} \right) \varepsilon.$$

Making $\nu \to \infty$, we see that, if S be any one of the limit points (§2.21) of S_ν, then $\left| S - \sum_{m=1}^{n} a_m \right| \le \varepsilon$. Therefore, since $|s - s_n| \le \varepsilon$, we have $|S - s| \le 2\varepsilon$. This inequality being true for *every* positive value of ε we infer, as in §2.21, that $S = s$; that is to say, S_ν has the unique limit s; this is the theorem which had to be proved.

Example 8.4.1 Frame a definition of 'uniform summability ($C1$) of a series of variable terms'.

Example 8.4.2 If $b_{n,\nu} \ge b_{n+1,\nu} \ge 0$ when $n < \nu$, and if, when n is *fixed*, $\lim_{\nu \to \infty} b_{n,\nu} = 1$ and if $\sum_{m=1}^{\infty} a_m = S$, then $\lim_{\nu \to \infty} \left\{ \sum_{n=1}^{\nu} a_n b_{n,\nu} \right\} = S$.

8.431 Cesàro's general method of summation

A series $\sum_{n=0}^{\infty} a_n$, is said to be ' summable (Cr)' if $\lim_{\nu \to \infty} \sum_{n=0}^{\nu} a_n b_{n,\nu}$ exists, where

$$b_{0,\nu} = 1, \quad b_{n,\nu} = \left\{ \left(1 + \frac{r}{\nu + 1 - n} \right) \left(1 + \frac{r}{\nu + 2 - n} \right) \cdots \left(1 + \frac{r}{\nu - 1} \right) \right\}^{-1}.$$

It follows from Example 8.4.2 that the *condition of consistency* is satisfied; in fact it can be proved [102, §122] that if a series is summable (Cr') it is also summable (Cr) when $r > r'$; the condition of consistency is the particular case of this result when $r = 0$.

8.44 The method of summation of Riesz [559]

A more extended method of *summing* a series than the preceding is by means of

$$\lim_{\nu \to \infty} \sum_{n=1}^{\nu} \left(1 - \frac{\lambda_n}{\lambda_\nu} \right)^r a_n,$$

in which λ_n is any real function of n which tends to infinity with n. A series for which this limit exists is said to be ' summable (Rr) with sum-function λ_n'.

8.5 Hardy's convergence theorem

This appears in Hardy [278]. For the proof here given, we are indebted to Mr. Littlewood.

Let $\sum_{n=1}^{\infty} a_n$ be a series which is summable ($C1$). Then if $a_n = O(1/n)$, the series $\sum_{n=1}^{\infty} a_n$ converges.

Let $s_n = a_1 + a_2 + \cdots + a_n$; then since $\sum_{n=1}^{\infty} a_n$ is summable ($C1$), we have $s_1 + s_2 + \cdots + s_n = n(s + o(1))$, where s is the sum ($C1$) of $\sum_{n=1}^{\infty} a_n$. Let $s_m - s = t_m$, $(m = 1, 2, \ldots, n)$, and let

$t_1 + t_2 + \cdots + t_n = \sigma_n$. With this notation, it is sufficient to shew that, if $|a_n| < Kn^{-1}$, where K is independent of n, and if $\sigma_n = n \cdot o(1)$, then $t_n \to 0$ as $n \to \infty$.

Suppose first that $a_1, a_2, \ldots$, are real. Then, if t_n does not tend to zero, there is some positive number h such that there are an unlimited number of the numbers t_n which satisfy *either* (i) $t_n > h$ or (ii) $t_n < -h$. We shall shew that either of these hypotheses implies a contradiction. Take the former[4], and choose n so that $t_n > h$. Then, when $r = 0, 1, 2, \ldots$, $|a_{n+r}| < K/n$.

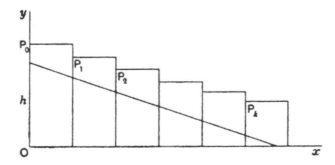

Now plot the points P_r whose coordinates are (r, t_{n+r}) in a Cartesian diagram. Since $t_{n+r+1} - t_{n+r} = a_{n+r+1}$, the slope of the line $P_r P_{r+1}$ is less than $\theta = \arctan(K/n)$. Therefore the points $P_0, P_1, P_2, \ldots$ lie above the line $y = h - x \tan \theta$. Let P_k be the last of the points $P_0, P_1, \ldots$ which lie on the left of $x = h \cot \theta$, so that $k \leq h \cot \theta$.

Draw rectangles as shewn in the figure. The area of these rectangles exceeds the area of the triangle bounded by $y = h - x \tan \theta$ and the axes; that is to say

$$\sigma_{n+k} - \sigma_{n-1} = t_n + t_{n+1} + \cdots + t_{n+k}$$
$$> \frac{1}{2} h^2 \cot \theta = \frac{1}{2} h^2 K^{-1} n.$$

But

$$|\sigma_{n+k} - \sigma_{n-1}| \leq |\sigma_{n+k}| + |\sigma_{n-1}|$$
$$= (n + k) \cdot o(1) + (n - 1) \cdot o(1)$$
$$= n \cdot o(1),$$

since $k \leq hnK^{-1}$, and h, K are independent of n. Therefore, for a set of values of n tending to infinity, $\frac{1}{2} h^2 K^{-1} n < n \cdot o(1)$, which is impossible since $\frac{1}{2} h^2 K^{-1}$ is *not* $o(1)$ as $n \to \infty$.

This is the contradiction obtained on the hypothesis that $\overline{\lim}\, t_n \geq h > 0$; therefore $\overline{\lim}\, t_n \leq 0$. Similarly, by taking the corresponding case in which $t_n \leq -h$, we arrive at the result $\underline{\lim}\, t_n \geq 0$. Therefore since $\overline{\lim}\, t_n \geq \underline{\lim}\, t_n$, we have $\overline{\lim}\, t_n = \underline{\lim}\, t_n = 0$, and so $t_n \to 0$. That is to say $s_n \to s$, *and so* $\sum\limits_{n=1}^{\infty} a_n$ *is convergent and its sum is* s.

If a_n be complex, we consider $\operatorname{Re} a_n$ and $\operatorname{Im} a_n$ separately, and find that $\sum\limits_{n=1}^{\infty} \operatorname{Re} a_n$ and

---

[4] The reader will see that the latter hypothesis involves a contradiction by using arguments of a precisely similar character to those which will be employed in dealing with the former hypothesis.

$\sum\limits_{n=1}^{\infty}$ Im a_n converge by the theorem just proved, and so $\sum\limits_{n=1}^{\infty} a_n$ converges. The reader will see in Chapter 9 that this result is of great importance in the modern theory of Fourier series.

Corollary 8.5.1 *If $a_n(\xi)$ be a function of ξ such that $\sum\limits_{n=1}^{\infty} a_n(\xi)$ is uniformly summable (C1) throughout a domain of values of ξ, and if $|a_n(\xi)| < K^{n-1}$, where K is independent of ξ, $\sum\limits_{n=1}^{\infty} a_n(\xi)$ converges uniformly throughout the domain.*

For, retaining the notation of the preceding section, if $t_n(\xi)$ does not tend to zero uniformly, we can find a positive number h independent of n and ξ such that an infinite sequence of values of n can be found for which $t_n(\xi_n) > h$ or $t_n(\xi_n) < -h$ for some point ξ_n of the domain[5]; the value of ξ_n depends on the value of n under consideration.

We then find, as in the original theorem, $\frac{1}{2}h^2 K^{-1} n < n \cdot o(1)$ for a set of values of n tending to infinity. The contradiction implied in the inequality shews[6] that h does not exist, and so $t_n(\xi) \to 0$ uniformly.

8.6 Miscellaneous examples

Example 8.1 Shew that $\displaystyle\int_0^\infty \frac{e^{-xt}}{1+t^2}\,dt \sim \frac{1}{x} - \frac{2\,!}{x^3} + \frac{4\,!}{x^6} - \cdots$ when x is real and positive.

Example 8.2 Discuss the representation of the function

$$f(x) = \int_{-\infty}^{0} \phi(t)e^{tx}\,dt$$

(where x is supposed real and positive, and ϕ is a function subject to certain general conditions) by means of the series

$$f(x) = \frac{\phi(0)}{x} - \frac{\phi'(0)}{x^2} + \frac{\phi''(0)}{x^3} - \cdots .$$

Shew that in certain cases (e.g. $\phi(t) = e^{at}$) the series is absolutely convergent, and represents $f(x)$ for large positive values of x; but that in certain other cases the series is the asymptotic expansion of $f(x)$.

Example 8.3 (Legendre [421, p. 340]) Shew that

$$e^z z^{-a} \int_z^\infty e^{-x} x^{a-1}\,dx \sim \frac{1}{z} + \frac{a-1}{z^2} + \frac{(a-1)(a-2)}{z^3} + \cdots$$

for large positive values of z.

---

[5] It is assumed that $a_n(\xi)$ is real; the extension to complex variables can be made as in the former theorem. If no such number h existed, $t_n(\xi)$ would tend to zero uniformly.

[6] It is essential to observe that the constants involved in the inequality do not depend on ξ_n. For if, say, K depended on ξ_n, K^{-1} would really be a function of n and might be $o(1)$ *qua* function of n, and the inequality would not imply a contradiction.

Example 8.4 (Schlömilch)   Shew that if, when $x > 0$,

$$f(x) = \int_0^\infty \left\{ \log u + \log \left(\frac{1}{1 - e^{-u}} \right) \right\} e^{-xu} \frac{du}{u},$$

then $f(x) \sim \dfrac{1}{2x} - \dfrac{B_1}{2^2 x^2} + \dfrac{B_2}{4^2 x^4} - \dfrac{B_3}{6^2 x^6} + \cdots$. Shew also that $f(x)$ can be expanded into an absolutely convergent series of the form

$$f(x) = \sum_{k-1}^\infty \frac{c_k}{(x + 1)(x + 2) \cdots (x + k)}.$$

Example 8.5 (Euler, Borel)   Shew that if the series $1 + 0 + 0 - 1 + 0 + 1 + 0 + 0 - 1 + \cdots$, in which two zeros precede each -1 and one zero precedes each $+1$, be *summed* by Cesàro's method, its sum is $\frac{3}{5}$.

Example 8.6   Shew that the series $1 - 2! + 4! - \cdots$ cannot be summed by Borel's method, but the series $1 + 0 - 2! + 0 + 4! + \cdots$ can be so summed.

9

Fourier Series and Trigonometric Series

9.1 Definition of Fourier series

Series of the type

$$\tfrac{1}{2}a_0 + (a_1 \cos x + b_1 \sin x) + (a_2 \cos 2x + b_2 \sin 2x) + \cdots = \tfrac{1}{2}a_0 + \sum_{n=1}^{\infty} (a_n \cos nx + b_n \sin nx),$$

where a_n, b_n are independent of x, are of great importance in many investigations. They are called *trigonometrical series*. (Throughout this chapter, except in §9.11, it is supposed that all the numbers involved are *real*.)

If there is a function $f(t)$ such that $\int_{-\pi}^{\pi} f(t)\,dt$ exists as a Riemann integral or as an improper integral which converges absolutely, and such that

$$\pi a_n = \int_{-\pi}^{\pi} f(t) \cos nt\,dt, \quad \pi b_n = \int_{-\pi}^{\pi} f(t) \sin nt\,dt,$$

then the trigonometrical series is called a *Fourier series*.

Trigonometrical series that are not Fourier series first appeared in analysis in connexion with the investigations of Daniel Bernoulli (1700–1782) on vibrating strings; d'Alembert had previously solved the equation of motion $\ddot{y} = a^2 \dfrac{d^2 y}{dx^2}$ in the form $y = \tfrac{1}{2}\{f(x+at) + f(x-at)\}$, where $y = f(x)$ is the initial shape of the string starting from rest; and Bernoulli shewed that a formal solution is

$$y = \sum_{n=1}^{\infty} b_n \sin \frac{n\pi x}{\ell} \cos \frac{n\pi at}{\ell},$$

the fixed ends of the string being $(0,0)$ and $(\ell, 0)$; and he asserted that this was the most general solution of the problem. This appeared to d'Alembert and Euler to be impossible, since such a series, having period 2ℓ, could not possibly represent such a function as[1] $cx(\ell - x)$ when $t = 0$. A controversy arose between these mathematicians, of which an account is given in Hobson [315].

Fourier, in his *Théorie de la Chaleur* [223] investigated a number of trigonometrical series and shewed that, in a large number of particular cases, a Fourier series *actually converged to the sum* $f(x)$. Poisson [531] attempted a general proof of this theorem. Two proofs were given by Cauchy [122] and [123, vol. 2, p. 341–376]. These proofs, which are based on the

---

[1] This function gives a simple form to the initial shape of the string.

theory of contour integration, are concerned with rather particular classes of functions and one is invalid. The second proof has been investigated by Harnack [283].

In 1829, Dirichlet [172] gave the first rigorous proof that, for a general class of functions, the Fourier series, defined as above, does converge to the sum $f(x)$. A modification of this proof was given later by Bonnet [82]. He employs the second mean-value theorem directly, while Dirichlet's original proof makes use of arguments precisely similar to those by which that theorem is proved. See §9.43.

The result of Dirichlet is that[2] if $f(t)$ is defined and bounded in the range $(-\pi, \pi)$ and if $f(t)$ has only a finite number of maxima and minima and a finite number of discontinuities in this range and, further, if $f(t)$ is defined by the equation $f(t + 2\pi) = f(t)$ outside the range $(-\pi, \pi)$, then, provided that

$$\pi a_n = \int_{-\pi}^{\pi} f(t) \cos nt \, dt, \quad \pi b_n = \int_{-\pi}^{\pi} f(t) \sin nt \, dt,$$

the series $\frac{1}{2} a_0 + \sum_{n=1}^{\infty} (a_n \cos nx + b_n \sin nx)$ converges to $\frac{1}{2}\{f(x+0) + f(x-0)\}$.

Later, Riemann and Cantor developed the theory of trigonometrical series generally, while still more recently Hurwitz, Fejér and others have investigated properties of Fourier series when the series does not necessarily converge. Thus Fejér has proved the remarkable theorem that a Fourier series (even if not convergent) is 'summable (C1)' at all points at which $f(x \pm 0)$ exist, and its sum (C1) is $\frac{1}{2}\{f(x + 0) + f(x - 0)\}$, provided that $\int_{-\pi}^{\pi} f(t) \, dt$ is an absolutely convergent integral. One of the investigations of the convergence of Fourier series which we shall give later (§9.42) is based on this result.

For a fuller account of investigations subsequent to Riemann, the reader is referred to Hobson [323], and to de la Vallée Poussin [639].

9.11 Nature of the region within which a trigonometrical series converges

Consider the series

$$\frac{1}{2} a_0 + \sum_{n=1}^{\infty} (a_n \cos nz + b_n \sin nz),$$

where z may be complex. If we write $e^{is} = \zeta$, the series becomes

$$\frac{1}{2} a_0 + \sum_{n=1}^{\infty} \left\{ \frac{1}{2}(a_n - ib_n)\zeta^n + \frac{1}{2}(a_n + ib_n)\zeta^{-n} \right\}.$$

This Laurent series will converge, if it converges at all, in a region in which $a \leq |\zeta| \leq b$, where a, b are positive constants. But, if $z = x + iy$, $|\zeta| = e^{-y}$, and so we get, as the region of convergence of the trigonometrical series, the strip in the z plane defined by the inequality

$$\log a \leq -y \leq \log b.$$

---

[2] The conditions postulated for $f(t)$ are known as *Dirichlet's conditions*; as will be seen in §§9.2, 9.42, they are unnecessarily stringent.

The case which is of the greatest importance in practice is that in which $a = b = 1$, and the strip consists of a single line, namely the real axis.

Example 9.1.1 Let

$$f(z) = \sin z - \frac{1}{2}\sin 2z + \frac{1}{3}\sin 3z - \frac{1}{4}\sin 4z + \cdots,$$

where $z = x + iy$. Writing this in the form

$$f(z) = -\frac{1}{2}i\left(e^{is} - \frac{1}{2}e^{2is} + \frac{1}{3}e^{3is} - \cdots\right) + \frac{1}{2}i\left(e^{-is} - \frac{1}{2}e^{-2is} + \frac{1}{3}e^{-3is} - \cdots\right)$$

we notice that the first series converges[3] only if $y \geq 0$, and the second only if $y \leq 0$. Writing x in place of z (x being real), we see that by Abel's theorem (§3.71),

$$f(x) = \lim_{r \to 1}\left(r \sin x - \frac{1}{2}r^2 \sin 2x + \frac{1}{3}r^3 \sin 3x - \cdots\right)$$

$$= \lim_{r \to 1}\left\{-\frac{1}{2}i\left(re^{ix} - \frac{1}{2}r^2 e^{2ix} + \frac{1}{3}r^3 e^{3ix} - \cdots\right)\right.$$

$$\left. + \frac{1}{2}i\left(re^{-ix} - \frac{1}{2}r^2 e^{-2ix} + \frac{1}{3}r^3 e^{-3ix} - \cdots\right)\right\}.$$

This is the limit of one of the values of

$$-\frac{1}{2}i\log\left(1 + re^{ix}\right) + \frac{1}{2}i\log\left(1 + re^{-ix}\right),$$

and as $r \to 1$ (if $-\pi < x < \pi$), this tends to $\frac{1}{2}x + k\pi$, where k is some integer.

Now $\sum_{n=1}^{\infty}\dfrac{(-1)^{n-1}\sin nx}{n}$ converges uniformly (Example 3.3.6) and is therefore continuous in the range $-\pi + \delta \leq x \leq \pi - \delta$, where δ is any positive constant. Since $\frac{1}{2}x$ is continuous, k has the same value wherever x lies in the range; and putting $x = 0$, we see that $k = 0$. *Therefore, when* $-\pi < x < \pi$, $f(x) = \frac{1}{2}x$. But, when $\pi < x < 3\pi$,

$$f(x) = f(x - 2\pi) = \frac{x - 2\pi}{2} = \frac{x}{2} - \pi,$$

and generally, if $(2n - 1)\pi < x < (2n + 1)\pi$, $f(x) = \frac{1}{2}x - n\pi$. We have thus arrived at an example in which $f(x)$ is not represented by a single analytical expression.

It must be observed that this phenomenon can only occur when the strip in which the Fourier series converges is a single line. For if the strip is not of zero breadth, the associated Laurent series converges in an annulus of non-zero breadth and represents an analytic function of ζ in that annulus; and, since ζ is an analytic function of z, the Fourier series represents an analytic function of z; such a series is given by

$$r \sin x - \frac{1}{2}r^2 \sin 2x + \frac{1}{3}r^3 \sin 3x - \cdots,$$

---

[3] The series *do converge* if $y = 0$, see Example 2.3.2.

where $0 < r < 1$; its sum is $\arctan\left(\dfrac{r \sin x}{1 + r \cos x}\right)$, the arctan always representing an angle between $\pm \frac{1}{2}\pi$.

Example 9.1.2   When $-\pi \le x \le \pi$,

$$\sum_{n=1}^{\infty} \frac{(-1)^{n-1} \cos nx}{n^2} = \frac{1}{12}\pi^2 - \frac{1}{4}x^2.$$

The series converges only when x is real; by §3.34 the convergence is then absolute and uniform. Since

$$\frac{1}{2}x = \sin x - \frac{1}{2}\sin 2x + \frac{1}{3}\sin 3x - \cdots \qquad (-\pi + \delta \le x \le \pi - \delta; \quad \delta > 0),$$

and this series converges uniformly, we may integrate term-by-term from 0 to x (§4.7), and consequently

$$\frac{1}{4}x^2 = \sum_{n=1}^{\infty} \frac{(-1)^{n-1}(1 - \cos nx)}{n^2} \qquad (-\pi + \delta \le x \le \pi - \delta).$$

That is to say, when $-\pi + \delta \le x \le \pi - \delta$,

$$C - \frac{1}{4}x^2 = \sum_{n=1}^{\infty} \frac{(-1)^{n-1} \cos nx}{n^2},$$

where C is a constant, at present undetermined.

But since the series on the right converges uniformly throughout the range $-\pi \le x \le \pi$, its sum is a continuous function of x in this extended range; and so, proceeding to the limit when $x \to \pm\pi$, we see that the last equation is still true when $x = \pm\pi$.

To determine C, integrate each side of the equation (§4.7) between the limits $-\pi, \pi$; and we get

$$2\pi C - \frac{1}{6}\pi^3 = 0.$$

Consequently

$$\frac{1}{12}\pi^2 - \frac{1}{4}x^2 = \sum_{n=1}^{\infty} \frac{(-1)^{n-1} \cos nx}{n^2} \qquad (-\pi \le x \le \pi).$$

Example 9.1.3   By writing $\pi - 2x$ for x in Example 9.1.2, shew that

$$\sum_{n=1}^{\infty} \frac{\sin^2 nx}{n^2} = \begin{cases} \frac{1}{2}x(\pi - x) & (0 \le x \le \pi), \\ \frac{1}{2}\{\pi|x| - x^2\} & (-\pi \le x \le \pi). \end{cases}$$

9.12 Values of the coefficients in terms of the sum of a trigonometrical series

Let the trigonometrical series $\frac{1}{2}c_0 + \sum\limits_{n=1}^{\infty} (c_n \cos nx + d_n \sin nx)$ be uniformly convergent in the range $(-\pi, \pi)$ and let its sum be $f(x)$. Using the obvious results

$$\int_{-\pi}^{\pi} \cos mx \cos nx \, dx = \begin{cases} 0 & (m \neq n) \\ \pi & (m = n \neq 0), \end{cases}$$

$$\int_{-\pi}^{\pi} \sin mx \sin nx \, dx = \begin{cases} 0 & (m \neq n) \\ \pi & (m = n \neq 0), \end{cases} \qquad \int_{-\pi}^{\pi} dx = 2\pi,$$

we find, on multiplying the equation $\frac{1}{2}c_0 + \sum\limits_{n=1}^{\infty} (c_n \cos nx + d_n \sin nx) = f(x)$ by[4] $\cos nx$; or by $\sin nx$ and integrating term-by-term (§4.7),

$$\pi c_n = \int_{-\pi}^{\pi} f(x) \cos nx \, dx, \qquad \pi d_n = \int_{-\pi}^{\pi} f(x) \sin nx \, dx.$$

These were given by Euler [203].

Corollary 9.1.1 *A trigonometrical series uniformly convergent in the range $(-\pi, \pi)$ is a Fourier series.*

Note Lebesgue [419, p. 124] has given a proof of a theorem communicated to him by Fatou that the trigonometrical series $\sum\limits_{n=2}^{\infty} \sin nx / \log n$, which converges for all real values of x (Example 2.3.1), is *not* a Fourier series.

9.2 On Dirichlet's conditions and Fourier's theorem

A theorem, of the type described in §9.1, concerning the expansibility of a function of a real variable into a trigonometrical series is usually described as *Fourier's theorem*. On account of the length and difficulty of a formal proof of the theorem (even when the function to be expanded is subjected to unnecessarily stringent conditions), we defer the proof until §9.42, §9.43. It is, however, convenient to state here certain *sufficient* conditions under which a function can be expanded into a trigonometrical series.

Let $f(t)$ be defined arbitrarily when $-\pi \leq t < \pi$ and defined [5] for all other real values of t by means of the equation $f(t + 2\pi) = f(t)$, so that $f(t)$ is a periodic function with period 2π.

Let $f(t)$ be such that $\int_{-\pi}^{\pi} f(t) \, dt$ exists; and if this is an improper integral, let it be absolutely convergent.

---

[4] Multiplying by these factors does not destroy the uniformity of the convergence.
[5] This definition frequently results in $f(t)$ not being expressible by a single analytical expression for all real values of t; cf. Example 9.1.1.

Let a_n, b_n be defined by the equations[6]

$$\pi a_n = \int_{-\pi}^{\pi} f(t) \cos nt \, dt, \quad \pi b_n = \int_{-\pi}^{\pi} f(t) \sin nt \, dt \quad (n = 0, 1, 2, \ldots).$$

Then, if x be an interior point of any interval (a, b) in which $f(t)$ has limited total fluctuation, the series

$$\frac{1}{2} a_0 + \sum_{n=1}^{\infty} (a_n \cos nx + b_n \sin nx)$$

is convergent, and its sum[7] *is $\frac{1}{2}\{f(x+0) + f(x-0)\}$. If $f(t)$ is continuous at $t = x$, this sum reduces to $f(x)$.*

This theorem will be assumed in §§9.21–9.32; these sections deal with theorems concerning Fourier series which are of some importance in practical applications. It should be stated here that every function which Applied Mathematicians need to expand into Fourier series satisfies the conditions just imposed on $f(t)$, so that the analysis given later in this chapter establishes the validity of all the expansions into Fourier series which are required in physical investigations.

The reader will observe that in the theorem just stated, $f(t)$ is subject to less stringent conditions than those contemplated by Dirichlet, and this decrease of stringency is of considerable practical importance. Thus, so simple a series as $\sum_{n=1}^{\infty} (-1)^{n-1} \frac{\cos nx}{n}$ is the expansion of the function[8] $\log |2 \cos \frac{1}{2} x|$; and this function does not satisfy Dirichlet's condition of boundedness at $\pm \pi$.

It is convenient to describe the series $\frac{1}{2} a_0 + \sum_{n=1}^{\infty} (a_n \cos nx + b_n \sin nx)$ as *the Fourier series associated with $f(t)$*. This description must, however, be taken as implying nothing concerning the convergence of the series in question.

9.21 The representation of a function by Fourier series for ranges other than $(-\pi, \pi)$

Consider a function $f(x)$ with an (absolutely) convergent integral, and with limited total fluctuation in the range $a \le x \le b$.

Write $x = \frac{1}{2}(a + b) - \frac{1}{2}(a - b)\pi^{-1} x', f(x) = F(x')$. Then it is known (§9.2) that

$$\frac{1}{2}\{F(x' + 0) + F(x' - 0)\} = \frac{1}{2} a_0 + \sum_{n=1}^{\infty} (a_n \cos nx' + b_n \sin nx'),$$

---

[6] The numbers a_n, b_n are called the *Fourier constants* of $f(t)$, and the symbols a_n, b_n will be used in this sense throughout §§9.2–9.5. It may be shewn that the convergence and absolute convergence of the integrals defining the Fourier constants are consequences of the convergence and absolute convergence of $\int_{-\pi}^{\pi} f(t) \, dt$; cf. §§2.32, 4.5.

[7] The limits $f(x \pm 0)$ exist, by Example 3.6.3.

[8] Example 9.6 at the end of the chapter.

and so

$$\frac{1}{2}\{f(x+0)+f(x-0)\} =$$

$$\frac{1}{2}a_0 + \sum_{n=1}^{\infty}\left\{a_n\cos\left(\frac{\pi n(2x-a-b)}{b-a}\right) + b_n\sin\left(\frac{\pi n(2x-a-b)}{b-a}\right)\right\},$$

where by an obvious transformation

$$\frac{1}{2}(b-a)a_n = \int_a^b f(x)\cos\left(\frac{\pi n(2x-a-b)}{b-a}\right)dx,$$

$$\frac{1}{2}(b-a)b_n = \int_a^b f(x)\sin\left(\frac{\pi n(2x-a-b)}{b-a}\right)dx.$$

9.22 The cosine series and the sine series

Let $f(x)$ be defined in the range $(0,\ell)$ and let it have an (absolutely) convergent integral and also let it have limited total fluctuation in that range. *Define $f(x)$ in the range $(-\ell,0)$ by the equation*

$$f(-x) = f(x).$$

Then

$$\frac{1}{2}\{f(x+0)+f(x-0)\} = \frac{1}{2}a_0 + \sum_{n=1}^{\infty}\left\{a_n\cos\frac{\pi nx}{\ell} + b_n\sin\frac{\pi nx}{\ell}\right\},$$

where, by §9.21,

$$\ell a_n = \int_{-\ell}^{\ell} f(t)\cos\frac{\pi nt}{\ell}\,dt = 2\int_0^{\ell} f(t)\cos\frac{\pi nt}{\ell}\,dt,$$

$$\ell b_n = \int_{-\ell}^{\ell} f(t)\sin\frac{\pi nt}{\ell}\,dt = 0,$$

so what when $-\ell \le x \le \ell$,

$$\frac{1}{2}\{f(x+0)+f(x-0)\} = \frac{1}{2}a_0 + \sum_{n=1}^{\infty} a_n\cos\frac{\pi nx}{\ell};$$

this is called the *cosine series*.

If, however, we define $f(x)$ in the range $(-\ell,0)$ by the equation

$$f(-x) = -f(-x),$$

we get, when $-\ell \le x \le \ell$,

$$\frac{1}{2}\{f(x+0)+f(x-0)\} = \sum_{n=1}^{\infty} b_n\sin\frac{\pi nx}{\ell},$$

where $\ell b_n = 2\int_0^{\ell} f(t)\sin\frac{\pi nt}{\ell}\,dt$; this is called the *sine series*.

Thus the series

$$\frac{1}{2}a_0 + \sum_{n=1}^{\infty} a_n \cos \frac{\pi n x}{\ell} + \sum_{n=1}^{\infty} b_n \sin \frac{\pi n x}{\ell},$$

where $\dfrac{\ell a_n}{2} = \displaystyle\int_0^\ell f(t) \cos \frac{\pi n t}{\ell}\, dt, \quad \dfrac{\ell b_n}{2} = \displaystyle\int_0^\ell f(t) \sin \frac{\pi n t}{\ell}\, dt$, *have the same sum when*
$0 \le x \le \ell$; but their sums are numerically equal and opposite in sign when $0 \ge x \ge -\ell$.

The cosine series was given by Clairaut [147] in a memoir dated July 9, 1757; the sine series was obtained between 1762 and 1765 by Lagrange [396, vol. I, p. 553].

Example 9.2.1 Expand $\frac{1}{2}(\pi - x)\sin x$ in a cosine series in the range $0 \le x \le \pi$.
Solution. We have, by the formula just obtained,

$$\frac{1}{2}(\pi - x)\sin x = \frac{1}{2}a_0 + \sum_{n=1}^{\infty} a_n \cos nx,$$

where

$$\frac{1}{2}\pi a_n = \int_0^\pi \frac{1}{2}(\pi - x)\sin x \cos nx\, dx.$$

But, integrating by parts, if $n \ne 1$,

$$\int_0^\pi 2(\pi - x)\sin x \cos nx\, dx = \int_0^\pi (\pi - x)\{\sin(n+1)x - \sin(n-1)x\}\, dx$$
$$= \left[(x - \pi)\left\{\frac{\cos(n+1)x}{n+1} - \frac{\cos(n+1)x}{n-1}\right\}\right]_0^\pi$$
$$- \int_0^\pi \left\{\frac{\cos(n+1)x}{n+1} - \frac{\cos(n-1)x}{n-1}\right\} dx$$
$$= \pi\left(\frac{1}{n+1} - \frac{1}{n-1}\right) = -\frac{2\pi}{(n+1)(n-1)}.$$

Whereas if $n = 1$, we get $\displaystyle\int_0^\pi 2(\pi - x)\sin x \cos x\, dx = \frac{1}{2}\pi$.
Therefore the required series is

$$\frac{1}{2} + \frac{1}{4}\cos x - \frac{1}{1\cdot 3}\cos 2x - \frac{1}{2\cdot 4}\cos 3x - \frac{1}{3\cdot 5}\cos 4x - \cdots$$

It will be observed that it is only for values of x between 0 and π that the sum of this series is proved to be $\frac{1}{2}(\pi - x)\sin x$; thus for instance when x has a value between 0 and $-\pi$, the sum of the series is not $\frac{1}{2}(\pi - x)\sin x$, but $-\frac{1}{2}(\pi + x)\sin x$; when x has a value between π and 2π, the sum of the series happens to be again $\frac{1}{2}(\pi - x)\sin x$, but this is a mere coincidence arising from the special function considered, and does not follow from the general theorem.

Example 9.2.2 Expand $\frac{1}{8}\pi x(\pi - x)$ in a sine series, valid when $0 \le x \le \pi$.
Answer. The series is $\sin x + \dfrac{\sin 3x}{3^3} + \dfrac{\sin 5x}{5^3} + \cdots$.

Example 9.2.3 Shew that, when $0 \leq x \leq \pi$,

$$\frac{1}{96}\pi(\pi - 2x)(\pi^2 + 2\pi x - 2x^2) = \cos x + \frac{\cos 3x}{3^4} + \frac{\cos 5x}{5^4} + \cdots .$$

Hint. Denoting the left-hand side by $f(x)$, we have, on integrating by parts and observing that $f'(0) = f'(\pi) = 0$,

$$\int_0^\pi f(x) \cos nx \, dx = \frac{1}{n} [f(x) \sin nx]_0^\pi - \frac{1}{n} \int_0^\pi f'(x) \sin nx \, dx$$

$$= \frac{1}{n^2} [f'(x) \cos nx]_0^\pi - \frac{1}{n^2} \int_0^\pi f''(x) \cos nx \, dx$$

$$= -\frac{1}{n^3} [f''(x) \sin nx]_0^\pi + \frac{1}{n^3} \int_0^\pi f'''(x) \sin nx \, dx$$

$$= -\frac{1}{n^4} [f'''(x) \cos nx]_0^\pi$$

$$= \frac{\pi}{4n^4} (1 - \cos n\pi).$$

Example 9.2.4 Shew that for values of x between 0 and π, e^{sx} can be expanded in the cosine series

$$\frac{2s}{\pi}(e^{s\pi} - 1)\left(\frac{1}{2s^2} + \frac{\cos 2x}{s^2 + 4} + \frac{\cos 4x}{s^2 + 16} + \cdots\right)$$

$$- \frac{2s}{\pi}(e^{s\pi} + 1)\left(\frac{\cos x}{s^2 + 1} + \frac{\cos 3x}{s^2 + 9} + \cdots\right),$$

and draw graphs of the function e^{sx} and of the sum of the series.

Example 9.2.5 Shew that for values of x between 0 and π, the function $\frac{1}{8}\pi(\pi - 2x)$ can be expanded in the cosine series

$$\cos x + \frac{\cos 3x}{3^2} + \frac{\cos 5x}{5^2} + \cdots,$$

and draw graphs of the function $\frac{1}{8}\pi(\pi - 2x)$ and of the sum of the series.

9.3 The nature of the coefficients in a Fourier series

The analysis of this section and of §9.31 is contained in Stokes' great memoir [608] (reproduced in [611, vol. I, pp. 236–313]).

Suppose that (as in the numerical examples which have been discussed) the interval $(-\pi, \pi)$ can be divided into a finite number of ranges $(-\pi, k_1), (k_1, k_2), \ldots, (k_n, \pi)$ such that throughout each range $f(x)$ and all its differential coefficients are continuous with limited total fluctuation and that they have limits on the right and on the left (§3.2) at the end points of these ranges.

Then

$$\pi a_m = \int_{-\pi}^{k_1} f(t) \cos mt \, dt + \int_{k_1}^{k_2} f(t) \cos mt \, dt + \cdots + \int_{k_n}^{\pi} f(t) \cos mt \, dt.$$

Integrating by parts we get

$$\pi a_m = \left[m^{-1}f(t)\sin mt\right]_{-\pi}^{k_1} + \left[m^{-1}f(t)\sin mt\right]_{k_1}^{k_2} + \cdots + \left[m^{-1}f(t)\sin mt\right]_{k_n}^{\pi}$$

$$- m^{-1}\int_{-\pi}^{k_1} f'(t)\sin mt\, dt - m^{-1}\int_{k_1}^{k_2} f'(t)\sin mt\, dt - \cdots - m^{-1}\int_{k_n}^{\pi} f'(t)\sin mt\, dt,$$

so that $a_m = \dfrac{A_m}{m} - \dfrac{b'_m}{m}$, where $\pi A_m = \sum\limits_{r=1}^{n} \sin mk_r\,[f(k_r - 0) - f(k_r + 0)]$, and b'_m is a Fourier

constant of $f'(x)$. Similarly $b_m = \dfrac{B_m}{m} + \dfrac{a'_m}{m}$, where

$$\pi B_m = -\sum_{r=1}^{n} \cos mk_r\,[f(k_r - 0) - f(k_r + 0)] - \cos m\pi\,[f(\pi - 0) - f(-\pi + 0)],$$

and a'_m is a Fourier constant of $f'(x)$. Similarly, we get

$$a'_m = \frac{A'_m}{m} - \frac{b''_m}{m}, \qquad b'_m = \frac{B'_m}{m} + \frac{a''_m}{m},$$

where a''_m, b''_m are the Fourier constants of $f''(x)$ and

$$\pi A'_m = \sum_{r=1}^{n} \sin mk_r\{f'(k_r - 0) - f'(k_r + 0)\},$$

$$\pi B'_m = -\sum_{r=1}^{n} \cos mk_r\{f'(k_r - 0) - f'(k_r + 0)\} - \cos m\pi\,\{f'(\pi - 0) - f'(-\pi + 0)\}.$$

Therefore

$$a_m = \frac{A_m}{m} - \frac{B'_m}{m^2} - \frac{a''_m}{m^2}, \qquad b_m = \frac{B_m}{m} + \frac{A'_m}{m^2} - \frac{b''_m}{m^2}.$$

Now as $m \to \infty$, we see that $A'_m = O(1)$, $B'_m = O(1)$, and, since the integrands involved in a''_m and b''_m are bounded, it is evident that $a''_m = O(1)$, $b''_m = O(1)$. Hence if $A_m = 0$, $B_m = 0$, the Fourier series for $f(x)$ converges absolutely and uniformly, by §3.34.

The necessary and sufficient conditions that $A_m = B_m = 0$ for all values of m are that

$$f(k_r - 0) = f(k_r + 0), \qquad f(\pi - 0) = f(-\pi + 0),$$

that is to say that[9] $f(x)$ should be continuous for *all* values of x.

9.31 *Differentiation of Fourier series*

The result of differentiating

$$\frac{1}{2}a_0 + \sum_{m=1}^{\infty} (a_m \cos mx + b_m \sin mx)$$

---

[9] Of course $f(x)$ is also subject to the conditions stated at the beginning of the section.

term by term is $\sum\limits_{m=1}^{\infty} \{mb_m \cos mx - ma_m \sin mx\}$. With the notation of §9.3, this is the same as

$$\frac{1}{2}a_0' + \sum_{m=1}^{\infty}(a_m' \cos mx + b_m' \sin mx),$$

provided that $A_m = B_m = 0$ and $\int_{-\pi}^{\pi} f'(x)dx = 0$; these conditions are satisfied if $f(x)$ is continuous for all values of x.

Consequently sufficient conditions for the legitimacy of differentiating a Fourier series term by term are that $f(x)$ should be continuous for *all* values of x and $f'(x)$ should have only a finite number of points of discontinuity in the range $(-\pi, \pi)$, both functions having limited total fluctuation throughout the range.

9.32 Determination of points of discontinuity

The expressions for a_m and b_m which have been found in §9.3 can frequently be applied in practical examples to determine the points at which the sum of a given Fourier series may be discontinuous. Thus, let it be required to determine the places at which the sum of the series

$$\sin x + \frac{1}{3}\sin 3x + \frac{1}{5}\sin 5x + \cdots$$

is discontinuous.

Assuming that the series is a Fourier series and not *any* trigonometrical series and observing that $a_m = 0$, $b_m = (2m)^{-1}(1 - \cos m\pi)$, we get on considering the formula found in §9.3,

$$A_m = 0, \quad B_m = \tfrac{1}{2} - \tfrac{1}{2}\cos m\pi, \quad a_m' = b_m' = 0.$$

Hence if $k_1, k_2, \ldots$ are the places at which the analytic character of the sum is broken, we have

$$0 = \pi A_m = \sin mk_1 \{f(k_1 - 0) - f(k_1 + 0)\} + \sin mk_2\{f(k_2 - 0) - f(k_2 + 0)\} + \cdots .$$

Since this is true for all values of m, the numbers $k_1, k_2, \ldots$ must be multiples of π; but there is only one even multiple of π in the range $-\pi < x \le \pi$, namely zero. So $k_1 = 0$, and $k_2, k_3, \ldots$ do not exist. Substituting $k_1 = 0$ in the equation $B_m = \tfrac{1}{2} - \tfrac{1}{2}\cos m\pi$, we have

$$\pi \left(\tfrac{1}{2} - \tfrac{1}{2}\cos m\pi\right) = -[\cos m\pi\{f(\pi - 0) - f(-\pi + 0)\} + f(-0) - f(+0)].$$

Since this is true for all values of m, we have

$$\tfrac{1}{2}\pi = f(+0) - f(-0), \quad \tfrac{1}{2}\pi = f(\pi - 0) - f(-\pi + 0).$$

This shews that, if the series is a Fourier series, $f(x)$ has discontinuities at the points $n\pi$ (n any integer) and since $a_m' = b_m' = 0$, we should expect to be constant in the open range $(-\pi, 0)$ and to be another constant in the open range $(0, \pi)$. In point of fact $f(x) = -\pi/4$ ($-\pi < x < 0$) and $\pi/4$ ($0 < x < \pi$).

9.4 Fejér's theorem

We now begin the discussion of the theory of Fourier series by proving the following theorem, due to Fejér [210], concerning the summability of the Fourier series associated with an arbitrary function $f(t)$:

Let $f(t)$ be a function of the real variable t, defined arbitrarily when $-\pi \leq t < \pi$, and defined by the equation $f(t + 2\pi) = f(t)$ for all other real values of t; and let $\int_{-\pi}^{\pi} f(t) \, dt$ exist and (if it is an improper integral) let it be absolutely convergent. Then the Fourier series associated with the function $f(t)$ is summable (C1) at all points x at which, the two limits $f(x \pm 0)$ exist. (See §8.43.) And its sum (C1) is

$$\frac{1}{2}\{f(x + 0) + f(x - 0)\}.$$

Let a_n, b_n, $(n = 0, 1, 2, \ldots)$ denote the Fourier constants (§9.2) of $f(t)$ and let

$$\frac{1}{2}a_0 = A_0, \qquad a_n \cos nx + b_n \sin nx = A_n(x), \qquad \sum_{n=0}^{m} A_n(x) = S_m(x).$$

Then we have to prove that

$$\lim_{m \to \infty} \frac{1}{m}\{A_0 + S_1(x) + S_2(x) + \cdots + S_{m-1}(x)\} = \frac{1}{2}\{f(x + 0) + f(x - 0)\},$$

provided that the limits on the right exist.

If we substitute for the Fourier constants their values in the form of integrals (§9.2), it is easy to verify that[10]

$$A_0 + \sum_{n=1}^{m-1} S_n(x) = mA_0 + (m - 1)A_1(x) + (m - 2)A_2(x) + \cdots + A_{m-1}(x)$$

$$= \frac{1}{\pi} \int_{-\pi}^{\pi} \left[\tfrac{1}{2}m + (m - 1)\cos(x - t) + (m - 2)\cos 2(x - t)\right.$$

$$\left. + \cdots + \cos(m - 1)(x - t)\right] f(t)dt$$

$$= \frac{1}{2\pi} \int_{-\pi}^{\pi} \frac{\sin^2 \frac{1}{2}m(x - t)}{\sin^2 \frac{1}{2}(x - t)} f(t)dt$$

$$= \frac{1}{2\pi} \int_{-\pi+x}^{\pi+x} \frac{\sin^2 \frac{1}{2}m(x - t)}{\sin^2 \frac{1}{2}(x - t)} f(t)dt,$$

the last step following from the periodicity of the integrand.

If now we bisect the path of integration and write $x \mp 2\theta$ in place of t in the two parts of

---

[10] It is obvious that, if we write λ for $e^t(x - t)$ in the second line, then

$$m + (m - 1)(\lambda + \lambda^{-1}) + (m - 2)(\lambda^2 + \lambda^{-2}) + \cdots + (\lambda^{m-1} + \lambda^{1-m})$$

$$= (1 - \lambda)^{-1}\{\lambda^{1-m} + \lambda^{2-m} + \cdots + \lambda^{-1} + 1 - \lambda - \lambda^2 - \cdots - \lambda^m\}$$

$$= (1 - \lambda)^{-2}\{\lambda^{1-m} - 2\lambda + \lambda^{m+1}\} = (\lambda^{\frac{1}{2}m} - \lambda^{-\frac{1}{2}m})^2/(\lambda^{\frac{1}{2}} - \lambda^{-\frac{1}{2}})^2.$$

the path, we get

$$A_0 + \sum_{n=1}^{m-1} S_n(x) = \frac{1}{\pi} \int_0^{\pi/2} \frac{\sin^2 m\theta}{\sin^2 \theta} f(x+2\theta)d\theta + \frac{1}{\pi} \int_0^{\pi/2} \frac{\sin^2 m\theta}{\sin^2 \theta} f(x-2\theta)d\theta.$$

Consequently it is sufficient to prove that, as $m \to \infty$, then

$$\frac{1}{m} \int_0^{\pi/2} \frac{\sin^2 m\theta}{\sin^2 \theta} f(x+2\theta)d\theta \to \frac{\pi}{2} f(x+0),$$

$$\frac{1}{m} \int_0^{\pi/2} \frac{\sin^2 m\theta}{\sin^2 \theta} f(x-2\theta)d\theta \to \frac{\pi}{2} f(x-0).$$

Now, if we integrate the equation

$$\frac{1}{2} \frac{\sin^2 m\theta}{\sin^2 \theta} = \frac{1}{2}m + (m-1)\cos 2\theta + \cdots + \cos 2(m-1)\theta,$$

we find that

$$\int_0^{\pi/2} \frac{\sin^2 m\theta}{\sin^2 \theta} d\theta = \frac{\pi m}{2},$$

and so we have to prove that

$$\frac{1}{m} \int_0^{\pi/2} \frac{\sin^2 m\theta}{\sin^2 \theta} \phi(\theta)d\theta \to 0 \quad \text{as } m \to \infty,$$

where $\phi(\theta)$ stands in turn for each of the two functions

$$f(x+2\theta) - f(x+0), \qquad f(x-2\theta) - f(x-0).$$

Now, given an arbitrary positive number ε, we can choose δ so that $|\phi(\theta)| < \varepsilon$ whenever $0 < \theta \le \frac{1}{2}\delta$, (on the assumption that $f(x\pm 0)$ exist). This choice of δ is obviously independent of m. Then

$$\left| \frac{1}{m} \int_0^{\pi/2} \frac{\sin^2 m\theta}{\sin^2 \theta} \phi(\theta)\, d\theta \right| \le \frac{1}{m} \int_0^{\delta/2} \frac{\sin^2 m\theta}{\sin^2 \theta} |\phi(\theta)|\, d\theta + \frac{1}{m} \int_{\delta/2}^{\pi/2} \frac{\sin^2 m\theta}{\sin^2 \theta} |\phi(\theta)|\, d\theta$$

$$< \frac{\varepsilon}{m} \int_0^{\delta/2} \frac{\sin^2 m\theta}{\sin^2 \theta}\, d\theta + \frac{1}{m \sin^2 (\delta/2)} \int_{\delta/2}^{\pi/2} |\phi(\theta)|\, d\theta$$

$$\le \frac{\varepsilon}{m} \int_0^{\pi/2} \frac{\sin^2 m\theta}{\sin^2 \theta}\, d\theta + \frac{1}{m \sin^2 (\delta/2)} \int_0^{\pi/2} |\phi(\theta)|\, d\theta$$

$$= \frac{\pi \varepsilon}{2} + \frac{1}{m \sin^2 (\delta/2)} \int_0^{\pi/2} |\phi(\theta)|\, d\theta.$$

Now the convergence of $\int_{-\pi}^{\pi} |f(t)|\, dt$ entails the convergence of $\int_0^{\pi/2} |\phi(\theta)|\, d\theta$, and so, given ε (and therefore δ), we can make

$$\frac{\pi m}{2} \varepsilon \sin^2 \frac{\delta}{2} > \int_0^{\pi/2} |\phi(\theta)|\, d\theta,$$

by taking m sufficiently large. Hence, by taking m sufficiently large, we can make

$$\left| \frac{1}{m} \int_0^{\pi/2} \frac{\sin^2 m\theta}{\sin^2 \theta} \phi(\theta) \, d\theta \right| < \pi\varepsilon,$$

where ε is an arbitrary positive number; that is to say, from the definition of a limit,

$$\lim_{m \to \infty} \frac{1}{m} \int_0^{\pi/2} \frac{\sin^2 m\theta}{\sin^2 \theta} \phi(\theta) \, d\theta = 0,$$

and so Fejér's theorem is established.

Corollary 9.4.1   *Let U and L be the upper and lower bounds of $f(t)$ in any interval (a, b) whose length does not exceed 2π, and let*

$$\int_{-\pi}^{\pi} |f(t)| \, dt = \pi A.$$

Then, if $a + \eta \leq x \leq b - \eta$, where η is any positive number, we have

$$U - \frac{1}{m} \left\{ A_0 + \sum_{n=1}^{m-1} S_n(x) \right\}$$

$$= \frac{1}{2m\pi} \left\{ \int_{-\pi+x}^{x-\eta} + \int_{x-\eta}^{x+\eta} + \int_{x+\eta}^{x+\eta} \right\} \frac{\sin^2 \frac{1}{2} m(x-t)}{\sin^2 \frac{1}{2}(x-t)} \{U - f(t)\} \, dt$$

$$\geq \frac{1}{2m\pi} \left\{ \int_{-\pi+x}^{x-\eta} + \int_{x+\eta}^{\pi+x} \right\} \frac{\sin^2 \frac{1}{2} m(x-t)}{\sin^2 \frac{1}{2}(x-t)} \{U - f(t)\} \, dt$$

$$\geq -\frac{1}{2m\pi} \left\{ \int_{-\pi+x}^{x-\eta} + \int_{x+\eta}^{\pi+x} \right\} \frac{|U| + |f(t)|}{\sin^2 \frac{1}{2}\eta} \, dt,$$

so that

$$\frac{1}{m} \left[A_0 + \sum_{n=1}^{m-1} S_n(x) \right] \leq U + \frac{|U| + \frac{1}{2} A}{m \sin^2(\eta/2)}.$$

Similarly

$$\frac{1}{m} \left[A_0 + \sum_{n=1}^{m-1} S_n(x) \right] \geq L - \frac{|L| + \frac{1}{2} A}{m \sin^2(\eta/2)}.$$

Corollary 9.4.2   *Let $f(t)$ be continuous in the interval $a \leq t \leq b$. Since continuity implies uniformity of continuity (§3.61), the choice of δ corresponding to any value of x in (a, b) is independent of x, and the upper bound of $|f(x \pm 0)|$, i.e. of $|f(x)|$, is also independent of x, so that*

$$\int_0^{\pi/2} |\phi(\theta)| \, d\theta = \int_0^{\pi/2} |f(x \pm 2\theta) - f(x \pm 0)| \, d\theta$$

$$\leq \frac{1}{2} \int_{-\pi}^{\pi} |f(t)| \, dt + \frac{1}{2} \pi |f(x \pm 0)|,$$

and the upper bound of the last expression is independent of x.

Hence the choice of m, which makes

$$\left| \frac{1}{m} \int_0^{\pi/2} \frac{\sin^2 m\theta}{\sin^2 \theta} \phi(\theta) \, d\theta \right| < \pi\varepsilon,$$

is independent of x, and consequently $\dfrac{1}{m} \left\{ A_0 + \displaystyle\sum_{n=1}^{m-1} S_n(x) \right\}$ *tends to the limit* $f(x)$, *as* $m \to \infty$,
uniformly throughout the interval $a \leq x \leq b$.

9.41 The Riemann–Lebesgue lemmas

In order to be able to apply Hardy's theorem; (§8.5) to deduce the convergence of Fourier series from Fejér's theorem, we need the two following lemmas:

(I) *Let* $\displaystyle\int_a^b \psi(\theta) \, d\theta$ *exist and (if it is an improper integral) let it be absolutely convergent.*

Then, as $\lambda \to \infty$, $\displaystyle\int_a^b \psi(\theta) \sin(\lambda\theta) \, d\theta$ *is* $O(1)$.

(II) *If, further,* $\psi(\theta)$ *has limited total fluctuation in the range* (a, b) *then, as* $\lambda \to \infty$,

$$\int_a^b \psi(\theta) \sin(\lambda\theta) \, d\theta \text{ is } \quad O(1/\lambda).$$

Of these results (I) was stated by W. R. Hamilton [270] and by Riemann [558, p. 241]. For Lebesgue's investigation see his [419, ch. III] in the case of bounded functions. The truth of (II) seems to have been well known before its importance was realised; it is a generalisation of a result established by Dirksen [180] and Stokes [608] (§9.3) in the case of functions with a continuous differential coefficient. The reader should observe that the analysis of this section remains valid when the sines are replaced throughout by cosines.

(I) It is convenient[11] to establish this lemma first in the case in which $\psi(\theta)$ is bounded in the range (a, b). In this case, let K be the upper bound of $|\psi(\theta)|$, and let ε be an arbitrary positive number. Divide the range (a, b) into n parts by the points $x_1, x_2, \ldots, x_{n-1}$, and form the sums S_n, s_n associated with the function $\psi(\theta)$ after the manner of §4.1. Take n so large that $S_n - s_n < \varepsilon$; this is possible since $\psi(\theta)$ is integrable.

In the interval (x_{r-1}, x_r) write $\psi(\theta) = \psi_r(x_{r-1}) + \omega_r(\theta)$, so that $|\omega_r(\theta)| \leq U_r - L_r$, where U_r and L_r are the upper and lower bounds of $\psi(\theta)$ in the interval (x_{r-1}, x_r). It is then clear

---

[11] For this proof we are indebted to Mr Hardy; it seems to be neater than the proofs given by other writers, e.g. de la Vallée Poussin [639, pp. 140–141].

that

$$\left| \int_a^b \psi(\theta) \, \sin(\lambda\theta) \, d\theta \right| = \left| \sum_{r=1}^n \psi_r(x_{r-1}) \int_{x_{r-1}}^{x_r} \sin(\lambda\theta) \, d\theta + \sum_{r=1}^n \int_{x_{r-1}}^n \omega_r(\theta) \sin(\lambda\theta) \, d\theta \right|$$

$$\leq \sum_{r=1}^n |\psi_r(x_{r-1})| \left| \int_{x_{r-1}}^{x_r} \sin(\lambda\theta) \, d\theta \right| + \sum_{r=1}^n \int_{x_{r-1}}^{x_r} |\omega_r(\theta)| \, d\theta$$

$$\leq nK \cdot (2/\lambda) + (S_n - s_n)$$

$$< (2nK/\lambda) + \varepsilon.$$

By taking λ large (n remaining fixed after ε has been chosen), the last expression may be made less than 2ε so that

$$\lim_{\lambda \to \infty} \int_a^b \psi(\theta) \sin(\lambda\theta) \, d\theta = 0,$$

and this is the result stated.

When $\psi(\theta)$ is unbounded, if it has an absolutely convergent integral, by §4.5, we may enclose the points at which it is unbounded in a finite number of intervals $\delta_1, \delta_2, \ldots, \delta_p$ (the *finiteness* of the number of intervals is assumed in the definition of an improper integral, §4.5) such that

$$\sum_{r=1}^p \int_{\delta_r} |\psi(\theta)| \, d\theta < \varepsilon.$$

If K denotes the upper bound of $|\psi(\theta)|$ for values of θ outside these intervals, and if $\gamma_1, \gamma_2, \ldots, \gamma_{p+1}$ denote the portions of the interval (a, b) which do not belong to $\delta_1, \delta_2, \ldots, \delta_p$ we may prove as before that

$$\left| \int_a^b \psi(\theta) \sin(\lambda\theta) \, d\theta \right| = \left| \sum_{r=1}^{p+1} \int_{\gamma_r} \psi(\theta) \sin(\lambda\theta) \, d\theta + \sum_{r=1}^p \int_{\delta_r} \psi(\theta) \sin(\lambda\theta) \, d\theta \right|$$

$$\leq \left| \sum_{r=1}^{p+1} \int_{\gamma_r} \psi(\theta) \sin(\lambda\theta) \, d\theta \right| + \sum_{r=1}^p \int_{\delta_r} |\psi(\theta) \sin(\lambda\theta)| \, d\theta$$

$$< (2nK/\lambda) + 2\varepsilon.$$

Now the choice of ε fixes n and K, so that the last expression may be made less than 3ε by taking λ sufficiently large. That is to say that, even if $\psi(\theta)$ be unbounded,

$$\lim_{\lambda \to \infty} \int_a^b \psi(\theta) \sin(\lambda\theta) \, d\theta = 0,$$

provided that $\psi(\theta)$ has an (improper) integral which is absolutely convergent. The first lemma is therefore completely proved.

(II) When $\psi(\theta)$ has limited total fluctuation in the range (a, b), by Example 3.6.2, we may write $\psi(\theta) = \chi_1(\theta) - \chi_2(\theta)$, where $\chi_1(\theta)$, $\chi_2(\theta)$ are positive increasing bounded functions. Then, by the second mean-value theorem (§4.14) a number ξ exists such that $a \leq \xi \leq b$

and

$$\left| \int_a^b \chi_1(\theta) \sin(\lambda\theta) \, d\theta \right| = \left| \chi_1(b) \int_\xi^b \sin(\lambda\theta) \, d\theta \right| \le 2\chi_1(b)/\lambda.$$

If we treat $\chi_2(\theta)$ in a similar manner, it follows that

$$\left| \int_a^b \psi(\theta) \sin(\lambda\theta) \, d\theta \right| \le \left| \int_a^b \chi_1(\theta) \sin(\lambda\theta) \, d\theta \right| + \left| \int_a^b \chi_2(\theta) \sin(\lambda\theta) \, d\theta \right|$$

$$\le 2(\chi_1(b) + \chi_2(b))/\lambda$$

$$= O(1/\lambda),$$

and the second lemma is established.

Corollary 9.4.3 *If $f(t)$ be such that $\displaystyle\int_{-\pi}^{\pi} f(t)$ exists and is an absolutely convergent integral, the Fourier constants a_n, b_n of $f(t)$ are $o(1)$ as $n \to \infty$; and if, further, $f(t)$ has limited total fluctuation in the range $(-\pi, \pi)$, the Fourier constants are $O(1/n)$.*

Note Of course these results are not sufficient to ensure the convergence of the Fourier series associated with $f(t)$; for a series, in which the terms are of the order of magnitude of the terms in the harmonic series (§2.3), is not necessarily convergent.

9.42 The proof of Fourier's theorem

We shall now prove the theorem enunciated in §9.2, namely:

Let $f(t)$ be a function defined arbitrarily when $-\pi \le t < \pi$, and defined by the equation $f(t+2\pi) = f(t)$ for all other real values of t; and let $\displaystyle\int_{-\pi}^{\pi} f(t) \, dt$ exist and (if it is an improper integral) let it be absolutely convergent. Let a_n, b_n be defined by the equations

$$\pi a_n = \int_{-\pi}^{\pi} f(t) \cos nt \, dt, \quad \pi b_n = \int_{-\pi}^{\pi} f(t) \sin nt \, dt.$$

Then, if x be an interior point of any interval (a,b) within which $f(t)$ has limited total fluctuation, the series

$$\frac{1}{2}a_0 + \sum_{n=1}^{\infty} a_n \cos nx + b_n \sin nx$$

is convergent and its sum is $\frac{1}{2}(f(x+0) + f(x-0))$.

It is convenient to give two proofs, one applicable to functions for which it is permissible to take the interval (a,b) to be the interval $(-\pi + x, \pi + x)$, the other applicable to functions for which it is not permissible.

(I) When the interval (a,b) may be taken to be $(-\pi + x, \pi + x)$, it follows from §9.41(II) that $a_n \cos nx + b_n \sin nx$ is as $O(1/n)$ as $n \to \infty$. Now by Fejér's theorem (§9.4) the series under consideration is summable (C1) and its sum (C1) is $\frac{1}{2}(f(x+0) + f(x-0))$. (The limits $f(x \pm 0)$ exist, by Example 3.6.3.)

Therefore, by Hardy's convergence theorem (§8.5), the series under consideration is *convergent* and its sum (by §8.43) is $\frac{1}{2}(f(x+0) + f(x-0))$.

(II) Even if it is not permissible to take the interval (a, b) to be the whole interval $(-\pi+x, \pi+x)$, it is possible, by hypothesis, to choose a positive number δ, less than π, such that $f(t)$ has limited total fluctuation in the interval $(x - \delta, x + \delta)$. We now define an auxiliary function $g(t)$, which is equal to $f(t)$ when $x - \delta \leq t \leq x + \delta$, and which is equal to zero throughout the rest of the interval $(-\pi + x, \pi + x)$; and $g(t + 2\pi)$ is to be equal to $g(t)$ for all real values of t.

Then $g(t)$ satisfies the conditions postulated for the functions under consideration in (I), namely that it has an integral which is absolutely convergent and it has limited total fluctuation in the interval $(-\pi + x, \pi + x)$; and so, if $a_n^{(1)}$, and $b_n^{(1)}$ denote the Fourier constants of $g(t)$, the arguments used in (I) prove that the Fourier series associated with $g(t)$, namely

$$\tfrac{1}{2}a_0^{(1)} + \sum_{n=1}^{\infty}(a_n^{(1)} \cos nx + b_n^{(1)} \sin nx),$$

is convergent and has the sum $\tfrac{1}{2}(g(x + 0) + g(x - 0))$, and this is equal to

$$\tfrac{1}{2}(f(x + 0) + f(x - 0)).$$

Now let $S_m(x)$ and $S_m^{(1)}(x)$ denote the sums of the first $m + 1$ terms of the Fourier series associated with $f(t)$ and $g(t)$ respectively. Then it is easily seen that

$$S_m(x) = \frac{1}{\pi} \int_{-\pi}^{\pi} \left\{ \frac{1}{2} + \cos(x - t) + \cos 2(x - t) + \cdots + \cos m(x - t) \right\} f(t)\, dt$$

$$= \frac{1}{2\pi} \int_{-\pi}^{\pi} \frac{\sin\left(m + \tfrac{1}{2}\right)(x - t)}{\sin \tfrac{1}{2}(x - t)} f(t)\, dt$$

$$= \frac{1}{2\pi} \int_{-\pi+x}^{\pi+x} \frac{\sin\left(m + \tfrac{1}{2}\right)(x - t)}{\sin \tfrac{1}{2}(x - t)} f(t)\, dt$$

$$= \frac{1}{\pi} \int_0^{\pi/2} \frac{\sin(2m + 1)\theta}{\sin \theta} f(x + 2\theta)\, d\theta + \frac{1}{\pi} \int_0^{\pi/2} \frac{\sin(2m + 1)\theta}{\sin \theta} f(x - 2\theta)\, d\theta,$$

by steps analogous to those given in §9.4.

In like manner

$$S_m^{(1)}(x) = \frac{1}{\pi} \int_0^{\pi/2} \frac{\sin(2m + 1)\theta}{\sin \theta} g(x + 2\theta)\, d\theta + \frac{1}{\pi} \int_0^{\pi/2} \frac{\sin(2m + 1)\theta}{\sin \theta} g(x - 2\theta)\, d\theta,$$

and so, using the definition of $g(t)$, we have

$$S_m(x) - S_m^{(1)}(x) = \frac{1}{\pi} \int_{\delta/2}^{\pi/2} \frac{\sin(2m + 1)\theta}{\sin \theta} f(x + 2\theta)\, d\theta$$

$$+ \frac{1}{\pi} \int_{\delta/2}^{\pi/2} \frac{\sin(2m + 1)\theta}{\sin \theta} f(x - 2\theta)\, d\theta.$$

Since $\operatorname{cosec} \theta$ is a continuous function in the range $\left(\tfrac{1}{2}\delta, \tfrac{1}{2}\pi\right)$, it follows that $f(x \pm 2\theta)\operatorname{cosec}\theta$ are integrable functions with absolutely convergent integrals; and so, by the Riemann–Lebesgue lemma of §9.41(I), *both the integrals on the right in the last equation tend to zero as $m \to \infty$*. That is to say $\lim\limits_{m \to \infty}\left(S_m(x) - S_m^{(1)}(x)\right) = 0$. Hence, since $\lim\limits_{m \to \infty} S_m^{(1)}(x) =$

$\frac{1}{2}\{f(x+0) + f(x-0)\}$, it follows also that

$$\lim_{m\to\infty} S_m(x) = \frac{1}{2}\left(f(x+0) + f(x-0)\right).$$

We have therefore proved that the Fourier series associated with $f(t)$, namely

$$\tfrac{1}{2}a_0 + \sum(a_n \cos nx + b_n \sin nx),$$

is convergent and its sum is $\frac{1}{2}\{f(x+0) + f(x-0)\}$.

9.43 The Dirichlet–Bonnet proof of Fourier's theorem

It is of some interest to prove directly the theorem of §9.42, without making use of the theory of summability; accordingly we now give a proof which is on the same general lines as the proofs due to Dirichlet and Bonnet.

As usual we denote the sum of the first $m+1$ terms of the Fourier series by $S_m(x)$, and then, by the analysis of §9.42, we have

$$S_m(x) = \frac{1}{\pi}\int_0^{\pi/2} \frac{\sin(2m+1)\theta}{\sin\theta} f(x+2\theta)\,d\theta + \frac{1}{\pi}\int_0^{\pi/2} \frac{\sin(2m+1)\theta}{\sin\theta} f(x-2\theta)\,d\theta.$$

Again, on integrating the equation

$$\frac{\sin(2m+1)\theta}{\sin\theta} = 1 + 2\cos 2\theta + 2\cos 4\theta + \cdots + 2\cos 2m\theta,$$

we have

$$\int_0^{\pi/2} \frac{\sin(2m+1)\theta}{\sin\theta}\,d\theta = \frac{\pi}{2},$$

so that

$$S_m(x) - \frac{1}{2}\{f(x+0) + f(x-0)\}$$
$$= \frac{1}{\pi}\int_0^{\pi/2} \frac{\sin(2m+1)\theta}{\sin\theta}\{f(x+2\theta) - f(x+0)\}\,d\theta$$
$$+ \frac{1}{\pi}\int_0^{\pi/2} \frac{\sin(2m+1)\theta}{\sin\theta}\{f(x-2\theta) - f(x-0)\}\,d\theta.$$

In order to prove that

$$\lim_{m\to\infty} S_m(x) = \tfrac{1}{2}\left(f(x+0) + f(x-0)\right),$$

it is therefore sufficient to prove that

$$\lim_{m\to\infty}\int_0^{\pi/2} \frac{\sin(2m+1)\theta}{\sin\theta}\phi(\theta)\,d\theta = 0,$$

where $\phi(\theta)$ stands in turn for each of the functions

$$f(x+2\theta) - f(x+0), \qquad f(x-2\theta) - f(x-0).$$

Now, by Example 3.6.4 $\theta\,\phi(\theta)/\sin\theta$ is a function with limited total fluctuation in an interval of which $\theta = 0$ is an end-point[12]; and so we may write

$$\frac{\theta\,\phi(\theta)}{\sin\theta} = \chi_1(\theta) - \chi_2(\theta),$$

where $\chi_1(\theta)$, $\chi_2(\theta)$ are bounded positive increasing functions of θ such that

$$\chi_1(+0) + \chi_2(+0) = 0.$$

Hence, given an arbitrary positive number ε, we can choose a positive number δ such that $0 \le \chi_1(\theta) < \varepsilon$, $0 \le \chi_2(\theta) < \varepsilon$ whenever $0 \le \theta \le \delta/2$.

We now obtain inequalities satisfied by the three integrals on the right of the obvious equation

$$\int_0^{\pi/2} \frac{\sin(2m+1)\theta}{\sin\theta}\phi(\theta)\,d\theta = \int_{\delta/2}^{\pi/2} \frac{\sin(2m+1)\theta}{\sin\theta}\phi(\theta)\,d\theta$$
$$+ \int_0^{\delta/2} \frac{\sin(2m+1)\theta}{\theta}\chi_1(\theta)\,d\theta - \int_0^{\delta/2} \frac{\sin(2m+1)\theta}{\theta}\chi_2(\theta)\,d\theta.$$

The modulus of the first integral can be made less than ε by taking m sufficiently large; this follows from §9.41(I) since $\phi(\theta)/\sin\theta$ has an integral which converges absolutely in the interval $(\frac{1}{2}\delta, \frac{1}{2}\pi)$.

Next, from the second mean-value theorem, it follows that there is a number ξ between 0 and δ such that

$$\left| \int_0^{\delta/2} \frac{\sin(2m+1)\theta}{\theta}\chi_1(\theta)\,d\theta \right| = \left| \chi_1\left(\frac{\delta}{2}\right) \int_{\xi/2}^{\delta/2} \frac{\sin(2m+1)\theta}{\theta}\,d\theta \right|$$
$$= \chi_1\left(\frac{\delta}{2}\right) \left| \int_{(m+\frac{1}{2})\xi}^{(m+\frac{1}{2})\delta} \frac{\sin u}{u}\,du \right|.$$

Since $\displaystyle\int^{\infty} \frac{\sin t}{t}\,dt$ is convergent, it follows that $\displaystyle\left| \int_{\beta}^{\infty} \frac{\sin u}{u}\,du \right|$ has an upper bound[13] B which is independent of β, and it is then clear that

$$\left| \int_0^{\delta/2} \frac{\sin(2m+1)\theta}{\theta}\chi_1(\theta)\,d\theta \right| \le 2B\chi_1\left(\frac{\delta}{2}\right) < 2B\varepsilon.$$

On treating the third integral in a similar manner, we see that we can make

$$\left| \int_0^{\pi/2} \frac{\sin(2m+1)\theta}{\sin\theta}\phi(\theta)\,d\theta \right| < (4B+1)\varepsilon$$

by taking m sufficiently large; *and so we have proved that*

$$\lim_{m\to\infty} \int_0^{\pi/2} \frac{\sin(2m+1)\theta}{\sin\theta}\phi(\theta)\,d\theta = 0.$$

---

[12]   The other end-point is $\theta = \frac{1}{2}(b-x)$ or $\theta = \frac{1}{2}(x-a)$, according as $\phi(\theta)$ represents one or other of the two functions.

[13]   The reader will find it interesting to prove that $B = \displaystyle\int_0^{\infty} \frac{\sin u}{u}\,du = \frac{\pi}{2}$.

But it has been seen that this is a sufficient condition for the limit of $S_m(x)$ to be $\frac{1}{2}(f(x+0) + f(x-0))$; and we have therefore established the convergence of a Fourier series in the circumstances enunciated in §9.42.

Note The reader should observe that in either proof of the convergence of a Fourier series the *second* mean-value theorem is required; but to prove the summability of the series, the *first* mean-value theorem is adequate. It should also be observed that, while restrictions are laid upon $f(t)$ throughout the range $(-\pi, \pi)$ in establishing the *summability* at any point x, the only additional restriction necessary to ensure *convergence* is a restriction on the behaviour of the function in the *immediate neighbourhood* of the point x. The fact that the convergence depends only on the behaviour of the function in the immediate neighbourhood of x (provided that the function has an integral which is absolutely convergent) was noticed by Riemann and has been emphasised by Lebesgue [418, p. 60].

It is obvious that the condition (due to Jordan [360]) that x should be an interior point of an interval in which $f(t)$ has limited total fluctuation is merely a *sufficient* condition for the convergence of the Fourier series; and it may be replaced by any condition which makes

$$\lim_{m \to \infty} \int_0^{\pi/2} \frac{\sin(2m+1)\theta}{\sin \theta} \phi(\theta) \, d\theta = 0.$$

Jordan's condition is, however, a natural modification of the Dirichlet condition that the function $f(t)$ should have only a finite number of maxima and minima, and it does not increase the difficulty of the proof.

Another condition with the same effect is due to Dini [170], namely that, if

$$\Phi(\theta) = \frac{1}{\theta} [f(x+2\theta) + f(x-2\theta) - f(x+0) - f(x-0)],$$

then $\int_0^a \Phi(\theta) d\theta$ should converge absolutely for some positive value of a. If the condition is satisfied, given ε we can find δ so that $\int_0^{\delta/2} |\Phi(\theta)| \, d\theta < \varepsilon$, and then

$$\left| \int_0^{\delta/2} \frac{\sin(2m+1)\theta}{\sin \theta} \theta \Phi(\theta) d\theta \right| < \frac{\pi \varepsilon}{2};$$

the proof that $\left| \int_{\delta/2}^{\pi/2} \frac{\sin(2m+1)\theta}{\sin \theta} \phi(\theta) d\theta \right| < \varepsilon$ for sufficiently large values of m follows from the Riemann–Lebesgue lemma.

A more stringent condition than Dini's is due to Lipschitz [440], namely $|\phi(\theta)| < C\theta^k$, where C and k are positive and independent of θ. For other conditions due to Lebesgue and to de la Vallée Poussin, see the latter's [639, II, pp. 149–150]. It should be noticed that Jordan's condition differs in character from Dini's condition; the latter is a condition that the series may converge *at a point*, the former that the series may converge *throughout an interval*.

9.44 The uniformity of the convergence of Fourier series

Let $f(t)$ satisfy the conditions enunciated in §9.42, and further let it be *continuous* (in addition to having limited total fluctuation) in an interval (a, b). Then the Fourier series associated with $f(t)$ converges uniformly *to the sum* $f(x)$ *at all points* x *for which* $a + \delta \leq x \leq b - \delta$, *where* δ *is any positive number.*

Let $h(t)$ be an auxiliary function defined to be equal to $f(t)$ when $a \leq t \leq b$ and equal to zero for other values of t in the range $(-\pi, \pi)$, and let a_n, b_n denote the Fourier constants of $h(t)$. Also let $S_m^{(2)}(x)$ denote the sum of the first $m + 1$ terms of the Fourier series associated with $h(t)$.

Then, by Corollary 9.4.2, it follows that $\frac{1}{2}a_0 + \sum\limits_{n=1}^{\infty}(a_n \cos nx + b_n \sin nx)$ is *uniformly summable throughout the interval* $(a + \delta, b - \delta)$; and since

$$|a_n \cos nx + b_n \sin nx| \leq (a_n^2 + b_n^2)^{1/2},$$

which is independent of x and which, by §9.41(II), is $O(1/n)$, it follows from Corollary 8.5.1 that

$$\frac{1}{2}a_0 + \sum_{n=1}^{\infty}(a_n \cos nx + b_n \sin nx)$$

converges uniformly to the sum $h(x)$, which is equal to $f(x)$.

Now, as in §9.42,

$$S_m(x) - S_m^{(2)}(x) = \frac{1}{\pi}\int_{\frac{1}{2}(b-x)}^{\pi/2} \frac{\sin(2m+1)\theta}{\sin\theta} f(x + 2\theta)\, d\theta$$

$$+ \frac{1}{\pi}\int_{\frac{1}{2}(x-a)}^{\pi/2} \frac{\sin(2m+1)\theta}{\sin\theta} f(x - 2\theta)\, d\theta.$$

As in §9.41 we choose an arbitrary positive number ε and then enclose the points at which $f(t)$ is unbounded in a set of intervals $\delta_1, \delta_2, \ldots, \delta_p$ such that

$$\sum_{r=1}^{p}\int_{\delta_r} |f(t)|\, dt < \varepsilon. \tag{9.1}$$

If K be the upper bound of $|f(t)|$ outside these intervals, we then have, as in §9.41,

$$|S_m(x) - S_m^{(2)}(x)| < \left(\frac{2nK}{2m+1} + 2\varepsilon\right)\operatorname{cosec}\delta,$$

where the choice of n depends only on a and b and the form of the function $f(t)$. Hence, by a choice of m *independent* of x we can make $|S_m(x) - S_m^{(2)}(x)|$ arbitrarily small; so that $S_m(x) - S_m^{(2)}(x)$ tends uniformly to zero. Since $S_m^{(2)}(x) \to f(x)$ uniformly, it is then obvious that $S_m(x) \to f(x)$ uniformly; and this is the result to be proved.

Note It must be observed that no general statement can be made about uniformity or absoluteness of convergence of Fourier series. Thus the series of Example 9.1.1 converges uniformly except near $x = (2n + 1)\pi$ but converges absolutely only when $x = n\pi$, whereas the series of Example 9.1.2 converges uniformly and absolutely for all real values of x.

Example 9.4.1   If $\phi(\theta)$ satisfies suitable conditions in the range $(0, \pi)$, shew that

$$\lim_{m \to \infty} \int_0^\pi \frac{\sin(2m+1)\theta}{\sin \theta} \phi(\theta)\, d\theta = \lim_{m \to \infty} \int_0^{\pi/2} \frac{\sin(2m+1)\theta}{\sin \theta} \phi(\theta)\, d\theta$$

$$+ \lim_{m \to \infty} \int_0^{\pi/2} \frac{\sin(2m+1)\theta}{\sin \theta} \phi(\pi - \theta)\, d\theta$$

$$= \frac{\pi}{2}\left(\phi(+0) + \phi(\pi - 0)\right).$$

Example 9.4.2 (Math. Trip. 1894)   Prove that, if $a > 0$,

$$\lim_{n \to \infty} \int_0^\infty \frac{\sin(2n+1)\theta}{\sin \theta} e^{-a\theta}\, d\theta = \frac{\pi}{2} \coth \frac{\pi a}{2}.$$

Hint. Shew that

$$\int_0^\infty \frac{\sin(2n+1)\theta}{\sin \theta} e^{-a\theta}\, d\theta = \lim_{m \to \infty} \int_0^{m\pi} \frac{\sin(2n+1)\theta}{\sin \theta} e^{-a\theta}\, d\theta$$

$$= \lim_{m \to \infty} \int_0^\pi \frac{\sin(2n+1)\theta}{\sin \theta}\left\{ e^{-a\theta} + e^{-a(\theta+\pi)} + \cdots + e^{-a(\theta+m\pi)} \right\} d\theta$$

$$= \int_0^\pi \frac{\sin(2n+1)\theta}{\sin \theta} \frac{e^{-a\theta}\, d\theta}{1 - e^{-a\pi}},$$

and use Example 9.4.1.

Example 9.4.3   Discuss the uniformity of the convergence of Fourier series by means of the Dirichlet–Bonnet integrals, without making use of the theory of summability.

9.5 The Hurwitz–Liapounoff theorem concerning Fourier constants

This appears in Hurwitz [328]. Liapounoff discovered the theorem in 1896 and published it in [434]. See also Stekloff [601].

Let $f(x)$ *be bounded in the interval* $(-\pi, \pi)$ *and let* $\int_{-\pi}^\pi f(x)\, dx$ *exist, so that the Fourier constants* a_n, b_n *of* $f(x)$ *exist. Then the series*

$$\frac{1}{2}a_0^2 + \sum_{n=1}^\infty (a_n^2 + b_n^2)$$

is convergent and its sum is

$$\frac{1}{\pi} \int_{-\pi}^\pi \{f(x)\}^2\, dx.$$

This integral exists by Example 4.1.3. A proof of the theorem has been given by de la Vallée Poussin, in which the sole restrictions on $f(x)$ are that the (improper) integrals of $f(x)$ and $\{f(x)\}^2$ exist in the interval $(-\pi, \pi)$. See [639, II, pp. 165–166].

It will first be shewn that, with the notation of §9.4,

$$\lim_{m \to \infty} \int_{-\pi}^\pi \left\{ f(x) - \frac{1}{m} \sum_{n=0}^{m-1} S_n(x) \right\}^2 dx = 0.$$

Divide the interval $(-\pi, \pi)$ into $4r$ parts, each of length δ; let the upper and lower bounds of $f(x)$ in the interval $\{(2p-1)\delta - \pi, (2p+3)\delta - \pi\}$ be U_p, L_p, and let the upper bound of $|f(x)|$ in the interval $(-\pi, \pi)$ be K. Then, by Corollary 9.4.1

$$\left| f(x) - \frac{1}{m} \sum_{n=0}^{m-1} S_n(x) \right| < U_p - L_p + 2K / \left(m \sin^2 \tfrac{1}{2}\delta \right)$$

$$< 2K \left[1 + 1/\left(m \sin^2 \tfrac{1}{2}\delta \right) \right],$$

when x lies between $2p\delta$ and $(2p+2)\delta$.

Consequently, by the first mean-value theorem,

$$\int_{-\pi}^{\pi} \left\{ f(x) - \frac{1}{m} \sum_{n=0}^{m-1} S_n(x) \right\}^2 dx$$

$$< 2K \left\{ 1 + \frac{1}{m \sin^2 \tfrac{1}{2}\delta} \right\} \left\{ 2\delta \sum_{p=0}^{2r-1} (U_p - L_p) + \frac{4Kr}{m \sin^2 \tfrac{1}{2}\delta} \right\}.$$

Since $f(x)$ satisfies the Riemann condition of integrability (§4.12), it follows that both $4\delta \sum_{p=0}^{r-1} (U_{2p} - L_{2p})$ and $4\delta \sum_{p=0}^{r-1} (U_{2p+1} - L_{2p+1})$ can be made arbitrarily small by giving r a sufficiently large value. When r (and therefore also δ) has been given such a value, we may choose m_1, so large that $r/\{m_1 \sin^2 \tfrac{1}{2}\delta\}$ is arbitrarily small. That is to say, we can make the expression on the right of the last inequality arbitrarily small by giving m any value greater than a determinate value m_1. Hence the expression on the left of the inequality tends to zero as $m \to \infty$.

But evidently

$$\int_{-\pi}^{\pi} \left\{ f(x) - \frac{1}{m} \sum_{n=0}^{m-1} S_n(x) \right\}^2 dx = \int_{-\pi}^{\pi} \left\{ f(x) - \sum_{n=0}^{m-1} \frac{m-n}{m} A_n(x) \right\}^2 dx$$

$$= \int_{-\pi}^{\pi} \left\{ f(x) - \sum_{n=0}^{m-1} A_n(x) + \sum_{n=0}^{m-1} \frac{n}{m} A_n(x) \right\}^2 dx$$

$$= \int_{-\pi}^{\pi} \left\{ f(x) - \sum_{n=0}^{m-1} A_n(x) \right\}^2 dx + \int_{-\pi}^{\pi} \left\{ \sum_{n=0}^{m-1} \frac{n}{m} A_n(x) \right\}^2 dx$$

$$+ 2 \int_{-\pi}^{\pi} \left\{ f(x) - \sum_{n=0}^{m-1} A_n(x) \right\} \left\{ \sum_{n=0}^{m-1} A_n(x) \right\} dx$$

$$= \int_{-\pi}^{\pi} \left\{ f(x) - \sum_{n=0}^{m-1} A_n(x) \right\}^2 dx + \frac{\pi}{m^2} \sum_{n=0}^{m-1} n^2 (a_n^2 + b_n^2),$$

since

$$\int_{-\pi}^{\pi} f(x) A_r(x)\, dx = \int_{-\pi}^{\pi} \left\{ \sum_{n=0}^{m-1} A_n(x) \right\} A_r(x)\, dx$$

when $r = 0, 1, 2, \ldots, m - 1$.

Since the original integral tends to zero and since it has been proved equal to the sum of two positive expressions, it follows that each of these expressions tends to zero; that is to say

$$\int_{-\pi}^{\pi} \left\{ f(x) - \sum_{n=0}^{m-1} A_n(x) \right\}^2 dx \to 0.$$

Now the expression on the left is equal to

$$\int_{-\pi}^{\pi} \{f(x)\}^2 dx - 2 \int_{-\pi}^{\pi} \left\{ f(x) - \sum_{n=0}^{m-1} A_n(x) \right\} \left\{ \sum_{n=0}^{m-1} A_n(x) \right\} dx$$

$$- \int_{-\pi}^{\pi} \left(\sum_{n=0}^{m-1} A_n(x) \right)^2 dx$$

$$= \int_{-\pi}^{\pi} \{f(x)\}^2 dx - \int_{-\pi}^{\pi} \left(\sum_{n=0}^{m-1} A_n(x) \right)^2 dx$$

$$= \int_{-\pi}^{\pi} \{f(x)\}^2 dx - \pi \left\{ \frac{1}{2} a_0^2 + \sum_{n=1}^{m-1} (a_n^2 + b_n^2) \right\},$$

so that, as $m \to \infty$,

$$\int_{-\pi}^{\pi} \{f(x)\}^2 dx - \pi \left(\frac{1}{2} a_0^2 + \sum_{n=0}^{m-1} (a_n^2 + b_n^2) \right) \to 0.$$

This is the theorem stated.

For the following corollary, Parseval assumed, of course, the permissibility of integrating the trigonometrical series term-by-term.

Corollary 9.5.1 (Parseval [516]) *If $f(x)$, $F(x)$ both satisfy the conditions laid on $f(x)$ at the beginning of this section, and if A_n, B_n be the Fourier constants of $F(x)$, it follows by subtracting the pair of equations which may be combined in the one form*

$$\int_{-\pi}^{\pi} \{f(x) \pm F(x)\}^2 dx = \pi \left[\frac{1}{2} (a_0 \pm A_0)^2 + \sum_{n=1}^{\infty} \{(a_0 \pm A_n)^2 + (b_n \pm B_n)^2\} \right]$$

that

$$\int_{-\pi}^{\pi} f(x) F(x) dx = \pi \left\{ \frac{1}{2} a_0 A_0 + \sum_{n=1}^{\infty} (a_n A_n + b_n B_n) \right\}.$$

9.6 Riemann's theory of trigonometrical series

The theory of Dirichlet concerning Fourier series is devoted to series which represent given functions. Important advances in the theory were made by Riemann, who considered properties of functions defined by a series of the type[14] $\frac{1}{2} a_0 + \sum_{n=1}^{\infty} (a_n \cos nx + b_n \sin nx)$, where

[14] Throughout §§9.6–9.632 the letters a_n, b_n do not necessarily denote Fourier constants.

it is assumed that $\lim_{n\to\infty}(a_n\cos nx + b_n\sin nx) = 0$. We shall give the propositions leading up to Riemann's theorem that if two trigonometrical series converge and are equal at all points of the range $(-\pi,\pi)$ with the possible exception of a finite number of points, corresponding coefficients in the two series are equal. The proof given is due to G. Cantor [113, 114].

9.61 Riemann's associated function

Let the sum of the series

$$\frac{1}{2}a_0 + \sum_{n=1}^{\infty}(a_n\cos nx + b_n\sin nx) = A_0 + \sum_{n=1}^{\infty}A_n(x),$$

at any point x where it converges, be denoted by $f(x)$. Let

$$F(x) = \frac{1}{2}A_0 x^2 - \sum_{n=1}^{\infty}\frac{A_n(x)}{n^2}.$$

Then, if the series defining $f(x)$ converges at all points of any finite interval, the series defining $F(x)$ converges for all real values of x.

To obtain this result we need the following lemma due to Cantor[15].

Lemma (Cantor) *If $\lim_{n\to\infty} A_n(x) = 0$ for all values of x such that $a \le x \le b$, then $a_n \to 0$, $b_n \to 0$.*

For take two points x, $x + \delta$ of the interval. Then, given ε, we can find n_0 (the value of n_0 depends on x and on δ) such that, when $n > n_0$

$$|a_n\cos nx + b_n\sin nx|< \varepsilon, \quad |a_n\cos n(x+\delta) + b_n\sin n(x+\delta)| < \varepsilon.$$

Therefore

$$|\cos n\delta(a_n\cos nx + b_n\sin nx) + \sin n\delta(-a_n\sin nx + b_n\cos nx)| < \varepsilon.$$

Since $|\cos n\delta(a_n\cos nx + b_n\sin nx)| < \varepsilon$, it follows that $|\sin n\delta(-a_n\sin nx + b_n\cos nx)| < 2\varepsilon$, and it is obvious that $|\sin n\delta(a_n\cos nx + b_n\sin nx)| < 2\varepsilon$. Therefore, squaring and adding,

$$(a_n^2 + b_n^2)^{1/2}|\sin n\delta| < 2\varepsilon\sqrt{2}.$$

Now suppose that a_n, b_n have not the unique limit 0; it will be shewn that this hypothesis involves a contradiction. For, by this hypothesis, *some* positive number ε_0 exists such that there is an unending increasing sequence $n_1, n_2, \ldots$ of values of n, for which

$$(a_n^2 + b_2^2)^{1/2} > 4\varepsilon_0.$$

Now let the range of values of δ be called the interval I_1 of length L_1 on the real axis. Take n_1' the smallest of the integers n_r such that $n_1'L_1 > 2\pi$; then $\sin n_1'y$ goes through all its phases in the interval I_1; call I_2 that sub-interval[16] of I_1 in which $\sin n_1'y > 1/\sqrt{2}$; its length

[15] Riemann appears to have regarded this result as obvious. The proof here given is a modification of Cantor's proof [114, 115].
[16] If there is more than one such sub-interval, take that which lies on the left.

is $\pi/(2n_1') = L_2$. Next take n_2' the smallest of the integers $n_r(> n_1')$ such that $n_2'L_2 > 2\pi$, so that $\sin n_2'y$ goes through all its phases in the interval I_2; call I_3 that sub-interval of I_2 in which $\sin n_2'y > 1/\sqrt{2}$; its length is $\pi/(2n_2') = L_3$. We thus get a sequence of decreasing intervals $I_1, I_2, \ldots$ each contained in all the previous ones. It is obvious from the definition of an irrational number that there is a certain point a which is not outside any of these intervals, and $\sin na \geq 1/\sqrt{2}$ when $n = n_1', n_2', \ldots (n_{r+1}' > n_r')$.

For these values of n, $(a_n^2 + b_n^2)^{1/2} \sin na > 2\varepsilon_0 \sqrt{2}$. But it has been shewn that corresponding to given numbers a and ε we can find n_0 such that when $n > n_0$, $(a_n^2 + b_n^2)^{1/2}(\sin na) < 2\varepsilon\sqrt{2}$; since some values of n_r' are greater than n_0, the required contradiction has been obtained, because we may take $\varepsilon < \varepsilon_0$; therefore $a_n \to 0$, $b_n \to 0$.

Assuming that the series defining $f(x)$ converges at all points of a certain interval of the real axis, we have just seen that $a_n \to 0$, $b_n \to 0$. Then, for all real values of x, $|a_n \cos nx + b_n \sin nx| \leq (a_n^2 + b_n^2)^{1/2} \to 0$, and so, by §3.34, the series $\frac{1}{2}A_0x^2 - \sum_{n=1}^{\infty} A_n(x)/n^2 = F(x)$ converges absolutely and uniformly for all real values of x; therefore (see §3.32) $F(x)$ is continuous for all real values of x.

9.62 Properties of Riemann's associated function; Riemann's first lemma

It is now possible to prove Riemann's first lemma *that if*

$$G(x, \alpha) = \frac{F(x + 2\alpha) + F(x - 2\alpha) - 2F(x)}{4\alpha^2}$$

then $\lim_{a \to 0} G(x, \alpha) = f(x)$, *provided that* $\sum_{n=0}^{\infty} A_n(x)$ *converges for the value of x under consideration.*

Since the series defining $F(x)$, $F(x \pm 2\alpha)$ converge absolutely, we may rearrange them; and, observing that

$$\cos n(x + 2\alpha) \cos n(x - 2\alpha) - 2\cos nx = -4\sin^2 n\alpha \cos nx,$$
$$\sin n(x + 2\alpha) + \sin n(x - 2\alpha) - 2\sin nx = -4\sin^2 n\alpha \sin nx,$$

it is evident that

$$G(x, \alpha) = A_0 + \sum_{n=1}^{\infty} \left(\frac{\sin n\alpha}{n\alpha}\right)^2 A_n(x).$$

It will now be shewn that this series converges uniformly with regard to α for all values of α, provided that $\sum_{n=1}^{\infty} A_n(x)$ converges. The result required is then an immediate consequence of §3.32: for, if $f_n(\alpha) = \left(\frac{\sin n\alpha}{n\alpha}\right)^2$, $(\alpha \neq 0)$, and $f_n(0) = 1$, then $f_n(\alpha)$ is continuous for all values of α, and so $G(x, \alpha)$ is a continuous function of α, therefore, by §3.2, $G(x, 0) = \lim_{a \to 0} G(x, \alpha)$.

To prove that the series defining $G(x, \alpha)$ converges uniformly, we employ the test given in Example 3.3.7. The expression corresponding to $\omega_n(x)$ is $f_n(\alpha)$, and it is obvious that $|f_n(\alpha)| \leq 1$; it is therefore sufficient to shew that $\sum_{n=1}^{x} |f_{n+1}(\alpha) - f_n(\alpha)| < K$, where K is independent of α.

In fact, since $x^{-1} \sin x$ decreases as x increases from 0 to π, if s be the integer such that $s|\alpha| \le \pi < (s+1)|\alpha|$, when $\alpha \ne 0$ we have

$$\sum_{n=1}^{s-1} |f_{n+1}(\alpha) - f_n(\alpha)| = \sum_{n=1}^{s-1} (f_n(\alpha) - f_{n+1}(\alpha)) = \frac{\sin^2 \alpha}{\alpha^2} - \frac{\sin^2 s\alpha}{s^2\alpha^2}.$$

Also

$$\sum_{n=s+1}^{\infty} |f_{n+1}(\alpha) - f_n(\alpha)| = \sum_{n=s+1}^{\infty} \left| \left\{ \frac{\sin^2 n\alpha}{\alpha^2} \left(\frac{1}{n^2} - \frac{1}{(n+1)^2} \right) \right\} + \frac{\sin^2 n\alpha - \sin^2(n+1)\alpha}{(n+1)^2\alpha^2} \right|$$

$$\le \sum_{n=s+1}^{\infty} \frac{1}{\alpha^2} \left(\frac{1}{n^2} - \frac{1}{(n+1)^2} \right) + \sum_{n=s+1}^{\infty} \frac{|\sin^2 n\alpha - \sin^2(n+1)\alpha|}{(n+1)^2\alpha^2}$$

$$\le \frac{1}{(s+1)^2\alpha^2} + \sum_{n=s+1}^{\infty} \frac{|\sin \alpha \sin(2n+1)\alpha|}{(n^2+1)^2\alpha^2}$$

$$\le \frac{1}{(s+1)^2\alpha^2} + \frac{|\sin \alpha|}{\alpha^2} \sum_{n=s+1}^{\infty} \frac{1}{(n+1)^2}$$

$$\le \frac{1}{\pi^2} + \frac{|\sin \alpha|}{\alpha^2} \int_{s}^{\infty} \frac{dx}{(x+1)^2}$$

$$\le \frac{1}{\pi^2} + \frac{1}{(s+1)|\alpha|}.$$

Therefore

$$\sum_{n=1}^{\infty} |f_{n+1}(\alpha) - f_n(\alpha)| \le \frac{\sin^2 \alpha}{\alpha^2} - \frac{\sin^2 s\alpha}{s^2\alpha^2} + \left(\frac{\sin^2 s\alpha}{s^2\alpha^2} + \frac{\sin^2(s+1)\alpha}{(s+1)^2\alpha^2} \right) + \frac{1}{\pi^2} + \frac{1}{\pi}$$

$$\le 1 + \frac{1}{\pi} + \frac{2}{\pi^2}.$$

Since this expression is independent of α, the result required has been obtained (this inequality is obviously true when $\alpha = 0$).

Hence, if $\sum_{n=0}^{\infty} A_n(x)$ converges, the series defining $G(x, \alpha)$ converges uniformly with respect to α for all values of α, and, as stated above,

$$\lim_{\alpha \to 0} G(x, \alpha) = G(x, 0) = A_0 + \sum_{n=1}^{\infty} A_n(x) = f(x).$$

Example (Riemann)   If

$$H(x, \alpha, \beta) = \frac{F(x + \alpha + \beta) - F(x + \alpha - \beta) - F(x - \alpha + \beta) + F(x - \alpha - \beta)}{4\alpha\beta}$$

shew that $H(x, \alpha, \beta) \to f(x)$ when $f(x)$ converges if $\alpha, \beta \to 0$ in such a way that α/β and β/α remain finite.

9.621 Riemann's second lemma

With the notation of §9.6 and §9.62, if $a_n, b_n \to 0$, then

$$\lim_{\alpha \to 0} \frac{F(x + 2\alpha) + F(x - 2\alpha) - 2F(x)}{4\alpha} = 0$$

for all values of x.

For

$$\tfrac{1}{4}\alpha^{-1} F(x + 2\alpha) + F(x - 2\alpha) - 2F(x) = A_0 \alpha + \sum_{n=1}^{\infty} \frac{\sin^2 n\alpha}{n^2 \alpha} A_n(x);$$

but by Example 9.1.3 if $\alpha > 0$,

$$\sum_{n=1}^{\infty} \frac{\sin^2 n\alpha}{n^2 \alpha} = \frac{1}{2}(\pi - \alpha);$$

and so, since

$$A_0(x)\alpha + \sum_{n=1}^{\infty} \frac{\sin^2 n\alpha}{n^2 \alpha} A_n(x) = A_0(x)\alpha + \frac{1}{2}(\pi - \alpha)A_1(x)$$

$$+ \sum_{n=1}^{\infty} \left\{ \frac{1}{2}(\pi - \alpha) - \sum_{m=1}^{n} \frac{\sin^2 m\alpha}{m^2 \alpha} \right\} \{A_{n+1}(x) - A_n(x)\},$$

it follows from Example 3.3.7, that this series converges uniformly with regard to α for all values of α greater than, or equal to, zero[17].

But

$$\lim_{\alpha \to +0} \frac{1}{4}\alpha^{-1} \{F(x + 2\alpha) + F(x - 2\alpha) - 2F(x)\}$$

$$= \lim_{\alpha \to 0^+} \left[A_0(x)\alpha + \frac{1}{2}(\pi - \alpha)A_1(x) + \sum_{n=1}^{\infty} g_n(\alpha) \{A_{n+1}(x) - A_n(x)\} \right]$$

and this limit is the value of the function when $\alpha = 0$, by §3.32; and this value is zero since $\lim_{n \to \infty} A_n(x) = 0$. By symmetry we see that $\lim_{\alpha \to 0^+} = \lim_{\alpha \to 0^-}$.

9.63 Riemann's theorem on trigonometrical series

Two trigonometrical series which converge and are equal at all points of the range $(-\pi, \pi)$, with the possible exception of a finite number of points, must have corresponding coefficients equal. The proof we give is due to G. Cantor [113].

An immediate deduction from this theorem is that a function of the type considered in §9.42 cannot be represented by any trigonometrical series in the range $(-\pi, \pi)$ other than the Fourier series. This fact was first noticed by Du Bois Reymond.

---

[17] If we define $g_n(\alpha)$ by the equations $g_n(\alpha) = (\pi - \alpha)/2 - \sum_{m=1}^{n} \frac{\sin^2 m\alpha}{m^2 \alpha}$ (with $\alpha \neq 0$), and $g_n(0) = \pi/2$, then $g_n(\alpha)$ is continuous when $\alpha \geq 0$, and $g_{n+1}(\alpha) \leq g_n(\alpha)$.

We observe that it is certainly possible to have other expansions of (say) the form

$$a_0 + \sum_{m=1}^{\infty}(a_m \cos \tfrac{1}{2}mx + \beta_m \sin \tfrac{1}{2}mx),$$

which represent $f(x)$ between $-\pi$ and π; for write $x = 2\xi$, and consider a function $\phi(\xi)$, which is such that $\phi(\xi) = f(2\xi)$ when $-\pi/2 < \xi < \pi/2$, and $\phi(\xi) = g(\xi)$ when $-\pi < \xi < -\pi/2$, and when $\pi/2 < \xi < \pi$, where $g(\xi)$ is any function satisfying the conditions of §9.43. Then if we expand $\phi(\xi)$ in a Fourier series of the form

$$a_0 + \sum_{m=0}^{\infty}(a_m \cos m\xi + \beta_m \sin m\xi),$$

this expansion represents $f(x)$ when $-\pi < x < \pi$; and clearly by choosing the function $g(\xi)$ in different ways an unlimited number of such expansions can be obtained.

The question now at issue is, whether other series proceeding in sines and cosines of *integral* multiples of x exist, which differ from Fourier's expansion and yet represent $f(x)$ between $-\pi$ and π.

If possible, let there be two trigonometrical series satisfying the given conditions, and let their difference be the trigonometrical series

$$A_0 + \sum_{n=1}^{\infty} A_n(x) = f(x).$$

Then $f(x) = 0$ at all points of the range $(-\pi, \pi)$ with a finite number of exceptions; let ξ_1, ξ_2 be a consecutive pair of these exceptional points, and let $F(x)$ be Riemann's associated function. We proceed to establish a lemma concerning the value of $F(x)$ when $\xi_1 < x < \xi_2$.

9.631 Schwartz' lemma

Quoted by G. Cantor [113]. *In the range $\xi_1 < x < \xi_2$, $F(x)$ is a linear function of x, if $f(x) = 0$ in this range.*

For if $\theta = 1$ or if $\theta = -1$

$$\phi(x) = \theta \left[F(x) - F(\xi_1) - \frac{x - \xi_1}{\xi_2 - \xi_1}\{F(\xi_2) - F(\xi_1)\}\right] - \frac{1}{2}h^2(x - \xi_1)(\xi_2 - x)$$

is a continuous function of x in the range $\xi_1 \le x \le \xi_2$ and $\phi(\xi_1) = \phi(\xi_2) = 0$.

If the first term of $\phi(x)$ is not zero throughout the range[18] there will be some point $x = c$ at which it is not zero. Choose the sign of θ so that the first term is positive at c, and then choose h so small that $\phi(c)$ is still positive. Since $\phi(x)$ is continuous it attains its upper bound (§3.62), and this upper bound is positive since $\phi(c) > 0$. Let $\phi(x)$ attain its upper bound at c_1, so that $c_1 \ne \xi_1, c_1 \ne \xi_2$. Then, by Riemann's first lemma,

$$\lim_{a \to 0} \frac{\phi(c_1 + a) + \phi(c_1 - a) - 2\phi(c_1)}{a^2} = h^2.$$

But $\phi(c_1 + a) \le \phi(c_1), \phi(c_1 - a) \le \phi(c_1)$, so this limit must be negative or zero. Hence, by supposing that the first term of $\phi(x)$ is not everywhere zero in the range (ξ_1, ξ_2), we have

---

[18] If it is zero throughout the range, $F(x)$ is a linear function of x.

arrived at a contradiction. Therefore it is zero; and consequently $F(x)$ is a linear function of x in the range $\xi_1 < x < \xi_2$. The lemma is therefore proved.

9.632 *Proof of Riemann's theorem*

We see that, in the circumstances under consideration, the curve $y = F(x)$ represents a series of segments of straight lines, the beginning and end of each line corresponding to an exceptional point; and as $F(x)$, being uniformly convergent is a continuous function of x, these lines must be connected.

But, by Riemann's second lemma, even if ξ be an exceptional point,

$$\lim_{a \to 0} \frac{F(\xi + \alpha) + F(\xi - \alpha) - 2F(\xi)}{\alpha} = 0.$$

Now the fraction involved in this limit is the difference of the slopes of the two segments which meet at that point whose abscissa is ξ; therefore the two segments are continuous in direction, so the equation $y = F(x)$ represents a single line. If then we write $F(x) = cx + c'$, it follows that c and c' have the same values for *all* values of x. Thus

$$\frac{1}{2}A_0 x^2 - cx - c' = \sum_{n=1}^{\infty} \frac{A_n(x)}{n^2},$$

the right-hand side of this equation being periodic, with period 2π.

The left-hand side of this equation must therefore be periodic, with period 2π. Hence $A_0 = 0$, $c = 0$, and $-c' = \sum_{n=1}^{\infty} \frac{A_n(x)}{n^2}$. Now the right-hand side of this equation converges uniformly, so we can multiply by $\cos nx$ or by $\sin nx$ and integrate. This process gives

$$\frac{\pi a_n}{n^2} = c' \int_{-\pi}^{\pi} \cos nx \, dx = 0,$$

$$\frac{\pi b_n}{n^2} = -c' \int_{-\pi}^{\pi} \sin nx \, dx = 0.$$

Therefore all the coefficients vanish, and therefore the two trigonometrical series whose difference is $A_0 + \sum_{n=1}^{\infty} A_n(x)$ have corresponding coefficients equal. This is the result stated in §9.63.

9.7 Fourier's representation of a function by an integral

This appears in Fourier [223]. For recent work on Fourier's integral and the modern theory of 'Fourier transforms', see Titchmarsh [629, 630].

It follows from §9.43 that, if $f(x)$ be continuous except at a finite number of discontinuities and if it have limited total fluctuation in the range $(-\infty, \infty)$, then, if x be any *internal* point of the range $(-\alpha, \beta)$,

$$\lim_{m \to \infty} \int_{-\alpha}^{\beta} \frac{\sin(2m + 1)(t - x)}{(t - x)} f(t) \, dt = \lim_{\theta \to 0} \frac{\pi}{2} \frac{\sin \theta}{\theta} \{f(x + 2\theta) + f(x - 2\theta)\}.$$

Now let λ be any real number, and choose the integer m so that $\lambda = 2m + 1 + 2\eta$ where

$0 \leq \eta < 1$. Then

$$\int_{-\alpha}^{\beta} [\sin \lambda(t - x) - \sin(2m + 1)(t - x)] \frac{f(t)}{t - x} \, dt$$

$$= \int_{-\alpha}^{\beta} 2 \cos[(2m + 1 + \eta)(t - x)] \sin \eta(t - x) \frac{f(t)}{t - x} \, dt$$

$$\to 0,$$

as $m \to \infty$ by §9.41, since $(t - x)^{-1} f(t) \sin \eta(t - x)$ has limited total fluctuation.

Consequently, from the proof of the Riemann–Lebesgue lemma of §9.41, it is obvious that if $\int_0^\infty |f(t)| \, dt$ and $\int_{-\infty}^0 |f(t)| \, dt$ converge, then[19]

$$\lim_{\lambda \to \infty} \int_{-\infty}^\infty \frac{\sin \lambda (t - x)}{(t - x)} f(t) \, dt = \frac{1}{2}\pi \{f(x + 0) + f(x - 0)\},$$

and so

$$\lim_{\lambda \to \infty} \int_{-\infty}^\infty \left\{\int_0^\lambda \cos u (t - x) \, du\right\} f(t) \, dt = \frac{\pi}{2} \{f(x + 0) + f(x - 0)\}.$$

To obtain Fourier's result, we must reverse the order of integration in this repeated integral. For any given value of λ and any arbitrary value of ε, there exists a number β such that

$$\int_\beta^\infty |f(t)| \, dt < \frac{\varepsilon}{2\lambda};$$

writing $\cos u(t - x) \cdot f(t) = \phi(t, u)$, we have[20]

$$\left|\int_0^\infty \left\{\int_0^\lambda \phi(t, u) \, du\right\} dt - \int_0^\lambda \left\{\int_0^\infty \phi(t, u) \, dt\right\} du\right|$$

$$= \left|\int_0^\beta \left\{\int_0^\lambda \phi(t, u) \, du\right\} dt + \int_\beta^\infty \left\{\int_0^\lambda \phi(t, u)du\right\} dt\right.$$

$$\left. - \int_0^\lambda \left\{\int_0^\beta \phi(t, u) \, dt\right\} du - \int_0^\lambda \left\{\int_\beta^\theta \phi(t, u) \, dt\right\} du\right|$$

$$= \left|\int_\beta^\infty \left\{\int_0^\lambda \phi(t, u) \, du\right\} dt - \int_0^\lambda \left\{\int_\beta^\infty \phi(t, u) \, dt\right\} du\right|$$

$$< \int_\beta^\infty \left\{\int_0^\lambda |\phi(t, u)| \, du\right\} dt + \int_0^\lambda \int_\beta^\infty |\phi(t, u)| \, dt \, du$$

$$< 2\lambda \int_\beta^\infty |f(t)| \, dt < \varepsilon.$$

Since this is true for all values of ε, no matter how small, we infer that

$$\int_0^\infty \int_0^\lambda = \int_0^\lambda \int_0^\infty; \qquad \text{similarly} \qquad \int_0^{-\infty} \int_0^\lambda = \int_0^\lambda \int_0^{-\infty}.$$

[19] $\int_{-\infty}^\infty$ means the double limit $\lim_{\rho \to \infty, \sigma \to \infty} \int_{-\rho}^\sigma$. If this limit exists, it is equal to $\lim_{\rho \to \infty} \int_{-\rho}^\rho$.

[20] The equation $\int_0^\beta \int_0^\lambda = \int_0^\lambda \int_0^\beta$ is easily justified by §4.3, by considering the ranges within which $f(x)$ is continuous.

Hence

$$\frac{1}{2}\pi \left\{f(x+0) + f(x-0)\right\} = \lim_{\lambda \to \infty} \int_0^\lambda \int_{-\infty}^\infty \cos u\,(t-x)f(t)\,dt\,du$$

$$= \int_0^\infty \int_{-\infty}^\infty \cos u\,(t-x)f(t)\,dt\,du.$$

This result is known as *Fourier's integral theorem*. For a proof of the theorem when $f(x)$ is subject to less stringent restrictions, see Hobson [316, pp. 492–493]. The reader should observe that, although $\lim_{\lambda \to \infty} \int_{-\infty}^\infty \int_0^\lambda$ exists, the repeated integral $\int_{-\infty}^\infty \left\{\int_0^\infty \sin u(t-x)\,du\right\} f(t)\,dt$ does not.

Example 9.7.1 (Rayleigh)   Verify Fourier's integral theorem directly (i) for the function $f(x) = (a^2 + x^2)^{-1/2}$, (ii) for the function defined by the equations

$$f(x) = 1,\ (-1 < x < 1); \qquad f(x) = 0,\ (|x| > 1).$$

9.8 Miscellaneous examples

Example 9.1   Obtain the expansions

(a) $\dfrac{1 - r\cos z}{1 - 2r\cos z + r^2} = 1 + r\cos z + r^2 \cos 2z + \cdots,$

(b) $\dfrac{1}{2}\log(1 - 2r\cos z + r^2) = -r\cos z - \dfrac{1}{2}r^2 \cos 2z - \dfrac{1}{3}r^3 \cos 3z - \cdots,$

(c) $\arctan \dfrac{r\sin z}{1 - r\cos z} = r\sin z + \dfrac{1}{2}r^2 \sin 2z + \dfrac{1}{3}r^3 \sin 3z + \cdots,$

(d) $\dfrac{1}{2}\arctan \dfrac{2r\sin z}{1 - r^2} = r\sin z + \dfrac{1}{3}r^3 \sin 3z + \dfrac{1}{5}r^5 \sin 5z + \cdots,$

and shew that, when $|r| < 1$, they are convergent for all values of z in certain strips parallel to the real axis in the z-plane.

Example 9.2 (Jesus, 1902)   Expand x^3 and x in Fourier sine series valid when $-\pi < x < \pi$; and hence find the value of the sum of the series

$$\sin x - \frac{1}{2^3}\sin 2x - \frac{1}{3^3}\sin 3x - \frac{1}{4^3}\sin 4x + \cdots,$$

for all values of x.

Example 9.3 (Pembroke, 1907)   Shew that the function of x represented by

$$\sum_{n=1}^\infty n^{-1}\sin nx \sin^2 na,$$

is constant $(0 < x < 2a)$ and zero $(2a < x < \pi)$, and draw a graph of the function.

Example 9.4 (Peterhouse, 1906)   Find the cosine series representing $f(x)$ where

$$f(x) = \begin{cases} \sin x + \cos x & (0 < x \le \tfrac{1}{2}\pi), \\ \sin x - \cos x & (\tfrac{1}{2}\pi \le x < \pi). \end{cases}$$

Example 9.5 (Trinity, 1895)   Shew that

$$\sin \pi x + \frac{\sin 3\pi x}{3} + \frac{\sin 5\pi x}{5} + \frac{\sin 7\pi x}{7} + \cdots = \frac{1}{4}\pi[x],$$

where $[x]$ denotes $+1$ or -1 according as the integer next inferior to x is even or uneven, and is zero if x is an integer.

Example 9.6   Shew that the expansions

$$\log \left| 2 \cos \frac{1}{2}x \right| = \cos x - \frac{1}{2} \cos 2x + \frac{1}{3} \cos 3x \cdots$$

and

$$\log \left| 2 \sin \frac{1}{2}x \right| = -\cos x - \frac{1}{2} \cos 2x - \frac{1}{3} \cos 3x \cdots$$

are valid for all real values of x, except multiples of π.

Example 9.7 (Trinity, 1898)   Obtain the expansion

$$\sum_{m=0}^{\infty} \frac{(-1)^m \cos mx}{(m+1)(m+2)}$$

$$= (\cos x + \cos 2x) \log \left(2 \cos \frac{x}{2} \right) + \frac{x}{2}(\sin 2x + \sin x) - \cos x,$$

and find the range of values of x for which it is applicable.

Example 9.8 (Trinity, 1895)   Prove that, if $0 < x < 2\pi$, then

$$\frac{\sin x}{a^2 + 1^2} + \frac{2 \sin 2x}{a^2 + 2^2} + \frac{3 \sin 3x}{a^2 + 3^2} + \cdots = \frac{\pi}{2} \frac{\sinh a(\pi - x)}{\sinh a\pi}.$$

Example 9.9   Shew that between the values $-\pi$ and $+\pi$ of x the following expansions hold:

$$\sin mx = \frac{2}{\pi} \sin m\pi \left(\frac{\sin x}{1^2 - m^2} - \frac{2 \sin 2x}{2^2 - m^2} + \frac{3 \sin 3x}{3^2 - m^2} - \cdots \right),$$

$$\cos mx = \frac{2}{\pi} \sin m\pi \left(\frac{1}{2m} + \frac{m \cos x}{1^2 - m^2} - \frac{m \cos 2x}{2^2 + m^2} + \frac{m \cos 3x}{3^2 - m^2} - \cdots \right),$$

$$\frac{e^{mx} + e^{-mx}}{e^{m\pi} - e^{-m\pi}} = \frac{2}{\pi} \left(\frac{1}{2m} - \frac{m \cos x}{1^2 + m^2} + \frac{m \cos 2x}{2^2 + m^2} - \frac{m \cos 3x}{3^2 + m^2} + \cdots \right).$$

Example 9.10 (Berger)   Let x be a real variable between 0 and 1, and let $n \geq 3$ be an odd number. Shew that

$$(-1)^s = \frac{1}{n} + \frac{2}{\pi} \sum_{m=1}^{\infty} \frac{1}{m} \tan \frac{m\pi}{n} \cos 2m\pi x,$$

if x is not a multiple of $1/n$, where s is the greatest integer contained in nx but

$$0 = \frac{1}{n} + \frac{2}{\pi} \sum_{m=1}^{\infty} \frac{1}{m} \tan \frac{m\pi}{n} \cos 2m\pi x$$

if x is an integer multiple of $1/n$.

Example 9.11 (Trinity, 1901)   Shew that the sum of the series

$$\frac{1}{3} + \frac{4}{\pi} \sum_{m=1}^{\infty} m^{-1} \sin\left(\frac{2m\pi}{3}\right) \cos 2m\pi x$$

is 1 when $0 < x < \frac{1}{3}$, and when $\frac{2}{3} < x < 1$, and is -1 when $\frac{1}{3} < x < \frac{2}{3}$.

Example 9.12 (Math. Trip. 1896)   If

$$\frac{a e^{ax}}{e^a - 1} = \sum_{n=0}^{\infty} \frac{a^n V_n(x)}{n!},$$

shew that, when $-1 < x < 1$,

$$\cos 2\pi x + \frac{\cos 4\pi x}{2^{2n}} + \frac{\cos 6\pi x}{3^{2n}} + \cdots = (-1)^{n-1} \frac{2^{2n-1}\pi^{2n}}{(2n)!} V_{2n}(x),$$

$$\sin 2\pi x + \frac{\sin 4\pi x}{2^{2n+1}} + \frac{\sin 6\pi x}{3^{2n+1}} + \cdots = (-1)^{n+1} \frac{2^{2n}\pi^{2n+1}}{(2n+1)!} V_{2n+1}(x).$$

Example 9.13 (Trinity, 1894)   If m is an integer, shew that, for all real values of x,

$$\cos^{2m} x = 2 \frac{1 \cdot 3 \cdot 5 \cdots (2m-1)}{2 \cdot 4 \cdot 6 \cdots 2m} \left\{ \frac{1}{2} + \frac{m}{m+1} \cos 2x + \frac{m(m-1)}{(m+1)(m+2)} \cos 4x \right.$$

$$\left. + \frac{m(m-1)(m-2)}{(m+1)(m+2)(m+3)} \cos 6x + \cdots \right\},$$

$$\left|\cos^{2m-1} x\right| = \frac{4}{\pi} \frac{2 \cdot 4 \cdot 6 \cdots (2m-2)}{1 \cdot 3 \cdot 5 \cdots (2m-1)} \left\{ \frac{1}{2} + \frac{2m-1}{2m+1} \cos 2x \right.$$

$$\left. + \frac{(2m-1)(2m-3)}{(2m+1)(2m+3)} \cos 4x + \cdots \right\}.$$

Example 9.14   A point moves in a straight line with a velocity which is initially u, and which receives constant increments, each equal to u, at equal intervals τ. Prove that the velocity at any time t after the beginning of the motion is

$$\frac{u}{2} + \frac{ut}{\tau} + \frac{u}{\pi} \sum_{m=1}^{\infty} \frac{1}{m} \sin \frac{2m\pi t}{\tau},$$

and that the distance traversed is

$$\frac{ut}{2\tau}(t + \tau) + \frac{u\tau}{12} - \frac{u\tau}{2\pi^2} \sum_{m=1}^{\infty} \frac{1}{m^2} \cos \frac{2m\pi t}{\tau}.$$

Example 9.15 (Math. Trip. 1893)   If

$$f(x) = \sum_{n=1}^{\infty} \frac{\sin(6n-3)x}{2n-1} - 2 \sum_{n=1}^{\infty} \frac{\sin(2n-1)x}{2n-1}$$

$$+ \frac{3\sqrt{3}}{\pi} \left\{ \sin x - \frac{\sin 5x}{5^2} + \frac{\sin 7x}{7^2} - \frac{\sin 11x}{11^2} + \cdots \right\},$$

shew that $f(+0) = f(\pi - 0) = -\frac{\pi}{4}$, and $f(\frac{\pi}{3} + 0) - f(\frac{\pi}{3} - 0) = -\frac{\pi}{2}$, $f(\frac{2\pi}{3} + 0) - f(\frac{2\pi}{3} - 0) = \frac{\pi}{2}$. Observing that the last series is

$$\frac{6}{\pi} \sum_{n=1}^{\infty} \frac{\sin \frac{(2n-1)\pi}{3} \sin(2n-1)x}{(2n-1)^2},$$

draw the graph of $f(x)$.

Example 9.16 (Trinity, 1908)　Shew that, when $0 < x < \pi$,

$$f(x) = \frac{2\sqrt{3}}{3} \left(\cos x - \frac{1}{5} \cos 5x + \frac{1}{7} \cos 7x - \frac{1}{11} \cos 11x + \cdots \right)$$

$$= \sin 2x + \frac{1}{2} \sin 4x + \frac{1}{4} \sin 8x + \frac{1}{5} \sin 10x + \cdots$$

where

$$f(x) = \begin{cases} \frac{1}{3}\pi & 0 < x < \frac{1}{3}\pi, \\ 0 & \frac{1}{3}\pi < x < \frac{2}{3}\pi, \\ -\frac{1}{3}\pi & \frac{2}{3}\pi < x < \pi. \end{cases}$$

Find the sum of each series when $x = 0$, $\frac{1}{3}\pi$, $\frac{2}{3}\pi$, π, and for all other values of x.

Example 9.17 (Math. Trip. 1895)　Prove that the locus represented by

$$\sum_{n=1}^{\infty} \frac{(-1)^{n-1}}{n^2} \sin nx \sin ny = 0$$

is two systems of lines at right angles, dividing the coordinate plane into squares of area π^2.

Example 9.18 (Trinity, 1903)　Shew that the equation

$$\sum_{n=1}^{\infty} \frac{(-1)^{n-1} \sin ny \cos nx}{n^3} = 0$$

represents the lines $y = \pm m\pi$, $(m = 0, 1, 2, \ldots)$ together with a set of arcs of ellipses whose semi-axes are π and $\pi/\sqrt{3}$, the arcs being placed in squares of area $2\pi^2$. Draw a diagram of the locus.

Example 9.19 (Math. Trip. 1904)　Shew that, if the point (x, y, z) lies inside the octahedron bounded by the planes $\pm x \pm y \pm z = \pi$, then

$$\sum_{n=1}^{\infty} (-1)^{n-1} \frac{\sin nx \sin ny \sin nz}{n^3} = \frac{1}{2} xyz.$$

Example 9.20 (Pembroke, 1902)　Circles of radius a are drawn having their centres at the alternate angular points of a regular hexagon of side a. Shew that the equation of the trefoil formed by the outer arcs of the circles can be put in the form

$$\frac{\pi r}{6\sqrt{3}a} = \frac{1}{2} + \frac{1}{2 \cdot 4} \cos 3\theta - \frac{1}{5 \cdot 7} \cos 6\theta + \frac{1}{8 \cdot 10} \cos 9\theta - \cdots,$$

the initial line being taken to pass through the centre of one of the circles.

Example 9.21 (Jesus, 1908)   Draw the graph represented by

$$\frac{r}{a} = 1 + \frac{2m}{\pi} \sin \frac{\pi}{m} \left\{ \frac{1}{2} + \sum_{n=1}^{\infty} \frac{(-1)^n \cos nm\theta}{1 - (nm)^2} \right\},$$

where m is an integer.

Example 9.22 (Trinity, 1905)   With each vertex of a regular hexagon of side $2a$ as centre, the arc of a circle of radius $2a$ lying within the hexagon is drawn. Shew that the equation of the figure formed by the six arcs is

$$\frac{\pi r}{4a} = 6 - 3\sqrt{3} + 2 \sum_{n=1}^{\infty} \frac{\left\{(-1)^{n-1}6 + 3\sqrt{3}\right\}}{(6n-1)(6n+1)} \cos 6n\theta,$$

the prime vector bisecting a petal.

Example 9.23 (Trinity, 1894)   Shew that, if $c > 0$,

$$\lim_{n\to\infty} \int_0^\infty e^{-cx} \cot x \sin(2n+1)x \, dx = \frac{1}{2}\pi \tanh \frac{1}{2}c\pi.$$

Example 9.24 (King's, 1901)   Shew that

$$\lim_{n\to\infty} \int_0^\infty \frac{\sin(2n+1)x}{\sin x} \frac{dx}{1+x^2} = \frac{1}{2}\pi \coth 1.$$

Example 9.25 (Math. Trip. 1905)   Shew that, when $-1 < x < 1$ and a is real,

$$\lim_{n\to\infty} \int_0^\infty \frac{\sin(2n+1)\theta \sin(1+x)\theta}{\sin\theta} \frac{\theta}{a^2 + \theta^2} d\theta = -\frac{\pi}{2} \frac{\sinh ax}{\sinh a}.$$

Example 9.26 (Math. Trip. 1898)   Assuming the possibility of expanding $f(x)$ in a uniformly convergent series of the form $\sum_k A_k \sin kx$, where k is a root of the equation $k \cos ak + b \sin ak = 0$ and the summation is extended to all positive roots of this equation, determine the constants A_k.

Example 9.27 (Beau)   If $f(x) = \frac{1}{2}a_0 + \sum_{n=1}^{\infty} (a_n \cos nx + b_n \sin nx)$ is a Fourier series, shew that, if $f(x)$ satisfies certain general conditions,

$$a_n = \frac{4}{\pi} \text{ P.V.} \int_0^\infty f(t) \cos nt \tan \frac{t}{2} \frac{dt}{t}, \qquad b_n = \frac{4}{\pi} \int_0^\infty f(t) \sin nt \tan \frac{t}{2} \frac{dt}{t},$$

where P.V. means principal value.

Example 9.28   If $S_n(x) = 2 \sum_{r=1}^{n} (-1)^{r-1} \frac{\sin rx}{r}$ prove that the highest maximum of $S_n(x)$ in the interval $(0, \pi)$ is at $x = \frac{n\pi}{n+1}$ and prove that, as $n \to \infty$,

$$S_n\left(\frac{n\pi}{n+1}\right) \to 2 \int_0^\pi \frac{\sin t}{t} dt.$$

Deduce that, as $n \to \infty$, the shape of the curve $y = S_n(x)$ in the interval $(0, \pi)$ tends to approximate to the shape of the curve formed by the line $y = x$, $0 \le x \le \pi$, together with the line $x = \pi$, $0 \le y \le G$, where

$$G = 2 \int_0^\pi \frac{\sin t}{t}\, dt.$$

Note   The fact that $G = 3.704 \cdots > \pi$ is known as *Gibbs' phenomenon*; see [243]. The phenomenon is characteristic of a Fourier series in the neighbourhood of a point of ordinary discontinuity of the function which it represents. For a full discussion of the phenomenon, which was discovered by Wilbraham [680], see Carslaw [119, Chapter 9].

10

Linear Differential Equations

10.1 Linear differential equations

The analysis contained in this chapter is mainly theoretical; it consists, for the most part, of existence theorems. It is assumed that the reader has some knowledge of practical methods of solving differential equations; these methods are given in works exclusively devoted to the subject, such as Forsyth [221, 222].

In some of the later chapters of this work, we shall be concerned with the investigation of extensive and important classes of functions which satisfy linear differential equations of the second order. Accordingly, it is desirable that we should now establish some general results concerning solutions of such differential equations.

The standard form of the linear differential equation of the second order will be taken to be

$$\frac{d^2u}{dz^2} + p(z)\frac{du}{dz} + q(z)u = 0, \qquad (10.1)$$

and it will be assumed that there is a domain S in which both $p(z)$, $q(z)$ are analytic except at a finite number of poles.

Any point of S at which $p(z)$, $q(z)$ are both analytic will be called an *ordinary point* of the equation; other points of S will be called *singular points*.

10.2 Solution of a differential equation valid in the vicinity of an ordinary point

This method is applicable only to equations of the second order. For a method applicable to equations of any order, see Forsyth [221].

Let b be an ordinary point of the differential equation, and let S_b be the domain formed by a circle of radius r_b, whose centre is b, and its interior, the radius of the circle being such that every point of S_b is a point of S, and is an ordinary point of the equation. Let z be a variable point of S_b.

In the equation write $u = v \exp\left\{-\frac{1}{2}\int_b^z p(\zeta)\,d\zeta\right\}$, and it becomes

$$\frac{d^2v}{dz^2} + J(z)v = 0, \qquad (10.2)$$

where $J(z) = q(z) - \frac{1}{2}\frac{dp(z)}{dz} - \frac{1}{4}\{p(z)\}^2$. It is easily seen (§5.22) that an ordinary point of equation (10.1) is also an ordinary point of equation (10.2).

Now consider the sequence of functions $v_n(z)$, analytic in S_b, defined by the equations

$$v_0(z) = a_0 + a_1(z - b),$$

$$v_n(z) = \int_b^z (\zeta - z) J(\zeta) v_{n-1}(\zeta) \, d\zeta, \qquad (n = 1, 2, 3, \ldots)$$

where a_0, a_1 are arbitrary constants.

Let M, μ be the upper bounds of $|J(z)|$ and $|v_0(z)|$ in the domain S_b. *Then at all points of this domain*

$$|v_n(z)| \le \frac{\mu M^n}{n!} |z - b|^{2n}. \qquad (10.3)$$

For this inequality is true when $n = 0$; if it is true when $n = 0, 1, \ldots, m - 1$, we have, by taking the path of integration to be a straight line,

$$
\begin{aligned}
|v_m(z)| &= \left| \int_b^z (\zeta - z) J(\zeta) v_{m-1}(\zeta) \, d\zeta \right| \\
&\le \frac{1}{(m-1)!} \int_b^z |\zeta - z| \, |J(\zeta)| \mu M^{m-1} |\zeta - b|^{2m-2} \, |d\zeta| \\
&\le \frac{\mu M^m}{(m-1)!} |z - b| \int_0^{|z-b|} t^{2m-2} \, dt \\
&\le \frac{\mu M^m}{m!} |z - b|^{2m}.
\end{aligned}
$$

Therefore, by induction, the inequality holds for all values of n.

Also, since $|v_n(z)| \le \frac{\mu M^n}{n!} r_b^{2n}$ when z is in S_b and $\sum_{n=0}^\infty \frac{\mu M^n}{n!} r_b^{2n}$ converges, it follows (§3.34) that $v(z) = \sum_{n=0}^\infty v_n(z)$ is a series of analytic functions uniformly convergent in S_b; while, from the definition of $v_n(z)$,

$$\frac{d}{dz} v_n(z) = -\int_b^z J(\zeta) v_{n-1}(\zeta) \, d\zeta, \qquad (n = 1, 2, 3, \ldots)$$

$$\frac{d^2}{dz^2} v_n(z) = -J(z) v_{n-1}(z);$$

hence it follows (§5.3) that

$$
\begin{aligned}
\frac{d^2 v(z)}{dz^2} &= \frac{d^2 v_0(z)}{dz^2} + \sum_{n=1}^\infty \frac{d^2 v_n(z)}{dz^2} \\
&= -J(z) v(z).
\end{aligned}
$$

Therefore $v(z)$ is a function of z, analytic in S_b, which satisfies the differential equation

$$\frac{d^2 v(z)}{dz^2} + J(z) v(z) = 0,$$

and, from the value obtained for $\dfrac{d}{dz} v_n(z)$, *it is evident that*

$$v(b) = a_0, \qquad v'(b) = \left\{ \frac{d}{dz} v(z) \right\}_{z=b} = a_1,$$

where a_0, a_1 are arbitrary.

10.21 Uniqueness of the solution

If there were two analytic solutions of the equation for v, say $v_1(z)$ and $v_2(z)$, such that $v_1(b) = v_2(b) = a_0$, and $v_1'(b) = v_2'(b) = a_1$, then, writing $w(z) = v_1(z) - v_2(z)$, we should have

$$\frac{d^2w(z)}{dz^2} + J(z)w(z) = 0.$$

Differentiating this equation $n - 2$ times and putting $z = b$, we get

$$w^{(n)}(b) + J(b)w^{(n-2)}(b) + \binom{n-2}{1}J'(b)w^{(n-3)}(b) + \cdots + J^{(n-2)}(b)w(b) = 0.$$

Putting $n = 2, 3, 4, \ldots$ in succession, we see that all the differential coefficients of $w(z)$ vanish at b; and so, by Taylor's theorem, $w(z) = 0$; that is to say the two solutions $v_1(z)$, $v_2(z)$ are identical.

Writing $u(z) = v(z) \exp\left\{-\frac{1}{2}\int_b^z p(\zeta)\,d\zeta\right\}$, we infer without difficulty that $u(z)$ is the only analytic solution of (10.1) such that $u(b) = A_0$, $u'(b) = A_1$, where

$$A_0 = a_0, \qquad A_1 = a_1 - \tfrac{1}{2}p(b)\,a_0.$$

Now that we know that a solution of (10.1) exists which is analytic in S_b and such that $u(b)$, $u'(b)$ have the arbitrary values A_0, A_1, the simplest method of obtaining the solution in the form of a Taylor's series is to assume

$$u(z) = \sum_{n=0}^{\infty} A_n(z - b)^n,$$

substitute this series in the differential equation and equate coefficients of successive powers of $z - b$ to zero (§3.73) to determine in order the values of $A_2, A_3, \ldots$ in terms of A_0, A_1.

Note In practice, in carrying out this process of substitution, the reader will find it much more simple to have the equation 'cleared of fractions' rather than in the canonical form (10.1) of §10.1. Thus the equations in Examples 10.2.1 and 10.2.2 below should be treated in the form in which they stand; the factors $1 - z^2$, $(z - 2)$, $(z - 3)$ should *not* be divided out. The same remark applies to the examples of §§10.3 and 10.32.

From the general theory of analytic continuation (§5.5) it follows that the solution obtained is analytic at all points of S except at singularities of the differential equation. The solution however is *not*, in general, *analytic throughout* S (see the footnote after Corollary 5.2.2), except at these points, as it may not be one-valued; i.e., it may not return to the same value when z describes a circuit surrounding one or more singularities of the equation.

The property that the solution of a linear differential equation is analytic except at singularities of the coefficients of the equation is common to linear equations of all orders.

When two particular solutions of an equation of the second order are not constant multiples of each other, they are said to form a *fundamental system*.

Example 10.2.1   Shew that the equation

$$(1 - z^2)u'' - 2zu' + \tfrac{3}{4}u = 0$$

has the fundamental system of solutions

$$u_1 = 1 - \frac{3}{8}z^2 - \frac{21}{128}z^4 - \cdots, \qquad u_2 = z + \frac{5}{24}z^3 + \frac{15}{128}z^5 + \cdots.$$

Determine the general coefficient in each series, and shew that the radius of convergence of each series is 1.

Example 10.2.2   Discuss the equation

$$(z - 2)(z - 3)u'' - (2z - 5)u' + 2u = 0$$

in a manner similar to that of Example 10.2.1.

10.3 Points which are regular for a differential equation

Suppose that a point c of S is such that, although $p(z)$ or $q(z)$ or both have poles at c, the poles are of such orders that $(z - c)p(z)$, $(z - c)^2 q(z)$ are analytic at c. Such a point is called a *regular point*[1] for the differential equation. Any poles of $p(z)$ or of $q(z)$ which are not of this nature are called *irregular points*. The reason for making the distinction will become apparent in the course of this section.

If c be a regular point, the equation may be written[2]

$$(z - c)^2 \frac{d^2 u}{dz^2} + (z - c)P(z - c)\frac{du}{dz} + Q(z - c)u = 0,$$

where $P(z - c)$, $Q(z - c)$ are analytic at c; hence, by Taylor's theorem,

$$P(z - c) = p_0 + p_1(z - c) + p_2(z - c)^2 + \cdots,$$
$$Q(z - c) = q_0 + q_1(z - c) + q_2(z - c)^2 + \cdots,$$

where $p_0, p_1, \ldots, q_0, q_1, \ldots$ are constants; and these series converge in the domain S_c formed by a circle of radius r (centre c) and its interior, where r is so small that c is the only singular point of the equation which is in S_c.

Let us assume as a *formal* solution of the equation

$$u = (z - c)^\alpha \left[1 + \sum_{n=1}^{\infty} a_n(z - c)^n \right],$$

where $\alpha, a_1, a_2, \ldots$ are constants to be determined.

Substituting in the differential equation (assuming that the term-by-term differentiations

---

[1] The name 'regular point' is due to Thomé [624]. Fuchs had previously used the phrase 'point of determinateness'.

[2] Frobenius calls this the normal form of the equation.

and multiplications of series are legitimate) we get

$$(z - c)^\alpha \left[\alpha(\alpha - 1) + \sum_{n=1}^{\infty} a_n(\alpha + n)(\alpha + n - 1)(z - c)^n \right]$$

$$+ (z - c)^\alpha P(z - c) \left[\alpha + \sum_{n=1}^{\infty} a_n(\alpha + n)(z - c)^n \right]$$

$$+ (z - c)^\alpha Q(z - c) \left[1 + \sum_{n=1}^{\infty} a_n(z - c)^n \right] = 0.$$

Substituting the series for $P(z - c)$, $Q(z - c)$, multiplying out and equating to zero the coefficients of successive powers of $z - c$, we obtain the following sequence of equations:

$$\alpha^2 + (p_0 - 1)\alpha + q_0 = 0,$$

$$a_1\{(\alpha + 1)^2 + (p_0 - 1)(\alpha + 1) + q_0\} + \alpha p_1 + q_1 = 0,$$

$$a_2\{(\alpha + 2)^2 + (p_0 - 1)(\alpha + 2) + q_0\} + a_1\{(\alpha + 1)p_1 + q_1\} + \alpha p_2 + q_2 = 0,$$

$$\vdots$$

$$a_n\{(\alpha + n)^2 + (p_0 - 1)(\alpha + n) + q_0\}$$

$$+ \sum_{m=1}^{n-1} a_{n-m}\{(\alpha + n - m)p_m + q_m\} + \alpha p_n + q_n = 0.$$

The first of these equations, called the *indicial equation* (the name is due to Cayley [139]), determines two values (which may, however, be equal) for α. The reader will easily convince himself that if c had been an *irregular* point, the indicial equation would have been (at most) of the first degree; and he will now appreciate the distinction made between regular and irregular singular points.

Let $\alpha = \rho_1$, $\alpha = \rho_2$ be the roots of

$$F(\alpha) \equiv \alpha^2 + (p_0 - 1)\alpha + q_0 = 0;$$

(these roots are called the *exponents* of the indicial equation) then the succeeding equations (when α has been chosen) determine $a_1, a_2, \ldots$, in order, uniquely, provided that $F(\alpha + n)$ does not vanish when $n = 1, 2, 3, \ldots$; that is to say, if $\alpha = \rho_1$, that ρ_2 is not one of the numbers $\rho_1 + 1$, $\rho_1 + 2, \ldots$; and, if $\alpha = \rho_2$, that ρ_1 is not one of the numbers $\rho_2 + 1$, $\rho_2 + 2, \ldots$.

Hence, if the difference of the exponents is not zero, or an integer, it is always possible to obtain two distinct series which formally satisfy the equation.

Example 10.3.1 Shew that, if m is not zero or an integer, the equation

$$u'' + \left(\frac{\frac{1}{4} - m^2}{z^2} - \frac{1}{4} \right) u = 0$$

is formally satisfied by two series whose leading terms are

$$z^{1/2+m} \left\{ 1 + \frac{z^2}{16(1 + m)} + \cdots \right\}, \quad z^{1/2-m} \left\{ 1 + \frac{z^2}{16(1 - m)} + \cdots \right\};$$

determine the coefficient of the general term in each series, and shew that the series converge for all values of z.

10.31 Convergence of the expansion of §10.3

If the exponents ρ_1, ρ_2 are not equal, let ρ_1 be that one whose real part is not inferior to the real part of the other, and let $\rho_1 - \rho_2 = s$; then

$$F(\rho_1 + n) = n(s + n).$$

Now, by §5.23, we can find a positive number M such that

$$|p_n| < Mr^{-n}, \quad |q_n| < Mr^{-n}, \quad |\rho_1 p_n + q_n| < Mr^{-n},$$

where M is independent of n; it is convenient to take $M \geq 1$.

Taking $\alpha = \rho_1$, we see that

$$|a_1| \frac{|\rho_1 p_1 + q_1|}{|F(\rho_1 + 1)|} < \frac{M}{r|s+1|} < \frac{M}{r},$$

since $|s+1| \geq 1$.

If now we assume $|a_n| < M^n r^{-n}$ when $n = 1, 2, \ldots, m-1$, we get

$$|a_m| = \left| \frac{\sum_{t=1}^{m-1} a_{m-t} \{(\rho_1 + m - t)p_1 + q_t\} + \rho_1 p_m + q_m}{F(\rho_1 + m)} \right|$$

$$\leq \frac{\sum_{t=1}^{m-1} |a_{m-t}| \cdot |\rho_1 p_1 + q_t| + |\rho_1 p_m + q_m| + \sum_{t=1}^{m-1} (m-t)|a_{m-t}||p_t|}{m|s+m|}$$

$$< \frac{mM^m r^{-m} + \left\{ \sum_{t=1}^{m-1} (m-t) \right\} M^m r^{-m}}{m^2 |1 + sm^{-1}|}.$$

Since $|1 + sm^{-1}| \geq 1$, because Re s is not negative, we get

$$|a_m| < \frac{m+1}{2m} M^m r^{-m} < M^m r^{-m},$$

and so, by induction, $|a_n| < M^n r^{-n}$ for all values of n.

If the values of the coefficients corresponding to the exponent ρ_2 be $a_1', a_2', \ldots$ we should obtain, by a similar induction,

$$|a_n'| < M^n k^n r^{-n},$$

where k is the upper bound of $|1-s|^{-1}, |1 - \frac{1}{2}s|^{-1}, |1 - \frac{1}{3}s|^{-1}, \ldots$; this bound exists when s is not a positive integer.

We have thus obtained two formal series

$$w_1(z) = (z - c)^{\rho_1} \left[1 + \sum_{n=1}^{\infty} a_n (z - c)^n \right],$$

$$w_2(z) = (z - c)^{\rho_2} \left[1 + \sum_{n=1}^{\infty} a'_n (z - c)^n \right].$$

The first, however, is a uniformly convergent series of analytic functions when $|z - c| < rM^{-1}$, as is also the second when $|z - c| < rM^{-1}k^{-1}$, provided in each case that $\arg(z - c)$ is restricted in such a way that the series are one-valued; consequently, the formal substitution of these series into the left-hand side of the differential equation is justified, and each of the series is a solution of the equation; provided always that $\rho_1 - \rho_2$ is not a positive integer or zero. (If $\rho_1 - \rho_2$ is a positive integer, k does not exist; if $\rho_1 = \rho_2$, the two solutions are the same.)

With this exception, we have therefore obtained a fundamental system of solutions valid in the vicinity of a regular singular point. And by the theory of analytic continuation, we see that if all the singularities in S of the equation are regular points, each member of a pair of fundamental solutions is analytic at all points of S except at the singularities of the equation, which are branch-points of the solution.

10.32 Derivation of a second solution in the case when the difference of the exponents is an integer or zero

In the case when $\rho_1 - \rho_2 = s$ is a positive integer or zero, the solution $w_2(z)$ found in §10.31 may break down or coincide with $w_1(z)$. (The coefficient a'_s may be indeterminate or it may be infinite; in the former case $w_2(z)$ will be a solution containing two arbitrary constants a'_0 and a'_s; the series of which a'_s is a factor will be a constant multiple of $w_1(z)$.) If we write $u = w_1(z)\zeta$, the equation to determine ζ is

$$(z - c)^2 \frac{d^2\zeta}{dz^2} + \left\{ 2(z - c)^2 \frac{w'_1(z)}{w_1(z)} + (z - c)P(z - c) \right\} \frac{d\zeta}{dz} = 0,$$

of which the general solution is

$$\zeta = A + B \int^z \frac{1}{\{w_1(z)\}^2} \exp\left\{ -\int^z \frac{P(z - c)}{z - c} dz \right\} dz$$

$$= A + B \int^z \frac{(z - c)^{-p_0}}{\{w_1(z)\}^2} \exp\left\{ -p_1(z - c) - \frac{1}{2}p_2(z - c)^2 - \cdots \right\} dz$$

$$= A + B \int^z (z - c)^{-p_0 - 2\rho_1} g(z) \, dz,$$

where A, B are arbitrary constants and $g(z)$ is analytic throughout the interior of any circle whose centre is c, which does not contain any singularities of $P(z - c)$ or singularities or zeros of $(z - c)^{-\rho_1} w_1(z)$; also $g(c) = 1$.

Let $g(z) = 1 + \sum\limits_{n=1}^{\infty} g_n(z-c)^n$. Then, if $s \neq 0$,

$$\zeta = A + B \int^z \left\{ 1 + \sum_{n=1}^{\infty} g_n(z-c)^n \right\} (z-c)^{-s-1} dz$$

$$= A + B \left[-\frac{1}{s}(z-c)^{-s} - \sum_{n=1}^{s-1} \frac{g_n}{s-n}(z-c)^{n-s} + g_s \log(z-c) \right.$$

$$\left. + \sum_{n=s+1}^{\infty} \frac{g_n}{n-s}(z-c)^{n-s} \right].$$

Therefore the general solution of the differential equation, which is analytic at all points of C (c excepted), is $Aw_1(z) + B\left[g_s w_1(z) \log (z-c) + \overline{w}(z)\right]$, where, by §2.53,

$$\overline{w}(z) = (z-c)^{p_2} \left\{ -\frac{1}{s} + \sum_{n=1}^{\infty} h_n (z-c)^n \right\},$$

the coefficients h_n being constants.

When $s = 0$, the corresponding form of the solution is

$$Aw_1(z) + B\left[w_1(z) \log(z-c) + (z-c)^{p_2} \sum_{n=1}^{\infty} h_n(z-c)^n \right].$$

The statement made at the end of §10.31 is now seen to hold in the exceptional case when s is zero or a positive integer.

In the special case when $g_s = 0$, the second solution does not involve a logarithm.

The solutions obtained, which are valid in the vicinity of a regular point of the equation, are called *regular integrals*.

Integrals of an equation valid near a regular point c may be obtained practically by first obtaining $w_1(z)$, and then determining the coefficients in a function $\overline{w}_1(z) = \sum\limits_{n=1}^{\infty} b_n(z-c)^{p_2+n}$, by substituting $w_1(z) \log(z-c) + \overline{w}_1(z)$ in the left-hand side of the equation and equating to zero the coefficients of the various powers of $z - c$ in the resulting expression. An alternative method due to Frobenius [227] is given by Forsyth [221, pp. 243–258].

Example 10.3.2   Shew that integrals of the equation

$$\frac{d^2u}{dz^2} + \frac{1}{z}\frac{du}{dz} - m^2 u = 0$$

regular near $z = 0$ are

$$w_1(z) = 1 + \sum_{n=1}^{\infty} \frac{m^{2n} z^{2n}}{2^{2n} n!^2}$$

and

$$w_1(z) \log z - \sum_{n=1}^{\infty} \frac{m^{2n} z^{2n}}{2^{2n} n!^2} \left(\frac{1}{1} + \frac{1}{2} + \cdots + \frac{1}{n} \right).$$

Verify that these series converge for all values of z.

Example 10.3.3   Shew that integrals of the equation

$$z(z-1)\frac{d^2u}{dz^2} + (2z-1)\frac{du}{dz} + \frac{1}{4}u = 0$$

regular near $z = 0$ are

$$w_1(z) = 1 + \sum_{n=1}^{\infty} \left(\frac{1\cdot 3\cdots(2n-1)}{2\cdot 4\cdots 2n}\right)^2 z^n$$

and

$$w_1(z)\log z + 4\sum_{n=1}^{\infty} \left(\frac{1\cdot 3\cdots(2n-1)}{2\cdot 4\cdots 2n}\right)^2 \left(\frac{1}{1} - \frac{1}{2} + \frac{1}{3} - \cdots - \frac{1}{2n}\right) z^n.$$

Verify that these series converge when $|z| < 1$ and obtain integrals regular near $z = 1$.

Example 10.3.4   Shew that the hypergeometric equation

$$z(1-z)\frac{d^2u}{dz^2} + \{c - (a+b+1)z\}\frac{du}{dz} - abu = 0$$

is satisfied by the hypergeometric series of §2.38. Obtain the complete solution of the equation when $c = 1$.

10.4 Solutions valid for large values of |z|

Let $z = 1/z_1$; then a solution of the differential is said to be valid for *large values of* $|z|$ if it is valid for sufficiently small values of $|z_1|$; and it is said that 'the point at infinity is an ordinary (or regular or irregular) point of the equation' when the point $z_1 = 0$ is an ordinary (or regular or irregular) point of the equation when it has been transformed so that z_1 is the independent variable.

Since

$$\frac{d^2u}{dz^2} + p(z)\frac{du}{dz} + q(z)u \equiv z_1^4\frac{d^2u}{dz_1^2} + \left\{2z_1^3 - z_1^2p\left(\frac{1}{z_1}\right)\right\}\frac{du}{dz_1} + q\left(\frac{1}{z_1}\right)u,$$

we see that the conditions that the point $z = \infty$ should be (i) an ordinary point, (ii) a regular point, are (i) that $2z - z^2p(z)$, $z^4q(z)$ should be analytic at infinity (§5.62) and (ii) that $zp(z)$, $z^2q(z)$ should be analytic at infinity.

Example 10.4.1   Shew that every point (including infinity) is either an ordinary point or a regular point for each of the equations

$$z(1-z)\frac{d^2u}{dz^2} + \{c - (a+b+1)z\}\frac{du}{dz} - abu = 0,$$

$$(1-z^2)\frac{d^2u}{dz^2} - 2z\frac{du}{dz} + n(n+1)u = 0,$$

where a, b, c, n are constants.

Example 10.4.2 Shew that every point except infinity is either an ordinary point or a regular point for the equation

$$z^2 \frac{d^2u}{dz^2} + z\frac{du}{dz} + (z^2 - n^2)u = 0,$$

where n is a constant.

Example 10.4.3 Shew that the equation

$$(1 - z^2)\frac{d^2u}{dz^2} - 2z\frac{du}{dz} + 6u = 0$$

has the two solutions

$$z^2 - \frac{1}{3}, \quad \frac{1}{z^3} + \frac{3\cdot 4}{2\cdot 7}\frac{1}{z^5} + \frac{3\cdot 4\cdot 5\cdot 6}{2\cdot 4\cdot 7\cdot 9}\frac{1}{z^7} + \cdots,$$

the latter converging when $|z| > 1$.

10.5 Irregular singularities and confluence

Near a point which is not a regular point of linear differential equations, an equation of the second order cannot have two regular integrals, for the indicial equation is at most of the first degree; there may be one regular integral or there may be none. We shall see later (e.g. §16.3) what is the nature of the solution near such points in some simple cases. A general investigation of such solutions is beyond the scope of this book. Some elementary investigations are given in Forsyth's [221]. Complete investigations are given in his *Theory of Differential Equations* [218].

It frequently happens that a differential equation may be derived from another differential equation by making two or more singularities of the latter tend to coincidence. Such a limiting process is called *confluence*; and the former equation is called a *confluent form* of the latter. It will be seen in §10.6 that the singularities of the former equation may be of a more complicated nature than those of the latter.

10.6 The differential equations of mathematical physics

The most general differential equation of the second order which has every point except a_1, a_2, a_3, a_4 and ∞ as an ordinary point, these five points being regular points with exponents α_r, β_r at a_r $(r = 1, 2, 3, 4)$ and exponents μ_1, μ_2 at ∞, may be verified[3] to be

$$\frac{d^2u}{dz^2} + \left\{ \sum_{r=1}^{4} \frac{1 - \alpha_r - \beta_r}{z - a_r} \right\}\frac{du}{dz} + \left\{ \sum_{r=1}^{4} \frac{\alpha_r\beta_r}{(z - a_r)^2} + \frac{Az^2 + 2Bz + C}{\prod_{r=1}^{4}(z - a_r)} \right\}u = 0,$$

---

[3] The coefficients of $\frac{du}{dz}$ and u must be rational or they would have an essential singularity at some point; the denominators of $p(z), q(z)$ must be $\prod_{r=1}^{4}(z - a_r), \prod_{r=1}^{4}(z - a_r)^2$ respectively; putting $p(z)$ and $q(z)$ into partial fractions and remembering that $p(z) = O(z^{-1}), q(z) = O(z^{-2})$ as $|z| \to \infty$, we obtain the required result without difficulty.

where A is such that μ_1 and μ_2 are the roots of

$$\mu^2 + \mu \left\{ \sum_{r=1}^{4} (\alpha_r + \beta_r) - 3 \right\} + \sum_{r=1}^{4} \alpha_r \beta_r + A = 0,$$

and B, C are constants. (It will be observed that μ_1, μ_2 are connected by the relation $\mu_1 + \mu_2 + \sum_{r=1}^{4} (\alpha_r + \beta_r) = 3$.)

The remarkable theorem has been proved by Klein [376] (see also [373]) and Bôcher [78] that all the linear differential equations which occur in certain branches of Mathematical Physics are confluent forms of the special equation of this type in which *the difference of the two exponents at each singularity is* $\frac{1}{2}$; a brief investigation of these forms will now be given.

If we put $\beta_r = \alpha_r + \frac{1}{2}$, $(r = 1, 2, 3, 4)$ and write ζ in place of z, the last written equation becomes

$$\frac{d^2u}{d\zeta^2} + \left\{ \sum_{r=1}^{4} \frac{\frac{1}{2} - 2\alpha_r}{\zeta - a_r} \right\} \frac{du}{d\zeta} + \left\{ \sum_{r=1}^{4} \frac{\alpha_r \left(\alpha_r + \frac{1}{2}\right)}{(\zeta - a_r)^2} + \frac{A\zeta^2 + 2B\zeta + C}{\prod_{r=1}^{4} (\zeta - a_r)} \right\} u = 0,$$

where (on account of the condition $\mu_2 - \mu_1 = \frac{1}{2}$)

$$A = \left(\sum_{r=1}^{4} \alpha_r \right)^2 - \sum_{r=1}^{4} \alpha_r^2 - \frac{3}{2} \sum_{r=1}^{4} \alpha_r + \frac{3}{16}.$$

This differential equation is called the *generalised Lamé equation*.

It is evident, on writing $a_1 = a_2$ throughout the equation, that the confluence of the two singularities a_1, a_2 yields a singularity at which the exponents α, β are given by the equations

$$\alpha + \beta = 2(\alpha_1 + \alpha_2), \quad \alpha\beta = \alpha_1 \left(\alpha_1 + \frac{1}{2}\right) + \alpha_2 \left(\alpha_2 + \frac{1}{2}\right) + D,$$

where

$$D = \frac{Aa_1^2 + 2Ba_1 + C}{(a_1 - a_3)(a_1 - a_4)}.$$

Therefore the exponent-difference at the confluent singularity *is not* $\frac{1}{2}$, *but it may have any assigned value by suitable choice of B and C*. In like manner, by the confluence of three or more singularities, we can obtain one irregular singularity.

By suitable confluences of the five singularities at our disposal, we can obtain six types of equations, which may be classified according to (a) the number of their singularities with exponent-difference $\frac{1}{2}$, (b) the number of their other regular singularities, (c) the number of their irregular singularities, by means of the following scheme, which is easily seen to be exhaustive:

	(a)	(b)	(c)	
(I)	3	1	0	Lamé
(II)	2	0	1	Mathieu
(III)	1	2	0	Legendre
(IV)	0	1	1	Bessel
(V)	1	0	1	Weber, Hermite
(VI)	0	0	1	Stokes

For instance the arrangement (a) 3, (b) 0, (c) 1, is inadmissible as it would necessitate *six* initial singularities. The last equation of this type was considered by Stokes [609] in his researches on Diffraction; it is, however, easily transformed into a particular case of Bessel.

These equations are usually known by the names of the mathematicians in the last column. Speaking generally, the later an equation comes in this scheme, the more simple are the properties of its solution. The solutions of (II)–(VI) are discussed in Chapters 15–19 of this work, and of (I) in Chapter 23. For properties of equations of type (I), see the works of Klein [376] and Forsyth [218]; also Todhunter [631]. The derivation of the standard forms of the equations from the generalised Lamé equation is indicated by the following examples:

Example 10.6.1   Obtain Lamé's equation

$$\frac{d^2u}{d\zeta^2} + \left\{ \sum_{r=1}^{3} \frac{\frac{1}{2}}{(\zeta - a_r)} \right\} \frac{du}{d\zeta} - \frac{\{n(n+1)\,\zeta + h\}\,u}{4 \prod_{r=1}^{3}(\zeta - a_r)} = 0,$$

(where h and n are constants) by taking

$$a_1 = a_2 = a_3 = a_4 = 0, \quad 8B = n(n+1)a_4, \quad 4C = ha_4,$$

and making $a_4 \to \infty$.

Example 10.6.2   Obtain the equation

$$\frac{d^2u}{d\zeta^2} + \left(\frac{\frac{1}{2}}{\zeta} + \frac{\frac{1}{2}}{\zeta - 1} \right) \frac{du}{d\zeta} - \frac{(a - 16q + 32q\zeta)u}{4\zeta(\zeta - 1)} = 0,$$

(where a and q are constants) by taking $a_1 = 0$, $a_2 = 1$, and making $a_3 = a_4 \to \infty$. Derive Mathieu's equation (§19.1)

$$\frac{d^2u}{dz^2} + (a + 16q \cos 2z)\,u = 0$$

by the substitution $\zeta = \cos^2 z$.

Example 10.6.3   Obtain the equation

$$\frac{d^2u}{d\zeta^2} + \left\{ \frac{\frac{1}{2}}{\zeta} + \frac{1}{\zeta - 1} \right\} \frac{du}{d\zeta} + \frac{1}{4} \left\{ \frac{n(n+1)}{\zeta} - \frac{m^2}{\zeta - 1} \right\} \frac{u}{\zeta(\zeta - 1)} = 0,$$

by taking

$$a_1 = a_2 = 1, \quad a_3 = a_4 = 0, \quad a_1 = a_2 = a_3 = 0, \quad a_4 = \frac{1}{4}.$$

Derive Legendre's equation (§15.13 and §15.5)

$$(1 - z^2) \frac{d^2 u}{dz^2} - 2z \frac{du}{dz} + \left\{ n(n+1) - \frac{m^2}{1-z^2} \right\} u = 0$$

by the substitution $\zeta = z^{-2}$.

Example 10.6.4 By taking $a_1 = a_2 = 0$, $a_1 = a_2 = a_3 = a_4 = 0$, and making $a_3 = a_4 \to \infty$, obtain the equation

$$\zeta^2 \frac{d^2 u}{d\zeta^2} + \zeta \frac{du}{d\zeta} + \frac{1}{4} (\zeta - n^2) u = 0.$$

Derive Bessel's equation (§17.11)

$$z^2 \frac{d^2 u}{dz^2} + z \frac{du}{dz} + (z^2 - n^2) u = 0$$

by the substitution $\zeta = z^2$.

Example 10.6.5 By taking $a_1 = 0$, $a_1 = a_2 = a_3 = a_4 = 0$, and making $a_2 = a_3 = a_4 \to \infty$, obtain the equation

$$\zeta \frac{d^2 u}{d\zeta^2} + \frac{1}{2} \frac{du}{d\zeta} + \frac{1}{4} \left(n + \frac{1}{2} - \frac{1}{4} \zeta \right) u = 0.$$

Derive Weber's equation (§16.5)

$$\frac{d^2 u}{dz^2} + \left(n + \frac{1}{2} - \frac{1}{4} z^2 \right) u = 0$$

by the substitution $\zeta = z^2$.

Example 10.6.6 By taking $a_r = 0$, and making $a_r \to \infty$ $(r = 1, 2, 3, 4)$, obtain the equation

$$\frac{d^2 u}{d\zeta^2} + (B_1 \zeta + C_1) u = 0.$$

By taking

$$u = (B_1 \zeta + C_1)^{\frac{1}{2}} v, \quad B_1 \zeta + C_1 = \left(\tfrac{3}{2} B_1 z \right)^{\frac{2}{3}},$$

shew that

$$z^2 \frac{d^2 v}{dz^2} + z \frac{dv}{dz} + \left(z^2 - \frac{1}{9} \right) v = 0.$$

Example 10.6.7 Shew that the general form of the generalised Lamé equation is unaltered (i) by any homographic change of independent variable such that ∞ is a singular point of the transformed equation, (ii) by any change of dependent variable of the type $u = (z - a_r)^\lambda v$.

Example 10.6.8 Deduce from Example 10.6.7 that the various confluent forms of the generalised Lamé equation may always be reduced to the forms given in Examples 10.6.1–10.6.6. (Note that a suitable homographic change of variable will transform any three distinct points into the points $0, 1, \infty$.)

10.7 Linear differential equations with three singularities

Let

$$\frac{d^2u}{dz^2} + p(z)\frac{du}{dz} + q(z)u = 0 \tag{10.4}$$

have three, and only three singularities, a, b, c; let these points be regular points, the exponents thereat being α, α'; β, β'; γ, γ'. The point at infinity is to be an ordinary point.

Then $p(z)$ is a rational function with simple poles at a, b, c, its residues at these poles being $1 - \alpha - \alpha'$, $1 - \beta - \beta'$, $1 - \gamma - \gamma'$; and as $z \to \infty$, $p(z) - 2z^{-1}$ is $O(z^{-2})$. Therefore

$$p(z) = \frac{1 - \alpha - \alpha'}{z - a} + \frac{1 - \beta - \beta'}{z - b} + \frac{1 - \gamma - \gamma'}{z - c}$$

and $\alpha + \alpha' + \beta + \beta' + \gamma + \gamma' = 1$. This relation must be satisfied by the exponents.

In a similar manner

$$q(z) = \left\{ \frac{\alpha\alpha'(a - b)(a - c)}{z - a} + \frac{\beta\beta'(b - c)(b - a)}{z - b} + \frac{\gamma\gamma'(c - a)(c - b)}{z - c} \right\}$$

$$\times \frac{1}{(z - a)(z - b)(z - c)},$$

and hence the differential equation is

$$\frac{d^2u}{dz^2} + \left\{ \frac{1 - \alpha - \alpha'}{z - a} + \frac{1 - \beta - \beta'}{z - b} + \frac{1 - \gamma - \gamma'}{z - c} \right\} \frac{du}{dz}$$

$$+ \left\{ \frac{\alpha\alpha'(a - b)(a - c)}{z - a} + \frac{\beta\beta'(b - c)\,(b - a)}{z - b} + \frac{\gamma\gamma'(c - a)(c - b)}{z - c} \right\}$$

$$\times \frac{u}{(z - a)(z - b)(z - c)} = 0.$$

This equation was first given by Papperitz [515].

To express the fact that u satisfies an equation of this type (which will be called Riemann's *P*-equation [556]; it will be seen from this memoir that, although Riemann did not apparently construct the equation, he must have inferred its existence from the hypergeometric equation) Riemann wrote

$$u = P \left\{ \begin{matrix} a & b & c & \\ \alpha & \beta & \gamma & z \\ \alpha' & \beta' & \gamma' & \end{matrix} \right\}.$$

The singular points of the equation are placed in the first row with the corresponding exponents directly beneath them, and the independent variable is placed in the fourth column.

Example 10.7.1   Shew that the hypergeometric equation

$$z(1 - z)\frac{d^2u}{dz^2} + \{c - (a + b + 1)z\}\frac{du}{dz} - abu = 0$$

is defined by the scheme

$$P \left\{ \begin{matrix} 0 & \infty & 1 & \\ 0 & a & 0 & z \\ 1 - c & b & c - a - b & \end{matrix} \right\}.$$

10.71 Transformations of Riemann's P-equation

The two transformations which are typified by the equations

(I)

$$\left(\frac{z-a}{z-b}\right)^k \left(\frac{z-c}{z-b}\right)^l P\left\{\begin{matrix} a & b & c \\ \alpha & \beta & \gamma & z \\ \alpha' & \beta' & \gamma' \end{matrix}\right\} = P\left\{\begin{matrix} a & b & c \\ \alpha+k & \beta-k-l & \gamma+l & z \\ \alpha'+k & \beta'-k-l & \gamma'+l \end{matrix}\right\},$$

(II)

$$P\left\{\begin{matrix} a & b & c \\ \alpha & \beta & \gamma & z \\ \alpha' & \beta' & \gamma' \end{matrix}\right\} = P\left\{\begin{matrix} a_1 & b_1 & c_1 \\ \alpha & \beta & \gamma & z_1 \\ \alpha' & \beta' & \gamma' \end{matrix}\right\}$$

(where z_1, a_1, b_1, c_1 are derived from z, a, b, c by the same homographic transformation) are of great importance. They may be derived by direct transformation of the differential equation of Papperitz and Riemann by suitable changes in the dependent and independent variables respectively; but the truth of the results of the transformations may be seen intuitively when we consider that Riemann's P-equation is determined *uniquely* by a knowledge of the three singularities and their exponents, and (I) that if

$$u = P\left\{\begin{matrix} a & b & c \\ \alpha & \beta & \gamma & z \\ \alpha' & \beta' & \gamma' \end{matrix}\right\},$$

then $u_1 = \left(\dfrac{z-a}{z-b}\right)^k \left(\dfrac{z-c}{z-b}\right)^l u$ satisfies a differential equation of the second order with the same three singular points and exponents $\alpha+k, \alpha'+k; \beta-k-l, \beta'-k-l; \gamma+l, \gamma'+l$; and that the sum of the exponents is 1.

Also (II) if we write $z = \dfrac{Az_1+B}{Cz_1+D}$, the equation in z_1 is a linear equation of the second order with singularities at the points derived from a, b, c by this homographic transformation, and exponents $\alpha, \alpha'; \beta, \beta'; \gamma, \gamma'$ thereat.

10.72 The connexion of Riemann's P-equation with the hypergeometric equation

By means of the results of §10.71 it follows that

$$P\left\{\begin{matrix} a & b & c \\ \alpha & \beta & \gamma & z \\ \alpha' & \beta' & \gamma' \end{matrix}\right\} = \left(\frac{z-a}{z-b}\right)^\alpha \left(\frac{z-c}{z-b}\right)^\gamma P\left\{\begin{matrix} a & b & c \\ 0 & \beta+\alpha+\gamma & 0 & z \\ \alpha'-\alpha & \beta'+\alpha+\gamma & \gamma'-\gamma \end{matrix}\right\}$$

$$= \left(\frac{z-a}{z-b}\right)^\alpha \left(\frac{z-c}{z-b}\right)^\gamma P\left\{\begin{matrix} 0 & \infty & 1 \\ 0 & \beta+\alpha+\gamma & 0 & x \\ \alpha'-\alpha & \beta'+\alpha+\gamma & \gamma'-\gamma \end{matrix}\right\},$$

where

$$x = \frac{(z-a)(c-b)}{(z-b)(c-a)}.$$

Hence, by Example 10.7.1, the solution of Riemann's P-equation can always be obtained in terms of the solution of the hypergeometric equation whose elements a, b, c, x are $\alpha + \beta + \gamma$, $\alpha + \beta' + \gamma$, $1 + \alpha - \alpha'$, $(z - a)(c - b)/(z - b)(c - a)$ respectively.

10.8 Linear differential equations with two singularities

If, in §10.7, we make the point c an ordinary point, we must have $1 - \gamma - \gamma' = 0$, $\gamma\gamma' = 0$ and $\dfrac{\alpha\alpha'(a - b)(a - c)}{z - a} + \dfrac{\beta\beta'(b - c)(b - a)}{z - b}$ must be divisible by $z - c$, in order that $p(z)$ and $q(z)$ may be analytic at c.

Hence $\alpha + \alpha' + \beta + \beta' = 0$, $\alpha\alpha' = \beta\beta'$, and the equation is

$$\frac{d^2u}{dz^2} + \left\{ \frac{1 - \alpha - \alpha'}{z - a} + \frac{1 + \alpha + \alpha'}{z - b} \right\} \frac{du}{dz} + \frac{\alpha\alpha'(a - b)^2 u}{(z - a)^2 (z - b)^2} = 0,$$

of which the solution is

$$u = A \left(\frac{z - a}{z - b} \right)^{\alpha} + B \left(\frac{z - a}{z - b} \right)^{\alpha'};$$

that is to say, the solution involves elementary functions only.

When $\alpha = \alpha'$, the solution is

$$u = A \left(\frac{z - a}{z - b} \right)^{\alpha} + B_1 \left(\frac{z - a}{z - b} \right)^{\alpha} \log \left(\frac{z - a}{z - b} \right).$$

10.9 Miscellaneous examples

Example 10.1   Shew that two solutions of the equation

$$\frac{d^2u}{dz^2} + zu = 0$$

are $z - \frac{1}{12}z^4 + \cdots$, and $1 - \frac{1}{6}z^3 + \cdots$, and investigate the region of convergence of these series.

Example 10.2   Obtain integrals of the equation

$$\frac{d^2u}{dz^2} + \frac{1 - z^2}{4z^2} u = 0,$$

regular near $z = 0$, in the form

$$u_1 = z^{1/2} \left\{ 1 + \frac{z^2}{16} + \frac{z^4}{1024} + \cdots \right\},$$

$$u_2 = u_1 \log z - \frac{z^{3/2}}{16} + \cdots.$$

Example 10.3   Shew that the equation

$$\frac{d^2u}{dz^2} + \left(n + \frac{1}{2} - \frac{1}{4}z^2 \right) u = 0$$

has the solutions

$$1 - \frac{2n+1}{4} z^2 + \frac{4n^2 + 4n + 3}{96} z^4 - \cdots,$$

$$z - \frac{2n+1}{12} z^3 + \frac{4n^2 + 4n + 7}{480} z^5 - \cdots,$$

and that these series converge for all values of z.

Example 10.4 (Klein)   Shew that the equation

$$\frac{d^2u}{dz^2} + \left\{ \sum_{r=1}^{n} \frac{1 - \alpha_r - \beta_r}{z - a_r} \right\} \frac{du}{dz} + \left\{ \sum_{r=1}^{n} \frac{\alpha_r \beta_r}{(z - a_r)^2} + \sum_{r=1}^{n} \frac{D_r}{z - a_r} \right\} u = 0,$$

where

$$\sum_{r=1}^{n} (\alpha_r + \beta_r) = n - 2, \quad \sum_{r=1}^{n} D_r = 0, \quad \sum_{r=1}^{n} (a_r D_r + \alpha_r \beta_r) = 0,$$

$$\sum_{r=1}^{n} (a_r^2 D_r + 2a_r \alpha_r \beta_r) = 0,$$

is the most general equation for which all points (including ∞), except $a_1, a_2, \ldots, a_n$, are ordinary points, and the points a_r are regular points with exponents α_r, β_r respectively.

Example 10.5 (Riemann)   Shew that, if $\beta + \gamma + \beta' + \gamma' = \frac{1}{2}$ then

$$P \left\{ \begin{matrix} 0 & \infty & 1 & \\ 0 & \beta & \gamma & z^2 \\ \frac{1}{2} & \beta' & \gamma' & \end{matrix} \right\} = P \left\{ \begin{matrix} -1 & \infty & 1 & \\ \gamma & 2\beta & \gamma & z \\ \gamma' & 2\beta' & \gamma' & \end{matrix} \right\}.$$

The differential equation in each case is

$$\frac{d^2u}{dz^2} + \frac{2z(1 - \gamma - \gamma')}{z^2 - 1} \frac{du}{dz} + \left(\beta\beta' + \frac{\gamma\gamma'}{z^2 - 1} \right) \frac{4u}{z^2 - 1} = 0.$$

Example 10.6 (Riemann)   Shew that, if $\gamma + \gamma' = \frac{1}{3}$ and if ω, ω^2 are the complex cube roots of unity, then

$$P \left\{ \begin{matrix} 0 & \infty & 1 & \\ 0 & 0 & \gamma & z^3 \\ \frac{1}{3} & \frac{1}{3} & \gamma' & \end{matrix} \right\} = P \left\{ \begin{matrix} 1 & \omega & \omega^2 & \\ \gamma & \gamma & \gamma & z \\ \gamma' & \gamma' & \gamma' & \end{matrix} \right\}.$$

The differential equation in each case is

$$\frac{d^2u}{dz^2} + \frac{2z^2}{z^3 - 1} \frac{du}{dz} + \frac{9\gamma\gamma' zu}{(z^3 - 1)^2} = 0.$$

Example 10.7 (Halm)   Shew that the equation

$$(1 - z^2) \frac{d^2u}{dz^2} - (2a + 1) z \frac{du}{dz} + n(n + 2a) u = 0$$

is defined by the scheme

$$P\left\{\begin{matrix} 1 & \infty & -1 \\ 0 & -n & 0 \\ \frac{1}{2} - a & n + 2a & \frac{1}{2} - a \end{matrix} \; z\right\},$$

and that the equation

$$(1 + \zeta^2)^2 \frac{d^2 u}{d\zeta^2} + n(n + 2)u = 0$$

may be obtained from it by taking $a = 1$ and changing the independent variable.

Example 10.8 (Cunningham)   Discuss the solutions of the equation

$$z \frac{d^2 u}{dz^2} + (z + 1 + m) \frac{du}{dz} + \left(n + 1 + \frac{1}{2}m\right) u = 0$$

valid near $z = 0$ and those valid near $z = \infty$.

Example 10.9 (Curzon)   Discuss the solutions of the equation

$$\frac{d^2 u}{dz^2} + \frac{2\mu}{z} \frac{du}{dz} - 2z \frac{du}{dz} + 2(v - \mu)u = 0$$

valid near $z = 0$ and those valid near $z = \infty$. Consider the following special cases:

$$\text{(i)} \quad \mu = -\frac{3}{2}; \qquad \text{(ii)} \quad \mu = \frac{1}{2}; \qquad \text{(iii)} \quad \mu + v = 3.$$

Example 10.10 (Lindemann; see §19.5)   Prove that the equation

$$z(1 - z) \frac{d^2 u}{dz^2} + \frac{1}{2}(1 - 2z) \frac{du}{dz} + (az + b)u = 0$$

has two particular integrals the product of which is a single-valued transcendental function. Under what circumstances are these two particular integrals coincident? If their product be $F(z)$, prove that the particular integrals are

$$u_1, u_2 = \sqrt{F(z)} \exp\left\{ \pm C \int^z \frac{dz}{F(z)\sqrt{z(1 - z)}} \right\},$$

where C is a determinate constant.

Example 10.11 (Math. Trip. 1912)   Prove that the general linear differential equation of the third order, whose singularities are 0, 1, ∞, which has all its integrals regular near each singularity (the exponents at each singularity being 1, 1, -1), is

$$\frac{d^3 u}{dz^3} + \left\{ \frac{2}{z} + \frac{2}{z - 1} \right\} \frac{d^2 u}{dz^2} - \left\{ \frac{1}{z^2} - \frac{3}{z(z - 1)} + \frac{1}{(z - 1)^2} \right\} \frac{du}{dz}$$
$$+ \left\{ \frac{1}{z^3} - \frac{3\cos^2 a}{z^2(z - 1)} - \frac{3\sin^2 a}{z(z - 1)^2} + \frac{1}{(z - 1)^3} \right\} u = 0,$$

where a may have any constant value.

11

Integral Equations

11.1 Definition of an integral equation

An integral equation is one which involves an unknown function under the sign of integration; and the process of determining the unknown function is called solving the equation. Except in the case of Fourier's integral (§9.7) we practically *always* need *continuous* solutions of integral equations.

The introduction of integral equations into analysis is due to Laplace (1782) who considered the equations

$$f(x) = \int e^{xt} \phi(t) \, dt, \qquad g(x) = \int t^{x-1} \phi(t) \, dt$$

(where in each case ϕ represents the unknown function), in connexion with the solution of differential equations. The first integral equation of which a solution was obtained, was Fourier's equation

$$f(x) = \int_{-\infty}^{\infty} \cos(xt) \phi(t) \, dt,$$

of which, in certain circumstances, a solution is[1]

$$\phi(x) = \frac{2}{\pi} \int_0^{\infty} \cos(ux) f(u) \, du,$$

$f(x)$ being an even function of x, since $\cos(xt)$ is an even function.

Later, Abel [6] was led to an integral equation in connexion with a mechanical problem and obtained two solutions of it; after this, Liouville investigated an integral equation which arose in the course of his researches on differential equations and discovered an important method for solving integral equations[2], which will be discussed in §11.4.

In recent years, the subject of integral equations has become of some importance in various branches of Mathematics; such equations (in physical problems) frequently involve repeated integrals and the investigation of them naturally presents greater difficulties than do those elementary equations which will be treated in this chapter.

To render the analysis as easy as possible, we shall suppose throughout that the constants a, b and the variables x, y, ξ are real and further that $a \leq x, y, \xi \leq b$; also that the given function, $K(x, y)$, which occurs under the integral sign in the majority of equations considered, is a real function of x and y and either (i) it is a continuous function of both variables in the range

---

[1] If this value of ϕ be substituted in the equation we obtain a result which is, effectively, that of §9.7.
[2] The numerical computation of solutions of integral equations has been investigated by Whittaker [677].

$(a \leq x \leq b, a \leq y \leq b)$, or (ii) it is a continuous function of both variables in the range $a \leq y \leq x \leq b$ and $K(x, y) = 0$ when $y > x$; in the latter case $K(x, y)$ has its discontinuities regularly distributed, and in either case it is easily proved that, if $f(y)$ is continuous when $a \leq y \leq b$, $\int_a^b f(y)K(x, y)\, dy$ is a continuous function of x when $a \leq x \leq b$.

Bôcher [80] in his important work on integral equations, always considers the more general case in which $K(x, y)$ has discontinuities *regularly distributed*, i.e. the discontinuities are of the nature described in Example 4.11. The reader will see from that example that the results of this chapter can almost all be generalised in this way. To make this chapter more simple we shall not consider such generalisations.

11.11 An algebraical lemma

The algebraical result which will now be obtained is of great importance in Fredholm's theory of integral equations.

Let (x_1, y_1, z_1), (x_2, y_2, z_2), (x_3, y_3, z_3) be three points at unit distance from the origin. The greatest (numerical) value of the volume of the parallelepiped, of which the lines joining the origin to these points are conterminous edges, is $+1$, the edges then being perpendicular. Therefore, if $x_r^2 + y_r^2 + z_r^2 = 1$ $(r = 1, 2, 3)$, the upper and lower bounds of the determinant

$$\begin{vmatrix} x_1 & y_1 & z_1 \\ x_2 & y_2 & z_2 \\ x_3 & y_3 & z_3 \end{vmatrix}$$

are ± 1.

A lemma due to Hadamard [265] generalises this result. Let

$$D = \begin{vmatrix} a_{11} & a_{12} & \cdots & a_{1n} \\ a_{21} & a_{22} & \cdots & a_{2n} \\ \vdots & \vdots & \vdots & \vdots \\ a_{n1} & a_{n2} & \cdots & a_{nn} \end{vmatrix}$$

where a_{mr} is real and $\sum_{r=1}^{n} a_{mr}^2 = 1$ $(m = 1, 2, \ldots, n)$; let A_{mr} be the cofactor of a_{mr} in D and let Δ be the determinant whose elements are A_{mr}, so that, by a well-known theorem (see Burnside and Panton [111, vol. 2, p. 40]), $\Delta = D^{n-1}$.

Since D is a continuous function of its elements, and is obviously bounded, the ordinary theory of maxima and minima is applicable, and if we consider variations in a_{1r} $(r = 1, 2, \ldots, n)$ only, D is stationary for such variations if $\sum_{r=1}^{n} \frac{\partial D}{\partial a_{1r}} \delta a_{1r} = 0$, where $\delta a_{1r}, \ldots$ are variations subject to the sole condition $\sum_{r=1}^{n} a_{1r} \delta a_{1r} = 0$; therefore[3]

$$A_{1r} = \frac{\partial D}{\partial a_{1r}} = \lambda a_{1r},$$

---

[3] By the ordinary theory of undetermined multipliers.

but $\sum\limits_{r=1}^{n} a_{1r} A_{1r} = D$, and so $\lambda \sum a_{1r}^{2} = D$; therefore $A_{1r} = D a_{1r}$.

Considering variations in the other elements of D, we see that D is stationary for variations in all elements when $A_{mr} = D a_{mr}$ $(m = 1, 2, \ldots, n; r = 1, 2, \ldots, n)$. Consequently $\Delta = D^n \cdot D$, and so $D^{n+1} = D^{n-1}$. Hence the maximum and minimum values of D are ± 1.

Corollary 11.1.1 *If a_{mr} be real and subject only to the condition $|a_{mr}| < M$, since*

$$\sum_{r=1}^{n} \left(\frac{a_{mr}}{n^{1/2} M} \right)^2 \leq 1,$$

we easily see that the maximum value of $|D|$ is $(n^{1/2} M)^n = n^{n/2} M^n$.

11.2 Fredholm's equation and its tentative solution

Fredholm's first paper on the subject appeared in [224]. His researches are also given in [225].

An important integral equation of a general type is

$$\phi(x) = f(x) + \lambda \int_a^b K(x, \xi)\, \phi(\xi)\, d\xi,$$

where $f(x)$ is a given continuous function, λ is a parameter (in general complex) and $K(x, \xi)$ is subject to the conditions laid down in §11.1. $K(x, \xi)$ is called the *nucleus*, or the *kernel* of the equation. The reader will observe that if $K(x, \xi) = 0$ $(\xi > x)$, the equation may be written

$$\phi(x) = f(x) + \lambda \int_a^x K(x, \xi)\, \phi(\xi)\, d\xi.$$

This is called an equation with *variable upper limit*.

This integral equation is known as *Fredholm's equation* or *the integral equation of the second kind* (see §11.3). It was observed by Volterra that an equation of this type could be regarded as a limiting form of a system of linear equations. Fredholm's investigation involved the tentative carrying out of a similar limiting process, and justifying it by the reasoning given below in §11.21. Hilbert [305] justified the limiting process directly.

We now proceed to write down the system of linear equations in question, and shall then investigate Fredholm's method of justifying the passage to the limit.

The integral equation is the limiting form (when $\delta \to 0$) of the equation

$$\phi(x) = f(x) + \lambda \sum_{q=1}^{n} K(x, x_q)\, \phi(x_q)\, \delta,$$

where $x_q - x_{q-1} = \delta$, $x_0 = a$, $x_n = b$.

Since this equation is to be true when $a \leq x \leq b$, it is true when x takes the values $x_1, x_2, \ldots, x_n$; and so

$$-\lambda\delta \sum_{q=1}^{n} K(x_p, x_q)\, \phi(x_q) + \phi(x_p) = f(x_p) \qquad (p = 1, 2, \ldots, n).$$

This system of equations for $\phi(x_p)$, $(p = 1, 2, \ldots, n)$ has a unique solution if the determinant formed by the coefficients of $\phi(x_p)$ does not vanish. This determinant is

$$D_n(\lambda) = \begin{vmatrix} 1 - \lambda\delta K(x_1, x_1) & -\lambda\delta K(x_1, x_2) & \cdots & -\lambda\delta K(x_1, x_n) \\ -\lambda\delta K(x_2, x_1) & 1 - \lambda\delta K(x_2, x_2) & \cdots & -\lambda\delta K(x_2, x_n) \\ \vdots & \vdots & \vdots & \vdots \\ -\lambda\delta K(x_n, x_1) & -\lambda\delta K(x_n, x_2) & \cdots & 1 - \lambda\delta K(x_n, x_n) \end{vmatrix}$$

$$= 1 - \lambda \sum_{p=1}^{n} \delta K(x_p, x_p) + \frac{\lambda^2}{2!} \sum_{p,q=1}^{n} \delta^2 \begin{vmatrix} K(x_p, x_p) & K(x_p, x_q) \\ K(x_q, x_p) & K(x_q, x_q) \end{vmatrix}$$

$$- \frac{\lambda^3}{3!} \sum_{p,q,r=1}^{n} \delta^3 \begin{vmatrix} K(x_p, x_p) & K(x_p, x_q) & K(x_p, x_r) \\ K(x_q, x_p) & K(x_q, x_q) & K(x_q, x_r) \\ K(x_r, x_p) & K(x_r, x_q) & K(x_r, x_r) \end{vmatrix} + \cdots$$

on expanding[4] in powers of λ. Making $\delta \to 0$, $n \to \infty$, and writing the summations as integrations, we are thus led to consider the series

$$D(\lambda) = 1 - \lambda \int_a^b K(\xi_1, \xi_1)\, d\xi_1 + \frac{\lambda^2}{2!} \int_a^b \int_a^b \begin{vmatrix} K(\xi_1, \xi_1) & K(\xi_1, \xi_2) \\ K(\xi_2, \xi_1) & K(\xi_2, \xi_2) \end{vmatrix} d\xi_1\, d\xi_2 - \cdots.$$

Further, if $D_n(x_\mu, x_\nu)$ is the cofactor of the term in $D_n(\lambda)$ which involves $K(x_\nu, x_\mu)$, the solution of the system of linear equations is

$$\phi(x_\mu) = \frac{f(x_1)D_n(x_\mu, x_1) + f(x_2)D_n(x_\mu, x_2) + \cdots + f(x_n)D_n(x_\mu, x_n)}{D_n(\lambda)}.$$

Now it is easily seen that the appropriate limiting form to be considered in association with $D_n(x_\mu, x_\mu)$ is $D(\lambda)$; also that, if $\mu \neq \nu$,

$$D_n(x_\mu, x_\nu) = \lambda\delta \left\{ K(x_\mu, x_\nu) - \lambda\delta \sum_{p=1}^{n} \begin{vmatrix} K(x_\mu, x_\nu) & K(x_\mu, x_p) \\ K(x_p, x_\nu) & K(x_p, x_p) \end{vmatrix} \right.$$

$$\left. + \frac{1}{2!}\lambda^2\delta^2 \sum_{p,q=1}^{n} \begin{vmatrix} K(x_\mu, x_\nu) & K(x_\mu, x_p) & K(x_\mu, x_q) \\ K(x_p, x_\nu) & K(x_p, x_p) & K(x_p, x_q) \\ K(x_q, x_\nu) & K(x_q, x_p) & K(x_q, x_q) \end{vmatrix} - \cdots \right\}.$$

So that the limiting form for $\delta^{-1}D(x_\mu, x_\nu)$ to be considered is

$$D(x_\mu, x_\nu; \lambda) = \lambda K(x_\mu, x_\nu) - \lambda^2 \int_a^b \begin{vmatrix} K(x_\mu, x_\nu) & K(x_\mu, \xi_1) \\ K(\xi_1, x_\nu) & K(\xi_1, \xi_1) \end{vmatrix} d\xi_1$$

$$+ \frac{1}{2!}\lambda^3 \int_a^b \int_a^b \begin{vmatrix} K(x_\mu, x_\nu) & K(x_\mu, \xi_1) & K(x_\mu, \xi_2) \\ K(\xi_1, x_\nu) & K(\xi_1, \xi_1) & K(\xi_1, \xi_2) \\ K(\xi_2, x_\nu) & K(\xi_2, \xi_1) & K(\xi_2, \xi_2) \end{vmatrix} d\xi_1 d\xi_2 - \cdots.$$

(The law of formation of successive terms is obvious from those written down.)

---

[4] The factorials appear because each determinant of s rows and columns occurs $s!$ times as $p, q, \ldots$ take *all* the values $1, 2, \ldots, n$, whereas it appears only once in the original determinant for $D_n(\lambda)$.

Consequently we are led to consider the possibility of the equation

$$\phi(x) = f(x) + \frac{1}{D(\lambda)} \int_a^b D(x, \xi; \lambda) f(\xi)\, d\xi$$

giving the solution of the integral equation.

Example 11.2.1 Shew that, in the case of the equation

$$\phi(x) = x + \lambda \int_0^1 xy\phi(y)\, dy,$$

we have

$$D(\lambda) = 1 - \tfrac{1}{3}\lambda, \quad D(x, y; \lambda) = \lambda xy$$

and a solution is

$$\phi(x) = \frac{3x}{3 - \lambda}.$$

Example 11.2.2 Shew that, in the case of the equation

$$\phi(x) = x + \lambda \int_0^1 (xy + y^2)\phi(y)\, dy,$$

we have

$$D(\lambda) = 1 - \tfrac{2}{3}\lambda - \tfrac{1}{72}\lambda^2,$$
$$D(x, y; \lambda) = \lambda(xy + y^2) + \lambda^2 \left(\tfrac{1}{2}xy^2 - \tfrac{1}{3}xy - \tfrac{1}{3}y^2 + \tfrac{1}{4}y \right),$$

and obtain a solution of the equation.

11.21 Investigation of Fredholm's solution

So far the construction of the solution has been purely tentative; we now start *ab initio* and verify that we actually do get a solution of the equation; to do this we consider the two functions $D(\lambda), D(x; y\lambda)$ arrived at in §11.2.

We write the series, by which $D(\lambda)$ was defined in §11.2, in the form $1 + \sum\limits_{n=1}^{\infty} \frac{a_n \lambda^n}{n!}$ so that

$$a_n = (-1)^n \int_a^b \int_a^b \cdots \int_a^b \begin{vmatrix} K(\xi_1, \xi_1) & K(\xi_1, \xi_2) & \cdots & K(\xi_1, \xi_n) \\ K(\xi_2, \xi_1) & K(\xi_2, \xi_2) & \cdots & K(\xi_2, \xi_n) \\ \vdots & \vdots & \vdots & \vdots \\ K(\xi_n, \xi_1) & K(\xi_n, \xi_2) & \cdots & K(\xi_n, \xi_n) \end{vmatrix} d\xi_1\, d\xi_2 \cdots d\xi_n;$$

since $K(x, y)$ is continuous and therefore bounded, we have $|K(x, y)| < M$, where M is independent of x and y; since $K(x, y)$ is real, we may employ Hadamard's lemma (§11.11) and we see at once that

$$|a_n| < n^{n/2} M^n (b - a)^n.$$

Write $n^{n/2} M^n (b - a)^n = n! b_n$; then

$$\lim_{n \to \infty} \frac{b_{n+1}}{b_n} = \lim_{n \to \infty} \frac{(b - a)M}{(n + 1)^{1/2}} \left(1 + \frac{1}{n} \right)^{n/2} = 0,$$

since $(1 + 1/n)^n \to e$.

The series $\sum\limits_{n=1}^{\infty} b_n \lambda^n$ is therefore absolutely convergent for all values of λ; and so (§2.34) the

series $1 + \sum\limits_{n=1}^{\infty} \frac{a_n \lambda^n}{n!}$ converges for all values of λ and therefore (§5.64) represents an integral function of λ.

Now write the series for $D(x, y; \lambda)$ in the form $\sum\limits_{n=0}^{\infty} \frac{v_n(x,y)\lambda^{n+1}}{n!}$. Then, by Hadamard's lemma (§11.11),

$$|v_{n-1}(x, y)| < n^{n/2} M^n (b - a)^{n-1},$$

and hence $\left| \dfrac{v_n(x, y)}{n!} \right| < c_n$ where c_n is independent of x and y and $\sum\limits_{n=0}^{\infty} c_n \lambda^{n+1}$ is absolutely convergent.

Therefore $D(x, y; \lambda)$ is an integral function of λ and the series for $D(x, y; \lambda) - \lambda K(x, y)$ is a uniformly convergent (§3.34) series of continuous[5] functions of x and y when $a \leq x \leq b$, $a \leq y \leq b$.

Now pick out the coefficient of $K(x, y)$ in $D(x, y; \lambda)$; and we get

$$D(x, y; \lambda) = \lambda D(\lambda) K(x, y) + \sum_{n=1}^{\infty} (-1)^n \lambda^{n+1} \frac{Q_n(x, y)}{n!},$$

where

$$Q_n(x, y) = \int_a^b \int_a^b \cdots \int_a^b \begin{vmatrix} 0 & K(x, \xi_1) & K(x, \xi_2) & \cdots & K(x, \xi_n) \\ K(\xi_1, y) & K(\xi_1, \xi_1) & K(\xi_1, \xi_2) & \cdots & K(\xi_1, \xi_n) \\ \vdots & \vdots & \vdots & \vdots & \vdots \\ K(\xi_n, y) & K(\xi_n, \xi_1) & K(\xi_n, \xi_2) & \cdots & K(\xi_n, \xi_n) \end{vmatrix} d\xi_1 \cdots d\xi_n.$$

Expanding in minors of the first column, we get $Q_n(x, y)$ equal to the integral of the sum of n determinants; writing $\xi_1, \xi_2, \ldots, \xi_{m-1}, \xi, \xi_m, \ldots, \xi_{n-1}$ in place of $\xi_1, \xi_2, \ldots, \xi_n$ in the mth of them, we see that the integrals of all the determinants[6] are equal and so

$$Q_n(x, y) = -n \int_a^b \int_a^b \cdots \int_a^b K(\xi, y) P_n \, d\xi \, d\xi_1 \cdots d\xi_{n-1},$$

where

$$P_n = \begin{vmatrix} K(x, \xi) & K(x, \xi_1) & \cdots & K(x, \xi_{n-1}) \\ K(\xi_1, \xi) & K(\xi_1, \xi_1) & \cdots & K(\xi_1, \xi_{n-1}) \\ \vdots & \vdots & \vdots & \vdots \\ K(\xi_{n-1}, \xi) & K(\xi_{n-1}, \xi_1) & \cdots & K(\xi_{n-1}, \xi) \end{vmatrix}.$$

It follows at once that

$$D(x; y; \lambda) = \lambda D(\lambda) K(x, y) + \lambda \int_a^b D(x, \xi; \lambda) K(\xi, y) \, d\xi.$$

---

[5] It is easy to verify that every term (except possibly the first) of the series for $D(x, y; \lambda)$ is a continuous function under either hypothesis (i) or hypothesis (ii) of §11.1.

[6] The order of integration is immaterial (§4.3).

Now take the equation

$$\phi(\xi) = f(\xi) + \lambda \int_a^b K(\xi, y)\phi(y)\, dy,$$

multiply by $D(x, \xi; \lambda)$ and integrate, and we get

$$\int_a^b f(\xi)D(x, \xi; \lambda)\, d\xi$$

$$= \int_a^b \phi(\xi)D(x, \xi; \lambda)\, d\xi - \lambda \int_a^b \int_a^b D(x; \xi; \lambda)K(\xi, y)\phi(y)\, dy\, d\xi,$$

the integrations in the repeated integral being in either order.

That is to say

$$\int_a^b f(\xi)D(x, \xi; \lambda)\, d\xi$$

$$= \int_a^b \phi(\xi)D(x, \xi; \lambda)\, d\xi - \int_a^b [D(x, y; \lambda) - \lambda D(\lambda)K(x, y)]\, \phi(y)\, dy$$

$$= \lambda D(\lambda) \int_a^b K(x, y)\phi(y)\, dy$$

$$= D(\lambda)[\phi(x) - f(x)],$$

in virtue of the given equation.

Therefore if $D(\lambda) \neq 0$ and if Fredholm's equation has a solution it can be none other than

$$\phi(x) = f(x) + \int_a^b f(\xi)\frac{D(x, \xi; \lambda)}{D(\lambda)}\, d\xi;$$

and, by actual substitution of this value of $\phi(x)$ in the integral equation, we see that it actually is a solution. This is, therefore, the unique continuous solution of the equation if $D(\lambda) \neq 0$.

Corollary 11.2.1 *If we put $f(x) \equiv 0$, the 'homogeneous' equation*

$$\phi(x) = \lambda \int_a^b K(x, \xi)\phi(\xi)\, d\xi$$

has no continuous solution except $\phi(x) = 0$, unless $D(\lambda) = 0$.

Example 11.2.3 By expanding the determinant involved in $Q_n(x, y)$ in minors of its first row, shew that

$$D(x, y; \lambda) = \lambda D(\lambda)K(x, y) + \lambda \int_a^b K(x, \xi)D(\xi, y; \lambda)\, d\xi.$$

Example 11.2.4 By using the formulae

$$D(\lambda) = 1 + \sum_{n=1}^\infty \frac{a_n \lambda^n}{n!}, \qquad D(x, y; \lambda) = \lambda D(\lambda)K(x, y) + \sum_{n=1}^\infty (-1)^n \frac{\lambda^{n+1} Q_n(x, y)}{n!},$$

shew that

$$\int_a^b D(\xi, \xi; \lambda)\, d\xi = -\lambda \frac{dD(\lambda)}{d\lambda}.$$

Example 11.2.5   If

$$K(x, y) = \begin{cases} 1 & \text{if } y \le x, \\ 0 & \text{if } y > x; \end{cases}$$

shew that $D(\lambda) = \exp\{-(b - a)\lambda\}$.

Example 11.2.6   Shew that, if $K(x, y) = f_1(x)f_2(y)$, and if

$$\int_a^b f_1(x)f_2(x)\, dx = A,$$

then

$$D(\lambda) = 1 - A\lambda, \qquad D(x, y; \lambda) = \lambda f_1(x)f_2(y),$$

and the solution of the corresponding integral equation is

$$\phi(x) = f(x) + \frac{\lambda f_1(x)}{1 - A\lambda} \int_a^b f(\xi)f_2(\xi)\, d\xi.$$

Example 11.2.7   Shew that, if

$$K(x, y) = f_1(x)g_1(y) + f_2(x)g_2(y),$$

then $D(\lambda)$ and $D(x, y; \lambda)$ are quadratic in λ, and, more generally, if

$$K(x, y) = \sum_{m=1}^n f_m(x)g_m(y),$$

then $D(\lambda)$ and $D(x, y, \lambda)$ are polynomials of degree n in λ.

11.22 Volterra's reciprocal functions

Two functions $K(x, y)$, $k(x, y; \lambda)$ are said to be *reciprocal* if they are bounded in the ranges $a \le x, y \le b$, if any discontinuities they may have are regularly distributed (§11.1, footnote on Bôcher's work), and if

$$K(x, y) + k(x, y; \lambda) = \lambda \int_a^b k(x, \xi; \lambda)K(\xi, y)\, d\xi.$$

We observe that, since the right-hand side is continuous (by Example 4.11), the sum of two reciprocal functions is continuous.

Also, a function $K(x, y)$ can only have one reciprocal if $D(\lambda) \ne 0$; for if there were two, their difference $k_1(x, y)$ would be a continuous solution of the homogeneous equation

$$k_1(x, y; \lambda) = \lambda \int_a^b k_1(x, \xi; \lambda)K(\xi, y)\, d\xi,$$

(where x is to be regarded as a parameter), and by Corollary 11.2.1, the only continuous solution of this equation is zero.

By the use of reciprocal functions, Volterra has obtained an elegant reciprocal relation between pairs of equations of Fredholm's type.

We first observe, from the relation

$$D(x, y; \lambda) = \lambda D(\lambda) K(x, y) + \lambda \int_a^b D(x, \xi, \lambda) K(\xi, y) \, d\xi,$$

proved in §11.21, that the value of $k(x, y; \lambda)$ is

$$-\frac{D(x, y; \lambda)}{\lambda D(\lambda)},$$

and from Example 11.2.3, the equation

$$k(x, y; \lambda) + K(x, y) = \lambda \int_a^b K(x, \xi) k(\xi, y; \lambda) \, d\xi$$

is evidently true.

Then, if we take the integral equation

$$\phi(x) = f(x) + \lambda \int_a^b K(x, \xi) \phi(\xi) \, d\xi,$$

when $a \le x \le b$, we have, on multiplying the equation

$$\phi(\xi) = f(\xi) + \lambda \int_a^b K(\xi, \xi_1) \phi(\xi_1) \, d\xi_1$$

by $k(x, \xi; \lambda)$ and integrating,

$$\int_a^b k(x, \xi; \lambda) \phi(\xi) \, d\xi = \int_a^b k(x, \xi; \lambda) f(\xi) \, d\xi$$
$$+ \lambda \int_a^b \int_a^b k(x, \xi, \lambda) K(\xi, \xi_1) \phi(\xi_1) \, d\xi_1 \, d\xi.$$

Reversing the order of integration[7] in the repeated integral and making use of the relation defining reciprocal functions, we get

$$\int_a^b k(x, \xi; \lambda) \phi(\xi) \, d\xi = \int_a^b k(x, \xi; \lambda) f(\xi) \, d\xi$$
$$+ \int_a^b \{ K(x, \xi) + k(x, \xi_1; \lambda) \} \phi(\xi_1) \, d\xi_1$$

and so

$$\lambda \int_a^b k(x, \xi; \lambda) f(\xi) \, d\xi = -\lambda \int_a^b K(x, \xi_1) \phi(\xi_1) \, d\xi_1$$
$$= -\phi(x) + f(x).$$

Hence $f(x) = \phi(x) + \lambda \int_a^b k(x, \xi; \lambda) f(\xi) \, d\xi$; similarly, from this equation we can derive the equation

$$\phi(x) = f(x) + \lambda \int_a^b K(x, \xi) \phi(\xi) \, d\xi,$$

---

[7] The reader will have no difficulty in extending the result of §4.3 to the integral under consideration.

so that either of these equations with reciprocal nuclei may be regarded as the solution of the other.

11.23 Homogeneous integral equations

The equation

$$\phi(x) = \lambda \int_a^b K(x,\xi)\phi(\xi)\,d\xi \tag{11.1}$$

is called a homogeneous integral equation. We have seen (Corollary 11.2.1) that the only continuous solution of the homogeneous equation, when $D(\lambda) \neq 0$, is $\phi(x) = 0$.

The roots of the equation $D(\lambda) = 0$ are therefore of considerable importance in the theory of the integral equation. They are called the *characteristic numbers of the nucleus*.

It will now be shewn that, when $D(\lambda) = 0$, a solution which is not identically zero can be obtained.

It will be proved in §11.51 that, if $K(x,y) \equiv K(y,x)$, the equation $D(\lambda) = 0$ has at least one root. Let $\lambda = \lambda_0$ be a root m times repeated of the equation $D(\lambda) = 0$. Since $D(\lambda)$ is an integral function, we may expand it into the convergent series

$$D(\lambda) = c_m(\lambda - \lambda_0)^m + c_{m+1}(\lambda - \lambda_0)^{m+1} + \cdots \qquad (m > 0;\ c_m \neq 0).$$

Similarly, since $D(x,y;\lambda)$ is an integral function of λ, there exists a Taylor series of the form

$$D(x,y;\lambda) = \frac{g_\ell(x,y)}{\ell!}(\lambda - \lambda_0)^\ell + \frac{g_{\ell+1}(x,y)}{(\ell+1)!}(\lambda - \lambda_0)^{\ell+1} + \cdots \qquad (\ell \geq 0;\ g_\ell \not\equiv 0);$$

by §3.34 it is easily verified that the series defining $g_n(x,y)$, $(n = \ell, \ell + 1, \ldots)$ converges absolutely and uniformly when $a \leq x \leq b, a \leq y \leq b$, and thence that the series for $D(x,y;\lambda)$ converges absolutely and uniformly in the same domain of values of x and y.

But, by Example 11.2.4,

$$\int_a^b D(\xi,\xi;\lambda)\,d\xi = -\lambda \frac{dD(\lambda)}{d\lambda};$$

now the right-hand side has a zero of order $m - 1$ at λ_0, while the left-hand side has a zero of order at least ℓ, *and so we have $m - 1 \geq \ell$.*

Substituting the series just given for $D(\lambda)$ and $D(x,y;\lambda)$ in the result of Example 11.2.3, viz.

$$D(x,y;\lambda) = \lambda D(\lambda)K(x,y) + \lambda \int_a^b K(x,\xi)D(\xi,y;\lambda)\,d\xi,$$

dividing by $(\lambda - \lambda_0)^\ell$ and making $\lambda \to \lambda_0$, we get

$$g_\ell(x,y) = \lambda_0 \int_a^b K(x,\xi)g_\ell(\xi,y)\,d\xi.$$

Hence if y have any constant value, $g_\ell(x,y)$ satisfies the homogeneous integral equation, and any linear combination of such solutions, obtained by giving y various values, is a solution.

Corollary 11.2.2 *The equation*

$$\phi(x) = f(x) + \lambda_0 \int_a^b K(x,\xi)\phi(\xi)\,d\xi$$

has no solution or an infinite number. For, if $\phi(x)$ is a solution, so is $\phi(x) + \sum_y c_y g_\ell(x,y)$, where c_y may be any function of y.

Example 11.2.8   Shew that solutions of

$$\phi(x) = \lambda \int_{-\pi}^{\pi} \cos^n(x - \xi)\phi(\xi)\,d\xi$$

are $\phi(x) = \cos(n-2r)x$, and $\phi(x) = \sin(n-2r)x$; where r assumes all positive integral values (zero included) not exceeding $\frac{1}{2}n$.

Example 11.2.9   Shew that

$$\phi(x) = \lambda \int_{-\pi}^{\pi} \cos^n(x + \xi)\phi(\xi)\,d\xi$$

has the same solutions as those given in Example 11.2.8, and shew that the corresponding values of λ give all the roots of $D(\lambda) = 0$.

11.3 Integral equations of the first and second kinds

Fredholm's equation is sometimes called an *integral equation of the second kind*; while the equation

$$f(x) = \lambda \int_a^b K(x,\xi)\phi(\xi)\,d\xi$$

is called the *integral equation of the first kind*.

In the case when $K(x,\xi) = 0$ if $\xi > x$, we may write the equations of the first and second kinds in the respective forms

$$f(x) = \lambda \int_a^x K(x,\xi)\phi(\xi)\,d\xi,$$

$$\phi(x) = f(x) + \lambda \int_a^x K(x,\xi)\phi(\xi)\,d\xi.$$

These are described as equations with *variable upper limits*.

11.31 Volterra's equation

The equation of the first kind with variable upper limit is frequently known as Volterra's equation. The problem of solving it has been reduced by that writer to the solution of Fredholm's equation.

Assuming that $K(x,\xi)$ is a continuous function of both variables when $\xi \leq x$, we have

$$f(x) = \lambda \int_a^x K(x,\xi)\phi(\xi)\,d\xi.$$

The right-hand side has a differential coefficient (see Example 4.2.1) if $\dfrac{\partial K}{\partial x}$ exists and is continuous, and so

$$f'(x) = \lambda K(x,x)\phi(x) + \lambda \int_a^x \frac{\partial K}{\partial x}\phi(\xi)\,d\xi.$$

This is an equation of Fredholm's type. If we denote its solution by $\phi(x)$, we get on integrating from a to x,

$$f(x) - f(a) = \lambda \int_a^x K(x,\xi)\phi(\xi)\,d\xi,$$

and so the solution of the Fredholm's equation gives a solution of Volterra's equation if $f(a) = 0$.

The solution of the equation of the first kind with constant upper limit can frequently be obtained in the form of a series. See Example 11.6. A solution valid under fewer restrictions is given by Bôcher.

11.4 The Liouville–Neumann method of successive substitutions

This appears in Liouville [438]. K. Neumann's investigations were later (1870); see [488].

A method of solving the equation

$$\phi(x) = f(x) + \lambda \int_a^b K(x,\xi)\phi(\xi)\,d\xi,$$

which is of historical importance, is due to Liouville.

It consists in continually substituting the value of $\phi(x)$ given by the right-hand side in the expression $\phi(\xi)$ which occurs on the right-hand side.

This procedure gives the series

$$S(x) = f(x) + \lambda \int_a^b K(x,\xi)f(\xi)\,d\xi$$

$$+ \sum_{m=2}^{\infty} \lambda^m \int_a^b K(x,\xi_1) \int_a^b K(\xi_1,\xi_2) \cdots \int_a^b K(\xi_{m-1},\xi_m)f(\xi_m)\,d\xi_m \cdots d\xi_1.$$

Since $|K(x,y)|$ and $|f(x)|$ are bounded, let their upper bounds be M, M'. Then the modulus of the general term of the series does not exceed $|\lambda|^m M^m M'(b-a)^m$. The series for $S(x)$ therefore converges uniformly when $|\lambda| < M^{-1}(b-a)^{-1}$; and, by actual substitution, it satisfies the integral equation.

If $K(x,y) = 0$ when $y > x$, we find by induction that the modulus of the general term in the series for $S(x)$ does not exceed

$$|\lambda|^m M^m M'(x-a)^m/m! \le |\lambda|^m M^m M'(b-a)^m/m!,$$

and so the series converges uniformly for *all* values of λ; and we infer that in this case Fredholm's solution is an integral function of λ.

It is obvious from the form of the solution that when $|\lambda| < M^{-1}(b-a)^{-1}$, the reciprocal function $k(x,\xi;\lambda)$ may be written in the form

$$k(x,\xi;\lambda) = -K(x,\xi)$$
$$-\sum_{m=2}^{\infty} \lambda^{m-1} \int_a^b K(x,\xi_1) \int_a^b K(\xi_1,\xi_2) \cdots \int_a^b K(\xi_{m-1},\xi)\, d\xi_{m-1}\, d\xi_{m-2} \cdots d\xi_1,$$

for with this definition of $k(x,\xi;\lambda)$, we see that

$$S(x) = f(x) - \lambda \int_a^b k(x,\xi;\lambda)f(\xi)\, d\xi,$$

so that $k(x,\xi;\lambda)$ is a reciprocal function, and by §11.22 there is *only one* reciprocal function if $D(\lambda) \neq 0$.

Write

$$K(x,\xi) = K_1(x,\xi), \qquad \int_a^b K(x,\xi')K_n(\xi',\xi)\, d\xi' = K_{n+1}(x,\xi),$$

and then we have

$$-K(x,\xi;\lambda) = \sum_{m=0}^{\infty} \lambda^m K_{m+1}(x,\xi),$$

while

$$\int_a^b K_m(x,\xi')K_n(\xi',\xi)\, d\xi' = K_{m+n}(x,\xi),$$

as may be seen at once on writing each side as an $(m+n-1)$-tuple integral.

The functions $K_m(x,\xi)$ are called *iterated* functions.

11.5 Symmetric nuclei

Let $K_1(x,y) \equiv K_1(y,x)$; then the nucleus $K(x,y)$ is said to be *symmetric*. The iterated functions of such a nucleus are also symmetric, i.e. $K_n(x,y) = K_n(y,x)$ for all values of n; for, if $K_n(x,y)$ is symmetric, then

$$K_{n+1}(x,y) = \int_a^b K_1(x,\xi)K_n(\xi,y)\, d\xi = \int_a^b K_1(\xi,x)K_n(y,\xi)\, d\xi$$
$$= \int_a^b K_n(y,\xi)K_1(\xi,x)\, d\xi$$
$$= K_{n+1}(y,x),$$

and the required result follows by induction.

Also, none of the iterated functions are identically zero; for, if possible, let $K_p(x,y) \equiv 0$; let n be chosen so that $2^{n-1} < p \leq 2^n$, and, since $K_p(x,y) \equiv 0$, it follows that $K_{2^n}(x,y) \equiv 0$, from the recurrence formula. But then

$$0 = K_{2^n}(x,x) = \int_a^b K_{2^{n-1}}(x,\xi)K_{2^{n-1}}(\xi,x)\,d\xi$$

$$= \int_a^b \{K_{2^{n-1}}(x,\xi)\}^2\,d\xi,$$

and so $K_{2^{n-1}}(x,\xi) \equiv 0$; continuing this argument, we find ultimately that $K_1(x,y) \equiv 0$, and the integral equation is trivial.

11.51 Schmidt's theorem that, if the nucleus is symmetric, the equation $D(\lambda) = 0$ has at least one root

The proof given is due to Kneser [380]. To prove this theorem, let

$$U_n = \int_a^b K_n(x,x)\,dx,$$

so that, when $|\lambda| < M^{-1}(b-a)^{-1}$, we have, by Example 11.2.4 and §11.4,

$$-\frac{1}{D(\lambda)}\frac{dD(\lambda)}{d\lambda} = \sum_{n=1}^{\infty} U_n \lambda^{n-1}.$$

Now since

$$\int_a^b \int_a^b (\mu K_{n+1}(x,\xi) + K_{n-1}(x,\xi))^2\,d\xi\,dx \geq 0$$

for all real values of μ, we have $\mu^2 U_{2n+2} + 2\mu U_{2n} + U_{2n-2} \geq 0$, and so $U_{2n+2}U_{2n-2} \geq U_{2n^2}$, $U_{2n-2} > 0$. Therefore $U_2, U_4, \ldots$ are all positive, and if $U_4/U_2 = \nu$, it follows, by induction from the inequality $U_{2n+2}U_{2n-2} \geq U_{2n^2}$, that $U_{2n+2}/U_{2n} \geq \nu^n$. Therefore when $|\lambda^2| \geq \nu^{-1}$, the terms of $\sum_{n=1}^{\infty} U_n \lambda^{n-1}$ do not tend to zero; and so, by §5.4, the function $\dfrac{1}{D(\lambda)}\dfrac{dD(\lambda)}{d\lambda}$ has a singularity inside or on the circle $|\lambda| = \nu^{-\frac{1}{2}}$; but since $D(\lambda)$ is an integral function, the only possible singularities of $\dfrac{1}{D(\lambda)}\dfrac{dD(\lambda)}{d\lambda}$ are at zeros of $D(\lambda)$; therefore $D(\lambda)$ has a zero inside or on the circle $|\lambda| = \nu^{-\frac{1}{2}}$.

Note By §11.21, $D(\lambda)$ is either an integral function or else a mere polynomial; in the latter case, it has a zero by Example 6.3.1; the point of the theorem is that in the former case $D(\lambda)$ cannot be such a function as e^{λ^2}, which has no zeros.

11.6 Orthogonal functions

The real continuous functions $\phi_1(x)$, $\phi_2(x)$, ... are said to be orthogonal and normal[8] for the range (a, b) if

$$\int_a^b \phi_m(x)\phi_n(x)\, dx = \begin{cases} 0 & (m \neq n), \\ 1 & (m = n). \end{cases}$$

If we are given n real continuous linearly independent functions $u_1(x), \ldots, u_n(x)$, we can form n linear combinations of them which are orthogonal.

For suppose we can construct $m - 1$ orthogonal functions $\phi_1, \ldots, \phi_{m-1}$ such that ϕ_p is a linear combination of $u_1, u_2, \ldots, u_p$ (where $p = 1, 2, \ldots, m - 1$); we shall now shew how to construct the function ϕ_m such that $\phi_1, \phi_2, \ldots, \phi_m$ are all normal and orthogonal.

Let $_1\phi_m(x) = c_{1,m}\phi_1(x) + c_{2,m}\phi_2(x) + \cdots + c_{m-1}\phi_{m-1}(x) + u_m(x)$, so that $_1\phi_m$ is a function of $u_1, u_2, \ldots, u_m$. Then, multiplying by ϕ_p and integrating,

$$\int_a^b {}_1\phi_m(x)\phi_p(x)\, dx = c_{p,m} + \int_a^b u_m(x)\phi_p(x)\, dx \qquad (p < m).$$

Hence $\int_a^b {}_1\phi_m(x)\phi_p(x)\, dx = 0$ if $c_{p,m} = -\int_a^b u_m(x)\phi_p(x)\, dx$; a function $_1\phi_m(x)$, orthogonal to $\phi_1(x), \phi_2(x), \ldots, \phi_{m-1}(x)$, is therefore constructed.

Now choose α so that $\alpha^2 \int_a^b \{_1\phi_m(x)\}^2\, dx = 1$; and take $\phi_m(x) = \alpha\, (_1\phi_m(x))$. Then

$$\int_a^b \phi_m(x)\phi_p(x)\, dx = \begin{cases} 0 & (p < m), \\ 1 & (p = m). \end{cases}$$

We can thus obtain the functions $\phi_1, \phi_2, \ldots$ in order.

The members of a finite set of orthogonal functions are linearly independent. For, if

$$\alpha_1\phi_1(x) + \alpha_2\phi_2(x) + \cdots + \alpha_n\phi_n(x) \equiv 0,$$

we should get, on multiplying by $\phi_p(x)$ and integrating, $\alpha_p = 0$; therefore all the coefficients α_p vanish and the relation is nugatory.

It is obvious that $\pi^{-1/2} \cos mx$, $\pi^{-1/2} \sin mx$ form a set of normal orthogonal functions for the range $(-\pi, \pi)$.

Example 11.6.1 From the functions $1, x, x^2, \ldots$ construct the following set of functions which are orthogonal (but not normal) for the range $(-1, 1)$;

$$1,\ x,\ x^2,\ -\frac{1}{3}x^3 - \frac{3}{5}x,\ x^4 - 2x^2 + \frac{3}{35}, \ldots.$$

Example 11.6.2 From the functions $1, x, x^2, \ldots$ construct a set of functions $f_0(x), f_1(x)$, $f_2(x), \ldots$ which are orthogonal (but not normal) for the range (a, b); where

$$f_n(x) = \frac{d^n}{dx^n} \left\{ (x - a)^n (x - b)^n \right\}.$$

A similar investigation is given in §15.14.

[8] They are said to be orthogonal if the first equation only is satisfied; the systematic study of such functions is due to Murphy [481, 483].

11.61 *The connexion of orthogonal functions with homogeneous integral equations*

Consider the homogeneous equation

$$\phi(x) = \lambda_0 \int_a^b \phi(\xi)K(x,\xi)\,d\xi,$$

where λ_0 is a *real* characteristic number for $K(x,\xi)$. It will be seen immediately that the characteristic numbers of a symmetric nucleus are all real. We have already seen how solutions of it may be constructed; let n linearly independent solutions be taken and construct from them n orthogonal and normal functions $\phi_1, \phi_2, \ldots, \phi_n$.

Then, since the functions ϕ_m are orthogonal and normal,

$$\int_a^b \left[\sum_{m=1}^n \phi_m(y) \int_a^b K(x,\xi)\phi_m(\xi)\,d\xi \right]^2 dy$$

$$= \sum_{m=1}^n \int_a^b \left[\phi_m(y) \int_a^b K(x,\xi)\phi_m(\xi)\,d\xi \right]^2 dy,$$

and it is easily seen that the expression on the right may be written in the form

$$\sum_{m=1}^n \left[\int_a^b K(x,\xi)\phi_m(\xi)\,d\xi \right]^2$$

on performing the integration with regard to y; and this is the same as

$$\sum_{m=1}^n \int_a^b K(x,y)\phi_m(y)\,dy \int_a^b K(x,\xi)\phi_m(\xi)\,d\xi.$$

Therefore, if we write K for $K(x,y)$ and Λ for

$$\sum_{m=1}^n \phi_m(y) \int_a^b K(x,\xi)\phi_m(\xi)\,d\xi,$$

we have $\int_a^b \Lambda^2\,dy = \int_a^b K\Lambda\,dy$, and so

$$\int_a^b \Lambda^2\,dy = \int_a^b K^2\,dy - \int_a^b (K - \Lambda)^2\,dy. \tag{11.2}$$

Therefore

$$\int_a^b \left[\sum_{m=1}^n \frac{\phi_m(y)\phi_m(x)}{\lambda_0} \right]^2 dy \le \int_a^b [K(x,y)]^2\,dy,$$

and so

$$\lambda_0^{-2} \sum_{m=1}^n [\phi_m(x)]^2 \le \int_a^b [K(x,y)]^2\,dy.$$

Integrating, we get

$$n \le \lambda_0^2 \int_a^b \int_a^b [K(x,y)]^2\,dy\,dx.$$

This formula gives an upper limit to the number, n, of orthogonal functions corresponding to any characteristic number λ_0.

These n orthogonal functions are called *characteristic functions* (or *auto-functions*) corresponding to λ_0.

Now let $\phi^{(0)}(x)$, $\phi^{(1)}(x)$ be characteristic functions corresponding to *different* characteristic numbers λ_0, λ_1. Then

$$\phi^{(0)}(x)\phi^{(1)}(x) = \lambda_1 \int_a^b K(x,\xi)\phi^{(0)}(x)\phi^{(1)}(\xi)\,d\xi,$$

and so

$$\int_a^b \phi^{(0)}(x)\phi^{(1)}(x)\,dx = \lambda_1 \int_a^b \int_a^b K(x,\xi)\phi^{(0)}(x)\phi^{(1)}(\xi)\,d\xi\,dx \qquad (11.3)$$

and similarly

$$\int_a^b \phi^{(0)}(x)\phi^{(1)}(x)\,dx = \lambda_0 \int_a^b \int_a^b K(x,\xi)\phi^{(0)}(\xi)\phi^{(1)}(x)\,d\xi\,dx$$

$$= \lambda_0 \int_a^b \int_a^b K(\xi,x)\phi^{(0)}(x)\phi^{(1)}(\xi)\,dx\,d\xi, \qquad (11.4)$$

on interchanging x and ξ.

We infer from (11.3) and (11.4) that if $\lambda_1 \neq \lambda_0$, *and if* $K(x,\xi) = K(\xi,x)$,

$$\int_a^b \phi^{(0)}(x)\phi^{(1)}(x)\,dx = 0,$$

and so the functions $\phi^{(0)}(x)$, $\phi^{(1)}(x)$ are *mutually* orthogonal.

If therefore the *nucleus be symmetric* and if, corresponding to each characteristic number, we construct the complete system of orthogonal functions, *all* the functions so obtained will be orthogonal.

Further, if the nucleus be symmetric *all the characteristic numbers are real*; for if λ_0, λ_1 be conjugate complex roots and if[9] $u_0(x) = v(x) + iw(x)$ be a solution for the characteristic number λ_0, then $u_1(x) = v(x) - iw(x)$ is a solution for the characteristic number λ_1; replacing $\phi^{(0)}(x)$, $\phi^{(1)}(x)$ in the equation

$$\int_a^b \phi^{(0)}(x)\phi^{(1)}(x)\,dx = 0$$

by $v(x) + iw(x)$, $v(x) - iw(x)$ (which is obviously permissible), we get

$$\int_a^b \left[(v(x))^2 + (w(x))^2\right]\,dx = 0,$$

which implies $v(x) \equiv w(x) \equiv 0$, so that the integral equation has no solution except zero corresponding to the characteristic numbers λ_0, λ_1; this is contrary to §11.23; hence, if the nucleus be symmetric, the characteristic numbers are real.

---

[9] $v(x)$ and $w(x)$ being real.

11.7 The development of a symmetric nucleus

This investigation is due to Schmidt, the result to Hilbert.

Let $\phi_1(x), \phi_2(x), \phi_3(x), \ldots$ be a complete set of orthogonal functions satisfying the homogeneous integral equation with symmetric nucleus

$$\phi(x) = \lambda \int_a^b K(x, \xi) \phi(\xi) \, d\xi,$$

the corresponding characteristic numbers being[10] $\lambda_1, \lambda_2, \lambda_3, \ldots$.

Now *suppose*[11] that the series $\sum\limits_{n=1}^{\infty} \dfrac{\phi_n(x)\, \phi_n(y)}{\lambda_n}$ is uniformly convergent when $a \leq x \leq b$, $a \leq y \leq b$. *Then it will be shewn that*

$$K(x, y) = \sum_{n=1}^{\infty} \frac{\phi_n(x)\phi_n(y)}{\lambda_n}.$$

For consider the symmetric nucleus

$$H(x, y) = K(x, y) - \sum_{n=1}^{\infty} \frac{\phi_n(x)\phi_n(y)}{\lambda_n}.$$

If this nucleus is not identically zero, it will possess (§11.51) at least one characteristic number μ.

Let $\psi(x)$ be any solution of the equation

$$\psi(x) = \mu \int_a^b H(x, \xi)\psi(\xi) \, d\xi,$$

which does not vanish identically. Multiply by $\phi_n(x)$ and integrate and we get

$$\int_a^b \psi(x)\phi_n(x)\, dx = \mu \int_a^b \int_a^b \left\{ K(x, \xi) - \sum_{m=1}^{\infty} \frac{\phi_m(x)\phi_m(\xi)}{\lambda_m} \right\} \psi(\xi)\phi_n(x)\, dx\, d\xi;$$

since the series converges uniformly, we may integrate term by term and get

$$\int_a^b \psi(x)\phi_n(x)\, dx = \frac{\mu}{\lambda_n} \int_a^b \psi(\xi)\phi_n(\xi)\, d\xi - \frac{\mu}{\lambda_n} \int_a^b \phi_n(\xi)\psi(\xi)\, d\xi = 0.$$

Therefore $\psi(x)$ is orthogonal to $\phi_1(x), \phi_2(x), \ldots$; and so taking the equation

$$\psi(x) = \mu \int_a^b \left\{ K(x, \xi) - \sum_{n=1}^{\infty} \frac{\phi_n(x)\phi_n(\xi)}{\lambda_n} \right\} \psi(\xi)\, d\xi,$$

we have

$$\psi(x) = \mu \int_a^b K(x, \xi)\psi(\xi)\, d\xi.$$

---

[10] These numbers are not all different if there is more than one orthogonal function to each characteristic number.

[11] The supposition is, of course, a matter for verification with any particular equation.

Therefore μ is a characteristic number of $K(x, y)$, and so $\psi(x)$ must be a linear combination of the (finite number of) functions $\phi_n(x)$ corresponding to this number; let

$$\psi(x) = \sum_m a_m \phi_m(x).$$

Multiply by $\phi_m(x)$ and integrate; then since $\psi(x)$ is orthogonal to all the functions $\phi_n(x)$, we see that $a_m = 0$, so, contrary to hypothesis, $\psi(x) \equiv 0$. The contradiction implies that the nucleus $H(x, y)$ must be identically zero; that is to say, $K(x, y)$ can be expanded in the given series, if it is uniformly convergent.

Example 11.7.1 Shew that, if λ_0 be a characteristic number, the equation

$$\phi(x) = f(x) + \lambda_0 \int_a^b K(x, \xi) \phi(\xi) \, d\xi$$

certainly has no solution when the nucleus is symmetric, unless $f(x)$ is orthogonal to all the characteristic functions corresponding to λ_0.

11.71 The solution of Fredholm's equation by a series

Retaining the notation of §11.7, consider the integral equation

$$\Phi(x) = f(x) + \lambda \int_a^b K(x, \xi) \Phi(\xi) \, d\xi,$$

where $K(x, \xi)$ is symmetric.

If we *assume* that $\Phi(\xi)$ can be expanded into a uniformly convergent series $\sum_{n=1}^{\infty} a_n \phi_n(\xi)$, we have

$$\sum_{n=1}^{\infty} a_n \phi_n(x) = f(x) + \sum_{n=1}^{\infty} \frac{\lambda}{\lambda_n} a_n \phi_n(x),$$

so that $f(x)$ can be expanded in the series

$$\sum_{n=1}^{\infty} a_n \frac{\lambda_n - \lambda}{\lambda_n} \phi_n(x).$$

Hence *if the function $f(x)$ can be expanded into the convergent series $\sum_{n=1}^{\infty} b_n \phi_n(x)$, then the series*

$$\sum_{n=1}^{\infty} \frac{b_n \lambda_n}{\lambda_n - \lambda} \phi_n(x),$$

if it converges uniformly in the range (a, b), is the solution of Fredholm's equation.

To determine the coefficients b_n we observe that $\sum_{n=1}^{\infty} b_n \phi_n(x)$ converges uniformly by §3.35; then, multiplying by $\phi_n(x)$ and integrating, we get

$$b_n = \int_a^b \phi_n(x) f(x) \, dx.$$

Since the numbers λ_n are all real we may arrange them in two sets, one negative the other positive, the members in each set being in order of magnitude; then, when $|\lambda_n| > \lambda$, it is evident that $\lambda_n/(\lambda_n - \lambda)$ is a monotonic sequence in the case of either set.

11.8 Solution of Abel's integral equation

This equation is of the form

$$f(x) = \int_a^x \frac{u(\xi)}{(x - \xi)^\mu}\, d\xi \qquad (0 < \mu < 1, \quad a \le x \le b),$$

where $f'(x)$ is continuous and $f(a) = 0$; we proceed to find a continuous solution $u(x)$.

Let $\phi(x) = \int_a^x u(\xi)\, d\xi$, and take the formula (this follows from Example 6.2.14, by writing $(z - x)/(x - \xi)$ in place of x)

$$\frac{\pi}{\sin \mu\pi} = \int_\xi^z \frac{dx}{(z - x)^{1-\mu}(x - \xi)^\mu},$$

multiply by $u(\xi)$ and integrate, and we get, on using Dirichlet's formula (Corollary 4.5.1),

$$\frac{\pi}{\sin \mu\pi}\{\phi(z) - \phi(a)\} = \int_a^z d\xi \int_\xi^z \frac{u(\xi)\, dx}{(z - x)^{1-\mu}(x - \xi)^\mu}$$

$$= \int_a^z dx \int_a^z \frac{u(\xi)\, d\xi}{(z - x)^{1-\mu}(x - \xi)^\mu}$$

$$= \int_a^z \frac{f(x)\, dx}{(z - x)^{1-\mu}}.$$

Since the original expression has a continuous derivate, so has the final one; therefore the continuous solution, *if it exists*, can be none other than

$$u(z) = \frac{\sin \mu\pi}{\pi} \frac{d}{dz} \int_a^z \frac{f(x)\, dx}{(z - x)^{1-\mu}};$$

and it can be verified by substitution[12] that this function actually *is* a solution.

11.81 Schlömilch's integral equation

This comes from [580]. The reader will easily see that this is reducible to a case of Volterra's equation with a discontinuous nucleus.

Let $f(x)$ *have a continuous differential coefficient when* $-\pi \le x \le \pi$. *Then the equation*

$$f(x) = \frac{2}{\pi} \int_0^{\pi/2} \phi(x \sin \theta)\, d\theta$$

has one solution with a continuous differential coefficient when $-\pi \le x \le \pi$, *namely*

$$\phi(x) = f(0) + x \int_0^{\pi/2} f'(x \sin \theta)\, d\theta.$$

[12] For the details we refer to Bôcher's tract [80].

From §4.2 it follows that

$$f'(x) = \frac{2}{\pi} \int_0^{\pi/2} \sin\theta\, \phi'(x\sin\theta)\, d\theta$$

(so that we have $\phi(0) = f(0)$, $\phi'(0) = \frac{\pi}{2} f'(0)$).

Write $x\sin\psi$ for x, and we have on multiplying by x and integrating

$$x \int_0^{\pi/2} f'(x\sin\psi)\, d\psi = \frac{2x}{\pi} \int_0^{\pi/2} \left\{ \int_0^{\pi/2} \sin\theta\, \phi'(x\sin\theta\sin\psi)\, d\theta \right\} d\psi.$$

Change the order of integration in the repeated integral (§4.3) and take a new variable χ in place of ψ, defined by the equation $\sin\chi = \sin\theta\sin\psi$. Then

$$x \int_0^{\pi/2} f'(x\sin\psi)\, d\psi = \frac{2x}{\pi} \int_0^{\pi/2} \left\{ \int_0^{\theta} \frac{\phi'(x\sin\chi)\cos\chi\, d\chi}{\cos\psi} \right\} d\theta.$$

Changing the order of integration again (§4.51),

$$x \int_0^{\pi/2} f'(x\sin\psi)\, d\psi = \frac{2x}{\pi} \int_0^{\pi/2} \left\{ \int_\chi^{\pi/2} \frac{\phi'(x\sin\chi)\cos\chi\sin\theta}{\sqrt{\sin^2\theta - \sin^2\chi}}\, d\theta \right\} d\chi.$$

But

$$\int_\chi^{\pi/2} \frac{\sin\theta\, d\theta}{\sqrt{\cos^2\chi - \cos^2\theta}} = \left[-\arcsin\left(\frac{\cos\theta}{\cos\chi}\right) \right]_\chi^{\pi/2} = \frac{\pi}{2},$$

and so

$$x \int_0^{\pi/2} f'(x\sin\psi)\, d\psi = x \int_0^{\pi/2} \phi'(x\sin\chi)\cos\chi\, d\chi$$
$$= \phi(x) - \phi(0).$$

Since $\phi(0) = f(0)$, we must have

$$\phi(x) = f(0) + x \int_0^{\pi/2} f'(x\sin\psi)\, d\psi;$$

and it can be verified by substitution that this function actually is a solution.

11.9 Miscellaneous examples

Example 11.1 (Abel)   Shew that if the time of descent of a particle down a smooth curve to its lowest point is independent of the starting-point (the particle starting from rest) the curve is a cycloid.

Example 11.2   Shew that, if $f(x)$ is continuous, the solution of

$$\phi(x) = f(x) + \lambda \int_0^\infty \cos(2xs)\phi(s)\, ds$$

is

$$\phi(x) = \frac{1}{1 - \frac{1}{4}\lambda^2\pi} \left(f(x) + \lambda \int_0^\infty f(s)\cos(2xs)\, ds \right),$$

assuming the legitimacy of a certain change of order of integration.

Example 11.3 (A. Milne)   Shew that the Weber–Hermite functions

$$D_n(x) = (-1)^n \, e^{\frac{1}{4}x^2} \frac{d^n}{dx^n} \left(e^{-\frac{1}{2}x^2} \right)$$

satisfy

$$\phi(x) = \lambda \int_{-\infty}^{\infty} e^{\frac{1}{2}isx} \phi(s) \, ds$$

for the characteristic values of λ.

Example 11.4 (Whittaker; see §19.21, [673])   Shew that even periodic solutions (with period 2π) of the differential equation

$$\frac{d^2\phi(x)}{dx^2} + (a^2 + k^2 \cos^2 x)\phi(x) = 0$$

satisfy the integral equation

$$\phi(x) = \lambda \int_{-\pi}^{\pi} e^{k \cos x \cos s} \phi(s) \, ds.$$

Example 11.5   Shew that the characteristic functions of the equation

$$\phi(x) = \lambda \int_{-\pi}^{\pi} \left\{ \frac{1}{4\pi}(x-y)^2 - \frac{1}{2}|x-y| \right\} \phi(y) \, dy$$

are $\phi(x) = \cos mx$, $\sin mx$, where $\lambda = m^2$ and m is any integer.

Example 11.6 (Bôcher)   Shew that

$$\phi(x) = \int_0^x \xi^{x-\xi} \, \phi(\xi) \, d\xi$$

has the discontinuous solution $\phi(x) = kx^{x-1}$.

Example 11.7   Shew that a solution of the integral equation with a symmetric nucleus

$$f(x) = \int_a^b K(x,\xi)\phi(\xi) \, d\xi \quad \text{is} \quad \phi(x) = \sum_{n=1}^{\infty} a_n \lambda_n \phi_n(x),$$

provided that this series converges uniformly, where λ_n, $\phi_n(x)$ are the characteristic numbers and functions of $K(x,\xi)$ and $\sum_{n=1}^{\infty} a_n \phi_n(x)$ is the expansion of $f(x)$.

Example 11.8   Shew that, if $|h| < 1$, the characteristic functions of the equation

$$\phi(x) = \frac{\lambda}{2\pi} \int_{-\pi}^{\pi} \frac{1-h^2}{1 - 2h\cos(\xi - x) + h^2} \phi(\xi) \, d\xi$$

are 1, $\cos mx$, $\sin mx$, the corresponding characteristic numbers being 1, $1/h^m$, $1/h^m$, where m takes all positive integral values.

Part II

The Transcendental Functions

12

The Gamma-Function

12.1 Definitions of the Gamma-function. The Weierstrassian product

Historically, the Gamma-function $\Gamma(z)$ was first defined by Euler as the limit of a product (§12.11) from which can be derived the infinite integral $\int_0^\infty t^{z-1} e^{-t} \, dt$. The notation $\Gamma(z)$ was introduced by Legendre in 1814. But in developing the theory of the function, it is more convenient to define it by means of an infinite product of Weierstrass' canonical form.

Consider the product

$$z e^{\gamma z} \prod_{n=1}^\infty \left\{ \left(1 + \frac{z}{n} \right) e^{-\frac{z}{n}} \right\},$$

where

$$\gamma = \lim_{m \to \infty} \left\{ \tfrac{1}{1} + \tfrac{1}{2} + \cdots + \tfrac{1}{m} - \log m \right\} \sim 0.5772157 \cdots . \tag{12.1}$$

(The constant γ is known as Euler's constant, or the Euler–Mascheroni constant.) To prove that it exists we observe that, if

$$u_n = \int_0^1 \frac{t \, dt}{n(n+t)} = \frac{1}{n} - \log \frac{n+1}{n},$$

u_n is positive and less than $\int_0^1 \frac{dt}{n^2} = \frac{1}{n^2}$; therefore $\sum_{n=1}^\infty u_n$ converges, and

$$\lim_{m \to \infty} \left\{ \frac{1}{1} + \frac{1}{2} + \cdots + \frac{1}{m} - \log m \right\} = \lim_{m \to \infty} \left\{ \sum_{n=1}^m u_n + \log \frac{m+1}{m} \right\} = \sum_{n=1}^\infty u_n.$$

The value of γ has been calculated by J. C. Adams to 260 places of decimals [10].

The product under consideration represents an analytic function of z, for all values of z; for, if N be an integer such that $|z| \leq \tfrac{1}{2} N$, we have[1] if $n > N$,

$$
\left| \log \left(1 + \frac{z}{n} \right) - \frac{z}{n} \right| = \left| -\frac{1}{2} \frac{z^2}{n^2} + \frac{1}{3} \frac{z^3}{n^3} - \cdots \right|
$$

$$
\leq \frac{|z|^2}{n^2} \left\{ 1 + \left| \frac{z}{n} \right| + \left| \frac{z^2}{n^2} \right| + \cdots \right\}
$$

$$
\leq \frac{1}{4} \frac{N^2}{n^2} \left\{ 1 + \frac{1}{2} + \frac{1}{2^2} + \cdots \right\} \leq \frac{1}{2} \frac{N^2}{n^2}.
$$

---

[1] Taking the principal value of $\log(1 + z/n)$.

243

Since the series $\sum\limits_{n=N+1}^{\infty} \dfrac{N^2}{2n^2}$ converges, it follows that, when $|z| \leq \frac{1}{2} N$,

$$\sum_{n=N+1}^{\infty} \{\log (1 + z/n) - z/n\}$$

is an absolutely and uniformly convergent series of analytic functions, and so it is an analytic function (§5.3); consequently its exponential

$$\prod_{n=N+1}^{\infty} \{(1 + z/n) e^{-z/n}\}$$

is an analytic function and $z e^{\gamma z} \prod\limits_{n=1}^{\infty} \{(1 + z/n) e^{-z/n}\}$ is an analytic function when $|z| \leq \frac{1}{2} N$, where N is any integer; that is to say, the product is analytic for all finite values of z.

The Gamma-function was defined by Weierstrass [659]. This formula for $\Gamma(z)$ had been obtained from Euler's formula (§12.11) in 1848 by F. W. Newman [493] by the equation

$$\frac{1}{\Gamma(z)} = z e^{\gamma z} \prod_{n=1}^{\infty} \left\{\left(1 + \frac{z}{n}\right) e^{-z/n}\right\} ;$$

from this equation *it is apparent that $\Gamma(z)$ is analytic except at the points $z = 0, -1, -2$, where it has simple poles.*

Note   Proofs have been published by Hölder [325], Moore [472] and Barnes [44] of a theorem known to Weierstrass that the Gamma-function does not satisfy any differential equation with rational coefficients.

Example 12.1.1   Prove that

$$\Gamma(1) = 1, \quad \Gamma'(1) = -\gamma,$$

where γ is Euler's constant. *Hint.* Justify differentiating logarithmically the equation

$$\frac{1}{\Gamma(z)} = z e^{\gamma z} \prod_{1}^{\infty} \left\{\left(1 + \frac{z}{n}\right) e^{-z/n}\right\}$$

by §4.7, and put $z = 1$ after the differentiations have been performed.

Example 12.1.2   Shew that

$$1 + \frac{1}{2} + \frac{1}{3} + \cdots + \frac{1}{n} = \int_0^1 \frac{1 - (1 - t)^n}{t} \, dt,$$

and hence that Euler's constant γ is given by

$$\lim_{n \to \infty} \left[\int_0^1 \left\{1 - \left(1 - \frac{t}{n}\right)^n\right\} \frac{dt}{t} - \int_1^n \left(1 - \frac{t}{n}\right)^n \frac{dt}{t} \right].$$

The reader will see later (Example 11.6.2) that this limit may be written

$$\int_0^1 (1 - e^{-t}) \frac{dt}{t} - \int_1^\infty \frac{e^{-t} \, dt}{t}.$$

Example 12.1.3   Shew that

$$\prod_{n=1}^{\infty} \left[\left(1 - \frac{x}{z+n} \right) e^{x/n} \right] = \frac{e^{\gamma x}\, \Gamma(z+1)}{\Gamma(z-x+1)}.$$

12.11 Euler's formula for the Gamma-function

By the definition of an infinite product we have

$$\frac{1}{\Gamma(z)} = z \left[\lim_{m\to\infty} e^{\left(1+\frac{1}{2}+\cdots+\frac{1}{m}-\log m\right)z} \right] \left[\lim_{m\to\infty} \prod_{n=1}^{m} \left\{ \left(1+\frac{z}{n} \right) e^{-\frac{z}{n}} \right\} \right]$$

$$= z \lim_{m\to\infty} \left[e^{\left(1+\frac{1}{2}+\cdots+\frac{1}{m}-\log m\right)z} \prod_{n=1}^{m} \left\{ \left(1+\frac{z}{n} \right) e^{-\frac{z}{n}} \right\} \right]$$

$$= z \lim_{m\to\infty} \left[m^{-z} \prod_{n=1}^{m} \left(1+\frac{z}{n} \right) \right]$$

$$= z \lim_{m\to\infty} \left[\prod_{n=1}^{m-1} \left(1+\frac{1}{n} \right)^{-z} \prod_{n=1}^{m} \left(1+\frac{z}{n} \right) \right]$$

$$= z \lim_{m\to\infty} \left[\prod_{n=1}^{m} \left\{ \left(1+\frac{z}{n} \right) \left(1+\frac{1}{n} \right)^{-z} \right\} \left(1+\frac{1}{m} \right)^{z} \right].$$

Hence

$$\Gamma(z) = \frac{1}{z} \prod_{n=1}^{\infty} \left\{ \left(1+\frac{1}{n} \right)^{z} \left(1+\frac{z}{n} \right)^{-1} \right\}. \tag{12.2}$$

This formula is due to Euler. It was given in 1729 in a letter to Goldbach, printed in Fuss [231]. It is valid except when $z = 0, -1, -2, \ldots$.

Example 12.1.4 (Euler)   Prove that

$$\Gamma(z) = \lim_{n\to\infty} \frac{1 \cdot 2 \cdot \, \cdots \, \cdot (n-1)}{z\,(z+1)\cdots(z+n-1)} n^{z}.$$

12.12 The difference equation satisfied by the Gamma-function

We shall now shew that the function $\Gamma(z)$ satisfies the difference equation

$$\Gamma(z+1) = z\Gamma(z).$$

For, by Euler's formula, if z is not a negative integer

$$\frac{\Gamma(z+1)}{\Gamma(z)} = \frac{1}{z+1}\left[\lim_{m\to\infty}\prod_{n=1}^{m}\frac{\left(1+\frac{1}{n}\right)^{z+1}}{1+\frac{z+1}{n}}\right]\left[\frac{1}{z}\lim_{m\to\infty}\prod_{n=1}^{m}\frac{\left(1+\frac{1}{n}\right)^{z}}{1+\frac{z}{n}}\right]^{-1}$$

$$= \frac{z}{z+1}\lim_{m\to\infty}\prod_{n=1}^{m}\left\{\frac{\left(1+\frac{1}{n}\right)(z+n)}{z+n+1}\right\}$$

$$= z\lim_{m\to\infty}\frac{m+1}{z+m+1} = z.$$

This is one of the most important properties of the Gamma-function. Since $\Gamma(1) = 1$, it follows that, if z is a positive integer, $\Gamma(z) = (z-1)!$.

Example 12.1.5 Prove that

$$\frac{1}{\Gamma(z+1)} + \frac{1}{\Gamma(z+2)} + \frac{1}{\Gamma(z+3)} + \cdots$$

$$= \frac{e}{\Gamma(z)}\left(\frac{1}{z} - \frac{1}{1!}\frac{1}{z+1} + \frac{1}{2!}\frac{1}{z+2} - \cdots\right).$$

Hint. Consider the expression

$$\frac{1}{z} + \frac{1}{z(z+1)} + \frac{1}{z(z+1)(z+2)} + \cdots + \frac{1}{z(z+1)\cdots(z+m)}.$$

It can be expressed in partial fractions in the form $\sum_{n=0}^{m}\frac{a_n}{z+n}$, where

$$a_n = \frac{(-1)^n}{n!}\left\{1 + \frac{1}{1!} + \frac{1}{2!} + \cdots + \frac{1}{(m-n)!}\right\} = \frac{(-1)^n}{n!}\left\{e - \sum_{r=m-n+1}^{\infty}\frac{1}{r!}\right\}.$$

Noting that $\sum_{r=m-n+1}^{\infty}displaystyle\frac{1}{r!} < \frac{e}{(m-n+1)!}$, prove that

$$\sum_{n=0}^{m}\frac{(-1)^n}{n!}\frac{1}{z+n}\left\{\sum_{r=m-n+1}^{\infty}\frac{1}{r!}\right\} \to 0$$

as $m \to \infty$ when z is not a negative integer.

12.13 *The evaluation of a general class of infinite products*

By means of the Gamma-function, it is possible to evaluate the general class of infinite products of the form $\prod_{n=1}^{\infty} u_n$, where u_n is any rational function of the index n.

For, resolving u_n into its factors, we can write the product in the form

$$\prod_{n=1}^{\infty}\left\{\frac{A(n-a_1)(n-a_2)\cdots(n-a_k)}{(n-b_1)\cdots(n-b_l)}\right\};$$

and it is supposed that no factor in the denominator vanishes.

In order that this product may converge, the number of factors in the numerator must clearly be the same as the number of factors in the denominator, and also $A = 1$ for, otherwise, the general factor of the product would not tend to the value unity as n tends to infinity.

We have therefore $k = l$, and, denoting the product by P, we may write

$$P = \prod_{n=1}^{\infty} \left\{ \frac{(n - a_1) \cdots (n - a_k)}{(n - b_1) \cdots (n - b_k)} \right\}.$$

The general term in this product can be written

$$\left(1 - \frac{a_1}{n}\right) \cdots \left(1 - \frac{a_k}{n}\right) \left(1 - \frac{b_1}{n}\right)^{-1} \cdots \left(1 - \frac{b_k}{n}\right)^{-1}$$

$$= 1 - \frac{a_1 + a_2 + \cdots + a_k - b_1 - \cdots - b_k}{n} + A_n,$$

where A_n is $O(n^{-2})$ when n is large. In order that the infinite product may be absolutely convergent, it is therefore necessary further (§2.7) that

$$a_1 + \cdots + a_k - b_1 - \cdots - b_k = 0.$$

We can therefore introduce the factor

$$\exp \left\{ n^{-1} \left(a_1 + \cdots + a_k - b_1 - \cdots - b_k \right) \right\}$$

into the general factor of the product, without altering its value; and thus we have

$$P = \prod_{n=1}^{\infty} \left\{ \frac{\left(1 - \frac{a_1}{n}\right) e^{a_1/n} \left(1 - \frac{a_2}{n}\right) e^{a_2/n} \cdots \left(1 - \frac{a_k}{n}\right) e^{a_k/n}}{\left(1 - \frac{b_1}{n}\right) e^{b_1/n} \left(1 - \frac{b_2}{n}\right) e^{b_2/n} \cdots \left(1 - \frac{b_k}{n}\right) e^{b_k/n}} \right\}.$$

But it is obvious from the Weierstrassian definition of the Gamma-function that

$$\prod_{n=1}^{\infty} \left\{ \left(1 - \frac{z}{n}\right) e^{z/n} \right\} = \frac{1}{-z\Gamma(-z)e^{-\gamma z}},$$

and so

$$P = \frac{b_1 \Gamma(-b_1) b_2 \Gamma(-b_2) \cdots b_k \Gamma(-b_k)}{a_1 \Gamma(-a_1) \cdots a_k \Gamma(-a_k)} = \prod_{m=1}^{k} \frac{\Gamma(1 - b_m)}{\Gamma(1 - a_m)},$$

a formula which expresses the general infinite product P in terms of the Gamma-function.

Example 12.1.6 Prove that

$$\prod_{s=1}^{\infty} \frac{s(a + b + s)}{(a + s)(b + s)} = \frac{\Gamma(a + 1)\Gamma(b + 1)}{\Gamma(a + b + 1)}.$$

Example 12.1.7 Shew that, if $a = \cos(2\pi/n) + i \sin(2\pi/n)$, then

$$x \left(1 - \frac{x}{1^n}\right) \left(1 - \frac{x}{2^n}\right) \cdots = \left(-\Gamma(-x^{\frac{1}{n}})\Gamma(-ax^{\frac{1}{n}}) \cdots \Gamma(-a^{n-1}x^{\frac{1}{n}})\right)^{-1}.$$

12.14 Connexion between the Gamma-function and the circular functions

We now proceed to establish another most important property of the Gamma-function, expressed by the equation

$$\Gamma(z)\Gamma(1-z) = \frac{\pi}{\sin \pi z}.$$

We have, by the definition of Weierstrass (§12.1),

$$\Gamma(z)\Gamma(-z) = -\frac{1}{z^2}\prod_{n=1}^{\infty}\left\{\left(1+\frac{z}{n}\right)e^{-z/n}\right\}^{-1}\prod_{n=1}^{\infty}\left\{\left(1-\frac{z}{n}\right)e^{z/n}\right\}^{-1}$$

$$= -\frac{\pi}{z\sin \pi z},$$

by Example 7.5.1. Since, by §12.12, $\Gamma(1-z) = -z\Gamma(-z)$ we have the result stated.

Corollary 12.1.1 *If we assign to z the value $\frac{1}{2}$, this formula gives*

$$\{\Gamma(\tfrac{1}{2})\}^2 = \pi;$$

since, by the formula of Weierstrass, $\Gamma(\frac{1}{2})$ is positive, we have

$$\Gamma\left(\tfrac{1}{2}\right) = \sqrt{\pi}.$$

Corollary 12.1.2 *If $\psi(z) = \Gamma'(z)/\Gamma(z)$, then $\psi(1-z) - \psi(z) = \pi \cot \pi z$.*

12.15 The multiplication-theorem of Gauss and Legendre

This appears in [236, p. 149]. The case in which $n = 2$ was given by Legendre.

We shall next obtain the result

$$\Gamma(z)\Gamma\left(z+\frac{1}{n}\right)\Gamma\left(z+\frac{2}{n}\right)\cdots\Gamma\left(z+\frac{n-1}{n}\right) = (2\pi)^{\frac{1}{2}(n-1)}n^{\frac{1}{2}-nz}\Gamma(nz).$$

For let

$$\phi(z) = \frac{n^{nz}\Gamma(z)\Gamma\left(z+\frac{1}{n}\right)\cdots\Gamma\left(z+\frac{n-1}{n}\right)}{n\Gamma(nz)}.$$

Then we have, by Euler's formula (Example 12.1.4),

$$\phi(z) = \frac{n^{nz}\prod_{r=0}^{n-1}\lim_{m\to\infty}\dfrac{1\cdot 2\cdots(m-1)\cdot m^{z+r/n}}{\left(z+\frac{r}{n}\right)\left(z+\frac{r}{n}+1\right)\cdots\left(z+\frac{r}{n}+m-1\right)}}{n\lim_{m\to\infty}\dfrac{1\cdot 2\cdots(nm-1)\cdot(nm)^{nz}}{nz(nz+1)\cdots(nz+nm-1)}}$$

$$= n^{nz-1}\lim_{m\to\infty}\frac{\{(m-1)!\}^n m^{nz+\frac{1}{2}(n-1)}n^{mn}}{(nm-1)!(nm)^{nz}}$$

$$= \lim_{m\to\infty}\frac{\{(m-1)!\}^n m^{\frac{1}{2}(n-1)}n^{mn-1}}{(nm-1)!}.$$

It is evident from this last equation that $\phi(z)$ is independent of z. Thus $\phi(z)$ is equal to the value which it has when $z = \frac{1}{n}$; and so

$$\phi(z) = \Gamma\left(\frac{1}{n}\right)\Gamma\left(\frac{2}{n}\right)\cdots\Gamma\left(\frac{n-1}{n}\right).$$

Therefore

$$\{\phi(z)\}^2 = \prod_{r=1}^{n-1}\left\{\Gamma\left(\frac{r}{n}\right)\Gamma\left(1 - \frac{r}{n}\right)\right\}$$

$$= \frac{\pi^{n-1}}{\sin\frac{\pi}{n}\sin\frac{2\pi}{n}\cdots\sin\frac{(n-1)\pi}{n}} = \frac{(2\pi)^{n-1}}{n}.$$

Thus, since $\phi(n^{-1})$ is positive, $\phi(z) = (2\pi)^{(n-1)/2}n^{-1/2}$, i.e.

$$\Gamma(z)\Gamma\left(z + \frac{1}{n}\right)\cdots\Gamma\left(z + \frac{n-1}{n}\right) = n^{\frac{1}{2}-nz}(2\pi)^{\frac{1}{2}(n-1)}\Gamma(nz). \tag{12.3}$$

Corollary 12.1.3 *Taking $n = 2$, we have*

$$2^{2z-1}\Gamma(z)\Gamma\left(z + \frac{1}{2}\right) = \pi^{1/2}\Gamma(2z).$$

This is called the duplication formula..

Example 12.1.8 If $B(p,q) = \dfrac{\Gamma(p)\Gamma(q)}{\Gamma(p+q)}$, shew that

$$B(np,nq) = n^{-nq}\frac{B(p,q)B\left(p + \frac{1}{n},q\right)\cdots B\left(p + \frac{n-1}{n},q\right)}{B(q,q)B(2q,q)\cdots B((n-1)q,q)}.$$

12.16 Expansion for the logarithmic derivates of the Gamma-function

We have

$$\{\Gamma(z+1)\}^{-1} = e^{\gamma z}\prod_{n=1}^{\infty}\left\{\left(1 + \frac{z}{n}\right)e^{-z/n}\right\}.$$

Differentiating logarithmically (§4.7), this gives

$$\frac{d\log\Gamma(z+1)}{dz} = -\gamma + \frac{z}{1(z+1)} + \frac{z}{2(z+2)} + \frac{z}{3(z+3)} + \cdots.$$

Therefore, since $\log\Gamma(z+1) = \log z + \log\Gamma(z)$, we have

$$\frac{d}{dz}\log\Gamma(z) = -\gamma - \frac{1}{z} + z\sum_{n=1}^{\infty}\frac{1}{n(z+n)}.$$

Differentiating again,

$$\frac{d^2}{dz^2}\log\Gamma(z+1) = \frac{d}{dz}\left\{\frac{z}{1(z+1)} + \frac{z}{2(z+2)} + \cdots\right\}$$

$$= \frac{1}{(z+1)^2} + \frac{1}{(z+2)^2} + \cdots. \tag{12.4}$$

These expansions are occasionally used in applications of the theory.

12.2 Euler's expression of $\Gamma(z)$ as an infinite integral

The infinite integral $\displaystyle\int_0^\infty e^{-t}t^{z-1}\,dt$ represents an analytic function of z when the real part of z is positive (if the real part of z is not positive the integral does not converge on account of the singularity of the integrand at $t = 0$, §5.32); it is called *the Eulerian Integral of the Second Kind*. The name was given by Legendre; see §12.4 for the Eulerian Integral of the First Kind. It will now be shewn that, when Re $z > 0$, the integral is equal to $\Gamma(z)$. Denoting the real part of z by x, we have $x > 0$. Now, if[2]

$$\Pi(z,n) = \int_0^n \left(1 - \frac{t}{n}\right)^n t^{z-1}\,dt,$$

we have $\displaystyle\Pi(z,n) = n^z\int_0^1 (1 - \tau)^n \tau^{z-1}\,d\tau$, if we write $t = n\tau$; it is easily shewn by repeated integrations by parts that, when $x > 0$ and n is a positive integer,

$$\int_0^1 (1-\tau)^n\tau^{z-1}d\tau = \left[\frac{1}{z}\tau^z(1-\tau)^n\right]_0^1 + \frac{n}{z}\int_0^1 (1-\tau)^{n-1}\tau^z d\tau$$

$$\vdots$$

$$= \frac{n(n-1)\cdots 1}{z(z+1)\cdots(z+n-1)}\int_0^1 \tau^{z+n-1}\,d\tau,$$

and so $\displaystyle\Pi(z,n) = \frac{1\cdot 2\cdots n}{z(z+1)\cdots(z+n)}n^z$. Hence, by Example 12.1.4, $\Pi(z,n) \to \Gamma(z)$ as $n \to \infty$.
  Consequently

$$\Gamma(z) = \lim_{n\to\infty}\int_0^n \left(1 - \frac{t}{n}\right)^n t^{z-1}dt.$$

And so, if $\displaystyle\Gamma_1(z) = \int_0^\infty e^{-t}t^{z-1}dt$, we have

$$\Gamma_1(z) - \Gamma(z) = \lim_{n\to\infty}\left[\int_0^n \left\{e^{-t} - \left(1 - \frac{t}{n}\right)^n\right\}t^{z-1}dt + \int_n^\infty e^{-t}t^{z-1}dt\right].$$

Now $\displaystyle\lim_{n\to\infty}\int_n^\infty e^{-t}t^{z-1}dt = 0$, since $\displaystyle\int_0^\infty e^{-t}t^{z-1}dt$ converges. To shew that zero is the limit of the first of the two integrals in the formula for $\Gamma_1(z) - \Gamma(z)$ we observe that

$$0 \le e^{-t} - \left(1 - \frac{t}{n}\right)^n \le n^{-1}t^2 e^{-t}.$$

Hint. To establish these inequalities, we proceed as follows: when $0 \le y < 1$,

$$1 + y \le e^y \le (1 - y)^{-1},$$

from the series for e^y and $(1 - y)^{-1}$. Writing t/n for y, we have

$$\left(1 + \frac{t}{n}\right)^{-n} \ge e^{-t} \ge \left(1 - \frac{t}{n}\right)^n,$$

---

[2] The many-valued function t^{z-1} is made precise by the equation $t^{z-1} = e^{(z-1)\log t}$, $\log t$ being purely real.

and so

$$0 \le e^{-t} - \left(1 - \frac{t}{n}\right)^n = e^{-t}\left\{1 - e^t\left(1 - \frac{t}{n}\right)^n\right\} \le e^{-t}\left\{1 - \left(1 - \frac{t^2}{n^2}\right)^n\right\}.$$

Now, if $0 \le a \le 1$, $(1 - a)^n \ge 1 - na$ by induction when $na < 1$ and obviously when $na \ge 1$; and, writing t^2/n^2 for a, we get

$$1 - \left(1 - \frac{t^2}{n^2}\right)^n \le \frac{t^2}{n}$$

and so

$$0 \le e^{-t} - \left(1 - \frac{t}{n}\right)^n \le e^{-t}t^2/n,$$

which is the required result. This analysis is a modification of that given by Schlömilch [583, Vol. 2, p. 243]. A simple method of obtaining a less precise inequality (which is sufficient for the object required) is given by Bromwich [102, p. 459].

From the inequalities, it follows at once that

$$\left|\int_0^n \left\{e^{-t} - \left(1 - \frac{t}{n}\right)^n\right\} t^{z-1} dt\right| \le \int_0^n n^{-1} e^{-t} t^{x+1} dt$$

$$< n^{-1} \int_0^\infty e^{-t} t^{x+1} dt \to 0,$$

as $n \to \infty$, since the last integral converges.

Consequently $\Gamma_1(z) = \Gamma(z)$ when the integral, by which $\Gamma_1(z)$ is defined, converges; that is to say, when the real part of z is positive,

$$\Gamma(z) = \int_0^\infty e^{-t} t^{z-1} \, dt.$$

And so, when the real part of z is positive, $\Gamma(z)$ may be defined either by this integral or by the Weierstrassian product.

Example 12.2.1 Prove that, when Re(z) is positive,

$$\Gamma(z) = \int_0^1 \left(\log \frac{1}{x}\right)^{z-1} dx.$$

Example 12.2.2 Prove that, if Re(z) > 0 and Re(s) > 0,

$$\int_0^\infty e^{-zx} x^{s-1} \, dx = \frac{\Gamma(s)}{z^s}.$$

Example 12.2.3 Prove that, if Re(z) > 0 and Re(s) > 1,

$$\frac{1}{(z+1)^s} + \frac{1}{(z+2)^s} + \frac{1}{(z+3)^s} + \cdots = \frac{1}{\Gamma(s)} \int_0^\infty \frac{e^{-xz} x^{s-1} \, dx}{e^x - 1}$$

Example 12.2.4 From Example 12.1.2 by using the inequality

$$0 \le e^{-t} - \left(1 - \frac{t}{n}\right)^n \le \frac{t^2 e^{-t}}{n}$$

deduce that

$$\gamma = \int_0^1 \frac{1 - e^{-t} - e^{-1/t}}{t} \, dt.$$

12.21 Extension of the infinite integral to the case in which the argument of the Gamma-function is negative

The formula of the last article is no longer applicable when the real part of z is negative. Cauchy [123, volume 2, pp. 91–92] and Saalschütz [568, 569] have shewn, however, that, for negative arguments, an analogous theorem exists. This can be obtained in the following way.

Consider the function

$$\Gamma_2(z) = \int_0^\infty t^{z-1} \left(e^{-t} - 1 + t - \frac{t^2}{2!} + \cdots + (-1)^{k+1} \frac{t^k}{k!} \right) dt,$$

where k is the integer so chosen that $-k > x > -k - 1$, x being the real part of z.

By partial integration we have, when $z < -1$,

$$\Gamma_2(z) = \left[\frac{t^z}{z} \left(e^{-t} - 1 + t - \frac{t^2}{2!} + \cdots + (-1)^{k+1} \frac{t^k}{k!} \right) \right]_0^\infty$$
$$+ \frac{1}{z} \int_0^\infty t^z \left(e^{-t} - 1 + t - \cdots + (-1)^k \frac{t^{k-1}}{(k-1)!} \right) dt.$$

The integrated part tends to zero at each limit, since $x + k$ is negative and $x + k + 1$ is positive: so we have

$$\Gamma_2(z) = \frac{\Gamma_2(z+1)}{z}.$$

The same proof applies when x lies between 0 and -1, and leads to the result $\Gamma(z+1) = z\Gamma_2(z)$ $(0 > x > -1)$. The last equation shews that, between the values 0 and -1 of x,

$$\Gamma_2(z) = \Gamma(z).$$

The preceding equation then shews that $\Gamma_2(z)$ is the same as $\Gamma(z)$ for all negative values of $\operatorname{Re}(z)$ less than -1. Thus, for all negative values of $\operatorname{Re}(z)$, we have the result of Cauchy and Saalschütz

$$\Gamma(z) = \int_0^\infty t^{s-1} \left(e^{-t} - 1 + t - \frac{t^2}{2!} + \cdots + (-1)^{k+1} \frac{t^k}{k!} \right) dt,$$

where k is the integer next less than $-\operatorname{Re}(z)$.

Example 12.2.5 (Saalschütz)   If a function $P(\mu)$ be such that for positive values of μ we have

$$P(\mu) = \int_0^1 x^{\mu-1} e^{-x} \, dx,$$

and if for negative values of μ we define $P_1(\mu)$ by the equation

$$P_1(\mu) = \int_0^1 x^{\mu-1} \left(e^{-x} - 1 + x - \cdots + (-1)^{k+1} \frac{x^k}{k!} \right) dx,$$

where k is the integer next less than $-\mu$, shew that

$$P_1(\mu) = P(\mu) - \frac{1}{\mu} + \frac{1}{1!(\mu+1)} - \cdots + (-1)^{k-1}\frac{1}{k!(\mu+k)}.$$

12.22 Hankel's expression of $\Gamma(z)$ as a contour integral

The integrals obtained for $\Gamma(z)$ in §§12.2, 12.21 are members of a large class of definite integrals by which the Gamma-function can be defined. The most general integral of the class in question is due to Hankel [272]; this integral will now be investigated.

Let D be a contour which starts from a point ρ on the real axis, encircles the origin once counter-clockwise and returns to ρ. Consider $\int_D (-t)^{z-1} e^{-t}\,dt$, when Re $z > 0$ and z is not an integer. The many-valued function $(-t)^{z-1}$ is to be made definite by the convention that $(-t)^{z-1} = e^{(z-1)\log(-t)}$ and $\log(-t)$ is purely real when t is on the negative part of the real axis, so that $-\pi \le \arg(-t) \le \pi$ on D.

The integrand is not analytic inside D, but, by Corollary 5.2.1, the path of integration may be deformed (without affecting the value of the integral) into the path of integration which starts from ρ, proceeds along the real axis to δ, describes a circle of radius δ counter-clockwise round the origin and returns to ρ along the real axis.

On the real axis in the first part of this new path we have $\arg(-t) = -\pi$, so that $(-t)^{z-1} = e^{-i\pi(z-1)}\,t^{z-1}$(where $\log t$ is purely real); and on the last part of the new path $(-t)^{z-1} = e^{i\pi(z-1)}\,t^{z-1}$. On the circle we write $-t = \delta e^{i\theta}$; then we get

$$\int_D (-t)^{z-1} e^{-t}\,dt = \int_\rho^\delta e^{-i\pi(z-1)} t^{z-1} e^{-t}\,dt + \int_{-\pi}^\pi (\delta e^{i\theta})^{z-1} e^{\delta(\cos\theta + i\sin\theta)}\delta e^{i\theta} i\,d\theta$$

$$+ \int_\delta^\rho e^{i\pi(z-1)} t^{s-1} e^{-t}\,dt$$

$$= 2i \sin(\pi z)\int_\delta^\rho t^{z-1} e^{-t}\,dt + i\delta^z \int_{-\pi}^\pi e^{iz\theta + \delta(\cos\theta + i\sin\theta)}\,d\theta.$$

This is true for all positive values of $\delta \le \rho$; now make $\delta \to 0$; then $\delta^z \to 0$ and

$$\int_{-\pi}^\pi e^{iz\theta + \delta(\cos\theta + i\sin\theta)}\,d\theta \to \int_{-\pi}^\pi e^{iz\theta}\,d\theta$$

since the integrand tends to its limit uniformly. *We consequently infer that*

$$\int_D (-t)^{z-1} e^{-t}\,dt = -2i\sin(\pi z)\int_0^\rho t^{z-1} e^{-t}\,dt.$$

This is true for all positive values of ρ; make $\rho \to \infty$, and let C be the limit of the contour D. Then

$$\int_C (-t)^{z-1} e^{-t}\,dt = -2i\,\sin(\pi z)\int_0^\infty t^{z-1} e^{-t}\,dt.$$

Therefore

$$\Gamma(z) = -\frac{1}{2i\sin\pi z}\int_C (-t)^{z-1} e^{-t}\,dt. \tag{12.5}$$

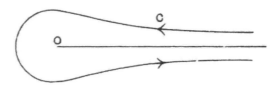

Now, since the contour C does not pass through the point $t = 0$, there is no need longer to stipulate that the real part of z is positive; and $\int_C (-t)^{z-1} e^{-t} dt$ is a one-valued analytic function of z for all values of z. Hence, by §5.5, *the equation, just proved when the real part of z is positive, persists for all values of z with the exception of the values $0, \pm 1, \pm 2, \ldots$.*

Consequently, for all except integer values of z,

$$\Gamma(z) = -\frac{1}{2i \sin \pi z} \int_C (-t)^{z-1} e^{-t} dt.$$

This is Hankel's formula; if we write $1 - z$ for z and make use of §12.14, we get the further result that

$$\frac{1}{\Gamma(z)} = \frac{i}{2\pi} \int_C (-t)^{-z} e^{-t} dt.$$

We shall write $\int_\infty^{(0+)}$ for $\int_C$, meaning thereby that the path of integration starts at 'infinity' on the real axis, encircles the origin in the positive direction and returns to the starting point.

Example 12.2.6 Shew that, if Re $z > 0$ and if a be any positive constant, $\int (-t)^{-z} e^{-t} dt$ tends to zero as $\rho \to \infty$, when the path of integration is either of the quadrants of circles of radius $\rho + a$ with centres at $-a$, the end points of one quadrant being ρ and $-a + i(\rho + a)$, and of the other ρ and $-a - i(\rho + a)$.

Deduce that

$$\lim_{\rho \to \infty} \int_{-a+i\rho}^{-a-i\rho} (-t)^{-z} e^{-t} dt = \lim_{\rho \to \infty} \int_C (-t)^{-z} e^{-t} dt,$$

and hence, by writing $t = -a - iu$, shew that

$$\frac{1}{\Gamma(z)} = \frac{1}{2\pi} \int_{-\infty}^\infty e^{a+iu}(a + iu)^{-z} du.$$

This formula was given by Laplace [410, p. 134], and it is substantially equivalent to Hankel's formula involving a contour integral.

Example 12.2.7 (Bourguet, L. [96, 97]) By taking $a = 1$, and putting $t = -1 + i \tan \theta$ in Example 12.2.6, shew that

$$\frac{1}{\Gamma(z)} = \frac{e}{\pi} \int_0^{\frac{1}{2}\pi} \cos(\tan \theta - z\theta) \cos^{z-2} \theta \, d\theta.$$

Example 12.2.8 By taking as contour of integration a parabola whose focus is the origin, shew that, if $a > 0$, then

$$\Gamma(z) = \frac{2a^z e^a}{\sin \pi z} \int_0^\infty e^{-at^2}(1 + t^2)^{z-\frac{1}{2}} \cos \{2at + (2z - 1) \arctan t\} \, dt.$$

Example 12.2.9 (St John's, 1902)   Investigate the values of x for which the integral

$$\frac{2}{\pi} \int_0^\infty t^{x-1} \sin t \, dt$$

converges; for such values of x express it in terms of Gamma-functions, and thence shew that it is equal to

$$e^{-\gamma x} \prod_{n=1}^\infty \left\{ \left(1 - \frac{x}{2n}\right) e^{x/(2n)} \right\} \Big/ \prod_{n=1}^\infty \left\{ \left(1 + \frac{x}{2n-1}\right) e^{-x/(2n-1)} \right\}.$$

Example 12.2.10 (St John's, 1902)   Prove that $\displaystyle\int_0^\infty (\log t)^m \frac{\sin t}{t} \, dt$ converges when $m > 0$, and, by means of Example 12.2.9, evaluate it when $m = 1$ and when $m = 2$.

12.3 Gauss' expression for the logarithmic derivate of the Gamma-function as an infinite integral

We shall now express the function $\dfrac{d}{dz} \log \Gamma(z) = \dfrac{\Gamma'(z)}{\Gamma(z)}$ as an infinite integral when $\operatorname{Re} z > 0$; the function in question is frequently written $\psi(z)$. (The results appear in [236, p. 159].) We first need a new formula for γ.

Take the formula in Example 11.6.2

$$\gamma = \int_0^1 \frac{1 - e^{-t}}{t} \, dt - \int_0^\infty \frac{e^{-t}}{t} \, dt = \lim_{\delta \to 0} \left\{ \int_\delta^1 \frac{dt}{t} - \int_\delta^\infty \frac{e^{-t}}{t} \, dt \right\}$$

$$= \lim_{\delta \to 0} \left\{ \int_\Delta^1 \frac{dt}{t} - \int_\delta^\infty \frac{e^{-t}}{t} \, dt \right\},$$

where $\Delta = 1 - e^\delta$, since

$$\int_\Delta^\delta \frac{dt}{t} = \log \frac{\delta}{1 - e^{-\delta}} \to 0 \quad \text{as} \quad \delta \to 0.$$

Writing $t = 1 - e^{-u}$ in the first of these integrals and then replacing u by t we have

$$\gamma = \lim_{\infty \to 0} \left\{ \int_\delta^\infty \frac{e^{-t}}{1 - e^{-t}} \, dt - \int_\delta^\infty \frac{e^{-t}}{t} \, dt \right\} = \int_0^\infty \left\{ \frac{1}{1 - e^{-t}} - \frac{1}{t} \right\} e^{-t} \, dt.$$

This is the formula for γ which was required.

To get Gauss' formula, take the equation (§12.16)

$$\frac{\Gamma'(z)}{\Gamma(z)} = -\gamma - \frac{1}{z} + \lim_{n \to \infty} \sum_{m=1}^n \left(\frac{1}{m} - \frac{1}{z+m} \right),$$

and write $\dfrac{1}{z+m} = \displaystyle\int_0^\infty e^{-t(z+m)} dt$; this is permissible when $m = 0, 1, 2, \dots$ if $\operatorname{Re} z > 0$. It follows that

$$\frac{\Gamma'(z)}{\Gamma(z)} = -\gamma - \int_0^\infty e^{-zt}\,dt + \lim_{n\to\infty}\int_0^\infty \sum_{m=1}^n (e^{-mt} - e^{-(m+z)t})\,dt$$

$$= -\gamma + \lim_{n\to\infty}\int_0^\infty \frac{e^{-t} - e^{-zt} - e^{-(n+1)t} + e^{-(z+n+1)t}}{1 - e^{-t}}\,dt$$

$$= \int_0^\infty \left(\frac{e^{-t}}{t} - \frac{e^{-zt}}{1 - e^{-t}}\right)dt - \lim_{n\to\infty}\int_0^\infty \frac{1 - e^{-zt}}{1 - e^{-t}} e^{-(n+1)t}\,dt.$$

Now, when $0 < t \le 1$, $\left|\dfrac{1 - e^{-zt}}{1 - e^{-t}}\right|$ is a bounded function of t whose limit as $t \to 0$ is finite; and when $t \ge 1$,

$$\left|\frac{1 - e^{-zt}}{1 - e^{-t}}\right| < \frac{1 + |e^{-zt}|}{1 - e^{-t}} < \frac{2}{1 - e^{-t}}.$$

Therefore we can find a number K independent of t such that, on the path of integration,

$$\left|\frac{1 - e^{-zt}}{1 - e^{-t}}\right| < K;$$

and so

$$\left|\int_0^\infty \frac{1 - e^{-zt}}{1 - e^{-t}} e^{(n+1)t}\,dt\right| < K\int_0^\infty e^{-(n+1)t}\,dt = \frac{K}{n+1} \to 0 \quad \text{as} \quad n \to \infty.$$

We have thus proved the formula

$$\psi(z) = \frac{d}{dz}\log\Gamma(z) = \int_0^\infty \left|\frac{e^{-t}}{t} - \frac{e^{-zt}}{1 - e^{-t}}\right|dt,$$

which is Gauss' expression of $\psi(z)$ as an infinite integral. It may be remarked that this is the first integral which we have encountered connected with the Gamma-function in which the integrand is a single-valued function.

Writing $t = \log(1 + x)$ in Gauss' result, we get, if $\Delta = e^\delta - 1$,

$$\frac{\Gamma'(z)}{\Gamma(z)} = \lim_{\delta\to 0}\int_\delta^\infty \left\{\frac{e^{-t}}{t} - \frac{e^{-zt}}{1 - e^{-t}}\right\}dt$$

$$= \lim_{\delta\to 0}\left\{\int_\delta^\infty \frac{e^{-t}}{t}\,dt - \int_\Delta^\infty \frac{dx}{x(1+x)^z}\right\}$$

$$= \lim_{\delta\to 0}\left\{\int_\Delta^\infty \frac{e^{-t}}{t}\,dt - \int_\Delta^\infty \frac{dx}{x(1+x)^z}\right\},$$

since

$$0 < \int_\delta^\Delta \frac{e^{-t}}{t}\,dt < \int_\delta^\Delta \frac{dt}{t} = \log\frac{e^\delta - 1}{\delta} \to 0$$

as $\delta \to 0$. Hence

$$\frac{\Gamma'(z)}{\Gamma(z)} = \lim_{\Delta\to 0}\int_\Delta^\infty \left\{e^{-x} - \frac{1}{(1+x)^z}\right\}\frac{dx}{x},$$

so that

$$\Gamma'(z) = \Gamma(z) \int_0^\infty \left\{ e^{-x} - \frac{1}{(1+x)^z} \right\} \frac{dx}{x},$$

an equation due to Dirichlet [178, p. 275].

Example 12.3.1 (Gauss)   Prove that, if Re $z > 0$,

$$\psi(z) = \int_0^1 \left\{ \frac{1}{-\log t} - \frac{t^{z-1}}{1-t} \right\} dt.$$

Example 12.3.2   (Dirichlet). Shew that

$$\gamma = \int_0^\infty \left\{ (1+t)^{-1} - e^{-t} \right\} t^{-1} dt.$$

12.31 Binet's first expression for $\log \Gamma(z)$ in terms of an infinite integral

Binet [73] has given two expressions for $\log \Gamma(z)$ which are of great importance as shewing the way in which $\log \Gamma(z)$ behaves as $|z| \to \infty$. To obtain the first of these expressions, we observe that, when the real part of z is positive,

$$\frac{\Gamma'(z+1)}{\Gamma(z+1)} = \int_0^\infty \left\{ \frac{e^{-t}}{t} - \frac{e^{-tz}}{e^t - 1} \right\} dt,$$

writing $z + 1$ for z in §12.3. Now, by Example 6.2.13, we have

$$\log z = \int_0^\infty \frac{e^{-t} - e^{-tz}}{t} dt,$$

and so, since $\dfrac{1}{z} = \displaystyle\int_0^\infty e^{-tz} dt$, we have

$$\frac{d}{dz} \log \Gamma(z+1) = \frac{1}{2z} + \log z - \int_0^\infty \left\{ \frac{1}{2} - \frac{1}{t} + \frac{1}{e^t - 1} \right\} e^{-tz} dt.$$

The integrand in the last integral is continuous as $t \to 0$; and since $\dfrac{1}{2} - \dfrac{1}{t} + \dfrac{1}{e^t - 1}$ is bounded as $t \to \infty$, it follows without difficulty that the integral converges uniformly when the real part of z is positive; we may consequently integrate from 1 to z under the sign of integration (§4.44) and we get[3]

$$\log \Gamma(z+1) = (z + \tfrac{1}{2}) \log z - z + 1 + \int_0^\infty \left\{ \frac{1}{2} - \frac{1}{t} + \frac{1}{e^t - 1} \right\} \frac{e^{-tz} - e^{-t}}{t} dt.$$

Since $\left\{ \dfrac{1}{2} - \dfrac{1}{t} + \dfrac{1}{e^t - 1} \right\}$ is continuous as $t \to 0$ by §7.2, and since

$$\log \Gamma(z+1) = \log z + \log \Gamma(z),$$

---

[3] $\log \Gamma(z + 1)$ means the sum of the principal values of the logarithms in the factors of the Weierstrassian product.

we have

$$\log \Gamma(z) = \left(z - \tfrac{1}{2}\right) \log z - z + 1 + \int_0^\infty \left\{\frac{1}{2} - \frac{1}{t} + \frac{1}{e^t - 1}\right\} \frac{e^{-tz}}{t} \, dt$$

$$- \int_0^\infty \left\{\frac{1}{2} - \frac{1}{t} + \frac{1}{e^t - 1}\right\} \frac{e^{-t}}{t} \, dt.$$

To evaluate the second of these integrals, let[4]

$$\int_0^\infty \left(\frac{1}{2} - \frac{1}{t} + \frac{1}{e^t - 1}\right) \frac{e^{-t}}{t} \, dt = I, \quad \int_0^\infty \left(\frac{1}{2} - \frac{1}{t} + \frac{1}{e^t - 1}\right) \frac{e^{-\frac{1}{2}t}}{t} dt = J;$$

so that, taking $z = \tfrac{1}{2}$ in the last expression for $\log \Gamma(z)$, we get

$$\frac{1}{2} \log \pi = \frac{1}{2} + J - I.$$

Also, since

$$I = \int_0^\infty \left(\frac{1}{2} - \frac{2}{t} + \frac{1}{e^{t/2} - 1}\right) \frac{e^{-t/2}}{t} \, dt,$$

we have

$$J - I = \int_0^\infty \left(\frac{1}{t} - \frac{e^{t/2}}{e^t - 1}\right) \frac{e^{-t/2} \, dt}{t}$$

$$= \int_0^\infty \left(\frac{e^{-t/2}}{t} - \frac{1}{e^t - 1}\right) \frac{dt}{t}.$$

And so

$$J = \int_0^\infty \left\{\frac{e^{-t/2}}{t} - \frac{1}{e^t - 1} + \frac{1}{2} e^{-t} - \frac{e^{-t}}{t} + \frac{e^{-t}}{e^t - 1}\right\} \frac{dt}{t}$$

$$= \int_0^\infty \left\{\frac{e^{-t/2} - e^{-t}}{t} - \frac{1}{2} e^{-t}\right\} \frac{dt}{t}$$

$$= \int_0^\infty \left\{-\frac{d}{dt} \left(\frac{e^{-t/2} - e^{-t}}{t}\right) - \frac{\frac{1}{2} e^{-t/2} - e^{-t}}{t} - \frac{e^{-t}}{2t}\right\} dt$$

$$= \left[-\frac{e^{-t/2} - e^{-t}}{t}\right]_0^\infty + \frac{1}{2} \int_0^\infty \frac{e^{-t} - e^{-t/2}}{t} \, dt$$

$$= \tfrac{1}{2} + \tfrac{1}{2} \log \tfrac{1}{2}.$$

Consequently $I = 1 - \tfrac{1}{2} \log(2\pi)$.

We therefore have Binet's result that, when $\mathrm{Re}\, z > 0$,

$$\log \Gamma(z) = \left(z - \tfrac{1}{2}\right) \log z - z + \tfrac{1}{2} \log(2\pi) + \int_0^\infty \left(\frac{1}{2} - \frac{1}{t} + \frac{1}{e^t - 1}\right) \frac{e^{-tz}}{t} \, dt.$$

If $z = x + iy$, we see that, if the upper bound of

$$\left| \left(\frac{1}{2} - \frac{1}{t} + \frac{1}{e^t - 1}\right) \frac{1}{t} \right|$$

---

[4] This artifice is due to Pringsheim [538].

for real values of t is K, then

$$\left| \log \Gamma(z) - \left(z - \frac{1}{2} \right) \log z + z - \frac{1}{2} \log(2\pi) \right| < K \int_0^\infty e^{-tz}\, dt$$
$$= Kx^{-1},$$

so that, when x is large, the terms $\left(z - \frac{1}{2} \right) \log z - z + \frac{1}{2} \log(2\pi)$ furnish an approximate expression for $\log \Gamma(z)$.

Example 12.3.3 (Malmstén) Prove that, when $\mathrm{Re}(z) > 0$,

$$\log \Gamma(z) = \int_0^\infty \left\{ \frac{e^{-zt} - e^{-t}}{1 - e^{-t}} + (z - 1)e^{-t} \right\} \frac{dt}{t}.$$

Example 12.3.4 (Féaux) Prove that, when $\mathrm{Re}(z) > 0$,

$$\log \Gamma(z) = \int_0^\infty \left\{ (z - 1)e^{-t} + \frac{(1 + t)^{-z} - (1 + t)^{-1}}{\log(1 + t)} \right\} \frac{dt}{t}.$$

Example 12.3.5 (Kummer) From the formula of §12.14 shew that, if $0 < x < 1$,

$$2 \log \Gamma(x) - \log \pi + \log \sin \pi x = \int_0^\infty \left\{ \frac{\sinh(\frac{1}{2} - x)t}{\sinh \frac{1}{2}t} - (1 - 2x)e^{-t} \right\} \frac{dt}{t}.$$

Example 12.3.6 (Kummer [390]) By expanding $\sinh(\frac{1}{2} - x)t$ and $1 - 2x$ in Fourier sine series, shew from Example 12.3.5 that, if $0 < x < 1$,

$$\log \Gamma(x) = \frac{1}{2} \log \pi - \frac{1}{2} \log \sin \pi x + 2 \sum_{n=1}^\infty a_n \sin 2n\pi x,$$

where

$$a_n = \int_0^\infty \left\{ \frac{2n\pi}{t^2 + 4n^2\pi^2} - \frac{e^{-t}}{2n\pi} \right\} \frac{dt}{t}.$$

Deduce from Example 12.3.2, that

$$a_n = \frac{1}{2n\pi} (\gamma + \log 2\pi + \log n).$$

12.32 Binet's second expression for $\log \Gamma(z)$ in terms of an infinite integral

Consider the application of Example 7.7 to the equation (12.4),

$$\frac{d^2}{dz^2} \log \Gamma(z) = \sum_{n=0}^\infty \frac{1}{(z + n)^2}.$$

The conditions there stated as sufficient for the transformation of a series into integrals are obviously satisfied by the function $\phi(\zeta) = 1/(z + \zeta)^2$, if $\mathrm{Re}\, z > 0$; and we have

$$\frac{d^2}{dz^2} \log \Gamma(z) = \frac{1}{2z^2} + \int_0^\infty \frac{d\xi}{(z + \xi)^2} - 2 \int_0^\infty \frac{q(t, z)}{e^{2\pi} - 1}\, dt + 2 \lim_{n \to \infty} \int_0^\infty \frac{q(t, z + n)}{e^{2\pi t} - 1}\, dt,$$

where $2iq(t) = \dfrac{1}{(z + it)^2} - \dfrac{1}{(z - it)^2}$. Since $|q(t, z + n)|$ is easily seen to be less than $K_1 t/n$, where K_1 is independent of t and n, it follows that the limit of the last integral is zero. Hence

$$\frac{d^2}{dz^2} \log \Gamma(z) = \frac{1}{2z^2} + \frac{1}{z} + \int_0^\infty \frac{4tz}{(z^2 + t^2)^2} \frac{dt}{e^{2\pi t} - 1}. \tag{12.6}$$

Since $\left| \dfrac{2z}{z^2 + t^2} \right|$ does not exceed K (where K depends only on δ) when the real part of z exceeds δ, the integral converges uniformly and we may integrate under the integral sign §4.44 from 1 to z. We get

$$\frac{d}{dz} \log \Gamma(z) = \frac{1}{2z} + \log z + C - 2 \int_0^\infty \frac{t \, dt}{(z^2 + t^2)(e^{2\pi t} - 1)},$$

where C is a constant. Integrating again,

$$\log \Gamma(z) = \left(z - \frac{1}{2} \right) \log z + (C - 1)z + C' + 2 \int_0^\infty \frac{\arctan(t/z)}{e^{2\pi t} - 1} \, dt,$$

where C' is a constant.

Now, if z is real, $0 \le \arctan t/z \le t/z$, and so

$$\left| \log \Gamma(z) - \left(z - \frac{1}{2} \right) \log z - (C - 1) z - C' \right| < \frac{2}{z} \int_0^\infty \frac{t}{e^{2\pi t} - 1} \, dt.$$

But it has been shewn in §12.31 that

$$\left| \log \Gamma(z) - \left(z - \frac{1}{2} \right) \log z + z - \frac{1}{2} \log(2\pi) \right| \to 0,$$

as $z \to \infty$ through real values. Comparing these results we see that $C = 0$,

$$C' = \frac{1}{2} \log(2\pi).$$

Hence for all values of z whose real part is positive,

$$\log \Gamma(z) = \left(z - \frac{1}{2} \right) \log z - z + \frac{1}{2} \log(2\pi) + 2 \int_0^\infty \frac{\arctan(t/z)}{e^{2\pi t} - 1} \, dt,$$

where $\arctan u$ is defined by the equation

$$\arctan u = \int_0^u \frac{dt}{1 + t^2},$$

in which the path of integration is a straight line. This is Binet's second expression for $\log \Gamma(z)$.

Example 12.3.7 Justify differentiating with regard to z under the sign of integration, so as to get the equation

$$\frac{\Gamma'(z)}{\Gamma(z)} = \log z - \frac{1}{2z} - 2 \int_0^\infty \frac{t \, dt}{(t^2 + z^2)(e^{2\pi t} - 1)}.$$

12.33 The asymptotic expansion of the logarithms of the Gamma-function

We can now obtain an expansion which represents the function $\log \Gamma(z)$ asymptotically (§8.2) for large values of $|z|$, and which is used in the calculation of the Gamma-function.

Let us assume that, if $z = x + iy$, then $x \geq \delta > 0$; and we have, by Binet's second formula,

$$\log \Gamma(z) = \left(z - \tfrac{1}{2}\right) \log z - z + \tfrac{1}{2} \log(2\pi) + \phi(z),$$

where

$$\phi(z) = 2 \int_0^\infty \frac{\arctan(t/z)}{e^{2\pi t} - 1} dt. \tag{12.7}$$

Now

$$\arctan(t/z) = \frac{t}{z} - \frac{1}{3}\frac{t^3}{z^3} + \frac{1}{5}\frac{t^5}{z^5} - \cdots + \frac{(-1)^{n-1}}{2n-1}\frac{t^{2n-1}}{z^{2n-1}} + \frac{(-1)^n}{z^{n-1}} \int_0^t \frac{u^{2n} du}{u^2 + z^2}.$$

Substituting and remembering (§7.2) that

$$\int_0^\infty \frac{t^{2n-1} dt}{e^{2\pi t} - 1} = \frac{B_n}{4n},$$

where $B_1, B_2, \ldots$ are Bernoulli's numbers, we have

$$\phi(z) = \sum_{r=1}^n \frac{(-1)^{r-1} B_r}{2r(2r-1)z^{2r-1}} + \frac{2(-1)^n}{z^{2n-1}} \int_0^\infty \left\{ \int_0^t \frac{u^{2n} du}{u^2 + z^2} \right\} \frac{dt}{e^{2\pi t} - 1}.$$

Let the upper bound[5] of $\left| \dfrac{z^2}{u^2 + z^2} \right|$ for positive values of u be K_z. Then

$$\left| \int_0^\infty \left\{ \int_0^t \frac{u^{2n} du}{u^2 + z^2} \right\} \frac{dt}{e^{2\pi t} - 1} \right| \leq K_z |z|^{-2} \int_0^\infty \left\{ \int_0^t u^{2n} du \right\} \frac{dt}{e^{2\pi t} - 1}$$

$$\leq \frac{K_z B_{n+1}}{4(n+1)(2n+1)|z|^2}.$$

Hence

$$\left| \frac{2(-1)^n}{z^{2n-1}} \int_0^\infty \left\{ \int_0^t \frac{u^{2n} du}{u^2 + z^2} \right\} \frac{dt}{e^{2\pi t-1}} \right| < \frac{K_z B_{n+1}}{2(n+1)(2n+1)|z|^{2n+1}},$$

and it is obvious that this tends to zero uniformly as $|z| \to \infty$ if $|\arg z| \leq \tfrac{1}{2}\pi - \Delta$, where $\tfrac{1}{4}\pi > \Delta > 0$, so that $K_z \leq \operatorname{cosec} 2\Delta$.

Also it is clear that if $|\arg z| \leq \tfrac{1}{4}\pi$ (so that $K_z = 1$) the error in taking the first n terms of the series

$$\sum_{r=1}^\infty \frac{(-1)^{r-1} B_r}{2r(2r-1)} \frac{1}{z^{2r-1}}$$

as an approximation to $\phi(z)$ is numerically less than the $(n+1)$th term.

[5] K_s^{-2} is the lower bound of $\dfrac{\{u^2 + (x^2 - y^2)\}^2 + 4x^2 y^2}{(x^2 + y^2)^2}$ and is consequently equal to $\dfrac{4x^2 y^2}{(x^2 + y^2)^2}$ or 1 as $x^2 < y^2$ or $x^2 \geq y^2$.

Since, if $|\arg z| \le \frac{1}{2}\pi - \Delta$,

$$\left| z^{2n-1}\left\{ \phi(z) - \sum_{r=1}^{n} \frac{(-1)^{r-1}B_r}{2r\,(2r-1)} \right\} \right| < \frac{B_{n+1}}{2(n+1)(2n+1)\sin^2(2\Delta)}\,|z|^{-2}$$

$$\to 0,$$

as $z \to \infty$, it is clear that

$$\frac{B_1}{1\cdot 2\cdot z} - \frac{B_2}{3\cdot 4\cdot z^3} + \frac{B_3}{5\cdot 6\cdot z^5} - \cdots$$

is the asymptotic expansion of $\phi(z)$ (§8.2). (The development is asymptotic; for if it converged when $|z| \ge \rho$, by §2.6 we could find K, such that $B_n < (2n-1)2nK\rho^{2n}$; and then the series $\sum_{n=1}^{\infty} \frac{(-1)^{n-1}B_n t^{2n}}{(2n)!}$ would define an integral function; this is contrary to §7.2.)

We see therefore that the series

$$\left(z - \tfrac{1}{2}\right)\log z - z + \tfrac{1}{2}\log(2\pi) + \sum_{r=1}^{\infty} \frac{(-1)^{r-1}B_r}{2r\,(2r-1)z^{2r-1}}$$

is the asymptotic expansion of $\log\Gamma(z)$ when $|\arg z| \le \pi/2 - \Delta$.

This is generally known as *Stirling's series*. In §13.6 it will be established over the extended range $|\arg z| \le \pi - \Delta$. In particular when z is positive ($= x$), we have

$$0 < 2\int_0^{\infty} \left\{ \int_0^t \frac{u^{2n}\,du}{u^2+x^2} \right\} \frac{dt}{e^{2\pi t}-1} < \frac{B_{n+1}}{2(n+1)(2n+1)x^2}.$$

Hence, when $x > 0$, the value of $\phi(x)$ always lies between the sum of n terms and the sum of $n + 1$ terms of the series for all values of n.

In particular $0 < \phi(x) < \frac{B_1}{2x}$, so that $\phi(x) = \frac{\theta}{12x}$ where $0 < \theta < 1$. Hence

$$\Gamma(x) = x^{x-1/2}e^{-x}(2\pi)^{\frac{1}{2}}e^{\theta/(12x)}.$$

Also, taking the exponential of Stirling's series, we get

$$\Gamma(x) = e^{-x}\,x^{x-1/2}(2\pi)^{1/2}\left\{ 1 + \frac{1}{12x} + \frac{1}{288x^2} - \frac{139}{51840x^3} - \frac{571}{2488320x^4} + O\left(\frac{1}{x^5}\right) \right\}.$$

This is an *asymptotic formula for the Gamma-function*. In conjunction with the formula $\Gamma(x+1) = x\Gamma(x)$, it is very useful for the purpose of computing the numerical value of the function for real values of x.

Tables of the function $\log_{10}\Gamma(x)$, correct to 12 decimal places, for values of x between 1 and 2, were constructed in this way by Legendre, and published in [421, vol. 2, p. 85], and his *Traité des fonctions elliptiques* [422, p. 489].

It may be observed that $\Gamma(x)$ has one minimum for positive values of x, when $x = 1.4616321\cdots$, the value of $\log_{10}\Gamma(x)$ then being $\bar{1}.9472391\cdots$.

Example 12.3.8 (Binet)   Obtain the expansion, convergent when $\mathrm{Re}(z) > 0$,

$$\log_e \Gamma(z) = \left(z - \tfrac{1}{2}\right)\log_e z - z + \tfrac{1}{2}\log_e(2\pi) + J(z),$$

where

$$J(z) = \frac{1}{2} \left\{ \frac{c_1}{z+1} + \frac{c_2}{2(z+1)(z+2)} + \frac{c_3}{3(z+1)(z+2)(z+3)} + \cdots \right\},$$

in which

$$c_1 = \frac{1}{6}, \quad c_2 = \frac{1}{3}, \quad c_3 = \frac{59}{90}, \quad c_4 = \frac{227}{60},$$

and generally

$$c_n = \int_0^1 (x+1)(x+2) \cdots (x+n-1)(2x-1)x \, dx.$$

12.4 The Eulerian integral of the first kind

The name *Eulerian Integral of the First Kind* was given by Legendre to the integral

$$B(p,q) = \int_0^1 x^{p-1} (1-x)^{q-1} \, dx,$$

which was first studied by Euler [200] and Legendre [421, vol. 1, p. 221]. In this integral, the real parts of p and q are supposed to be positive; and x^{p-1}, $(1-x)^{q-1}$ are to be understood to mean those values of $e^{(p-1)\log x}$ and $e^{(q-1)\log(1-x)}$ which correspond to the real determinations of the logarithms.

With these stipulations, it is easily seen that $B(p,q)$ exists, as a (possibly improper) integral (see formula (2) in Example 4.5.1).

We have, on writing $(1-x)$ for x,

$$B(p,q) = B(q,p).$$

Also, integrating by parts,

$$\int_0^1 x^{p-1}(1-x)^q \, dx = \left[\frac{x^p(1-x)^q}{p} \right]_0^1 + \frac{q}{p} \int_0^1 x^p (1-x)^{q-1} \, dx,$$

so that $B(p, q+1) = \dfrac{q}{p} B(p+1, q)$.

Example 12.4.1   Shew that

$$B(p,q) = B(p+1,q) + B(p,q+1).$$

Example 12.4.2   Deduce from Example 12.4.1 that

$$B(p, q+1) = \frac{q}{p+q} B(p,q).$$

Example 12.4.3   Prove that if n is a positive integer,

$$B(p, n+1) = \frac{1 \cdot 2 \cdots n}{p(p+1) \cdots (p+n)}.$$

Example 12.4.4   Prove that

$$B(x,y) = \int_0^\infty \frac{a^{x-1}}{(1+a)^{x+y}} \, da.$$

Example 12.4.5   Prove that

$$\Gamma(z) = \lim_{n\to\infty} n^z B(z,n).$$

12.41 Expression of the Eulerian integral of the first kind in terms of the Gamma-function

We shall now establish the important theorem that

$$B(m,n) = \frac{\Gamma(m)\Gamma(n)}{\Gamma(m+n)}.$$

First let the real parts of m and n exceed $\frac{1}{2}$; then

$$\Gamma(m)\,\Gamma(n) = \int_0^\infty e^{-x}x^{m-1}\,dx \times \int_0^\infty e^{-y}y^{n-1}\,dy.$$

On writing x^2 for x and y^2 for y, this gives

$$\Gamma(m)\,\Gamma(n) = 4\lim_{R\to\infty}\int_0^R e^{-x^2}x^{2m-1}\,dx \times \int_0^R e^{-y^2}y^{2n-1}\,dy$$

$$= 4\lim_{R\to\infty}\int_0^R\int_0^R e^{-(x^2+y^2)}x^{2m-1}y^{2n-1}\,dx\,dy.$$

Now, for the values of m and n under consideration, the integrand is continuous over the range of integration, and so the integral may be considered as a double integral taken over a square S_R. Calling the integrand $f(x,y)$, and calling Q_R the quadrant with centre at the origin and radius R, we have, if T_R be the part of S_R outside Q_R,

$$\left|\iint_{S_R} f(x,y)\,dx\,dy - \iint_{Q_R} f(x,y)\,dx\,dy\right| = \left|\iint_{T_R} f(x,y)\,dx\,dy\right|$$

$$\le \iint_{T_R} |f(x,y)|\,dx\,dy$$

$$\le \iint_{S_R} |f(x,y)|\,dx\,dy - \iint_{S_{R/2}} |f(x,y)|\,dx\,dy$$

$$\to 0 \text{ as } R\to\infty,$$

since $\iint_{S_R} |f(x,y)|\,dx\,dy$ converges to a limit, namely

$$2\int_0^\infty e^{-x^2}x^{2m-1}\,dx \times 2\int_0^\infty e^{-y}y^{2n-1}\,dy.$$

Therefore

$$\lim_{R\to\infty}\iint_{S_R} f(x,y)\,dx\,dy = \lim_{R\to\infty}\iint_{Q_R} f(x,y)\,dx\,dy.$$

Changing to polar[6] coordinates $(x = r \cos \theta, y = r \sin \theta)$, we have

$$\int \int_{Q_R} f(x, y)\, dx\, dy = \int_0^R \int_0^{\pi/2} e^{-r^2} (r \cos \theta)^{2m-1} (r \sin \theta)^{2n-1} r\, dr\, d\theta.$$

Hence

$$\Gamma(m)\Gamma(n) = 4 \int_0^\infty e^{-r^2} r^{2(m+n)-1}\, dr \int_0^{\pi/2} \cos^{2m-1} \theta \sin^{2n-1} \theta\, d\theta$$

$$= 2\Gamma(m + n) \int_0^{\pi/2} \cos^{2m-1} \theta \sin^{2n-1} \theta\, d\theta.$$

Writing $\cos^2 \theta = u$ we at once get

$$\Gamma(m)\Gamma(n) = \Gamma(m + n) \cdot B(m, n).$$

This has only been provided when the real parts of m and n exceed $1/2$; but it can obviously be deduced when these are less than $1/2$ by Example 12.4.2. This result, discovered by Euler, connects the Eulerian Integral of the First Kind with the Gamma-function.

Example 12.4.6   Shew that

$$\int_{-1}^1 (1 + x)^{p-1}(1 - x)^{q-1}\, dx = 2^{p+q-1} \frac{\Gamma(p)\Gamma(q)}{\Gamma(p + q)}.$$

Example 12.4.7 (Jesus, 1901)   Shew that, if

$$f(x, y) = \frac{1}{x} - y\frac{1}{x + 1} + \frac{y(y - 1)}{2!}\frac{1}{x + 2} - \frac{y(y - 1)(y - 2)}{3!}\frac{1}{x + 3} + \cdots,$$

then

$$f(x, y) = f(y + 1, x - 1),$$

where x and y have such values that the series are convergent.

Example 12.4.8 (Math. Trip. 1894)   Prove that

$$\int_0^1 \int_0^1 f(xy)(1 - x)^{\mu-1} y^\mu (1 - y)^{\nu-1}\, dx\, dy = \frac{\Gamma(\mu)\Gamma(\nu)}{\Gamma(\mu + \nu)} \int_0^1 f(z)(1 - z)^{\mu+\nu-1}\, dz.$$

12.42 Evaluation of trigonometrical integrals in terms of the Gamma-function

We can now evaluate the integral $\int_0^{\pi/2} \cos^{m-1} x \sin^{n-1} x\, dx$, where m and n are not restricted to be integers, but have their real parts positive.

For, writing $\cos^2 x = t$, we have, as in §12.41,

$$\int_0^{\pi/2} \cos^{m-1} x \sin^{n-1} x\, dx = \frac{\Gamma(\frac{m}{2})\Gamma(\frac{n}{2})}{2\Gamma(\frac{m+n}{2})}.$$

---

[6] It is easily provided by the methods of §4.11 that the areas $A_{m,\mu}$ of §4.3 need not be rectangles provided only that their greatest diameters can be made arbitrarily small by taking the number of areas sufficiency large; so the areas may be taken to be the regions bounded by radii vectors and circular arcs.

The well-known elementary formulae for the cases in which m and n are integers can be at once derived from this result.

Example 12.4.9 (Trinity, 1898)   Prove that, when $|k| < 1$,

$$\int_0^{\pi/2} \frac{\cos^m \theta \sin^n \theta \, d\theta}{(1 - k \sin^2 \theta)^{1/2}} = \frac{\Gamma(\frac{m+1}{2})\Gamma(\frac{n+1}{2})}{\Gamma(\frac{m+n+1}{2})\sqrt{\pi}} \int_0^{\pi/2} \frac{\cos^{m+n} \theta \, d\theta}{(1 - k \sin^2 \theta)^{(n+1)/2}}.$$

12.43 Pochhammer's extension of the Eulerian integral of the first kind

This appears in [527]. The use of the double circuit integrals of this section seems to be due to Jordan [363].

We have seen in §12.22 that it is possible to replace the second Eulerian integral for $\Gamma(z)$ by a contour integral which converges for all values of z. A similar process has been carried out by Pochhammer for Eulerian integrals of the first kind.

Let P be any point on the real axis between 0 and 1; consider the integral

$$e^{-\pi i(\alpha+\beta)} \int_P^{(1+,0+,1-,0-)} t^{\alpha-1}(1 - t)^{\beta-1} \, dt = \varepsilon(\alpha, \beta).$$

The notation employed is that introduced at the end of §12.22 and means that the path of integration starts from P, encircles the point 1 in the positive (counter-clockwise) direction and returns to P, then encircles the origin in the positive direction and returns to P, and so on.

At the starting-point the arguments of t and $1 - t$ are both zero; after the circuit $(1+)$ they are 0 and 2π; after the circuit $(0+)$ they are 2π and 2π; after the circuit $(1-)$ they are 2π and 0 and after the circuit $(0-)$ they are both zero, so that the final value of the integrand is the same as the initial value.

It is easily seen that, since the path of integration may be deformed in any way so long as it does not pass over the branch points 0, 1 of the integrand, the path may be taken to be that shewn in the figure, wherein the four parallel lines are supposed to coincide with the real axis.

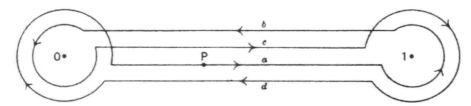

If the real parts of α and β are positive the integrals round the circles tend to zero as the radii of the circles tend to zero (the reader ought to have no difficulty in proving this); the integrands on the paths marked a, b, c, d are

$$t^{a-1}(1 - t)^{\beta-1}, \quad t^{a-1}(1 - t)^{\beta-1}e^{2\pi i(\beta-1)},$$
$$t^{a-1}e^{2\pi i(\alpha-1)}(1 - t)^{\beta-1}e^{2\pi i(\beta-1)}, \quad t^{a-1}e^{2\pi i(\alpha-1)}(1 - t)^{\beta-1}$$

respectively, the arguments of t and $1 - t$ *now being zero in each case*.

Hence we may write $\varepsilon(\alpha,\beta)$ as the sum of four (possibly improper) integrals, thus:

$$\varepsilon(\alpha,\beta) = e^{\pi i(\alpha+\beta)}\left[\int_0^1 t^{\alpha-1}(1-t)^{\beta-1}\,dt + \int_1^0 t^{\alpha-1}(1-t)^{\beta-1}e^{2\pi i\beta}\,dt\right.$$

$$\left. + \int_0^1 t^{\alpha-1}(1-t)^{\beta-1}e^{2\pi i(\alpha+\beta)}\,dt + \int_1^0 t^{\alpha-1}(1-t)^{\beta-1}e^{2\pi i\alpha}\,dt\right].$$

Hence

$$\varepsilon(\alpha,\beta) = e^{\pi i(\alpha+\beta)}(1-e^{2\pi i\alpha})(1-e^{2\pi i\beta})\int_0^1 t^{\alpha-1}(1-t)^{\beta-1}\,dt$$

$$= -4\sin(\alpha\pi)\sin(\beta\pi)\frac{\Gamma(\alpha)\Gamma(\beta)}{\Gamma(\alpha+\beta)}$$

$$= \frac{-4\pi^2}{\Gamma(1-\alpha)\Gamma(1-\beta)\Gamma(\alpha+\beta)}.$$

Now $\varepsilon(\alpha,\beta)$ and this last expression are analytic functions of α and of β for *all* values of α and β. So, by the theory of analytic continuation, this equality, proved when the real parts of α and β are positive, holds for all values of α and β. *Hence for all values of α and β we have proved that*

$$\varepsilon(\alpha,\beta) = \frac{-4\pi^2}{\Gamma(1-\alpha)\Gamma(1-\beta)\Gamma(\alpha+\beta)}.$$

12.5 Dirichlet's integral

This material appears in [178, pp. 375, 391]. We shall now shew how the repeated integral

$$I = \int\int\cdots\int f(t_1+t_2+\cdots+t_n)t_1^{\alpha_1-1}t_2^{\alpha_2-1}\cdots t_n^{\alpha_n-1}\,dt_1\,dt_2\cdots dt_n$$

may be reduced to a simple integral, where f is continuous, $\alpha_r > 0$ $(r = 1,2,\ldots,n)$ and the integration is extended over all positive values of the variables such that $t_1+t_2+\cdots+t_n \le 1$.
To simplify

$$\int_0^{1-\lambda}\int_0^{1-\lambda-T} f(t+T+\lambda)t^{\alpha-1}T^{\beta-1}\,dt\,dT$$

(where we have written t,T,α,β for $t_1,t_2,\alpha_1,\alpha_2$ and λ for $t_3+t_4+\cdots+t_n$), put $t = T(1-v)/v$; the integral becomes (if $\lambda \ne 0$)

$$\int_0^{1-\lambda}\int_{T/(1-\lambda)}^1 f(\lambda+T/v)(1-v)^{\alpha-1}v^{-\alpha-1}T^{\alpha+\beta-1}\,dv\,dT.$$

Changing the order of integration (§4.51), the integral becomes

$$\int_0^1\int_0^{(1-\lambda)v} f(\lambda+T/v)(1-v)^{\alpha-1}v^{-\alpha-1}T^{\alpha+\beta-1}\,dT\,dv.$$

Putting $T = v\tau_2$, the integral becomes

$$\int_0^1 \int_0^{1-\lambda} f(\lambda + \tau_2)(1-v)^{\alpha-1} v^{\beta-1} \tau_2^{\alpha+\beta-1} d\tau_2\, dv = \frac{\Gamma(\alpha)\Gamma(\beta)}{\Gamma(\alpha+\beta)} \int_0^{1-\lambda} f(\lambda + \tau_2)\tau_2^{\alpha+\beta-1}\, d\tau_2.$$

Hence

$$I = \frac{\Gamma(\alpha_1)\Gamma(\alpha_2)}{\Gamma(\alpha_1+\alpha_2)} \int \int \cdots \int f(\tau_2 + t_3 + \cdots + t_n)\tau_2^{\alpha_1+\alpha_2-1} t_3^{\alpha_3-1} \cdots t_n^{\alpha_n-1}\, d\tau_2\, dt_3 \cdots dt_n,$$

the integration being extended over all positive values of the variables such that $\tau_2 + t_3 + \cdots + t_n \le 1$.

Continually reducing in this way we get

$$I = \frac{\Gamma(\alpha_1)\Gamma(\alpha_2)\cdots\Gamma(\alpha_n)}{\Gamma(\alpha_1+\alpha_2+\cdots+\alpha_n)} \int_0^1 f(\tau)\tau^{\Sigma\alpha-1}\, d\tau,$$

which is Dirichlet's result.

Example 12.5.1 (Dirichlet)   Reduce

$$\int \int \int f\left[\left(\frac{x}{a}\right)^\alpha + \left(\frac{y}{b}\right)^\beta + \left(\frac{z}{c}\right)^\gamma\right] x^{p-1} y^{q-1} z^{r-1}\, dx\, dy\, dz$$

to a simple integral; the range of integration being extended over all positive values of the variables such that

$$\left(\frac{x}{a}\right)^\alpha + \left(\frac{y}{b}\right)^\beta + \left(\frac{z}{c}\right)^\gamma \le 1,$$

it being assumed that $a, b, c, \alpha, \beta, \gamma, p, q, r$ are positive.

Example 12.5.2 (Pembroke, 1907)   Evaluate $\int\int x^p y^q\, dx\, dy$, where m and n are positive and $x \ge 0$, $y \ge 0$, $x^m + y^n \le 1$.

Example 12.5.3   Shew that the moment of inertia of a homogeneous ellipsoid of unit density, taken about the axis of z, is

$$\frac{4\pi}{15}(a^2 + b^2)abc,$$

where a, b, c are the semi-axes.

Example 12.5.4   Shew that the area of the hypocycloid $x^{2/3} + y^{2/3} = \ell^{2/3}$ is $3\pi\ell^2/8$.

12.6 Miscellaneous examples

Example 12.1 (Trinity, 1897)   Shew that

$$(1-z)\left(1+\frac{z}{2}\right)\left(1-\frac{z}{3}\right)\left(1+\frac{z}{4}\right)\cdots = \frac{\sqrt{\pi}}{\Gamma(1+\frac{1}{2}z)\Gamma(\frac{1}{2}-\frac{1}{2}z)}.$$

Example 12.2 (Trinity, 1885)   Shew that

$$\lim_{n\to\infty} \frac{1}{1+x}\frac{1}{1+\frac{1}{2}x}\frac{1}{1+\frac{1}{3}x}\cdots\frac{1}{1+\frac{1}{n}x} n^x = \Gamma(x+1).$$

Example 12.3 (Jesus, 1903)   Prove that

$$\frac{\Gamma'(1)}{\Gamma(1)} - \frac{\Gamma'(\frac{1}{2})}{\Gamma(\frac{1}{2})} = 2\log 2.$$

Example 12.4 (Trinity, 1891)   Shew that

$$\frac{[\Gamma(\frac{1}{4})]^4}{16\pi^2} = \frac{3^2}{3^2-1} \cdot \frac{5^2-1}{5^2} \cdot \frac{7^2}{7^2-1} \cdot \frac{9^2-1}{9^2} \cdot \frac{11^2}{11^2-1} \cdots$$

Example 12.5 (Trinity, 1905)   Shew that

$$\prod_{n=0}^{\infty} \left[\frac{(n-\alpha)(n+\beta+\gamma)}{(n+\beta)(n+\gamma)} \left(1 + \frac{\alpha}{n+1}\right) \right] = -\frac{1}{\pi}\sin(\pi\alpha)B(\beta,\gamma).$$

Example 12.6 (Peterhouse, 1906)   Shew that

$$\prod_{k=1}^{\infty} \Gamma\left(\frac{k}{3}\right) = \frac{640}{3^6}\left(\frac{\pi}{\sqrt{3}}\right)^3.$$

Example 12.7 (Trinity, 1904)   Shew that, if $z = i\zeta$ where ζ is real, then

$$|\Gamma(z)| = \sqrt{\frac{\pi}{\zeta\sinh(\pi\zeta)}}.$$

Example 12.8 (Math. Trip. 1897)   When x is positive, shew that

$$\frac{\Gamma(x)\Gamma(\frac{1}{2})}{\Gamma(x+\frac{1}{2})} = \sum_{n=0}^{\infty} \frac{(2n)!}{2^{2n}n!^2}\frac{1}{x+n}.$$

(This and some other examples are most easily proved by the result of §14.11.)

Example 12.9   If a is positive, shew that

$$\frac{\Gamma(z)\Gamma(a+1)}{\Gamma(z+a)} = \sum_{n=0}^{\infty} \frac{(-1)^n(a)(a-1)(a-2)\cdots(a-n)}{n!}\frac{1}{z+n}.$$

Example 12.10   If $x > 0$ and

$$P(x) = \int_0^1 e^{-t}t^{x-1}\,dt,$$

shew that

$$P(x) = \frac{1}{x} - \frac{1}{1!}\frac{1}{x+1} + \frac{1}{2!}\frac{1}{x+2} - \frac{1}{3!}\frac{1}{x+3} + \cdots,$$

and

$$P(x+1) = xP(x) - 1/e.$$

Example 12.11 (Euler)   Shew that if $\lambda > 0$, $x > 0$, $-\pi/2 < a < \pi/2$, then

$$\int_0^\infty t^{x-1}e^{-\lambda t\cos a}\cos(\lambda t\sin a)\,dt = \lambda^{-x}\Gamma(x)\cos ax,$$

$$\int_0^\infty t^{x-1}e^{-\lambda t\cos a}\sin(\lambda t\sin a)\,dt = \lambda^{-x}\Gamma(x)\sin ax.$$

Example 12.12 (Euler)   Prove that, if $b > 0$, then, when $0 < z < 2$,

$$\int_0^\infty \frac{\sin bx}{x^z} \, dx = \frac{\pi}{2} \frac{b^{z-1}}{\Gamma(z)} \operatorname{cosec} \left(\frac{\pi z}{2} \right),$$

and, when $0 < z < 1$,

$$\int_0^\infty \frac{\cos bx}{x^z} \, dx = \frac{\pi}{2} \frac{b^{z-1}}{\Gamma(z)} \sec \left(\frac{\pi z}{2} \right).$$

Example 12.13 (Peterhouse, 1895)   If $0 < n < 1$, prove that

$$\int_0^\infty (1 + x)^{n-1} \cos x \, dx = \Gamma(n) \left\{ \cos \left(\frac{n\pi}{2} - 1 \right) - \frac{1}{\Gamma(n+1)} + \frac{1}{\Gamma(n+3)} - \cdots \right\}.$$

Example 12.14 (Bourguet [97])   By taking as contour of integration a parabola with its vertex at the origin, derive from the formula

$$\Gamma(a) = -\frac{1}{2i \sin a\pi} \int_\infty^{(0+)} (-z)^{a-1} e^{-z} \, dz$$

the result

$$\Gamma(a) = \frac{1}{2 \sin a\pi} \int_0^\infty e^{-x^2} x^{a-1} (1 + x^2)^{a/2} [3 \sin \{x + a \operatorname{arccot}(-x)\}$$

$$+ \sin \{x + (a - 2) \operatorname{arccot}(-x)\}] \, dx,$$

the arccot denoting an obtuse angle.

Example 12.15 (Math. Trip. 1907)   Shew that, if $\operatorname{Re} a_n > 0$ and $\sum\limits_{n=1}^\infty 1/a_n^2$ is convergent, then

$$\prod_{n=1}^\infty \left[\frac{\Gamma(a_n)}{\Gamma(z + a_n)} \exp \left\{ \sum_{n=1}^m \frac{2^s}{s!} \psi^{(s)}(a_n) \right\} \right]$$

is convergent when $m > 2$, where $\psi^{(s)}(z) = \dfrac{d^s}{dz^s} \log \Gamma(z)$.

Example 12.16 (Legendre)   Prove that

$$\frac{d \log \Gamma(z)}{dz} = \int_0^\infty \frac{e^{-a} - e^{-za}}{1 - e^{-a}} \, da - \gamma$$

$$= \int_0^\infty \left[(1 + a)^{-1} - (1 + a)^{-s} \right] \frac{da}{a} - \gamma$$

$$= \int_0^1 \frac{x^{s-1} - 1}{x - 1} \, dx - \gamma.$$

Example 12.17 (Binet)   Prove that, when $\operatorname{Re}(z) > 0$,

$$\log \Gamma(z) = \int_0^1 \left\{ \frac{x^s - x}{x - 1} - x(z - 1) \right\} \frac{dx}{x \log x}.$$

Example 12.18   Prove that, for all values of z except negative real values,

$$\log \Gamma(z) = \left(z - \frac{1}{2}\right) \log z - z + \frac{1}{2} \log(2\pi) + \frac{1}{2}\left\{\frac{1}{2 \cdot 3} \sum_{r=1}^{\infty} \frac{1}{(z+r)^2}\right.$$

$$\left. + \frac{2}{3 \cdot 4} \sum_{r=1}^{\infty} \frac{1}{(z+r)^3} + \frac{3}{4 \cdot 5} \sum_{r=1}^{\infty} \frac{1}{(z+r)^4} + \cdots \right\}.$$

Example 12.19   Prove that, when $\text{Re}(z) > 0$,

$$\frac{d}{dz} \log \Gamma(z) = \log z - \int_0^1 \frac{x^{s-1}}{(1-x)\log x}[1 - x + \log x]\, dx.$$

Example 12.20   Prove that, when $\text{Re}(z) > 0$,

$$\frac{d^2}{dz^2} \log \Gamma(z) = \int_0^{\infty} \frac{xe^{-xs}\, dx}{1 - e^{-x}}.$$

Example 12.21 (Raabe, [546])   If $\displaystyle\int_s^{s+1} \log \Gamma(t)\, dt = u$, shew that

$$\frac{du}{dz} = \log z,$$

and deduce from §12.33 that, for all values of z except negative real values,

$$u = z \log z - z + \tfrac{1}{2}\log(2\pi).$$

Example 12.22 (Bourguet)   Prove that, for all values of z except negative real values,

$$\log \Gamma(z) = \left(z - \tfrac{1}{2}\right)\log z - z + \tfrac{1}{2}\log(2\pi) + \sum_{n=1}^{\infty} \int_0^{\infty} \frac{dx}{x+z}\frac{\sin 2n\pi x}{n\pi}.$$

This result is attributed to Bourguet by Stieltjes [606].

Example 12.23 (Binet)   Prove that

$$B(p,p)B\left(p + \tfrac{1}{2}, p + \tfrac{1}{2}\right) = \frac{\pi}{2^{4p-1}p}.$$

Example 12.24   Prove that, when $-t < r < t$,

$$B(t+r, t-r) = \frac{1}{4^{t-1}}\int_0^{\infty} \frac{\cosh(2ru)\, du}{\cosh^{2t} u}.$$

Example 12.25   Prove that, when $q > 1$,

$$B(p,q) + B(p+1,q) + B(p+2,q) + \cdots = B(p, q-1).$$

Example 12.26   Prove that, when $p - a > 0$,

$$\frac{B(p-a,q)}{B(p,q)} = 1 + \frac{aq}{p+q} + \frac{a(a+1)q(q+1)}{1 \cdot 2 \cdot (p+q)(p+q+1)} + \cdots.$$

Example 12.27 (Euler)   Prove that

$$B(p,q)B(p+q,r) = B(q,r)B(q+r,p).$$

Example 12.28 (Trinity, 1908)   Shew that

$$\int_0^1 x^{a-1}(1-x)^{b-1}\frac{dx}{(x+p)^{a+b}} = \frac{\Gamma(a)\Gamma(b)}{\Gamma(a+b)}\frac{1}{(1+p)^a p^b},$$

if $a > 0$, $b > 0$, $p > 0$.

Example 12.29 (St John's, 1904)   Shew that, if $m > 0$, $n > 0$, then

$$\int_{-1}^1 \frac{(1+x)^{2m-1}(1-x)^{2n-1}}{(1+x^2)^{m+n}}\,dx = 2^{m+n-2}\frac{\Gamma(m)\Gamma(n)}{\Gamma(m+n)};$$

and deduce that, when a is real and not an integer multiple of $\pi/2$,

$$\int_{-\pi/2}^{\pi/2}\left(\frac{\cos\theta + \sin\theta}{\cos\theta - \sin\theta}\right)^{\cos 2a}\,d\theta = \frac{\pi}{2\sin(\pi\cos^2 a)}.$$

Example 12.30 (Kummer)   Shew that, if $\alpha > 0$, $\beta > 0$,

$$\int_0^1 \frac{t^{\alpha-1}}{1+t}\,dt = \frac{1}{2}\psi\left(\frac{\alpha+1}{2}\right) - \frac{1}{2}\psi\left(\frac{\alpha}{2}\right),$$

and

$$\int_0^1 \frac{t^{\alpha-1} - t^{\beta-1}}{(1+t)\log t}\,dt = \log\frac{\Gamma(\frac{\alpha+1}{2})\Gamma(\frac{\beta}{2})}{\Gamma(\frac{\alpha}{2})\Gamma(\frac{\beta+1}{2})}.$$

Example 12.31   Shew that, if $a > 0$, $a + b > 0$,

$$\int_0^1 \frac{x^{a-1}(1-x^b)}{1-x}\,dx = \lim_{\delta\to 0}\left\{\frac{\Gamma(a)\Gamma(\delta)}{\Gamma(a+\delta)} - \frac{\Gamma(a+b)\Gamma(\delta)}{\Gamma(a+b+\delta)}\right\}$$

$$= \psi(a+b) - \psi(a).$$

Deduce that, if in addition $a + c > 0$, $a + b + c > 0$,

$$\int_0^1 \frac{x^{a-1}(1-x^b)(1-x^c)}{(1-x)(-\log x)}\,dx = \log\frac{\Gamma(a)\Gamma(a+b+c)}{\Gamma(a+b)\Gamma(a+c)}.$$

Example 12.32   Shew that, if a, b, c be such that the integral converges,

$$\int_0^1 \frac{(1-x^a)(1-x^b)(1-x^c)}{(1-x)(-\log x)}\,dx$$

$$= \log\frac{\Gamma(b+c+1)\Gamma(c+a+1)\Gamma(a+b+1)}{\Gamma(a+1)\Gamma(b+1)\Gamma(c+1)\Gamma(a+b+c+1)}.$$

Example 12.33 (St John's, 1896)   By the substitution $\cos\theta = 1 - 2\tan\frac{1}{2}\phi$, shew that

$$\int_0^\pi \frac{d\theta}{(3-\cos\theta)^{1/2}} = \frac{[\Gamma(\frac{1}{4})]^2}{4\sqrt{\pi}}.$$

Example 12.34 (Clare, 1898)   Evaluate in terms of Gamma-functions the integral

$$\int_0^\infty \frac{\sin^p x}{x}\, dx,$$ when p is a fraction greater than unity whose numerator and denomina-

tor are both odd integers. Shew that the integral is

$$\frac{1}{2} \int_0^\pi \sin^p x \left[\frac{1}{x} + \sum_{n=1}^\infty (-1)^n \left(\frac{1}{x+n\pi} + \frac{1}{x-n\pi} \right) \right] dx.$$

Example 12.35 Shew that

$$\int_0^{\pi/2} \left(1 + \tfrac{1}{2}\sin^2 x \right)^{n-1/2} dx = \frac{n!}{2^{n+2}\sqrt{\pi}} \sum_{r=0}^n \frac{2^{3r}}{(2r)!(n-r)!} \left[\Gamma \left(\frac{2r+1}{4} \right) \right]^2.$$

Example 12.36 (Euler) Prove that

$$\log B(p,q) = \log \left(\frac{p+q}{pq} \right) + \int_0^1 \frac{(1-v^p)(1-v^q)}{(1-v)\log v}\, dv.$$

Example 12.37 (Binet) Prove that, if $p > 0$, $p + s > 0$, then

$$B(p,p+s) = \frac{B(p,p)}{2^s} \left\{ 1 + \frac{s(s-1)}{2(2p+1)} + \frac{s(s-1)(s-2)(s-3)}{2 \cdot 4(2p+1)(2p+3)} + \cdots \right\}.$$

Example 12.38 The curve $r^m = 2^{m-1}a^m \cos m\theta$ is composed of m equal closed loops. Shew that the length of the arc of half of one of the loops is

$$\frac{a}{m} \int_0^{\pi/2} \left(\tfrac{1}{2}\cos x \right)^{1/m-1} dx,$$

and hence that the total perimeter of the curve is

$$a \frac{\Gamma \left(\frac{1}{2m} \right)^2}{\Gamma \left(\frac{1}{m} \right)}.$$

Example 12.39 (Cauchy) Draw the straight line joining the points $\pm i$, and the semicircle of $|z| = 1$ which lies on the right of this line. Let C be the contour formed by indenting this figure at $-i$, 0, i. By considering

$$\int_C z^{p-q-1}(z + z^{-1})^{p+q-2}\, dz,$$

shew that, if $p + q > 1, q < 1$,

$$\int_0^{\pi/2} \cos^{p+q-2}\theta \cos(p-q)\theta\, d\theta = \frac{\pi}{(p+q-1)2^{p+q-1}B(p,q)}.$$

Prove that the result is true for all values of p and q such that $p + q > 1$.

Example 12.40 If s is positive (not necessarily integral), and $-\tfrac{1}{2}\pi \le x \le \tfrac{1}{2}\pi$, shew that

$$(\cos x)^s = \frac{1}{2^{s-1}} \frac{\Gamma(s+1)}{\{\Gamma(\tfrac{1}{2}s+1)\}^2} \left\{ \frac{1}{2} + \frac{s}{s+2}\cos 2x + \frac{s(s-2)}{(s+2)(s+4)}\cos 4x + \cdots \right\},$$

and draw graphs of the series and of the function $\cos^s x$.

Example 12.41 (Cauchy)   Obtain the expansion

$$(\cos x)^s =$$
$$\frac{a}{2^{s-1}}\Gamma(s+1)\left[\frac{\cos ax}{\Gamma(\tfrac{1}{2}s+\tfrac{1}{2}a+1)\Gamma(\tfrac{1}{2}s-\tfrac{1}{2}a+1)}+\frac{\cos 3ax}{\Gamma(\tfrac{1}{2}s+\tfrac{3}{2}a+1)\Gamma(\tfrac{1}{2}s-\tfrac{3}{2}a+1)}+\cdots\right],$$

and find the values of x for which it is applicable.

Example 12.42 (Binet)   Prove that, if $p > \tfrac{1}{2}$,

$$\Gamma(2p) = \frac{2^{2p-1}}{\sqrt{\pi}}\;\{\Gamma(p)\}^2\left[\frac{2p^2}{2p+1}\left\{1+\frac{1^2}{2(2p+3)}+\frac{1^2\cdot 3^2}{2\cdot 4\cdot(2p+3)(2p+5)}+\cdots\right\}\right]^{1/2}.$$

Example 12.43   Shew that, if $x < 0$, $x + z > 0$, then

$$\frac{\Gamma(-x)}{\Gamma(z)}\left\{\frac{-x}{z}+\frac{1}{2}\frac{(-x)(1-x)}{z(1+z)}+\frac{1}{3}\frac{(-x)(1-x)(2-x)}{z(1+z)(2+z)}+\cdots\right\}$$

$$=\frac{1}{\Gamma(x+z)}\int_0^1 t^{-x-1}\left\{-\log(1-t)\right\}(1-t)^{x+z-1}\,dt,$$

and deduce that, when $x + z > 0$,

$$\frac{d}{dz}\log\frac{\Gamma(z+x)}{\Gamma(z)}=\frac{x}{z}-\frac{1}{2}\frac{x(x-1)}{z(z+1)}+\frac{1}{3}\frac{x(x-1)(x-2)}{z(z+1)(z+2)}-\cdots.$$

Example 12.44 (Binet [73])   Using the result of Example 12.43 above, prove that

$$\log\Gamma(z+a) = \log\Gamma(z) + a\log z - \frac{a-a^2}{2z}$$

$$-\sum_{n=1}^{\infty}\frac{a\int_0^1 t(l-t)(2-t)\cdots(n-t)\,dt - \int_0^a t(l-t)(2-t)\cdots(n-t)\,dt}{(n+1)z(z+1)(z+2)\cdots(z+n)},$$

investigating the region of convergence of the series.

Example 12.45   Prove that, if $p > 0$, $q > 0$, then

$$B(p,q) = \frac{p^{p-\frac{1}{2}}q^{q-\frac{1}{2}}}{(p+q)^{p+q-\frac{1}{2}}}(2\pi)^{\frac{1}{2}}e^{M(p,q)},$$

where

$$M(p,q) = 2\rho\int_0^{\infty}\frac{1}{e^{2\pi t\rho}-1}\arctan\left\{\frac{(t^3+t)\rho^3}{pq(p+q)}\right\}dt,$$

and $\rho^2 = p^2 + q^2 + pq$.

Example 12.46   If $U = 2^{x/2}/\Gamma(1-\tfrac{1}{2}x)$, $V = 2^{x/2}/\Gamma(\tfrac{1}{2}-\tfrac{1}{2}x)$, and if the function $F(x)$ be defined by the equation

$$F(x) = \sqrt{\pi}\left(V\frac{dU}{dx}-U\frac{dV}{dx}\right),$$

shew:

(a) that $F(x)$ satisfies the equation

$$F(x + 1) = xF(x) + \frac{1}{\Gamma(1 - x)};$$

(b) that, for all positive integral values of x,

$$F(x) = \Gamma(x);$$

(c) that $F(x)$ is analytic for all finite values of x;
(d) that

$$F(x) = \frac{1}{\Gamma(1 - x)} \frac{d}{dx} \log \frac{\Gamma\left(\frac{1-x}{2}\right)}{\Gamma\left(1 - \frac{x}{2}\right)}.$$

Example 12.47 Expand $1/\Gamma(a)$ as a series of ascending powers of a.

Note Various evaluations of the coefficients in this expansion have been given by Bourguet [94]; Schlömilch [585, 586].

Example 12.48 (Alexeiwksky) Prove that the G-function, defined by the equation

$$G(z + 1) = (2\pi)^{\frac{1}{2}z} e^{-\frac{1}{2}z(z+1)-\frac{1}{2}\gamma z^2} \prod_{n=1}^{\infty} \left\{ \left(1 + \frac{z}{n}\right)^n e^{-z+z^2/(2n)} \right\}$$

is an integral function which satisfies the relations

$$G(z + 1) = \Gamma(z)G(z), \qquad G(1) = 1,$$
$$\frac{(n!)^n}{G(n + 1)} = 1^1 \cdot 2^2 \cdot 3^3 \cdots n^n.$$

The most important properties of the G-function are discussed in Barnes [42].

Example 12.49 Shew that

$$\frac{G'(z + 1)}{G(z + 1)} = \frac{1}{2} \log(2\pi) + \frac{1}{2} - z + z\frac{\Gamma'(z)}{\Gamma(z)},$$

and deduce that

$$\log \frac{G(1 - z)}{G(1 + z)} = \int_0^s \pi z \cot \pi z \, dz - z \log(2\pi).$$

Example 12.50 Shew that

$$\int_0^s \log \Gamma(t + 1) \, dt = \frac{1}{2}z \log(2\pi) - \frac{1}{2}z(z + 1) + z \log \Gamma(z + 1) - \log G(z + 1).$$

13

The Zeta-Function of Riemann

13.1 Definition of the zeta-function

Let $s = \sigma + it$ where σ and t are real[1]; then, if $\delta > 0$, the series

$$\zeta(s) = \sum_{n=1}^{\infty} \frac{1}{n^s}$$

is a uniformly convergent series of analytic functions (§§2.33, 3.34) in any domain in which $\sigma \geq 1 + \delta$; and consequently the series is an analytic function of s in such a domain. The function is called the *zeta-function*; although it was known to Euler [197], its most remarkable properties were not discovered before Riemann [557] who discussed it in his memoir on prime numbers; it has since proved to be of fundamental importance, not only in the Theory of Prime Numbers, but also in the higher theory of the Gamma-function and allied functions.

13.11 The generalised zeta-function

The definition of this function appears to be due to Hurwitz [327].

Many of the properties possessed by the zeta-function are particular cases of properties possessed by a more general function defined, when $\sigma \geq 1 + \delta$, by the equation

$$\zeta(s,a) = \sum_{n=0}^{\infty} \frac{1}{(a+n)^s}, \tag{13.1}$$

where a is a constant. For simplicity, we shall suppose that $0 < a \leq 1$, and then we take $\arg(a + n) = 0$. It is evident that $\zeta(s, 1) = \zeta(s)$. (When a has this range of values, the properties of the function are, in general, much simpler than the corresponding properties for other values of a. The results of §13.14 are true for all values of a (negative integer values excepted); and the results of §§13.12, 13.13, 13.2 are true when $\mathrm{Re}(a) > 0$.)

13.12 The expression of $\zeta(s,a)$ as an infinite integral

Since

$$(a+n)^{-s}\,\Gamma(s) = \int_0^{\infty} x^{s-1} e^{-(n+a)x}\,dx,$$

---

[1] The letters σ, t will be used in this sense throughout the chapter.

when $\arg x = 0$ and $\sigma > 0$ (and *a fortiori* when $\sigma \geq (1 + \delta)$), we have, when $\sigma \geq 1 + \delta$,

$$\Gamma(s)\,\zeta(s,a) = \lim_{N\to\infty} \sum_{n=0}^{N} \int_0^\infty x^{s-1} e^{-(n+a)x}\, dx$$

$$= \lim_{N\to\infty} \left\{ \int_0^\infty \frac{x^{s-1} e^{-ax}}{1 - e^{-x}}\, dx - \int_0^\infty \frac{x^{s-1}}{1 - e^{-x}} e^{-(N+1+a)x}\, dx \right\}.$$

Now, when $x \geq 0$, $e^x \geq 1 + x$, and so the modulus of the second of these integrals does not exceed

$$\int_0^\infty x^{\sigma-2} e^{-(N+a)x}\, dx = (N + a)^{1-\sigma}\Gamma(\sigma - 1),$$

which (when $\sigma \geq 1 + \delta$) tends to 0 as $N \to \infty$. Hence, when $\sigma \geq 1 + \delta$ and $\arg x = 0$,

$$\zeta(s,a) = \frac{1}{\Gamma(s)} \int_0^\infty \frac{x^{s-1} e^{-ax}}{1 - e^{-x}}\, dx;$$

this formula corresponds in some respects to Euler's integral for the Gamma-function.

13.13 The expression of $\zeta(s,a)$ as a contour integral

.

This was given by Riemann for the ordinary zeta-function.

When $\sigma \geq 1 + \delta$, consider

$$\int_\infty^{(0+)} \frac{(-z)^{s-1} e^{-az}}{1 - e^{-z}}\, dz,$$

the contour of integration being of Hankel's type (§12.22) and not containing the points $\pm 2n\pi i$ ($n = 1, 2, 3, \ldots$) which are poles of the integrand; it is supposed (as in §12.22) that $|\arg(-z)| \leq \pi$.

It is legitimate to modify the contour, precisely as in §12.22, when[2] $\sigma \geq 1 + \delta$; and we get

$$\int_\infty^{(0+)} \frac{(-z)^{s-1} e^{-az}}{1 - e^{-z}}\, dz = \left\{ e^{\pi i(s-1)} - e^{-\pi i(s-1)} \right\} \int_0^\infty \frac{x^{s-1} e^{-ax}}{1 - e^{-x}}\, dx.$$

Therefore

$$\zeta(s,a) = -\frac{\Gamma(1-s)}{2\pi i} \int_\infty^{(0+)} \frac{(-z)^{s-1} e^{-az}}{1 - e^{-z}}\, dz.$$

Now this last integral is a one-valued analytic function of s for *all* values of s. Hence the only possible singularities of $\zeta(s,a)$ are at the singularities of $\Gamma(1 - s)$, i.e. at the points $1, 2, 3, \ldots$, and, with the exception of these points, the integral affords a representation of $\zeta(s,a)$ valid over the whole plane. The result obtained corresponds to Hankel's integral for the Gamma-function. Also, we have seen that $\zeta(s,a)$ is analytic when $\sigma \geq 1 + \delta$, and so the only singularity of $\zeta(s,a)$ is at the point $s = 1$. Writing $s = 1$ in the integral, we get

$$\frac{1}{2\pi i} \int_\infty^{(0+)} \frac{e^{-az}\, dz}{1 - e^{-z}},$$

[2] If $\sigma \leq 1$, the integral taken along any straight line up to the origin does not converge.

which is the residue at $z = 0$ of the integrand, and this residue is 1. Hence

$$\lim_{s \to 1} \frac{\zeta(s, a)}{\Gamma(1 - s)} = -1.$$

Since $\Gamma(1 - s)$ has a single pole at $s = 1$ with residue -1, it follows that the only singularity of $\zeta(s, a)$ is a simple pole with residue $+1$ at $s = 1$.

Example 13.1.1   Shew that, when $\operatorname{Re}(s) > 0$,

$$(1 - 2^{1-s}) \zeta(s) = \frac{1}{1^s} - \frac{1}{2^s} + \frac{1}{3^s} - \frac{1}{4^s} + \cdots$$

$$= \frac{1}{\Gamma(s)} \int_0^\infty \frac{x^{s-1}}{e^x + 1} \, dx.$$

Example 13.1.2   Shew that, when $\operatorname{Re}(s) > 1$,

$$(2^s - 1) \zeta(s) = \zeta\left(s, \tfrac{1}{2}\right)$$

$$= \frac{2^s}{\Gamma(s)} \int_0^\infty \frac{x^{s-1} e^x}{e^{2x} - 1} \, dx.$$

Example 13.1.3   Shew that

$$\zeta(s) = -\frac{2^{1-s} \Gamma(1 - s)}{2\pi i (2^{1-s} - 1)} \int_\infty^{(0+)} \frac{(-z)^{s-1}}{e^s + 1} \, dz,$$

where the contour does not include any of the points $\pm \pi i, \pm 3\pi i, \pm 5\pi i, \ldots$.

13.14 Values of $\zeta(s, a)$ for special values of s

In the special case when s is an integer (positive or negative),

$$\frac{(-z)^{s-1} e^{-az}}{1 - e^{-z}}$$

is a one-valued function of z. We may consequently apply Cauchy's theorem, so that

$$\frac{1}{2\pi i} \int_\infty^{(0+)} \frac{(-z)^{s-1} e^{-az}}{1 - e^{-z}} \, dz$$

is the residue of the integrand at $z = 0$, that is to say, it is the coefficient of z^{-s} in $\dfrac{(-1)^{s-1} e^{-az}}{1 - e^{-z}}$.
To obtain this coefficient we differentiate the expansion (§7.2)

$$-z \frac{e^{-az} - 1}{e^{-z} - 1} = \sum_{n=1}^\infty \frac{(-1)^n \phi_n(a) z^n}{n!}$$

term-by-term with regard to a, where $\phi_n(a)$ denotes the Bernoullian polynomial. (This is obviously legitimate, by §4.7, when $|z| < 2\pi$, since $\dfrac{z^2 e^{-az}}{e^{-z} - 1}$ can be expanded into a power series in z uniformly convergent with respect to a.) Then

$$\frac{z^2 e^{-az}}{e^{-z} - 1} = \sum_{n=1}^\infty \frac{(-1)^n \phi_n'(a) z^n}{n!}. \tag{13.2}$$

Therefore if s is zero or a negative integer (= −m), we have

$$\zeta(-m, a) = -\frac{\phi'_{m+2}(a)}{(m+1)(m+2)}.$$

In the special case when $a = 1$, if $s = -m$, then $\zeta(s)$ is the coefficient of z^{1-s} in the expansion of $\frac{(-1)^s m! z}{e^z - 1}$. Hence, by §7.2

$$\zeta(-2m) = 0; \qquad \zeta(1 - 2m) = \frac{(-1)^m B_m}{2m} \quad (m = 1, 2, 3, \ldots); \qquad \zeta(0) = -\frac{1}{2}.$$

These equations give the value of $\zeta(s)$ when s is a negative integer or zero.

13.15 The formula of Hurwitz for $\zeta(s, a)$ when $\sigma < 0$

This appears in [327, p. 95]. Consider $-\frac{1}{2\pi i} \int_C \frac{(-z)^{s-1} e^{-az}}{1 - e^{-z}} dz$ taken round a contour C consisting of a (large) circle of radius $(2N+1)\pi$, (N an integer), starting at the point $(2N+1)\pi$ and encircling the origin in the positive direction, $\arg(-z)$ being zero at $z = -(2N+1)\pi$.

In the region between C and the contour $(2N\pi + \pi, 0+)$, of which the contour of §13.13 is the limiting form, $(-z)^{s-1} e^{-az} (1 - e^{-z})^{-1}$ is analytic and one-valued except at the simple poles $\pm 2\pi i, \pm 4\pi i, \ldots, \pm 2N\pi i$. Hence

$$\frac{1}{2\pi i} \int_C \frac{(-z)^{s-1} e^{-az}}{1 - e^{-z}} dz - \frac{1}{2\pi i} \int_{(2N+1)\pi}^{(0+)} \frac{(-z)^{s-1} e^{-az}}{1 - e^{-z}} dz = \sum_{n=1}^{N} (R_n + R'_n),$$

where R_n, R'_n are the residues of the integrand at $2n\pi i$, $-2n\pi i$ respectively. At the point at which $-z = 2n\pi e^{-\pi i/2}$, the residue is

$$(2n\pi)^{s-1} e^{-\frac{1}{2}\pi i(s-1)} e^{-2an\pi i},$$

and hence $R_n + R'_n = 2(2n\pi)^{s-1} \sin(s\pi/2 + 2\pi an)$. Hence

$$-\frac{1}{2\pi i} \int_{(2N+1)\pi}^{(0+)} \frac{(-z)^{s-1} e^{-az}}{1 - e^{-z}} dz$$

$$= \frac{2\sin(s\pi/2)}{(2\pi)^{1-s}} \sum_{n=1}^{N} \frac{\cos(2\pi an)}{n^{1-s}}$$

$$+ \frac{2\cos(s\pi/2)}{(2\pi)^{1-s}} \sum_{n=1}^{N} \frac{\sin(2\pi an)}{n^{1-s}} - \frac{1}{2\pi i} \int_C \frac{(-z)^{s-1} e^{-az}}{1 - e^{-z}} dz.$$

Now, since $0 < a \leq 1$, it is easy to see that we can find a number K independent of N such that $\left| e^{-az} (1 - e^{-z})^{-1} \right| < K$ when z is on C. Hence

$$\left| \frac{1}{2\pi i} \int_C \frac{(-z)^{s-1} e^{-az}}{1 - e^{-z}} dz \right| < \frac{K}{2\pi} \int_{-\pi}^{\pi} \left| [(2N+1)\pi]^s e^{si\theta} \right| d\theta$$

$$< K [(2N+1)\pi]^s e^{\pi|s|}$$

$$\to 0 \quad \text{as } N \to \infty \qquad \text{if } \sigma < 0.$$

Making $N \to \infty$, we obtain the result of Hurwitz that, if $\sigma < 0$,

$$\zeta(s,a) = \frac{2\Gamma(1-s)}{(2\pi)^{1-s}} \left[\sin\left(\frac{s\pi}{2}\right) \sum_{n=1}^{\infty} \frac{\cos(2\pi an)}{n^{1-s}} + \cos\left(\frac{s\pi}{2}\right) \sum_{n=1}^{\infty} \frac{\sin(2\pi an)}{n^{1-s}} \right],$$

each of these series being convergent.

13.151 Riemann's relation between $\zeta(s)$ and $\zeta(1-s)$

If we write $a = 1$ in the formula of Hurwitz given in §13.15, and employ §12.14, we get the remarkable result, due to Riemann, that

$$2^{1-s}\Gamma(s)\zeta(s)\cos\left(\frac{s\pi}{2}\right) = \pi^s \zeta(1-s). \tag{13.3}$$

Since both sides of this equation are analytic functions of s, save for isolated values of s at which they have poles, this equation, proved when $\sigma < 0$, persists (by §5.5) for all values of s save those isolated values.

Example 13.1.4 If m be a positive integer, shew that

$$\zeta(2m) = \frac{2^{2m-1}\pi^{2m}B_m}{(2m)!}.$$

Example 13.1.5 (Riemann)   Shew that $\Gamma(s/2)\pi^{-s/2}\zeta(s)$ is unaltered by replacing s by $1-s$.

Example 13.1.6   Deduce from Riemann's relation that the zeros of Hermite's integral for $\zeta(s)$ at $-2, -4, -6, \ldots$ are zeros of the first order.

13.2 Hermite's formula for $\zeta(s,a)$

This appears in [296]. Let us apply Plana's theorem (Example 7.7 in Chapter 7) to the function $\phi(z) = (a+z)^{-s}$, where $\arg(a+z)$ has its principal value.

Define the function $q(x,y)$ by the equation

$$q(x,y) = \frac{1}{2i}\left[(a+x+iy)^{-s} - (a+x-iy)^{-s}\right]$$

$$= -\left[(a+x)^2 + y^2\right]^{-s/2} \sin\left(s \arctan\frac{y}{x+a}\right).$$

Since[3] $\left|\arctan\dfrac{y}{x+a}\right|$ does not exceed the smaller of $\dfrac{\pi}{2}$ and $\dfrac{|y|}{x+a}$, we have

$$|q(x,y)| \le \left\{(a+x)^2 + y^2\right\}^{(1-\sigma)/2} \left|y^{-1}\sinh\left(\frac{\pi s}{2}\right)\right|,$$

$$\le \left\{(a+x)^2 + y^2\right\}^{-\sigma/2} \left|\sinh\frac{y|s|}{x+a}\right|.$$

Using the first result when $|y| > a$ and the second when $|y| < a$, it is evident that, if $\sigma > 0$, $\displaystyle\int_0^{\infty} q(x,y)\left(e^{2\pi y} - 1\right)^{-1} dy$ is convergent when $x \ge 0$ and tends to 0 as $x \to \infty$; also

---

[3] If $\xi > 0$, $\arctan \xi = \displaystyle\int_0^{\xi} \frac{dt}{1+t^2} < \int_0^{\infty} \frac{dt}{1+t^2}$; and $\arctan \xi < \displaystyle\int_0^{\xi} dt$.

$\int_0^\infty (a + x)^{-s} dx$ converges if $\sigma > 1$. Hence, if $\sigma > 1$, it is legitimate to make $x_2 \to \infty$ in the result contained in the example cited; and we have

$$\zeta(s, a) = \frac{1}{2}a^{-s} + \int_0^\infty (a + x)^{-s} dx + 2 \int_0^\infty (a^2 + y^2)^{-s/2} \left[\sin\left(s \arctan \frac{y}{a} \right) \right] \frac{dy}{e^{2\pi y} - 1}.$$

So

$$\zeta(s, a) = \frac{1}{2}a^{-s} + \frac{a^{1-s}}{s - 1} + 2 \int_0^\infty (a^2 + y^2)^{-s/2} \left[\sin\left(s \arctan \frac{y}{a} \right) \right] \frac{dy}{e^{2\pi y} - 1}.$$

This is Hermite's formula[4]; using the results that, if $y \geq 0$,

$$\arctan \frac{y}{a} \leq \frac{y}{a}, \quad \left(y < \tfrac{\pi a}{2}\right); \qquad \arctan \frac{y}{a} < \frac{\pi}{2} \quad \left(y > \tfrac{\pi a}{2}\right),$$

we see that the integral involved in the formula converges for all values of s. Further, the integral defines an analytic function of s for all values of s.

To prove this, it is sufficient (§5.31) to shew that the integral obtained by differentiating under the sign of integration converges uniformly; that is to say we have to prove that

$$\int_0^\infty \left[-\frac{1}{2} \log(a^2 + y^2)(a^2 + y^2)^{-s/2} \sin\left(s \arctan \frac{y}{a} \right) \right] \frac{dy}{e^{2\pi y} - 1}$$

$$+ \int_0^\infty \left[(a^2 + y^2)^{-s/2} \arctan \frac{y}{a} \cos\left(s \arctan \frac{y}{a} \right) \right] \frac{dy}{e^{2\pi y} - 1}$$

converges uniformly with respect to s in any domain of values of s. Now when $|s| \leq \Delta$, where Δ is any positive number, we have

$$\left| (a^2 + y^2)^{-s/2} \arctan \frac{y}{a} \cos\left(s \arctan \frac{y}{a} \right) \right| < (a^2 + y^2)^{\Delta/2} \frac{y}{a} \cosh\left(\tfrac{1}{2}\pi\Delta\right);$$

since

$$\frac{\Delta}{a} \int_0^\infty (a^2 + y^2)^{\Delta/2} \frac{y \, dy}{e^{2\pi y} - 1}$$

converges, uniform convergence of the second integral is justified using de la Vallée Poussin's test in (I) of §4.431.

By dividing the path of integration of the first integral into two parts $(0, \tfrac{1}{2}\pi a)$, $(\tfrac{1}{2}\pi a, \infty)$ and using the results

$$\left| \sin\left(s \arctan \frac{y}{a} \right) \right| < \sinh \frac{\Delta y}{a}, \left| \sin\left(s \arctan \frac{y}{a} \right) \right| < \sinh \frac{1}{2}\pi\Delta$$

in the respective parts, we can similarly shew that the first integral converges uniformly.

Consequently Hermite's formula is valid (§5.5) for all values of s, and it is legitimate to differentiate under the sign of integration, and the differentiated integral is a continuous function of s.

---

[4] The corresponding formula when $a = 1$ had been previously given by Jensen.

13.21 Deductions from Hermite's formula

Writing $s = 0$ in Hermite's formula, we see that $\zeta(0,a) = \frac{1}{2} - a$. Making $s \to 1$, from the uniformity of convergence of the integral involved in Hermite's formula we see that

$$\lim_{s\to 1}\left\{\zeta(s,a) - \frac{1}{s-1}\right\} = \lim_{s\to 1}\frac{a^{1-s}-1}{s-1} + \frac{1}{2a} + 2\int_0^\infty \frac{y\,dy}{(a^2+y^2)(e^{2\pi y}-1)}.$$

Hence, by the example of §12.32, we have

$$\lim_{s\to 1}\left(\zeta(s,a) - \frac{1}{s-1}\right) = -\frac{\Gamma'(a)}{\Gamma(a)}.$$

Further, differentiating the formula for $\zeta(s,a)$ and then making $s \to 0$ (this was justified in §13.2), we get

$$\left\{\frac{d}{ds}\zeta(s,a)\right\}_{s=0} = \lim_{s\to 0}\left[-\frac{1}{2}a^{-s}\log a - \frac{a^{1-s}\log a}{s-1} - \frac{a^{1-s}}{(s-1)^2}\right.$$

$$+ 2\int_0^\infty \left\{-\frac{1}{2}\log(a^2+y^2)\cdot(a^2+y^2)^{-\frac{1}{2}s}\sin\left(s\arctan\frac{y}{a}\right)\right.$$

$$\left.\left.+(a^2+y^2)^{-s/2}\arctan\frac{y}{a}\cos\left(s\arctan\frac{y}{a}\right)\right\}\frac{dy}{e^{2\pi y}-1}\right]$$

$$= \left(a-\frac{1}{2}\right)\log a - a + 2\int_0^\infty \frac{\arctan(y/a)}{e^{2\pi y}-1}dy.$$

Hence, by §12.32,

$$\left\{\frac{d}{ds}\zeta(s,a)\right\}_{s=0} = \log\Gamma(a) - \frac{1}{2}\log(2\pi).$$

These results had previously been obtained in a different manner by Lerch [425]. The formula for $\zeta(s,a)$ from which Lerch derived these results is given in a memoir published by the Academy of Sciences of Prague. A summary of his memoir is contained in [429].

Corollary 13.2.1 $\lim_{s\to 1}\left(\zeta(s) - \frac{1}{s-1}\right) = \gamma, \quad \zeta'(0) = -\frac{1}{2}\log(2\pi).$

13.3 Euler's product for $\zeta(s)$

Let $\sigma \geq 1 + \delta$; and let $2, 3, 5, \ldots, p, \ldots$ be the prime numbers in order. Then, subtracting the series for $2^{-s}\zeta(s)$ from the series for $\zeta(s)$, we get

$$\zeta(s)\cdot(1-2^{-s}) = \frac{1}{1^s} + \frac{1}{3^s} + \frac{1}{5^s} + \frac{1}{7^s} + \cdots,$$

all the terms of $\sum n^{-s}$ for which n is a multiple of 2 being omitted; then in like manner

$$\zeta(s)\cdot(1-2^{-s})(1-3^{-s}) = \frac{1}{1^s} + \frac{1}{5^s} + \frac{1}{7^s} + \cdots,$$

all the terms for which n is a multiple of 2 or 3 being omitted, and so on; so that

$$\zeta(s)\cdot(1-2^{-s})(1-3^{-s})\cdots(1-p^{-s}) = 1 + \sum{}' n^{-s},$$

the prime denoting that only those values of n (greater than p) which are prime to $2, 3, \ldots, p$ occur in the summation. Now, since the first term of $\sum'$ starts with the prime next greater than p,

$$\left| \sum' n^{-s} \right| \leq \sum' n^{-1-s} \leq \sum_{n=p+1}^{\infty} n^{-1-s} \to 0 \quad \text{as } p \to \infty.$$

Therefore if $\sigma \geq 1 + \delta$, the product $\zeta(s) \prod_p (1 - p^{-s})$ converges to 1, where the number p assumes the prime values $2, 3, 5, \ldots$ only. But the product $\prod_p (1 - p^{-s})$ converges when $\sigma \geq 1 + \delta$, for it consists of some of the factors of the absolutely convergent product $\prod_{n=2}^{\infty} (1 - n^{-s})$. Consequently we infer that $\zeta(s)$ *has no zeros at which $\sigma \geq 1 + \delta$;* for if it had any such zeros, $\prod_p (1 - p^{-s})$ would not converge at them. Therefore, if $\sigma \geq 1 + \delta$,

$$\prod_p \left(1 - \frac{1}{p^s} \right) = \frac{1}{\zeta(s)}.$$

This is Euler's result.

13.31 Riemann's hypothesis concerning the zeros of $\zeta(s)$

It has just been proved that $\zeta(s)$ has no zeros at which $\sigma > 1$. From the formula (13.3)

$$\zeta(s) = \frac{2^{s-1}\pi^s}{\Gamma(s)} \sec\left(\frac{s\pi}{2}\right) \zeta(1-s)$$

it is now apparent that the only zeros of $\zeta(s)$ for which $\sigma < 0$ are the zeros of $\sec(s\pi/2)/\Gamma(s)$, i.e. the points $s = -2, -4, \ldots$.

Hence all the zeros of $\zeta(s)$ except those at $-2, -4, \ldots$ lie in that strip of the domain of the complex variable s which is defined by $0 \leq \sigma \leq 1$.

It was conjectured by Riemann, but it has not yet been proved, that all the zeros of $\zeta(s)$ in this strip lie on the line $\sigma = \frac{1}{2}$; while it has quite recently been proved by Hardy [279] that an infinity of zeros of $\zeta(s)$ actually lie on $\sigma = \frac{1}{2}$. It is highly probable that Riemann's conjecture is correct, and the proof of it would have far-reaching consequences in the theory of Prime Numbers.

13.4 Riemann's integral for $\zeta(s)$

It is easy to see that, if $\sigma > 0$,

$$n^{-s}\Gamma\left(\frac{s}{2}\right)\pi^{-s/2} = \int_0^{\infty} e^{-n^2\pi x} x^{s/2-1}\, dx.$$

Hence, when $\sigma > 0$,

$$\zeta(s)\Gamma\left(\frac{s}{2}\right)\pi^{-s/2} = \lim_{N \to \infty} \int_0^{\infty} \sum_{n=1}^{N} e^{-n^2\pi x} x^{s/2-1}\, dx.$$

Now, if $\vartheta(x) = \sum\limits_{n=1}^{\infty} e^{-n^2 \pi x}$, since, by Example 6.17 of Chapter 6,

$$1 + 2\vartheta(x) = x^{-1/2}(1 + 2\vartheta(1/x)), \qquad (13.4)$$

we have $\lim\limits_{x\to 0} x^{1/2}\vartheta(x) = 1/2$; and hence $\int_0^\infty \vartheta(x)x^{s/2-1}\, dx$ converges when $\sigma > 1$.
Consequently, if $\sigma > 2$,

$$\zeta(s)\Gamma\left(\frac{s}{2}\right)\pi^{-s/2} = \lim_{N\to\infty}\left[\int_0^\infty \vartheta(x)x^{s/2-1}\, dx - \int_0^\infty \sum_{n=N+1}^\infty e^{-n^2\pi x}x^{s/2-1}\, dx\right].$$

Now, as in §13.12, the modulus of the last integral does not exceed

$$\int_0^\infty \left\{\sum_{n=N+1}^\infty e^{-n(N+1)\pi x}\right\}x^{\sigma/2-1}\, dx = \int_0^\infty \frac{e^{-(N+1)^2\pi x}x^{\sigma/2-1}}{1 - e^{-(N+1)\pi x}}\, dx$$

$$< \{\pi(N+1)\}^{-1}\int_0^\infty e^{-(N^2+2N)\pi x}x^{\sigma/2-2}\, dx$$

$$= \{\pi(N+1)\}^{-1}\{(N^2+2N)\pi\}^{1-\sigma/2}\Gamma(\sigma/2-1)$$

$$\to 0 \quad \text{as } N\to\infty, \qquad \text{since} \quad \sigma > 2.$$

Hence, when $\sigma > 2$,

$$\zeta(s)\Gamma\left(\frac{s}{2}\right)\pi^{-s/2} = \int_0^\infty \vartheta(x)x^{s/2-1}\, dx$$

$$= \int_0^1 \left\{-\tfrac{1}{2} + \tfrac{1}{2}x^{-\frac{1}{2}} + x^{-\frac{1}{2}}\vartheta(1/x)\right\}x^{s/2-1}\, dx + \int_1^\infty \vartheta(x)x^{s/2-1}\, dx$$

$$= -\frac{1}{s} + \frac{1}{s-1} + \int_\infty^1 x^{1/2}\vartheta(x)x^{-s/2+1}\left(-\frac{1}{x^2}\right)\, dx + \int_1^\infty \vartheta(x)x^{s/2-1}\, dx.$$

Consequently

$$\zeta(s)\Gamma\left(\frac{s}{2}\right)\pi^{-s/2} - \frac{1}{s(s-1)} = \int_1^\infty (x^{(1-s)/2} + x^{s/2})\frac{\vartheta(x)}{x}\, dx.$$

Now the integral on the right represents an analytic function of s for *all* values of s, by §5.32, since on the path of integration

$$\vartheta(x) < e^{-\pi x}\sum_{n=0}^x e^{-m\pi x} \le e^{-\pi x}(1 - e^{-\pi})^{-1}.$$

Consequently, by §5.5, the above equation, proved when $\sigma > 2$, persists for all values of s.

If now we put

$$s = \frac{1}{2} + it, \qquad \frac{1}{2}s(s-1)\zeta(s)\Gamma\left(\frac{s}{2}\right)\pi^{-s/2} = \xi(t),$$

we have

$$\xi(t) = \frac{1}{2} - (t^2 + \tfrac{1}{4})\int_1^\infty x^{-3/4}\vartheta(x)\cos\left(\tfrac{1}{2}t\log x\right)\, dx.$$

Since

$$\int_1^\infty x^{-3/4}\vartheta(x)\left\{\tfrac{1}{2}\log x\right\}^n \cos\left(\tfrac{1}{2}t\log x + \tfrac{1}{2}n\pi\right) dx$$

satisfies the test of Corollary 4.4.1, we may differentiate any number of times under the sign of integration, and then put $t = 0$. Hence, by Taylor's theorem, we have for all values[5] of t

$$\xi(t) = \sum_{n=0}^\infty a_{2n}t^{2n}; \tag{13.5}$$

by considering the last integral a_{2n} is obviously real. This result is fundamental in Riemann's researches.

13.5 Inequalities satisfied by $\zeta(s,a)$ when $\sigma > 0$

We shall now investigate the behaviour of $\zeta(s,a)$ as $t \to \pm\infty$, for given values of σ. When $\sigma > 1$, it is easy to see that, if N be any integer,

$$\zeta(s,a) = \sum_{n=0}^N \frac{1}{(a+n)^s} - \frac{1}{(1-s)(N+a)^{s-1}} - \sum_{n=N}^\infty f_n(s),$$

where

$$f_n(s) = \frac{1}{1-s}\left\{\frac{1}{(n+1+a)^{s-1}} - \frac{1}{(n+a)^{s-1}}\right\} - \frac{1}{(n+1+a)^s}$$

$$= s\int_n^{n+1} \frac{u-n}{(u+a)^{s+1}}\, du.$$

Now, when $\sigma > 0$,

$$|f_n(s)| \le |s|\int_n^{n+1} \frac{u-n}{(u+a)^{\sigma+1}}\, du$$

$$< |s|\int_n^{n+1} \frac{du}{(n+a)^{\sigma+1}}$$

$$= |s|\,(n+a)^{-\sigma-1}.$$

Therefore the series $\sum_{n=N}^\infty f_n(s)$ is a uniformly convergent series of analytic functions when $\sigma > 0$; so that $\sum_{n=N}^\infty f_n(s)$ is an analytic function when $\sigma > 0$; and consequently, by §5.5, the function $\zeta(s,a)$ may be defined when $\sigma > 0$ by the series

$$\zeta(s,a) = \sum_{n=0}^N \frac{1}{(a+n)^s} - \frac{1}{(1-s)(N+a)^{s-1}} - \sum_{n=N}^\infty f_n(s).$$

---

[5] In this particular piece of analysis it is convenient to regard t as a complex variable, defined by the equation $s = \tfrac{1}{2} + it$; and then $\xi(t)$ is an integral function of t.

Now let $\lfloor t \rfloor$ be the greatest integer in $|t|$; and take $N = \lfloor t \rfloor$. Then

$$|\zeta(s,a)| \leq \sum_{n=0}^{\lfloor t \rfloor} |(a+n)^{-s}| + |\{(1-s)^{-1}(\lfloor t \rfloor + a)^{1-s}\}| + \sum_{n=\lfloor t \rfloor}^{\infty} |s|(n+a)^{-\sigma-1}$$

$$< \sum_{n=0}^{\lfloor t \rfloor} (a+n)^{-\sigma} + |t|^{-1}(\lfloor t \rfloor + a)^{1-\sigma} + |s| \sum_{n=\lfloor t \rfloor}^{\infty} (n+a)^{-\sigma-1}.$$

Using the Maclaurin–Cauchy sum formula (§4.43), we get

$$|\zeta(s,a)| < a^{-\sigma} + \int_0^{\lfloor t \rfloor} (a+x)^{-\sigma}\,dx + |t|^{-1}(\lfloor t \rfloor + a)^{1-\sigma} + |s| \int_{\lfloor t \rfloor - 1}^{\infty} (x+a)^{-\sigma-1}\,dx.$$

Now when $\delta \leq \sigma \leq 1 - \delta$ where $\delta > 0$, we have

$$|\zeta(s,a)| < a^{-\sigma} + (1-\sigma)^{-1}\left\{(a+\lfloor t \rfloor)^{1-\sigma} - a^{1-\sigma}\right\} + |t|^{-1}(\lfloor t \rfloor + a)^{1-\sigma} + |s|\sigma^{-1}(\lfloor t \rfloor - 1 + a)^{-\sigma}.$$

Hence $\zeta(s,a) = O(|t|^{1-\sigma})$, the constant implied in the symbol O being independent of s. But, when $1 - \delta \leq \sigma \leq 1 + \delta$, we have

$$|\zeta(s,a)| = O(|t|^{1-\sigma}) + \int_0^{\lfloor t \rfloor} (a+x)^{-\sigma}\,dx$$

$$< O(|t|^{1-\sigma}) + \left\{a^{1-\sigma} + (a+t)^{1-\sigma}\right\} \int_0^{\lfloor t \rfloor} (a+x)^{-1}\,dx,$$

since $(a+x)^{-\sigma} \leq a^{1-\sigma}(a+x)^{-1}$ when $\sigma \geq 1$, and $(a+x)^{-\sigma} \leq (a+\lfloor t \rfloor)^{1-\sigma}(a+x)^{-1}$ when $\sigma \leq 1$, *and so*

$$\zeta(s,a) = O\left\{|t|^{1-\sigma} \log |t|\right\}.$$

When $\sigma \geq 1 + \delta$,

$$|\zeta(s,a)| \leq a^{-\sigma} + \sum_{n=1}^{\infty} (a+n)^{-1-\delta} = O(1).$$

13.51 *Inequalities satisfied by $\zeta(s,a)$ when $\sigma \leq 0$*

We next obtain inequalities of a similar nature when $\sigma \leq \delta$. In the case of the function $\zeta(s)$ we use Riemann's relation

$$\zeta(s) = 2^s \pi^{s-1} \Gamma(1-s)\,\zeta(1-s)\,\sin\left(\frac{s\pi}{2}\right).$$

Now, when $\sigma < 1 - \delta$, we have, by §12.33,

$$\Gamma(1-s) = O\left(e^{(\frac{1}{2}-s)\log(1-s)-(1-s)}\right)$$

and so

$$\zeta(s) = O\left[\exp\left\{\frac{\pi}{2}|t| + (\tfrac{1}{2} - \sigma - it)\log|1-s| + i\arctan\frac{t}{(1-\sigma)}\right\}\right]\zeta(1-s).$$

Since $\arctan t/(1-\sigma) = \pm\frac{1}{2}\pi + O(t^{-1})$, according as t is positive or negative, we see, from the results already obtained for $\zeta(s,a)$, that

$$\zeta(s) = O\left(|t|^{1/2-\sigma}\right)\zeta(1-s).$$

In the case of the function $\zeta(s,a)$, we have to use the formula of Hurwitz (§13.15) to obtain the generalisation of this result; we have, when $\sigma < 0$,

$$\zeta(s,a) = -i(2\pi)^{s-1}\Gamma(1-s)\left[e^{s\pi i/2}\zeta_a(1-s) - e^{-s\pi i/2}\,\zeta_{-a}(1-s)\right],$$

where

$$\zeta_a(1-s) = \sum_{n=1}^{\infty}\frac{e^{2n\pi i a}}{n^{1-s}}.$$

Hence

$$(1-e^{2\pi i a})\zeta_a(1-s) = e^{2\pi i a} + \sum_{n=2}^{N}e^{2n\pi i a}\left[n^{s-1} - (n-1)^{s-1}\right]$$

$$+(s-1)\sum_{n=N+1}^{\infty}e^{2n\pi i a}\int_{n-1}^{n}u^{s-2}\,du;$$

since the series on the right is a uniformly convergent series of analytic functions whenever $\sigma \le 1 - \delta$, this equation gives the continuation of $\zeta_a(1-s)$ over the range $0 \le \sigma \le 1 - \delta$; so that, whenever $\sigma \le 1 - \delta$, we have

$$|\sin\pi a\,\zeta_a(1-s)| \le 1 + \sum_{n=2}^{N}\left\{n^{\sigma-1} + (n-1)^{\sigma-1}\right\} + |s-1|\sum_{n=N+1}^{\infty}\int_{n-1}^{n}u^{\sigma-2}\,du.$$

Taking $N = \lfloor t \rfloor$, we obtain, as in §13.5,

$$\zeta_a(1-s) = O(|t|^{\sigma}) \qquad (\delta \le \sigma \le 1 - \delta)$$
$$= O(|t|^{\sigma}\log|t|) \qquad (-\delta \le \sigma < \delta),$$

and obviously

$$\zeta_a(1-s) = O(1) \qquad (\sigma < -\delta).$$

Consequently, whether a is unity or not, we have the results

$$\zeta(s,a) = O(|t|^{1/2-\sigma}) \qquad (\sigma \le \delta)$$
$$= O(|t|^{1/2}) \qquad (\delta \le \sigma \le 1 - \delta)$$
$$= O(|t|^{1/2}\log|t|) \qquad (-\delta \le \sigma \le \delta).$$

We may combine these results and those of §13.5, into the single formula

$$\zeta(s,a) = O(|t|^{\tau(\sigma)}\log|t|),$$

where[6]

$$
\tau(\sigma) = \begin{cases}
\frac{1}{2} - \sigma & \sigma \le 0, \\
\frac{1}{2} & 0 \le \sigma \le \frac{1}{2}, \\
1 - \sigma & \frac{1}{2} \le \sigma \le 1, \\
0 & \sigma \ge 1,
\end{cases}
$$

and the $\log |t|$ may be suppressed except when $-\delta \le \sigma \le \delta$ or when $1 - \delta \le \sigma \le 1 + \delta$.

13.6 The asymptotic expansion of $\log \Gamma(z + a)$

From Example 12.1.3 it follows that

$$
\left(1 + \frac{z}{a}\right) \prod_{n=1}^{\infty} \left\{\left(1 + \frac{z}{a + n}\right) e^{-z/n}\right\} = \frac{e^{-\gamma z} \Gamma(a)}{\Gamma(z + a)}.
$$

Now, the principal values of the logarithms being taken,

$$
\log\left(1 + \frac{z}{a}\right) + \log \prod_{n=1}^{\infty} \left\{\left(1 + \frac{z}{a + n}\right) e^{-z/n}\right\}
$$

$$
= \sum_{n=1}^{\infty} \left[\left(\frac{-az}{n(a + n)}\right) + \sum_{m=2}^{\infty} \frac{(-1)^{m-1}}{m} \frac{z^m}{(a + n)^m}\right] + \sum_{m=1}^{\infty} \frac{(-1)^{m-1}}{m} \frac{z^m}{a^m}.
$$

If $|z| < a$, the double series is absolutely convergent since

$$
\sum_{n=1}^{\infty} \left[\frac{a|z|}{n(a + n)} - \log\left(1 - \frac{|z|}{a + n}\right) + \frac{|z|}{a + n}\right]
$$

converges.

Consequently

$$
\log \frac{e^{-\gamma z} \Gamma(a)}{\Gamma(z + a)} = \frac{z}{a} - \sum_{n=1}^{\infty} \frac{az}{n(a + n)} + \sum_{m=2}^{\infty} \frac{(-1)^{m-1}}{m} z^m \zeta(m, a).
$$

Now consider $-\dfrac{1}{2\pi i} \displaystyle\int_C \dfrac{\pi z^s}{s \sin \pi s} \zeta(s, a) \, ds$, the contour of integration being similar to that of §12.22 enclosing the points $s = 2, 3, 4, \ldots$ but not the points $1, 0, -1, -2, \ldots$; the residue of the integrand at $s = m$ ($m \ge 2$) is $\dfrac{(-1)^m}{m} z^m \zeta(m, a)$; and since, as $\sigma \to \infty$ (where $s = \sigma + it$), $\zeta(s, a) = O(1)$, the integral converges if $|z| < 1$.

Consequently

$$
\log \frac{e^{-\gamma z} \Gamma(a)}{\Gamma(z + a)} = \frac{z}{a} - \sum_{n=1}^{\infty} \frac{az}{n(a + n)} - \frac{1}{2\pi i} \int_C \frac{\pi z^s}{s \sin \pi s} \zeta(s, a) \, ds.
$$

Hence

$$
\log \frac{\Gamma(a)}{\Gamma(z + a)} - z \frac{\Gamma'(a)}{\Gamma(a)} - \frac{1}{2\pi i} \int_C \frac{\pi z^s}{s \sin \pi s} \zeta(s, a) \, ds.
$$

---

[6] It can be proved that $\tau(\sigma)$ may be taken to be $\frac{1}{2}(1 - \sigma)$ when $0 \le \sigma \le 1$. See Landau [405, §237].

Now let D be a semicircle of (large) radius N with centre at $s = \frac{3}{2}$, the semicircle lying on the right of the line $\sigma = \frac{3}{2}$. On this semicircle $\zeta(s, a) = O(1)$, $|z^s| = |z|^\sigma e^{-t \arg s}$, and so the integrand is $O\{|z|^\sigma e^{-\pi|t| - t \arg z}\}$. The constants implied in the symbol O are independent of s and z throughout. Hence if $|z| < 1$ and $-\pi + \delta \le \arg z \le \pi - \delta$, where δ is positive, the integrand is $O\left(|z|^\sigma e^{-\delta|t|}\right)$, and hence

$$\int_D \frac{\pi z^s}{s \sin \pi s} \zeta(s, a) \, ds \to 0$$

as $N \to \infty$. It follows at once that, if $|\arg z| \le \pi - \delta$ and $|z| < 1$,

$$\log \frac{\Gamma(a)}{\Gamma(z + a)} = -z \frac{\Gamma'(a)}{\Gamma(a)} + \frac{1}{2\pi i} \int_{\frac{3}{2} - i\infty}^{\frac{3}{2} + i\infty} \frac{\pi z^s}{s \sin \pi s} \zeta(s, a) \, ds.$$

But this integral defines an analytic function of z for all values of $|z|$ if $|\arg z| \le \pi - \delta$. Hence, by §5.5 the above equation, proved when $|z| < 1$, persists for all values of $|z|$ when $|\arg z| \le \pi - \delta$.

Now consider $\int_{-n-\frac{1}{2} \pm iR}^{\frac{3}{2} \pm iR} \frac{\pi z^s}{s \sin \pi s} \zeta(s, a) \, ds$, where n is a fixed integer and R is going to tend to infinity. By §13.51, the integrand is $O\left(z^\sigma e^{-sR} R^{\tau(\sigma)}\right)$, where $-n - \frac{1}{2} \le \sigma \le \frac{3}{2}$; and hence if the upper signs be taken, or if the lower signs be taken, the integral tends to zero as $R \to \infty$. Therefore, by Cauchy's theorem,

$$\log \frac{\Gamma(a)}{\Gamma(z + a)} = -z \frac{\Gamma'(a)}{\Gamma(a)} + \frac{1}{2\pi i} \int_{-n-\frac{1}{2} - i\infty}^{-n-\frac{1}{2} + i\infty} \frac{\pi z^s}{s \sin \pi s} \zeta(s, a) \, ds + \sum_{m=-1}^{n} R_m,$$

where R_m is the residue of the integrand at $s = -m$.

Now, on the new path of integration

$$\left| \frac{\pi z^s}{s \sin \pi s} \zeta(s, a) \right| < K z^{-n-\frac{1}{2}} e^{-\delta|t| \tau(-n-\frac{1}{2})} |t|,$$

where K is independent of z and t, and $\tau(\sigma)$ is the function defined in §13.51. Consequently, since $\int_{-\infty}^{\infty} e^{-\delta|t|} |t|^{\tau(-n-\frac{1}{2})} \, dt$ converges, we have

$$\log \frac{\Gamma(a)}{\Gamma(z + a)} = -z \frac{\Gamma'(a)}{\Gamma(a)} + \sum_{m=-1}^{n} R_m + O(z^{-n-\frac{1}{2}}),$$

when $|z|$ is large.

Now, when m is a positive integer, $R_m = \dfrac{(-1)^m z^{-m} \zeta(-m, a)}{-m}$, and so by §13.14, $R_m = \dfrac{(-1)^m z^{-m} \phi'_{m+2}(a)}{m(m+1)(m+2)}$, where $\phi'_m(a)$ denotes the derivative of Bernoulli's polynomial. Also R_0 is the residue at $s = 0$ of

$$\frac{1}{s} \left(1 + \tfrac{1}{6}\pi^2 s^2 + \cdots\right) \left(1 + s \log z + \cdots\right) \left(\tfrac{1}{2} - a + s \, \zeta'(0, a) + \cdots\right),$$

and so

$$R_0 = \left(\tfrac{1}{2} - a\right) \log z + \zeta'(0, a)$$
$$= \left(\tfrac{1}{2} - a\right) \log z + \log \Gamma(a) - \tfrac{1}{2} \log(2\pi),$$

by §13.21.

And, using §13.21 and writing $s = S + 1$, R_{-1} is the residue at $S = 0$ of

$$-\frac{1}{S}(1 - S + S^2 - \cdots)\left(1 + \frac{\pi^2 S^2}{6} + \cdots\right) z(1 + S \log z + \cdots)\left(\frac{1}{S} - \frac{\Gamma'(a)}{\Gamma(a)} + \cdots\right).$$

Hence $R_{-1} = -z \log z + z\dfrac{\Gamma'(a)}{\Gamma(a)} + z$. Consequently, finally, if $|\arg z| \leq \pi - \delta$ and $|z|$ is large,

$$\log \Gamma(z + a) = \left(z + a - \tfrac{1}{2}\right) \log z - z + \tfrac{1}{2} \log(2\pi)$$
$$+ \sum_{m=1}^{n} \frac{(-1)^{m-1} \phi'_{m+2}(a)}{m(m+1)(m+2)z^m} + O(z^{-n-\frac{1}{2}}).$$

In the special case when $a = 1$, this reduces to the formula found previously in §12.33 for a more restricted range of values of arg z.

The asymptotic expansion just obtained is valid when a is not restricted by the inequality $0 < a \leq 1$; but the investigation of it involves the rather more elaborate methods which are necessary for obtaining inequalities satisfied by $\zeta(s, a)$ when a does not satisfy the inequality $0 < a \leq 1$. But if, in the formula just obtained, we write $a = 1$ and then put $z + a$ for z, it is easily seen that, when $|\arg(z + a)| \leq \pi - \delta$, we have

$$\log \Gamma(z + a + 1) = \left(z + a + \tfrac{1}{2}\right) \log(z + a) - z - a + \tfrac{1}{2} \log(2\pi) + o(1);$$

subtracting $\log(z + a)$ from each side, we easily see that when both

$$|\arg(z + a)| \leq \pi - \delta \quad \text{and} \quad |\arg z| \leq \pi - \delta,$$

we have the asymptotic formula

$$\log \Gamma(z + a) = \left(z + a - \tfrac{1}{2}\right) \log z - z + \tfrac{1}{2} \log(2\pi) + o(1),$$

where the expression which is $o(1)$ tends to zero as $|z| \to \infty$.

13.7 Miscellaneous examples

Example 13.1 (Jensen [359])   Shew that

$$(2^s - 1)\zeta(s) = \frac{2^{s-1}s}{s-1} + 2\int_0^\infty \left(\tfrac{1}{4} + y^2\right)^{-s/2} \sin(s \arctan 2y) \frac{dy}{e^{2\pi y} - 1}.$$

Example 13.2 (Jensen)   Shew that

$$\zeta(s) = \frac{2^{s-1}}{s-1} - 2^s \int_0^\infty (1 + y^2)^{-s/2} \sin(s \arctan y) \frac{dy}{e^{\pi y} + 1}.$$

Example 13.3 (Barnes)   Discuss the asymptotic expansion of $\log G(z + a)$, (Example 12.48) by aid of the generalised zeta-function.

Example 13.4 (Dirichlet [176])   Shew that, if $\sigma > 1$,

$$\log \zeta(s) = \sum_{p} \sum_{m=1}^{\infty} \frac{1}{m p^{ms}},$$

the summation extending over the prime numbers $p = 2, 3, 5, \ldots$.

Example 13.5   Shew that, if $\sigma > 1$,

$$-\frac{\zeta'(s)}{\zeta(s)} = \sum_{n-1}^{\infty} \frac{\Lambda(n)}{n^s},$$

where $\Lambda(n) = 0$ when n is not a power of a prime, and $\Lambda(n) = \log p$ when n is a power of a prime p.

Example 13.6 (Lerch [429])   Prove that

$$\int_0^{\infty} \frac{e^{-x^2}\, dx}{\left(1 + \frac{w^2}{4x^2}\right)^{s/2}} = \frac{\pi^{\frac{1}{2}}}{\Gamma(\frac{s}{2})} \int_0^{\infty} e^{-x^2 - wx} x^{s-1}\, dx.$$

Example 13.7 (Appell [28])   If

$$\phi(s, x) = \sum_{n=1}^{\infty} \frac{x^n}{n^s},$$

where $|x| < 1$, and $\operatorname{Re} s > 0$, shew that

$$\phi(s, x) = \frac{1}{\Gamma(s)} \int_0^{\infty} \frac{x z^{s-1}\, dz}{e^s - x}$$

and, if $s < 1$,

$$\lim_{x \to 1} (1 - x)^{1-s} \phi(s, x) = \Gamma(1 - s).$$

Example 13.8 (Lerch [426])   If x, a, and s be real, and $0 < a < 1$, and $s > 1$, and if

$$\phi(x, a, s) = \sum_{n=0}^{\infty} \frac{e^{2n\pi ix}}{(a + n)^s},$$

shew that

$$\phi(x, a, s) = \frac{1}{\Gamma(s)} \int_0^{\infty} \frac{e^{-as} z^{s-1}\, dz}{1 - e^{2\pi ix - z}}$$

and

$$\phi(x, a, 1 - s) =$$
$$\frac{\Gamma(s)}{(2\pi)^s} \times \left\{ e^{\pi i(s/2 - 2ax)} \phi(-a, x, s) + e^{\pi i \{-s/2 + 2a(1-x)\}} \phi(a, 1 - x, s) \right\}.$$

Example 13.9 (Hardy)   By evaluating the residues at the poles on the left of the straight line taken as contour, shew that, if $k > 0$, and $|\arg y| < \frac{\pi}{2}$,

$$e^{-y} = \frac{1}{2\pi i} \int_{k-i\infty}^{k+i\infty} \Gamma(u) y^{-u}\, du,$$

and deduce that, if $k > \frac{1}{2}$,

$$\frac{1}{2\pi i} \int_{k-i\infty}^{k+i\infty} \frac{\Gamma(u)}{(\pi x)^u} \zeta(2u)\, du = \vartheta(x),$$

and thence that, if a is an acute angle,

$$\int_0^\infty \frac{\cosh \frac{1}{2} at}{t^2 + \frac{1}{4}} \xi(t)\, dt = \pi \cos(a/4) - \tfrac{\pi}{2} e^{ia/4} \left(1 + 2\vartheta(e^{ia})\right).$$

Example 13.10 (Hardy)   By differentiating $2n$ times under the integral sign in the last result (Example 13.9) and then making $a \to \pi/2$, deduce from Example 6.17 of Chapter 6 that

$$\int_0^\infty \frac{\cosh \frac{1}{4}\pi t}{t^2 + \frac{1}{4}} t^{2n} \xi(t)\, dt = \frac{(-1)^n \pi}{2^{2n}} \cos \frac{\pi}{8}.$$

By taking n large, deduce that there is no number t_0 such that $\xi(t)$ is of fixed sign when $t > t_0$, and thence that $\zeta(s)$ has an infinity of zeros on the line $\sigma = \frac{1}{2}$.

Note   Hardy and Littlewood [281] have shewn that the number of zeros on the line $\sigma = \frac{1}{2}$ for which $0 < t < T$ is at least $O(T)$ as $T \to \infty$; if the Riemann hypothesis is true, the number is

$$\frac{1}{2\pi} T \log T - \frac{1 + \log 2\pi}{2\pi} T + O(\log T);$$

see Landau [405, p. 370].

14

The Hypergeometric Function

14.1 The hypergeometric series

We have already (§2.38) considered the *hypergeometric series*[1]

$$1 + \frac{a \cdot b}{1 \cdot c}z + \frac{a(a+1) \cdot b(b+1)}{1 \cdot 2 \cdot c(c+1)}z^2 + \frac{a(a+1)(a+2) \cdot b(b+1)(b+2)}{1 \cdot 2 \cdot 3 \cdot c(c+1)(c+2)}z^3 + \cdots$$

from the point of view of its convergence. It follows from §2.38 and §5.3 that the series defines a function which is analytic when $|z| < 1$.

It will appear later (§14.53) that this function has a branch point at $z = 1$ and that if a cut[2] (i.e. an impassable barrier) is made from $+1$ to $+\infty$ along the real axis, the function is analytic and one-valued throughout the cut plane. The function will be denoted by $F(a, b; c; z)$.

Many important functions employed in Analysis can be expressed by means of hypergeometric functions. Thus[3]

$$(1 + z)^n = F(-n, \beta; \beta; -z),$$
$$\log(1 + z) = zF(1, 1; 2; -z),$$
$$e^z = \lim_{\beta \to \infty} F(1, \beta; 1; z/\beta).$$

Example 14.1.1 Shew that

$$\frac{d}{dz} F(a, b; c; z) = \frac{ab}{c}F(a + 1, b + 1; c + 1; z).$$

14.11 *The value of* $F(a, b; c; 1)$ *when* $\mathrm{Re}(c - a - b) > 0$

This analysis is due to Gauss. A method more easy to remember but more difficult to justify is given in Example 14.6.2.

[1] The name was given by Wallis in 1655 to the series whose nth term is

$$a(a + b)(a + 2b) \cdots (a + (n - 1)b).$$

Euler used the term hypergeometric in this sense, the modern use of the term being apparently due to Kummer [388, 389].

[2] The plane of the variable z is said to be *cut* along a curve when it is convenient to consider only such variations in z which do not involve a passage across the curve in question; so that the cut may be regarded as an impassable barrier.

[3] It will be a good exercise for the reader to construct a rigorous proof of the third of these results.

293

The reader will easily verify, by considering the coefficients of x^n in the various series, that if $0 \le x < 1$, then

$$c\{c - 1 - (2c - a - b - 1)x\}F(a, b; c; x) + (c - a)(c - b)xF(a, b; c + 1; x)$$
$$= c(c - 1)(1 - x)F(a, b; c - 1; x)$$
$$= c(c - 1)\left(1 + \sum_{n=1}^{\infty}(u_n - u_{n-1})x^n\right),$$

where u_n is the coefficient of x^n in $F(a, b; c - 1; x)$.

Now make $x \to 1$. By §3.71, the right-hand side tends to zero if $1 + \sum_{n=1}^{\infty}(u_n - u_{n-1})$ converges to zero, i.e. if $u_n \to 0$, which is the case when $\operatorname{Re}(c - a - b) > 0$. Also, by §2.38 and §3.71, the left-hand side tends to

$$c(a + b - c)F(a, b; c; 1) + (c - a)(c - b)F(a, b; c + 1; 1)$$

under the same condition; and therefore

$$F(a, b; c; 1) = \frac{(c - a)(c - b)}{c(c - a - b)}F(a, b; c + 1; 1).$$

Repeating this process, we see that

$$F(a, b; c; 1) = \left\{\prod_{n=0}^{m-1}\frac{(c - a + n)(c - b + n)}{(c + n)(c - a - b + n)}\right\}F(a, b; c + m; 1)$$
$$= \left\{\lim_{m \to \infty}\prod_{n=0}^{m-1}\frac{(c - a + n)(c - b + n)}{(c + n)(c - a - b + n)}\right\}\lim_{m \to \infty}F(a, b; c + m; 1),$$

if these two limits exist.

But (§12.13) the former limit is $\dfrac{\Gamma(c)\Gamma(c - a - b)}{\Gamma(c - a)\Gamma(c - b)}$, if c is not a negative integer; and, if $u_n(a, b, c)$ be the coefficient of x^n in $F(a, b; c; x)$, and $m > |c|$, we have

$$|F(a, b; c + m) - 1| \le \sum_{n=1}^{\infty}|u_n(a, b, c + m)|$$
$$\le \sum_{n=1}^{\infty}u_n(|a|, |b|, m - |c|)$$
$$< \frac{|ab|}{m - |c|}\sum_{n=0}^{\infty}u_n(|a| + 1, |b| + 1, m + 1 - |c|).$$

Now the last series converges, when $m > |c| + |a| + |b| - 1$, and is a positive decreasing function of m; therefore, since $(m - |c|)^{-1} \to 0$, we have

$$\lim_{m \to \infty}F(a, b; c + m; 1) = 1;$$

and therefore, finally,

$$F(a, b; c; 1) = \frac{\Gamma(c)\Gamma(c - a - b)}{\Gamma(c - a)\Gamma(c - b)}.$$

14.2 The differential equation satisfied by $F(a, b; c; z)$

The reader will verify without difficulty, by the methods of §10.3 that the hypergeometric series is an integral valid near $z = 0$ of the *hypergeometric equation* (this equation was given by Gauss).

$$z(1 - z)\frac{d^2 u}{dz^2} + \{c - (a + b + 1)z\}\frac{du}{dz} - abu = 0;$$

from §10.3, it is apparent that every point is an 'ordinary point' of this equation, with the exception of 0, 1, ∞, and that these are 'regular points'.

Example 14.2.1 Shew that an integral of the equation

$$z\left(z\frac{d}{dz} + a\right)\left(z\frac{d}{dz} + b\right)u - \left(z\frac{d}{dz} - \alpha\right)\left(z\frac{d}{dz} - \beta\right)u = 0$$

is $z^a F(a + \alpha, b + \alpha; \alpha - \beta + 1; z)$.

14.3 Solutions of Riemann's P-equation by hypergeometric functions

In §10.72 it was observed that Riemann's differential equation[4]

$$\frac{d^2 u}{dz^2} + \left\{\frac{1 - \alpha - \alpha'}{z - a} + \frac{1 - \beta - \beta'}{z - b} + \frac{1 - \gamma - \gamma'}{z - c}\right\}\frac{du}{dz}$$
$$+ \left\{\frac{\alpha\alpha'(a - b)(a - c)}{z - a} + \frac{\beta\beta'(b - c)(b - a)}{(z - c)} + \frac{\gamma\gamma'(c - a)(c - b)}{z - c}\right\}$$
$$\times \frac{u}{(z - a)(z - b)(z - c)} = 0,$$

by a suitable change of variables, could be reduced to a hypergeometric equation; and, carrying out the change, we see that a solution of Riemann's equation is

$$\left(\frac{z - a}{z - b}\right)^\alpha \left(\frac{z - c}{z - b}\right)^\gamma F\left\{\alpha + \beta + \gamma, \alpha + \beta' + \gamma; 1 + \alpha - \alpha'; \frac{(z - a)(c - b)}{(z - b)(c - a)}\right\},$$

provided that $\alpha - \alpha'$ is not a negative integer; for simplicity, we shall, throughout this section, suppose that no one of the exponent differences $\alpha - \alpha'$, $\beta - \beta'$, $\gamma - \gamma'$ is zero or an integer, as (§10.32) in this exceptional case the general solution of the differential equation may involve logarithmic terms; the formulae in the exceptional case will be found in Lindelöf's memoir [435] to which the reader is referred. See also Klein's lithographed lectures [372].

Now if α be interchanged with α', or γ with γ', in this expression, it must still satisfy Riemann's equation, since the latter is unaffected by this change.

---

[4] The constants are subject to the condition $\alpha + \alpha' + \beta + \beta' + \gamma + \gamma' = 1$.

We thus obtain altogether four expressions, namely,

$$u_1 = \left(\frac{z-a}{z-b}\right)^{\alpha}\left(\frac{z-c}{z-b}\right)^{\gamma} F\left\{\alpha+\beta+\gamma, \alpha+\beta'+\gamma; 1+\alpha-\alpha'; \frac{(c-b)(z-a)}{(c-a)(z-b)}\right\},$$

$$u_2 = \left(\frac{z-a}{z-b}\right)^{\alpha'}\left(\frac{z-c}{z-b}\right)^{\gamma} F\left\{\alpha'+\beta+\gamma, \alpha'+\beta'+\gamma; 1+\alpha'-\alpha; \frac{(c-b)(z-a)}{(c-a)(z-b)}\right\},$$

$$u_3 = \left(\frac{z-a}{z-b}\right)^{\alpha}\left(\frac{z-c}{z-b}\right)^{\gamma'} F\left\{\alpha+\beta+\gamma', \alpha+\beta'+\gamma'; 1+\alpha-\alpha'; \frac{(c-b)(z-a)}{(c-a)(z-b)}\right\},$$

$$u_4 = \left(\frac{z-a}{z-b}\right)^{\alpha'}\left(\frac{z-c}{z-b}\right)^{\gamma'} F\left\{\alpha'+\beta+\gamma', \alpha'+\beta'+\gamma'; 1+\alpha'-\alpha; \frac{(c-b)(z-a)}{(c-a)(z-b)}\right\},$$

which are all solutions of the differential equation.

Moreover, the differential equation is unaltered if the triads (α, α', a), (β, β', b), (γ, γ', c) are interchanged in any manner. If therefore we make such changes in the above solutions, they will still be solutions of the differential equation.

There are five such changes possible, for we may write

$$\{b, c, a\}, \quad \{c, a, b\}, \quad \{a, c, b\}, \quad \{c, b, a\}, \quad \{b, a, c\}$$

in turn in place of $\{a, b, c\}$, with corresponding changes of α, α', β, β', γ, γ'.

We thus obtain $4 \times 5 = 20$ new expressions, which with the original four make altogether twenty-four series.

The twenty new solutions may be written down as follows:

$$u_5 = \left(\frac{z-b}{z-c}\right)^{\beta}\left(\frac{z-a}{z-c}\right)^{\alpha} F\left\{\beta+\gamma+\alpha, \beta+\gamma'+\alpha; 1+\beta-\beta'; \frac{(a-c)(z-b)}{(a-b)(z-c)}\right\},$$

$$u_6 = \left(\frac{z-b}{z-c}\right)^{\beta'}\left(\frac{z-a}{z-c}\right)^{\alpha} F\left\{\beta'+\gamma+\alpha, \beta'+\gamma'+\alpha; 1+\beta'-\beta; \frac{(a-c)(z-b)}{(a-b)(z-c)}\right\},$$

$$u_7 = \left(\frac{z-b}{z-c}\right)^{\beta}\left(\frac{z-a}{z-c}\right)^{\alpha'} F\left\{\beta+\gamma+\alpha', \beta+\gamma'+\alpha'; 1+\beta-\beta'; \frac{(a-c)(z-b)}{(a-b)(z-c)}\right\},$$

$$u_8 = \left(\frac{z-b}{z-c}\right)^{\beta'}\left(\frac{z-a}{z-c}\right)^{\alpha'} F\left\{\beta'+\gamma+\alpha', \beta'+\gamma'+\alpha'; 1+\beta'-\beta; \frac{(a-c)(z-b)}{(a-b)(z-c)}\right\},$$

$$u_9 = \left(\frac{z-c}{z-a}\right)^{\gamma}\left(\frac{z-b}{z-a}\right)^{\beta} F\left\{\gamma+\alpha+\beta, \gamma+\alpha'+\beta; 1+\gamma-\gamma'; \frac{(b-a)(z-c)}{(b-c)(z-a)}\right\},$$

$$u_{10} = \left(\frac{z-c}{z-a}\right)^{\gamma'}\left(\frac{z-b}{z-a}\right)^{\beta} F\left\{\gamma'+\alpha+\beta, \gamma'+\alpha'+\beta; 1+\gamma'-\gamma; \frac{(b-a)(z-c)}{(b-c)(z-a)}\right\},$$

$$u_{11} = \left(\frac{z-c}{z-a}\right)^{\gamma} \left(\frac{z-b}{z-a}\right)^{\beta'} F\left\{\gamma + \alpha + \beta', \gamma + \alpha' + \beta'; 1 + \gamma - \gamma'; \frac{(b-a)(z-c)}{(b-c)(z-a)}\right\},$$

$$u_{12} = \left(\frac{z-c}{z-a}\right)^{\gamma'} \left(\frac{z-b}{z-a}\right)^{\beta'} F\left\{\gamma' + \alpha + \beta', \gamma' + \alpha' + \beta'; 1 + \gamma' - \gamma; \frac{(b-a)(z-c)}{(b-c)(z-a)}\right\},$$

$$u_{13} = \left(\frac{z-a}{z-c}\right)^{\alpha} \left(\frac{z-b}{z-c}\right)^{\beta} F\left\{\alpha + \gamma + \beta, \alpha + \gamma' + \beta; 1 + \alpha - \alpha'; \frac{(b-c)(z-a)}{(b-a)(z-c)}\right\},$$

$$u_{14} = \left(\frac{z-a}{z-c}\right)^{\alpha'} \left(\frac{z-b}{z-c}\right)^{\beta} F\left\{\alpha' + \gamma + \beta, \alpha' + \gamma' + \beta; 1 + \alpha' - \alpha; \frac{(b-c)(z-a)}{(b-a)(z-c)}\right\},$$

$$u_{15} = \left(\frac{z-a}{z-c}\right)^{\alpha} \left(\frac{z-b}{z-c}\right)^{\beta'} F\left\{\alpha + \gamma + \beta', \alpha + \gamma' + \beta'; 1 + \alpha - \alpha'; \frac{(b-c)(z-a)}{(b-a)(z-c)}\right\},$$

$$u_{16} = \left(\frac{z-a}{z-c}\right)^{\alpha'} \left(\frac{z-b}{z-c}\right)^{\beta'} F\left\{\alpha' + \gamma + \beta', \alpha' + \gamma' + \beta'; 1 + \alpha' - \alpha; \frac{(b-c)(z-a)}{(b-a)(z-c)}\right\},$$

$$u_{17} = \left(\frac{z-c}{z-b}\right)^{\gamma} \left(\frac{z-a}{z-b}\right)^{\alpha} F\left\{\gamma + \beta + \alpha, \gamma + \beta' + \alpha; 1 + \gamma - \gamma'; \frac{(a-b)(z-c)}{(a-c)(z-b)}\right\},$$

$$u_{18} = \left(\frac{z-c}{z-b}\right)^{\gamma'} \left(\frac{z-a}{z-b}\right)^{\alpha} F\left\{\gamma' + \beta + \alpha, \gamma' + \beta' + \alpha; 1 + \gamma' - \gamma; \frac{(a-b)(z-c)}{(a-c)(z-b)}\right\},$$

$$u_{19} = \left(\frac{z-c}{z-b}\right)^{\gamma} \left(\frac{z-a}{z-b}\right)^{\alpha'} F\left\{\gamma + \beta + \alpha', \gamma + \beta' + \alpha'; 1 + \gamma - \gamma'; \frac{(a-b)(z-c)}{(a-c)(z-b)}\right\},$$

$$u_{20} = \left(\frac{z-c}{z-b}\right)^{\gamma'} \left(\frac{z-a}{z-b}\right)^{\alpha'} F\left\{\gamma' + \beta + \alpha', \gamma' + \beta' + \alpha'; 1 + \gamma' - \gamma; \frac{(a-b)(z-c)}{(a-c)(z-b)}\right\},$$

$$u_{21} = \left(\frac{z-b}{z-a}\right)^{\beta} \left(\frac{z-c}{z-a}\right)^{\gamma} F\left\{\beta + \alpha + \gamma, \beta + \alpha' + \gamma; 1 + \beta - \beta'; \frac{(c-a)(z-b)}{(c-b)(z-a)}\right\},$$

$$u_{22} = \left(\frac{z-b}{z-a}\right)^{\beta'} \left(\frac{z-c}{z-a}\right)^{\gamma} F\left\{\beta' + \alpha + \gamma, \beta' + \alpha' + \gamma; 1 + \beta' - \beta; \frac{(c-a)(z-b)}{(c-b)(z-a)}\right\},$$

$$u_{23} = \left(\frac{z-b}{z-a}\right)^{\beta} \left(\frac{z-c}{z-a}\right)^{\gamma'} F\left\{\beta + \alpha + \gamma', \beta + \alpha' + \gamma'; 1 + \beta - \beta'; \frac{(c-a)(z-b)}{(c-b)(z-a)}\right\},$$

$$u_{24} = \left(\frac{z-b}{z-a}\right)^{\beta'} \left(\frac{z-c}{z-a}\right)^{\gamma'} F\left\{\beta' + \alpha + \gamma', \beta' + \alpha' + \gamma'; 1 + \beta' - \beta; \frac{(c-a)(z-b)}{(c-b)(z-a)}\right\}.$$

By writing $0, 1-C, A, B, 0, C-A-B, x$ for $\alpha, \alpha', \beta, \beta', \gamma, \gamma', \dfrac{(z-a)(c-b)}{(z-b)(c-a)}$ respectively, we obtain 24 solutions of the hypergeometric equation satisfied by $F(A, B; C; x)$. The existence of these 24 solutions was first shewn by Kummer [388, 389]. They are obtained in a different manner in Forsyth [221, Chap. VI].

14.4 Relations between particular solutions of the hypergeometric equation

It has just been shewn that 24 expressions involving hypergeometric series are solutions of the hypergeometric equation; and, from the general theory of linear differential equations of the second order, it follows that, if any three have a common domain of existence, there must be a linear relation with constant coefficients connecting those three solutions.

If we simplify $u_1, u_2, u_3, u_4; u_{17}, u_{18}, u_{21}, u_{22}$ in the manner indicated at the end of §14.3, we obtain the following solutions of the hypergeometric equation with elements A, B, C, x:

$$y_1 = F(A, B; C; x),$$
$$y_2 = (-x)^{1-C} F(A - C + 1, B - C + 1; 2 - C; x),$$
$$y_3 = (1 - x)^{C-A-B} F(C - B, C - A; C; x),$$
$$y_4 = (-x)^{1-C} (1 - x)^{C-A-B} F(1 - B, 1 - A; 2 - C; x),$$
$$y_{17} = F(A, B; A + B - C + 1; 1 - x),$$
$$y_{18} = (1 - x)^{C-A-B} F(C - B, C - A; C - A - B + 1; 1 - x),$$
$$y_{21} = (-x)^{-B} F(A, A - C + 1; A - B + 1; x^{-1}),$$
$$y_{22} = (-x)^{-A} F(B, B - C + 1; B - A + 1; x^{-1}).$$

If $|\arg(1 - x)| < \pi$, it is easy to see from §2.53 that, when $|x| < 1$, the relations connecting y_1, y_2, y_3, y_4 must be $y_1 = y_3, y_2 = y_4$, by considering the form of the expansions near $x = 0$ of the series involved. In this manner we can group the functions $u_1, \ldots, u_{24}$ into six sets of four[5], viz.

$$u_1, u_3, u_{13}, u_{15}; \quad u_2, u_4, u_{14}, u_{16}; \quad u_5, u_7, u_{21}, u_{23};$$
$$u_6, u_8, u_{22}, u_{24}; \quad u_9, u_{11}, u_{17}, u_{19}; \quad u_{10}, u_{12}, u_{18}, u_{20},$$

such that members of the same set are constant multiples of one another throughout a suitably chosen domain.

In particular, we observe that u_1, u_3, u_{13}, u_{15} are constant multiples of a function which (by §5.4, §2.53) can be expanded in the form

$$(z - a)^a \left\{ 1 + \sum_{n=1}^{\infty} e_n (z - a)^n \right\}$$

when $|z - a|$ is sufficiently small; when $\arg(z - a)$ is so restricted that $(z - a)^a$ is one-valued,

---

[5] The special formula

$$F(A, 1; C; x) = \frac{1}{1 - x} F\left(C - A, 1; C; \frac{x}{x - 1}\right),$$

which is derivable from the relation connecting u_1 with u_{13}, was discovered in 1730 by Stirling [607, Prop. VII].

this solution of Riemann's equation is usually written $P^{(\alpha)}$. And $P^{(\alpha')}$; $P^{(\beta)}$, $P^{(\beta')}$; $P^{(\gamma)}$, $P^{(\gamma')}$ are defined in a similar manner when $|z - a|$, $|z - b|$, $|z - c|$ respectively are sufficiently small.

To obtain the relations which connect three members of separate sets of solutions is much more difficult. The relations have been obtained by elaborate transformations of the double circuit integrals which will be obtained later in §14.61; but a more simple and singularly elegant method has recently been discovered by Barnes; of his investigation we shall give a brief account.

14.5 Barnes' contour integrals for the hypergeometric function

This appears in [48]. References to previous work on similar topics by Pincherle, Mellin and Barnes are there given.

Consider

$$\frac{1}{2\pi i} \int_{-i\infty}^{i\infty} \frac{\Gamma(a + s)\Gamma(b + s)\Gamma(-s)}{\Gamma(c + s)}(-z)^s \, ds,$$

where $|\arg(-z)| < \pi$, and the path of integration is curved (if necessary) to ensure that the poles of $\Gamma(a + s)\Gamma(b + s)$, viz. $s = -a - n, -b - n$ ($n = 0, 1, 2, \ldots$) lie on the left of the path and the poles of $\Gamma(-s)$, viz. $s = 0, 1, 2, \ldots$, lie on the right of the path. It is assumed that a and b are such that the contour can be drawn, i.e. that a and b are not negative integers (in which case the hypergeometric series is merely a polynomial).

From §13.6 it follows that the integrand is

$$O\left(|s|^{a+b-c-1} \exp\{-\arg(-z)\operatorname{Im}(s) - \pi|\operatorname{Im}(s)|\}\right)$$

as $s \to \infty$ on the contour, and hence it is easily seen (§5.32) that the integrand is an analytic function of z throughout the domain defined by the inequality $|\arg z| \le \pi - \delta$, where δ is any positive number.

Now, taking note of the relation $\Gamma(-s)\Gamma(1 + s) = -\pi \operatorname{cosec} \pi s$, consider

$$\frac{1}{2\pi i} \int_C \frac{\Gamma(a + s)\Gamma(b + s)}{\Gamma(c + s)\Gamma(1 + s)} \frac{\pi(-z)^s}{\sin \pi s} \, ds,$$

where C is the semicircle of radius $N + \frac{1}{2}$ on the right of the imaginary axis with centre at the origin, and N is an integer. Now, by §13.6, we have

$$\frac{\Gamma(a + s)\Gamma(b + s)}{\Gamma(c + s)\Gamma(1 + s)} \frac{\pi(-z)^s}{\sin s\pi} = O(N^{a+b-c-1}) \cdot \frac{(-z)^s}{\sin \pi s}$$

as $N \to \infty$, the constant implied in the symbol O being independent of $\arg s$ when s is on the semicircle; and, if $s = \left(N + \frac{1}{2}\right)e^{i\theta}$ and $|z| < 1$, we have

$$(-z)^s \operatorname{cosec} \pi s = O\left[\exp\left\{\left(N + \tfrac{1}{2}\right)\cos\theta \log |z| - \left(N + \tfrac{1}{2}\right)\sin\theta \arg(-z)\right.\right.$$
$$\left.\left. - \left(N + \tfrac{1}{2}\right)\pi|\sin\theta|\right\}\right]$$
$$= O\left[\exp\left\{\left(N + \tfrac{1}{2}\right)\cos\theta \log |z| - \left(N + \tfrac{1}{2}\right)\delta|\sin\theta|\right\}\right]$$
$$= \begin{cases} O\left[\exp\left\{2^{-\frac{1}{2}}\left(N + \tfrac{1}{2}\right)\log|z|\right\}\right] & 0 \le |\theta| \le \tfrac{1}{4}\pi, \\ O\left[\exp\left\{-2^{-\frac{1}{2}}\delta\left(N + \tfrac{1}{2}\right)\right\}\right] & \tfrac{1}{4}\pi \le |\theta| \le \tfrac{1}{2}\pi. \end{cases}$$

Hence *if* $\log|z|$ *is negative* (i.e. $|z| < 1$), the integrand tends to zero sufficiently rapidly (for all values of θ under consideration) to ensure that $\int_C \to 0$ as $N \to \infty$.

Now

$$\int_{-i\infty}^{i\infty} - \left\{ \int_{-i\infty}^{-i(N+\frac{1}{2})} + \int_C + \int_{i(N+\frac{1}{2})}^{i\infty} \right\},$$

by Cauchy's theorem, is equal to $-2\pi i$ times the sum of the residues of the integrand at the points $s = 0, 1, \ldots, N$. Make $N \to \infty$, and the last three integrals tend to zero when $|\arg(-z)| \le \pi - \delta$, and $|z| < 1$, and so, in these circumstances,

$$\frac{1}{2\pi i} \int_{-i\infty}^{i\infty} \frac{\Gamma(a+s)\Gamma(b+s)\Gamma(-s)}{\Gamma(c+s)} (-z)^s \, ds = \lim_{N \to \infty} \sum_{n=0}^{N} \frac{\Gamma(a+n)\Gamma(b+n)}{\Gamma(c+n)n!} z^n,$$

the general term in this summation being the residue of the integrand at $s = n$.

Thus, an analytic function (namely the integral under consideration) exists throughout the domain defined by the inequality $|\arg z| < \pi$, and, when $|z| < 1$, this analytic function may be represented by the series

$$\sum_{n=0}^{\infty} \frac{\Gamma(a+n)\Gamma(b+n)}{\Gamma(c+n)n!} z^n.$$

The symbol $F(a, b; c; z)$ will, in future, be used to denote this function divided by $\Gamma(a)\Gamma(b)/\Gamma(c)$.

14.51 The continuation of the hypergeometric series

To obtain a representation of the function $F(a, b; c; z)$ in the form of series convergent when $|z| > 1$, we shall employ the integral obtained in §14.5. If D be the semicircle of radius ρ on the left of the imaginary axis with centre at the origin, it may be shewn[6] by the methods of §14.5 that

$$\frac{1}{2\pi i} \int_D \frac{\Gamma(a+s)\Gamma(b+s)\Gamma(-s)}{\Gamma(c+s)} (-z)^s \, ds \to 0$$

as $\rho \to \infty$, provided that $|\arg(-z)| < \pi$, $|z| > 1$ and $\rho \to \infty$ in such a way that the lower bound of the distance of D from poles of the integrand is a *positive* number (not zero).

Hence it can be proved (as in the corresponding work of §14.5) that, when $|\arg(-z)| < \pi$ and $|z| > 1$,

$$\frac{1}{2\pi i} \int \frac{\Gamma(a+s)\Gamma(b+s)\Gamma(-s)}{\Gamma(c+s)} (-z)^s \, ds$$

$$= \sum_{n=0}^{\infty} \frac{\Gamma(a+n)\Gamma(1-c+a+n)}{\Gamma(1+n)\Gamma(1-b+a+n)} \frac{\sin(c-a-n)\pi}{\cos n\pi \sin(b-a-n)\pi} (-z)^{-a-n}$$

$$+ \sum_{n=0}^{\infty} \frac{\Gamma(b+n)\Gamma(1-c+b+n)}{\Gamma(1+n)\Gamma(1-a+b+n)} \frac{\sin(c-b-n)\pi}{\cos n\pi \sin(a-b-n)\pi} (-z)^{-b-n},$$

---

[6] In considering the asymptotic expansion of the integrand when $|s|$ is large on the contour or on D, it is simplest to transform $\Gamma(a+s)$, $\Gamma(b+s)$, $\Gamma(c+s)$ by the relation of §12.14.

the expressions in these summations being the residues of the integrand at the points $s = -a - n$, $s = -b - n$ respectively.

It then follows at once on simplifying these series that the analytic continuation of the series, by which the hypergeometric function was originally defined, is given by the equation

$$\frac{\Gamma(a)\Gamma(b)}{\Gamma(c)}F(a,b;c;z) = \frac{\Gamma(a)\Gamma(b-a)}{\Gamma(c-a)}(-z)^{-a}F(a,1-c+a;1-b+a;z^{-1})$$
$$+ \frac{\Gamma(b)\Gamma(a-b)}{\Gamma(c-b)}(-z)^{-b}F(b,1-c+b;1-a+b;z^{-1}),$$

where $|\arg(-z)| < \pi$.

It is readily seen that each of the three terms in this equation is a solution of the hypergeometric equation (see §14.4).

This result has to be modified when $a - b$ is an integer or zero, as some of the poles of $\Gamma(a+s)\Gamma(b+s)$ are double poles, and the right-hand side then may involve logarithmic terms, in accordance with §14.3.

Corollary 14.5.1 *Putting $b = c$, we see that, if $|\arg(-z)| < \pi$,*

$$\Gamma(a)(1-z)^{-a} = \frac{1}{2\pi i}\int_{-i\infty}^{i\infty}\Gamma(a+s)\Gamma(-s)(-z)^s\,ds,$$

where $(1-z)^{-a} \to 1$ as $z \to 0$, and so the value of $|\arg(1-z)|$ which is less than π always has to be taken in this equation, in virtue of the cut (see §14.1) from 0 to $+\infty$ caused by the inequality $|\arg(-z)| < \pi$.

14.52 Barnes' lemma

Barnes' lemma states that, *if the path of integration is curved so that the poles of $\Gamma(\gamma - s)\Gamma(\delta - s)$ lie on the right of the path and the poles of $\Gamma(\alpha + s)\Gamma(\beta + s)$ lie on the left[7], then*

$$\frac{1}{2\pi i}\int_{-i\infty}^{i\infty}\Gamma(\alpha+s)\Gamma(\beta+s)\Gamma(\gamma-s)\Gamma(\delta-s)\,ds = \frac{\Gamma(\alpha+\gamma)\Gamma(\alpha+\delta)\Gamma(\beta+\gamma)\Gamma(\beta+\delta)}{\Gamma(\alpha+\beta+\gamma+\delta)}.$$

Write I for the expression on the left.

If C be defined to be the semicircle of radius ρ on the right of the imaginary axis with centre at the origin, and if $\rho \to \infty$ in such a way that the lower bound of the distance of C from the poles of $\Gamma(\gamma - s)\Gamma(\delta - s)$ is *positive* (not zero), it is readily seen that

$$\Gamma(\alpha+s)\Gamma(\beta+s)\Gamma(\gamma-s)\Gamma(\delta-s)$$
$$= \frac{\Gamma(\alpha+s)\Gamma(\beta+s)}{\Gamma(1-\gamma+s)\Gamma(1-\delta+s)}\pi^2\,\operatorname{cosec}\pi(\gamma-s)\,\operatorname{cosec}\pi(\delta-s)$$
$$= O[s^{\alpha+\beta+\gamma+\delta-2}\exp\{-2\pi\,|\mathrm{Im}(s)|\}],$$

as $|s| \to \infty$ on the imaginary axis or on C.

Hence the original integral converges; and $\int_C \to 0$ as $\rho \to \infty$, when $\mathrm{Re}(\alpha + \beta + \gamma + \delta - 1) < 0$. Thus, as in §14.5, the integral involved in I is $-2\pi i$ times

[7] It is supposed that $\alpha, \beta, \gamma, \delta$ are such that no pole of the first set coincides with any pole of the second set.

the sum of the residues of the integrand at the poles of $\Gamma(\gamma - s)\Gamma(\delta - s)$; evaluating these residues we get[8]

$$I = \sum_{n=0}^{\infty} \frac{\Gamma(a + \gamma + n)\Gamma(\beta + \gamma + n)}{\Gamma(n+1)\Gamma(1 + \gamma - \delta + n)} \frac{\pi}{\sin \pi(\delta - \gamma)}$$

$$+ \sum_{n=0}^{\infty} \frac{\Gamma(\alpha + \delta + n)\Gamma(\beta + \delta + n)}{\Gamma(n+1)\Gamma(1 + \delta - \gamma + n)} \frac{\pi}{\sin \pi(\gamma - \delta)}.$$

And so, using the result of §12.14 freely, by §14.11:

$$I = \frac{\pi}{\sin \pi(\gamma - \delta)} \left\{ \frac{\Gamma(\alpha + \delta)\Gamma(\beta + \delta)}{\Gamma(1 - \gamma + \delta)} F(\alpha + \delta, \beta + \delta; 1 - \gamma + \delta; 1) \right.$$

$$\left. - \frac{\Gamma(\alpha + \gamma)\Gamma(\beta + \gamma)}{\Gamma(1 - \delta + \gamma)} F(\alpha + \gamma, \beta + \gamma; 1 - \delta + \gamma; 1) \right\}$$

$$= \frac{\pi\Gamma(1 - \alpha - \beta - \gamma - \delta)}{\sin \pi(\gamma - \delta)} \left\{ \frac{\Gamma(\alpha + \delta)\Gamma(\beta + \delta)}{\Gamma(1 - \alpha - \gamma)\Gamma(1 - \beta - \gamma)} \right.$$

$$\left. - \frac{\Gamma(\alpha + \gamma)\Gamma(\beta + \gamma)}{\Gamma(1 - \alpha - \delta)\Gamma(1 - \beta - \delta)} \right\}$$

$$= \frac{\Gamma(\alpha + \gamma)\Gamma(\beta + \gamma)\Gamma(\alpha + \delta)\Gamma(\beta + \delta)}{\Gamma(\alpha + \beta + \gamma + \delta) \sin \pi(\alpha + \beta + \gamma + \delta) \sin \pi(\gamma - \delta)}$$

$$\{\sin \pi(\alpha + \gamma) \sin \pi(\beta + \gamma) - \sin \pi(\alpha + \delta) \sin \pi(\beta + \delta)\}.$$

But

$$2 \sin \pi(\alpha + \gamma) \sin \pi(\beta + \gamma) - 2 \sin \pi(\alpha + \delta) \sin \pi(\beta + \delta)$$

$$= \cos \pi(\alpha - \beta) - \cos \pi(\alpha + \beta + 2\gamma) - \cos \pi(\alpha - \beta) + \cos \pi(\alpha + \beta + 2\delta)$$

$$= 2 \sin \pi(\gamma - \delta) \sin \pi(\alpha + \beta + \gamma + \delta).$$

Therefore

$$I = \frac{\Gamma(\alpha + \gamma)\Gamma(\beta + \gamma)\Gamma(\alpha + \delta)\Gamma(\beta + \delta)}{\Gamma(\alpha + \beta + \gamma + \delta)}$$

which is the required result; it has, however, only been proved when $\text{Re}(\alpha + \beta + \gamma + \delta - 1) < 0$; but, by the theory of analytic continuation, it is true throughout the domain through which both sides of the equation are analytic functions of, say α, and hence it is true for all values of $\alpha, \beta, \gamma, \delta$ for which none of the poles of $\Gamma(\alpha + s)\Gamma(\beta + s)$, *qua* function of s, coincide with any of the poles of $\Gamma(\gamma - s)\Gamma(\delta - s)$.

Corollary 14.5.2   *Writing $s + k, \alpha - k, \beta - k, \gamma + k, \delta + k$ in place of $s, \alpha, \beta, \gamma, \delta$, we see that the result is still true when the limits of integration are $-k \pm i\infty$, where k is any real constant.*

---

[8] These two series converge (§2.38).

14.53 The connexion between hypergeometric functions of z and of $1 - z$

We have seen that, if $|\arg(-z)| < \pi$,

$$\frac{\Gamma(a)\Gamma(b)}{\Gamma(c)} F(a, b; c; z) = \frac{1}{2\pi i} \int_{-i\infty}^{i\infty} \frac{\Gamma(a + s)\Gamma(b + s)\Gamma(-s)}{\Gamma(c + s)} (-z)^s \, ds$$

$$= \frac{1}{2\pi i} \int_{-i\infty}^{i\infty} \left\{ \frac{1}{2\pi i} \int_{-k-i\infty}^{-k+i\infty} \Gamma(a + t)\Gamma(b + t)\Gamma(s - t)\Gamma(c - a - b - t) \, dt \right\}$$

$$\times \frac{\Gamma(-s)(-z)^s}{\Gamma(c - a)\Gamma(c - b)} \, ds,$$

by Barnes' lemma.

If k be so chosen that the lower bound of the distance between the s contour and the t contour is positive (not zero), it may be shewn that the order of the integrations[9] may be interchanged.

Carrying out the interchange, we see that if $\arg(1 - z)$ be given its principal value,

$$\frac{\Gamma(c - a)\Gamma(c - b)\Gamma(a)\Gamma(b)}{\Gamma(c)} F(a, b; c; z)$$

$$= \frac{1}{2\pi i} \int_{-k-i\infty}^{-k+i\infty} \Gamma(a + t)\Gamma(b + t)\Gamma(c - a - b - t)$$

$$\left\{ \frac{1}{2\pi i} \int_{-i\infty}^{i\infty} \Gamma(s - t)\Gamma(-s)(-z)^s \, ds \right\} dt$$

$$= \frac{1}{2\pi i} \int_{-k-i\infty}^{-k+i\infty} \Gamma(a + t)\Gamma(b + t)\Gamma(c - a - b - t)\Gamma(-t)(1 - z)^t \, dt.$$

Now, when $|\arg(1 - z)| < 2\pi$ and $|1 - z| < 1$, this last integral may be evaluated by the methods of Barnes' lemma (§14.52); and so we deduce that

$$\Gamma(c - a)\Gamma(c - b)\Gamma(a)\Gamma(b)F(a, b; c; z)$$

$$= \Gamma(c)\Gamma(a)\Gamma(b)\Gamma(c - a - b)F(a, b; a + b - c + 1; 1 - z)$$

$$+ \Gamma(c)\Gamma(c - a)\Gamma(c - b)\Gamma(a + b - c)(1 - z)^{c-a-b}$$

$$\times F(c - a, c - b; c - a - b + 1; 1 - z),$$

a result which shews the nature of the singularity of $F(a, b; c; z)$ at $z = 1$.

This result has to be modified if $c - a - b$ is an integer or zero, as then $\Gamma(a + t)\Gamma(b + t)\Gamma(c - a - b - t)\Gamma(-t)$ has double poles, and logarithmic terms may appear. With this exception, the result is valid when $|\arg(-z)| < \pi$, $|\arg 1 - z| < \pi$. Taking $|z| < 1$, we may make z tend to a real value, and we see that the result still holds for real values of z such that $0 < z < 1$.

14.6 Solution of Riemann's equation by a contour integral

We next proceed to establish a result relating to the expression of the hypergeometric function by means of contour integrals.

---

[9] Methods similar to those of §4.51 may be used, or it may be proved without much difficulty that conditions established by Bromwich [102, §177] are satisfied.

Let the dependent variable u in Riemann's P-equation (§10.7) be replaced by a new dependent variable I, defined by the relation

$$u = (z-a)^\alpha (z-b)^\beta (z-c)^\gamma I.$$

The differential equation satisfied by I is easily found to be

$$\frac{d^2 I}{dz^2} + \left\{ \frac{1+\alpha-\alpha'}{z-a} + \frac{1+\beta-\beta'}{z-b} + \frac{1+\gamma-\gamma'}{z-c} \right\} \frac{dI}{dz}$$
$$+ \frac{(\alpha+\beta+\gamma)\{(\alpha+\beta+\gamma+1)z + \sum a(\alpha+\beta'+\gamma'-1)\}}{(z-a)(z-b)(z-c)} I = 0,$$

which can be written in the form

$$Q(z)\frac{d^2 I}{dz^2} - \{(\lambda-2)Q'(z) + R(z)\} \frac{dI}{dz}$$
$$+ \{\tfrac{1}{2}(\lambda-2)(\lambda-1)Q''(z) + (\lambda-1)R'(z)\} I = 0,$$

where

$$\lambda = 1-\alpha-\beta-\gamma = \alpha'+\beta'+\gamma',$$
$$Q(z) = (z-a)(z-b)(z-c),$$
$$R(z) = \sum (\alpha'+\beta+\gamma)(z-b)(z-c).$$

It must be observed that the function I is not analytic at ∞, and consequently the above differential equation in I is not a case of the generalised hypergeometric equation.

We shall now shew that this differential equation can be satisfied by an integral of the form

$$I = \int_C (t-a)^{\alpha'+\beta+\gamma-1}(t-b)^{\alpha+\beta'+\gamma-1}(t-c)^{\alpha+\beta+\gamma'-1}(z-t)^{-\alpha-\beta-\gamma}\, dt,$$

provided that C, the contour of integration, is suitably chosen.

For, if we substitute this value of I in the differential equation, the condition[10] that the equation should be satisfied becomes

$$\int_C (t-a)^{\alpha'+\beta+\gamma-1}(t-b)^{\alpha+\beta'+\gamma-1}(t-c)^{\alpha+\beta+\gamma'-1}(z-t)^{-\alpha-\beta-\gamma-2}K\, dt = 0,$$

where

$$K = (\lambda-2)\left\{ Q(z) + (t-z)Q'(z) + \tfrac{1}{2}(t-z)^2 Q''(z) \right\}$$
$$+ (t-z)\{R(z) + (t-z)R'(z)\}$$
$$= (\lambda-2)\{Q(t) - (t-z)^3\} + (t-z)\{R(t) - (t-z)^2 \sum (\alpha'+\beta+\gamma)\}$$
$$= -(1+\alpha+\beta+\gamma)(t-a)(t-b)(t-c) + \sum (\alpha'+\beta+\gamma)(t-b)(t-c)(t-z).$$

It follows that the condition to be satisfied reduces to $\displaystyle\int_C \frac{dV}{dt}\, dt = 0$, where

$$V = (t-a)^{\alpha'+\beta+\gamma}(t-b)^{\alpha+\beta'+\gamma}(t-c)^{\alpha+\beta+\gamma'}(t-z)^{-(1+\alpha+\beta+\gamma)}.$$

---

[10] The differentiations under the sign of integration are legitimate (§4.2) if the path C does not depend on z and does not pass through the points a, b, c, z; if C be an infinite contour or if C passes through the points a, b, c or z, further conditions are necessary.

The integral I is therefore a solution of the differential equation, when C is such that V resumes its initial value after t has described C. Now

$$V = (t-a)^{\alpha'+\beta+\gamma-1}(t-b)^{\alpha+\beta'+\gamma-1}(t-c)^{\alpha+\beta+\gamma'-1}(z-t)^{-\alpha-\beta-\gamma}U,$$

where $U = (t-a)(t-b)(t-c)(z-t)^{-1}$. Now U is a one-valued function of t; hence, if C be a closed contour, it must be such that the integrand in the integral I resumes its original value after t has described the contour.

Hence finally any integral of the type

$$(z-a)^{\alpha}(z-b)^{\beta}(z-c)^{\gamma}\int_C (t-a)^{\beta+\gamma+\alpha'-1}(t-b)^{\gamma+\alpha+\beta'-1}$$

$$\times (t-c)^{\alpha+\beta+\gamma'-1}(z-t)^{-\alpha-\beta-\gamma}\,dt,$$

where C is either a closed contour in the t-plane such that the integrand resumes its initial value after t has described it, or else is a simple curve such that V has the same value at its termini, is a solution of the differential equation of the general hypergeometric function.

Note The reader is referred to the memoirs of Pochhammer [527], and Hobson [314], for an account of the methods by which integrals of this type are transformed so as to give rise to the relations of §14.51 and §14.53.

Example 14.6.1 To deduce a real definite integral which, in certain circumstances, represents the hypergeometric series.

The hypergeometric series $F(a,b;c;z)$ is, as already shewn, a solution of the differential equation defined by the scheme

$$P\left\{\begin{array}{ccc} 0 & \infty & 1 \\ 0 & a & 0 \\ 1-c & b & c-a-b \end{array}\;z\right\}.$$

If in the integral

$$(z-a)^{\alpha}\left(1-\frac{z}{b}\right)^{\beta}(z-c)^{\gamma}\int_C (t-a)^{\beta+\gamma+\alpha'-1}\left(1-\frac{t}{b}\right)^{\gamma+\alpha+\beta'-1}$$

$$\times (t-c)^{\alpha+\beta+\gamma'-1}(t-z)^{-\alpha-\beta-\gamma}\,dt,$$

which is a constant multiple of that just obtained, we make $b \to \infty$ (without paying attention to the validity of this process), we are led to consider

$$\int_C t^{a-c}(t-1)^{c-b-1}(t-z)^{-a}\,dt.$$

Now the limiting form of V in question is

$$t^{1-c+a}(t-1)^{c-b}(t-z)^{-1-a},$$

and this tends to zero at $t=1$ and $t=\infty$, provided $\mathrm{Re}(c) > \mathrm{Re}(b) > 0$. We accordingly consider $\displaystyle\int_1^{\infty} t^{a-c}(t-1)^{c-b-1}(t-z)^{-a}\,dt$, where z is not[11] positive and greater than 1. In this

---

[11] This ensures that the point $t=1/z$ is not on the path of integration.

integral, write $t = u^{-1}$; the integral becomes

$$\int_0^1 u^{b-1}(1-u)^{c-b-1}(1-uz)^{-a}\,du.$$

We are therefore led to expect that this integral may be a solution of the differential equation for the hypergeometric series.

The reader will easily see that if $\mathrm{Re}(c) > \mathrm{Re}(b) > 0$, and if $\arg u = \arg(1 - u) = 0$, while the branch of $1 - uz$ is specified by the fact that $(1 - uz)^{-a} \to 1$ as $u \to 0$, the integral just found is

$$\frac{\Gamma(b)\Gamma(c-b)}{\Gamma(c)}\,F(a,b;c;z).$$

This can be proved by expanding[12] $(1 - uz)^{-a}$ in ascending powers of z when $|z| < 1$ and using §12.41.

Example 14.6.2   Deduce the result of (§14.11) from the preceding example.

14.61 Determination of an integral which represents $P^{(\alpha)}$

We shall now shew how an integral which represents the particular solution $P^{(\alpha)}$ (§14.3) of the hypergeometric differential equation can be found.

We have seen (§14.6) that the integral

$$I = (z - a)^\alpha (z - b)^\beta (z - c)^\gamma \int_C (t - a)^{\beta+\gamma+\alpha'-1} (t - b)^{\gamma+\alpha+\beta'-1}$$
$$\times (t - c)^{\alpha+\beta+\gamma'-1} (t - z)^{-\alpha-\beta-\gamma}\,dt$$

satisfies the differential equation of the hypergeometric function, provided C is a closed contour such that the integrand resumes its initial value after t has described C. Now the singularities of this integrand in the t-plane are the points a, b, c, z; and after describing the double circuit contour (§12.43) symbolised by $(b+, c+, b-, c-)$ the integrand returns to its original value.

Now, if z lie in a circle whose centre is a, the circle not containing either of the points b and c, we can choose the path of integration so that t is outside this circle, and so $|z - a| < |t - a|$ for all points t on the path.

Now choose $\arg(z - a)$ to be numerically less than π and $\arg(z - b)$, $\arg(z - c)$ so that they reduce to $\arg(a - b)$, $\arg(a - c)$ when $z \to a$, the values of $\arg(a - b)$, $\arg(a - c)$ being fixed. Now fix $\arg(t - a)$, $\arg(t - b)$, $\arg(t - c)$ at the point N at which the path of integration starts and ends; also choose $\arg(t - z)$ to reduce to $\arg(t - a)$ when $z \to a$.

Then

$$(z - b)^\beta = (a - b)^\beta \left\{ 1 + \beta \left(\frac{z - a}{a - b} \right) + \cdots \right\},$$
$$(z - c)^\gamma = (a - c)^\gamma \left\{ 1 + \gamma \left(\frac{z - a}{a - c} \right) + \cdots \right\},$$

---

[12] The justification of this process by (§4.7) is left to the reader.

and since we can expand $(t-z)^{-a-\beta-\gamma}$ into an absolutely and uniformly convergent series

$$(t-a)^{-a-\beta-\gamma}\left\{1-(\alpha+\beta+\gamma)\frac{a-z}{t-a}+\cdots\right\},$$

we may expand the integral into a series which converges absolutely.

Multiplying up the absolutely convergent series, we get a series of integer powers of $z-a$ multiplied by $(z-a)^{\alpha}$. Consequently we must have

$$I = (a-b)^{\beta}(a-c)^{\gamma}P^{(\alpha)}\int_{N}^{(b+,c+,b-,c-)}(t-a)^{\beta+\gamma+\alpha'-1}$$
$$\times(t-b)^{\gamma+\alpha+\beta'-1}(t-c)^{\alpha+\beta+\gamma'-1}\,dt.$$

We can define $P^{(\alpha')}, P^{(\beta)}, P^{(\beta')}, P^{(\gamma)}, P^{(\gamma')}$ by double circuit integrals in a similar manner.

14.7 Relations between contiguous hypergeometric functions

Let $P(z)$ be a solution of Riemann's equation with argument z, singularities a, b, c, and exponents α, α', β, β', γ, γ'. Further, let $P(z)$ be a constant multiple of one of the six functions $P^{(\alpha)}, P^{(\alpha')}, P^{(\beta)}, P^{(\beta')}, P^{(\gamma)}, P^{(\gamma')}$. Let $P_{\ell+1,m-1}(z)$ denote the function which is obtained by replacing two of the exponents, ℓ and m, in $P(z)$ by $\ell+1$ and $m-1$ respectively. Such functions $P_{\ell+1,m-1}(z)$ are said to be *contiguous* to $P(z)$. There are $6\times5 = 30$ contiguous functions, since ℓ and m may be any two of the six exponents.

It was first shewn by Riemann [556][13] that the function $P(z)$ and any two of its contiguous functions are connected by a linear relation, the coefficients in which are polynomials in z. There will clearly be $\frac{1}{2}\times30\times29 = 435$ of these relations. To shew how to obtain them, we shall take $P(z)$ in the form

$$P(z) = (z-a)^{\alpha}(z-b)^{\beta}(z-c)^{\gamma}\int_{C}(t-a)^{\beta+\gamma+\alpha'-1}(t-b)^{\gamma+\alpha+\beta'-1}$$
$$\times(t-c)^{\alpha+\beta+\gamma'-1}(t-z)^{\alpha-\beta-\gamma}\,dt,$$

where C is a double circuit contour of the type considered in (§14.61).

First, since the integral round C of the differential of any function which resumes its initial value after t has described C is zero, we have

$$0 = \int_{C}\frac{d}{dt}\left\{(t-a)^{\alpha'+\beta+\gamma}(t-b)^{\alpha+\beta'+\gamma-1}(t-c)^{\alpha+\beta+\gamma'-1}(t-z)^{-\alpha-\beta-\gamma}\right\}\,dt.$$

On performing the differentiation by differentiating each factor in turn, we get

$$(\alpha'+\beta+\gamma)P + (\alpha+\beta'+\gamma-1)P_{\alpha'+1,\beta'-1} + (\alpha+\beta+\gamma'-1)P_{\alpha'+1,\gamma'-1}$$
$$= \frac{(\alpha+\beta+\gamma)}{z-b}P_{\beta+1,\gamma'-1}.$$

Considerations of symmetry shew that the right-hand side of this equation can be replaced by

$$\frac{(\alpha+\beta+\gamma)}{z-c}P_{\beta'-1,\gamma+1}.$$

[13] Gauss had previously obtained 15 relations between contiguous hypergeometric functions.

These, together with the analogous formulae obtained by cyclical interchange[14] of (a, α, α') with (b, β, β') and (c, γ, γ'), are six linear relations connecting the hypergeometric function P with the twelve contiguous functions

$$P_{\alpha+1,\beta'-1}, \quad P_{\beta+1,\gamma'-1}, \quad P_{\gamma+1,\alpha'-1}, P_{\alpha+1,\gamma'-1}, \quad P_{\beta+1,\alpha'-1}, \quad P_{\gamma+1,\beta'-1},$$

$$P_{\alpha'+1,\beta'-1}, \quad P_{\alpha'+1,\gamma'-1}, \quad P_{\beta'+1,\gamma'-1}, P_{\beta'+1,\alpha'-1}, \quad P_{\gamma'+1,\alpha'-1}, \quad P_{\gamma'+1,\beta'-1}.$$

Next, writing $t - a = (t - b) + (b - a)$, and using[15] $P_{\alpha'-1}$ to denote the result of writing $\alpha' - 1$ for α' in P, we have

$$P = P_{\alpha'-1,\beta'+1} + (b - a)P_{\alpha'-1}.$$

Similarly $P = P_{\alpha'-1,\gamma'+1} + (c - a)P_{\alpha'-1}$. Eliminating $P_{\alpha'-1}$ from these equations, we have

$$(c - b)P + (a - c)P_{\alpha'-1,\beta'+1} + (b - a)P_{\alpha'-1,\gamma'+1} = 0.$$

This and the analogous formulae are three more linear relations connecting P with the last six of the twelve contiguous functions written above.

Next, writing $(t - z) = (t - a) - (z - a)$, we readily find the relation

$$P = \frac{1}{z - b}P_{\beta+1,\gamma'-1} - (z - a)^{\alpha+1}(z - b)^{\beta}(z - c)^{\gamma}$$

$$\times \int_C (t - a)^{\beta+\gamma+\alpha'-1}(t - b)^{\gamma+\alpha+\beta'-1}(t - c)^{\alpha+\beta+\gamma'-1}(t - z)^{-\alpha-\beta-\gamma-1}\, dt,$$

which gives the equations

$$(z - a)^{-1}\left\{P - (z - b)^{-1}P_{\beta+1,\gamma'-1}\right\} = (z - b)^{-1}\left\{P - (z - c)^{-1}P_{\gamma+1,\alpha'-1}\right\}$$

$$= (z - c)^{-1}\left\{P - (z - a)^{-1}P_{\alpha+1,\beta'-1}\right\}.$$

These are two more linear equations between P and the above twelve contiguous functions.

We have therefore now altogether found eleven linear relations between P and these twelve functions, the coefficients in these relations being rational functions of z. Hence each of these functions can be expressed linearly in terms of P and some selected one of them; that is, *between P and any two of the above functions there exists a linear relation*. The coefficients in this relation will be rational functions of z, and therefore will become polynomials in z when the relation is multiplied throughout by the least common multiple of their denominators.

The theorem is therefore proved, so far as the above twelve contiguous functions are concerned. It can, without difficulty, be extended so as to be established for the rest of the thirty contiguous functions.

Corollary 14.7.1 *If functions be derived from P by replacing the exponents α, α', β, β', γ, γ' by $\alpha + p$, $\alpha' + q$, $\beta + r$, $\beta' + s$, $\gamma + t$, $\gamma' + u$, where p, q, r, s, t, u are integers satisfying the relation*

$$p + q + r + s + t + u = 0,$$

then between P and any two such functions there exists a linear relation, the coefficients in which are polynomials in z.

---

[14] The interchange is to be made only in the integrands; the contour C is to remain, unaltered.

[15] $P_{\alpha'-1}$ is not a function of Riemann's type since the sum of its exponents at a, b, c is not unity.

This result can be obtained by connecting P with the two functions by a chain of intermediate contiguous functions, writing down the linear relations which connect them with P and the two functions, and from these relations eliminating the intermediate contiguous functions.

Many theorems which will be established subsequently, e.g. the recurrence formulae for the Legendre functions (§15.21), are really cases of the theorem of this article.

14.8 Miscellaneous examples

Example 14.1   Shew that

$$F(a,b+1;c;z) - F(a,b;c;z) = \frac{az}{c}F(a+1,b+1;c+1;z).$$

Example 14.2   Shew that if a is a negative integer while β and γ are not integers, then the ratio $F(\alpha,\beta;\alpha+\beta+1-\gamma;1-x)/F(\alpha,\beta;\gamma;x)$ is independent of x, and find its value.

Example 14.3   If $P(z)$ be a hypergeometric function, express its derivatives $\dfrac{dP}{dz}$ and $\dfrac{d^2P}{dz^2}$ linearly in terms of P and contiguous functions, and hence find the linear relation between P, $\dfrac{dP}{dz}$, and $\dfrac{d^2P}{dz^2}$, i.e. verify that P satisfies the hypergeometric differential equation.

Example 14.4   Shew that $F\left\{\frac{1}{4},\frac{1}{4};1;4z(1-z)\right\}$ satisfies the hypergeometric equation satisfied by $F(\frac{1}{2},\frac{1}{2};1;z)$. Shew that, in the left-hand half of the lemniscate $|z(1-z)| = \frac{1}{4}$, these two functions are equal; and in the right-hand half of the lemniscate, the former function is equal to $F(\frac{1}{2},\frac{1}{2};1;1-z)$.

Example 14.5 (Gauss)   If $F_{a+} = F(a+1,b;c;x)$, $F_{a-} = F(a-1,b;c;x)$ determine the 15 linear relations with polynomial coefficients which connect $F(a,b;c;x)$ with pairs of the six functions F_{a+}, F_{a-}, F_{b+}, F_{b-}, F_{c+}, F_{c-}.

Example 14.6   Shew that the hypergeometric equation

$$x(x-1)\frac{d^2y}{dx^2} - \{\gamma - (\alpha+\beta+1)x\}\frac{dy}{dx} + \alpha\beta y = 0$$

is satisfied by the two integrals (supposed convergent)

$$\int_0^1 z^{\beta-1}(1-z)^{\gamma-\beta-1}(1-xz)^{-\alpha}\,dz$$

and

$$\int_0^1 z^{\beta-1}(1-z)^{\alpha-\gamma}\{1-(1-x)z\}^{-\alpha}\,dz.$$

Example 14.7 (Math. Trip. 1896)   Shew that, for values of x between 0 and 1, the solution of the equation

$$x(1-x)\frac{d^2y}{dx^2} + \frac{1}{2}(\alpha+\beta+1)(1-2x)\frac{dy}{dx} - \alpha\beta y = 0$$

is

$$AF\left(\frac{\alpha}{2},\frac{\beta}{2};\frac{1}{2};(1-2x)^2\right)+B(1-2x)F\left(\frac{\alpha+1}{2},\frac{\beta+1}{2};\frac{3}{2}(1-2x)^2\right),$$

where A, B are arbitrary constants and $F(\alpha,\beta;\gamma;x)$ represents the hypergeometric series.

Example 14.8 (Hardy)   Shew that

$$\lim_{x\to 1^-}\left[F(\alpha,\beta;\gamma;x)-\right.$$

$$\left.\sum_{n=0}^{k}(-1)^n\,\frac{\Gamma(\alpha+\beta-\gamma-n)\Gamma(\gamma-\alpha+n)\Gamma(\gamma-\beta+n)\Gamma(\gamma)}{n\,!\Gamma(\gamma-\alpha)\Gamma(\gamma-\beta)\Gamma(\alpha)\Gamma(\beta)}(1-x)^{n+\gamma-\alpha-\beta}\right]$$

$$=\frac{\Gamma(\gamma-\alpha-\beta)\Gamma(\gamma)}{\Gamma(\gamma-\alpha)\Gamma(\gamma-\beta)}$$

where k is the integer such that $k\leq \mathrm{Re}(\alpha+\beta-\gamma)<k+1$. (This specifies the manner in which the hypergeometric function becomes infinite when $x\to 1^-$ provided that $\alpha+\beta-\gamma$ is not an integer.)

Example 14.9 (M. J. M. Hill [308])   Shew that, when $\mathrm{Re}(\gamma-\alpha-\beta)<0$, then

$$S_n\left/\frac{\Gamma(\gamma)n^{\alpha+\beta-\gamma}}{(\alpha+\beta-\gamma)\Gamma(\alpha)\Gamma(\beta)}\right. \to 1 \qquad \text{as}\quad n\to\infty,$$

where S_n denotes the sum of the first n terms of the series for $F(\alpha,\beta;\gamma;1)$.

Example 14.10 (Appell [29])   Shew that, if y_1, y_2 be independent solutions of

$$\frac{d^2y}{dx^2}+P\frac{dy}{dx}+Qy=0,$$

then the general solution of

$$\frac{d^3z}{dx^3}+3P\frac{d^2z}{dx^2}+\left\{2P^2+\frac{dP}{dx}+4Q\right\}\frac{dz}{dx}+\left\{4PQ+2\frac{dQ}{dx}\right\}z=0$$

is $z=Ay_1{}^2+By_1y_2+cy_2{}^2$, where A,B,C are constants.

Example 14.11 (Clausen [148])   Deduce from Example 14.10 above that, if $a+b+\frac{1}{2}=c$,

$$(F(a,b;c;x))^2=\frac{\Gamma(c)\Gamma(2c-1)}{\Gamma(2a)\Gamma(2b)\Gamma(a+b)}\sum_{n=0}^{\infty}\frac{\Gamma(2a+n)\Gamma(a+b+n)\Gamma(2b+n)}{n!\Gamma(c+n)\Gamma(2c-1+n)}x^n.$$

Example 14.12 (Kummer)   Shew that, if $|x|<\frac{1}{2}$ and $|x(1-x)|<\frac{1}{4}$,

$$F\left(2\alpha,2\beta;\alpha+\beta+\frac{1}{2};x\right)=F\left(\alpha,\beta;\alpha+\beta+\frac{1}{2};4x(1-x)\right).$$

Example 14.13   Deduce from Example 14.12 above that

$$F\left(2\alpha,2\beta;\alpha+\beta+\frac{1}{2};\frac{1}{2}\right)=\frac{\Gamma(\alpha+\beta+\frac{1}{2})\Gamma(\frac{1}{2})}{\Gamma(\alpha+\frac{1}{2})\Gamma(\beta+\frac{1}{2})}.$$

Example 14.14 (Watson [647]) Shew that, if $\omega = e^{2\pi i/3}$ and $\text{Re}(a) < 1$,

$$F(a, 3a - 1; 2a; -\omega^2) = 3^{3(\alpha-1)/2} \exp\left[\frac{\pi i(3a - 1)}{6}\right] \frac{\Gamma(2a)\Gamma(a - \frac{1}{3})}{\Gamma(3a - 1)\Gamma(\frac{2}{3})},$$

$$F(a, 3a - 1; 2a; -\omega) = 3^{3(\alpha-1)/2} \exp\left[-\frac{\pi i(3a - 1)}{6}\right] \frac{\Gamma(2a)\Gamma(a - \frac{1}{3})}{\Gamma(3a - 1)\Gamma(\frac{2}{3})}.$$

Example 14.15 (Heymann [302]) Shew that

$$F\left(-\frac{1}{2}n, -\frac{1}{2}n + \frac{1}{2}; n + \frac{3}{2}; -\frac{1}{3}\right) = \left(\frac{8}{9}\right)^n \frac{\Gamma(\frac{4}{3})\Gamma(n + \frac{3}{2})}{\Gamma(\frac{3}{2})\Gamma(n + \frac{4}{3})}.$$

Example 14.16 (Cayley [133]. See also Orr [510]) If

$$(1 - x)^{\alpha+\beta-\gamma} F(2\alpha, 2\beta; 2\gamma; x) = 1 + Bx + Cx^2 + Dx^3 + \cdots,$$

shew that

$$F(\alpha, \beta; \gamma + \tfrac{1}{2}; x) \, F(\gamma - \alpha, \gamma - \beta; \gamma + \tfrac{1}{2}; x)$$

$$= 1 + \frac{\gamma}{\gamma + \frac{1}{2}} Bx + \frac{\gamma(\gamma + 1)}{(\gamma + \frac{1}{2})(\gamma + \frac{3}{2})} Cx^2$$

$$+ \frac{\gamma(\gamma + 1)(\gamma + 2)}{(\gamma + \frac{1}{2})(\gamma + \frac{3}{2})(\gamma + \frac{5}{2})} Dx^3 + \cdots.$$

Example 14.17 (Le Vavasseur) If the function $F(\alpha, \beta, \beta', \gamma; x, y)$ be defined by the equation

$$F(\alpha, \beta, \beta', \gamma; x, y) = \frac{\Gamma(\gamma)}{\Gamma(\alpha)\Gamma(\gamma - \alpha)} \int_0^1 u^{\alpha-1}(1 - u)^{\gamma-\alpha-1}(1 - ux)^{-\beta}(1 - uy)^{-\beta'} \, du,$$

then shew that between F and any three of its eight contiguous functions

$$F(\alpha \pm 1), F(\beta \pm 1), F(\beta' \pm 1), F(\gamma \pm 1),$$

there exists a homogeneous linear equation, whose coefficients are polynomials in x and y.

Example 14.18 (Math. Trip. 1893) If $\gamma - \alpha - \beta < 0$, shew that, as $x \to 1^-$,

$$F(\alpha, \beta; \gamma; x) \left/ \left\{\frac{\Gamma(\gamma)\Gamma(\alpha + \beta - \gamma)}{\Gamma(\alpha)\Gamma(\beta)}(1 - x)^{\gamma-\alpha-\beta}\right\} \right. \to 1,$$

and that, if $\gamma - \alpha - \beta = 0$, the corresponding approximate formula is

$$F(\alpha, \beta; \gamma; x) \left/ \left\{\frac{\Gamma(\alpha + \beta)}{\Gamma(\alpha)\Gamma(\beta)} \log \frac{1}{1 - x}\right\} \right. \to 1.$$

Example 14.19 (Pochhammer) Shew that, when $|x| < 1$,

$$\int_c^{(x^+, 0^+, x^-, 0^-)} x^{1-\gamma}(v - x)^{\gamma-\alpha-1} v^{\alpha-1}(1 - v)^{-\beta} \, dv$$

$$= -4e^{\pi i \gamma} \sin \pi\alpha \sin \pi(\gamma - \alpha) \cdot \frac{\Gamma(\gamma - \alpha)\Gamma(\alpha)}{\Gamma(\gamma)} F(\alpha, \beta; \gamma; x),$$

where c denotes a point on the straight line joining the points 0, x, the initial arguments of $v - x$ and of v are the same as that of x, and $\arg(1 - v) \to 0$ as $v \to 0$.

Example 14.20 (Barnes)   If, when $|\arg(1-x)| < 2\pi$,

$$K(x) = \frac{1}{2\pi i} \int_{-i\infty}^{i\infty} \left[\Gamma(-s)\Gamma\left(\tfrac{1}{2}+s\right)\right]^2 (1-x)^s \, ds,$$

and, when $|\arg x| < 2\pi$,

$$K'(x) = \frac{1}{2\pi i} \int_{-i\infty}^{i\infty} \left[\Gamma(-s)\Gamma\left(\tfrac{1}{2}+s\right)\right]^2 x^s \, ds,$$

by changing the variable s in the integral or otherwise, obtain the following relations:

$$
\begin{aligned}
K(x) &= K'(1-x), && \text{if } |\arg(1-x)| < \pi, \\
K(1-x) &= K'(x), && \text{if } |\arg x| < \pi, \\
K(x) &= (1-x)^{-1/2} K\left(\frac{x}{x-1}\right), && \text{if } |\arg(1-x)| < \pi, \\
K(1-x) &= x^{-1/2} K\left(\frac{x-1}{x}\right), && \text{if } |\arg x| < \pi, \\
K'(x) &= x^{-1/2} K'(1/x), && \text{if } |\arg x| < \pi, \\
K'(1-x) &= (1-x)^{-1/2} K'\left(\frac{1}{1-x}\right), && \text{if } |\arg(1-x)| < \pi.
\end{aligned}
$$

Example 14.21 (Barnes)   With the notation of the preceding example, obtain the following results

$$2K(x) = \sum_{n=0}^{\infty} \left[\frac{\Gamma\left(\tfrac{1}{2}+n\right)}{n!}\right]^2 x^n,$$

$$2\pi K'(x) = -\sum_{n=0}^{\infty} \left[\frac{\Gamma\left(\tfrac{1}{2}+n\right)}{n!}\right]^2 x^n$$

$$\times \left[\log x - 4\log 2 + 4\left(\frac{1}{1} - \frac{1}{2} + \cdots - \frac{1}{2n}\right)\right],$$

when $|x| < 1, |\arg x| < \pi$; and

$$K(x) = \mp i (-x)^{-\frac{1}{2}} K(1/x) + (-x)^{-\frac{1}{2}} K'(1/x),$$

when $|\arg(-x)| < \pi$, the ambiguous sign being the same as the sign of $\mathrm{Im}(x)$.

Example 14.22 (Appell [30])   Hypergeometric series in two variables are defined by the equations

$$F_1(\alpha; \beta, \beta'; \gamma; x, y) = \sum_{m,n} \frac{\alpha_{m+n}\beta_m\beta'_n}{m!n!\gamma_{m+n}} x^m y^n,$$

$$F_2(\alpha; \beta, \beta'; \gamma, \gamma'; x, y) = \sum_{m,n} \frac{\alpha_{m+n}\beta_m\beta'_n}{m!n!\gamma_m\gamma'_n} x^m y^n,$$

$$F_3(\alpha, \alpha', \beta, \beta'; \gamma; x, y) = \sum_{m,n} \frac{\alpha_m\alpha'_n\beta_m\beta'_n}{m!n!\gamma_{m+n}} x^m y^n,$$

$$F_4(\alpha, \beta; \gamma, \gamma'; x, y) = \sum_{m,n} \frac{\alpha_{m+n}\beta_{m+n}}{m!n!\gamma_m\gamma'_n} x^m y^n,$$

where $\alpha_m = \alpha(\alpha + 1) \cdots (\alpha + m - l)$, and $\sum_{m,n}$ means $\sum_{m=0}^{\infty} \sum_{n=0}^{\infty}$.

Obtain the differential equations

$$x(1-x)\frac{\partial^2 F_1}{\partial x^2} + y(1-x)\frac{\partial^2 F_1}{\partial x \partial y} + \{\gamma - (\alpha+\beta+1)x\}\frac{\partial F_1}{\partial x} - \beta y \frac{\partial F_1}{\partial y} - \alpha\beta F_1 = 0,$$

$$x(1-x)\frac{\partial^2 F_2}{\partial x^2} - xy\frac{\partial^2 F_2}{\partial x \partial y} + \{\gamma - (\alpha+\beta+1)x\}\frac{\partial F_2}{\partial x} - \beta y \frac{\partial F_2}{\partial y} - \alpha\beta F_2 = 0,$$

$$x(1-x)\frac{\partial^2 F_3}{\partial x^2} + y\frac{\partial^2 F_3}{\partial x \partial y} + \{\gamma - (\alpha+\beta+1)x\}\frac{\partial F_3}{\partial x} - \alpha\beta F_3 = 0,$$

$$x(1-x)\frac{\partial^2 F_4}{\partial x^2} - 2xy\frac{\partial^2 F_4}{\partial x \partial y} - y^2\frac{\partial^2 F_4}{\partial y^2} + \{\gamma - (\alpha+\beta+1)x\}\frac{\partial F_4}{\partial x}$$

$$- (\alpha+\beta+1)y\frac{\partial F_4}{\partial y} - \alpha\beta F_4 = 0,$$

and four similar equations, derived from these by interchanging x with y and α, β, γ with α', β', γ' when α', β', γ' occur in the corresponding series.

Example 14.23 (Hermite [292]) If α is negative, and if

$$\alpha = -\nu + a,$$

where ν is an integer and α is positive, shew that

$$\frac{\Gamma(x)\Gamma(\alpha)}{\Gamma(x+\alpha)} = \sum_{n=1}^{\infty} \left\{ \frac{R_n}{x+n} + G_n(x) \right\},$$

where

$$R_n = \frac{(-1)^n(\alpha-1)(\alpha-2)\cdots(\alpha-n)}{n!} G(-n),$$

$$G(x) = \left(1 + \frac{x}{\alpha-1}\right)\left(1 + \frac{x}{\alpha-2}\right)\cdots\left(1 + \frac{x}{\alpha-\nu}\right)$$

$$G_n(x) = \frac{G(x) - G(-n)}{x+n}.$$

Example 14.24   When $\alpha < 1$, shew that

$$\frac{\Gamma(x)\Gamma(\alpha - x)}{\Gamma(\alpha)} = \sum_{n=1}^{\infty} \frac{R_n}{x + n} - \sum_{n=1}^{\infty} \frac{R_n}{x - \alpha - n},$$

where

$$R_n = \frac{(-1)^n \alpha (\alpha + 1) \cdots (\alpha + n - 1)}{n!}.$$

Example 14.25 (Hermite [292])   When $\alpha > 1$, and v and a are respectively the integral and fractional parts of α, shew that

$$\frac{\Gamma(x)\Gamma(\alpha - x)}{\Gamma(\alpha)} = \sum_{n=1}^{\infty} \frac{G(x)\rho_n}{x + n} - \sum_{n=1}^{\infty} \frac{G(x)\rho_{v+n}}{x - a - n}$$

$$- G(x) \left[\frac{\rho_0}{x - a} + \frac{\rho_1}{x - a - 1} + \cdots + \frac{\rho_{v-1}}{x - a - v + 1} \right],$$

where

$$G(x) = \left(1 - \frac{x}{a}\right)\left(1 - \frac{x}{a + 1}\right) \cdots \left(1 - \frac{x}{a + v - 1}\right)$$

and

$$\rho_n = \frac{(-1)^n a(a + 1) \cdots (a + n - 1)}{n!}.$$

Example 14.26 (Saalschütz [570])   (A number of similar results are given by Dougall [187].) If

$$f_n(x, y, v) =$$

$$1 - \binom{n}{1}\frac{x(y + v + n - 1)}{y(x + v)} + \binom{n}{2}\frac{x(x + 1)(y + v + n - 1)(y + v + n)}{y(y + 1)(x + v)(x + v + 1)} - \cdots,$$

where n is a positive integer and $\binom{n}{1}, \binom{n}{2}, \ldots$ are binomial coefficients, shew that

$$f_n(x, y, v) = \frac{\Gamma(y)\Gamma(y - x + n)\Gamma(x + v)\Gamma(v + n)}{\Gamma(y - x)\Gamma(y + n)\Gamma(v)\Gamma(x + v + n)}.$$

Example 14.27 (Dixon [183])   If

$$F(\alpha, \beta, \gamma; \delta, \varepsilon; x) = 1 + \frac{\alpha\beta\gamma}{\delta\varepsilon + 1}x + \frac{\alpha(\alpha + 1)\beta(\beta + 1)\gamma(\gamma + 1)}{\delta(\delta + 1)\varepsilon(\varepsilon + 1)1 \cdot 2}x^2 + \cdots,$$

shew that, when $\mathrm{Re}(\delta + \varepsilon - \frac{3}{2}\alpha - 1) > 0$, then

$$F(\alpha, \alpha - \delta + 1, \alpha - \varepsilon + 1; \delta, \varepsilon; 1)$$

$$= 2^{-\alpha} \frac{\Gamma(\frac{1}{2})\Gamma(\delta)\Gamma(\varepsilon)\Gamma(\delta + \varepsilon - \frac{3}{2}\alpha - 1)}{\Gamma(\delta - \frac{1}{2}\alpha)\Gamma(\varepsilon - \frac{1}{2}\alpha)\Gamma(\frac{1}{2} + \frac{1}{2}\alpha)\Gamma(\delta + \varepsilon - \alpha - 1)}.$$

Example 14.28 (Morley [475])   Shew that, if $\mathrm{Re}(\alpha) < \frac{2}{3}$, then

$$1 + \sum_{n=1}^{\infty} \left\{ \frac{\alpha(\alpha + 1) \cdots (\alpha + n - 1)}{n!} \right\}^3 = \cos\left(\frac{\pi\alpha}{2}\right) \frac{\Gamma(1 - \frac{3}{2}\alpha)}{\left[\Gamma(1 - \frac{1}{2}\alpha)\right]^3}.$$

Example 14.29 (Dixon [184])   If

$$\int_0^1 \int_0^1 x^{i-1}(1-x)^{j-1}y^{l-1}(1-y)^{k-1}(1-xy)^{m-j-k}\,dx\,dy = B(i,j,k,l,m),$$

shew, by integrating with respect to x, and also with respect to y, that $B(i,j,k,l,m)$ is a symmetric function of $i+j, j+k, k+l, l+m, m+i$. Deduce that

$$\frac{F(\alpha,\beta,\gamma;\delta,\varepsilon;1)}{\Gamma(\delta)\Gamma(\varepsilon)\Gamma(\delta+\varepsilon-\alpha-\beta-\gamma)}$$

is a symmetric function of $\delta, \varepsilon, \delta+\varepsilon-\alpha-\beta, \delta+\varepsilon-\beta-\gamma, \delta+\varepsilon-\gamma-\alpha$. For a proof of a special case by Barnes, see [50].

Example 14.30   If

$$F_n = F(-n,\alpha+n;\gamma;x)$$
$$= \frac{x^{1-\gamma}(1-x)^{\gamma-\alpha}}{\gamma(\gamma+1)\cdots(\gamma+n-1)}\frac{d^n}{dx^n}\{x^{\gamma+n-1}(1-x)^{\alpha+n-\gamma}\},$$

shew that, when n is a large positive integer, and $0 < x < 1$,

$$F_n = \frac{\Gamma(\gamma)}{n^{\gamma-\frac{1}{2}}\sqrt{\pi}}(\sin\phi)^{\frac{1}{2}-\gamma}(\cos\phi)^{\gamma-\alpha-\frac{1}{2}}\cos\{(2n+a)\phi - \frac{\pi}{4}(2\gamma-1)\} + O\left(\frac{1}{n^{\gamma+\frac{1}{2}}}\right),$$

where $x = \sin^2\phi$.

Note   This result is contained in the great memoir by Darboux [163, 164]. For a systematic development of hypergeometric functions in which one (or more) of the constants is large, see [652].

15

Legendre Functions

15.1 Definition of Legendre polynomials

Consider the expression $(1 - 2zh + h^2)^{-\frac{1}{2}}$; when $|2zh - h^2| < 1$, it can be expanded in a series of ascending powers of $2zh - h^2$. If, in addition, $|2zh| + |h|^2 < 1$, these powers can be multiplied out and the resulting series rearranged in any manner (§2.52) since the expansion of $[1 - \{|2zh| + |h|^2\}]^{-\frac{1}{2}}$ in powers of $|2zh| + |h|^2$ then converges absolutely. In particular, if we rearrange in powers of h, we get

$$(1 - 2zh + h^2)^{-\frac{1}{2}} = P_0(z) + hP_1(z) + h^2 P_2(z) + h^3 P_3(z) + \cdots,$$

where

$$P_0(z) = 1, \quad P_1(z) = z, \quad P_2(z) = \frac{1}{2}(3z^2 - 1), \quad P_3(z) = \frac{1}{2}(5z^3 - 3z),$$

$$P_4(z) = \frac{1}{8}(35z^4 - 30z^2 + 3), \quad P_5(z) = \frac{1}{8}(63z^5 - 70z^3 + 15z),$$

and generally

$$P_n(z) = \frac{(2n)!}{2^n (n!)^2} \left\{ z^n - \frac{n(n-1)}{2(2n-1)} z^{n-2} + \frac{n(n-1)(n-2)(n-3)}{2 \cdot 4 \cdot (2n-1)(2n-3)} z^{n-4} - \cdots \right\}$$

$$= \sum_{r=0}^{m} (-1)^r \frac{(2n-2r)!}{2^n r! (n-r)! (n-2r)!} z^{n-2r},$$

where $m = \frac{1}{2}n$ or $\frac{1}{2}(n-1)$, whichever is an integer.

If a, b and δ be positive constants, b being so small that $2ab + b^2 \leq 1 - \delta$, the expansion of $(1 - 2zh + h^2)^{-\frac{1}{2}}$ converges uniformly with respect to z and h when $|z| \leq a, |h| \leq b$.

The expressions $P_0(z), P_1(z), \ldots$, which are clearly all polynomials in z, are known as *Legendre polynomials*, $P_n(z)$ being called the *Legendre polynomial of degree n*. Other names are *Legendre coefficients* and *Zonal Harmonics*. They were introduced into analysis in 1784 by Legendre [420].

It will appear later (§15.2) that these polynomials are particular cases of a more extensive class of functions known as *Legendre functions*.

Example 15.1.1 By giving z special values in the expression $(1 - 2zh + h^2)^{-\frac{1}{2}}$, shew that

$$P_n(1) = 1, \quad P_n(-1) = (-1)^n,$$

$$P_{2n+1}(0) = 0, \quad P_{2n}(0) = (-1)^n \frac{1 \cdot 3 \cdots (2n-1)}{2 \cdot 4 \cdots (2n)}.$$

Example 15.1.2 (Legendre)   From the expansion

$$(1 - 2h\cos\theta + h^2)^{-\frac{1}{2}} = \left(1 + \frac{1}{2}he^{i\theta} + \frac{1\cdot 3}{2\cdot 4}h^2 e^{2i\theta} + \cdots\right)$$
$$\times \left(1 + \frac{1}{2}he^{-i\theta} + \frac{1\cdot 3}{2\cdot 4}h^2 e^{-2i\theta} + \cdots\right),$$

shew that

$$P_n(\cos\theta) = \frac{1\cdot 3\cdots(2n-1)}{2\cdot 4\cdots(2n)}\left\{2\cos n\theta + \frac{1\cdot(2n)}{2\cdot(2n-1)}2\cos(n-2)\theta\right.$$
$$\left. + \frac{1\cdot 3\cdot(2n)\cdot(2n-2)}{2\cdot 4\cdot(2n-1)(2n-3)}2\cos(n-4)\theta + \cdots\right\}.$$

Deduce that, if θ be a real angle,

$$|P_n(\cos\theta)| \leq \frac{1\cdot 3\cdots(2n-1)}{2\cdot 4\cdots 2n}\left\{2 + \frac{1\cdot(2n)}{2\cdot(2n-1)}\cdot 2 + \frac{1\cdot 3\cdot(2n)(2n-2)}{2\cdot 4\cdot(2n-1)(2n-3)}\cdot 2 + \cdots\right\}$$
$$= P_n(1),$$

so that $|P_n(\cos\theta)| \leq 1$.

Example 15.1.3 (Clare, 1905)   Shew that, when $z = -\frac{1}{2}$,

$$P_n = P_0 P_{2n} - P_1 P_{2n-1} + P_2 P_{2n-2} - \cdots + P_{2n}P_0.$$

15.11 Rodrigues' formula for the Legendre polynomials [561]

It is evident that, when n is an integer,

$$\frac{d^n}{dz^n}(z^2 - 1)^n = \frac{d^n}{dz^n}\left\{\sum_{r=0}^{n}(-1)^r\frac{n!}{r!(n-r)!}z^{2n-2r}\right\}$$
$$= \sum_{r=0}^{m}(-1)^r\frac{n!}{r!(n-r)!}\frac{(2n-2r)!}{(n-2r)!}z^{n-2r},$$

where $m = \frac{1}{2}n$ or $\frac{1}{2}(n-1)$, the coefficients of negative powers of z vanishing. From the general formula for $P_n(z)$ it follows at once that

$$P_n(z) = \frac{1}{2^n n!}\frac{d^n}{dz^n}(z^2 - 1)^n;$$

this result is known as Rodrigues' formula.

Example 15.1.4   Shew that $P_n(z) = 0$ has n real roots, all lying between ± 1.

15.12 Schläfli's integral for $P_n(z)$ [579]

From the result of §15.11 combined with §5.22, it follows at once that

$$P_n(z) = \frac{1}{2\pi i}\int_C \frac{(t^2 - 1)^n}{2^n(t - z)^{n+1}}\,dt,$$

where C is a contour which encircles the point z once counter-clockwise; this result is called *Schläfli's integral formula* for the Legendre polynomials.

15.13 Legendre's differential equation

We shall now prove that the function $u = P_n(z)$ is a solution of the differential equation

$$(1 - z^2)\frac{d^2u}{dz^2} - 2z\frac{du}{dz} + n(n+1)u = 0,$$

which is called *Legendre's differential equation for functions of degree n.*

For, substituting Schläfli's integral in the left-hand side, we have, by §5.22,

$$(1 - z^2)\frac{d^2 P_n(z)}{dz^2} - 2z\frac{dP_n(z)}{dz} + n(n+1)P_n(z)$$

$$= \frac{(n+1)}{2\pi i}\int_C \frac{(t^2-1)^n}{2^n(t-z)^{n+3}}\{-(n+2)(t^2-1) + 2(n+1)t(t-z)\}\, dt$$

$$= \frac{(n+1)}{2\pi i \cdot 2^n}\int_C \frac{d}{dt}\left\{\frac{(t^2-1)^{n+1}}{(t-z)^{n+2}}\right\}\, dt,$$

and this integral is zero, since $(t^2-1)^{n+1}(t-z)^{-n-2}$ resumes its original value after describing C when n is an integer. The Legendre polynomial therefore satisfies the differential equation. The result just obtained can be written in the form

$$\frac{d}{dz}\left\{(1-z^2)\frac{dP_n(z)}{dz}\right\} + n(n+1)P_n(z) = 0.$$

Note It will be observed that Legendre's equation is a particular case of Riemann's equation, defined by the scheme

$$P\left\{\begin{matrix} -1 & \infty & 1 \\ 0 & n+1 & 0 \\ 0 & -n & 0 \end{matrix}\; z\right\}.$$

Example 15.1.5 Shew that the equation satisfied by $\dfrac{d^r P_n(z)}{dz^r}$ is defined by the scheme

$$P\left\{\begin{matrix} -1 & -\infty & 1 \\ -r & n+r+1 & -r \\ 0 & -n+r & 0 \end{matrix}\; z\right\}.$$

Example 15.1.6 If $z^2 = \eta$, shew that Legendre's differential equation takes the form

$$\frac{d^2y}{d\eta^2} + \left\{\frac{1}{2\eta} - \frac{1}{1-\eta}\right\}\frac{dy}{d\eta} + \frac{n(n+1)y}{4\eta(1-\eta)} = 0.$$

Shew that this is a hypergeometric equation.

Example 15.1.7 Deduce Schläfli's integral for the Legendre functions, as a limiting case of the general hypergeometric integral of §14.6. *Hint.* Since Legendre's equation is given by the scheme

$$P\left\{\begin{matrix} -1 & \infty & 1 \\ 0 & n+1 & 0 & z \\ 0 & -n & 0 \end{matrix}\right\},$$

the integral suggested is

$$\lim_{b\to\infty}\left(1-\frac{z}{b}\right)^{n+1}\int_C (t+1)^n(t-1)^n \lim_{b\to\infty}\left(1-\frac{t}{b}\right)^{-n}(t-z)^{-n-1}\, dt$$

$$= \int_C (t^2-1)^n(t-z)^{-n-1}\, dt,$$

taken round a contour C such that the integrand resumes its initial value after describing it; and this gives Schläfli's integral.

15.14 The integral properties of the Legendre polynomials

We shall now shew that

$$\int_{-1}^{1} P_m(z)P_n(z)\, dz = \begin{cases} 0 & (m \neq n), \\ \dfrac{2}{2n+1} & (m=n). \end{cases}$$

These two results were given by Legendre in 1784 and 1789.

Let $\{u\}_r$ denote $\dfrac{d^r u}{dz^r}$; then, if $r \le n$, $\{(z^2-1)^n\}_r$ is divisible by $(z^2-1)^{n-r}$; and so, if $r < n$, $\{(z^2-1)^n\}_r$ vanishes when $z = 1$ and when $z = -1$.

Now, of the two numbers m, n, let m be that one which is equal to or greater than the other. Then, integrating by parts continually,

$$\int_{-1}^{1}\left\{(z^2-1)^m\right\}_m \left\{(z^2-1)^n\right\}_n dz = \left[\left\{(z^2-1)^m\right\}_{m-1}\left\{(z^2-1)^n\right\}_n\right]_{-1}^{1} -$$

$$\int_{-1}^{1}\left\{(z^2-1)^m\right\}_{m-1}\left\{(z^2-1)^n\right\}_{n+1} dz$$

$$\vdots$$

$$= (-1)^m \int_{-1}^{1}(z^2-1)^m \left\{(z^2-1)^n\right\}_{n+m} dz,$$

since $\left\{(z^2-1)^m\right\}_{m-1}, \left\{(z^2-1)^m\right\}_{m-2}, \ldots$ vanish at both limits.

Now, when $m > n$, $\left\{(z^2-1)^n\right\}_{m+n} = 0$, since differential coefficients of $(z^2-1)^n$ of order higher than $2n$ vanish; and so, *when m is greater than n*, it follows from Rodrigues' formula that

$$\int_{-1}^{1} P_m(z)P_n(z)\, dz = 0.$$

When $m = n$, we have, by the transformations just obtained,

$$\int_{-1}^{1} \left\{ (z^2 - 1)^n \right\}_n \left\{ (z^2 - 1)^n \right\}_n \, dz = (-1)^n \int_{-1}^{1} (z^2 - 1)^n \frac{d^{2n}}{dz^{2n}} (z^2 - 1)^n \, dz$$

$$= (2n)! \int_{-1}^{1} (1 - z^2)^n \, dz$$

$$= 2 \cdot (2n)! \int_{0}^{1} (1 - z^2)^n \, dz$$

$$= 2 \cdot (2n)! \int_{0}^{\frac{1}{2}\pi} \sin^{2n+1} \theta \, d\theta$$

$$= 2 \cdot (2n)! \frac{2 \cdot 4 \cdots (2n)}{3 \cdot 5 \cdots (2n + 1)},$$

where $\cos \theta$ has been written for z in the integral; hence, by Rodrigues' formula,

$$\int_{-1}^{1} \left\{ P_n(z) \right\}^2 \, dz = \frac{2 \cdot (2n)!}{(2^n n!)^2} \frac{(2^n; n!)^2}{(2n + 1)!} = \frac{2}{2n + 1}.$$

We have therefore obtained both the required results.

Note   It follows that, in the language of Chapter 11, the functions $\left(n + \frac{1}{2} \right)^{1/2} P_n(z)$ are normal orthogonal functions for the interval $(-1, 1)$.

Example 15.1.8 (Clare, 1908)   Shew that, if $x > 0$,

$$\int_{-1}^{1} (\cosh 2x - z)^{-\frac{1}{2}} P_n(z) \, dz = \sqrt{2} \left(n + \frac{1}{2} \right)^{-1} e^{-(2n+1)x}.$$

Example 15.1.9 (Clare, 1902)   If $I = \int_{0}^{1} P_m(z) P_n(z) \, dz$, then

(i) $I = 1/(2n + 1)$ when $m = n$,
(ii) $I = 0$ when $m - n$ is even,
(iii) $I = \dfrac{(-1)^{\mu+\nu}}{2^{m+n-1}(n - m)(n + m + 1)} \dfrac{n! m!}{(\nu!)^2 (\mu!)^2}$ when $n = 2\nu + 1, m = 2\mu$.

15.2 Legendre functions

Hitherto we have supposed that the degree n of $P_n(z)$ is a positive integer; in fact, $P_n(z)$ has not been defined except when n is a positive integer. We shall now see how $P_n(z)$ can be defined for values of n which are not necessarily integers.

An analogy can be drawn from the theory of the Gamma-function. The expression $z!$ as ordinarily defined (viz. as $z(z - 1)(z - 2) \cdots 2 \cdot 1$) has a meaning only for positive integral values of z; but when the Gamma-function has been introduced, $z!$ can be defined to be $\Gamma(z + 1)$, and so a function $z!$ will exist for values of z which are not integers.

Referring to §15.13, we see that the differential equation

$$(1 - z^2) \frac{d^2 u}{dz^2} - 2z \frac{du}{dz} + n(n + 1)u = 0$$

is satisfied by the expression

$$u = \frac{1}{2\pi i} \int_C \frac{(t^2 - 1)^n}{2^n (t - z)^{n+1}} dt,$$

even when n is not a positive integer, provided that C is a contour such that $(t^2 - 1)^{n+1}(t - z)^{-n-2}$ resumes its original value after describing C.

Suppose then that n is no longer taken to be a positive integer. The function $(t^2 - 1)^{n+1}(t - z)^{-n-2}$ has three singularities, namely the points $t = 1, t = -1, t = z$; and it is clear that after describing a circuit round the point $t = 1$ counter-clockwise, the function resumes its original value multiplied by $e^{2\pi i(n+1)}$; while after describing a circuit round the point $t = z$ counter-clockwise, the function resumes its original value multiplied by $e^{2\pi i(-n-2)}$. If therefore C be a contour enclosing the points $t = 1$ and $t = z$, but not enclosing the point $t = -1$, then the function $(t^2 - 1)^{n+1}(t - z)^{-n-2}$ will resume its original value after t has described the contour C. Hence, *Legendre's differential equation for functions of degree n,*

$$(1 - z^2) \frac{d^2 u}{dz^2} - 2z \frac{du}{dz} + n(n + 1)u = 0,$$

is satisfied by the expression

$$u = \frac{1}{2\pi i} \int_A^{(1+,z+)} \frac{(t^2 - 1)^n}{2^n (t - z)^{n+1}} dt,$$

for all values of n; the many-valued functions will be specified precisely by taking A on the real axis on the right of the point $t = 1$ (and on the right of z if z be real), and by taking $\arg(t - 1) = \arg(t + 1) = 0$ and $|\arg(t - z)| < \pi$ at A.

This expression will be denoted by $P_n(z)$, and will be termed the Legendre function of degree n of the first kind.

We have thus defined a function $P_n(z)$, the definition being valid whether n is an integer or not.

Note The function $P_n(z)$ thus defined is not a one-valued function of z; for we might take two contours as shewn in the figure, and the integrals along them would not be the same; to

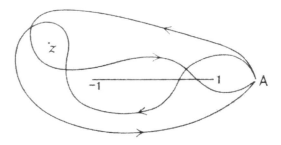

make the contour integral unique, make a cut in the t plane from -1 to $-\infty$ along the real axis; this involves making a similar cut in the z plane, for if the cut were not made, then, as z varied continuously across the negative part of the real axis, the contour would not vary continuously. It follows, by §5.31, that $P_n(z)$ is analytic throughout the cut plane.

15.21 The recurrence formulae

We proceed to establish a group of formulae (which are really particular cases of the relations between contiguous Riemann P-functions which were shewn to exist in §14.7) connecting Legendre functions of different degrees.

If C be the contour of §15.2, and writing $P_n'(z)$ for $\frac{d}{dz}P_n(z)$, we have

$$P_n(z) = \frac{1}{2^{n+1}\pi i}\int_C \frac{(t^2-1)^n}{(t-z)^{n+1}}\,dt; \qquad P_n'(z) = \frac{n+1}{2^{n+1}\pi i}\int_C \frac{(t^2-1)^n}{(t-z)^{n+2}}\,dt.$$

Now

$$\frac{d}{dt}\frac{(t^2-1)^{n+1}}{(t-z)^{n+1}} = \frac{2(n+1)t(t^2-1)^n}{(t-z)^{n+1}} - \frac{(n+1)(t^2-1)^{n+1}}{(t-z)^{n+2}},$$

and so, integrating,

$$0 = 2\int_C \frac{t(t^2-1)^n}{(t-z)^{n+1}}\,dt - \int_C \frac{(t^2-1)^{n+1}}{(t-z)^{n+2}}\,dt.$$

Therefore

$$\frac{1}{2^{n+1}\pi i}\int_C \frac{(t^2-1)^n}{(t-z)^n}\,dt = \frac{1}{2^{n+2}\pi i}\int_C \frac{(t^2-1)^{n+1}}{(t-z)^{n+2}}\,dt - \frac{z}{2^{n+1}\pi i}\int_C \frac{(t^2-1)^n}{(t-z)^{n+1}}\,dt.$$

Consequently

$$P_{n+1}(z) - zP_n(z) = \frac{1}{2^{n+1}\pi i}\int_C \frac{(t^2-1)^n}{(t-z)^n}\,dt. \tag{15.1}$$

Differentiating[1], we get

$$P_{n+1}'(z) - zP_n'(z) - P_n(z) = nP_n(z),$$

and so

$$P_{n+1}'(z) - zP_n'(z) = (n+1)P_n(z) \tag{15.2}$$

This is the first of the required formulae.

Next, expanding the equation

$$\int_C \frac{d}{dt}\left\{\frac{t(t^2-1)^n}{(t-z)^n}\right\}\,dt = 0,$$

we find that

$$\int_C \frac{(t^2-1)^n}{(t-z)^n}\,dt + 2n\int_C \frac{t^2(t^2-1)^{n-1}}{(t-z)^n}\,dt - n\int_C \frac{t(t^2-1)}{(t-z)^{n+1}}\,dt = 0.$$

Writing $(t^2-1)+1$ for t^2 and $(t-z)+z$ for t in this equation, we get

$$(n+1)\int_C \frac{(t^2-1)^n}{(t-z)^n}\,dt + 2n\int_C \frac{(t^2-1)^{n-1}}{(t-z)^n}\,dt - nz\int_C \frac{(t^2-1)^n}{(t-z)^{n+1}}\,dt = 0.$$

Using (15.1), we have at once

$$(n+1)\{P_{n+1}(z) - zP_n(z)\} + nP_{n-1}(z) - nzP_n(z) = 0.$$

---

[1] The process of differentiating under the sign of integration is readily justified by §4.2.

That is to say

$$(n + 1)P_{n+1}(z) - (2n + 1)zP_n(z) + nP_{n-1}(z) = 0, \tag{15.3}$$

a relation[2] connecting three Legendre functions of consecutive degrees. This is the second of the required formulae.

We can deduce the remaining formulae from (15.2) and (15.3) thus:
Differentiating (15.3), we have

$$(n + 1)\left\{P'_{n+1}(z) - zP'_n(z)\right\} - n\left\{zP'_n(z) - P'_{n-1}(z)\right\} - (2n + 1)P_n(z) = 0.$$

Using (15.2) to eliminate $P'_{n+1}(z)$, and then dividing by[3] n we get

$$zP'_n(z) - P'_{n-1}(z) = nP_n(z). \tag{15.4}$$

Adding (15.2) and (15.4) we get

$$P'_{n+1}(z) - P'_{n-1}(z) = (2n + 1)P_n(z). \tag{15.5}$$

Lastly, writing $n - 1$ for n in (15.2) and eliminating $P'_{n-1}(z)$ between the equation so obtained and (15.4), we have

$$(z^2 - 1)P'_n(z) = nzP_n(z) - nP_{n-1}(z). \tag{15.6}$$

The formulae (15.2), (15.3), (15.4), (15.5), (15.6), are called the *recurrence formulae*.

Note The above proof holds whether n is an integer or not, i.e. it is applicable to the general Legendre functions. Another proof which, however, only applies to the case when n is a positive integer (i.e. is only applicable to the Legendre polynomials) is as follows:

Write $V = (1 - 2hz + h^2)^{-\frac{1}{2}}$. Then, equating coefficients[4] of powers of h in the expansions on each side of the equation

$$(1 - 2hz + h^2)\frac{\partial V}{\partial h} = (z - h)V,$$

we have $nP_n(z) - (2n - 1)zP_{n-1}(z) + (n - 1)P_{n-2}(z) = 0$, which is the formula (15.3).

Similarly, equating coefficients of powers of h in the expansions on each side of the equation

$$h\frac{\partial V}{\partial h} = (z - h)\frac{\partial V}{\partial z},$$

we have

$$z\frac{dP_n(z)}{dz} - \frac{dP_{n-1}(z)}{dz} = nP_n(z),$$

which is the formula (15.4). The others can be deduced from these.

Example 15.2.1 (Hargreaves) Shew that, for all values of n,

$$\frac{d}{dz}\{z(P_n{}^2 + P_{n+1}^2) - 2P_n P_{n+1}\} = (2n + 3)P_{n+1}^2 - (2n + 1)P_n^2.$$

---

[2] This relation was given in substance by Lagrange [392] in a memoir on probability.
[3] If $n = 0$, we have $P_0(z) = 1$, $P_{-1}(z) = 1$, and the result (15.4) is true but trivial.
[4] The reader is recommended to justify these processes.

Example 15.2.2 (Trinity, 1900)   If

$$M_n(x) = \left[\left(\frac{d}{dz}\right)^n (ze^{xz} \operatorname{cosech} z)\right]_{z=0},$$

shew that

$$\frac{dM_n(x)}{dx} = nM_{n-1}(x) \quad \text{and} \quad \int_{-1}^{1} M_n(x)\,dx = 0.$$

Example 15.2.3 (Clare, 1898)   Prove that if m and n are integers such that $m \le n$, both being even or both odd,

$$\int_{-1}^{1} \frac{dP_m(z)}{dz}\frac{dP_n(z)}{dz}\,dz = m(m+1).$$

Example 15.2.4 (Math. Trip. 1897)   Prove that, if m, n are integers and $m \ge n$,

$$\int_{-1}^{1} \frac{d^2 P_m(z)}{dz^2}\frac{d^2 P_n(z)}{dz^2}\,dz = \frac{(n-1)n(n+1)(n+2)}{48}\{3m(m+1) - n(n+1) + 6\}$$
$$\times \{1 + (-1)^{n+m}\}.$$

15.211 The expression of any polynomial as a series of Legendre polynomials

Let $f_n(z)$ be a polynomial of degree n in z. Then it is always possible to choose $a_0, a_1, \ldots, a_n$ so that

$$f_n(z) \equiv a_0 P_0(z) + a_1 P_1(z) + \cdots + a_n P_n(z),$$

for, on equating coefficients of z^n, $z^{n-1}, \ldots$ on each side, we obtain equations which determine $a_n, a_{n-1}, \ldots$ uniquely in turn, in terms of the coefficients of powers of z in $f_n(z)$.

To determine $a_0, a_1, \ldots, a_n$ in the most simple manner, multiply the identity by $P_r(z)$, and integrate. Then, by §15.14,

$$\int_{-1}^{1} f_n(z)P_r(z)\,dz = \frac{2a_r}{2r+1},$$

when $r = 0, 1, 2, \ldots, n$; when $r > n$, the integral on the left vanishes.

Example 15.2.5 (Legendre [421])   Given $z^n = a_0 P_0(z) + a_1 P_1(z) + \cdots + a_n P_n(z)$, determine $a_0, a_1, \ldots, a_n$.
Hint. Equate coefficients of z^n on both sides; this gives

$$a_n = \frac{2^n (n!)^2}{(2n)!}.$$

Let $I_{n,m} = \int_{-1}^{1} z^n P_m(z)\,dz$, so that, by the result just given,

$$I_{m,m} = \frac{2^{m+1}(m!)^2}{(2m+1)!}.$$

Now when $n-m$ is odd, $I_{n,m}$ is the integral of an odd function with limits ± 1, and so vanishes; and $I_{n,m}$ also vanishes when $n - m$ is negative and even.

To evaluate $I_{n,m}$ when $n - m$ is a positive even integer, we have from Legendre's equation

$$m(m + 1) \int_{-1}^{1} z^n P_m(z) \, dz = - \int_{-1}^{1} z^n \frac{d}{dz} \{(1 - z^2) P'_m(z)\} \, dz$$

$$= - \left[z^n (1 - z^2) P'_m(z) \right]_{-1}^{1} + n \int_{-1}^{1} z^{n-1} (1 - z^2) P'_m(z) \, dz$$

$$= n \left[z^{n-1} (1 - z^2) P_m(z) \right]_{-1}^{1}$$

$$- n \int_{-1}^{1} \{(n - 1) z^{n-2} - (n + 1) z^n\} P_m(z) \, dz,$$

on integrating by parts twice; and so

$$m(m + 1) I_{n,m} = n(n + 1) I_{n,m} - n(n - 1) I_{n-2,m}.$$

Therefore

$$I_{n,m} = \frac{n(n - 1)}{(n - m)(n + m + 1)} I_{n-2,m}$$

$$= \frac{n(n - 1) \cdots (m + 1)}{(n - m)(n - 2 - m) \cdots 2 \cdot (n + m + 1)(n + m - 1) \cdots (2m + 3)} I_{m,m},$$

by carrying on the process of reduction.

Consequently

$$I_{n,m} = \frac{2^{m+1} n! (\tfrac{1}{2}n + \tfrac{1}{2}m)!}{(\tfrac{1}{2}n - \tfrac{1}{2}m)! (n + m + 1)!},$$

and so

$$a_m = \begin{cases} 0, & \text{when } n - m \text{ is odd or negative and} \\[2mm] \dfrac{(2m + 1) 2^m n! (\tfrac{1}{2}n + \tfrac{1}{2}m)!}{(\tfrac{1}{2}n - \tfrac{1}{2}m)! (n + m + 1)!} & \text{when } n - m \text{ is even and positive.} \end{cases} \quad (15.7)$$

Example 15.2.6 Express $\cos n\theta$ as a series of Legendre polynomials of $\cos \theta$ when n is an integer.

Example 15.2.7 (St John's, 1899) Evaluate the integrals

$$\int_{-1}^{1} z P_n(z) P_{n+1}(z) \, dz, \quad \int_{-1}^{1} z^2 P_n(z) P_{n+1}(z) \, dz.$$

Example 15.2.8 (Trinity, 1894) Shew that

$$\int_{-1}^{1} (1 - z^2) \{P'_n(z)\}^2 \, dz = \frac{2n(n + 1)}{2n + 1}.$$

Example 15.2.9 (St John's, 1898) Shew that

$$n P_n(\cos \theta) = \sum_{r=1}^{n} \cos r\theta P_{n-r}(\cos \theta).$$

Example 15.2.10 (Trinity, 1895)   If $u_n = \int_{-1}^{1} (1 - z^2)^n P_{2m}(z) \, dz$, where $m < n$, shew that

$$(n - m)(2n + 2m + 1)u_n = 2n^2 u_{n-1}.$$

15.22 *Murphy's expression of $P_n(z)$ as a hypergeometric function*

This appears in [482]. Murphy's result was obtained only for the Legendre polynomials.

Since (§15.13) Legendre's equation is a particular case of Riemann's equation, it is to be expected that a formula can be obtained giving $P_n(z)$ in terms of hypergeometric functions. To determine this formula, take the integral of §15.2 for the Legendre function and suppose that $|1 - z| < 2$; to fix the contour C, let δ be any constant such that $0 < \delta < 1$, and suppose that z is such that $|1 - z| \le 2(1 - \delta)$; and then take C to be the circle[5] $|1 - t| = 2 - \delta$. Since

$$\left| \frac{1 - z}{1 - t} \right| \le \frac{2 - 2\delta}{2 - \delta} < 1,$$

we may expand $(t - z)^{-n-1}$ into the uniformly convergent series[6]

$$(t - z)^{-n-1} = (t - 1)^{-n-1} \left\{ 1 + (n + 1)\frac{z - 1}{t - 1} + \frac{(n + 1)(n + 2)}{2!} \left(\frac{z - 1}{t - 1} \right)^2 + \cdots \right\}.$$

Substituting this result in Schläfli's integral, and integrating term-by-term (§4.7) we get

$$P_n(z) = \sum_{r=0}^{\infty} \frac{(z - 1)^r}{2^{n+1}\pi i} \frac{(n + 1)(n + 2) \cdots (n + r)}{r!} \int_A^{(1+, z+)} \frac{(t^2 - 1)^n}{(t - 1)^{n+1+r}} \, dt$$

$$= \sum_{r=0}^{\infty} \frac{(z - 1)^r (n + 1)(n + 2) \cdots (n + r)}{2^n (r!)^2} \left[\frac{d^r}{dt^r}(t + 1)^n \right]_{t=1},$$

by §5.22. Since $\arg(t + 1) = 0$ when $t = 1$, we get

$$\left[\frac{d^r}{dt^r}(t + 1)^n \right]_{t=1} = 2^{n-r} n(n - 1) \cdots (n - r + 1),$$

and so, when $|1 - z| \le 2(1 - \delta) < 2$, we have

$$P_n(z) = \sum_{r=0}^{\infty} \frac{(n + 1)(n + 2) \cdots (n + r) \cdot (-n)(1 - n) \cdots (r - 1 - n)}{(r!)^2} \left(\frac{1}{2} - \frac{1}{2}z \right)^r$$

$$= F\left(n + 1, -n; 1; \frac{1}{2} - \frac{1}{2}z \right).$$

This is the required expression; it supplies a reason §14.53 why the cut from -1 to $-\infty$ could not be avoided in §15.2.

Corollary 15.2.1   *From this result, it is obvious that, for all values of n,*

$$P_n(z) = P_{-n-1}(z).$$

---

[5] This circle contains the points $t = 1, t = z$.
[6] The series terminates if n be a negative integer.

Note When n is a positive integer, the result gives the Legendre polynomial as a polynomial in $1 - z$ with simple coefficients.

Example 15.2.11 (Trinity, 1907) Shew that, if m be a positive integer,

$$\left\{ \frac{d^{m+1} P_{m+n}(z)}{dz^{m+1}} \right\}_{z=1} = \frac{\Gamma(2m + n + 2)}{2^{m+1} (m + 1)! \Gamma(n)}.$$

Example 15.2.12 (Murphy) Shew that the Legendre polynomial $P_n(\cos \theta)$ is equal to

$$(-1)^n F \left(n + 1, -n; 1; \cos^2(\theta/2) \right),$$

and to

$$\cos^n (\theta/2) \times F \left(-n, -n; 1; \tan^2(\theta/2) \right).$$

15.23 Laplace's integrals for $P_n(z)$

This appears in Laplace's *Mécanique Céleste*, [409, Livre XI, Ch. 2.]. For the contour employed in this section, and for some others introduced later in the chapter, we are indebted to Mr J. Hodgkinson.

We shall next shew that, for all values of n and for certain values of z, the Legendre function $P_n(z)$ can be represented by the integral (called *Laplace's first integral*)

$$\frac{1}{\pi} \int_0^\pi \{z + (z^2 - 1)^{\frac{1}{2}} \cos \phi\}^n \, d\phi.$$

(A) *Proof applicable only to the Legendre polynomials.*

When n is a positive integer, we have, by §15.12,

$$P_n(z) = \frac{1}{2^{n+1} \pi i} \int_C \frac{(t^2 - 1)^n}{(t - z)^{(n+1)}} \, dt,$$

where C is any contour which encircles the point z counter-clockwise. Take C to be the circle with centre z and radius $|z^2 - 1|^{1/2}$, so that, on C, $t = z + (z^2 - 1)^{\frac{1}{2}} e^{i\phi}$, where ϕ may be taken to increase from $-\pi$ to π.

Making the substitution, we have, for *all* values of z,

$$P_n(z) = \frac{1}{2^{n+1} \pi i} \int_{-\pi}^\pi \left(\frac{\{z - 1 + (z^2 - 1)^{\frac{1}{2}} e^{i\phi}\}\{z + 1 + (z^2 - 1)^{\frac{1}{2}} e^{i\phi}\}}{(z^2 - 1)^{\frac{1}{2}} e^{i\phi}} \right)^n i \, d\phi$$

$$= \frac{1}{2\pi} \int_{-\pi}^\pi \{z + (z^2 - 1)^{\frac{1}{2}} \cos \phi\}^n \, d\phi$$

$$= \frac{1}{\pi} \int_0^\pi \{z + (z^2 - 1)^{\frac{1}{2}} \cos \phi\}^n \, d\phi,$$

since the integrand is an even function of ϕ. The choice of the branch of the two-valued function $(z^2 - 1)^{\frac{1}{2}}$ is obviously a matter of indifference.

(B) *Proof applicable to the Legendre functions, where n is unrestricted.*

Make the same substitution as in (A) in Schläfli's integral defining $P_n(z)$; it is, however, necessary in addition to verify that $t = 1$ is inside the contour and $t = -1$ outside it, and it is also necessary that we should specify the branch of $\{z + (z^2 - 1)^{\frac{1}{2}} \cos \phi\}^n$, which is now a many-valued function of ϕ.

The conditions that $t = 1, t = -1$ should be inside and outside C respectively are that the distances of z from these points should be less and greater than $|z^2 - 1|^{\frac{1}{2}}$. These conditions are both satisfied if $|z - 1| < |z + 1|$, which gives $\text{Re}(z) > 0$, and so (giving $\arg z$ its principal value) *we must have* $|\arg z| < \frac{1}{2}\pi$.

Therefore

$$P_n(z) = \frac{1}{2\pi} \int_{-\pi}^{\pi} \{z + (z^2 - 1)^{\frac{1}{2}} \cos \phi\}^n \, d\phi,$$

where the value of $\arg\{z + (z^2 - 1)^{\frac{1}{2}} \cos \phi\}$ is specified by the fact that it, being equal to $\arg(t^2 - 1) - \arg(t - z)$, is numerically less than π when t is on the real axis and on the right of z (see §15.2).

Now as ϕ increases from $-\pi$ to π, $z + (z^2 - 1)^{\frac{1}{2}} \cos \phi$ describes a straight line in the Argand diagram going from $z - (z^2 - 1)^{\frac{1}{2}}$ to $z + (z^2 - 1)^{\frac{1}{2}}$ and back again; and since this line does not pass through the origin[7], $\arg\{z + (z^2 - 1)^{\frac{1}{2}} \cos \phi\}$ does not change by so much as π on the range of integration.

Now suppose that the branch of $\{z + (z^2 - 1)^{\frac{1}{2}} \cos \phi\}^n$ which has to be taken is such that it reduces to $z^n e^{2k\pi i} n$ (where k is an integer) when $\phi = \frac{1}{2}\pi$. Then

$$P_n(z) = \frac{e^{2nk\pi i}}{2\pi} \int_{-\pi}^{\pi} \{z + (z^2 - 1)^{\frac{1}{2}} \cos \phi\}^n \, d\phi,$$

where now that branch of the many-valued function is taken which is equal to z^n when $\phi = \frac{1}{2}\pi$. Now make $z \to 1$ by a path which avoids the zeros of $P_n(z)$; since $P_n(z)$ and the integral are analytic functions of z when $|\arg z| < \frac{1}{2}\pi$, k does not change as z describes the path. And so we get $e^{2nk\pi i} = 1$. Therefore, when $|\arg z| < \frac{1}{2}\pi$ and n is unrestricted,

$$P_n(z) = \frac{1}{2\pi} \int_{-\pi}^{\pi} \{z + (z^2 - 1)^{\frac{1}{2}} \cos \phi\}^n \, d\phi,$$

where $\arg\{z + (z^2 - 1)^{\frac{1}{2}} \cos \phi\}$ is to be taken equal to $\arg z$ when $\phi = \frac{1}{2}\pi$. This expression for $P_n(z)$, which may, again, obviously be written

$$\frac{1}{\pi} \int_0^{\pi} \{z + (z^2 - 1)^{\frac{1}{2}} \cos \phi\}^n \, d\phi,$$

is known as *Laplace's first integral* for $P_n(z)$.

Corollary 15.2.2 *From Corollary 15.2.1 it is evident that, when* $|\arg z| < \frac{\pi}{2}$,

$$P_n(z) = \frac{1}{n} \int_0^{\pi} \frac{d\phi}{\{z + (z^2 - 1)^{\frac{1}{2}} \cos \phi\}^{n+1}},$$

a result, due to Jacobi [351], known as Laplace's second integral for $P_n(z)$.

---

[7] It only does so if z is a pure imaginary; and such values of z have been excluded.

Example 15.2.13   Obtain Laplace's first integral by considering

$$\sum_{n=0}^{\infty} h^n \int_0^{\pi} \{z + (z^2 - 1)^{\frac{1}{2}} \cos \phi\}^n \, d\phi,$$

and using Example 6.2.1.

Example 15.2.14   Shew, by direct differentiation, that Laplace's integral is a solution of Legendre's equation.

Example 15.2.15 (Binet)   If $s < 1, |h| < 1$ and

$$(1 - 2h \cos \theta + h^2)^{-s} = \sum_{n=0}^{\infty} b_n \cos n\theta,$$

shew that

$$b_n = \frac{2 \sin s\pi}{\pi} \int_0^1 \frac{h^n x^{n+s-1} \, dx}{(1 - x)^s (1 - xh^2)^s}.$$

Example 15.2.16   When $z > 1$, deduce Laplace's second integral from his first integral by the substitution

$$\{z - (z^2 - 1)^{\frac{1}{2}} \cos \theta\}\{z + (z^2 - 1)^{\frac{1}{2}} \cos \phi\} = 1.$$

Example 15.2.17   By expanding in powers of $\cos \phi$, shew that for a certain range of values of z,

$$\frac{1}{\pi} \int_0^{\pi} \{z + (z^2 - 1)^{\frac{1}{2}} \cos \phi\}^n \, d\phi = z^n F \left(-\frac{n}{2}, \frac{1 - n}{2}; 1; 1 - z^{-2} \right).$$

Example 15.2.18   Shew that Legendre's equation is defined by the scheme

$$P \left\{ \begin{matrix} 0 & \infty & 1 \\ -\frac{1}{2}n & \frac{1}{2} + \frac{1}{2}n & 0 & \xi \\ \frac{1}{2} + \frac{1}{2}n & -\frac{1}{2}n & 0 \end{matrix} \right\},$$

where $z = \frac{1}{2}(\xi^{\frac{1}{2}} + \xi^{-\frac{1}{2}})$.

15.231 The Mehler–Dirichlet integral for $P_n(z)$

This comes from Dirichlet [174] and Mehler [465].

Another expression for the Legendre function as a definite integral may be obtained in the following way: For all values of n, we have, by the preceding theorem,

$$P_n(z) = \frac{1}{\pi} \int_0^{\pi} \{z + (z^2 - 1)^{1/2} \cos \phi\}^n \, d\phi.$$

In this integral, replace the variable ϕ by a new variable h, defined by the equation $h = z + (z^2 - 1)^{1/2} \cos \phi$, and we get

$$P_n(z) = \frac{i}{\pi} \int_{z-(z^2-1)^{1/2}}^{z+(z^2-1)^{1/2}} h^n (1 - 2hz + h^2)^{-1/2} \, dh;$$

the path of integration is a straight line, arg h is determined by the fact that $h = z$ when $\phi = \frac{1}{2}\pi$, and $(1 - 2hz + h^2)^{-1/2} = -i\,(z^2 - 1)^{1/2}\sin\phi$.

Now let $z = \cos\theta$; then

$$P_n(\cos\theta) = \frac{i}{\pi}\int_{e^{-i\theta}}^{e^{i\theta}} h^n(1 - 2hz + h^2)^{-1/2}\,dh.$$

Now (θ being restricted so that $-\frac{\pi}{2} < \theta < \frac{\pi}{2}$ when n is not a positive integer) the path of integration may be deformed[8] into that arc of the circle $|h| = 1$ which passes through $h = 1$, and joins the points $h = e^{-i\theta}$, $h = e^{i\theta}$, since the integrand is analytic throughout the region between this arc and its chord[9].

Writing $h = e^{i\phi}$ we get

$$P_n(\cos\theta) = \frac{1}{\pi}\int_{-\theta}^{\theta} \frac{e^{(n+1/2)i\phi}\,d\phi}{(2\cos\phi - 2\cos\theta)^{1/2}},$$

and so

$$P_n(\cos\theta) = \frac{2}{\pi}\int_0^{\theta} \frac{\cos(n + \frac{1}{2})\phi\,d\phi}{\{2(\cos\phi - \cos\theta)\}^{1/2}};$$

it is easy to see that the positive value of the square root is to be taken. This is known as *Mehler's simplified form of Dirichlet's integral*. The result is valid for all values of n.

Example 15.2.19   Prove that, when n is a positive integer,

$$P_n(\cos\theta) = \frac{2}{\pi}\int_{\theta}^{\pi} \frac{\sin(n + \frac{1}{2})\phi\,d\phi}{\{2(\cos\theta - \cos\phi)\}^{1/2}}.$$

(Write $\pi - \theta$ for θ and $\pi - \phi$ for ϕ in the result just obtained.)

Example 15.2.20   Prove that

$$P_n(\cos\theta) = \frac{1}{2\pi i}\int \frac{h^n\,dh}{(h^2 - 2h\cos\theta + 1)^{1/2}},$$

the integral being taken along a closed path which encircles the two points $h = e^{\pm i\theta}$, and a suitable meaning being assigned to the radical.

Note   Hence (or otherwise) prove that, if θ lie between $\frac{1}{6}\pi$ and $\frac{5}{6}\pi$,

$$P_n(\cos\theta) = \frac{4}{\pi}\frac{2\cdot 4\cdots 2n}{3\cdot 5\cdots(2n+1)}\left\{\begin{array}{l} \dfrac{\cos(n\theta + \phi)}{(2\sin\theta)^{\frac{1}{2}}} + \dfrac{1^2}{2(2n+3)}\dfrac{\cos(n\theta + 3\phi)}{(2\sin\theta)^{\frac{3}{2}}} \\[2ex] + \dfrac{1^2\cdot 3^2}{2\cdot 4\cdot(2n+3)(2n+5)}\dfrac{\cos(n\theta + 5\phi)}{(2\sin\theta)^{\frac{5}{2}}} + \cdots\cdots \end{array}\right\},$$

where ϕ denotes $\frac{1}{2}\theta - \frac{1}{4}\pi$.

Shew also that the first few terms of the series give an approximate value of $P_n(\cos\theta)$ for

---

[8]   If θ be complex and Re $\cos\theta > 0$ the deformation of the contour presents slightly greater difficulties. The reader will easily modify the analysis given to cover this case.

[9]   The integrand is not analytic at the ends of the arc but behaves like $(h - e^{\pm i\theta})^{-1/2}$ near them; but if the region be indented §6.23 at $e^{\pm i\theta}$ and the radii of the indentations be made to tend to zero, we see that the deformation is legitimate.

all values of θ between 0 and π which are not nearly equal to either 0 or π. And explain how this theorem may be used to approximate to the roots of the equation $P_n(\cos\theta) = 0$. (See Heine [287, vol. I, p. 171]; Darboux [161].)

15.3 Legendre functions of the second kind

We have hitherto considered only one solution of Legendre's equation, namely $P_n(z)$. We proceed to find a second solution.

We have seen (§15.2) that Legendre's equation is satisfied by

$$\int (t^2 - 1)^n (t - z)^{-n-1}\, dt,$$

taken round any contour such that the integrand returns to its initial value after describing it. Let D be a figure-of-eight contour formed in the following way: let z be not a real number between ± 1; draw an ellipse in the t-plane with the points ± 1 as foci, the ellipse being so small that the point $t = z$ is outside. Let A be the end of the major axis of the ellipse on the right of $t = 1$. Let the contour D start from A and describe the circuits $(1^-, -1^+)$, returning to A (cf. §12.43), and lying wholly inside the ellipse. Let $|\arg z| \le \pi$ and let $|\arg(z-t)| \to \arg z$ as $t \to 0$ on the contour. Let $\arg(t + 1) = \arg(t - 1) = 0$ at A.

Then a solution of Legendre's equation valid in the plane (cut along the real axis from 1 to $-\infty$) is

$$Q_n(z) = \frac{1}{4i \sin n\pi} \int_D \frac{(t^2 - 1)^n\, dt}{2^n (z - t)^{n+1}},$$

if n is not an integer.

When $\mathrm{Re}(n + 1) > 0$, we may deform the path of integration as in §12.43, and get

$$Q_n(z) = \frac{1}{2^{n+1}} \int_{-1}^{1} (1 - t^2)^n (z - t)^{-n-1}\, dt$$

(where $\arg(1 - t) = \arg(1 + t) = 0$); this will be taken as the definition of $Q_n(z)$ when n is a positive integer or zero. When n is a negative integer ($= -m - 1$) Legendre's differential equation for functions of degree n is identical with that for functions of degree m, and accordingly we shall take the two fundamental solutions to be $P_m(z)$, $Q_m(z)$.

We call $Q_n(z)$ *the Legendre function of degree n of the second kind*.

15.31 Expansion of $Q_n(z)$ as a power series

We now proceed to express the Legendre function of the second kind as a power series in z^{-1}. We have, when the real part of $n + 1$ is positive,

$$Q_n(z) = \frac{1}{2^{n+1}} \int_{-1}^{1} (1 - t^2)^n (z - t)^{-n-1}\, dt.$$

Suppose that $|z| > 1$. Then the integrand can be expanded in a series uniformly convergent with regard to t, so that

$$Q_n(z) = \frac{1}{2^{n+1} z^{n+1}} \int_{-1}^{1} (1 - t^2)^n \left(1 - \frac{t}{z}\right)^{-n-1} dt$$

$$= \frac{1}{2^{n+1} z^{n+1}} \int_{-1}^{1} (1 - t^2)^n \left\{1 + \sum_{r=1}^{\infty} \left(\frac{t}{z}\right)^r \frac{(n+1)(n+2)\cdots(n+r)}{r!}\right\} dt$$

$$= \frac{1}{2^n z^{n+1}} \left[\int_0^1 (1 - t^2)^n dt + \sum_{s=1}^{\infty} \frac{(n+1)\cdots(n+2s)}{2s! z^{2s}} \int_0^1 (1 - t^2)^n t^{2s} dt\right],$$

where $r = 2s$, the integrals arising from odd values of r vanishing. Writing $t^2 = u$ we get without difficulty, from §12.41,

$$Q_n(z) = \frac{\pi^{\frac{1}{2}} \Gamma(n+1)}{2^{n+1} \Gamma(n + \frac{3}{2})} \frac{1}{z^{n+1}} F\left(\frac{1}{2}n + \frac{1}{2}, \frac{1}{2}n + 1; n + \frac{3}{2}; z^{-2}\right).$$

The proof given above applies only when the real part of $(n + 1)$ is positive (see §4.5); but a similar process can be applied to the integral

$$Q_n(z) = \frac{1}{4i \sin n\pi} \int_D \frac{1}{2n} (t^2 - 1)^n (z - t)^{-n-1} dt,$$

the coefficients being evaluated by writing $\int_D (t^2 - 1)^n t^r \, dt$ in the form

$$e^{n\pi i} \int_0^{(1-)} (1 - t^2)^n t^r \, dt + e^{n\pi i} \int_0^{(-1+)} (1 - t^2)^n t^r \, dt;$$

and then, writing $t^2 = u$ and using §12.43, the same result is reached, so that the formula

$$Q_n(z) = \frac{\pi^{\frac{1}{2}}}{2^{n+1}} \frac{\Gamma(n+1)}{\Gamma(n + \frac{3}{2})} \frac{1}{z^{n+1}} F\left(\frac{1}{2}n + \frac{1}{2}, \frac{1}{2}n + 1; n + \frac{3}{2}; \frac{1}{z^2}\right)$$

is true for unrestricted values of n (negative integer values excepted) and for all values[10] of z, such that $|z| > 1$, $|\arg z| < \pi$.

Example 15.3.1   Shew that, when n is a positive integer,

$$Q_n(z) = \frac{(-2)^n n!}{(2n)!} \frac{d^n}{dz^n} \left\{(z^2 - 1)^n \int_z^{\infty} (v^2 - 1)^{-n-1} dv\right\}.$$

It is easily verified that Legendre's equation can be derived from the equation

$$(1 - z^2)\frac{d^2 w}{dz^2} + 2(n - 1)z\frac{dw}{dz} + 2nw = 0,$$

by differentiating n times and writing $u = \dfrac{d^n w}{dz^n}$. Two independent solutions of this equation are found to be

$$(z^2 - 1)^n \quad \text{and} \quad (z^2 - 1)^n \int_z^{\infty} (v^2 - 1)^{-n-1} dv.$$

---

[10]   When n is a positive integer it is unnecessary to restrict the value of arg z.

It follows that

$$\frac{d^n}{dz^n}\left\{(z^2-1)^n\int_z^\infty (v^2-1)^{-n-1}\,dv\right\}$$

is a solution of Legendre's equation. As this expression, when expanded in ascending powers of z^{-1}, commences with a term in z^{-n-1}, it must be a constant multiple[11] of $Q_n(z)$; and on comparing the coefficient of z^{-n-1} in this expression with the coefficient of z^{-n-1} in the expansion of $Q_n(z)$, as found above, we obtain the required result.

Example 15.3.2 Shew that, when n is a positive integer, the Legendre function of the second kind can be expressed by the formula

$$Q_n(z) = 2^n n! \int_z^\infty \int_v^\infty \int_v^\infty \cdots \int_v^\infty (v^2-1)^{-n-1}\,(dv)^{n+1}.$$

Example 15.3.3 Shew that, when n is a positive integer,

$$Q_n(z) = \sum_{t=0}^{n} \frac{2^n \cdot n!}{t!(n-t)!}(-z)^{n-t}\int_z^\infty v^t(v^2-1)^{-n-1}\,dv.$$

This result can be obtained by applying the general integration theorem

$$\int_z^\infty \int_v^\infty \int_v^\infty \cdots \int_v^\infty f(v)\,(dv)^{n+1} = \sum_{t=0}^{n} \frac{(-z)^{n-t}}{t!(n-t)!}\int_z^\infty v^t f(v)\,dv$$

to the preceding result.

15.32 *The recurrence formulae for* $Q_n(z)$

The functions $P_n(z)$ and $Q_n(z)$ have been defined by means of integrals of precisely the same form, namely

$$\int (t^2-1)^n(t-z)^{-n-1}\,dt,$$

taken round different contours.

It follows that the general proof of the recurrence formulae for $P_n(z)$, given in §15.21, is equally applicable to the function $Q_n(z)$; and hence that *the Legendre function of the second kind satisfies the recurrence formulae*

$$Q'_{n+1}(z) - zQ'_n(z) = (n+1)Q_n(z),$$
$$(n+1)Q_{n+1}(z) - (2n+1)zQ_n(z) + nQ_{n-1}(z) = 0,$$
$$zQ'_n(z) - Q'_{n-1}(z) = nQ_n(z),$$
$$Q'_{n+1}(z) - Q'_{n-1}(z) = (2n+1)Q_n(z),$$
$$(z^2-1)Q'_n(z) = nzQ_n(z) - nQ_{n-1}(z).$$

Example 15.3.4 Shew that

$$Q_0(z) = \frac{1}{2}\log\frac{z+1}{z-1}, \qquad Q_1(z) = \frac{1}{2}z\log\frac{z+1}{z-1} - 1,$$

----

[11] $P_n(z)$ contains *positive* powers of z when n is an integer.

and deduce that

$$Q_2(z) = \frac{1}{2} P_2(z) \log \frac{z+1}{z-1} - \frac{3}{2} z$$

and that

$$\frac{Q_n(z)}{P_n(z)} = \frac{1}{2} \log \frac{z+1}{z-1} - \cfrac{1}{z - \cfrac{1^2}{3z - \cfrac{2^2}{5z - \cfrac{3^2}{7z - \cdots - \cfrac{(n-1)^2}{(2n-1)z}}}}}.$$

Example 15.3.5 Shew by the recurrence formulae that, when n is a positive integer[12],

$$\frac{1}{2} P_n(z) \log \left(\frac{z+1}{z-1} \right) - Q_n(z) = f_{n-1}(z),$$

where $f_{n-1}(z)$ consists of the positive (and zero) powers of z in the expansion of $\frac{1}{2} P_n(z) \log \left(\frac{z+1}{z-1} \right)$ in descending powers of z.

Note   This example shews the nature of the singularities of $Q_n(z)$ at ± 1, when n is an integer, which make the cut from -1 to $+1$ necessary. For the connexion of the result with the theory of continued fractions, see Gauss [233], and Frobenius [226]; the formulae of Example 15.3.4 are due to them.

15.33 The Laplacian integral for Legendre functions of the second kind

This formula was first given by Heine [287, p. 147].

It will now be proved that, when $\mathrm{Re}(n+1) > 0$,

$$Q_n(z) = \int_0^\infty \left\{ z + (z^2 - 1)^{\frac{1}{2}} \cosh \theta \right\}^{-n-1} d\theta,$$

where $\arg\{z + (z^2 - 1)^{\frac{1}{2}} \cosh \theta\}$ has its principal value when $\theta = 0$, if n be not an integer.
First suppose that $z > 1$. In the integral of §15.3, viz.

$$Q_n(z) = \frac{1}{2^{n+1}} \int_{-1}^1 (1 - t^2)^n (z - t)^{-n-1} \, dt,$$

write

$$t = \frac{e^\theta (z+1)^{1/2} - (z-1)^{1/2}}{e^\theta (z+1)^{1/2} + (z-1)^{1/2}},$$

---

[12] If $-1 < z < 1$, it is apparent from these formulae that $Q_n(z + 0i) - Q_n(z - 0i) = -\pi i P_n(z)$. It is convenient to *define* $Q_n(z)$ for such values of z to be $\frac{1}{2} Q_n(z + 0i) + \frac{1}{2} Q_n(z - 0i)$. The reader will observe that this function satisfies Legendre's equation for real values of z.

so that the range $(-1, 1)$ of real values of t corresponds to the range $(-\infty, \infty)$ of real values of θ. It then follows (as in (15.1)) by straightforward substitution that

$$Q_n(z) = \frac{1}{2} \int_{-\infty}^{\infty} \left\{ z + (z^2 - 1)^{\frac{1}{2}} \cosh \theta \right\}^{-n-1} d\theta$$

$$= \int_{0}^{\infty} \left\{ z + (z^2 - 1)^{\frac{1}{2}} \cosh \theta \right\}^{-n-1} d\theta,$$

since the integrand is an even function of θ.

Note To prove the result for values of z not comprised in the range of real values greater than 1, we observe that the branch points of the integrand, *qua* function of z, are at the points ± 1 and at points where $z + (z^2 - 1)^{1/2} \cosh \theta$ vanishes; the latter are the points at which $z = \pm \coth \theta$. Hence $Q_n(z)$ and

$$\int_{0}^{\infty} \{ z + (z^2 - 1)^{\frac{1}{2}} \cosh \theta \}^{-n-1} d\theta$$

are both analytic[13] at all points of the plane when cut along the line joining the points $z = \pm 1$. By the theory of analytic continuation the equation proved for positive values of $z - 1$ persists for all values of z in the cut plane, provided that $\arg \{ z + (z^2 - 1)^{\frac{1}{2}} \cosh \theta \}$ is given a suitable value, namely that one which reduces to zero when $z - 1$ is positive. The integrand is one-valued in the cut plane [and so is $Q_n(z)$] when n is a positive integer; but $\arg \{ z + (z^2 - 1)^{\frac{1}{2}} \cosh \theta \}$ increases by 2π as $\arg z$ does so, and therefore if n be not a positive integer, a further cut has to be made from $z = -1$ to $z = -\infty$. These cuts give the necessary limitations on the value of z; and the cut when n is not an integer ensures that $\arg \{ z + (z^2 - 1)^{\frac{1}{2}} \} = 2 \arg \{ (z + 1)^{\frac{1}{2}} + (z - 1)^{\frac{1}{2}} \}$ has its principal value.

Example 15.3.6 Obtain this result for complex values of z by taking the path of integration to be a certain circular arc before making the substitution

$$t = \frac{e^\theta (z + 1)^{\frac{1}{2}} - (z - 1)^{\frac{1}{2}}}{e^\theta (z + 1)^{\frac{1}{2}} + (z - 1)^{\frac{1}{2}}},$$

where θ is real.

Example 15.3.7 (Trinity, 1893) Shew that, if $z > 1$ and $\coth a = z$,

$$Q_n(z) = \int_{0}^{a} \{ z - (z^2 - 1)^{\frac{1}{2}} \cosh u \}^n du,$$

where $\arg \{ z - (z^2 - 1)^{\frac{1}{2}} \cosh u \} = 0$.

15.34 Neumann's formula for $Q_n(z)$, when n is an integer

This appears in F. Neumann [490]. When n is a positive integer, and z is not a real number between 1 and -1, the function $Q_n(z)$ is expressed in terms of the Legendre function of the first kind by the relation

$$Q_n(z) = \frac{1}{2} \int_{-1}^{1} P_n(y) \frac{dy}{z - y},$$

---

[13] It is easy to shew that the integral has a unique derivative in the cut plane.

which we shall now establish.

When $|z| > 1$ we can expand the integrand in the uniformly convergent series

$$P_n(y) \sum_{m=0}^{\infty} \frac{y^m}{z^{m+1}}.$$

Consequently

$$\frac{1}{2} \int_{-1}^{1} P_n(y) \frac{dy}{z-y} = \frac{1}{2} \sum_{m=0}^{\infty} z^{-m-1} \int_{-1}^{1} y^m P_n(y)\, dy.$$

The integrals for which $m - n$ is odd or negative vanish (15.7); and so

$$\frac{1}{2} \int_{-1}^{1} P_n(y) \frac{dy}{z-y} = \frac{1}{2} \sum_{m=0}^{\infty} z^{-n-2m-1} \int_{-1}^{1} y^{n+2m} P_n(y)\, dy$$

$$= \frac{1}{2} \sum_{m=0}^{\infty} z^{-n-2m-1} \frac{2^{n+1}(n+2m)!(n+m)!}{m!(2n+2m+1)!}$$

$$= \frac{2^n(n!)^2}{(2n+1)!} z^{-n-1} F\left(\frac{n}{2} + \frac{1}{2}, \frac{n}{2} + 1; n + \frac{3}{2}; z^{-2}\right)$$

$$= Q_n(z),$$

by §15.31. The theorem is thus established for the case in which $|z| > 1$. Since each side of the equation

$$Q_n(z) = \frac{1}{2} \int_{-1}^{1} P_n(y) \frac{dy}{z-y}$$

represents an analytic function, even when $|z|$ is not greater than unity, provided that z is not a real number between -1 and $+1$, it follows that, with this exception, the result is true (§5.5) for all values of z.

The reader should notice that Neumann's formula apparently expresses $Q_n(z)$ as a one-valued function of z, whereas it is known to be many-valued (Example 15.3.4). The reason for the apparent discrepancy is that Neumann's formula has been established when the z-plane is cut from -1 to $+1$, and $Q_n(z)$ is *one-valued in the cut plane*.

Example 15.3.8   Shew that, when $-1 \le \operatorname{Re} z \le 1$, $|Q_n(z)| \le |\operatorname{Im} z|^{-1}$; and that for other values of z, $|Q_n(z)|$ does not exceed the larger of $|z-1|^{-1}$ and $|z+1|^{-1}$.

Example 15.3.9   Shew that, when n is a positive integer, $Q_n(z)$ is the coefficient of h^n in the expansion of

$$(1 - 2hz + h^2)^{-1/2} \operatorname{arccosh}\left\{\frac{h-z}{(z^2-1)^{1/2}}\right\}.$$

Hint. For $|h|$ sufficiently small,

$$\sum_{n=0}^{\infty} h^n Q_n(z) = \sum_{n=0}^{\infty} \frac{h^n}{2} \int_{-1}^{1} \frac{P_n(y)\, dy}{z-y} = \frac{1}{2} \int_{-1}^{1} \frac{(1 - 2hy + h^2)^{-1/2}\, dy}{(z-y)}$$

$$= (1 - 2hz + h^2)^{-1/2} \operatorname{arccosh}\left\{\frac{h-z}{(z^2-1)^{1/2}}\right\}.$$

This result has been investigated by Heine [287, vol. I, p. 134] and Laurent [411].

15.4 Heine's development of $(t - z)^{-1}$ as a series of Legendre polynomials in z

This appears in [286]. We shall now obtain an expansion which will serve as the basis of a general class of expansions involving Legendre polynomials. The reader will readily prove by induction from the recurrence formulae

$$(2m + 1)tQ_m(t) - (m + 1)Q_{m+1}(t) - mQ_{m-1}(t) = 0,$$
$$(2m + 1)zP_m(z) - (m + 1)P_{m+1}(z) - mP_{m-1}(z) = 0,$$

that

$$\frac{1}{t - z} = \sum_{m=0}^{n}(2m + 1)P_m(z)Q_m(t) + \frac{n + 1}{t - z}\{P_{n+1}(z)Q_n(t) - P_n(z)Q_{n+1}(t)\}.$$

Using Laplace's integrals, we have

$$P_{n+1}(z)Q_n(t) - P_n(z)Q_{n+1}(t) = \frac{1}{\pi}\int_0^{\pi}\int_0^{\infty}\frac{\{z + (z^2 - 1)^{\frac{1}{2}}\cos\phi\}^n}{\{t + (t^2 - 1)^{\frac{1}{2}}\cosh u\}^{n+1}}$$
$$\times [z + (z^2 - 1)^{\frac{1}{2}}\cos\phi - \{t + (t^2 - 1)^{\frac{1}{2}}\cosh u\}^{-1}]\,d\phi\,du.$$

Now consider

$$\left|\frac{z + (z^2 - 1)^{\frac{1}{2}}\cos\phi}{t + (t^2 - 1)^{\frac{1}{2}}\cosh u}\right|.$$

Let $\cosh a$, $\sinh a$ be the semi-major axes of the ellipses with foci ± 1 which pass through z and t respectively. Let θ be the eccentric angle of z; then

$$z = \cosh(a + i\theta),$$
$$|z \pm (z^2 - 1)^{\frac{1}{2}}\cos\phi| = |\cosh(a + i\theta) \pm \sinh(a + i\theta)\cos\phi|$$
$$= \{\cosh^2 a - \sin^2\theta + (\cosh^2 a - \cos^2\theta)$$
$$\cos^2\phi \pm 2\sinh a\cosh a\cos\phi\}^{\frac{1}{2}}.$$

This is a maximum for real values of ϕ when $\cos\phi = \mp 1$; and hence

$$|z \pm (z^2 - 1)^{\frac{1}{2}}\cos\phi|^2 \leq 2\cosh^2 a - 1 + 2\cosh a(\cosh^2 a - 1)^{\frac{1}{2}} = \exp(2\alpha).$$

Similarly $|t + (t^2 - 1)^{\frac{1}{2}}\cosh u| \leq \exp a$. Therefore

$$|P_{n+1}(z)Q_n(t) - P_n(z)Q_{n+1}(t)| \leq \pi^{-1}\exp\{n(a - \alpha)\}\int_0^{\pi}\int_0^{\infty} V\,d\phi\,du,$$

where

$$|V| = \left|\frac{z + (z^2 - 1)^{\frac{1}{2}}\cos\phi}{t + (t^2 - 1)^{\frac{1}{2}}\cosh u}\right| + |\{t + (t^2 - 1)^{\frac{1}{2}}\cosh u\}|^{-2}.$$

Therefore $|P_{n+1}(z)Q_n(t) - P_n(z)Q_{n+1}(t)| \to 0$, as $n \to \infty$, provided $a < \alpha$. And further, if t varies, α remaining constant, it is easy to see that the upper bound of $\int_0^{\pi}\int_0^{\infty} V\,d\phi\,du$ is independent of t, and so $P_{n+1}(z)Q_n(t) - P_n(z)Q_{n+1}(t)$ tends to zero uniformly with regard to t.

Hence *if the point z is in the interior of the ellipse which passes through the point t and has the points ± 1 for its foci, then the expansion*

$$\frac{1}{t-z} = \sum_{n=0}^{\infty} (2n+1)P_n(z)Q_n(t)$$

is valid; and if t be a variable point on an ellipse with foci ± 1 such that z is a fixed point inside it, the expansion converges uniformly with regard to t.

15.41 Neumann's expansion of an arbitrary function in a series of Legendre polynomials

This comes from Neumann [485]. See also Thomé [623]. Neumann also gives an expansion, in Legendre functions of both kinds, valid in the annulus bounded by two ellipses.

We proceed now to discuss the expansion of a function in a series of Legendre polynomials. The expansion is of special interest, as it stands next in simplicity to Taylor's series, among expansions in series of polynomials.

Let $f(z)$ be any function which is analytic inside and on an ellipse C, whose foci are the points $z = \pm 1$. We shall shew that

$$f(z) = a_0 P_0(z) + a_1 P_1(z) + a_2 P_2(z) + a_3 P_3(z) + \cdots,$$

where $a_0, a_1, a_2, \ldots$ are independent of z, this expansion being valid for all points z in the interior of the ellipse C. Let t be any point on the circumference of the ellipse.

Then, since $\sum_{n=0}^{\infty} (2n+1)P_n(z)Q_n(t)$ converges uniformly with regard to t,

$$f(z) = \frac{1}{2\pi i} \int_C \frac{f(t)\,dt}{t-z} = \frac{1}{2\pi i} \sum_{n=0}^{\infty} \int_C (2n+1)P_n(z)Q_n(t)f(t)\,dt$$

$$= \sum_{n=0}^{\infty} a_n P_n(z),$$

where

$$a_n = \frac{2n+1}{2\pi i} \int_C f(t)Q_n(t)\,dt.$$

This is the required expansion; since $\sum_{n=0}^{\infty} (2n+1)P_n(z)Q_n(t)$ may be proved[14] to converge uniformly with regard to z when z lies in any domain C' lying wholly inside C, the expansion converges uniformly throughout C'. Another form for a_n can therefore be obtained by integrating, as in §15.211, so that

$$a_n = \left(n + \tfrac{1}{2}\right) \int_{-1}^{1} f(x)P_n(x)\,dx.$$

---

[14] The proof is similar to the proof in §15.4 that convergence is uniform with regard to t.

A form of this equation which is frequently useful is

$$a_n = \frac{n + \frac{1}{2}}{2^n n!} \int_{-1}^{1} f^{(n)}(x) \cdot (1 - x^2)^n \, dx,$$

which is obtained by substituting for $P_n(x)$ from Rodrigues' formula and integrating by parts.

The theorem which bears the same relation to Neumann's expansion as Fourier's theorem bears to the expansion of §9.11 is as follows:

Let $f(t)$ be defined when $-1 \le t \le 1$, and let the integral of $(1 - t^2)^{-1/4} f(t)$ exist and be absolutely convergent; also let

$$a_n = \left(n + \tfrac{1}{2}\right) \int_{-1}^{1} f(t) P_n(t) \, dt.$$

Then $\sum a_n P_n(x)$ is convergent and has the sum $\frac{1}{2}\{f(x + 0) + f(x - 0)\}$ at any point x, for which $-1 < x < 1$, if any condition of the type stated at the end of §9.43 is satisfied.

For a proof, the reader is referred to memoirs by Hobson [317, 319] and Burkhardt [108].

Example 15.4.1   Shew that, if $\rho \ge 1$ be the radius of convergence of the series $\sum c_n z^n$, then $\sum c_n P_n(z)$ converges inside an ellipse whose semi-axes are $\frac{1}{2}(\rho + \rho^{-1})$ and $\frac{1}{2}(\rho - \rho^{-1})$.

Example 15.4.2   If $z = \left(\dfrac{y - 1}{y + 1}\right)^{\frac{1}{2}}$, $k^2 = \dfrac{(x - 1)(y + 1)}{(x + 1)(y - 1)}$, where $y > x > 1$, prove that

$$\int^{1} \frac{dz}{\{(1 - z^2)(1 - k^2 z^2)\}^{\frac{1}{2}}} = \{(x + 1)(y - 1)\}^{\frac{1}{2}} \sum_{n=0}^{\infty} P_n(x) Q_n(y).$$

Hint. Substitute Laplace's integrals on the right and integrate with regard to ϕ.

Example 15.4.3 (Frobenius [226])   Shew that

$$\frac{1}{2(y - x)} \log \frac{(x + 1)(y - 1)}{(x - 1)(y + 1)} = \sum_{n=0}^{\infty} (2n + 1) Q_n(x) Q_n(y).$$

15.5 Ferrers' associated Legendre functions $P_n^m(z)$ and $Q_n^m(z)$

We shall now introduce a more extended class of Legendre functions.

If m be a positive integer and $-1 < z < 1$, n being unrestricted (when n is a positive integer it is unnecessary to restrict the value of arg z), the functions

$$P_n^m(z) = (1 - z^2)^{m/3} \frac{d^m P_n(z)}{dz^m}, \qquad Q_n^m(z) = (1 - z^2)^{m/2} \frac{d^m Q_n(z)}{dz^m}$$

will be called Ferrers' associated Legendre functions of degree n and order m of the first and second kinds respectively. (Ferrers writes $T_n^m(z)$ for $P_n^m(z)$.)

It may be shewn that these functions satisfy a differential equation analogous to Legendre's equation.

For, differentiate Legendre's equation

$$(1 - z^2) \frac{d^2 y}{dz^2} - 2z \frac{dy}{dz} + n(n + 1)y = 0$$

m times and write v for $\dfrac{d^m y}{dz^m}$. We obtain the equation

$$(1 - z^2)\frac{d^2 v}{dz^2} - 2z(m + 1)\frac{dv}{dz} + (n - m)(n + m + 1)v = 0.$$

Write $w = (1 - z^2)^{m/2}v$, and we get

$$(1 - z^2)\frac{d^2 w}{dz^2} - 2z\frac{dw}{dz} + \left\{ n(n + 1) - \frac{m^2}{1 - z^2} \right\} w = 0.$$

This is the differential equation satisfied by $P_n^m(z)$ and $Q_n^m(z)$.

Note From the definitions given above, several expressions for the associated Legendre functions may be obtained. Thus, from Schläfli's formula we have

$$P_n^m(z) = \frac{(n + 1)(n + 2) \cdots (n + m)}{2^{n+1} \pi i}(1 - z^2)^{m/2} \int_A^{(1+,z+)} (t^2 - 1)^n (t - z)^{-n-m-1}\, dt,$$

where the contour does not enclose the point $t = -1$. Further, when n is a positive integer, we have, by Rodrigues' formula,

$$P_n^m(z) = \frac{(1 - z^2)^{m/2}}{2^n n!} \frac{d^{n+m}(z^2 - 1)^n}{dz^{n+m}}.$$

Example 15.5.1 (Olbricht)   Shew that Legendre's associated equation is defined by the scheme

$$P \left\{ \begin{matrix} 0 & \infty & 1 & \\ \frac{1}{2}m & n + 1 & \frac{1}{2}m & \frac{1}{2} - \frac{1}{2}z \\ -\frac{1}{2}m & -n & -\frac{1}{2}m & \end{matrix} \right\}.$$

15.51 The integral properties of the associated Legendre functions

The generalisation of the theorem of §15.14 is the following: When n, r, m are positive integers and $n > m$, $r > m$, then

$$\int_{-1}^1 P_n^m(z)P_r^m(z)\, dz = \begin{cases} 0 & (r \neq n), \\ \dfrac{2}{2n + 1}\dfrac{(n + m)!}{(n - m)!} & (r = n). \end{cases}$$

To obtain the first result, multiply the differential equations for $P_n^m(z)$, $P_r^m(z)$ by $P_r^m(z)$, $P_n^m(z)$ respectively and subtract; this gives

$$\frac{d}{dz}\left[(1 - z^2)\left\{ P_r^m(z)\frac{dP_n^m(z)}{dz} - P_n^m(z)\frac{dP_r^m(z)}{dz} \right\} \right]$$
$$+ (n - r)(n + r + 1)P_r^m(z)P_n^m(z) = 0.$$

On integrating between the limits -1, $+1$, the result follows when n and r are unequal, since the expression in square brackets vanishes at each limit. To obtain the second result, we observe that

$$P_n^{m+1}(z) = (1 - z^2)^{1/2}\frac{dP_n^m(z)}{dz} + mz(1 - z^2)^{-1/2}P_n^m(z);$$

squaring and integrating, we get

$$\int_{-1}^{1} \left[P_n^{m+1}(z) \right]^2 \, dz = \int_{-1}^{1} \left[(1 - z^2) \left\{ \frac{dP_n^m(z)}{dz} \right\}^2 + 2mz P_n^m(z) \frac{dP_n^m(z)}{dz} \right.$$

$$\left. + \frac{m^2 z^2}{1 - z^2} \{P_n^m(z)\}^2 \right] dz$$

$$= -\int_{-1}^{1} P_n^m(z) \frac{d}{dz} \left\{ (1 - z^2) \frac{dP_n^m(z)}{dz} \right\} \, dz - m \int_{-1}^{1} \{P_n^m(z)\}^2 \, dz$$

$$+ \int_{-1}^{1} \frac{m^2 z^2}{1 - z^2} \{P_n^m(z)\}^2 \, dz,$$

on integrating the first two terms in the first lines on the right by parts. If now we use the differential equation for $P_n^m(z)$ to simplify the first integral in the second line, we at once get

$$\int_{-1}^{1} \{P_n^{m+1}(z)\}^2 \, dz = (n - m)(n + m + 1) \int_{-1}^{1} \{P_n^m(z)\}^2 \, dz.$$

By repeated applications of this result we get

$$\int_{-1}^{1} \{P_n^m(z)\}^2 \, dz = (n - m + 1)(n - m + 2) \cdots n$$

$$\times (n + m)(n + m - 1) \cdots (n + 1) \int_{-1}^{1} \{P_n(z)\}^2 \, dz,$$

and so

$$\int_{-1}^{1} \{P_n^m(z)\}^2 \, dz = \frac{2}{2n + 1} \frac{(n + m)!}{(n - m)!}. \tag{15.8}$$

15.6 Hobson's definition of the associated Legendre functions

So far it has been taken for granted that the function $(1 - z^2)^{m/2}$ which occurs in Ferrers' definition of the associated functions is purely real; and since, in the more elementary physical applications of Legendre functions, it usually happens that $-1 < z < 1$, no complications arise. But as we wish to consider the associated functions as functions of a complex variable, it is undesirable to introduce an additional cut in the z-plane by giving $\arg(1 - z)$ its principal value.

Accordingly, in future, when z is not a real number such that $-1 < z < 1$, we shall follow Hobson in defining the associated functions by the equations

$$P_n^m(z) = (z^2 - 1)^{m/2} \frac{d^m P_n(z)}{dz^m} \qquad Q_n^m(z) = (z^2 - 1)^{m/2} \frac{d^m Q_n(z)}{dz^m},$$

where m is a positive integer, n is unrestricted and $\arg z$, $\arg(z + 1)$, $\arg(z - 1)$ have their principal values.

When m is unrestricted, $P_n^m(z)$ is defined by Hobson to be

$$\frac{1}{\Gamma(1 - m)} \left(\frac{z + 1}{z - 1} \right)^{m/2} F\left(-n, n + 1; 1 - m; \frac{1}{2} - \frac{1}{2}z \right);$$

and Barnes has given a definition of $Q_n^m(z)$ from which the formula

$$Q_n^m(z) = \frac{\sin(n+m)\pi}{\sin n\pi} \frac{\Gamma(n+m+1)\Gamma\left(\frac{1}{2}\right)}{2^{n+1}\Gamma\left(n+\frac{3}{2}\right)} \frac{(z^2-1)^{m/2}}{z^{n+m+1}}$$

$$\times F\left(\frac{n}{2}+\frac{m}{2}+1, \frac{n}{2}+\frac{m}{2}+\frac{1}{2}; n+\frac{3}{2}; z^{-2}\right)$$

may be obtained.

Throughout this work we shall take m to be a positive integer.

15.61 Expression of $P_n^m(z)$ as an integral of Laplace's type

If we make the necessary modification in the Schläfli integral of §15.5, in accordance with the definition of §15.6, we have

$$P_n^m(z) = \frac{(n+1)(n+2)\cdots(n+m)}{2^{n+1}\pi i}(z^2-1)^{m/2}\int_A^{(1+,z+)}(t^2-1)^n(t-z)^{-n-m-1}\,dt.$$

Write $t = z + (z^2-1)^{1/2}e^{i\phi}$, as in §15.23; then

$$P_n^m(z) = \frac{(n+1)(n+2)\cdots(n+m)}{2\pi}(z^2-1)^{m/2}\int_\alpha^{2\pi+\alpha}\frac{\{z+(z^2-1)^{1/2}\cos\phi\}^n}{\{(z^2-1)^{1/2}e^{i\phi}\}^m}\,d\phi,$$

where α is the value of ϕ when t is at A, so that $|\arg(z^2-1)^{1/2}+\alpha| < \pi$.

Now, as in §15.23, the integrand is a one-valued periodic function of the real variable ϕ with period 2π, and so

$$P_n^m(z) = \frac{(n+1)(n+2)\cdots(n+m)}{2\pi}\int_{-\pi}^{\pi}\{z+(z^2-1)^{1/2}\cos\phi\}^n e^{-mi\phi}\,d\phi.$$

Since $\{z+(z^2-1)^{1/2}\cos\phi\}^n$ is an even function of ϕ, we get, on dividing the range of integration into the parts $(-\pi,0)$ and $(0,\pi)$,

$$P_n^m(z) = \frac{(n+1)(n+2)\cdots(n+m)}{\pi}\int_0^{\pi}\{z+(z^2-1)^{1/2}\cos\phi\}^n\cos m\phi\,d\phi.$$

The ranges of validity of this formula, which is due to Heine (according as n is or is not an integer), are precisely those of the formula of §15.23.

Example 15.6.1   Shew that, if $|\arg z| < \frac{1}{2}\pi$,

$$P_n^m(z) = (-1)^m\frac{n(n-1)\cdots(n-m+1)}{\pi}\int_0^{\pi}\frac{\cos m\phi\,d\phi}{\{z+(z^2-1)^{1/2}\cos\phi\}^{n+1}},$$

where the many-valued functions are specified as in §15.23.

15.7 The addition-theorem for the Legendre polynomials

This appears in Legendre [421, vol. II, p. 262–269]. An investigation of the theorem based on physical reasoning will be given subsequently (§18.4).

Let $z = xx' - (x^2 - 1)^{1/2}(x'^2 - 1)^{1/2} \cos \omega$, where x, x', ω are unrestricted complex numbers. Then we shall shew that

$$P_n(z) = P_n(x)P_n(x') + 2 \sum_{m=1}^{n} (-1)^m \frac{(n-m)!}{(n+m)!} P_n^m(x)P_n^m(x') \cos m\omega.$$

First let $\text{Re}(x') > 0$, so that

$$\left| \frac{x + (x^2 - 1)^{1/2} \cos(\omega - \phi)}{x' + (x'^2 - 1)^{1/2} \cos \phi} \right|$$

is a bounded function of ϕ in the range $0 < \phi < 2\pi$. If M be its upper bound and if $|h| < M^{-1}$, then

$$\sum_{n=0}^{\infty} h^n \frac{\{x + (x^2 - 1)^{1/2} \cos(\omega - \phi)\}^n}{\{x' + (x'^2 - 1)^{1/2} \cos \phi\}^{n+1}}$$

converges uniformly with regard to ϕ, and so (§4.7)

$$\sum_{n=0}^{\infty} h^n \int_{-\pi}^{\pi} \frac{\{x + (x^2 - 1)^{1/2} \cos(\omega - \phi)\}^n}{\{x' + (x'^2 - 1)^{1/2} \cos \phi\}^{n+1}} \, d\phi$$

$$= \int_{-\pi}^{\pi} \sum_{n=0}^{\infty} \frac{h^n \{x + (x^2 - 1)^{1/2} \cos(\omega - \phi)\}^n}{\{x' + (x'^2 - 1)^{1/2} \cos \phi\}^{n+1}} \, d\phi$$

$$= \int_{-\pi}^{\pi} \frac{d\phi}{x' + (x'^2 - 1)^{1/2} \cos \phi - h\{x + (x^2 - 1)^{1/2} \cos(\omega - \phi)\}}.$$

Now, by a slight modification of Example 6.2.1 it follows that

$$\int_{-\pi}^{\pi} \frac{d\phi}{A + B \cos \phi + C \sin \phi} = \frac{2\pi}{(A^2 - B^2 - C^2)^{1/2}},$$

where that value of the radical is taken which makes

$$|A - (A^2 - B^2 - C^2)^{1/2}| < |(B^2 + C^2)^{1/2}|.$$

Therefore

$$\int_{-\pi}^{\pi} \frac{d\phi}{x' + (x'^2 - 1)^{1/2} \cos \phi - h\{x + (x^2 - 1)^{1/2} \cos(\omega - \phi)\}}$$

$$= \frac{2\pi}{[(x' - hx)^2 - \{(x^2 - 1)^{1/2} - h(x^2 - 1)^{1/2} \cos \omega\}^2 - \{h(x^2 - 1)^{1/2} \sin \omega\}^2]^{1/2}}$$

$$= \frac{2\pi}{(1 - 2hz + h^2)^{1/2}};$$

and when $h \to 0$, this expression has to tend to $2\pi P_0(x')$ by §15.23. Expanding in powers of h and equating coefficients, we get

$$P_n(z) = \frac{1}{2\pi} \int_{-\pi}^{\pi} \frac{\{x + (x^2 - 1)^{1/2} \cos(\omega - \phi)\}^n}{\{x' + (x'^2 - 1)^{1/2} \cos \phi \}^{n+1}} \, d\phi.$$

Now $P_n(z)$ is a polynomial of degree n in $\cos \omega$, and can consequently be expressed in the form $\dfrac{1}{2}A_0 + \displaystyle\sum_{m=1}^{n} A_m \cos m\omega$, where the coefficients $A_0, A_1, \ldots, A_n$ are independent of ω; to determine them, we use Fourier's rule (§9.12), and we get

$$
\begin{aligned}
A_m &= \frac{1}{\pi} \int_{-\pi}^{\pi} P_n(z) \cos m\omega \, d\omega \\
&= \frac{1}{2\pi^2} \int_{-\pi}^{\pi} \left[\int_{-\pi}^{\pi} \frac{\{x + (x^2 - 1)^{1/2} \cos(\omega - \phi)\}^n \cos m\omega}{\{x' + (x'^2 - 1)^{1/2} \cos \phi\}^{n+1}} \, d\phi \right] d\omega \\
&= \frac{1}{2\pi^2} \int_{-\pi}^{\pi} \left[\int_{-\pi}^{\pi} \frac{\{x + (x^2 - 1)^{1/2} \cos(\omega - \phi)\}^n \cos m\omega}{\{x' + (x'^2 - 1)^{1/2} \cos \phi\}^{n+1}} \, d\omega \right] d\phi \\
&= \frac{1}{2\pi^2} \int_{-\pi}^{\pi} \left[\int_{-\pi}^{\pi} \frac{\{x + (x^2 - 1)^{1/2} \cos \psi\}^n \cos m(\phi + \psi)}{\{x' + (x'^2 - 1)^{1/2} \cos \phi\}^{n+1}} \, d\psi \right] d\phi,
\end{aligned}
$$

on changing the order of integration, writing $\omega = \phi + \psi$ and changing the limits for ψ from $\pm\pi - \phi$ to $\pm\pi$.

Now

$$
\int_{-\pi}^{\pi} \{x + (x^2 - 1)^{1/2} \cos \psi\}^n \sin m\psi \, d\psi = 0,
$$

since the integrand is an odd function; and so, by §15.61,

$$
\begin{aligned}
A_m &= \frac{n!}{\pi(n+m)!} \int_{-\pi}^{\pi} \frac{\cos m\phi \, P_n^m(x)}{\{x' + (x'^2 - 1)^{1/2} \cos \phi\}^{n+1}} \, d\phi \\
&= 2(-1)^m \frac{(n-m)!}{(n+m)!} P_n^m(x) P_n^m(x').
\end{aligned}
$$

Therefore, when $|\arg z'| < \tfrac{1}{2}\pi$,

$$
P_n(z) = P_n(x) P_n(x') + 2 \sum_{m=1}^{n} (-1)^m \frac{(n-m)!}{(n+m)!} P_n^m(x) P_n^m(x') \cos m\omega.
$$

But this is a mere algebraical identity in x, x' and $\cos \omega$ (since n is a positive integer) and so is true independently of the sign of $\mathrm{Re}(x')$. The result stated has therefore been proved. The corresponding theorem with Ferrers' definition is

$$
P_n\{xx' + (1 - x^2)^{1/2}(1 - x'^2)^{1/2} \cos \omega\}
$$

$$
= P_n(x) P_n(x') + 2 \sum_{m=1}^{n} \frac{(n-m)!}{(n+m)!} P_n^m(x) P_n^m(x') \cos m\omega.
$$

15.71 The addition theorem for the Legendre functions

Let x, x' be two constants, real or complex, whose arguments are numerically less than $\tfrac{1}{2}\pi$; and let $(x \pm 1)^{1/2}$, $(x' \pm 1)^{1/2}$ be given their principal values; let ω be real and let

$$
z = xx' - (x^2 - 1)^{1/2}, \qquad (x'^2 - 1)^{1/2} \cos \omega.
$$

Then we shall shew that, if $|\arg z| < \tfrac{1}{2}\pi$ for all values of the real variable ω, and n be not a positive integer,

$$P_n(z) = P_n(x)P_n(x') + 2\sum_{m=1}^{\infty}(-1)^m \frac{\Gamma(n-m+1)}{\Gamma(n+m+1)}P_n^m(x)P_n^m(x')\cos m\omega.$$

Let $\cosh\alpha$, $\cosh\alpha'$ be the semi-major axes of the ellipses with foci ± 1 passing through x, x' respectively. Let β, β' be the eccentric angles of x, x' on these ellipses so that

$$-\frac{\pi}{2} < \beta < \frac{\pi}{2}, \qquad -\frac{\pi}{2} < \beta' < \frac{\pi}{2}.$$

Let $\alpha + i\beta = \xi$, $\alpha' + i\beta' = \xi'$, so that $x = \cosh xi$, $x' = \cosh\xi'$. Now as ω passes through all real values, $\mathrm{Re}(z)$ oscillates between

$$\mathrm{Re}(xx') \pm \mathrm{Re}(x^2 - 1)^{1/2}(x'^2 - 1)^{1/2} = \cosh(\alpha \pm \alpha')\cos(\beta \pm \beta'),$$

so that it is necessary that $\beta \pm \beta'$ be acute angles positive or negative.

Now take Schläfli's integral

$$P_n(z) = \frac{1}{2^{n+1}\pi i}\int_A^{(1+,z+)} \frac{(t^2-1)^n}{(t-z)^{n+1}}\,dt,$$

and write

$$t = \frac{e^{i\phi}\{e^{-i\omega}\sinh\xi\cosh\frac{1}{2}\xi' - \cosh\xi\sinh\frac{1}{2}\xi'\} + \cosh\frac{1}{2}\xi' - e^{i\omega}\sinh\xi\sinh\frac{1}{2}\xi'}{\cosh\frac{1}{2}\xi' + e^{i\phi}\sinh\frac{1}{2}\xi'}.$$

The path of t, as ϕ increases from $-\pi$ to π, may be shewn to be a circle; and the reader will verify that

$$t - 1 = \frac{2\{e^{i(\phi-\omega)}\cosh\frac{1}{2}\xi + \sinh\frac{1}{2}\xi\}\{\sinh\frac{1}{2}\xi\cosh\frac{1}{2}\xi' - e^{i\omega}\cosh\frac{1}{2}\xi\sinh\frac{1}{2}\xi'\}}{\cosh\frac{1}{2}\xi' + e^{i\phi}\sinh\frac{1}{2}\xi'},$$

$$t + 1 = \frac{2\{e^{i(\phi-\omega)}\sinh\frac{1}{2}\xi + \cosh\frac{1}{2}\xi\}\{\cosh\frac{1}{2}\xi\cosh\frac{1}{2}\xi' - e^{i\omega}\sinh\frac{1}{2}\xi\sinh\frac{1}{2}\xi'\}}{\cosh\frac{1}{2}\xi' + e^{i\phi}\sinh\frac{1}{2}\xi'},$$

$$t - z =$$
$$\frac{\{e^{i\phi}\cosh\frac{1}{2}\xi' + \sinh\xi'\}\{e^{i\omega}\sinh\frac{1}{2}\xi\sinh^2\frac{1}{2}\xi' + e^{-i\omega}\sinh\xi\cosh^2\frac{1}{2}\xi' - \cosh\xi + \sinh\xi'\}}{\cosh\frac{1}{2}\xi' + e^{i\phi}\sinh\frac{1}{2}\xi'}.$$

Since[15] $|\cosh\frac{1}{2}\xi'| > |\sinh\frac{1}{2}\xi'|$, the argument of the denominators does not change when ϕ increases by 2π; for similar reasons, the arguments of the first and third numerators increase by 2π, and the argument of the second does not change; therefore the circle contains the points $t = 1, t = z$, and not $t = -1$, so it is a possible contour.

Making these substitutions it is readily found that

$$P_n(z) = \frac{1}{2\pi}\int_{-\pi}^{\pi} \frac{\{x + (x^2-1)^{1/2}\cos(\omega - \phi)\}^n}{\{x' + (x'^2-1)^{1/2}\cos\phi\}^{n+1}}\,d\phi,$$

and the rest of the work follows the course of §15.7 except that the general form of Fourier's theorem has to be employed.

---

[15] This follows from the fact that $\cos\beta' > 0$.

Example 15.7.1 (Heine [287], Neumann [489])   Shew that, if n be a positive integer,

$$Q_n\{xx' + (x^2 - 1)^{1/2}(x'^2 - 1)^{1/2} \cos \omega\}$$

$$= Q_n(x)P_n(x') + 2 \sum_{m=1}^{\infty} Q_n^m(x)P_n^{-m}(x') \cos m\omega,$$

when ω is real, $\mathrm{Re}(x') \geq 0$, and $|(x' - 1)(x + 1)| < |(x - 1)(x' + 1)|$.

15.8 The function $C_n^\nu(z)$

A function connected with the associated Legendre function $P_n^m(z)$ is the function $C_n^\nu(z)$, which for integral values of n is defined to be the coefficient of h^n in the expansion of $(1 - 2hz + h^2)^{-\nu}$ in ascending powers of h. This function has been studied by Gegenbauer [240].

It is easily seen that $C_n^\nu(z)$ satisfies the differential equation

$$\frac{d^2y}{dz^2} + \frac{(2\nu + 1)z}{z^2 - 1}\frac{dy}{dz} - \frac{n(n + 2\nu)}{z^2 - 1}y = 0.$$

For all values of n and ν, it may be shewn that we can define a function, satisfying this equation, by a contour integral of the form

$$(1 - z^2)^{1/2-\nu} \int_C \frac{(1 - t^2)^{n+\nu-1/2}}{(t - z)^{n+1}} \, dt,$$

where C is the contour of §15.2; this corresponds to Schläfli's integral.

The reader will easily prove the following results:

(I) When n is a integer

$$C_n^\nu(z) = \frac{(-2)^n \nu(\nu + 1) \cdots (\nu + n - 1)}{n!(2n + 2\nu - 1)(2n + 2\nu - 2) \cdots (n + 2\nu)}(1 - z^2)^{1/2-\nu}\frac{d^n}{dz^n}\{(1 - z^2)^{n+\nu-\frac{1}{2}}\};$$

since $P_n(z) = C_n^{1/2}(z)$, Rodrigues' formula is a particular case of this result.

(II) When r is an integer,

$$C_{n-r}^{r+\frac{1}{2}}(z) = \frac{1}{(2r - 1)(2r - 3) \cdots 3 \cdot 1}\frac{d^r}{dz^r}P_n(z),$$

whence

$$C_{n-r}^{r+\frac{1}{2}}(z) = \frac{(z^2 - 1)^{-r/3}}{(2r - 1) \cdot (2r - 3) \cdots 3 \cdot 1}P_n^r(z).$$

The last equation gives the connexion between the functions $C_n^\nu(z)$ and $P_n^r(z)$.

(III) Modifications of the recurrence formulae for $P_n(z)$ are the following:

$$C_{n-1}^{v+1}(z) - C_{n-2}^{v+1}(z) - \frac{n}{2v}C_n^v(z) = 0,$$

$$C_n^{v+1}(z) - zC_{n-1}^{v+1}(z) = \frac{n+2v}{2v}C_n^v(z),$$

$$\frac{dC_n^v(z)}{dz} = 2vC_{n-1}^{v+1}(z),$$

$$nC_n^v(z) = (n-1+2v)zC_{n-1}^v(z) - 2v(1-z^2)C_{n-2}^{v-1}(z).$$

15.9 Miscellaneous examples

The functions involved in Examples 15.1–15.30 are Legendre *polynomials*.

Example 15.1 (Math. Trip. 1898) Prove that when n is a positive integer,

$$P_n(z) = \sum_0^n \frac{(n+p)!(-1)^p}{(n-p)!p!^2 2^{p+1}}\{(1-z)^p + (-1)^n(1+z)^p\}.$$

Example 15.2 (Math. Trip. 1896) Prove that $\int_{-1}^1 z(1-z^2)\frac{dP_n}{dz}\frac{dP_m}{dz}dz$ is zero unless $m - n = \pm 1$, and determine its value in these cases.

Example 15.3 (Math. Trip. 1899) Shew (by induction or otherwise) that when n is a positive integer,

$$(2n+1)\int_z^1 P_n^2(z)\,dz =$$
$$1 - zP_n^2 - 2z(P_1^2 + P_2^2 + \cdots + P_{n-1}^2) + 2(P_1P_2 + P_2P_3 + \cdots + P_{n-1}P_n).$$

Example 15.4 (Clare, 1906) Shew that

$$zP_n'(z) = nP_n(z) + (2n-3)P_{n-2}(z) + (2n-7)P_{n-4}(z) + \cdots .$$

Example 15.5 (Math. Trip. 1904) Shew that

$$z^2 P_n''(z) = n(n-1)P_n(z) + \sum_{r=1}^p (2n-4r+1)\{r(2n-2r+1) - 2\}P_{n-2r}(z),$$

where $p = \frac{1}{2}n$ or $\frac{1}{2}(n-1)$.

Example 15.6 (Trin. Coll. Dublin) Shew that the Legendre polynomial satisfies the relation

$$(z^2 - 1)^2\frac{d^2 P_n}{dz^2} = n(n-1)(n+1)(n+2)\int_1^z dz \int_1^z P_n(z)\,dz.$$

Example 15.7 (Peterhouse, 1905) Shew that

$$\int_0^1 z^2 P_{n+1}(z)P_{n-1}(z)\,dz = \frac{n(n+1)}{(2n-1)(2n+1)(2n+3)}.$$

Example 15.8 (Peterhouse, 1907)   Shew that the values of

$$\int_{-1}^{1} (1 - z^2)^2 P_m'''(z) P_n'(z)\, dz$$

are as follows:

1. $8n(n + 1)$ when $m - n$ is positive and even,
2. $-2n(n^2 - 1)(n - 2)/(2n + 1)$ when $m = n$,
3. 0 for other values of m and n.

Example 15.9 (Math. Trip. 1907)   Shew that

$$\sin^n \theta P_n(\sin \theta) = \sum_{r=0}^{n} (-1)^r \frac{n!}{r!(n - r)!} \cos^r \theta P_r(\cos \theta).$$

Example 15.10 (Clare, 1903)   Shew, by evaluating $\int_0^{\pi} P_n(\cos \theta)\, d\theta$ (Example 15.3.1), and then integrating by parts, that

$$\int_{-1}^{1} P_n(\mu) \arcsin \mu\, d\mu = \begin{cases} 0 & \text{when } n \text{ is even} \\ \pi \left\{ \dfrac{1 \cdot 3 \cdots (n - 2)}{2 \cdot 4 \cdots (n + 1)} \right\}^2 & \text{when } n \text{ is odd.} \end{cases}$$

Example 15.11 (Adams [9])   If m and n be positive integers, and $m \le n$, shew by induction that

$$P_m(z) P_n(z) = \sum_{r=0}^{m} \frac{A_{m-r} A_r A_{n-r}}{A_{n+m-r}} \left(\frac{2n + 2m - 4r + 1}{2n + 2m - 2r + 1} \right) P_{n+m-2r}(z),$$

where $A_m = \dfrac{1 \cdot 3 \cdot 5 \cdots (2m - 1)}{m!}$.

Example 15.12   By expanding in ascending powers of u shew that

$$P_n(z) = \frac{(-1)^n}{n!} \frac{d^n}{dz^n} (u^2 + z^2)^{-1/2},$$

where u^2 is to be replaced by $(1 - z^2)$ after the differentiation has been performed.

Example 15.13 (Heun [299])   Shew that $P_n(z)$ can be expressed as a constant multiple of a determinant in which all elements parallel to the auxiliary diagonal are equal (i.e. all elements are equal for which the sum of the row-index and column-index is the same); the determinant containing n rows, and its elements being

$$z, \quad -\frac{1}{3}, \quad \frac{1}{3}z, \quad -\frac{1}{5}, \quad \frac{1}{5}z, \quad \ldots, \quad \frac{1}{2n - 1}z.$$

Example 15.14 (Silva)   Shew that, if the path of integration passes above $t = 1$,

$$P_n(z) = \frac{2}{\pi i} \int_0^{\infty} \frac{\{z(1 - t^2) - 2t(1 - z^2)^{1/2}\}^n}{(1 - t^2)^{n+1}}\, dt.$$

Example 15.15 (Math. Trip. 1893) By writing $\cot \theta' = \cot \theta - h \csc \theta$ and expanding $\sin \theta'$ in powers of h by Taylor's theorem, shew that

$$P_n(\cos \theta) = \frac{(-1)^n}{n!} \csc^{n+1} \theta \frac{d^n(\sin \theta)}{d(\cot \theta)^n}.$$

Example 15.16 (Glaisher [247]) By considering $\sum_{n=0}^{\infty} h^n P_n(z)$, shew that

$$P_n(z) = \frac{1}{n!\sqrt{\pi}} \int_{-\infty}^{\infty} e^{-(1-z^2)t^2} \left(-\frac{d}{dz}\right)^n e^{-z^2 t^2} \, dt.$$

Example 15.17 (Math. Trip. 1894) The equation of a nearly spherical surface of revolution is

$$r = 1 + \alpha\{P_1(\cos \theta) + P_3(\cos \theta) + \cdots + P_{2n-1}(\cos \theta)\},$$

where α is small; shew that if α^2 be neglected the radius of curvature of the meridian is

$$1 + \alpha \sum_{m=0}^{n-1} \{n(4m+3) - (m+1)(8m+3)\} P_{2m+1}(\cos \theta).$$

Example 15.18 (Trinity, 1894) The equation of a nearly spherical surface of revolution is

$$r = \alpha \{1 + \varepsilon P_n(\cos \theta)\},$$

where ε is small. Shew that if ε^3 be neglected, its area is

$$4\pi\alpha^2 \left\{1 + \frac{1}{2} \varepsilon^2 \frac{n^2 + n + 2}{2n+1}\right\}.$$

Example 15.19 (Routh [564]) Shew that, if k is an integer and

$$(1 - 2hz + h^2)^{-k/2} = \sum_{n=0}^{\infty} \alpha_n P_n(z),$$

then

$$\alpha_n = \frac{h^n}{(1-h^2)^{k-2}} \frac{2^{\frac{1}{2}(k-3)}(2n+1)}{1 \cdot 3 \cdot 5 \cdots (k-2)} \left(h^2 \frac{\partial}{\partial x} + \frac{\partial}{\partial y}\right)^{\frac{1}{2}(k-3)} x^{-n+k/2-2} y^{n+k/2-2},$$

where x and y are to be replaced by unity after the differentiations have been performed.

Example 15.20 (Catalan) Shew that

$$\int_{-1}^{1} \frac{1}{z - x} \{P_n(x)P_{n-1}(z) - P_{n-1}(x)P_n(z)\} \, dx = -\frac{2}{n},$$

$$\sum_{n=1}^{\infty} \frac{1}{2n+1} \frac{d}{dz} \left[P_n(z) \left(\frac{1}{n} P_{n-1}(z) + \frac{1}{n+1} P_{n+1}(z)\right)\right] = -1.$$

Example 15.21 Let $x^2 + y^2 + z^2 = r^2$, $z = \mu r$, the numbers involved being real, so that $-1 < \mu < 1$. Shew that

$$P_n(\mu) = \frac{(-1)^n r^{n+1}}{n!} \frac{\partial^n}{\partial z^n} \left(\frac{1}{r}\right),$$

where r is to be treated as a function of the independent variables x, y, z in performing the differentiations.

Example 15.22   With the notation of Example 15.3.4, shew that

$$Q_n(\mu) = \frac{(-1)^n r^{n+1}}{n!} \frac{\partial^n}{\partial z^n} \left\{ \frac{1}{2r} \log\left(\frac{r-z}{r+z} \right) \right\},$$

$$(n+1)P_n(\mu) + \mu P_n'(\mu) = \frac{(-1)^n r^{n+3}}{n!} \frac{\partial^n}{\partial z^n} \left(\frac{1}{r^3} \right).$$

Example 15.23   Shew that, if $|h|$ and $|z|$ are sufficiently small,

$$\frac{1 - h^2}{(1 - 2hz + h^2)^{3/2}} = \sum_{n=0}^{\infty} (2n+1)h^n P_n(z).$$

Example 15.24 (Math. Trip. 1894)   Prove that

$$P_{n+1}(z)Q_{n-1}(z) - P_{n-1}(z)Q_{n+1}(z) = \frac{2n+1}{n(n+1)} z.$$

Example 15.25 (Bauer)   If the arbitrary function $f(x)$ can be expanded in the series

$$f(x) = \sum_{n=0}^{\infty} \alpha_n P_n(x),$$

converging uniformly in a domain which includes the point $x = 1$, shew that the expansion of the integral of this function is

$$\int_1^x f(x)\, dx = -\alpha_0 - \frac{1}{3}\alpha_1 + \sum_{n=1}^{\infty} \left(\frac{\alpha_{n-1}}{2n-1} - \frac{\alpha_{n+1}}{2n+3} \right) P_n(x).$$

Example 15.26 (Bauer [58])   Determine the coefficients in Neumann's expansion of $e^{\alpha z}$ in a series of Legendre polynomials.

Example 15.27 (Catalan)   Deduce from Example 15.25 that

$$\arcsin z = \frac{\pi}{2} \sum_0^{\infty} \left\{ \frac{1 \cdot 3 \cdot 5 \cdots (2n-1)}{2 \cdot 4 \cdot 6 \cdots 2n} \right\}^2 \{P_{2n+1}(z) - P_{2n-1}(z)\}.$$

Example 15.28 (Schläfli; Hermite [293])   Shew that

$$Q_n(z) = \frac{1}{2} \log\left(\frac{z+1}{z-1} \right) \cdot P_n(z) - \{P_{n-1}(z)P_0(z) + \frac{1}{2}P_{n-2}(z)P_1(z)$$

$$+ \frac{1}{3}P_{n-3}(z)P_2(z) + \cdots + \frac{1}{n}P_0(z)P_{n-1}(z)\}.$$

Example 15.29 (Math. Trip. 1898)   Shew that

$$Q_n(z) = \frac{1}{2^n n!} \frac{d^n}{dz^n} \left\{ (z^2 - 1)^n \log \frac{z+1}{z-1} \right\} - \frac{1}{2}P_n(z) \log \frac{z+1}{z-1}.$$

Prove also that

$$Q_n(z) = \frac{1}{2}P_n(z) \log \frac{z+1}{z-1} - f_{n-1}(z),$$

where

$$f_{n-1}(z) = \frac{2n-1}{1 \cdot n} P_{n-1}(z) + \frac{2n-5}{3(n-1)} P_{n-3}(z) + \frac{2n-9}{5(n-2)} P_{n-5}(z) + \cdots$$

$$= \left\{ \begin{array}{l} k_n + (k_n - 1)\frac{n(n+1)}{1^2}\left(\frac{z-1}{2}\right) + \left(k_n - 1 - \frac{1}{2}\right)\frac{n(n-1)(n+1)(n+2)}{1^2 2^2}\left(\frac{z-1}{2}\right)^2 \\[2mm] + \left(k_n - 1 - \frac{1}{2} - \frac{1}{3}\right)\frac{n(n-1)(n-2)(n+1)(n+2)(n+3)}{1^2 2^2 3^2}\left(\frac{z-1}{2}\right)^3 + \cdots \end{array} \right\},$$

where $k_n = 1 + \frac{1}{2} + \frac{1}{3} + \cdots + \frac{1}{n}$.

The first of these expressions for $f_{n-1}(z)$ was given by Christoffel [145] and he also gives a generalisation of Example 15.28; the second was given by Stieltjes [298, p. 59].

Example 15.30 Shew that the complete solution of Legendre's differential equation is

$$y = A P_n(z) + B P_n(z) \int_z^\infty \frac{dt}{(t^2 - 1)\{P_n(t)\}^2},$$

the path of integration being the straight line which when produced backwards passes through the point $t = 0$.

Example 15.31 (Schläfli)   Shew that

$$\{z + (z^2 - 1)^{1/2}\}^\alpha = \sum_{m=0}^{\infty} B_m Q_{2m-\alpha-1}(z),$$

where

$$B_m = -\frac{\alpha(\alpha - 2m + \frac{1}{2})}{2\pi} \frac{\Gamma(m - \frac{1}{2})\Gamma(m - \alpha - \frac{1}{2})}{m!\,\Gamma(m - \alpha + 1)}.$$

Example 15.32   Shew that, when $\mathrm{Re}(n + 1) > 0$,

$$Q_n(z) = \int_{z+(z^2-1)^{1/2}}^{\infty} \frac{h^{-n-1}\,dh}{(1 - 2hz + h^2)^{1/2}}, \quad \text{and}$$

$$Q_n(z) = \int_0^{z+(z^2-1)^{1/2}} \frac{h^n\,dh}{(1 - 2hz + h^2)^{1/2}}.$$

Example 15.33 (Hobson)   Shew that

$$Q_n^m(z) = e^{m\pi i} \frac{\Gamma(n+1)}{\Gamma(n-m+1)} \int_0^\infty \frac{\cosh mu\,du}{\{z + (z^2-1)^{1/2}\cosh u\}^{n+1}},$$

where $\mathrm{Re}(n+1) > m$.

Example 15.34   Obtain the expansion of $P_n(z)$ when $|\arg z| < \pi$ as a series of powers of $1/z$, when n is not an integer, namely

$$P_n(z) = \frac{\tan n\pi}{\pi}\{Q_n(z) - Q_{-n-1}(z)\}$$

$$= \frac{2^n \Gamma(n + \frac{1}{2})}{\Gamma(n+1)\Gamma(\frac{1}{2})} z^n F\left(\frac{1-n}{2}, -\frac{n}{2}; \frac{1}{2} - n, \frac{1}{z^2}\right)$$

$$+ \frac{2^{-n-1}\Gamma(-n - \frac{1}{2})}{\Gamma(-n)\Gamma(\frac{1}{2})} z^{-n-1} F\left(\frac{n}{2} + 1, \frac{n+1}{2}; n + \frac{3}{2}, \frac{1}{z^2}\right).$$

[This is most easily obtained by the method of §14.51.]

Example 15.35 (Olbricht)   Shew that the differential equation for the associated Legendre function $P_n^m(z)$ is defined by the schemes[16]

$$
P \left\{ \begin{matrix} 0 & \infty & 1 & \\ -\frac{1}{2}n & m & -\frac{1}{2}n & \frac{z+(z^2-1)^{1/2}}{z-(z^2-1)^{1/2}} \\ \frac{1}{2}n+\frac{1}{2} & -m & \frac{1}{2}n+\frac{1}{2} & \end{matrix} \right\}, \quad
P \left\{ \begin{matrix} 0 & \infty & 1 & \\ -\frac{1}{2}n & \frac{1}{2}m & 0 & \frac{1}{1-z^2} \\ \frac{1}{2}n+\frac{1}{2} & -\frac{1}{2}m & \frac{1}{2} & \end{matrix} \right\}.
$$

Example 15.36   Shew that the differential equation for $C_n^\nu(z)$ is defined by the scheme

$$
P \left\{ \begin{matrix} -1 & \infty & 1 & \\ -\frac{1}{2}-\nu & n+2\nu & \frac{1}{2}-\nu & z \\ 0 & -n & 0 & \end{matrix} \right\}.
$$

Example 15.37 (Math. Trip. 1896)   Prove that, if

$$
y_s = \frac{(2n+1)(2n+3)\cdots(2n+2s-1)}{n(n^2-1)(n^2-4)\cdots\{n^2-(s-1)^2\}(n+s)}(z^2-1)^s \frac{d^s P_n}{dz^s},
$$

then

$$
y_2 = P_{n+2} - \frac{2(2n+1)}{2n-1}P_n + \frac{2n+3}{2n-1}P_{n-2},
$$

$$
y_3 = P_{n+3} - \frac{3(2n+3)}{2n-1}P_{n+1} + \frac{3(2n+5)}{2n-3}P_{n-1} - \frac{(2n+3)(2n+5)}{(2n-1)(2n-3)}P_{n-3},
$$

and find the general formula.

Example 15.38 (Math. Trip. 1901)   Shew that

$$
P_n^m(\cos\theta) = \frac{2}{\sqrt{\pi}}\frac{\Gamma(n+m+1)}{\Gamma(n+\frac{3}{2})} \left[\frac{\cos\{(n+\frac{1}{2})\theta - \frac{1}{4}\pi + \frac{1}{2}m\pi\}}{(2\sin\theta)^{1/2}} \right.
$$

$$
+ \frac{(1^2-4m^2)}{2(2n+3)}\frac{\cos\{(n+\frac{3}{2})\theta - \frac{3}{4}\pi + \frac{1}{2}m\pi\}}{(2\sin\theta)^{3/2}}
$$

$$
\left. + \frac{(1^2-4m^2)(3^2-4m^2)}{2\cdot 4\cdot(2n+3)(2n+5)}\frac{\cos\{(n+\frac{5}{2})\theta - \frac{5}{4}\pi + \frac{1}{2}m\pi\}}{(2\sin\theta)^{5/2}} + \cdots \right],
$$

obtaining the ranges of values of m, n and θ for which it is valid.

Example 15.39 (Macdonald [445, 447])   Shew that the values of n, for which $P_n^{-m}(\cos\theta)$ vanishes, decrease as θ increases from 0 to π when m is positive; and that the number of real zeros of $P_n^{-m}(\cos\theta)$ for values of θ between $-\pi$ and π is the greatest integer less than $n - m + 1$.

---

[16] See also Example 15.5.1.

Example 15.40 (Legendre) Obtain the formula

$$\frac{1}{2\pi} \int_{-\pi}^{\pi} \left[1 - 2h\{\cos\omega\cos\phi + \sin\omega\sin\phi\cos(\theta' - \theta)\} + h^2\right]^{-1/2} d\theta$$

$$= \sum_{n=0}^{\infty} h^n P_n(\cos\omega) P_n(\cos\phi).$$

Example 15.41 (Trinity, 1893) If $f(x) = x^2$ for $x \geq 0$, and $f(x) = -x^2$ for $x < 0$, shew that, if $f(x)$ can be expanded into a uniformly convergent series of Legendre polynomials in the range $(-1, 1)$, the expansion is

$$f(x) = \frac{3}{4}P_1(x) - \sum_{r=1}^{\infty}(-1)^r \frac{1 \cdot 3 \cdots (2r-3)}{4 \cdot 6 \cdot 8 \cdots 2r} \frac{4r+3}{2r+4} P_{2r+1}(x).$$

Example 15.42 (Gegenbauer [241]) If $\dfrac{1}{(1 - 2hz + h^2)^\nu} = \sum_{n=0}^{\infty} h^n C_n^\nu(z)$, shew that

$$C_n^\nu \{xx_1 - (x^2 - 1)^{1/2}(x_1^2 - 1)^{1/2} \cos\phi\}$$

$$= \frac{\Gamma(2\nu - 1)}{\{\Gamma(\nu)\}^2} \sum_{\lambda=0}^{n}(-1)^\lambda \frac{4^\lambda \Gamma(n - \lambda + 1)\{\Gamma(\nu + \lambda)\}^2(2\nu + 2\lambda - 1)}{\Gamma(n + 2\nu + \lambda)}$$

$$\times (x^2 - 1)^{\frac{1}{2}\lambda}(x_1^2 - 1)^{\frac{1}{2}\lambda} C_{n-\lambda}^{\nu+\lambda}(x) C_{n-\lambda}^{\nu+\lambda}(x_1) C_\lambda^{\nu-\frac{1}{2}}(\cos\phi)$$

Example 15.43 (Pincherle [522]) If $\sigma_n(z) = \displaystyle\int_0^{e_1}(t^3 - 3tz + 1)^{-1/2} t^n \, dt$, where e_1 is the least root of $t^3 - 3tz + 1 = 0$, shew that

$$(2n + 1)\sigma_{n+1} - 3(2n - 1)z\sigma_{n-1} + 2(n - 1)\sigma_{n-2} = 0,$$

and

$$4(4z^3 - 1)\sigma_n''' + 144z^2\sigma_n'' - z(12n^2 - 24n - 291)\sigma_n' - (n - 3)(2n - 7)(2n + 5)\sigma_n = 0,$$

where $\sigma_n''' = \dfrac{d^3\sigma_n(z)}{dz^3}$, etc.

Example 15.44 (Pincherle [520]) If $(h^3 - 3hz + 1)^{-1/2} = \sum_{n=0}^{\infty} R_n(z)h^n$, shew that

$$2(n + 1)R_{n+1} - 3(2n + 1)zR_n + (2n - 1)R_{n-2} = 0,$$
$$nR_n + R_{n-2}' - zR_n' = 0,$$

and

$$4(4z^3 - 1)R_n''' + 96z^2 R_n'' - z(12n^2 + 24n - 91)R_n' - n(2n + 3)(2n + 9)R_n = 0,$$

where $R_n''' = \dfrac{d^3 R_n}{dz^3}$, etc.

Example 15.45 (Schendel [575]) If $A_n(x) = \dfrac{1}{2^n n!(x - 1)} \dfrac{d^n}{dx^n}\{(x^2 - 1)^n(x - 1)\}$, obtain the recurrence formula

$$(n + 1)(2n - 1)A_n(x) - \{(4n^2 - 1)x + 1\}A_{n-1}(x) + (n - 1)(2n + 1)A_{n-2}(x) = 0.$$

Example 15.46 If n is not negative and m is a positive integer, shew that the equation

$$(x^2 - 1)\frac{d^2y}{dx^2} + (2n + 2)x\frac{dy}{dx} = m(m + 2n + 1)y$$

has the two solutions

$$K_m(x) = (x^2 - 1)^{-n}\frac{d^m}{dx^m}(x^2 - 1)^{m+n}, \qquad L_m(x) = (x^2 - 1)^{-n}\int_{-1}^{1}\frac{(t^2 - 1)^n}{x - t}K_m(t)\,dt,$$

when x is not a real number such that $-1 \leq x \leq 1$.

Example 15.47 (Clare, 1901) Prove that

$$\{1 - hx - (1 - 2hx + h^2)^{1/2}\}^m = m(x^2 - 1)^m \sum_{n=m}^{\infty} \frac{h^{n+m}}{(n + m)!}\frac{1}{n}\frac{d^{n+m}}{dx^{n+m}}\left(\frac{x^2 - 1}{2}\right)^n.$$

Example 15.48 (Trinity, 1905) If $F_{\alpha,n}(x) = \sum_{m=0}^{\infty} \frac{(m + \alpha)^n}{m!}x^m$, shew that

$$F_{\alpha,n}(x) = \left\{\frac{d^n}{dt^n}(e^{\alpha t + xs^t})\right\}_{t=0} = e^x P_n(x, \alpha),$$

where $P_n(x, \alpha)$ is a polynomial of degree n in x; and deduce that

$$P_{n+1}(x, \alpha) = (x + \alpha)P_n(x, \alpha) + x\frac{d}{dx}P_n(x, \alpha).$$

Example 15.49 (Léauté) If $F_n(x)$ be the coefficient of z^n in the expansion of

$$\frac{2hz}{e^{hz} - e^{-kz}}e^{xz}$$

in ascending powers of z, so that

$$F_0(x) = 1, \quad F_1(x) = x, \quad F_2(x) = \frac{3x^2 - h^2}{6}, \quad \ldots,$$

shew that:

1. $F_n(x)$ is a homogeneous polynomial of degree n in x and h;
2. $\dfrac{dF_n(x)}{dx} = F_{n-1}(x)$ for $n \geq 1$;
3. $\displaystyle\int_{-k}^{k} F_n(x)\,dx = 0$ for $n \geq 1$;
4. If $y = \alpha_0 F_0(x) + \alpha_1 F_1(x) + \alpha_2 F_2(x) + \cdots$, where $\alpha_0, \alpha_1, \alpha_2, \ldots$ are real constants, then the mean-value of $\dfrac{d^r y}{dx^r}$ in the interval from $x = -h$ to $x = +h$ is α_r.

Example 15.50 (Appell) If $F_n(x)$ be defined as in the preceding example, shew that, when $-h < x < h$,

$$F_{2m}(x) = (-1)^m\frac{2h^{2m}}{\pi^{2m}}\left(\cos\frac{\pi x}{h} - \frac{1}{2^{2m}}\cos\frac{2\pi x}{h} + \frac{1}{3^{2m}}\cos\frac{3\pi x}{h} + \cdots\right),$$

$$F_{2m+1}(x) = (-1)^m\frac{2h^{2m+1}}{\pi^{2m+1}}\left(\sin\frac{\pi x}{h} - \frac{1}{2^{2m+1}}\sin\frac{2\pi x}{h} + \frac{1}{3^{2m+1}}\sin\frac{3\pi x}{h} + \cdots\right).$$

16

The Confluent Hypergeometric Function

16.1 The confluence of two singularities of Riemann's equation

We have seen (§10.8) that the linear differential equation with two regular singularities only can be integrated in terms of elementary functions; while the solution of the linear differential equation with three regular singularities is substantially the topic of Chapter 14. As the next type in order of complexity, we shall consider a modified form of the differential equation which is obtained from Riemann's equation by the confluence of two of the singularities. This confluence gives an equation with an irregular singularity (corresponding to the confluent singularities of Riemann's equation) and a regular singularity corresponding to the third singularity of Riemann's equation.

The confluent equation is obtained by making $c \to \infty$ in the equation defined by the scheme

$$P \left\{ \begin{matrix} 0 & \infty & c \\ \frac{1}{2} + m & -c & c - k & z \\ \frac{1}{2} - m & 0 & k \end{matrix} \right\}.$$

The equation in question is readily found to be

$$\frac{d^2 u}{dz^2} + \frac{du}{dz} + \left(\frac{k}{z} + \frac{\frac{1}{4} - m^2}{z^2} \right) u = 0. \tag{16.1}$$

We modify this equation by writing $u = e^{-\frac{1}{2}z} W_{k,m}(z)$ and obtain as the equation[1]

$$\frac{d^2 W}{dz^2} + \left\{ -\frac{1}{4} + \frac{k}{z} + \frac{\frac{1}{4} - m^2}{z^2} \right\} W = 0. \tag{16.2}$$

The reader will verify that the singularities of this equation are at 0 and ∞, the former being regular and the latter irregular; and when $2m$ is *not an integer*, two integrals of equation (16.2) which are regular near 0 and valid for all finite values of z are given by the series

$$M_{k,m}(z) = z^{1/2+m} e^{-\frac{1}{2}z} \left\{ 1 + \frac{\frac{1}{2} + m - k}{1!(2m+1)} z + \frac{(\frac{1}{2} + m - k)(\frac{3}{2} + m - k)}{2!(2m+1)(2m+2)} z^2 + \cdots \right\},$$

$$M_{k,-m}(z) = z^{1/2-m} e^{-\frac{1}{2}z} \left\{ 1 + \frac{\frac{1}{2} - m - k}{1!(1-2m)} z + \frac{(\frac{1}{2} - m - k)(\frac{3}{2} - m - k)}{2!(1-2m)(2-2m)} z^3 + \cdots \right\}.$$

---

[1] This equation was given by Whittaker [671], for $W_{k,m}(z)$.

These series obviously form a fundamental system of solutions.

Note Series of the type above have been considered by Kummer [389, p. 139] and more recently by Jacobsthal [355] and Barnes [49]; the special series in which $k = 0$ had been investigated by Lagrange in 1762–1765 [396, vol. I, p. 480]. In the notation of Kummer, modified by Barnes, they would be written $_1F_1\left\{\frac{1}{2} \pm m - k; \pm 2m + 1; z\right\}$; the reason for discussing solutions of equation (16.2) rather than those of the equation $z\dfrac{d^2y}{dz^2} - (z - \rho)\dfrac{dy}{dz} - ay = 0$, of which $_1F_1(a; \rho; z)$ is a solution, is the greater appearance of symmetry in the formulae, together with a simplicity in the equations giving various functions of Applied Mathematics (see §16.2) in terms of solutions of equation (16.2).

16.11 Kummer's formulae

(I) We shall now shew that, if $2m$ is not a negative integer, then

$$z^{-1/2-m}M_{k,m}(z) = (-z)^{-1/2-m}M_{-k,m}(-z);$$

that is to say,

$$e^{-z}\left\{1 + \frac{\frac{1}{2} + m - k}{1!(2m+1)}z + \frac{(\frac{1}{2} + m - k)(\frac{3}{2} + m - k)}{2!(2m+1)(2m+2)}z^2 + \cdots\right\}$$

$$= 1 - \frac{\frac{1}{2} + m + k}{1!(2m+1)}z + \frac{(\frac{1}{2} + m + k)(\frac{3}{2} + m + k)}{2!(2m+1)(2m+2)}z^2 - \cdots.$$

For, replacing e^{-z} by its expansion in powers of z, the coefficient of z^n in the product of absolutely convergent series on the left is

$$\frac{(-1)^n}{n!}F\left(\tfrac{1}{2} + m - k, -n; 2m+1; 1\right) = \frac{(-1)^n}{n!}\frac{\Gamma(2m+1)\Gamma(m + \frac{1}{2} + k + n)}{\Gamma(m + \frac{1}{2} + k)\Gamma(2m + 1 + n)},$$

by §14.11, and this is the coefficient of z^n on the right (the result is still true when $m + \frac{1}{2} + k$ is a negative integer, by a slight modification of the analysis of §14.11); we have thus obtained the required result. This will be called *Kummer's first formula*.

(II) The equation

$$M_{0,m}(z) = z^{1/2+m}\left\{1 + \sum_{p=1}^{\infty}\frac{z^{2p}}{2^{4p}p!(m+1)(m+2)\cdots(m+p)}\right\},$$

valid when $2m$ is not a negative integer, will be called *Kummer's second formula*.

To prove it we observe that the coefficient of $z^{n+m+1/2}$ in the product

$$2^{m+1/2}e^{-z/2}{}_1F_1(m + \tfrac{1}{2}; 2m+1; z),$$

of which the second and third factors possess absolutely convergent expansions, is (§3.73)

$$\frac{(\frac{1}{2} + m)(\frac{3}{2} + m)\cdots(n - m + \frac{1}{2})}{n!(2m+1)(2m+2)\cdots(2m+n)}F\left(-n, -2m - n; -n + \tfrac{1}{2} - m; \tfrac{1}{2}\right)$$

$$= \frac{(\frac{1}{2} + m)(\frac{3}{2} + m)\cdots(n - m + \frac{1}{2})}{n!(2m+1)(2m+2)\cdots(2m+n)}F\left(-\tfrac{1}{2}n, -m - \tfrac{1}{2}n; -n + \tfrac{1}{2} - m; 1\right),$$

by Kummer's relation (see Chapter 14, Examples 14.12 and 14.13)

$$F(2\alpha, 2\beta; \alpha + \beta + \tfrac{1}{2}; x) = F\{\alpha, \beta; \alpha + \beta + \tfrac{1}{2}; 4x(1 - x)\},$$

valid when $0 \le x \le \tfrac{1}{2}$; and so the coefficient of $z^{n+m+1/2}$ (by §14.11) is

$$\frac{(\tfrac{1}{2} + m)(\tfrac{3}{2} + m) \cdots (n - m + \tfrac{1}{2})}{n!(2m + 1)(2m + 2) \cdots (2m + n)} \frac{\Gamma(-n + \tfrac{1}{2} - m)\Gamma(\tfrac{1}{2})}{\Gamma(\tfrac{1}{2} - m - \tfrac{1}{2}n)\Gamma(\tfrac{1}{2} - \tfrac{1}{2}n)}$$

$$= \frac{\Gamma(\tfrac{1}{2} - m)\Gamma(\tfrac{1}{2})}{n!(2m + 1)(2m + 2) \cdots (2m + n)\Gamma(\tfrac{1}{2} - m - \tfrac{1}{2}n)\Gamma(\tfrac{1}{2} - \tfrac{1}{2}n)},$$

and when n is odd this vanishes; for even values of n $(= 2p)$ it is

$$\frac{\Gamma(\tfrac{1}{2} - m)(-\tfrac{1}{2})(-\tfrac{3}{2}) \cdots (\tfrac{1}{2} - p)}{(2p)!2^{2p}(m + \tfrac{1}{2})(m + \tfrac{3}{2}) \cdots (m + p - \tfrac{1}{2})(m + 1)(m + 2) \cdots (m + p)\Gamma(\tfrac{1}{2} - m - p)}$$

$$= \frac{1 \cdot 3 \cdots (2p - 1)}{(2p)! \, 2^{3p}(m + 1)(m + 2) \cdots (m + p)} = \frac{1}{2^{4p} \cdot p!(m + 1)(m + 2) \cdots (m + p)}.$$

16.12 Definition of the function $W_{k,m}(z)$

The function $W_{k,m}(z)$ was defined by means of an integral in this manner by Whittaker [671]. The solutions $M_{k,\pm m}(z)$ of equation (16.2) of §16.1 are not, however, the most convenient to take as the standard solutions, on account of the disappearance of one of them when $2m$ is an integer.

The integral obtained by confluence from that of §14.6, when multiplied by a constant multiple of $e^{z/2}$, is[2]

$$W_{k,m}(z) = -\frac{1}{2\pi i}\Gamma\left(k + \tfrac{1}{2} - m\right)e^{-z/2}z^k \int_{\infty}^{(0+)} (-t)^{-k-\frac{1}{2}+m}\left(1 + t/z\right)^{k-\frac{1}{2}+m}e^{-t}\,dt.$$

It is supposed that arg z has its principal value and that the contour is so chosen that the point $t = -z$ is outside it. The integrand is rendered one-valued by taking $|\arg(-t)| \le \pi$ and taking that value of $\arg(1 + t/z)$ which tends to zero as $t \to 0$ by a path lying inside the contour. Under these circumstances it follows from §5.32 that the integral is an analytic function of z. To shew that it satisfies equation (16.2), write

$$v = \int_{\infty}^{(0+)} (-t)^{-k-\frac{1}{2}+m}(1 + t/z)^{k-\frac{1}{2}+m}e^{-t}\,dt;$$

and we have without difficulty[3]

$$\frac{d^2v}{dz^2} + \left(\frac{2k}{z} - 1\right)\frac{dv}{dz} + \frac{\tfrac{1}{4} - m^2 + k(k - 1)}{z^2}v$$

$$= -\frac{(k - \tfrac{1}{2} + m)}{z^2}\int_{\infty}^{(0+)} \frac{d}{dt}\left\{t^{-k+1/2+m}\left(1 + t/z\right)^{k-3/2+m}e^{-t}\right\}\,dt$$

$$= 0,$$

[2] A suitable contour has been chosen and the variable t of §14.6 replaced by $-t$.
[3] The differentiations under the sign of integration are legitimate by Corollary 4.4.1.

since the expression in braces tends to zero as $t \to +\infty$; and this is the condition that $e^{-z/2}z^k v$ should satisfy (16.2).

Accordingly the function $W_{k,m}(z)$ defined by the integral

$$-\frac{1}{2\pi i}\Gamma\left(k + \tfrac{1}{2} - m\right) e^{-\frac{1}{2}z}z^k \int_{\infty}^{(0+)} (-t)^{-k-\frac{1}{2}+m} (1 + t/z)^{k-\frac{1}{2}+m} e^{-t}\, dt$$

is a solution of the differential equation (16.2).

The formula for $W_{k,m}(z)$ becomes nugatory when $k - \tfrac{1}{2} - m$ is a negative integer. To overcome this difficulty, we observe that *whenever* $\mathrm{Re}\left(k - \tfrac{1}{2} - m\right) \le 0$ *and* $k - \tfrac{1}{2} - m$ *is not an integer*, we may transform the contour integral into an infinite integral, after the manner of §12.22; and so, when $\mathrm{Re}\left(k - \tfrac{1}{2} - m\right) \le 0$,

$$W_{k,m}(z) = \frac{e^{-\frac{1}{2}z}z^k}{\Gamma(\tfrac{1}{2} - k + m)} \int_0^{\infty} t^{-k-\frac{1}{2}+m} (1 + t/z)^{k-\frac{1}{2}+m} e^{-t}\, dt.$$

This formula suffices to define $W_{k,m}(z)$ in the critical cases when $m + \tfrac{1}{2} - k$ is a positive integer, and so $W_{k,m}(z)$ is defined for all values of k and m and all values of z except negative real values[4].

Example 16.1.1   Solve the equation

$$\frac{d^2u}{dz^2} + \left(a + \frac{b}{z} + \frac{c}{z^2}\right) u = 0$$

in terms of functions of the type $W_{k,m}(z)$, where a, b, c are any constants.

16.2 Expression of various functions by functions of the type $W_{k,m}(z)$

It has been shewn[5] that various functions employed in Applied Mathematics are expressible by means of the function $W_{k,m}(z)$; the following are a few examples:

(I) *The Error function*[6] which occurs in connexion with the theories of Probability, Errors of Observation, Refraction and Conduction of Heat is defined by the equation

$$\mathrm{Erfc}(x) = \int_x^{\infty} e^{-t^2}\, dt,$$

where x is real.

---

[4] When z is real and negative, $W_{k,m}(z)$ may be defined to be either $W_{k,m}(z + 0i)$ or $W_{k,m}(z - 0i)$, whichever is more convenient.

[5] Whittaker [671]; this paper contains a more complete account than is given here.

[6] This name is also applied to the function

$$\mathrm{Erf}(x) = \int_0^x e^{-t^2}\, dt = \frac{\sqrt{\pi}}{2} - \mathrm{Erfc}(x).$$

Writing $t = x^2(w^2 - 1)$ and then $w = s/x$ in the integral for $W_{-\frac{1}{4},\frac{1}{4}}(x^2)$, we get

$$W_{-\frac{1}{4},\frac{1}{4}}(x^2) = x^{-\frac{1}{2}}e^{-\frac{1}{2}x^2} \int_0^\infty \left(1 + t/x^2\right)^{-1/2} e^{-t}\, dt$$

$$= 2x^{\frac{3}{2}}e^{-\frac{1}{2}x^2} \int_1^\infty e^{x^2(1-w^2)}\, dw$$

$$= 2x^{\frac{1}{2}}e^{\frac{1}{2}x^2} \int_x^\infty e^{-s^2}\, ds,$$

and so the error function is given by the formula

$$\mathrm{Erfc}(x) = \tfrac{1}{2}x^{-\frac{1}{2}}e^{-\frac{1}{2}x^2}W_{-\frac{1}{4},\frac{1}{4}}(x^2).$$

Other integrals which occur in connexion with the theory of Conduction of Heat, e.g. $\int_a^b e^{-t^2 - x^2/t^2}\, dt$, can be expressed in terms of error functions, and so in terms of $W_{k,m}$ functions.

Example 16.2.1 Shew that the formula for the error function is true for complex values of x.

(II) *The Incomplete Gamma-function*, studied by Legendre and others[7], is defined by the equation

$$\gamma(n, x) = \int_0^x t^{n-1}e^{-t}\, dt.$$

By writing $t = s - x$ in the integral for $W_{\frac{1}{2}(n-1),\frac{1}{2}n}(x)$, the reader will verify that

$$\gamma(n, x) = \Gamma(n) - x^{\frac{1}{2}(n-1)}e^{-\frac{1}{2}x}W_{\frac{1}{2}(n-1),\frac{1}{2}n}(x).$$

(III) *The Logarithmic-integral function*, which has been discussed by Euler and others[8], is of considerable importance in the higher parts of the Theory of Prime Numbers; see Landau [405]. It is defined, when $|\arg(-\log z)| < \pi$, by the equation

$$\mathrm{li}(z) = \int_0^z \frac{dt}{\log t}.$$

On writing $s - \log z = u$ and then $u = -\log t$ in the integral for

$$W_{-\frac{1}{2},0}(-\log z),$$

it may be verified that

$$\mathrm{li}(z) = -(-\log z)^{-\frac{1}{2}}z^{\frac{1}{2}}W_{-\frac{1}{2},0}(-\log z).$$

It will appear later that Weber's Parabolic Cylinder functions (§16.5) and Bessel's Circular Cylinder functions (Chapter 17) are particular cases of the $W_{k,m}$ function. Other functions of like nature are given in the Miscellaneous Examples at the end of this chapter.

---

[7] Legendre [421, vol. I, p. 339]; Hočevar [324]; Schlömilch [584]; Prym [545].
[8] Euler [201]; Soldner [597]; Bessel [69]; Laguerre [397]; Stieltjes [605].

Note   The error function has been tabulated by Encke [195], and Burgess [107]. The logarithmic-integral function has been tabulated by Bessel and by Soldner. Jahnke & Emde [356], and Glaisher [253], should also be consulted.

16.3 The asymptotic expansion of $W_{k,m}(z)$, when $|z|$ is large

From the contour integral by which $W_{k,m}(z)$ was defined, it is possible to obtain an asymptotic expansion for $W_{k,m}(z)$ valid when $|\arg z| < \pi$. For this purpose, we employ the result given in Chapter 5, Example 5.6, that

$$\left(1 + \frac{t}{z}\right)^\lambda = 1 + \frac{\lambda}{1}\frac{t}{z} + \cdots + \frac{\lambda(\lambda - 1)\cdots(\lambda - n + 1)}{n!}\frac{t^n}{z^n} + R_n(t, z),$$

where

$$R_n(t, z) = \frac{\lambda(\lambda - 1)\cdots(\lambda - n)}{n!}\left(1 + \frac{t}{z}\right)^\lambda \int_0^{t/z} u^n(1 + u)^{-\lambda - 1}\, du.$$

Substituting this in the formula of §16.12 and integrating term-by-term, it follows from the result of §12.22 that

$$W_{k,m}(z) = e^{-\frac{1}{2}z}z^k \left\{ 1 + \frac{m^2 - (k - \frac{1}{2})^2}{1!z} + \frac{\{m^2 - (k - \frac{1}{2})^2\}\{m^2 - (k - \frac{3}{2})^2\}}{2!z^2} \right.$$

$$+ \cdots + \frac{\{m^2 - (k - \frac{1}{2})^2\}\{m^2 - (k - \frac{3}{2})^2\} \cdots \{m^2 - (k - n + \frac{1}{2})^2\}}{n!z^n}$$

$$\left. + \frac{1}{\Gamma(-k + \frac{1}{2} + m)} \int_0^\infty t^{-k-\frac{1}{2}+m} R_n(t, z)e^{-t}\, dt \right\}$$

provided that n be taken so large that $\operatorname{Re}\left(n - k - \frac{1}{2} + m\right) > 0$.

Now, if $|\arg z| \le \pi - \alpha$ and $|z| > 1$, then

$$\left. \begin{array}{l} 1 \le |(1 + t/z)| \le 1 + t \quad \operatorname{Re}(z) \ge 0 \\ |1 + t/z| \ge \sin\alpha \quad \operatorname{Re}(z) \le 0 \end{array} \right\},$$

and so[9]

$$|R_n(t, z)| \le \left|\frac{\lambda(\lambda - 1)\cdots(\lambda - n)}{n!}\right| \frac{(1 + t)^{|\lambda|}}{(\sin\alpha)^{|\lambda|}} \int_0^{|t/z|} u^n(1 + u)^{|\lambda|}\, du.$$

Therefore

$$|R_n(t, z)| < \left|\frac{\lambda(\lambda - 1)\cdots(\lambda - n)}{n!}\right| \frac{(1 + t)^{|\lambda|}}{(\sin\alpha)^{|\lambda|}} |t/z|^{n+1}(1 + t)^{|\lambda|}(n + 1)^{-1},$$

since $1 + u < 1 + t$. Therefore, when $|z| > 1$,

$$\left|\frac{1}{\Gamma(-k + \frac{1}{2} + m)} \int_0^\infty t^{-k-\frac{1}{2}+m} R_n(t, z)e^{-t}\, dt\right| = O\left\{\int_0^\infty t^{-k+\frac{1}{2}+m+n}(1 + t)^{2|\lambda|}|z|^{-n-1}e^{-t}\, dt\right\}$$

$$= O(z^{-n-1}),$$

---

[9]   It is supposed that λ is real; the inequality has to be slightly modified for complex values of λ.

since the integral converges. The constant implied in the symbol O is independent of arg z, but depends on α, and tends to infinity as $\alpha \to 0$. *That is to say, the asymptotic expansion of $W_{k,m}(z)$ is given by the formula*

$$W_{k,m}(z) \sim e^{-\frac{1}{2}z} z^k$$

$$\times \left\{ 1 + \sum_{n=1}^{\infty} \frac{\left\{ m^2 - (k - \frac{1}{2})^2 \right\} \left\{ m^2 - (k - \frac{3}{2})^2 \right\} \cdots \left\{ m^2 - (k - n + \frac{1}{2})^2 \right\}}{n! z^n} \right\}$$

for large values of $|z|$ when $|\arg z| \le \pi - \alpha < \pi$.

16.31 The second solution of the equation for $W_{k,m}(z)$

The differential equation (16.2) of §16.1 satisfied by $W_{k,m}(z)$ is unaltered if the signs of z and k are changed throughout. Hence, if $|\arg(-z)| < \pi$, $W_{-k,m}(-z)$ is a solution of the equation. Since, when $|\arg z| < \pi$,

$$W_{k,m}(z) = e^{-z/2} z^k \left\{ 1 + O\left(z^{-1}\right) \right\},$$

whereas, when $|\arg(-z)| < \pi$,

$$W_{-k,m}(-z) = e^{z/2}(-z)^{-k} \left\{ 1 + O\left(z^{-1}\right) \right\},$$

the ratio $W_{k,m}(z)/W_{-k,m}(-z)$ cannot be a constant, and so $W_{k,m}(z)$ and $W_{-k,m}(-z)$ form a fundamental system of solutions of the differential equation.

16.4 Contour integrals of the Mellin–Barnes type for $W_{k,m}(z)$

Consider now

$$I = \frac{e^{-z/2} z^k}{2\pi i} \int_{-i\infty}^{i\infty} \frac{\Gamma(s)\Gamma(-s - k - m + \frac{1}{2})\Gamma(-s - k + m + \frac{1}{2})}{\Gamma(-k - m + \frac{1}{2})\Gamma(-k + m + \frac{1}{2})} z^s \, ds, \qquad (16.3)$$

where $|\arg z| < \frac{3}{2}\pi$, and neither of the numbers $k \pm m + \frac{1}{2}$ is a positive integer or zero[10]; the contour has loops if necessary so that the poles of $\Gamma(s)$ and those of $\Gamma\left(-s - k - m + \frac{1}{2}\right) \times \Gamma\left(-s - k + m + \frac{1}{2}\right)$ are on opposite sides of it.

It is easily verified, by §13.6, that, as $s \to \infty$ on the contour,

$$\Gamma(s)\Gamma\left(-s - k - m + \frac{1}{2}\right) \Gamma\left(-s - k + m + \frac{1}{2}\right) = O(e^{-3\pi|s|/2}|s|^{-2k-1/2}),$$

and so the integral represents a function of z which is analytic at all points[11] in the domain $|\arg z| \le \frac{3}{2}\pi - \alpha < \frac{3}{2}\pi$.

Now choose N so that the poles of $\Gamma\left(-s - k - m + \frac{1}{2}\right) \Gamma\left(-s - k + m + \frac{1}{2}\right)$ are on the right of the line $\text{Re}(s) = -N - \frac{1}{2}$; and consider the integral taken round the rectangle whose corners are $\pm i\xi$, $-N - \frac{1}{2} \pm i\xi$, where ξ is positive[12] and large. The reader will verify that,

---

[10] In these cases the series of §16.3 terminates and $W_{k,m}(z)$ is a combination of elementary functions.
[11] The integral is rendered one-valued when $\text{Re}(z) < 0$ by specifying arg z.
[12] The line joining $\pm i\xi$ may have loops to avoid poles of the integrand as explained above.

when $|\arg z| \leq \frac{3}{2}\pi - \alpha$, the integrals $\displaystyle\int_{-i\xi}^{-N-\frac{1}{2}-i\xi}$ and $\displaystyle\int_{i\xi}^{-N-\frac{1}{2}+i\xi}$ tend to zero as $\xi \to \infty$; and so, by Cauchy's theorem,

$$\frac{e^{-z/2}z^k}{2\pi i} \int_{-i\infty}^{i\infty} \frac{\Gamma(s)\Gamma(-s-k-m+\frac{1}{2})\Gamma(-s-k+m+\frac{1}{2})}{\Gamma(-k-m+\frac{1}{2})\Gamma(-k+m+\frac{1}{2})} z^s \, ds$$

$$= e^{-\frac{1}{2}z}z^k \left\{ \sum_{n=0}^{N} R_n + \frac{1}{2\pi i} \int_{-N-\frac{1}{2}-i\infty}^{-N-\frac{1}{2}+i\infty} \frac{\Gamma(s)\Gamma(-s-k-m+\frac{1}{2})\Gamma(-s-k+m+\frac{1}{2})}{\Gamma(-k-m+\frac{1}{2})\Gamma(-k+m+\frac{1}{2})} z^s \, ds \right\},$$

where R_n is the residue of the integrand at $s = -n$.

Write $s = -N - \frac{1}{2} + it$, and the modulus of the last integrand is

$$|z|^{-N-\frac{1}{2}} O\left\{ e^{-\alpha|t|}|t|^{N-2k} \right\},$$

where the constant implied in the symbol O is independent of z. Since $\displaystyle\int^{\pm\infty} e^{-\alpha|t|}|t|^{N-2k} \, dt$ converges, we find that

$$I = e^{-z/2}z^k \left\{ \sum_{n=0}^{N} R_n + O(|z|^{-N-1/2}) \right\}.$$

But, on calculating the residue R_n, we get

$$R_n = \frac{\Gamma(n-k-m+\frac{1}{2})\Gamma(n-k+m+\frac{1}{2})}{n!\,\Gamma(-k-m+\frac{1}{2})\Gamma(-k+m+\frac{1}{2})}(-1)^n z^{-n}$$

$$= \frac{\{m^2 - (k-\frac{1}{2})^2\}\{m^2 - (k-\frac{3}{2})^2\} \cdots \{m^2 - (k-n+\frac{1}{2})^2\}}{n!\,z^n},$$

and so *I has the same asymptotic expansion as* $W_{k,m}(z)$.

Further, *I* satisfies the differential equation for $W_{k,m}(z)$; for, on substituting

$$\int_{-i\infty}^{i\infty} \Gamma(s)\Gamma\left(-s-k-m+\tfrac{1}{2}\right)\Gamma\left(-s-k+m+\tfrac{1}{2}\right) z^s \, ds$$

for v in the expression (given in §16.12)

$$z^2 \frac{d^2 v}{dz^2} + 2kz \frac{dv}{dz} + \left(k-m-\tfrac{1}{2}\right)\left(k+m-\tfrac{1}{2}\right) v - z^2 \frac{dv}{dz},$$

we get

$$\int_{-i\infty}^{i\infty} \Gamma(s)\Gamma\left(-s-k-m+\tfrac{3}{2}\right)\Gamma\left(-s-k+m+\tfrac{3}{2}\right) z^s \, ds$$

$$- \int_{-\infty i}^{\infty i} \Gamma(s+1)\Gamma\left(-s-k-m+\tfrac{1}{2}\right)\Gamma\left(-s-k+m+\tfrac{1}{2}\right) z^{s+1} \, ds$$

$$= \left(\int_{-i\infty}^{i\infty} - \int_{1-i\infty}^{1+i\infty} \right) \Gamma(s)\Gamma\left(-s-k-m+\tfrac{3}{2}\right)\Gamma\left(-s-k+m+\tfrac{3}{2}\right) z^s \, ds.$$

Since there are no poles of the last integrand between the contours, and since the integrand tends to zero as $|s| \to \infty$, s being between the contours, the expression under consideration vanishes, by Cauchy's theorem; and so I satisfies the equation for $W_{k,m}(z)$.

Therefore $I = AW_{k,m}(z) + BW_{-k,m}(-z)$, where A and B are constants. Making $|z| \to \infty$ when $\operatorname{Re}(z) > 0$ we see, from the asymptotic expansions obtained for I and $W_{\mp k,m}(\pm z)$, that $A = 1$, $B = 0$. Accordingly, by the theory of analytic continuation, the equality

$$I = W_{k,m}(z)$$

persists for all values of z such that $|\arg z| < \pi$; and, for values[13] of $\arg z$ such that $\pi \le |\arg z| < \frac{3}{2}\pi$, $W_{k,m}(z)$ may be *defined* to be the expression I.

Example 16.4.1   Shew that

$$W_{k,m}(z) = \frac{e^{-\frac{1}{2}z}}{2\pi i} \int_{-i\infty}^{i\infty} \frac{\Gamma(s-k)\Gamma(-s-m+\frac{1}{2})\Gamma(-s+m+\frac{1}{2})}{\Gamma(-k-m+\frac{1}{2})\Gamma(-k+m+\frac{1}{2})} z^s \, ds,$$

taken along a suitable contour.

Example 16.4.2   Obtain Barnes' integral for $W_{k,m}(z)$ by writing

$$\frac{1}{2\pi i} \int_{-i\infty}^{i\infty} \frac{\Gamma(s)\Gamma(-s-k-m+\frac{1}{2})}{\Gamma(-k-m+\frac{1}{2})} z^s t^{-s} \, ds$$

for $(1+t/z)^{k-\frac{1}{2}+m}$ in the integral of §16.12 and changing the order of integration.

16.41 Relations between $W_{k,m}(z)$ and $M_{k,\pm m}(z)$

If we take the expression

$$F(s) \equiv \Gamma(s)\Gamma\left(-s-k-m+\tfrac{1}{2}\right)\Gamma\left(-s-k+m+\tfrac{1}{2}\right)$$

which occurs in Barnes' integral for $W_{k,m}(z)$, and write it in the form

$$\frac{\pi^2 \Gamma(s)}{\Gamma(s+k+m+\frac{1}{2})\Gamma(s+k-m+\frac{1}{2})\cos(s+k+m)\pi \cos(s+k-m)\pi},$$

we see, by §13.6, that when $\operatorname{Re}(s) \ge 0$, we have, as $|s| \to \infty$,

$$F(s) = O\left[\exp\left\{(-s-\tfrac{1}{2}-2k)\log s + s\right\}\right] \sec(s+k+m)\pi \sec(s+k-m)\pi.$$

Hence, if $|\arg z| < \frac{3}{2}\pi$, $\int F(s)z^s \, ds$, taken round a semicircle on the right of the imaginary axis, tends to zero as the radius of the semicircle tends to infinity, provided the lower bound of the distance of the semicircle from the poles of the integrand is positive (not zero). Therefore

$$W_{k,m}(z) = -\frac{e^{-\frac{1}{2}z}z^k \cdot (\sum R')}{\Gamma(-k-m+\frac{1}{2})\Gamma(-k+m+\frac{1}{2})},$$

[13] It would have been possible, by modifying the path of integration in §16.3, to have shewn that that integral could be made to define an analytic function when $|\arg z| < 3\pi/2$. But the reader will see that it is unnecessary to do so, as Barnes' integral affords a simpler definition of the function.

where $\sum R'$ denotes the sum of the residues of $F(s)$ at its poles on the right of the contour (cf. §14.5) which occurs in equation (16.3) of §16.4.

Evaluating these residues we find without difficulty that, when $|\arg z| < \frac{3}{2}\pi$, and $2m$ is not an integer[14],

$$W_{k,m}(z) = \frac{\Gamma(-2m)}{\Gamma(\frac{1}{2} - m - k)} M_{k,m}(z) + \frac{\Gamma(2m)}{\Gamma(\frac{1}{2} + m - k)} M_{k,-m}(z).$$

Example 16.4.3 (Barnes)   Shew that, when $|\arg(-z)| < \frac{3}{2}\pi$ and $2m$ is not an integer,

$$W_{-k,m}(-z) = \frac{\Gamma(-2m)}{\Gamma(\frac{1}{2} - m + k)} M_{-k,m}(-z) + \frac{\Gamma(2m)}{\Gamma(\frac{1}{2} + m + k)} M_{-k,-m}(-z).$$

These results are given in the notation explained in §16.1.

Example 16.4.4   When $-\frac{1}{2}\pi < \arg z < \frac{3}{2}\pi$; and $-\frac{3}{2}\pi < \arg(-z) < \frac{1}{2}\pi$, shew that

$$M_{k,m}(z) = \frac{\Gamma(2m + 1)}{\Gamma(\frac{1}{2} + m - k)} e^{k\pi i} W_{-k,m}(-z) + \frac{\Gamma(2m + 1)}{\Gamma(\frac{1}{2} + m + k)} e^{(\frac{1}{2} + m + k)\pi i} W_{k,m}(z).$$

Example 16.4.5 (Barnes)   Obtain Kummer's first formula (§16.11) from the result

$$z^n e^{-z} = \frac{1}{2\pi i} \int_{-i\infty}^{i\infty} \Gamma(n - s) z^s \, ds.$$

16.5 The parabolic cylinder functions. Weber's equation

Consider the differential equation satisfied by $w = z^{-\frac{1}{2}} W_{k,-\frac{1}{4}}\left(\frac{1}{2}z^2\right)$; it is

$$\frac{1}{z}\frac{d}{dz}\left\{\frac{1}{z}\frac{d(wz^{\frac{1}{2}})}{dz}\right\} + \left\{-\frac{1}{4} + \frac{2k}{z^2} + \frac{\frac{3}{4}}{z^4}\right\} wz^{\frac{1}{2}} = 0;$$

this reduces to

$$\frac{d^2w}{dz^2} + \left\{2k - \frac{1}{4}z^2\right\} w = 0.$$

Therefore the function

$$D_n(z) = 2^{\frac{1}{2}n+\frac{1}{4}} z^{-\frac{1}{2}} W_{\frac{1}{2}n+\frac{1}{4},-\frac{1}{4}}\left(\frac{1}{2}z^2\right)$$

satisfies the differential equation

$$\frac{d^2 D_n(z)}{dz^2} + \left(n + \frac{1}{2} - \frac{1}{4}z^2\right) D_n(z) = 0.$$

Accordingly $D_n(z)$ is one of the functions associated with the parabolic cylinder in harmonic analysis (see Weber [655] and Whittaker [670]); the equation satisfied by it will be called Weber's equation.

---

[14] When $2m$ is an integer some of the poles are generally double poles, and their residues involve logarithms of z. The result has not been proved when $k - \frac{1}{2} \pm m$ is a positive integer or zero, but may be obtained for such values of k and m by comparing the terminating series for $W_{k,m}(z)$ with the series for $M_{k,\pm m}(z)$.

From §16.41, it follows that

$$D_n(z) = \frac{\Gamma(\frac{1}{2})2^{\frac{1}{2}n+\frac{1}{4}}z^{-\frac{1}{2}}}{\Gamma(\frac{1}{2}-\frac{1}{2}n)} M_{\frac{n}{2}+\frac{1}{4},-\frac{1}{4}}\left(\tfrac{1}{2}z^2\right) + \frac{\Gamma(-\frac{1}{2})2^{\frac{1}{2}n+\frac{1}{4}}z^{-\frac{1}{2}}}{\Gamma(-\frac{1}{2}n)} M_{\frac{n}{2}+\frac{1}{4},\frac{1}{4}}\left(\tfrac{1}{2}z^2\right)$$

when $|\arg z| < \frac{3}{4}\pi$. But

$$z^{-\frac{1}{2}}M_{\frac{n}{2}+\frac{1}{4},-\frac{1}{4}}\left(\tfrac{1}{2}z^2\right) = 2^{-\frac{1}{4}}e^{-\frac{1}{4}z^2}\,{}_1F_1\left(-\tfrac{n}{2};\tfrac{1}{2};\tfrac{1}{2}z^2\right),$$

$$z^{-\frac{1}{2}}M_{\frac{n}{2}+\frac{1}{4},\frac{1}{4}}\left(\tfrac{1}{2}z^2\right) = 2^{-\frac{3}{4}}ze^{-\frac{1}{4}z^2}\,{}_1F_1\left(\tfrac{1}{2}-\tfrac{n}{2};\tfrac{3}{2};\tfrac{1}{2}z^2\right),$$

and these are *one-valued* analytic functions of z throughout the z-plane. Accordingly $D_n(z)$ is a one-valued function of z throughout the z-plane; and, by §16.4, its asymptotic expansion when $|\arg z| < \frac{3}{4}\pi$ is

$$e^{-\frac{1}{4}z^2}z^n\left\{1 - \frac{n(n-1)}{2z^2} + \frac{n(n-1)(n-2)(n-3)}{2\cdot 4z^4} - \cdots\right\}.$$

16.51 The second solution of Weber's equation

Since Weber's equation is unaltered if we simultaneously replace n and z by $-n-1$ and $\pm iz$ respectively, it follows that $D_{-n-1}(iz)$ and $D_{-n-1}(-iz)$ are solutions of Weber's equation, as is also $D_n(-z)$.

It is obvious from the asymptotic expansions of $D_n(z)$ and $D_{-n-1}(ze^{\frac{1}{2}\pi i})$, valid in the range $-\frac{3}{4}\pi < \arg z < \frac{1}{4}\pi$, that the ratio of these two solutions is not a constant.

16.511 The relation between the functions $D_n(z), D_{-n-1}(\pm iz)$

From the theory of linear differential equations, a relation of the form

$$D_n(z) = aD_{-n-1}(iz) + bD_{-n-1}(-iz)$$

must hold when the ratio of the functions on the right is not a constant.

To obtain this relation, we observe that if the functions involved be expanded in ascending powers of z, the expansions are

$$\frac{\Gamma(\frac{1}{2})2^{\frac{n}{2}}}{\Gamma(\frac{1}{2}-\frac{n}{2})} + \frac{\Gamma(-\frac{1}{2})2^{\frac{n}{2}-\frac{1}{2}}}{\Gamma(-\frac{1}{2}n)}z + \cdots$$

and

$$a\left\{\frac{\Gamma(\frac{1}{2})2^{-\frac{1}{2}n-\frac{1}{2}}}{\Gamma(1+\frac{n}{2})} + \frac{\Gamma(-\frac{1}{2})2^{-\frac{n}{2}-1}}{\Gamma(\frac{1}{2}+\frac{n}{2})}iz + \cdots\right\} + b\left\{\frac{\Gamma(\frac{1}{2})2^{-\frac{1}{2}n-\frac{1}{2}}}{\Gamma(1+\frac{1}{2}n)} - \frac{\Gamma(-\frac{1}{2})2^{-\frac{n}{2}-1}}{\Gamma(\frac{1}{2}+\frac{n}{2})}iz + \cdots\right\}.$$

Comparing the first two terms we get

$$a = (2\pi)^{-\frac{1}{2}}\Gamma(n+1)\,e^{\frac{1}{2}n\pi i}, \quad b = (2\pi)^{-\frac{1}{2}}\Gamma(n+1)e^{-\frac{1}{2}n\pi i},$$

and so

$$D_n(z) = \frac{\Gamma(n+1)}{\sqrt{2\pi}}\left[e^{n\pi i/2}D_{-n-1}(iz) + e^{-n\pi i/2}D_{-n-1}(-iz)\right].$$

16.52 The general asymptotic expansion of $D_n(z)$

So far the asymptotic expansion of $D_n(z)$ for large values of z has only been given (§16.5) in the sector $|\arg z| < \frac{3}{4}\pi$. To obtain its form for values of $\arg z$ not comprised in this range we write $-iz$ for z and $-n-1$ for n in the formula of the preceding section, and get

$$D_n(z) = e^{n\pi i} D_n(-z) + \frac{\sqrt{2\pi}}{\Gamma(-n)} e^{\frac{1}{2}(n+1)\pi i} D_{-n-1}(-iz).$$

Now, if $\frac{5}{4}\pi > \arg z > \frac{1}{4}\pi$, we can assign to $-z$ and $-iz$ arguments between $\pm\frac{3}{4}\pi$; and $\arg(-z) = \arg z - \pi$, $\arg(-iz) = \arg z - \frac{1}{2}\pi$; and then, applying the asymptotic expansion of §16.5 to $D_n(-z)$ and $D_{-n-1}(-iz)$, we see that, if $\frac{3}{4}\pi > \arg z > \frac{1}{4}\pi$,

$$D_n(z) \sim e^{-\frac{1}{4}z^2} z^n \left\{ 1 - \frac{n(n-1)}{2z^2} + \frac{n(n-1)(n-2)(n-3)}{2 \cdot 4z^4} - \cdots \right\}$$

$$- \frac{\sqrt{2\pi}}{\Gamma(-n)} e^{n\pi i} e^{\frac{1}{4}z^2} z^{-n-1} \left\{ 1 + \frac{(n+1)(n+2)}{1 \cdot 2z^2} + \frac{(n+1)(n+2)(n+3)(n+4)}{2 \cdot 4z^4} + \cdots \right\}.$$

This formula is not inconsistent with that of §16.5 since in their common range of validity, viz. $\frac{1}{4}\pi < \arg z < \frac{3}{4}\pi$, $e^{\frac{1}{2}z^2} z^{-2n-1}$ is $o(z^{-m})$ for all positive values of m.

To obtain a formula valid in the range $-\frac{1}{4}\pi > \arg z > -\frac{5}{4}\pi$, we use the formula

$$D_n(z) = e^{-n\pi i} D_n(-z) + \frac{\sqrt{2\pi}}{\Gamma(-n)} e^{-\frac{1}{2}(n+1)\pi i} D_{-n-1}(iz),$$

and we get an asymptotic expansion which differs from that which has just been obtained only in containing $e^{-n\pi i}$ in place of $e^{n\pi i}$.

Since $D_n(z)$ is one-valued and one or other of the expansions obtained is valid for all values of $\arg z$ in the range $-\pi \leq \arg z \leq \pi$, the complete asymptotic expansion of $D_n(z)$ has been obtained.

16.6 A contour integral for $D_n(z)$

Consider $\displaystyle\int_\infty^{(0+)} e^{-zt-\frac{1}{2}t^2}(-t)^{-n-1}\, dt$, where $|\arg(-t)| \leq \pi$; it represents a one-valued analytic function of z throughout the z-plane (§5.32) and further

$$\left(\frac{d^2}{dz^2} - z\frac{d}{dz} + n \right) \int_\infty^{(0+)} e^{-zt-\frac{1}{2}t^2}(-t)^{-n-1}\, dt = \int_\infty^{(0+)} \frac{d}{dt}\{e^{-zt-\frac{1}{2}t^2}(-t)^{-n}\}\, dt = 0,$$

the differentiations under the sign of integration being easily justified; accordingly the integral satisfies the differential equation satisfied by $e^{\frac{1}{4}z^2} D_n(z)$; and therefore

$$e^{-\frac{1}{4}z^2} \int_\infty^{(0+)} e^{-zt-\frac{1}{2}t^2}(-t)^{-n-1}\, dt = aD_n(z) + bD_{-n-1}(iz),$$

where a and b are constants.

Now, if the expression on the right be called $E_n(z)$, we have

$$E_n(0) = \int_\infty^{(0+)} e^{-\frac{1}{2}t^2}(-t)^{-n-1}\, dt, \qquad E_n'(0) = \int_\infty^{(0+)} e^{-\frac{1}{2}t^2}(-t)^{-n}\, dt.$$

To evaluate these integrals, which are analytic functions of n, we suppose first that $\mathrm{Re}(n) < 0$; then, deforming the paths of integration, we get

$$E_n(0) = -2i\sin(n+1)\pi \int_0^\infty e^{-\frac{1}{2}t^2}t^{-n-1}\,dt$$

$$= 2^{-n/2}i\sin n\pi \int_0^\infty e^{-u}u^{-n/2-1}\,du$$

$$= 2^{-n/2}i\sin(n\pi)\Gamma\left(-\tfrac{n}{2}\right).$$

Similarly $E_n'(0) = -2^{(1-n)/2}i\sin(n\pi)\Gamma(\tfrac{1}{2}(1-n))$. Both sides of these equations being analytic functions of n, the equations are true for all values of n; and therefore

$$b = 0, \quad a = \frac{\Gamma(\tfrac{1}{2}-\tfrac{1}{2}n)}{\Gamma(\tfrac{1}{2})2^{n/2}}2^{-\frac{1}{2}n}i\sin(n\pi)\Gamma\left(-\tfrac{n}{2}\right) = 2i\Gamma(-n)\sin n\pi.$$

Therefore

$$D_n(z) = -\frac{\Gamma(n+1)}{2\pi i}e^{-\frac{1}{4}z^2}\int_\infty^{(0+)} e^{-zt-\frac{1}{2}t^2}(-t)^{-n-1}\,dt.$$

16.61 Recurrence formulae for $D_n(z)$

Form the equation

$$0 = \int_\infty^{(0+)} \frac{d}{dt}\left\{e^{-zt-\frac{1}{2}t^2}(-t)^{-n-1}\right\}\,dt$$

$$= \int_\infty^{(0+)} \left\{-z(-t)^{-n-1} + (-t)^{-n} + (n+1)(-t)^{-n-2}\right\}e^{-zt-\frac{1}{2}t^2}\,dt,$$

after using §16.6, we see that

$$D_{n+1}(z) - zD_n(z) + nD_{n-1}(z) = 0.$$

Further, by differentiating the integral of §16.6, it follows that

$$D_n'(z) + \tfrac{1}{2}zD_n(z) - nD_{n-1}(z) = 0.$$

Example 16.6.1 Obtain these results from the ascending power series of §16.5.

16.7 Properties of $D_n(z)$ when n is an integer

When n is an integral, we may write the integral of §16.6 in the form

$$D_n(z) = -\frac{n!e^{-\frac{1}{4}z^2}}{2\pi i}\int^{(0+)} \frac{e^{-zt-\frac{1}{2}t^2}}{(-t)^{n+1}}\,dt.$$

If now we write $t = v - z$, we get

$$D_n(z) = (-1)^n\frac{n!e^{\frac{1}{4}z^2}}{2\pi i}\int^{(z+)} \frac{e^{-\frac{1}{2}v^2}}{(v-z)^{n+1}}\,dv$$

$$= (-1)^n e^{\frac{1}{4}z^2}\frac{d^n}{dz^n}\left(e^{-\frac{1}{2}z^2}\right),$$

a result due to Hermite [289].

Also, if m and n be unequal integers, we see from the differential equations that

$$D_n(z)D_m''(z) - D_m(z)D_n''(z) + (m-n)D_m(z)D_n(z) = 0,$$

and so

$$(m-n)\int_{-\infty}^{\infty} D_m(z)D_n(z)\,dz = \left[D_n(z)D_m'(z) - D_m(z)D_n'(z)\right]_{-\infty}^{\infty}$$
$$= 0,$$

by the expansion of §16.5 in descending powers of z (which terminates and is valid for all values of arg z when n is a positive integer). Therefore if m and n are unequal positive integers

$$\int_{-\infty}^{\infty} D_m(z)D_n(z)\,dz = 0.$$

On the other hand, when $m = n$, we have

$$(n+1)\int_{-\infty}^{\infty} \{D_n(z)\}^2\,dz = \int_{-\infty}^{\infty} D_n(z)\left\{D_{n+1}'(z) + \tfrac{1}{2}zD_{n+1}(z)\right\}\,dz$$
$$= [D_n(z)D_{n+1}(z)]_{-\infty}^{\infty}$$
$$+ \int_{-\infty}^{\infty} \left\{\tfrac{1}{2}zD_n(z)D_{n+1}(z) - D_{n+1}(z)D_n'(z)\right\}\,dz$$
$$= \int_{-\infty}^{\infty} \{D_{n+1}(z)\}^2\,dz,$$

on using the recurrence formula, integrating by parts and then using the recurrence formula again. It follows by induction that

$$\int_{-\infty}^{\infty} \{D_n(z)\}^2\,dz = n!\int_{-\infty}^{\infty} \{D_0(z)\}^2\,dz$$
$$= n!\int_{-\infty}^{\infty} e^{-\frac{1}{2}z^2}\,dz$$
$$= (2\pi)^{\frac{1}{2}}n!,$$

by Corollary 12.1.1 and §12.2.

It follows at once that if, for a function $f(z)$, an expansion of the form

$$f(z) = a_0 D_0(z) + a_1 D_1(z) + \cdots + a_n D_n(z) + \cdots$$

exists, and if it is legitimate to integrate term-by-term between the limits $-\infty$ and ∞, then

$$a_n = \frac{1}{(2\pi)^{\frac{1}{2}}n!}\int_{-\infty}^{\infty} D_n(t)f(t)\,dt.$$

16.8 Miscellaneous examples

Example 16.1   Shew that, if the integral is convergent, then

$$M_{k,m}(z) = \frac{\Gamma(2m+1)z^{m+\frac{1}{2}}2^{-2m}}{\Gamma(\frac{1}{2}+m+k)\Gamma(\frac{1}{2}+m-k)}$$

$$\times \int_{-1}^{1} (1+u)^{-\frac{1}{2}+m-k}(1-u)^{-\frac{1}{2}+m+k}\, e^{\frac{1}{2}zu}\, du.$$

Example 16.2   Shew that

$$M_{k,m}(z) = z^{\frac{1}{2}+m}\, e^{-z/2} \lim_{\rho\to\infty} F\left(\tfrac{1}{2}+m-k, \tfrac{1}{2}+m-k+\rho; 2m+1; z/\rho\right).$$

Example 16.3   Obtain the recurrence formulae

$$W_{k,m}(z) = z^{\frac{1}{2}} W_{k-\frac{1}{2},m-\frac{1}{2}}(z) + \left(\tfrac{1}{2}-k+m\right) W_{k-1,m}(z),$$

$$W_{k,m}(z) = z^{\frac{1}{2}} W_{k-\frac{1}{2},m+\frac{1}{2}}(z) + \left(\tfrac{1}{2}-k-m\right) W_{k-1,m}(z),$$

$$zW_{k,m}'(z) = \left(k - \tfrac{1}{2}z\right) W_{k,m}(z) - \left\{m^2 - \left(k - \tfrac{1}{2}\right)^2\right\} W_{k-1,m}(z).$$

Example 16.4   Prove that $W_{k,m}(z)$ is the integral of an elementary function when either of the numbers $k - \frac{1}{2} \pm m$ is a negative integer.

Example 16.5   Shew that, by a suitable change of variables, the equation

$$(a_2 + b_2 x)\frac{d^2 y}{dx^2} + (a_1 + b_1 x)\frac{dy}{dx} + (a_0 + b_0 x)y = 0$$

can be brought to the form

$$\xi\frac{d^2\eta}{d\xi^2} + (c - \xi)\frac{d\eta}{d\xi} - a\eta = 0.$$

Derive this equation from the equation for $F(a, b; c; x)$ by writing $x = \xi/b$ and making $b \to \infty$.

Example 16.6   Shew that the cosine integral of Schlömilch and Besso [71], defined by the equation

$$\mathrm{Ci}(z) = \int_{z}^{\infty} \frac{\cos t}{t}\, dt,$$

is equal to

$$\frac{1}{2}z^{-\frac{1}{2}}\, e^{\frac{1}{2}iz+\frac{1}{4}\pi i} W_{-\frac{1}{2},0}(-iz) + \frac{1}{2}z^{-\frac{1}{2}}e^{-\frac{1}{2}iz-\frac{1}{4}\pi i} W_{-\frac{1}{2},0}(iz).$$

Shew also that Schlömilch's function, defined in [582] by the equations

$$S(v, z) = \int_{0}^{\infty} (1+t)^{-v}e^{-zt}\, dt = z^{v-1}e^{z} \int_{z}^{\infty} \frac{e^{-u}}{u^{v}}\, du,$$

is equal to $z^{v/2-1}e^{z/2}W_{-\frac{1}{2}v, \frac{1}{2}-\frac{1}{2}v}(z)$.

Example 16.7   Express in terms of $W_{k,m}$ functions the two functions

$$\mathrm{Si}(z) \equiv \int_0^z \frac{\sin t}{t}\, dt, \quad \mathrm{Ei}(z) \equiv \int_z^\infty \frac{e^{-t}}{t}\, dt.$$

The results of Examples 16.8, 16.9, 16.10 below were communicated to us by Mr. Bateman.

Example 16.8   Shew that Sonine's polynomial [598], defined by the equation

$$T_m^n(z) = \frac{z^n}{n!(m+n)!0!} - \frac{z^{n-1}}{(n-1)!(m+n-1)!1!}$$

$$+ \frac{z^{n-2}}{(n-2)!(m+n-2)!2!} - \cdots,$$

is equal to

$$\frac{1}{n!(m+n)!} z^{-\frac{1}{2}(m+1)} e^{z/2} W_{n+\frac{1}{2}m+\frac{1}{2},\frac{1}{2}m}(z).$$

Example 16.9   Shew that the function $\phi_m(z)$ defined by Lagrange [394] in 1762–1765 and by Abel [5, p. 284] as the coefficient of h^m in the expansion of $(1-h)^{-1} e^{-hz/(1-k)}$ is equal to

$$\frac{(-1)^m}{m!} z^{-1/2}\, e^{z/2} W_{m+\frac{1}{2},0}(z).$$

Example 16.10   Shew that the Pearson–Cunningham function [517, 158], $\omega_{n,m}(z)$, defined as

$$\frac{e^{-z}(-z)^{n-\frac{1}{2}m}}{\Gamma(n-\frac{1}{2}m+1)}\left\{ 1 - \frac{(n+\frac{1}{2}m)(n-\frac{1}{2}m)}{z} \right.$$

$$\left. + \frac{(n+\frac{1}{2}m)(n+\frac{1}{2}m-1)(n-\frac{1}{2}m)(n-\frac{1}{2}m-1)}{2!z^2} - \cdots \right\},$$

is equal to

$$\frac{(-1)^{n-\frac{1}{2}m}}{\Gamma(n-\frac{1}{2}m+1)} z^{-\frac{1}{2}(m+1)} e^{-z/2} W_{n+\frac{1}{2},\frac{1}{2}m}(z).$$

Example 16.11 (Whittaker)   Shew that, if $|\arg z| < \frac{1}{4}\pi$ and $|\arg(1+t)| \le \pi$,

$$D_n(z) = \frac{\Gamma(\frac{1}{2}n+1)}{2^{-\frac{1}{2}(n-1)}\pi i} \int_{-\infty}^{(-1+)} e^{\frac{1}{4}z^2 t}(1+t)^{-\frac{1}{2}n-1}(1-t)^{\frac{1}{2}(n-1)}\, dt.$$

Example 16.12   Shew that, if n be not a positive integer and if $|\arg z| < \frac{3}{4}\pi$, then

$$D_n(z) = \frac{1}{2\pi i}\, e^{-\frac{1}{4}z^2} \int_{-i\infty}^{i\infty} \frac{\Gamma(\frac{1}{2}t - \frac{1}{2}n)\Gamma(-t)}{\Gamma(-n)}(\sqrt{2})^{t-n-2} z^t\, dt,$$

and that this result holds for all values of arg z if the integral be $\displaystyle\int_\infty^{(0-)}$, the contours enclosing the poles of $\Gamma(-t)$ but not those of $\Gamma(\frac{t-n}{2})$.

Example 16.13 Shew that, if $|\arg a| < \frac{1}{2}\pi$,

$$\int_x^{(0+)} e^{(\frac{1}{4}-a)z^2} z^m D_n(z)\, dz = \frac{\pi^{3/2} 2^{n/2-m} e^{\pi i(m-\frac{1}{2})}}{\Gamma(-m)\Gamma(\frac{1}{2}m - \frac{1}{2}n + 1)a^{\frac{1}{2}(m+1)}}$$

$$\times F\left(-\frac{n}{2}, \frac{m+1}{2}; \frac{m-n}{2} + 1; 1 - \frac{1}{2a}\right).$$

Example 16.14 (Watson) Deduce from Example 16.13 that, if the integral is convergent, then

$$\int_0^x e^{-\frac{3}{4}z^2} z^m D_{m+1}(z)\, dz = \sqrt{2}^{-1-m}\Gamma(m+1)\sin\left((1-m)\tfrac{\pi}{4}\right).$$

Example 16.15 (Watson) Shew that, if n be a positive integer, and if

$$E_n(x) = \int_{-\infty}^{\infty} e^{-\frac{1}{4}z^2}(z-x)^{-1} D_n(z)\, dz,$$

then $E_n(x) = \pm i e^{\mp n\pi i}\sqrt{2\pi}\Gamma(n+1)e^{-\frac{1}{4}x^2} D_{-n-1}(\mp ix)$, the upper or lower signs being taken according as the imaginary part of x is positive or negative.

Example 16.16 (Adamoff) Shew that, if n be a positive integer,

$$D_n(x) = (-1)^\mu 2^{n+2}(2\pi)^{-\frac{1}{2}} e^{\frac{1}{4}x^2} \int_0^x u^n e^{-2u^2} \begin{Bmatrix} \cos \\ \sin \end{Bmatrix} (2xu)\, du,$$

where μ is $\frac{1}{2}n$ or $\frac{1}{2}(n-1)$, whichever is an integer, and the cosine or sine is taken as n is even or odd.

Example 16.17 (Adamoff) Shew that, if n be a positive integer,

$$D_n(x) = (-1)^\mu \sqrt{\frac{2}{\pi}} \sqrt{n}^{n+1} e^{\frac{1}{4}x^2} e^{-\frac{1}{2}n}(J_1 + J_2 - J_3),$$

where

$$J_1 = \int_{-\infty}^{\infty} e^{-n(v-1)^2} \begin{Bmatrix} \cos \\ \sin \end{Bmatrix} (xv\sqrt{n})\, dv,$$

$$J_2 = \int_0^{\infty} \sigma(v) \begin{Bmatrix} \cos \\ \sin \end{Bmatrix} (xv\sqrt{n})\, dv,$$

$$J_3 = \int_{-\infty}^0 e^{-n(v-1)^2} \begin{Bmatrix} \cos \\ \sin \end{Bmatrix} (xv\sqrt{n})\, dv,$$

and $\sigma(v) = e^{\frac{1}{2}n(1-v^2)}v^n - e^{-n(v-1)^2}$.

Example 16.18 (Adamoff) With the notation of the preceding examples, shew that, when x is real,

$$J_1 = \pi^{\frac{1}{2}} n^{-\frac{1}{2}} e^{-\frac{1}{4}x^2} \begin{Bmatrix} \cos \\ \sin \end{Bmatrix} (xv\sqrt{n})\, dv$$

while J_3 satisfies both the inequalities

$$|J_3| < \frac{2e^{-n}}{|x|\sqrt{n}}, \quad |J_3| < \left(\frac{\pi}{2n}\right)^{\frac{1}{2}} e^{-n}.$$

Shew also that as v increases from 0 to 1, $\sigma(v)$ decreases from 0 to a minimum at $v = 1 - h_1$ and then increases to 0 at $v = 1$; and as v increases from 1 to ∞, $\sigma(v)$ increases to a maximum at $1 + h_2$ and then decreases, its limit being zero; where

$$\frac{1}{2}\sqrt{\frac{3}{2n}} < h_1 < \sqrt{\frac{3}{2n}}, \quad \frac{1}{2}\sqrt{\frac{3}{2n}} < h_2 < \sqrt{\frac{3}{2n}},$$

and $|\sigma(1 - h_1)| < An^{-\frac{1}{2}}, \sigma(1 + h_2) < An^{-\frac{1}{2}}$, where $A = 0.0742\cdots$.

Example 16.19 (Adamoff) By employing the second mean-value theorem when necessary, shew that

$$D_n(x) = \sqrt{2}\sqrt{n}^n e^{-n/2}\left[\cos\left(x\sqrt{n} - \tfrac{1}{2}n\pi\right) + \frac{\omega_n(x)}{\sqrt{n}}\right],$$

where $\omega_n(x)$ satisfies both the inequalities

$$|\omega_n(x)| < \frac{3.35\cdots}{|x|\sqrt{\pi}}e^{\frac{1}{4}x^2}, \quad |\omega_n(0)| < \frac{1}{6\sqrt{n}},$$

when x is real and n is an integer greater than 2.

Example 16.20 (Milne) Shew that, if n be positive but otherwise unrestricted, and if m be a *positive integer* (or zero), then the equation in z

$$D_n(z) = 0$$

has m positive roots when $2m - 1 < n < 2m + 1$.

17

Bessel Functions

17.1 The Bessel coefficients

In this chapter we shall consider a class of functions known as *Bessel functions* or *cylindrical functions* which have many analogies with the Legendre functions of Chapter 15. Just as the Legendre functions proved to be particular forms of the hypergeometric function with three regular singularities, so the Bessel functions are particular forms of the confluent hypergeometric function with one regular and one irregular singularity. As in the case of the Legendre functions, we first introduce a certain set of the Bessel functions as coefficients in an expansion. This procedure is due to Schlömilch [581].

For all values of z and t ($t = 0$ excepted), the function

$$e^{\frac{1}{2}z\left(t-\frac{1}{t}\right)}$$

can be expanded by Laurent's theorem in a series of positive and negative powers of t. If the coefficient of t^n, where n is any integer positive or negative, be denoted by $J_n(z)$, it follows, from §5.6, that

$$J_n(z) = \frac{1}{2\pi i} \int^{(0+)} u^{-n-1} e^{\frac{1}{2}z(u-1/u)} \, du.$$

To express $J_n(z)$ as a power series in z, write $u = 2t/z$; then

$$J_n(z) = \frac{1}{2\pi i} \left(\frac{z}{2}\right)^n \int^{(0+)} t^{-n-1} \exp\left(t - z^2/4t\right) \, dt;$$

since the contour is any one which encircles the origin once counter-clockwise, we may take it to be the circle $|t| = 1$; as the integrand can be expanded in a series of powers of z uniformly convergent on this contour, it follows from §4.7 that

$$J_n(z) = \frac{1}{2\pi i} \sum_{r=0}^{\infty} \frac{(-1)^r}{r!} \left(\frac{z}{2}\right)^{n+2r} \int^{(0+)} t^{-n-r-1} e^t \, dt. \tag{17.1}$$

Now the residue of the integrand at $t = 0$ is $1/(n + r)!$ by §6.1, when $n + r$ is a positive integer or zero; when $n+r$ is a negative integer the residue is zero. Therefore, if n is a positive integer or zero,

$$J_n(z) = \sum_{r=0}^{\infty} \frac{(-1)^r (\frac{z}{2})^{n+2r}}{r!(n+r)!}$$

$$= \frac{z^n}{2^n n!} \left\{ 1 - \frac{z^2}{2^2 \cdot 1(n+1)} + \frac{z^4}{2^4 \cdot 1 \cdot 2(n+1)(n+2)} - \cdots \right\};$$

whereas, when n is a negative integer equal to $-m$,

$$J_n(z) = \sum_{r=m}^{\infty} \frac{(-1)^r \left(\frac{z}{2}\right)^{2r-m}}{r!(r-m)!} = \sum_{s=0}^{\infty} \frac{(-1)^{m+s}\left(\frac{1}{2}z\right)^{m+2s}}{(m+s)!s!},$$

and so $J_n(z) = (-1)^m J_m(z)$.

The function $J_n(z)$, which has now been defined for all integral values of n, positive and negative, is called the *Bessel coefficient* of order n; the series defining it converges for all values of z. We shall see later (§17.2) that Bessel coefficients are a particular case of a class of functions known as *Bessel functions*.

The series by which $J_n(z)$ is defined occurs in a memoir by Euler [206], on the vibrations of a stretched circular membrane, an investigation dealt with below in §18.51; it also occurs in a memoir by Lagrange [393] on elliptic motion. The earliest systematic study of the functions was made in 1824 by Bessel in [70]; special cases of Bessel coefficients had, however, appeared in researches published before 1769; the earliest of these is in a letter, dated Oct. 3, 1703, from Jakob Bernoulli to Leibniz [424], in which occurs a series which is now described as a Bessel function of order $\frac{1}{3}$; the Bessel coefficient of order zero occurs in 1732 in Daniel Bernoulli's memoir on the oscillations of heavy chains [65]. In reading some of the earlier papers on the subject, it should be remembered that the notation has changed, what was formerly written $J_n(z)$ being now written $J_n(2z)$.

Example 17.1.1 (Math. Trip. 1896)   Prove that if

$$\frac{2b(1+\theta^2)}{(1-2u\theta-\theta^2)^2 + 4b^2\theta^2} = A_1 + A_2\theta + A_3\theta^2 + \cdots,$$

then $e^{as}\sin bz = A_1 J_1(z) + A_2 J_2(z) + A_3 J_3(z) + \cdots$. *Hint.* For, if the contour D in the u-plane be a circle with centre $u = 0$ and radius large enough to include the zeros of the denominator, we have

$$e^{\frac{1}{2}z\left(u-\frac{1}{u}\right)} \frac{2b\left(\frac{1}{u^2}+\frac{1}{u^4}\right)}{\left(1-\frac{2a}{u}-\frac{1}{u^2}\right)^2 + \frac{4bz}{u^2}} = \sum_{n=1}^{\infty} e^{\frac{1}{2}z\left(u-\frac{1}{u}\right)} A_n u^{-n-1},$$

the series on the right converging uniformly on the contour; and so, using §4.7 and replacing the integrals by Bessel coefficients, we have

$$\frac{1}{2\pi i} \int_D e^{\frac{1}{2}z\left(u-\frac{1}{u}\right)} \frac{2b\left(\frac{1}{u^2}+\frac{1}{u^4}\right)}{\left(1-\frac{2a}{u}-\frac{1}{u^2}\right)^2 + \frac{4bz}{u^2}}\, du$$

$$= \frac{1}{2\pi i} \int_D e^{\frac{1}{2}z\left(u-\frac{1}{u}\right)} \left(\frac{A_1}{u^2} + \frac{A_2}{u^3} + \frac{A_3}{u^4} + \cdots\right) du$$

$$= A_1 J_1(z) + A_2 J_2(z) + A_3 J_3(z) + \cdots.$$

In the integral on the left write $\frac{1}{2}(u - u^{-1}) - a = t$, so that as u describes a circle of radius e^β, t describes an ellipse with semi-axes $\cosh\beta$ and $\sinh\beta$ with foci at $-a \pm i$; then we have

$$\sum_{n=1}^{\infty} A_n J_n(z) = \frac{1}{2\pi i} \int \frac{e^{s(t+a)} b\, dt}{t^2 + b^2},$$

the contour being the ellipse just specified, which contains the zeros of $t^2 + b^2$. Evaluating the integral by §6.1, we have the required result.

Example 17.1.2 (K. Neumann and Schläfli)   Shew that, when n is an integer,

$$J_n\,(y+z) = \sum_{m=-\infty}^{\infty} J_m(y)J_{n-m}(z).$$

Hint. Consider the expansion of each side of the equation

$$\exp\left\{\frac{1}{2}(y+z)\left(t - \frac{1}{t}\right)\right\} = \exp\left\{\frac{1}{2}y\left(t - \frac{1}{t}\right)\right\} \cdot \exp\left\{\frac{1}{2}z\left(t - \frac{1}{t}\right)\right\}.$$

Example 17.1.3   Shew that

$$e^{iz\cos\phi} = J_0(z) + 2i\cos\phi J_1(z) + 2i^2 \cos 2\phi J_2(z) + \cdots.$$

Example 17.1.4 (K. Neumann and E. Lommel)   Shew that if $r^2 = x^2 + y^2$

$$J_0(r) = J_0(x)J_0(y) - 2J_2(x)J_2(y) + 2J_4(x)J_4(y) - \cdots.$$

17.11 Bessel's differential equation

We have seen that, when n is an integer, the Bessel coefficient of order n is given by the formula

$$J_n(z) = \frac{1}{2\pi i}\left(\frac{z}{2}\right)^n \int^{(0+)} t^{-n-1} \exp\left(t - \frac{z^2}{4t}\right) dt.$$

From this formula we shall now shew that $J_n(z)$ is a solution of the linear differential equation

$$\frac{d^2 y}{dz^2} + \frac{1}{z}\frac{dy}{dz} + \left(1 - \frac{n^2}{z^2}\right)y = 0,$$

which is called Bessel's equation for functions of order n.

For we find on performing the differentiations (§4.2) that

$$\frac{d^2 J_n(z)}{dz^2} + \frac{1}{z}\frac{dJ_n(z)}{dz} + \left(1 - \frac{n^2}{z^2}\right)J_n(z)$$

$$= \frac{1}{2\pi i}\left(\frac{z}{2}\right)^n \int^{(0+)} t^{-n-1}\left\{1 - \frac{n+1}{t} + \frac{z^2}{4t^2}\right\}\exp\left(t - \frac{z^2}{4t}\right)dt$$

$$= -\frac{1}{2\pi i}\left(\frac{z}{2}\right)^n \int^{(0+)} \frac{d}{dt}\left\{t^{-n-1}\exp\left(t - \frac{z^2}{4t}\right)\right\}dt$$

$$= 0,$$

since $t^{-n-1}\exp(t - z^2/4t)$ is one-valued. *Thus we have proved that*

$$\frac{d^2 J_n(z)}{dz^2} + \frac{1}{z}\frac{dJ_n(z)}{dz} + \left(1 - \frac{n^2}{z^2}\right)J_n(z) = 0.$$

The reader will observe that $z = 0$ is a regular point and $z = \infty$ an irregular point, all other points being ordinary points of this equation.

Example 17.1.5 (St John's, 1899)   By differentiating the expansion

$$e^{\frac{1}{2}z\left(t-\frac{1}{t}\right)} = \sum_{n=-\infty}^{\infty} t^n J_n(z)$$

with regard to z and with regard to t, shew that the Bessel coefficients satisfy Bessel's equation.

Example 17.1.6   The function $P_n^m\left(1 - \frac{z^2}{2n^2}\right)$ satisfies the equation defined by the scheme

$$P\left\{\begin{matrix} 4n^2 & \infty & 0 \\ \frac{1}{2}m & n+1 & \frac{1}{2}m & z^2 \\ -\frac{1}{2}m & -n & -\frac{1}{2}m \end{matrix}\right\}.$$

Shew that $J_m(z)$ satisfies the confluent form of this equation obtained by making $n \to \infty$.

17.2 The solution of Bessel's equation when n is not necessarily an integer

We now proceed, after the manner of §15.2, to extend the definition of $J_n(z)$ to the case when n is any number, real or complex. It appears by methods similar to those of §17.11 that, for all values of n, the equation

$$\frac{d^2y}{dz^2} + \frac{1}{z}\frac{dy}{dz} + \left(1 - \frac{n^2}{z^2}\right)y = 0$$

is satisfied by an integral of the form

$$y = z^n \int_C t^{-n-1} \exp\left(t - \frac{z^2}{4t}\right) dt$$

provided that $t^{-n-1} \exp\left(t - \frac{z^2}{4t}\right)$ resumes its initial value after describing C and that differentiations under the sign of integration are justified.

Accordingly, we define $J_n(z)$ by the equation

$$J_n(z) = \frac{z^n}{2^{n+1}\pi i} \int_{-\infty}^{(0+)} t^{-n-1} \exp\left(t - \frac{z^2}{4t}\right) dt,$$

the expression being rendered precise by giving arg z its principal value and taking $|\arg t| \leq \pi$ on the contour.

To express this integral as a power series, we observe that it is an analytic function of z; and we may obtain the coefficients in the Taylor's series in powers of z by differentiating under the sign of integration (§§5.32 and 4.44). Hence we deduce that

$$J_n(z) = \frac{z^n}{2^{n+1}\pi i} \sum_{r=0}^{\infty} \frac{(-1)^r z^{2r}}{2^{2r} r!} \int_{-\infty}^{(0+)} e^t t^{-n-r-1} dt$$

$$= \sum_{r=0}^{\infty} \frac{(-1)^r z^{n+2r}}{2^{n+2r} r! \, \Gamma(n+r+1)},$$

by §12.22. This is the expansion in question.

Accordingly, for general values of n, we define the Bessel function $J_n(z)$ *by the equations*

$$J_n(z) = \frac{1}{2\pi i} \left(\frac{z}{2}\right)^n \int_{-\infty}^{(0+)} t^{-n-1} \exp\left(t - \frac{z^2}{4t}\right) dt$$

$$= \sum_{r=0}^{\infty} \frac{(-1)^r z^{n+2r}}{2^{n+2r} r! \, \Gamma(n+r+1)}.$$

This function reduces to a Bessel coefficient when n is an integer; it is sometimes called a Bessel function *of the first kind.*

The reader will observe that since Bessel's equation is unaltered by writing $-n$ for n, fundamental solutions are $J_n(z)$, $J_{-n}(z)$, except when n is an integer, in which case the solutions are not independent. With this exception *the general solution of Bessel's equation is*

$$\alpha J_n(z) + \beta J_{-n}(z),$$

where α and β are arbitrary constants.

A second solution of Bessel's equation when n is an integer will be given later (§17.6).

17.21 The recurrence formulae for the Bessel functions

As the Bessel function satisfies a confluent form of the hypergeometric equation, it is to be expected that recurrence formulae will exist, corresponding to the relations between contiguous hypergeometric functions indicated in §14.7.

To establish these relations for general values of n, real or complex, we have recourse to the result of §17.2. On writing the equation

$$0 = \int_{-\infty}^{(0+)} \frac{d}{dt} \left\{ t^{-n} \exp\left(t - \frac{z^2}{4t}\right) \right\} dt$$

at length, we have

$$0 = \int_{-\infty}^{(0+)} \left(t^{-n} + \frac{1}{4} z^2 t^{-n-2} - n t^{-n-1} \right) \exp\left(t - \frac{z^2}{4t}\right) dt$$

$$= 2\pi i \left\{ (2z^{-1})^{n-1} J_{n-1}(z) + \frac{1}{4} z^2 (2z^{-1})^{n+1} J_{n+1}(z) - n(2z^{-1})^n J_n(z) \right\},$$

and so

$$J_{n-1}(z) + J_{n+1}(z) = \frac{2n}{z} J_n(z). \tag{17.2}$$

Next we have, by §4.44,

$$\frac{d}{dz} \{z^{-n} J_n(z)\} = \frac{1}{2^{n+1}\pi i} \frac{d}{dz} \int_{-\infty}^{(0+)} t^{-n-1} \exp\left(t - \frac{z^2}{4t}\right) dt$$

$$= -\frac{z}{2^{n+2}\pi i} \int_{-\infty}^{(0+)} t^{-n-2} \exp\left(t - \frac{z^2}{4t}\right) dt$$

$$= -z^{-n} J_{n+1}(z),$$

and consequently, if primes denote differentiations with regard to z,

$$J_n'(z) = \frac{n}{z} J_n(z) - J_{n+1}(z). \tag{17.3}$$

From (17.2) and (17.3) it is easy to derive the other recurrence formulae

$$J_n'(z) = \frac{1}{2} \{J_{n-1}(z) - J_{n+1}(z)\} \tag{17.4}$$

and

$$J_n'(z) = J_{n-1}(z) - \frac{n}{z} J_n(z). \tag{17.5}$$

Example 17.2.1   Obtain these results from the power series for $J_n(z)$.

Example 17.2.2   Shew that $\dfrac{d}{dz} \{z^n J_n(z)\} = z^n J_{n-1}(z)$.

Example 17.2.3   Shew that $J_0'(z) = -J_1(z)$.

Example 17.2.4   Shew that

$$16 J_n^{(iv)}(z) = J_{n-4}(z) - 4J_{n-2}(z) + 6J_n(z) - 4J_{n+2}(z) + J_{n+4}(z).$$

Example 17.2.5   Shew that

$$J_2(z) - J_0(z) = 2J_0''(z).$$

Example 17.2.6   Shew that

$$J_2(z) = J_0''(z) - z^{-1} J_0'(z).$$

17.211 *Relation between two Bessel functions whose orders differ by an integer*

From the last article can be deduced an equation connecting any two Bessel functions whose orders differ by an integer, namely

$$z^{-n-r} J_{n+r}(z) = (-1)^r \frac{d^r}{(z\,dz)^r} \{z^{-n} J_n(z)\},$$

where n is unrestricted and r is any positive integer. This result follows at once by induction from formula (17.3), when it is written in the form

$$z^{-n-1} J_{n+1}(z) = -\frac{d}{z\,dz} \{z^{-n} J_n(z)\}.$$

17.212 *The connexion between $J_n(z)$ and $W_{k,m}$ functions*

The reader will verify without difficulty that, if in Bessel's equation we write $y = z^{-\frac{1}{2}} v$ and then write $z = x/2i$, we get

$$\frac{d^2 v}{dx^2} + \left(-\frac{1}{4} + \frac{\frac{1}{4} - n^2}{x^2} \right) v = 0,$$

which is the equation satisfied by $W_{0,n}(x)$; it follows that

$$J_n(z) = A z^{-\frac{1}{2}} M_{0,n}(2iz) + B z^{-\frac{1}{2}} M_{0,-n}(2iz).$$

Comparing the coefficients of $z^{\pm n}$ on each side we see that

$$J_n(z) = \frac{z^{-\frac{1}{2}}}{2^{2n+\frac{1}{2}} i^{n+\frac{1}{2}} \Gamma(n+1)} M_{0,n}(2iz),$$

except in the critical cases when $2n$ is a negative integer; when n is half of a negative odd integer, the result follows from Kummer's second formula (§16.11).

17.22 The zeros of Bessel functions whose order n is real

The relations of §17.21 enable us to deduce the interesting theorem that *between any two consecutive real zeros of $z^{-n} J_n(z)$, there lies one and only one zero $z^{-n} J_{n+1}(z)$*. Proofs of this theorem have been given by Bôcher [79], Gegenbauer [242] and Porter [535].

For, from relation (17.3) when written in the form

$$z^{-n} J_{n+1}(z) = -\frac{d}{dz} \{z^{-n} J_n(z)\},$$

it follows from Rolle's theorem that between each consecutive pair of zeros of $z^{-n} J_n(z)$ there is at least one zero of $z^{-n} J_{n+1}(z)$. Rolle's theorem is proved in [111, I. p. 157] for polynomials. It may be deduced for any functions with continuous differential coefficients by using the first mean-value theorem (§4.14).

Similarly, from relation (17.5) when written in the form

$$z^{n+1} J_n(z) = \frac{d}{dz} \{z^{n+1} J_{n+1}(z)\},$$

it follows that between each consecutive pair of zeros of $z^{n+1} J_{n+1}(z)$ there is at least one zero of $z^{n+1} J_n(z)$.

Further, $z^{-n} J_n(z)$ and $\frac{d}{dz} \{z^{-n} J_n(z)\}$ have no common zeros; for the former function satisfies the equation

$$z \frac{d^2 y}{dz^2} + (2n+1) \frac{dy}{dz} + zy = 0,$$

and it is easily verified by induction on differentiating this equation that if both y and $\dfrac{dy}{dz}$ vanish for any value of z, all differential coefficients of y vanish, and y is zero by §5.4.

The theorem required is now obvious except for the numerically smallest zeros $\pm\xi$ of $z^{-n} J_n(z)$, since (except for $z = 0$), $z^{-n} J_n(z)$ and $z^{n+1} J_n(z)$ have the same zeros. But $z = 0$ is a zero of $z^{-n} J_{n+1}(z)$, and if there were any other positive zero of $z^{-n} J_{n+1}(z)$, say ξ_1, which was less than ξ, then $z^{n+1} J_n(z)$ would have a zero between 0 and ξ_1, which contradicts the hypothesis that there were no zeros of $z^{n+1} J_n(z)$ between 0 and ξ.

The theorem is therefore proved.

Note   See also Examples 17.3.3 and 17.3.4, and Example 17.19 at the end of the chapter.

17.23 Bessel's integral for the Bessel coefficients

We shall next obtain an integral first given by Bessel in the particular case of the Bessel functions for which n is a positive integer; in some respects the result resembles Laplace's integrals given in §15.23 and §15.33 for the Legendre functions.

In the integral of §17.1, viz.

$$J_n(z) = \frac{1}{2\pi i} \int^{(0+)} u^{-n-1} e^{z(u-1/u)/2} \, du,$$

take the contour to be the circle $|u| = 1$ and write $u = e^{i\theta}$, so that

$$J_n(z) = \frac{1}{2\pi} \int_{-\pi}^{\pi} e^{-ni\theta + iz \sin \theta} \, d\theta.$$

Bisect the range of integration and in the former part write $-\theta$ for θ; we get

$$J_n(z) = \frac{1}{2\pi} \int_0^{\pi} e^{ni\theta - iz \sin \theta} \, d\theta + \frac{1}{2\pi} \int_0^{\pi} e^{-ni\theta + iz \sin \theta} \, d\theta,$$

and so

$$J_n(z) = \frac{1}{\pi} \int_0^{\pi} \cos(n\theta - z \sin \theta) \, d\theta,$$

which is the formula in question.

Example 17.2.7 Shew that, when z is real and n is an integer,

$$|J_n(z)| \le 1.$$

Example 17.2.8 Shew that, for all values of n (real or complex), the integral

$$y = \frac{1}{\pi} \int_0^{\pi} \cos(n\theta - z \sin \theta) \, d\theta$$

satisfies

$$\frac{d^2 y}{dz^2} + \frac{1}{z} \frac{dy}{dz} + \left(1 - \frac{n^2}{z^2}\right) y = \frac{\sin n\pi}{\pi} \left(\frac{1}{z} - \frac{n}{z^2}\right),$$

which reduces to Bessel's equation when n is an integer. *Hint*. It is easy to shew, by differentiating under the integral sign, that the expression on the left is equal to

$$-\frac{1}{\pi} \int_0^{\pi} \frac{d}{d\theta} \left\{ \left(\frac{n}{z^2} + \frac{\cos \theta}{z}\right) \sin(n\theta - z \sin \theta) \right\} d\theta.$$

17.231 The modification of Bessel's integral when n is not an integer

We shall now shew that, for general values of n,

$$J_n(z) = \frac{1}{\pi} \int_0^{\pi} \cos(n\theta - z \sin \theta) \, d\theta - \frac{\sin n\pi}{\pi} \int_0^{\infty} e^{-n\theta - z \sinh \theta} \, d\theta, \qquad (17.6)$$

when Re $z > 0$. This result is due to Schläfli, [577]. This obviously reduces to the result of §17.23 when n is an integer.

Taking the integral of §17.2, viz.

$$J_n(z) = \frac{z^n}{2^{n+1}\pi i} \int_{-\infty}^{(0+)} t^{-n-1} \exp\left(t - \frac{z^2}{4t}\right) dt,$$

and supposing that z is positive, we have, on writing $t = \frac{1}{2}uz$,

$$J_n(z) = \frac{1}{2\pi i} \int_{-\infty}^{(0+)} u^{-n-1} \exp\left\{\frac{1}{2}z\left(u - \frac{1}{u}\right)\right\} du.$$

But, if the contour be taken to be that of the figure consisting of the real axis from -1 to $-\infty$ taken twice and the circle $|u| = 1$, this integral represents an analytic function of z when $\mathrm{Re}\,(zu) < 0$ as $|u| \to \infty$ on the path, i.e. when $|\arg z| < \frac{1}{2}\pi$; and so, by the theory of analytic continuation, the formula (which has been proved by a direct transformation for *positive* values of z) is true whenever $\mathrm{Re}\,z > 0$. Hence

$$J_n(z) = \frac{1}{2\pi i} \left\{ \int_{-\infty}^{-1} + \int_C + \int_{-1}^{-\infty} \right\} u^{-n-1} \exp\left\{\frac{1}{2}z\left(u - \frac{1}{u}\right)\right\} du,$$

where C denotes the circle $|u| = 1$, and $\arg u = -\pi$ on the first path of integration while $\arg u = +\pi$ on the third path.

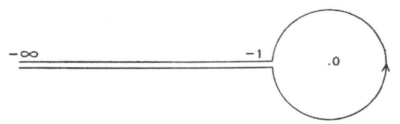

Writing $u = te^{\mp i\pi}$ in the first and third integrals respectively (so that in each case $\arg t = 0$), and $u = e^{i\theta}$ in the second, we have

$$J_n(z) = \frac{1}{2\pi} \int_{-\pi}^{\pi} e^{-ni\theta + iz\sin\theta} d\theta + \left\{\frac{e^{(n+1)\pi i}}{2\pi i} - \frac{e^{-(n+1)\pi i}}{2\pi i}\right\} \int_1^{\infty} t^{-n-1} e^{\frac{1}{2}z\left(-t+\frac{1}{t}\right)} dt.$$

Modifying the former of these integrals as in §17.23 and writing e^{θ} for t in the latter, we have at once

$$J_n(z) = \frac{1}{\pi} \int_0^{\pi} \cos(n\theta - z\sin\theta)\,d\theta + \frac{\sin(n+1)\pi}{\pi} \int_0^{\infty} e^{-n\theta - z\sinh\theta}\,d\theta,$$

which is the required result, when $|\arg z| < \frac{1}{2}\pi$.

When $|\arg z|$ lies between $\frac{1}{2}\pi$ and π, since $J_n(z) = e^{\pm n\pi i} J_n(-z)$, we have

$$J_n(z) = \frac{e^{\pm n\pi i}}{\pi} \left\{ \int_0^{\pi} \cos(n\theta + z\sin\theta)\,d\theta - \sin n\pi \int_0^{\infty} e^{-n\theta + z\sinh\theta}\,d\theta \right\}, \tag{17.7}$$

the upper or lower sign being taken as $\arg z > \frac{1}{2}\pi$ or $< -\frac{1}{2}\pi$.

When n is an integer (17.6) reduces at once to Bessel's integral, and (17.7) does so when we make use of the equation $J_n(z) = (-1)^n J_{-n}(z)$, which is true for integer values of n.

Equation (17.6), as already stated, is due to Schläfli [576, p. 148], and equation (17.7) was given by Sonine [598].

These trigonometric integrals for the Bessel functions may be regarded as corresponding to Laplace's integrals for the Legendre functions. For (see Example 17.1.6) $J_m(z)$ satisfies the confluent form (obtained by making $n \to \infty$) of the equation for $P_n^m(1 - z^2/2n^2)$.

But Laplace's integral for this function is a multiple of

$$\int_0^\pi \left[1 - \frac{z^2}{2n^2} + \left\{ \left(1 - \frac{z^2}{2n^2} \right)^2 - 1 \right\}^{1/2} \cos \phi \right]^n \cos m\phi \, d\phi$$

$$= \int_0^\pi \left\{ 1 + \frac{iz}{n} \cos \phi + O(n^{-2}) \right\}^n \cos m\phi \, d\phi.$$

The limit of the integrand as $n \to \infty$ is $e^{iz \cos \phi} \cos m\phi$, and this exhibits the similarity of Laplace's integral for $P_n^m(z)$ to the Bessel–Schläfli integral for $J_m(z)$.

Example 17.2.9 (Callandreau [112])   From the formula $J_0(x) = \dfrac{1}{2\pi} \displaystyle\int_{-\pi}^\pi e^{-ix \cos \phi} \, d\phi$, by a change of order of integration, shew that, when n is a positive integer and $\cos \theta > 0$,

$$P_n(\cos \theta) = \frac{1}{\Gamma(n+1)} \int_0^\infty e^{-x \cos \theta} J_0(x \sin \theta) x^n \, dx.$$

Example 17.2.10   Shew that, with Ferrers' definition of $P_n^m(\cos \theta)$,

$$P_n^m(\cos \theta) = \frac{1}{\Gamma(n - m + 1)} \int_0^\infty e^{-x \cos \theta} J_m(x \sin \theta) x^n \, dx$$

when n and m are positive integers and $\cos \theta > 0$. See Hobson [313].

17.24 Bessel functions whose order is half an odd integer

We have seen (§17.2) that when the order n of a Bessel function $J_n(z)$ is half an odd integer, the difference of the roots of the indicial equation at $z = 0$ is $2n$, which is an integer. We now shew that, in such cases, $J_n(z)$ is expressible in terms of elementary functions.

For

$$J_{1/2}(z) = \left(\frac{2z}{\pi} \right)^{1/2} \left\{ 1 - \frac{z^2}{2 \cdot 3} + \frac{z^4}{2 \cdot 3 \cdot 4 \cdot 5} - \cdots \right\} = \left(\frac{2}{\pi z} \right)^{1/2} \sin z,$$

and therefore (§17.211) if k is a positive integer

$$J_{k+1/2}(z) = \frac{(-1)^k (2z)^{k + \frac{1}{2}}}{\pi^{\frac{1}{2}}} \frac{d^k}{d(z^2)^k} \left(\frac{\sin z}{z} \right).$$

On differentiating out the expression on the right, we obtain the result that

$$J_{k+1/2}(z) = P_k \sin z + Q_k \cos z,$$

where P_k, Q_k are polynomials in $z^{-\frac{1}{2}}$.

Example 17.2.11   Shew that $J_{-\frac{1}{2}}(z) = \left(\dfrac{2}{\pi z} \right)^{1/2} \cos z$.

Example 17.2.12 Prove by induction that if k be an integer and $n = k + \frac{1}{2}$, then

$$
J_n(z) = \left(\frac{2}{\pi z}\right)^{1/2}
$$

$$
\times \left[\cos\left(z - \tfrac{1}{2}n\pi - \tfrac{1}{4}\pi\right) \left\{1 + \sum_{r=1} \frac{(-1)^r (4n^2 - 1^2)(4n^2 - 3^2) \cdots \{4n^2 - (4r-1)^2\}}{(2r)! \, 2^{6r} z^{2r}}\right\} \right.
$$

$$
\left. + \sin\left(z - \tfrac{1}{2}n\pi - \tfrac{1}{4}\pi\right) \sum_{r=1} \frac{(-1)^r (4n^2 - 1^2)(4n^2 - 3^2) \cdots \{4n^2 - (4r-3)^2\}}{(2r-1)! \, 2^{6r-3} z^{2r-1}} \right],
$$

the summations being continued as far as the terms with the vanishing factors in the numerators.

Example 17.2.13 Shew that $z^{k+\frac{1}{2}} \dfrac{d^k}{d(z^2)^k}\left(\dfrac{\cos z}{z}\right)$ is a solution of Bessel's equation for $J_{k+\frac{1}{2}}(z)$.

Example 17.2.14 (Lommel) Shew that the solution of $z^{m+\frac{1}{2}} \dfrac{d^{2m+1} y}{dz^{2m+1}} + y = 0$ is

$$
y = z^{\frac{1}{2}m+\frac{1}{4}} \sum_{p=0}^{2m} c_p \left\{ J_{-m-\frac{1}{2}}(2a_p z^{\frac{1}{2}}) + i J_{m+\frac{1}{2}}(2a_p z^{\frac{1}{2}}) \right\},
$$

where $c_0, c_1, \ldots, c_{2m}$ are arbitrary and $a_0, a_1, \ldots, a_{2m}$ are the roots of $a^{2m+1} = i$.

17.3 Hankel's contour integral for $J_n(z)$

This appears in [273]. Consider the integral

$$
y = z^n \int_A^{(1+,-1-)} (t^2 - 1)^{n-\frac{1}{2}} \cos(zt) \, dt,
$$

where A is a point on the right of the point $t = 1$, and

$$
\arg(t - 1) = \arg(t + 1) = 0
$$

at A; the contour may conveniently be regarded as being in the shape of a figure of eight.

We shall shew that this integral is a constant multiple of $J_n(z)$. It is easily seen that the integrand returns to its initial value after t has described the path of integration; for $(t-1)^{n-\frac{1}{2}}$ is multiplied by the factor $e^{(2n-1)\pi i}$ after the circuit $(1+)$ has been described, and $(t+1)^{n-\frac{1}{2}}$ is multiplied by the factor $e^{-(2n-1)\pi i}$ after the circuit $(-1-)$ has been described.

Since $\sum_{r=0}^{\infty} \frac{(-1)^r (zt)^{2r}}{(2r)!}(t^2-1)^{n-1/2}$ converges uniformly on the contour, we have

$$
y = \sum_{r=0}^{\infty} \frac{(-1)^r z^{n+2r}}{(2r)!} \int_A^{(1+,-1-)} t^{2r}(t^2-1)^{n-1/2} \, dt,
$$

(see §4.7). To evaluate these integrals, we observe firstly that they are analytic functions of n for all values of n, and secondly that, when $\mathrm{Re}\,(n + \frac{1}{2}) > 0$, we may deform the contour into the circles $|t - 1| = \delta$, $|t + 1| = \delta$ and the real axis joining the points $t = \pm(1 - \delta)$ taken twice, and then we may make $\delta \to 0$; the integrals round the circles tend to zero and, assigning to $t - 1$ and $t + 1$ their appropriate arguments on the modified path of integration, we get, if $\arg(1 - t^2) = 0$ and $t^2 = u$,

$$\int_A^{(1+,-1-)} t^{2\tau} (t^2 - 1)^{n-1/2} \, dt$$

$$= e^{(n-1/2)\pi i} \int_1^{-1} t^{2\tau} (1 - t^2)^{n-1/2} \, dt + e^{-(n-1/2)\pi i} \int_{-1}^1 t^{2\tau} (1 - t^2)^{n-1/2} \, dt$$

$$= -4i \sin\left(n - \tfrac{1}{2}\right) \pi \int_0^1 t^{2\tau} (1 - t^2)^{n-1/2} \, dt$$

$$= -2i \sin\left(n - \tfrac{1}{2}\right) \pi \int_0^1 u^{\tau-1/2} (1 - u)^{n-1/2} \, du$$

$$= 2i \sin\left(n + \tfrac{1}{2}\right) \pi \Gamma\left(r + \tfrac{1}{2}\right) \Gamma\left(n + \tfrac{1}{2}\right) / \Gamma(n + r + 1).$$

Since the initial and final expressions are analytic functions of n for all values of n, it follows from §5.5 that this equation, proved when $\operatorname{Re}\left(n + \tfrac{1}{2}\right) > 0$, is true for all values of n. Accordingly

$$y = \sum_{r=0}^{\infty} \frac{(-1)^r z^{n+2r} 2i \sin(n + \tfrac{1}{2}) \pi \Gamma(r + \tfrac{1}{2})\Gamma(n + \tfrac{1}{2})}{(2r)! \Gamma(n + r + 1)}$$

$$= 2^{n+1} i \sin\left(n + \tfrac{1}{2}\right) \pi \Gamma\left(n + \tfrac{1}{2}\right) \Gamma\left(\tfrac{1}{2}\right) J_n(z),$$

on reduction.

Accordingly, when $\left\{ \Gamma\left(\tfrac{1}{2} - n\right) \right\}^{-1} \neq 0$, *we have*

$$J_n(z) = \frac{\Gamma(\tfrac{1}{2} - n)(\tfrac{1}{2}z)^n}{2\pi i \Gamma\left(\tfrac{1}{2}\right)} \int_A^{(1+,-1-)} (t^2 - 1)^{n-1/2} \cos(zt) \, dt.$$

Corollary 17.3.1   *When* $\operatorname{Re}\left(n + \tfrac{1}{2}\right) > 0$, *we may deform the path of integration, and obtain the result*

$$J_n(z) = \frac{\left(\tfrac{z}{2}\right)^n}{\Gamma\left(n + \tfrac{1}{2}\right) \Gamma\left(\tfrac{1}{2}\right)} \int_{-1}^1 (1 - t^2)^{n-1/2} \cos(zt) \, dt$$

$$= \frac{2 \left(\tfrac{z}{2}\right)^n}{\Gamma\left(n + \tfrac{1}{2}\right) \Gamma\left(\tfrac{1}{2}\right)} \int_0^{\pi/2} \sin^{2n} \phi \cos(z \cos \phi) \, d\phi.$$

Example 17.3.1   Shew that, when $\operatorname{Re}\left(n + \tfrac{1}{2}\right) > 0$,

$$J_n(z) = \frac{\left(\tfrac{z}{2}\right)^n}{\Gamma\left(n + \tfrac{1}{2}\right) \Gamma\left(\tfrac{1}{2}\right)} \int_0^{\pi} e^{\pm iz \cos \phi} \sin^{2n} \phi \, d\phi.$$

Example 17.3.2   Obtain the result

$$J_n(z) = \frac{\left(\tfrac{z}{2}\right)^n}{\Gamma\left(n + \tfrac{1}{2}\right) \Gamma\left(\tfrac{1}{2}\right)} \int_0^{\pi} \cos(z \cos \phi) \sin^{2n} \phi \, d\phi,$$

when $\operatorname{Re}(n) > 0$, by expanding in powers of z and integrating (§4.7) term-by-term.

Example 17.3.3 Shew that when $-\frac{1}{2} < n < \frac{1}{2}$, $J_n(z)$ has an infinite number of real zeros. *Hint.* Let $z = \left(m + \frac{1}{2}\right)\pi$ where m is zero or a positive integer; then by the corollary above

$$J_n\left(m\pi + \tfrac{1}{2}\pi\right) = \frac{z^n}{2^{n-1}\Gamma\left(n + \frac{1}{2}\right)\Gamma\left(\frac{1}{2}\right)}\left\{\tfrac{1}{2}u_0 - u_1 + u_2 - \cdots + (-1)^m u_m\right\},$$

where

$$u_r = \left|\int_{\frac{2r-1}{2m+1}}^{\frac{2r+1}{2m+1}} (1 - t^2)^{n-1/2}\cos\left\{\left(m + \tfrac{1}{2}\right)\pi t\right\}\, dt\right|$$

$$= \int_0^{1/\left(m+\frac{1}{2}\right)} \left\{1 - \left(t + \frac{2r-1}{2m+1}\right)^2\right\}^{n-1/2}\sin\left\{\left(m + \tfrac{1}{2}\right)\pi t\right\}\, dt,$$

so, since $n - \frac{1}{2} < 0$, $u_m > u_{m-1} > u_{m-2} > \cdots$, and hence $J_n\left(m\pi + \frac{1}{2}\pi\right)$ has the sign of $(-1)^m$. *This method of proof, for $n = 0$, is due to Bessel.*

Example 17.3.4 Shew that if n be real, $J_n(z)$ has an infinite number of real zeros; and find an upper limit to the numerically smallest of them. *Hint.* Use Example 17.3.3 combined with §17.22.

17.4 Connexion between Bessel coefficients and Legendre functions

We shall now establish a result due to Heine [287], which renders precise the statement of §17.11, Example 17.1.6 concerning the expression of Bessel coefficients as limiting forms of hypergeometric functions. The apparently different result given in [287] is due to the difference between Heine's associated Legendre function and Ferrers' function.

When $|\arg(1 \pm z)| < \pi$, n is unrestricted and m is a positive integer, it follows by differentiating the formula of §15.22 that, with Ferrers' definition of $P_n^m(z)$,

$$P_n^m(z) = \frac{\Gamma(n + m + 1)}{2^m \cdot m!\, \Gamma(n - m + 1)}(1 - z)^{\frac{1}{2}m}(1 + z)^{\frac{1}{2}m}$$
$$\times F\left(-n + m, n + 1 + m; m + 1; \tfrac{1}{2} - \tfrac{1}{2}z\right),$$

and so, if $|\arg z| < \frac{1}{2}\pi$, $\left|\arg\left(1 - \frac{1}{4}z^2/n^2\right)\right| < \pi$, we have

$$P_n^m\left(1 - \frac{z^2}{2n^2}\right) = \frac{\Gamma(n + m + 1)z^m n^{-m}}{2^m \cdot m!\, \Gamma(n - m + 1)}\left(1 - \frac{z^2}{4n^2}\right)^{m/2}$$
$$\times F\left(-n + m, n + 1 + m; m + 1; \tfrac{1}{4}z^2 n^{-2}\right).$$

Now make $n \to +\infty$ (n being positive, but not necessarily integral), so that, if $\delta = n^{-1}$, $\delta \to 0$ continuously through positive values. Then

$$\frac{\Gamma(n + m + 1)n^{-m}}{\Gamma(n - m + 1)n^m} \to 1,$$

by §13.6, and

$$\left(1 - z^2/4n^2\right)^{m/2} \to 1.$$

Further, the $(r + 1)$th term of the hypergeometric series is

$$(-1)^r \frac{(1 - m\delta)(1 + m\delta + r\delta)\left(1 - (m+1)^2\delta^2\right)\left(1 - (m+2)^2\delta^2\right)\cdots\left(1 - (m+r-1)^2\delta^2\right)}{(m+1)(m+2)\cdots(m+r)\cdot r!}$$

$$\times \left(\frac{z}{2}\right)^{2r};$$

this is a continuous function of δ and the series of which this is the $(r + 1)$th term is easily seen to converge uniformly in a range of values of δ including the point $\delta = 0$; so, by §3.32, we have

$$\lim_{n\to\infty}\left[n^{-m}P_n^m\left(1 - \frac{z^2}{2n^2}\right)\right] = \frac{z^m}{2^m\cdot m!}\sum_{r=0}^{\infty}\frac{(-1)^r\left(\frac{z}{2}\right)^r}{(m+1)(m+2)\cdots(m+r)\,r!}$$

$$= J_m(z),$$

which is the relation required.

Example 17.4.1   Shew that

$$\lim_{n\to\infty}\left[n^{-m}P_n^m\left(\cos\frac{z}{n}\right)\right] = J_m(z).$$

The special case of this when $m = 0$ was given by Mehler [464]; see also [465].

Example 17.4.2   Shew that Bessel's equation is the confluent form of the equations defined by the schemes

$$P\left\{\begin{matrix} 0 & \infty & c & \\ n & ic & \frac{1}{2} + ic & z \\ -n & -ic & \frac{1}{2} - ic & \end{matrix}\right\}, e^{is}P\left\{\begin{matrix} 0 & \infty & c & \\ n & \frac{1}{2} & 0 & z \\ -n & \frac{3}{2} - 2ic & 2ic - 1 & \end{matrix}\right\},$$

$$P\left\{\begin{matrix} 0 & \infty & c^2 & \\ \frac{1}{2}n & \frac{1}{2}(c - n) & 0 & z^2 \\ -\frac{1}{2}n & -\frac{1}{2}(c + n) & n + 1 & \end{matrix}\right\},$$

the confluence being obtained by making $c \to \infty$.

17.5 Asymptotic series for $J_n(z)$ when $|z|$ is large

We have seen (§17.212) that

$$J_n(z) = \frac{z^{-1/2}}{2^{2n+1/2}\,e^{\frac{1}{2}(n+\frac{1}{2})\pi i}\,\Gamma(n+1)}M_{0,n}(2iz),$$

where it is supposed that $|\arg z| < \pi$, $-\frac{1}{2}\pi < \arg(2iz) < \frac{3}{2}\pi$. But for this range of values of z

$$M_{0,n}(2iz) = \frac{\Gamma(2n+1)}{\Gamma\left(\frac{1}{2} + n\right)}e^{(n+\frac{1}{2})\pi i}\,W_{0,n}(2iz) + \frac{\Gamma(2n+1)}{\Gamma\left(\frac{1}{2} + n\right)}W_{0,n}(-2iz)$$

by Example 16.4.4, if $-\frac{3}{2}\pi < \arg(-2iz) < \frac{1}{2}\pi$; and so, when $|\arg z| < \pi$,

$$J_n(z) = \frac{1}{(2\pi z)^{1/2}}\left\{e^{\frac{1}{2}(n+\frac{1}{2})\pi i}W_{0,n}(2iz) + e^{-\frac{1}{2}(n+\frac{1}{2})\pi i}W_{0,n}(-2iz)\right\}.$$

But, for the values of z under consideration, the asymptotic expansion of $W_{0,n}(\pm 2iz)$ is

$$e^{\mp iz}\left\{1 \pm \frac{(4n^2 - 1^2)}{8iz} + \frac{(4n^2 - 1^2)(4n^2 - 3^2)}{2!(8iz)^2} \pm \cdots \right.$$

$$\left. + \frac{(\pm 1)^r \left\{4n^2 - 1^2\right\}\left\{4n^2 - 3^2\right\}\cdots\left\{4n^2 - (2r-1)^2\right\}}{r!(8iz)^r} + O(z^{-r})\right\},$$

and therefore, combining the series, the asymptotic expansion of $J_n(z)$, when $|z|$ is large and $|\arg z| < \pi$, is

$$J_n(z) \sim \left(\frac{2}{\pi z}\right)^{\frac{1}{2}}\left[\cos\left(z - \frac{n\pi}{2} - \frac{\pi}{4}\right)\right.$$

$$\times \left\{1 + \sum_{r=1}^{\infty} \frac{(-1)^r \left\{4n^2 - 1^2\right\}\left\{4n^2 - 3^2\right\}\cdots\left\{4n^2 - (4r-1)^2\right\}}{(2r)!2^{6r}z^{2r}}\right\}$$

$$\left. + \sin\left(z - \frac{n\pi}{2} - \frac{\pi}{4}\right)\sum_{r=1}^{\infty} \frac{(-1)^r\left\{4n^2 - 1^2\right\}\left\{4n^2 - 3^2\right\}\cdots\left\{4n^2 - (4r-3)^2\right\}}{(2r-1)!2^{6r-3}z^{2r-1}}\right]$$

$$= \left(\frac{2}{\pi z}\right)^{\frac{1}{2}}\left[\cos\left(z - \frac{n\pi}{2} - \frac{\pi}{4}\right)\cdot U_n(z) - \sin\left(z - \frac{n\pi}{2} - \frac{\pi}{4}\right)\cdot V_n(z)\right],$$

where $U_n(z)$, $-V_n(z)$ have been written in place of the series.

The reader will observe that if n is half an odd integer these series terminate and give the result of Example 17.2.12.

Even when z is not very large, the value of $J_n(z)$ can be computed with great accuracy from this formula. Thus, for all positive values of z greater than 8, the first three terms of the asymptotic expansion give the value of $J_0(z)$ and $J_1(z)$ to six places of decimals.

This asymptotic expansion was given by Poisson [530, p. 350] (for $n = 0$) and by Jacobi (for $n = 0$) and by Jacobi [353, p. 94] (for general integral values of n) for real values of z. Complex values of z were considered by Hankel [273] and several subsequent writers. The method of obtaining the expansion here given is due to Barnes [49]. Asymptotic expansions for $J_n(z)$ when the order n is large have been given by Debye [168] and Nicholson [495].

An approximate formula for $J_n(nx)$ when n is large and $0 < x < 1$, namely

$$\frac{x^n \exp\left\{n\sqrt{1 - x^2}\right\}}{(2\pi n)^{1/2}(1 - x^2)^{1/4}\left\{1 + \sqrt{1 - x^2}\right\}^n},$$

was obtained by Carlini in 1817 in a memoir reprinted in Jacobi [354, vol. VII, pp. 189–245]. The formula was also investigated by Laplace [409, vol. V, 1827] in 1827, on the hypothesis that x is purely imaginary. A more extended account of researches on Bessel functions of large order is given in [676].

Example 17.5.1 By suitably modifying Hankel's contour integral (§17.3), shew that, when

$|\arg z| < \frac{1}{2}\pi$ and $\mathrm{Re}\left(n + \frac{1}{2}\right) > 0$,

$$J_n(z) = \frac{1}{\Gamma\left(n+\frac{1}{2}\right)(2\pi z)^{\frac{1}{2}}} \left[e^{i(z-\frac{1}{2}n\pi-\frac{1}{4}\pi)} \int_0^\infty e^{-u} u^{n-\frac{1}{2}} \left(1 + \frac{iu}{2z}\right)^{n-\frac{1}{2}} du \right.$$

$$\left. + e^{-i(z-\frac{1}{2}n\pi-\frac{1}{4}\pi)} \int_0^\infty e^{-u} u^{n-\frac{1}{2}} \left(1 - \frac{iu}{2z}\right)^{n-\frac{1}{2}} du \right];$$

and deduce the asymptotic expansion of $J_n(z)$ when $|z|$ is large and $|\arg z| < \frac{1}{2}\pi$. *Hint.* Take the contour to be the rectangle whose corners are ± 1, $\pm 1 + iN$, the rectangle being indented at ± 1, and make $N \to \infty$; the integrand being $(1 - t^2)^{n-\frac{1}{2}} e^{ist}$.

Example 17.5.2 Shew that, when $|\arg z| < \frac{\pi}{2}$ and $\mathrm{Re}\left(n + \frac{1}{2}\right) > 0$,

$$J_n(z) = \frac{2^{n+1} z^n}{\Gamma\left(n+\frac{1}{2}\right) \pi^{\frac{1}{2}}} \int_0^{\pi/2} e^{-2z \cot \phi} \cos^{n-\frac{1}{2}} \phi \, \mathrm{cosec}^{2n+1} \phi \sin\left\{z - \left(n - \frac{1}{2}\right)\phi\right\} d\phi.$$

Hint. Write $u = 2z \cot \phi$ in the preceding example.

Example 17.5.3 (Schafheitlin, [573])   Shew that, if $|\arg z| < \frac{1}{2}\pi$ and $\mathrm{Re}\left(n + \frac{1}{2}\right) > 0$, then

$$A e^{iz} z^n \int_0^\infty v^{n-\frac{1}{2}} (1 + iv)^{n-\frac{1}{2}} e^{-2vz} \, dv + B e^{-iz} z^n \int_0^\infty v^{n-\frac{1}{2}} (1 - iv)^{n-\frac{1}{2}} e^{-2vz} \, dv$$

is a solution of Bessel's equation. Further, determine A and B so that this may represent $J_n(z)$.

17.6 The second solution of Bessel's equation when the order is an integer

We have seen in §17.2 that, when the order n of Bessel's differential equation is not an integer, the general solution of the equation is

$$\alpha J_n(z) + \beta J_{-n}(z),$$

where α and β are arbitrary constants. When, however, n is an integer, we have seen that

$$J_n(z) = (-1)^n J_{-n}(z),$$

and consequently the two solutions $J_n(z)$ and $J_{-n}(z)$ are not really distinct. We therefore require in this case to find another particular solution of the differential equation, distinct from $J_n(z)$, in order to have the general solution.

We shall now consider the function

$$Y_n(z) = 2\pi e^{n\pi i} \frac{J_n(z) \cos n\pi - J_{-n}(z)}{\sin 2n\pi},$$

which is a solution of Bessel's equation when $2n$ is not an integer. The introduction of this function $Y_n(z)$ is due to Hankel [273, p. 472]. When n is an integer, $Y_n(z)$ is defined by the limiting form of this equation, namely

$$Y_n(z) = \lim_{\varepsilon \to 0} 2\pi e^{(n+\varepsilon)\pi i} \frac{J_{n+\varepsilon}(z)\cos(n\pi + \varepsilon\pi) - J_{-n-\varepsilon}(z)}{\sin 2(n+\varepsilon)\pi}$$

$$= \lim_{\varepsilon \to 0} \frac{2\pi e^{n\pi i}}{\sin 2\varepsilon\pi} \{(-1)^n J_{n+\varepsilon}(z) - J_{-n-\varepsilon}(z)\}$$

$$= \lim_{\varepsilon \to 0} \varepsilon^{-1} \{J_{n+\varepsilon}(z) - (-1)^n J_{-n-\varepsilon}(z)\}.$$

To express $Y_n(z)$ in terms of $W_{k,m}$ functions, we have recourse to the result of §17.5, which gives

$$Y_n(z) = \lim_{e \to 0} \frac{\varepsilon^{-1}}{(2\pi z)^{\frac{1}{2}}} \left[\left\{ e^{\frac{1}{2}(n+\varepsilon+\frac{1}{2})\pi i} W_{0,n+\varepsilon}(2iz) + e^{-\frac{1}{2}(n+\varepsilon+\frac{1}{2})\pi i} W_{0,n+\varepsilon}(-2iz) \right\} \right.$$

$$\left. -(-1)^n \left\{ e^{\frac{1}{2}(-n-\varepsilon+\frac{1}{2})\pi i} W_{0,n+\varepsilon}(2iz) + e^{-\frac{1}{2}(-n-\varepsilon+\frac{1}{2})\pi i} W_{0,n+\varepsilon}(-2iz) \right\} \right],$$

remembering that $W_{k,m} = W_{k,-m}$. Hence, since[1] $\lim_{e \to 0} W_{0,n+\varepsilon}(2iz) = W_{0,n}(2iz)$, we have

$$Y_n(z) = \left(\frac{\pi}{2z}\right)^{\frac{1}{2}} \left\{ e^{(\frac{1}{2}n+\frac{3}{4})\pi i} W_{0,n}(2iz) + e^{-(\frac{1}{2}n+\frac{3}{4})\pi i} W_{0,n}(-2iz) \right\}.$$

This function (n being an integer) is obviously a solution of Bessel's equation; it is called a *Bessel function of the second kind*.

Another function (also called a function of the second kind) was first used by Weber [657, p. 148] and by Schläfli [578, p. 17]. It is defined by the equation

$$Y_n(z) = \frac{J_n(z)\cos n\pi - J_{-n}(z)}{\sin n\pi} = \frac{Y_n(z)\cos n\pi}{\pi e^{n\pi i}},$$

or by the limits of these expressions when n is an integer. This function which exists for *all* values of n is taken as the canonical function of the second kind by Nielsen [500], and formulae involving it are generally (but not always) simpler than the corresponding formulae involving Hankel's function.

The asymptotic expansion for $Y_n(z)$, corresponding to that of §17.5 for $J_n(z)$, is that, when $|\arg z| < \pi$ and n is an integer,

$$Y_n(z) \sim \left(\frac{2}{\pi z}\right)^{1/2} \left[\sin\left(z - \tfrac{1}{2}n\pi - \tfrac{1}{4}\pi\right) U_n(z) + \cos\left(z - \tfrac{1}{2}n\pi - \tfrac{1}{4}\pi\right) V_n(z) \right],$$

where $U_n(z)$ and $V_n(z)$ are the asymptotic expansions defined in §17.5, their leading terms being 1 and $(4n^2 - 1)/8z$ respectively.

Example 17.6.1 (Hankel)   Prove that

$$Y_n(z) = \frac{dJ_n(z)}{dn} - (-1)^n \frac{dJ_{-n}(z)}{dn},$$

where n is made an integer after differentiation.

Example 17.6.2   Shew that if $Y_n(z)$ be defined by the equation of Example 17.6.1, it is a solution of Bessel's equation when n is an integer.

---

[1] This is most easily seen from the uniformity of the convergence with regard to ε of Barnes' contour integral (§16.4) for $W_{0,n+\varepsilon}(2iz)$.

17.61 The ascending series for $Y_n(z)$

The series of §17.6 is convenient for calculating $Y_n(z)$ when $|z|$ is large. To obtain a convenient series for small values of $|z|$, we observe that, since the ascending series for $J_{\pm(n+\varepsilon)}(z)$ are uniformly convergent series of analytic functions[2] of ε, each term may be expanded in powers of ε and this double series may then be arranged in powers of ε (§§5.3, 5.4).

Accordingly, to obtain $Y_n(z)$, we have to sum the coefficients of the first power of ε in the terms of the series

$$\sum_{r=0}^{\infty} \frac{(-1)^r (\frac{1}{2}z)^{n+2r+\varepsilon}}{r! \Gamma(n+\varepsilon+r+1)} - (-1)^n \sum_{r=0}^{\infty} \frac{(-1)^r (\frac{1}{2}z)^{-n+2r-\varepsilon}}{r! \Gamma(-n-\varepsilon+r+1)}.$$

Now, if s be a positive integer or zero and t a negative integer, the following expansions in powers of ε are valid:

$$\left(\frac{z}{2}\right)^{n+\varepsilon+2r} = \left(\frac{z}{2}\right)^{n+2r} \left\{1 + \varepsilon \log\left(\frac{1}{2}z\right) + \cdots\right\},$$

$$\frac{1}{\Gamma(s+\varepsilon+1)} = \frac{1}{\Gamma(s+1)} \left\{1 - \varepsilon \frac{\Gamma'(s+1)}{\Gamma(s+1)} + \cdots\right\}$$

$$= \frac{1}{\Gamma(s+1)} \left\{1 - \varepsilon\left(-\gamma + \sum_{m=1}^{s} m^{-1}\right) + \cdots\right\},$$

$$\frac{1}{\Gamma(t+\varepsilon+1)} = -\frac{\sin(t+\varepsilon)\pi}{\pi} \Gamma(-t-\varepsilon) = (-1)^{t+1} \varepsilon \Gamma(-t) + \cdots,$$

where γ is Euler's constant (§12.1).

Accordingly, picking out the coefficient of ε, we see that

$$Y_n(z) = \log\left(\frac{z}{2}\right) \left[\sum_{r=0}^{\infty} \frac{(-1)^r (\frac{z}{2})^{n+2r}}{r! \Gamma(n+r+1)} + (-1)^n \sum_{r=0}^{\infty} \frac{(-1)^r (\frac{z}{2})^{-n+2r}}{r! \Gamma(-n+r+1)}\right]$$

$$+ \sum_{r=0}^{\infty} \frac{(-1)^r (\frac{z}{2})^{n+2r}}{r! \Gamma(n+r+1)} \left(\gamma - \sum_{m=1}^{n+r} \frac{1}{m}\right)$$

$$+ (-1)^n \sum_{r=n}^{\infty} \frac{(-1)^r (\frac{z}{2})^{-n+2r}}{r! \Gamma(-n+r+1)} \left(\gamma - \sum_{m=1}^{n-r} \frac{1}{m}\right)$$

$$+ (-1)^n \sum_{r=0}^{n-1} \frac{(-1)^r (\frac{z}{2})^{-n+2r}}{r!} (-1)^{r-n+1} \Gamma(n-r),$$

and so

$$Y_n(z) = \sum_{r=0}^{\infty} \frac{(-1)^r (\frac{z}{2})^{n+2r}}{r! (n+r)!} \left\{2 \log\left(\frac{z}{2}\right) + 2\gamma - \sum_{m=1}^{n+r} \frac{1}{m} - \sum_{m=1}^{r} \frac{1}{m}\right\}$$

$$- \sum_{r=0}^{n-1} \frac{(\frac{z}{2})^{-n+2r} (n-r-1)!}{r!}.$$

---

[2] The proof of this is left to the reader.

When n is an integer, fundamental solutions[3] of Bessel's equations, regular near $z = 0$, are $J_n(z)$ and $Y_n(z)$ or $Y_n(z)$.

Karl Neumann [486, p. 41] took as the second solution the function $Y^{(n)}(z)$ defined by the equation

$$Y^{(n)}(z) = \frac{1}{2} Y_n(z) + J_n(z)(\log 2 - \gamma);$$

but $Y_n(z)$ and $Y_n(z)$ are more useful for physical applications.

Example 17.6.3 Shew that the function $Y_n(z)$ satisfies the recurrence formulae

$$n Y_n(z) = \frac{1}{2} z \left(Y_{n+1}(z) + Y_{n-1}(z) \right) \quad \text{and} \quad Y_n'(z) = \frac{1}{2} \left(Y_{n-1}(z) - Y_{n+1}(z) \right).$$

Shew also that Hankel's function $Y_n(z)$ and Neumann's function $Y^{(n)}(z)$ satisfy the same recurrence formulae. *Note.* These are the same as the recurrence formulae satisfied by $J_n(z)$.

Example 17.6.4 (Schläfli [576]) Shew that, when $|\arg z| < \frac{1}{2}\pi$,

$$\pi Y_n(z) = \int_0^\pi \sin(z \sin\theta - n\theta)\, d\theta - \int_0^\infty e^{-z \sinh\theta} \left\{ e^{n\theta} + (-1)^n e^{-n\theta} \right\} d\theta.$$

Example 17.6.5 Shew that

$$Y^{(0)}(z) = J_0(z) \log z + 2 \left(J_2(z) - \frac{1}{2} J_4(z) + \frac{1}{3} J_6(z) - \cdots \right).$$

17.7 Bessel functions with purely imaginary argument

The function[4]

$$I_n(z) = i^{-n} J_n(iz) = \sum_{r=0}^\infty \frac{(\frac{z}{2})^{n+2r}}{r!\,(n+r)!}$$

is of frequent occurrence in various branches of applied mathematics; in these applications z is usually positive.

The reader should have no difficulty in obtaining the following formulae:

(i) $I_{n-1}(z) - I_{n+1}(z) = \frac{2n}{z} I_n(z).$

(ii) $\frac{d}{dz} \left\{ z^n I_n(z) \right\} = z^n I_{n-1}(z).$

(iii) $\frac{d}{dz} \left\{ z^{-n} I_n(z) \right\} = z^{-n} I_{n+1}(z).$

(iv) $\dfrac{d^2 I_n(z)}{dz^2} + \dfrac{1}{z} \dfrac{d I_n(z)}{dz} - \left(1 + \dfrac{n^2}{z^2} \right) I_n(z) = 0.$

(v) When $\operatorname{Re}\left(n + \frac{1}{2} \right) > 0$, $I_n(z) = \dfrac{z^n}{2^n \Gamma(\frac{1}{2}) \Gamma(n + \frac{1}{2})} \displaystyle\int_0^\pi \cosh(z \cos\phi) \sin^{2n}\phi\, d\phi.$

---

[3] Euler [202, pp. 187, 233] gave a second solution (involving a logarithm) of the equation in the special cases $n = 0, n = 1$.

[4] This notation was introduced by Basset [51, p. 17]; in 1886 he had defined $I_n(z)$ as $i^n J_n(iz)$; [52].

(vi)   When $-\frac{3}{2}\pi < \arg z < \frac{1}{2}\pi$, the asymptotic expansion of $I_n(z)$ is

$$I_n(z) \sim \frac{e^z}{(2\pi z)^{1/2}}\left[1 + \sum_{r=1}^{\infty}(-1)^r\frac{\{4n^2-1^2\}\{4n^2-3^2\}\cdots\{4n^2-(2r-1)^2\}}{r!\,2^{3r}\,z^r}\right]$$
$$+ \frac{e^{-(n+\frac{1}{2})\pi i}e^{-z}}{(2\pi z)^{1/2}}\left[1 + \sum_{r=1}^{\infty}\frac{\{4n^2-1^2\}\{4n^2-3^2\}\cdots\{4n^2-(2r-1)^2\}}{r!\,2^{3r}\,z^r}\right],$$

the second series being negligible when $|\arg z| < \frac{1}{2}\pi$. The result is easily seen to be valid over the extended range $-\frac{3}{2}\pi < \arg z < \frac{3}{2}\pi$ if we write $e^{\pm(n+\frac{1}{2})\pi i}$ for $e^{-(n+\frac{1}{2})\pi i}$, the upper or lower sign being taken according as $\arg z$ is positive or negative.

17.71 Modified Bessel functions of the second kind

When n is a positive integer or zero, $I_{-n}(z) = I_n(z)$; to obtain a second solution of the modified Bessel equation (iv) above, we define[5] the function $K_n(z)$ for all values of n by the equation

$$K_n(z) = \left(\frac{\pi}{2z}\right)^{1/2}\cos n\pi\, W_{0,n}(2z),$$

so that $K_n(z) = \frac{\pi}{2}\left(I_{-n}(z) - I_n(z)\right)\cot n\pi$.

Whether n be an integer or not, this function is a solution of the modified Bessel equation, and when $|\arg z| < \frac{3}{2}\pi$ it possesses the asymptotic expansion

$$K_n(z) \sim \left(\frac{\pi}{2z}\right)^{1/2}e^{-z}\cos(n\pi)\left[1 + \sum_{r=1}^{\infty}\frac{\{4n^2-1^2\}\{4n^2-3^2\}\cdots\{4n^2-(2r-1)^2\}}{r!\,2^{3r}\,z^r}\right]$$

for large values of $|z|$.

When n is an integer, $K_n(z)$ is defined by the equation

$$K_n(z) = \lim_{\varepsilon\to 0}\frac{\pi}{2}\{I_{-n-\varepsilon}(z) - I_{n+\varepsilon}(z)\}\cot\pi\varepsilon,$$

which gives (cf. §17.61)

$$K_n(z) = -\sum_{r=0}^{\infty}\frac{(\frac{z}{2})^{n+2r}}{r!(n+r)!}\left\{\log\frac{z}{2} + \gamma - \frac{1}{2}\sum_{m=1}^{n+r}\frac{1}{m} - \frac{1}{2}\sum_{m=1}^{r}\frac{1}{m}\right\}$$
$$+ \frac{1}{2}\sum_{r=0}^{n-1}\left(\frac{z}{2}\right)^{-n+2r}\frac{(-1)^{n-r}(n-r-1)!}{r!}$$

---

[5] The notation $K_n(z)$ was used by Basset [52, p. 11] to denote a function which differed from the function now defined by the omission of the factor $\cos n\pi$, and Basset's notation has since been used by various writers, notably Macdonald. The object of the insertion of the factor is to make $I_n(z)$ and $K_n(z)$ satisfy the same recurrence formulae. Subsequently Basset [51, p. 19] used the notation $K_n(z)$ to denote a slightly different function, but the latter usage has not been followed by other writers. The definition of $K_n(z)$ for *integral* values of n which is given here is due to Gray and Mathews [259, p. 68], and is now common (see Example 17.40), but the corresponding definition for non-integral values has the serious disadvantage that the function vanishes identically when $2n$ is an odd integer. The function was considered by Riemann [555] and Hankel [273, p. 498].

as an ascending series.

Example 17.7.1   Shew that $K_n(z)$ satisfies the same recurrence formulae as $I_n(z)$.

17.8 Neumann's expansion of an analytic function in a series of Bessel coefficients

We shall now consider the expansion of an arbitrary function $f(z)$, analytic in a domain including the origin, in a series of Bessel coefficients, in the form

$$f(z) = a_0 J_0(z) + a_1 J_1(z) + a_2 J_2(z) + \cdots,$$

where $\alpha_0, \alpha_1, \alpha_2, \ldots$ are independent of z. This appears in Neumann [487] and Kapteyn [365].

Assuming the possibility of expansions of this type, let us first consider the expansion of $1/(t - z)$; let it be

$$\frac{1}{t - z} = O_0(t) J_0(z) + 2 O_1(t) J_1(z) + 2 O_2(t) J_2(z) + \cdots,$$

where the functions $O_n(t)$ are independent of z.

We shall now determine conditions which $O_n(t)$ must satisfy if the series on the right is to be a uniformly convergent series of analytic functions; by these conditions $O_n(t)$ will be determined, and it will then be shewn that, if $O_n(t)$ is so determined, then the series on the right actually converges to the sum $1/(t - z)$ when $|z| < |t|$.

Since $\left(\dfrac{\partial}{\partial t} + \dfrac{\partial}{\partial z} \right) \dfrac{1}{t - z} = 0$, we have

$$O_0'(t) J_0(z) + 2 \sum_{n=1}^{\infty} O_n'(t) J_n(z) + O_0(t) J_0'(z) + 2 \sum_{n=1}^{\infty} O_n(t) J_n'(z) \equiv 0,$$

so that, on replacing $2 J_n'(z)$ by $J_{n-1}(z) - J_{n+1}(z)$, we find

$$\{O_0'(t) + O_1(t)\} J_0(z) + \sum_{n=1}^{\infty} \{2 O_n'(t) + O_{n+1}(t) - O_{n-1}(t)\} J_n(z) = 0.$$

Accordingly the successive functions $O_1(t)$, $O_2(t)$, $O_3(t)$, . . . are determined by the recurrence formulae

$$O_1(t) = -O_0{}'(t), \qquad O_{n+1}(t) = O_{n-1}(t) - 2 O_n'(t),$$

and, putting $z = 0$ in the original expansion, we see that $O_0(t)$ *is to be defined by the equation* $O_0(t) = 1/t$. These formulae shew without difficulty that $O_n(t)$ is a polynomial of degree n in $1/t$.

We shall next prove by induction that $O_n(t)$, so defined, is equal to

$$\frac{1}{2} \int_0^{\infty} e^{-tu} \left[\left\{ u + \sqrt{u^2 + 1} \right\}^n + \left\{ u - \sqrt{u^2 + 1} \right\}^n \right] du$$

when $\mathrm{Re}\,(t) > 0$. For the expression is obviously equal to $O_0(t)$ or $O_1(t)$ when n is equal to 0 or 1 respectively; and

$$\frac{1}{2}\int_0^\infty e^{-tu}\left\{u \pm \sqrt{u^2+1}\right\}^{n-1} du - \frac{d}{dt}\int_0^\infty e^{-tu}\left\{u \pm \sqrt{u^2+1}\right\}^n du$$

$$= \frac{1}{2}\int_0^\infty e^{-tu}\left\{u \pm \sqrt{u^2+1}\right\}^{n-1}\left\{1 + 2u^2 \pm 2u\sqrt{u^2+1}\right\} du$$

$$= \frac{1}{2}\int_0^\infty e^{-tu}\left\{u \pm \sqrt{u^2+1}\right\}^{n+1} du,$$

whence the induction is obvious.

Writing $u = \sinh\theta$, we see that, according as n is even or odd, see [321, §§79, 264],

$$\frac{1}{2}\left[\left\{u + \sqrt{u^2+1}\right\}^n + \left\{u - \sqrt{u^2+1}\right\}^n\right] = \begin{Bmatrix} \cosh \\ \sinh \end{Bmatrix} n\theta$$

$$= 2^{n-1}\left\{\sinh^n\theta + \frac{n(n-1)}{2(2n-2)}\sinh^{n-2}\theta\right.$$

$$\left. + \frac{n(n-1)(n-2)(n-3)}{2\cdot 4(2n-2)(2n-4)}\sinh^{n-4}\theta + \cdots\right\},$$

and hence, when $\mathrm{Re}\,(t) > 0$, we have on integration,

$$O_n(t) = \frac{2^{n-1}n!}{t^{n+1}}\left\{1 + \frac{t^2}{2\,(2n-2)} + \frac{t^4}{2\cdot 4\,(2n-2)\,(2n-4)} + \cdots\right\},$$

the series terminating with the term in t^n or t^{n-1}; now, whether $\mathrm{Re}(t)$ be positive or not, $O_n(t)$ is defined as a polynomial in $1/t$; and so the expansion obtained for $O_n(t)$ is the value of $O_n(t)$ for *all* values of t.

Example 17.8.1   Shew that, for all values of t,

$$O_n(t) = \frac{1}{2t^{n+1}}\int_0^\infty e^{-x}\left[\left\{x + \sqrt{x^2+t^2}\right\}^n + \left\{x - \sqrt{x^2+t^2}\right\}^n\right] dx,$$

and verify that the expression on the right satisfies the recurrence formulae for $O_n(t)$.

17.81 Proof of Neumann's expansion

The method of §17.8 merely determined the coefficients in Neumann's expansion of $1/(t-z)$, on the hypothesis that the expansion existed and that the rearrangements were legitimate. To obtain a proof of the validity of the expansion, we observe that

$$J_n(z) = \frac{(\frac{z}{2})^n}{n!}\{1 + \theta_n\}, \qquad O_n(t) = \frac{2^{n-1}n!}{t^{n+1}}\{1 + \phi_n\},$$

where $\theta_n \to 0$, $\phi_n \to 0$ as $n \to \infty$, when z and t are fixed. Hence the series

$$O_0(t)J_0(z) + 2\sum_{n=1}^\infty O_n(t)J_n(z) \equiv F(z,t)$$

is comparable with the geometrical progression whose general term is z^n/t^{n+1}, and this progression is absolutely convergent when $|z| < |t|$, and so the expansion for $F(z,t)$ is absolutely convergent (§2.34) in the same circumstances.

Again if $|z| \leq r$, $|t| \geq R$, where $r < R$, the series is comparable with the geometrical progression whose general term is r^n/R^{n+1}, and so the expansion for $F(z,t)$ converges uniformly throughout the domains $|z| \leq r$ and $|t| \geq R$ by §3.34. Hence, by §5.3, term-by-term differentiations are permissible, and so

$$\left(\frac{\partial}{\partial t} + \frac{\partial}{\partial z}\right) F(z,t) = O_0'(t) J_0(z) + 2\sum_{n=1}^{\infty} O_n'(t)J_n(z) + O_0(t) J_0'(z) + 2\sum_{n=1}^{\infty} O_n(t) J_n'(z)$$

$$= \{O_0'(t) + O_1(t)\} J_0(z) + \sum_{n=1}^{\infty} \{2O_n'(t) + O_{n+1}(t) - O_{n-1}(t)\} J_n(z)$$

$$= 0,$$

by the recurrence formulae.

Since $\left(\dfrac{\partial}{\partial t} + \dfrac{\partial}{\partial z}\right) F(z,t) = 0$, it follows that $F(z,t)$ is expressible as a function of $t - z$; and since $F(0,t) = O_0(t) = 1/t$, it is clear that $F(z,t) = 1/(t - z)$. It is therefore proved that

$$\frac{1}{t-z} = O_0(t) J_0(z) + 2\sum_{n=1}^{\infty} O_n(t) J_n(z), \tag{17.8}$$

provided that $|z| < |t|$.

Hence, if $f(z)$ be analytic when $|z| \leq r$, we have, when $|z| < r$,

$$f(z) = \frac{1}{2\pi i} \int \frac{f(t)}{t-z} \, dt$$

$$= \frac{1}{2\pi i} \int f(t) \left\{ O_0(t)J_0(z) + 2\sum_{n=1}^{\infty} O_n(t)J_n(z) \right\} dt$$

$$= J_0(z)f(0) + \frac{1}{\pi i} \sum_{n=1}^{\infty} J_n(z) \int O_n(t)f(t) \, dt,$$

by §4.7, the paths of integration being the circle $|t| = r$; and this establishes the validity of Neumann's expansion when $|z| < r$ and $f(z)$ is analytic when $|z| \leq r$.

Example 17.8.2 (K. Neumann)   Shew that

$$\cos z = J_0(z) - 2J_2(z) + 2J_4(z) - \cdots,$$
$$\sin z = 2J_1(z) - 2J_3(z) + 2J_5(z) - \cdots.$$

Example 17.8.3 (K. Neumann)   Shew that

$$\left(\frac{z}{2}\right)^n = \sum_{r=0}^{\infty} \frac{(n+2r)(n+r-1)!}{r!} J_{n+2r}(z).$$

Example 17.8.4 (W. Kapteyn)   Shew that, when $|z| < |t|$,

$$O_0(t)J_0(z) + 2\sum_{n=1}^{\infty} O_n(t)J_n(z) = \sum_{n=-\infty}^{\infty} J_n(z) \int_0^{\infty} t^{-n-1} e^{-x} \left\{ x + \sqrt{x^2 + t^2} \right\}^n dx$$

$$= \int_0^{\infty} \frac{e^{-x}}{t^{n+1}} \sum_{n=-\infty}^{\infty} J_n(z) \left\{ x + \sqrt{x^2 + t^2} \right\}^n dx$$

$$= \frac{1}{t} \int_0^{\infty} \exp\left(\frac{zx}{t} - x \right) dx$$

$$= \frac{1}{t - z}.$$

17.82 Schlömilch's expansion of an arbitrary function in a series of Bessel coefficients of order zero

Schlömilch [581] (see also Chapman [142]) has given an expansion of quite a different character from that of Neumann. His result may be stated thus:

Any function $f(x)$, which has a continuous differential coefficient with limited total fluctuation for all values of x in the closed range $(0, \pi)$, may be expanded in the series

$$f(x) = a_0 + a_1 J_0(x) + a_2 J_0(2x) + a_3 J_0(3x) + \cdots,$$

valid in this range, where

$$a_0 = f(0) + \frac{1}{\pi} \int_0^{\pi} u \int_0^{\pi/2} f'(u \sin \theta) \, d\theta \, du,$$

$$a_n = \frac{2}{\pi} \int_0^{\pi} u \cos nu \int_0^{\pi/2} f'(u \sin \theta) \, d\theta \, du \qquad (n > 0).$$

Schlömilch's proof is substantially as follows:

Let $F(x)$ be the continuous solution of the integral equation

$$f(x) = \frac{2}{\pi} \int_0^{\pi/2} F(x \sin \phi) \, d\phi.$$

Then (§11.81)

$$F(x) = f(0) + x \int_0^{\pi/2} f'(x \sin \theta) \, d\theta.$$

In order to obtain Schlömilch's expansion, it is merely necessary to apply Fourier's theorem to the function $F(x \sin \phi)$. We thus have

$$f(x) = \frac{2}{\pi} \int_0^{\pi/2} d\phi \left\{ \frac{1}{\pi} \int_0^{\pi} F(u) \, du + \frac{2}{\pi} \sum_{n=1}^{\infty} \int_0^{\pi} \cos nu \cos(nx \sin \phi) F(u) \, du \right\}$$

$$= \frac{1}{\pi} \int_0^{\pi} F(u) \, du + \frac{2}{\pi} \sum_{n=1}^{\infty} \int_0^{\pi} \cos nu \, F(u) J_0(nx) \, du,$$

the interchange of summation and integration being permissible by §4.7 and §9.44.

In this equation, replace $F(u)$ by its value in terms of $f(u)$. Thus we have

$$f(x) = \frac{1}{\pi} \int_0^\pi \left\{ f(0) + u \int_0^{\pi/2} f'(u \sin \theta)\, d\theta \right\} du$$

$$+ \frac{2}{\pi} \sum_{n=1}^\infty J_0(nx) \int_0^\infty \cos nu \left\{ f(0) + u \int_0^{\pi/2} f'(u \sin \theta)\, d\theta \right\} du,$$

which gives Schlömilch's expansion.

Example 17.8.5 (Math. Trip. 1895)   Shew that, if $0 \le x \le \pi$, the expression

$$\frac{\pi^2}{4} - 2 \left\{ J_0(x) + \frac{1}{9} J_0(3x) + \frac{1}{25} J_0(5x) + \cdots \right\}$$

is equal to x; but that, if $\pi \le x \le 2\pi$, its value is

$$x + 2\pi \arccos(\pi/x) - 2\sqrt{x^2 - \pi^2},$$

where $\arccos(\pi/x)$ is taken between 0 and $\frac{\pi}{3}$. Find the value of the expression when x lies between 2π and 3π.

17.9 Tabulation of Bessel functions

Hansen used the asymptotic expansion (§17.5) to calculate tables of $J_n(x)$ which are given by Lommel in [441]. Meissel [466] tabulated $J_0(x)$ and $J_1(x)$ to 12 places of decimals from $x = 0$ to $x = 15.5$, while the *British Assoc. Report* (1909), p. 33, gives tables by which $J_n(x)$ and $Y_n(x)$ may be calculated when $x > 10$. Tables of $J_{\frac{1}{3}}(x)$, $J_{\frac{2}{3}}(x)$, $J_{-\frac{1}{3}}(x)$, $J_{-\frac{2}{3}}(x)$ are given by Dinnik [171].

Tables of the second solution of Bessel's equation have been given by the following writers: B. A. Smith [594] (see also [469]), Aldis [16], Airey [14]. The functions $I_n(x)$ have been tabulated in the *British Assoc. Reports*, (1889) p. 28, (1893) p. 223, (1896) p. 98, (1907) p. 94; also by Aldis [15], by Isherwood [338] and by E. Anding [18].

Tables of $J_n(x\sqrt{i})$, a function employed in the theory of alternating currents in wires, have been given in the *British Assoc. Reports*, 1889, 1893, 1896 and 1912; by Kelvin [627], by Aldis [16] and by Savidge [572]. Formulae for computing the zeros of $J_0(z)$ were given by Stokes [610] and the 40 smallest zeros were tabulated by Willson and Peirce [682]. The roots of an equation involving Bessel functions were computed by Kalähne [364]. A number of tables connected with Bessel functions are given in *British Assoc. Reports*, 1910–1914, and also by Jahnke & Emde [356].

17.10 Miscellaneous examples

Example 17.1 (K. Neumann)   Shew that

$$\cos(z \sin \theta) = J_0(z) + 2J_2(z) \cos 2\theta + 2J_4(z) \cos 4\theta + \cdots,$$

$$\sin(z \sin \theta) = 2J_1(z) \sin \theta + 2J_3(z) \sin 3\theta + 2J_5(z) \sin 5\theta + \cdots$$

Example 17.2   By expanding each side of the equations of Example 17.1 in powers of $\sin \theta$, express z^n as a series of Bessel coefficients.

Example 17.3   By multiplying the expansions for $\exp\left\{\frac{z}{2}(t - 1/t)\right\}$ and $\exp\left\{-\frac{z}{2}(t - 1/t)\right\}$ and considering the terms independent of t, shew that

$$\{J_0(z)\}^2 + 2\{J_1(z)\}^2 + 2\{J_2(z)\}^2 + 2\{J_3(z)\}^2 + \cdots = 1.$$

Deduce that, for the Bessel coefficients, $|J_0(z)| \leq 1$, $|J_n(z)| \leq 1/\sqrt{2}$, for $n \geq 1$, when z is real.

Example 17.4 (Bourget [94])   If $J_m^k(z) = \dfrac{1}{\pi}\displaystyle\int_0^\pi 2^k \cos^k u \cos(mu - z \sin u)\, du$ (this function reduces to a Bessel coefficient when k is zero and m an integer), shew that

$$J_m^k(z) = \sum_{p=0}^\infty \frac{1}{p!}\left(\frac{z}{2}\right)^p N_{-m,k,p},$$

where $N_{-m,k,p}$ is the *Cauchy number* defined by the equation

$$N_{-m,k,p} = \frac{1}{2\pi}\int_{-\pi}^\pi e^{-miu}(e^{iu} + e^{-iu})^k(e^{iu} - e^{-iu})^p\, du.$$

Shew further that;

$$J_m^k(z) = J_{m-1}^{k-1}(z) + J_{m+1}^{k-1}(z),$$

and $zJ_m^{k+2}(z) = 2mJ_m^{k+1}(z) - 2(k+1)\left\{J_{m-1}^k(z) - J_{m+1}^k(z)\right\}.$

Example 17.5 (Bourget)   If v and M are connected by the equations

$$M = E - e\sin E, \qquad \cos v = \frac{\cos E - e}{1 - e\cos E}, \qquad \text{where } |e| < 1,$$

shew that

$$v = M + 2(1 - e^2)^{\frac{1}{2}}\sum_{m=1}^\infty \sum_{k=0}^\infty \left(\frac{1}{2}e\right)^k J_m{}^k(me)\frac{1}{m}\sin mM,$$

where $J_m^k(z)$ is defined as in Example 17.4.

Example 17.6 (Math. Trip. 1893)   Prove that, if m and n are integers,

$$P_n^m(\cos\theta) = \frac{c_n^m}{r^n}J_m\left((x^2 + y^2)^{\frac{1}{2}}\frac{\partial}{\partial z}\right)z^n,$$

where $z = r\cos\theta$, $x^2 + y^2 = r^2\sin^2\theta$, and c_n^m is independent of z.

Example 17.7   Shew that the solution of the differential equation

$$\frac{d^2y}{dz^2} - \frac{\phi'}{\phi}\frac{dy}{dz} + \left\{\frac{1}{4}\left(\frac{\phi'}{\phi}\right)^2 - \frac{1}{2}\frac{d}{dz}\left(\frac{\phi'}{\phi}\right) - \frac{1}{4}\left(\frac{\psi''}{\psi'}\right)^2 + \frac{1}{2}\frac{d}{dz}\left(\frac{\psi''}{\psi'}\right) + \left(\psi^2 - v^2 + \frac{1}{4}\right)\left(\frac{\psi'}{\psi}\right)^2\right\}y$$

$$= 0,$$

where ϕ and ψ are arbitrary functions of z, is

$$y = \left(\frac{\phi\psi}{\psi'}\right)^{\frac{1}{2}}\{AJ_v(\psi) + BJ_{-v}(\psi)\}.$$

Example 17.8 (Trinity, 1908)   Shew that

$$J_1(x) + J_3(x) + J_5(x) + \cdots = \frac{1}{2}\left[J_0(x) + \int_0^x \{J_0(t) + J_1(t)\}\, dt - 1 \right].$$

Example 17.9 (Schläfli [576] and Schönholzer [588])   Shew that

$$J_\mu(z)J_\nu(z) = \sum_{n=0}^{\infty} \frac{(-1)^n \Gamma(\mu + \nu + 2n + 1)(\frac{z}{2})^{\mu+\nu+2n}}{n!\, \Gamma(\mu + n + 1)\Gamma(\nu + n + 1)\Gamma(\mu + \nu + n + 1)}$$

for all values of μ and ν.

Example 17.10 (Math. Trip. 1899)   Shew that, if n is a positive integer and $m + 2n + 1$ is positive,

$$(m-1)\int_0^x x^m J_{n+1}(x)J_{n-1}(x)\, dx = x^{m+1}\{J_{n+1}(x)J_{n-1}(x) - J_n^2(x)\} + (m+1)\int_0^x x^m J_n^2(x)\, dx.$$

Example 17.11   Shew that

$$J_3(z) + 3\frac{dJ_0(z)}{dz} + 4\frac{d^3 J_0(z)}{dz^3} = 0.$$

Example 17.12   Shew that

$$\frac{J_{n+1}(z)}{J_n(z)} = \cfrac{\frac{1}{2}z/n(n+1)}{1 - \cfrac{(\frac{1}{2}z)^2/(n+1)(n+2)}{1 - \cfrac{(\frac{1}{2}z)^2/(n+2)(n+3)}{1 - \cdots}}}.$$

Example 17.13 (Lommel)   Shew that

$$J_{-n}(z)\, J_{n-1}(z) + J_{-n+1}(z)J_n(z) = \frac{2\sin n\pi}{\pi z}.$$

Example 17.14   If $\dfrac{J_{n+1}(z)}{zJ_n(z)}$ be denoted by $Q_n(z)$, shew that

$$\frac{dQ_n(z)}{dz} = \frac{1}{z} - \frac{2(n+1)}{z}Q_n(z) + z\{Q_n(z)\}^2.$$

Example 17.15 (K. Neumann)   Shew that, if $R^2 = r^2 + r_1^2 - 2rr_1\cos\theta$ and $r_1 > r > 0$,

$$J_0(R) = J_0(r)J_0(r_1) + 2\sum_{n=1}^{\infty} J_n(r)\, J_n(r_1)\, \cos n\theta,$$

$$Y_0(R) = J_0(r)\, Y_0(r_1) + 2\sum_{n=1}^{\infty} J_n(r)Y_n(r_1)\, \cos n\theta.$$

Example 17.16 (K. Neumann)   Shew that, if $\operatorname{Re}\left(n + \frac{1}{2}\right) > 0$,

$$\int_0^{\frac{1}{2}\pi} J_{2n}\left(2z\cos\theta\right) d\theta = \frac{\pi}{2}\{J_n(z)\}^2.$$

Example 17.17 (Math. Trip. 1896)   Shew how to express $z^{2n} J_{2n}(z)$ in the form $AJ_2(z) + BJ_0(z)$, where A and B are polynomials in z; and prove that

$$J_4(\sqrt{6}) + 3J_0(\sqrt{6}) = 0, \qquad 3J_6(\sqrt{30}) + 5J_2(\sqrt{30}) = 0.$$

Example 17.18   Shew that, if $\alpha \neq \beta$ and $n > -1$,

$$(\alpha^2 - \beta^2) \int_0^x x J_n(\alpha x) J_n(\beta x)\, dx = x \left\{ J_n(\alpha x) \frac{d}{dx} J_n(\beta x) - J_n(\beta x) \frac{d}{dx} J_n(\alpha x) \right\},$$

$$2\alpha^2 \int_0^x x \{ J_n(\alpha x) \}^2\, dx = (\alpha^2 x^2 - n^2) \{ J_n(ax) \}^2 + \left\{ x \frac{d}{dx} J_n(\alpha x) \right\}^2.$$

Example 17.19 (Lommel [441])   Prove that, if $n > -1$, and $J_n(\alpha) = J_n(\beta) = 0$ while $\alpha \neq \beta$,

$$\int_0^1 x J_n(\alpha x) J_n(\beta x)\, dx = 0, \quad \text{and} \quad \int_0^1 x \{ J_n(\alpha x) \}^2\, dx = \frac{1}{2} \{ J_{n+1}(\alpha) \}^2.$$

Hence prove that, when $n > -1$, the roots of $J_n(x) = 0$, other than zero, are all real and unequal. *Hint.* If α could be complex, take β to be the conjugate complex.

Example 17.20   Let $x^{1/2} f(x)$ have an absolutely convergent integral in the range $0 \leq x \leq 1$; let H be a real constant and let $n \geq 0$. Then, if $k_1, k_2, \ldots$ denote the positive roots of the equation

$$k^{-n} \{ k\, J_n'(k) + H\, J_n(k) \} = 0,$$

shew that, at any point x for which $0 < x < 1$ and $f(x)$ satisfies one of the conditions of §9.43, $f(x)$ can be expanded in the form

$$f(x) = \sum_{r=1}^{\infty} A_r\, J_n(k_r x),$$

where

$$A_r = \left[\int_0^1 x \{ J_n(k_r x) \}^2\, dx \right]^{-1} \int_0^1 x f(x)\, J_n(k_r x)\, dx.$$

In the special case when $H = -n$, k_1 is to be taken to be zero, the equation determining $k_1, k_2, \ldots$ being $J_{n+1}(k) = 0$, and the first term of the expansion is $A_0 x^n$ where

$$A_0 = (2n + 2) \int_0^1 x^{n+1} f(x)\, dx.$$

Discuss, in particular, the case when H is infinite, so that $J_n(k) = 0$, shewing that

$$A_r = 2 \{ J_{n+1}(k_r) \}^{-2} \int_0^1 x f(x)\, J_n(k_r x)\, dx.$$

Note   This result is due to Hobson [318]; see also W. H. Young [686]. The formal expansion was given with H infinite (when $n = 0$) by Fourier and (for general values of n) by Lommel; proofs were given by Hankel and Schläfli. The formula when $H = -n$ was given incorrectly by Dini [170], the term $A_0 x^n$ being printed as A_0, and this error was not corrected by Nielsen. See Bridgeman [98] and Chree [144]. The expansion is usually called the *Fourier–Bessel expansion*.

Example 17.21 (Clare, 1900)   Prove that, if the expansion

$$\alpha^2 - x^2 = A_1 J_0(\lambda_1 x) + A_2 J_0(\lambda_2 x) + \cdots$$

exists as a uniformly convergent series when $-\alpha \le x \le \alpha$, where $\lambda_1, \lambda_2, \ldots$ are the positive roots of $J_0(\lambda\alpha) = 0$, then $A_n = 8\{\alpha\lambda_n{}^3 J_1(\lambda_n\alpha)\}^{-1}$.

Example 17.22 (Math. Trip. 1906)   If $k_1, k_2, \ldots$ are the positive roots of $J_n(k\alpha) = 0$, and if

$$x^{n+2} = \sum_{r=1}^{\infty} A_r J_n(k_r x),$$

this series converging uniformly when $0 \le x \le \alpha$, then

$$A_r = \frac{2a^{n-1}}{k_r{}^2}(4n + 4 - \alpha^2 k_r{}^2) \div \frac{dJ_n(k_r\alpha)}{da}.$$

Example 17.23 (Sonine [598])   Shew that

$$J_n(x) = \frac{x^{n-m}}{2^{n-m-1}\,\Gamma(n-m)} \int_0^{\frac{\pi}{2}} J_m(x \sin\theta) \cos^{2n-2m-1}\theta \, \sin^{m+1}\theta \, d\theta$$

when $n > m > -1$.

Example 17.24 (Nicholson [497])   Shew that, if $\sigma > 0$,

$$\int_0^{\infty} \cos(t^3 - \sigma t)\, dt = \frac{\pi\sigma^{1/2}}{3\sqrt{3}} \left\{ J_{1/3}\left(\frac{2\sigma^{3/2}}{3^{3/2}} \right) + J_{-1/3}\left(\frac{2\sigma^{3/2}}{3^{3/2}} \right) \right\}.$$

Example 17.25 (Math. Trip. 1904)   If m be a positive integer and $u > 0$, deduce from Bessel's integral formula that

$$\int_0^{\infty} e^{-x \sinh u} J_m(x)\, dx = e^{-mu} \operatorname{sech} u.$$

Example 17.26 (Sonine [598])   Prove that, when $x > 0$,

$$J_0(x) = \frac{2}{\pi} \int_0^{\infty} \sin(x \cosh t)\, dt, \qquad Y_0(x) = -\frac{2}{\pi} \int_0^{\infty} \cos(x \cosh t)\, dt.$$

Hint. Take the contour of §17.1 to be the imaginary axis indented at the origin and a semicircle on the left of this line.

Example 17.27 (Weber [656])   Shew that

$$\int_0^{\infty} x^{-1} J_0(xt) \sin x \, dx = \begin{cases} \dfrac{\pi}{2} & 0 < t < 1 \\ \operatorname{arccosec} t & t > 1 \end{cases}$$

and that

$$\int_0^{\infty} x^{-1} J_1(xt) \sin x \, dx = \begin{cases} t^{-1}\{1 - (1 - t^2)^{1/2}\} & 0 < t < 1 \\ t^{-1} & t > 1 \end{cases}$$

Example 17.28 (Poisson [531]; see also Stokes [609])   Shew that

$$u = \int_0^\pi e^{nr\cos\theta}\{A + B\log(r\sin^2\theta)\}\,d\theta$$

is the solution of

$$\frac{d^2u}{dr^2} + \frac{1}{r}\frac{du}{dr} - n^2u = 0.$$

Example 17.29 (Math. Trip. 1901)   Prove that no relation of the form

$$\sum_{s=0}^k N_s J_{n+s}(x) = 0$$

can exist for rational values of N_s, n and x except relations which are satisfied when the Bessel functions are replaced by arbitrary solutions of the recurrence formula of (17.2). *Hint.* Express the left-hand side in terms of $J_n(x)$ and $J_{n+1}(x)$, and shew by Example 17.12 that $J_{n+1}(x)/J_n(x)$ is irrational when n and x are rational.

Example 17.30 (Hargreave [282]; Macdonald [444])   Prove that, when Re $(n) > -\frac{1}{2}$,

$$J_n(z) = \frac{z^n}{2^{n-1}\Gamma(n+\frac{1}{2})\Gamma(\frac{1}{2})}\left(1 + \frac{d^2}{dz^2}\right)^{n-1/2}\left(\frac{\sin z}{z}\right),$$

$$-Y_n(z) = \frac{z^n}{2^{n-1}\Gamma(n+\frac{1}{2})\Gamma(\frac{1}{2})}\left(1 + \frac{d^2}{dz^2}\right)^{n-1/2}\left(\frac{\cos z}{z}\right).$$

Here,

$$\left(1 + \frac{d^2}{dz^2}\right)^{n-1/2}$$

means

$$1 + \frac{n-\frac{1}{2}}{1!}\frac{d^2}{dz^2} + \frac{(n-\frac{1}{2})(n-\frac{3}{2})}{2!}\frac{d^4}{dz^4} + \cdots.$$

Hint. Write $\dfrac{e^{iz}}{z} = \displaystyle\int_{i\infty}^1 ie^{izt}\,dt.$

Example 17.31 (Hobson)   Shew that, when Re $(m + \frac{1}{2}) > 0$,

$$\left(\frac{2}{\pi}\right)^{\frac{1}{2}}\int_0^{\frac{\pi}{2}} J_m(z\sin\theta)\sin^{m+1}\theta\,d\theta = z^{-\frac{1}{2}}J_{m+\frac{1}{2}}(z).$$

Example 17.32 (Weber [654]; Math. Trip. 1898)   Shew that, if $2n + 1 > m > -1$,

$$\int_0^\infty x^{-n+m}J_n(ax)\,dx = 2^{-n+m}a^{n-m-1}\frac{\Gamma(\frac{1}{2}m+\frac{1}{2})}{\Gamma(n-\frac{1}{2}m+\frac{1}{2})}.$$

Example 17.33 (Lommel)   Shew that

$$\frac{z}{\pi} = \sum_{p=0}^\infty \frac{2p+1}{2}\{J_{p+\frac{1}{2}}(z)\}^2.$$

Example 17.34 (Math. Trip. 1894)   In the equation

$$\frac{d^2 y}{dz^2} + \frac{1}{z}\frac{dy}{dz} + \left(1 + \frac{n^2}{z^2} +\right) y = 0,$$

n is real; shew that a solution is given by

$$\cos(n \log z) - \sum_{m=1}^{\infty} \frac{(-1)^m z^{2m} \cos(u_m - n \log z)}{2^{2m} m! \, (1 + n^2)^{\frac{1}{2}} (4 + n^2)^{\frac{1}{2}} \cdots (m^2 + n^2)^{\frac{1}{2}}},$$

where u_m denotes $\sum\limits_{r=1}^{m} \arctan(n/r)$.

Example 17.35 (Cauchy [128]; Nicholson [496])   Shew that, when *n* is large

$$J_n(n) = 2^{-\frac{2}{3}} 3^{-\frac{1}{6}} \pi^{-1} \Gamma\left(\frac{1}{3}\right) n^{-\frac{1}{3}} + O(n^{-1}).$$

Example 17.36 (Mehler [464])   Shew that

$$K_0(x) = \int_0^{\infty} \frac{t J_0(tx)}{1 + t^2}\, dt.$$

Example 17.37 (Math. Trip. 1900)   Shew that

$$e^{\lambda \cos \theta} = 2^{n-1} \Gamma(n) \sum_{k=0}^{\infty} (n + k) C_k^n (\cos \theta) \lambda^{-n} I_{n+k}(\lambda).$$

Example 17.38 (Sonine [598])   Shew that, if

$$W = \int_0^{\infty} J_m(ax) J_m(bx) J_m(cx) x^{1-m}\, dx,$$

a, b, c being positive, and *m* is a positive integer or zero, then

$$W = \begin{cases} 0 & \text{if } (a-b)^2 > c^2, \\ \dfrac{a^{-m} b^{-m} c^{-m}}{2^{3m-1} \pi^{\frac{1}{2}} \Gamma\left(m+\frac{1}{2}\right)} \{2 \sum b^2 c^2 - \sum a^4\}^{m-\frac{1}{2}} & \text{if } (a+b)^2 > c^2 > (a-b)^2, \\ 0 & \text{if } (a+b)^2 > c^2. \end{cases}$$

Example 17.39 (Macdonald [448])   Shew that, if $n > -1$, $m > -\frac{1}{2}$ and

$$W = \int_0^{\infty} J_n(ax) J_n(bx) J_m(cx) x^{1-m}\, dx,$$

a, b, c being positive, then

$$W = \begin{cases} 0 & \text{if } (a-b)^2 > c^2, \\ (2\pi)^{-\frac{1}{2}} a^{m-1} b^{m-1} c^{-m} (1 - \mu^2)^{\frac{1}{4}(2m-1)} P_{n-\frac{1}{2}}^{\frac{1}{2}-m}(\mu) & \text{if } (a+b)^2 > c^2 \\ & \qquad > (a-b)^2, \\ \left(\frac{1}{2}\pi\right)^{-\frac{1}{2}} a^{m-1} b^{m-1} c^{-m} \frac{\sin(m-n)\pi}{\pi} e^{(m-\frac{1}{2})\pi i} (\mu_1^2 - 1)^{\frac{1}{4}(2m-1)} Q_{n-\frac{1}{2}}^{\frac{1}{2}-m}(\mu_1) & \text{if } c^2 > (a+b)^2, \end{cases}$$

where $\mu = (a^2 + b^2 - c^2)/2ab$, and $\mu_1 = -\mu$.

Example 17.40 (Math. Trip. 1898; Basset [52])   Shew that, if $\mathrm{Re}\,(m + \frac{1}{2}) > 0$,

$$I_m(z) = \frac{z^m}{2^m \Gamma(m + \frac{1}{2})\Gamma(\frac{1}{2})} \int_0^\pi \cosh(z \cos \phi) \sin^{2m} \phi\, d\phi,$$

and, if $|\arg z| < \frac{1}{2}\pi$,

$$K_m(z) = \frac{z^m \Gamma\left(\frac{1}{2}\right) \cos m\pi}{2^m \Gamma(m + \frac{1}{2})} \int_0^\infty e^{-z \cosh \phi} \sinh^{2m} \phi\, d\phi.$$

Prove also that

$$K_m(z) = \frac{(2z)^m}{\sqrt{\pi}} \Gamma\left(m + \tfrac{1}{2}\right) \cos m\pi \int_0^\infty (u^2 + z^2)^{-m-\frac{1}{2}} \cos u\, du.$$

Hint. The first integral may be obtained by expanding in powers of z and integrating term-by-term. To obtain the second, consider

$$z^m \int_\infty^{(1+, -1+1)} e^{-st} (t^2 - 1)^{m-\frac{1}{2}}\, dt,$$

where initially $\arg(t - 1) = \arg(t + 1) = 0$. Take $|t| > 1$ on the contour, expand $(t^2 - 1)^{m-\frac{1}{2}}$ in descending powers of t, and integrate term-by-term. The result is

$$2ie^{2m\pi i} \sin(2m\pi)\Gamma(2m)2^{-m}\Gamma(1 - m)I_{-m}(z).$$

Also, deforming the contour by flattening it, the integral becomes

$$2ie^{2m\pi i} z^m \sin 2m\pi \int_1^\infty e^{-st}(t^2 - 1)^{m-\frac{1}{2}}\, dt + 2ie^{2m\pi i} z^m \cos m\pi \int_{-1}^1 e^{-st}(1 - t^2)^{m-\frac{1}{2}}\, dt;$$

and consequently

$$I_{-m}(z) - I_m(z) = \frac{2^{1-m} \sin(m\pi)z^m}{\Gamma\left(\frac{1}{2}\right)\Gamma(m + \frac{1}{2})} \int_1^\infty e^{-st}(t^2 - 1)^{m-\frac{1}{2}}\, dt.$$

Example 17.41 (K. Neumann)   Shew that $O_n(z)$ satisfies the differential equation

$$\frac{d^2 O_n(z)}{dz^2} + \frac{3}{z}\frac{dO_n(z)}{dz} + \left\{1 - \frac{n^2 - 1}{2}\right\} O_n(z) = g_n,$$

where

$$g_n = \begin{cases} z^{-1} & \text{if } n \text{ is even} \\ nz^{-2} & \text{if } n \text{ is odd.} \end{cases}$$

Example 17.42 (K. Neumann)   If $f(z)$ be analytic throughout the ring-shaped region bounded by the circles c, C whose centres are at the origin, establish the expansion

$$f(z) = \frac{1}{2}\alpha_0 J_0(z) + \alpha_1 J_1(z) + \alpha_2 J_2(z) + \cdots + \frac{1}{2}\beta_0 O_0(z) + \beta_1 O_1(z) + \beta_2 O_2(z) + \cdots,$$

where

$$\alpha_n = \frac{1}{\pi i} \int_C f(t)O_n(t)\, dt, \qquad \beta_n = \frac{1}{\pi i} \int_C f(t)J_n(t)\, dt.$$

Example 17.43 (Math. Trip. 1905)   Shew that, if x and y are positive,

$$\int_0^\infty \frac{e^{-\beta x}}{\beta} J_0(ky)k \, dk = \frac{e^{-ir}}{r},$$

where $r = \sqrt{x^2 + y^2}$ and $\beta = \sqrt{k^2 - 1}$ or $i\sqrt{1 - k^2}$, according as $k > 1$ or $k < 1$.

Example 17.44   Shew that, with suitable restrictions on n and on the form of the function $f(x)$,

$$f(x) = \int_0^\infty J_n(tx)t\left\{\int_0^\infty f(x')J_n(tx')x' \, dx'\right\} dt.$$

Note   A proof with an historical account of this important theorem is given by Nielsen [500, p. 360–363]. It is due to Hankel, but (in view of the result of §9.7) it is often called the *Fourier–Bessel integral*.

Example 17.45 (K. Neumann)   If C be any closed contour, and m and n are integers, shew that

$$\int_C J_m(z)J_n(z) \, dz = \int_C O_m(z)O_n(z) \, dz = \int_C J_m(z)O_n(z) \, dz = 0,$$

unless C contains the origin and $m = n$; in which case the first two integrals are still zero, but the third is equal to πi (or $2\pi i$, if $m = 0$) if C encircles the origin once counter-clockwise.

Example 17.46 (K. Neumann)   Shew that, if

$$a_{p,q} = \frac{(-1)^p}{p! \, q!},$$

and if n be a positive integer, then

$$z^{-2n} = \sum_{m=1}^n a_{n-m,n+m-1}O_{2m-1}(z),$$

while

$$z^{1-2n} = a_{n-1,n-1}O_0(z) + 2\sum_{m=1}^{n-1} a_{m-1,n+m-1}O_{2m}(z).$$

Example 17.47 (K. Neumann)   If

$$\Omega_n(y) = \sum_{m=0}^n \frac{2^{2m}(m!)^2}{(2m)!} \frac{n^2(n^2 - 1^2)(n^2 - 2^2)\cdots(n^2 - (m-1)^2)}{y^{2m+2}},$$

shew that

$$(y^2 - x^2)^{-1} = \Omega_0(y)J_0^2(x) + 2\sum_{n=1}^\infty \Omega_n(y)J_n^2(x)$$

when the series on the right converges.

Example 17.48 (Macdonald [446])   Shew that, if $c > 0$, $\text{Re}\,(n) > -1$, and $\text{Re}\,(a \pm b)^2 > 0$, then

$$J_n(a)J_n(b) = \frac{1}{2\pi i}\int_{c-\infty i}^{c+\infty i} t^{-1}\exp\{(t^2 - a^2 - b^2)/(2t)\} I_n(ab/t) \, dt.$$

Example 17.49 (Gegenbauer [239])   Deduce from Example 17.48, or otherwise prove, that

$$(a^2 + b^2 - 2ab\cos\theta)^{-\frac{1}{2}n} J_n\{(a^2 + b^2 - 2ab\cos\theta)^{\frac{1}{2}}\}$$

$$= 2^n \Gamma(n) \sum_{m=0}^{\infty} (m+n) a^{-n} b^{-n} J_{m+n}(a) J_{m+n}(b) C_m^n(\cos\theta).$$

Example 17.50 (Schafheitlin [573]; Math. Trip. 1903)   Shew that

$$y = \int_C J_m(t) J_n\left(tz^{1/2}\right) t^{k-1}\, dt$$

satisfies the equation

$$\frac{d^2y}{dz^2} + \left(\frac{1}{z} + \frac{k}{z-1}\right)\frac{dy}{dz} + \left(k^2 - m^2 + \frac{n^2}{z}\right)\frac{y}{4z(z-1)} = 0$$

if

$$kt^k J_m(t) J_n\left(tz^{1/2}\right) - t^{k+1} J_m'(t) J_n\left(tz^{1/2}\right) + z^{1/2} t^{k+1} J_m(t) J_m'\left(tz^{1/2}\right)$$

resumes its initial value after describing the contour. Deduce that, when $0 < z < 1$,

$$\int_0^\infty J_{a-\beta}(t) J_{\gamma-1}\left(tz^{\frac{1}{2}}\right) t^{a+\beta-\gamma}\, dt = \frac{\Gamma(a) z^{\frac{1}{2}(\gamma-1)}}{2\gamma^{-a-\beta}\Gamma(1-\beta)\Gamma(\gamma)} F(a, \beta; \gamma; z).$$

18

The Equations of Mathematical Physics

18.1 The differential equations of mathematical physics

The functions which have been introduced in the preceding chapters are of importance in the applications of mathematics to physical investigations. Such applications are outside the province of this book; but most of them depend essentially on the fact that, by means of these functions, it is possible to construct solutions of certain partial differential equations, of which the following are among the most important:

(I) Laplace's equation

$$\frac{\partial^2 V}{\partial x^2} + \frac{\partial^2 V}{\partial y^2} + \frac{\partial^2 V}{\partial z^2} = 0,$$

which was originally introduced in a memoir [408, p. 252] on Saturn's rings.

If (x, y, z) be the rectangular coordinates of any point in space, this equation is satisfied by the following functions which occur in various branches of mathematical physics:

(i) The gravitational potential in regions not occupied by attracting matter.
(ii) The electrostatic potential in a uniform dielectric, in the theory of electrostatics.
(iii) The magnetic potential in free aether, in the theory of magnetostatics.
(iv) The electric potential, in the theory of the steady flow of electric currents in solid conductors.
(v) The temperature, in the theory of thermal equilibrium in solids.
(vi) The velocity potential at points of a homogeneous liquid moving irrotationally, in hydrodynamical problems.

Notwithstanding the physical differences of these theories, the mathematical investigations are much the same for all of them: thus, the problem of thermal equilibrium in a solid when the points of its surface are maintained at given temperatures is mathematically identical with the problem of determining the electric intensity in a region when the points of its boundary are maintained at given potentials.

(II) The equation of wave motions

$$\frac{\partial^2 V}{\partial x^2} + \frac{\partial^2 V}{\partial y^2} + \frac{\partial^2 V}{\partial z^2} = \frac{1}{c^2}\frac{\partial^2 V}{\partial t^2}.$$

This equation is of general occurrence in investigations of undulatory disturbances propagated with velocity c independent of the wave length; for example, in the theory of electric waves and the electro-magnetic theory of light, it is the equation satisfied by each component of the electric or magnetic vector; in the theory of elastic vibrations, it is the equation satisfied by each component of the displacement; and in the theory of sound, it is the equation satisfied by the velocity potential in a perfect gas.

(III) The equation of conduction of heat

$$\frac{\partial^2 V}{\partial x^2} + \frac{\partial^2 V}{\partial y^2} + \frac{\partial^2 V}{\partial z^2} = \frac{1}{k}\frac{\partial V}{\partial t}.$$

This is the equation satisfied by the temperature at a point of a homogeneous isotropic body; the constant k is proportional to the heat conductivity of the body and inversely proportional to its specific heat and density.

(IV) Two-dimensional wave motion

A particular case of the preceding equation, when the variable z is absent, is

$$\frac{\partial^2 V}{\partial x^2} + \frac{\partial^2 V}{\partial y^2} = \frac{1}{c^2}\frac{\partial^2 V}{\partial t^2}.$$

This is the equation satisfied by the displacement in the theory of transverse vibrations of a membrane; the equation also occurs in the theory of wave motion in two dimensions.

(V) The equation of telegraphy

$$LK\frac{\partial^2 V}{\partial t^2} + KR\frac{\partial V}{\partial t} = \frac{\partial^2 V}{\partial x^2}.$$

This is the equation satisfied by the potential in a telegraph cable when the inductance L, the capacity K, and the resistance R per unit length are taken into account.

It would not be possible, within the limits of this chapter, to attempt an exhaustive account of the theories of these and the other differential equations of mathematical physics; but, by considering selected typical cases, we shall expound some of the principal methods employed, with special reference to the uses of the transcendental functions.

18.2 Boundary conditions

A problem which arises very frequently is the determination, for one of the equations of §18.1, of a solution which is subject to certain boundary conditions; thus we may desire to find the temperature at any point inside a homogeneous isotropic conducting solid in thermal equilibrium when the points of its outer surface are maintained at given temperatures. This amounts to finding a solution of Laplace's equation at points inside a given surface, when the value of the solution at points on the surface is given.

A more complicated problem of a similar nature occurs in discussing small oscillations of a liquid in a basin, the liquid being exposed to the atmosphere; in this problem we are given, effectively, the velocity potential at points of the free surface and the normal derivate of the velocity potential where the liquid is in contact with the basin.

The nature of the boundary conditions, necessary to determine a solution uniquely, varies very much with the form of differential equation considered, even in the case of equations which, at first sight, seem very much alike. Thus a solution of the equation

$$\frac{\partial^2 V}{\partial x^2} + \frac{\partial^2 V}{\partial y^2} = 0$$

(which occurs in the problem of thermal equilibrium in a conducting cylinder) is uniquely determined at points inside a closed curve in the x–y-plane by a knowledge of the value of V at points on the curve; but in the case of the equation

$$\frac{\partial^2 V}{\partial x^2} - \frac{1}{c^2}\frac{\partial^2 V}{\partial t^2} = 0$$

(which effectively only differs from the former in a change of sign), occurring in connexion with transverse vibrations of a stretched string, where V denotes the displacement at time t at distance x from the end of the string, it is physically evident that a solution is determined uniquely only if *both* V and $\dfrac{\partial V}{\partial t}$ are given for all values of x such that $0 \le x \le l$, when $t = 0$ (where l denotes the length of the string).

Physical intuitions will usually indicate the nature of the boundary conditions which are necessary to determine a solution of a differential equation uniquely; but the existence theorems, which are necessary from the point of view of the pure mathematician, are usually very tedious and difficult (see Forsyth [220, §§216–220], where an apparently simple problem is discussed).

18.3 A general solution of Laplace's equation

It is possible to construct a general solution of Laplace's equation in the form of a definite integral (see Whittaker [672]). This solution can be employed to solve various problems involving boundary conditions.

Let $V(x, y, z)$ be a solution of Laplace's equation which can be expanded into a power series in three variables valid for points of (x, y, z) sufficiently near a given point (x_0, y_0, z_0). Accordingly we write

$$x = x_0 + X, \qquad y = y_0 + Y, \qquad z = z_0 + Z;$$

and we assume the expansion

$$V = a_0 + a_1 X + b_1 Y + c_1 Z + a_2 X^2 + b_2 Y^2 + c_2 Z^2 + 2d_2 YZ + 2e_2 ZX + 2f_2 XY + \cdots,$$

it being supposed that this series is absolutely convergent whenever

$$|X|^2 + |Y|^2 + |Z|^2 \le a,$$

where a is some positive constant (the functions of applied mathematics satisfy this condition). If this expansion exists, V is said to be analytic at (x_0, y_0, z_0). It can be proved by the methods of §§3.7, 4.7 that the series converges uniformly throughout the domain indicated and may be differentiated term-by-term with regard to X, Y or Z any number of times at points inside the domain.

If we substitute the expansion in Laplace's equation, which may be written

$$\frac{\partial^2 V}{\partial X^2} + \frac{\partial^2 V}{\partial Y^2} + \frac{\partial^2 V}{\partial Z^2} = 0,$$

and equate to zero (§3.73) the coefficients of the various powers of X, Y and Z, we get an infinite set of linear relations between the coefficients, of which

$$a_2 + b_2 + c_2 = 0$$

may be taken as typical.

There are $\frac{1}{2} n(n-1)$ of these relations[1] between the $\frac{1}{2}(n+2)(n+1)$ coefficients of the terms of degree n in the expansion of V, so that there are only $\frac{1}{2}(n+2)(n+1) - \frac{1}{2} n(n-1) = 2n+1$ independent coefficients in the terms of degree n in V. Hence the terms of degree n in V must be a linear combination of $2n+1$ linearly independent particular solutions of Laplace's equation, these solutions being each of degree n in X, Y and Z.

To find a set of such solutions, consider $(Z + iX \cos u + iY \sin u)^n$; it is a solution of Laplace's equation which may be expanded in a series of sines and cosines of multiples of u, thus:

$$\sum_{m=0}^{n} g_m(X, Y, Z) \cos mu + \sum_{m=1}^{n} h_m(X, Y, Z) \sin mu,$$

the functions $g_m(X, Y, Z)$ and $h_m(X, Y, Z)$ being independent of u. The highest power of Z in $g_m(X, Y, Z)$ and $h_m(X, Y, Z)$ is Z^{n-m} and the former function is an even function of Y, the latter an odd function, hence the functions are linearly independent. They therefore form a set of $2n+1$ functions of the type sought.

Now by Fourier's rule[2] (§9.12)

$$\pi g_m(X, Y, Z) = \int_{-\pi}^{\pi} (Z + iX \cos u + iY \sin u)^n \cos mu \, du,$$

$$\pi h_m(X, Y, Z) = \int_{-\pi}^{\pi} (Z + iX \cos u + iY \sin u)^n \sin mu \, du,$$

and so any linear combination of the $2n+1$ solutions can be written in the form

$$\int_{-\pi}^{\pi} (Z + iX \cos u + iY \sin u)^n f_n(u) \, du,$$

where $f_n(u)$ is a rational function of e^{iu}.

Now it is readily verified that, if the terms of degree n in the expression assumed for V

---

[1] If $a_{r,s,t}$ (where $r + s + t = n$) be the coefficient of $X^r Y^s Z^t$ in V, and if the terms of degree $n-2$ in

$$\frac{\partial^2 V}{\partial X^2} + \frac{\partial^2 V}{\partial Y^2} + \frac{\partial^2 V}{\partial Z^2}$$

be arranged primarily in powers of X and secondarily in powers of Y, the coefficient $a_{r,s,t}$ does not occur in any term after $X^{r-2} Y^s Z^t$ (or $X^r Y^{s-2} Z^t$ if $r = 0$ or 1), and hence the relations are all linearly independent.
[2] 2π must be written for π in the coefficient of $g_0(X, Y, Z)$.

be written in this form, the series of terms under the integral sign converges uniformly if $|X|^2 + |Y|^2 + |Z|^2$ be sufficiently small, and so (§4.7) we may write

$$V = \int_{-\pi}^{\pi} \sum_{n=0}^{\infty} (Z + iX \cos u + iY \sin u)^n f_n(u) du.$$

But any expression of this form may be written

$$V = \int_{-\pi}^{\pi} F(Z + iX \cos u + iY \sin u, u) \, du,$$

where F is a function such that differentiations with regard to X, Y or Z under the sign of integration are permissible. And, conversely, if F be any function of this type, V is a solution of Laplace's equation.

This result may be written

$$V = \int_{-\pi}^{\pi} f(z + ix \cos u + iy \sin u, u) \, du,$$

on absorbing the terms $-z_0 - ix_0 \cos u - iy_0 \sin u$ into the second variable; and, if differentiations under the sign of integration are permissible, this gives a general solution of Laplace's equation; that is to say, every solution of Laplace's equation which is analytic throughout the interior of some sphere is expressible by an integral of the form given.

This result is the three-dimensional analogue of the theorem that

$$V = j(x + iy) + g(x - iy)$$

is the general solution of

$$\frac{\partial^2 V}{\partial x^2} + \frac{\partial^2 V}{\partial y^2} = 0.$$

Remark 18.3.1 A distinction has to be drawn between the primitive of an ordinary differential equation and general integrals of a partial differential equation of order higher than the first. For a discussion of general integrals of such equations, see Forsyth [219, chapter 12].

Two apparently distinct primitives are always directly transformable into one another by means of suitable relations between the constants; thus in the case of $\frac{d^2 y}{dx^2} + y = 0$, we can obtain the primitive $C \sin(x + \varepsilon)$ from $A \cos x + B \sin x$ by defining C and ε by the equations $C \sin \varepsilon = A$, $C \cos \varepsilon = B$. On the other hand, every solution of Laplace's equation is expressible in each of the forms

$$\int_{-\pi}^{\pi} f(x \cos t + y \sin t + iz, t) \, dt, \qquad \int_{-\pi}^{\pi} g(y \cos u + z \sin u + ix, u) \, du;$$

but if these are known to be the same solution, there appears to be no general analytical relation, connecting the functions f and g, which will directly transform one form of the solution into the other.

Example 18.3.1 Shew that the potential of a particle of unit mass at (a, b, c) is

$$\frac{1}{2\pi} \int_{-\pi}^{\pi} \frac{du}{(z - c) + i(x - a) \cos u + i(y - b) \sin u}$$

at all points for which $z > c$.

Example 18.3.2   Shew that a general solution of Laplace's equation of zero degree in x, y, z is

$$\int_{-\pi}^{\pi} \log(x \cos t + y \sin t + iz)g(t)\, dt,$$

if $\int_{-\pi}^{\pi} g(t)\, dt = 0$. Express the solutions $\dfrac{x}{z+r}$ and $\log \dfrac{r+z}{r-z}$ in this form, where $r^2 = x^2 + y^2 + z^2$.

Example 18.3.3   Shew that, in the case of the equation

$$p^{1/2} + q^{1/2} = x + y$$

(where $p = \partial z/\partial x$, $q = \partial z/\partial y$), integrals of Charpit's subsidiary equations (see Forsyth [221, chapter 9]), are

(i) $p^{1/2} - x = y - q^{1/2} = a$,

(ii) $p = q + a^2$.

Deduce that the corresponding general integrals are derived from

(i) $z = \frac{1}{3}(x + a)^3 + \frac{1}{3}(y - a)^3 + F(a)$, $0 = (x + a)^2 - (y - a)^2 + F'(a)$;

(ii) $4z = \frac{1}{3}(x + y)^3 + 2a^2(x - y) - a^4(x + y)^{-1} + G(a)$, $0 = 4a(x - y) - 4a^3(x + y)^{-1} + G'(a)$

and thence obtain a differential equation determining the function $G(a)$ in terms of the function $F(a)$ when the two general integrals are the same.

18.31 Solutions of Laplace's equation involving Legendre functions

If an expansion for V, of the form assumed in §18.3, exists when

$$x_0 = y_0 = z_0 = 0,$$

we have seen that we can express V as a series of expressions of the type

$$\int_{-\pi}^{\pi} (z + ix \cos u + iy \sin u)^n \cos mu\, du,$$

$$\int_{-\pi}^{\pi} (z + ix \cos u + iy \sin u)^n \sin mu\, du,$$

where n and m are integers such that $0 \le m \le n$. We shall now examine these expressions more closely.

If we take polar coordinates, defined by the equations

$$x = r \sin\theta \cos\phi, \qquad y = r \sin\theta \sin\phi, \qquad z = r \cos\theta,$$

we have

$$\int_{-\pi}^{\pi} (z + ix \cos u + iy \sin u)^n \cos mu \, du$$

$$= r^n \int_{-\pi}^{\pi} \{\cos\theta + i \sin\theta \cos(u - \phi)\}^n \cos mu \, du$$

$$= r^n \int_{-\pi-\phi}^{\pi-\phi} \{\cos\theta + i \sin\theta \cos\psi\}^n \cos m(\phi + \psi) \, d\psi$$

$$= r^n \int_{-\pi}^{\pi} \{\cos\theta + i \sin\theta \cos\psi\}^n \cos m(\phi + \psi) \, d\psi$$

$$= r^n \cos m\phi \int_{-\pi}^{\pi} \{\cos\theta + i \sin\theta \cos\psi\}^n \cos m\psi \, d\psi,$$

since the integrand is a periodic function of ψ and

$$(\cos\theta + i \sin\theta \cos\psi)^n \sin m\psi$$

is an odd function of ψ. Therefore (§15.61), with Ferrers' definition of the associated Legendre function,

$$\int_{-\pi}^{\pi} (z + ix \cos u + iy \sin u)^n \cos mu \, du = \frac{2\pi i^m n!}{(n+m)!} r^n P_n^m(\cos\theta) \cos m\phi.$$

Similarly

$$\int_{-\pi}^{\pi} (z + ix \cos u + iy \sin u)^n \sin mu \, du = \frac{2\pi i^m \cdot n!}{(n+m)!} r^n P_n{}^m(\cos\theta) \sin m\phi.$$

Therefore $r^n P_n^m(\cos\theta) \cos m\phi$ and $r^n P_n^m(\cos\theta) \sin m\phi$ are polynomials in x, y, z and are particular solutions of Laplace's equation. Further, by §18.3, every solution of Laplace's equation, which is analytic near the origin, can be expressed in the form

$$V = \sum_{n=0}^{\infty} r^n \left\{ A_n P_n(\cos\theta) + \sum_{m=1}^{n} (A_n^{(m)} \cos m\phi + B_n^{(m)} \sin m\phi) P_n^m(\cos\theta) \right\}.$$

Any expression of the form

$$A_n P_n(\cos\theta) + \sum_{m=1}^{n} (A_n^{(m)} \cos m\phi + B_n^{(m)} \sin m\phi) P_n^m(\cos\theta),$$

where n is a positive integer, is called a *surface harmonic* of degree n; a surface harmonic of degree n multiplied by r^n is called a *solid harmonic* (or a *spherical harmonic*) of degree n.

The curves on a unit sphere (with centre at the origin) on which $P_n(\cos\theta)$ vanishes are n parallels of latitude which divide the surface of the sphere into zones, and so $P_n(\cos\theta)$ is called (see §15.1) a *zonal harmonic*; and the curves on which $\begin{Bmatrix} \cos \\ \sin \end{Bmatrix} m\phi \cdot P_n^m(\cos\theta)$ vanishes are $n - m$ parallels of latitude and $2m$ meridians, which divide the surface of the sphere into quadrangles whose angles are right angles, and so these functions are called *tesseral harmonics*.

A solid harmonic of degree n is evidently a homogeneous polynomial of degree n in x, y, z and it satisfies Laplace's equation.

It is evident that, if a change of rectangular coordinates[3] is made by rotating the axes about the origin, a solid harmonic (or a surface harmonic) of degree n transforms into a solid harmonic (or a surface harmonic) of degree n in the new coordinates.

Spherical harmonics were investigated with the aid of Cartesian coordinates by W. Thomson in 1862, see [626] and Thomson and Tait [628, pp. 171–218]; they were also investigated independently in the same manner at about the same time by Clebsch [149].

Example 18.3.4 If coordinates r, θ, ϕ are defined by the equations

$$x = r \cos \theta, \quad y = (r^2 - 1)^{1/2} \sin \theta \cos \phi, \quad z = (r^2 - 1)^{1/2} \sin \theta \sin \phi,$$

shew that $P_n^m(r) P_n^m(\cos \theta) \cos m\phi$ satisfies Laplace's equation.

18.4 The solution of Laplace's equation which satisfies assigned boundary conditions at the surface of a sphere

We have seen (§18.31) that any solution of Laplace's equation which is analytic near the origin can be expanded in the form

$$V(r, \theta, \phi) =$$

$$\sum_{n=0}^{\infty} r^n \left\{ A_n P_n(\cos \theta) + \sum_{m=1}^{n} (A_n^{(m)} \cos m\phi + B_n^{(m)} \sin m\phi) P_n^m(\cos \theta) \right\};$$

and, from §3.7, it is evident that if it converges for a given value of r, say a, for all values of θ and ϕ such that $0 \le \theta \le \pi$, $-\pi \le \phi \le \pi$, it converges absolutely and uniformly when $r < a$.

To determine the constants, we must know the boundary conditions which V must satisfy. A boundary condition of frequent occurrence is that V is a given bounded integrable function of θ and ϕ, say $f(\theta, \phi)$, on the surface of a given sphere, which we take to have radius a, and V is analytic at points inside this sphere.

We then have to determine the coefficients $A_n, A_n^{(m)}, B_n^{(m)}$ from the equation

$$f(\theta, \phi) = \sum_{n=0}^{\infty} a^n \left\{ A_n P_n(\cos \theta) + \sum_{m=1}^{n} (A_n^{(m)} \cos m\phi + B_n^{(m)} \sin m\phi) P_n^m(\cos \theta) \right\}.$$

Assuming that this series converges uniformly throughout the domain $0 \le \theta \le \pi$, $-\pi \le \phi \le \pi$, (this is usually the case in physical problems), multiplying by

$$P_n^m(\cos \theta) \begin{Bmatrix} \cos \\ \sin \end{Bmatrix} m\phi,$$

integrating term-by-term (§4.7) and using the results of §§15.14 and 15.51 on the integral

---

[3] Laplace's operator $\dfrac{\partial^2}{\partial x^2} + \dfrac{\partial^2}{\partial y^2} + \dfrac{\partial^2}{\partial z^2}$ is invariant for changes of rectangular axes.

properties of Legendre functions, we find that

$$\int_{-\pi}^{\pi} \int_{0}^{\pi} f(\theta', \phi') P_n^m(\cos \theta') \cos m\phi' \sin \theta' \, d\theta' \, d\phi' = \pi a^n \frac{2}{2n+1} \frac{(n+m)!}{(n-m)!} A_n^{(m)},$$

$$\int_{-\pi}^{\pi} \int_{0}^{\pi} f(\theta', \phi') P_n^m(\cos \theta') \sin m\phi' \sin \theta' \, d\theta' \, d\phi' = \pi a^n \frac{2}{2n+1} \cdot \frac{(n+m)!}{(n-m)!} B_n^{(m)},$$

$$\int_{-\pi}^{\pi} \int_{0}^{\pi} f(\theta', \phi') P_n(\cos \theta') \sin \theta' \, d\theta' \, d\phi' = 2\pi a^n \frac{2}{2n+1} A_n.$$

Therefore, when $r < a$,

$$V(r, \theta, \phi) = \sum_{n=0}^{\infty} \frac{2n+1}{4\pi} \left(\frac{r}{a}\right)^n \int_{-\pi}^{\pi} \int_{0}^{\pi} f(\theta', \phi') \left\{ P_n(\cos \theta) P_n(\cos \theta') \right.$$

$$\left. + 2 \sum_{m=1}^{n} \frac{(n-m)!}{(n+m)!} P_n{}^m(\cos \theta) P_n^m(\cos \theta') \cos m(\phi - \phi') \right\} \sin \theta' \, d\theta' \, d\phi'.$$

The series which is here integrated term-by-term converges uniformly when $r < a$, since the expression under the integral sign is a bounded function of $\theta, \theta', \phi, \phi'$, and so (§4.7)

$$4\pi V(r, \theta, \phi) = \int_{-\pi}^{\pi} \int_{0}^{\pi} f(\theta', \phi') \sum_{n=0}^{\infty} (2n+1) \left(\frac{r}{a}\right)^n \left\{ P_n(\cos \theta) P_n(\cos \theta') \right.$$

$$\left. + 2 \sum_{m=1}^{n} \frac{(n-m)!}{(n+m)!} P_n^m(\cos \theta) P_n^m(\cos \theta') \cos m(\phi - \phi') \right\} \sin \theta' \, d\theta' \, d\phi'.$$

Now suppose that we take the line (θ, ϕ) as a new polar axis and let (θ_1', ϕ_1') be the new coordinates of the line whose old coordinates were (θ', ϕ'); we consequently have to replace $P_n(\cos \theta)$ by 1 and $P_n^m(\cos \theta)$ by zero; and so we get

$$4\pi V(r, \theta, \phi) = \int_{-\pi}^{\pi} \int_{0}^{\pi} f(\theta', \phi') \sum_{n=0}^{\infty} (2n+1) \left(\frac{r}{a}\right)^n P_n(\cos \theta_1') \sin \theta_1' \, d\theta_1' \, d\phi_1'$$

$$= \int_{-\pi}^{\pi} \int_{0}^{\pi} f(\theta', \phi') \sum_{n=0}^{\infty} (2n+1) \left(\frac{r}{a}\right)^n P_n(\cos \theta_1') \sin \theta' \, d\theta' \, d\phi'.$$

If, in this formula, we make use of the result of Example 15.23 of Chapter 15 we get

$$4\pi V(r, \theta, \phi) = \int_{-\pi}^{\pi} \int_{0}^{\pi} f(\theta', \phi') \frac{a(a^2 - r^2) \sin \theta' d\theta' d\phi'}{(r^2 - 2ar \cos \theta_1' + a^2)^{\frac{3}{2}}},$$

and so

$$V(r, \theta, \phi) = \frac{a(a^2 - r^2)}{4\pi} \int_{-\pi}^{\pi} \int_{0}^{\pi} \frac{f(\theta', \phi') \sin \theta' d\theta' d\phi'}{[r^2 - 2ar\{\cos \theta \cos \theta' + \sin \theta \sin \theta' \cos(\phi - \phi')\} + a^2]^{\frac{3}{2}}}.$$

In this compact formula the Legendre functions have ceased to appear explicitly.

The last formula can be obtained by the theory of *Green's functions*. For properties of such functions the reader is referred to Thomson and Tait [628].

Remark 18.4.1   From the integrals for $V(r, \theta, \phi)$ involving Legendre functions of $\cos \theta_1'$ and of $\cos \theta$, $\cos \theta'$ respectively, we can obtain a new proof of the addition theorem for the Legendre polynomial.

For let

$$\chi_n(\theta', \phi') = P_n(\cos \theta_1') - \left\{ P_n(\cos \theta) P_n(\cos \theta') \right.$$

$$\left. + 2 \sum_{m=1}^{n} \frac{(n-m)!}{(n+m)!} P_n{}^m(\cos \theta) P_n{}^m(\cos \theta') \cos m(\phi - \phi') \right\},$$

and we get, on comparing the two formulae for $V(r, \theta, \phi)$,

$$0 = \int_{-\pi}^{\pi} \int_{0}^{\pi} f(\theta', \phi') \sum_{n=0}^{\infty} (2n+1) \left(\frac{r}{a} \right)^n \chi_n(\theta', \phi') \sin \theta' \, d\theta' \, d\phi'.$$

If we take $f(\theta', \phi')$ to be a surface harmonic of degree n, the term involving r^n is the only one which occurs in the integrated series; and in particular, if we take $f(\theta', \phi') = \chi_n(\theta', \phi')$, we get

$$\int_{-\pi}^{\pi} \int_{0}^{\pi} \{\chi_n(\theta', \phi')\}^2 \sin \theta' \, d\theta' \, d\phi' = 0.$$

Since the integrand is continuous and is not negative it must be zero; and so $\chi_n(\theta', \phi') \equiv 0$; that is to say we have proved the formula

$$P_n(\cos \theta_1') = P_n(\cos \theta) P_n(\cos \theta') + 2 \sum_{m=1}^{n} \frac{(n-m)!}{(n+m)!} P_n^m(\cos \theta) P_n^m(\cos \theta') \cos m(\phi - \phi'),$$

wherein it is obvious that

$$\cos \theta_1' = \cos \theta \cos \theta' + \sin \theta \sin \theta' \cos(\phi - \phi'),$$

from geometrical considerations.

We have thus obtained a physical proof of a theorem proved elsewhere (§15.7) by purely analytical reasoning. (The absence of the factor $(-1)^m$ which occurs in §15.7 is due to the fact that the functions now employed are Ferrers' associated functions.)

Example 18.4.1   Find the solution of Laplace's equation analytic inside the sphere $r = 1$ which has the value $\sin 3\theta \cos \phi$ at the surface of the sphere.
Solution. $\frac{8}{15} r^3 P_3^1(\cos \theta) \cos \phi - \frac{1}{5} r P_1^1(\cos \theta) \cos \phi$.

Example 18.4.2   Let $f_n(r, \theta, \phi)$ be equal to a homogeneous polynomial of degree n in x, y, z. Shew that

$$\int_{-\pi}^{\pi} \int_{0}^{\pi} f_n(a, \theta, \phi) P_n\{\cos \theta \cos \theta' + \sin \theta \sin \theta' \cos(\phi - \phi')\} a^2 \sin \theta \, d\theta \, d\phi$$

$$= \frac{4\pi a^2}{2n+1} f_n(a, \theta', \phi').$$

Hint. Take the direction (θ', ϕ') as a new polar axis.

18.5 Solutions of Laplace's equation which involve Bessel coefficients

A particular case of the result of §18.3 is that

$$\int_{-\pi}^{\pi} e^{k(z+ix\cos u+iy\sin u)} \cos mu\, du$$

is a solution of Laplace's equation, k being any constant and m being any integer.

Taking cylindrical-polar coordinates (ρ, ϕ, z) defined by the equations

$$x = \rho\cos\phi, \quad y = \rho\sin\phi,$$

the above solution becomes

$$e^{kz}\int_{-\pi}^{\pi} e^{ik\rho\cos(u-\phi)} \cos mu\, du = e^{kz}\int_{-\pi}^{\pi} e^{ik\rho\cos v} \cos m(v+\phi)\cdot dv$$

$$= 2e^{kz}\int_{0}^{\pi} e^{ik\rho\cos v} \cos mv\cos m\phi\, dv$$

$$= 2e^{kz}\cos(m\phi)\int_{0}^{\pi} e^{ik\rho\cos v} \cos mv\, dv,$$

and so, using Example 17.1.3 *we see that* $2\pi i^m e^{kz}\cos(m\phi)J_m(k\rho)$ *is a solution of Laplace's equation analytic near the origin.*

Similarly, from the expression

$$\int_{-\pi}^{\pi} e^{k(z+ix\cos u+iy\sin u)} \sin mu\, du,$$

where m is an integer, *we deduce that* $2\pi i^m e^{kz} \sin(m\phi)J_m(k\rho)$ *is a solution of Laplace's equation.*

18.51 The periods of vibration of a uniform membrane

This is based upon Euler [205], Poisson [533] and Bourget [95]. For a detailed discussion of vibrations of membranes, see also Rayleigh [550]. The equation satisfied by the displacement V at time t of a point (x, y) of a uniform plane membrane vibrating harmonically is

$$\frac{\partial^2 V}{\partial x^2} + \frac{\partial^2 V}{\partial y^2} = \frac{1}{c^2}\frac{\partial^2 V}{\partial t^2},$$

where c is a constant depending on the tension and density of the membrane. The equation can be reduced to Laplace's equation by the change of variable given by $z = cti$. It follows, from §18.5, that expressions of the form

$$J_m(k\rho)\begin{Bmatrix}\sin\\\cos\end{Bmatrix}(m\phi)\begin{Bmatrix}\sin\\\cos\end{Bmatrix}(ckt)$$

satisfy the equation of motion of the membrane.

Take as a particular case a drum, that is to say a membrane with a fixed circular boundary of radius R. Then one possible type of vibration is given by the equation

$$V = J_m(k\rho)\cos m\phi\cos ckt,$$

provided that $V = 0$ when $\rho = R$; so that we have to choose k to satisfy the equation

$$J_m(kR) = 0.$$

This equation to determine k has an infinite number of real roots (Example 17.3.3), $k_1, k_2, k_3, \ldots$ say. A possible type of vibration is then given by

$$V = J_m(k_r \rho) \cos m\phi \cos ck_r t \qquad (r = 1, 2, 3, \ldots).$$

This is a periodic motion with period $2\pi/(ck_r)$; and so the calculation of the periods depends essentially on calculating the zeros of Bessel coefficients (see §17.9).

Example 18.5.1   The equation of motion of air in a circular cylinder vibrating perpendicularly to the axis OZ of the cylinder is

$$\frac{\partial^2 V}{\partial x^2} + \frac{\partial^2 V}{\partial y^2} = \frac{1}{c^2} \frac{\partial^2 V}{\partial t^2},$$

V denoting the velocity potential. If the cylinder have radius R, the boundary condition is that $\dfrac{\partial V}{\partial \rho} = 0$ when $\rho = R$. Shew that the determination of the free periods depends on finding the zeros of $J'_m(\zeta) = 0$.

18.6 A general solution of the equation of wave motions

It may be shewn[4] by the methods of §18.3 that *a general solution of the equation of wave motions*

$$\frac{\partial^2 V}{\partial x^2} + \frac{\partial^2 V}{\partial y^2} + \frac{\partial^2 V}{\partial z^2} = \frac{1}{c^2} \frac{\partial^2 V}{\partial t^2}$$

is

$$V = \int_{-\pi}^{\pi} \int_{-\pi}^{\pi} f(x \sin u \cos v + y \sin u \sin v + z \cos u + ct, u, v) \, du \, dv,$$

where f is a function (of three variables) of the type considered in §18.3.

Regarding an integral as a limit of a sum, we see that a physical interpretation of this equation is that the velocity potential V is produced by a number of plane waves, the disturbance represented by the element

$$f(x \sin u \cos v + y \sin u \sin v + z \cos u + ct, u, v) \, \delta u \, \delta v$$

being propagated in the direction $(\sin u \cos v, \sin u \sin v, \cos u)$ with velocity c. *The solution therefore represents an aggregate of plane waves travelling in all directions with velocity c.*

18.61 Solutions of the equation of wave motions which involve Bessel functions

We shall now obtain a class of particular solutions of the equation of wave motions, useful for the solution of certain special problems.

In physical investigations, it is desirable to have the time occurring by means of a factor

---

[4] See the paper previously cited [672, p. 342–345] or [646].

$\sin ckt$ or $\cos ckt$, where k is constant. This suggests that we should consider solutions of the type

$$V = \int_{-\pi}^{\pi} \int_{0}^{\pi} e^{ik(x \sin u \cos v + y \sin u \sin v + z \cos u + ct)} f(u, v) \, du \, dv.$$

Physically this means that we consider motions in which all the elementary waves have the same period.

Now let the polar coordinates of (x, y, z) be (r, θ, ϕ) and let (ω, ψ) be the polar coordinates of the direction (u, v) referred to new axes such that the polar axis is the direction (θ, ϕ), and the plane $\psi = 0$ passes through OZ; so that

$$\cos \omega = \cos \theta \cos u + \sin \theta \sin u \cos(\phi - v),$$

$$\sin u \sin(\phi - v) = \sin \omega \sin \psi.$$

Also, take the arbitrary function $f(u, v)$ to be $S_n(u, v) \sin u$, where S_n denotes a surface harmonic in u, v of degree n; so that we may write

$$S_n(u, v) = \bar{S}_n(\theta, \phi; \omega, \psi),$$

where (§18.31) $\bar{S}_n$ is a surface harmonic in ω, ψ of degree n. We thus get

$$V = e^{ikct} \int_{-\pi}^{\pi} \int_{0}^{\pi} e^{ikr \cos \omega} \bar{S}_n(\theta, \phi; \omega, \psi) \sin \omega \, d\omega \, d\psi.$$

Now we may write (§18.31)

$$\bar{S}_n(\theta, \phi; \omega, \psi) = A_n(\theta, \phi) P_n(\cos \omega)$$

$$+ \sum_{m=1}^{n} \left\{ A_n^{(m)}(\theta, \phi) \cos m\psi + B_n^{(m)}(\theta, \phi) \sin m\psi \right\} P_n^m(\cos \omega),$$

where $A_n(\theta, \phi)$, $A_n^{(m)}(\theta, \phi)$ and $B_n^{(m)}(\theta, \phi)$ are independent of ψ and ω.

Performing the integration with respect to ψ, we get

$$V = 2\pi e^{ikct} A_n(\theta, \phi) \int_{0}^{\pi} e^{ikr \cos \omega} P_n(\cos \omega) \sin \omega \, d\omega$$

$$= 2\pi e^{ikct} A_n(\theta, \phi) \int_{-1}^{1} e^{ikr\mu} P_n(\mu) \, d\mu$$

$$= 2\pi e^{ikct} A_n(\theta, \phi) \int_{-1}^{1} e^{ikr\mu} \frac{1}{2^n n!} \frac{d^n}{d\mu^n} (\mu^2 - 1)^n \, d\mu,$$

by Rodrigues' formula (§15.11); on integrating by parts n times and using Hankel's integral (Corollary 17.3.1), we obtain the equation

$$V = \frac{2\pi}{2^n n} e^{ikct} A_n(\theta, \phi)(ikr)^n \int_{-1}^{1} e^{ikr\mu} (1 - \mu^2)^n \, d\mu$$

$$= (2\pi)^{\frac{3}{2}} i^n e^{ikct} (kr)^{-\frac{1}{2}} J_{n+\frac{1}{2}}(kr) A_n(\theta, \phi),$$

and so V is a constant multiple of $e^{ikct} r^{-\frac{1}{2}} J_{n+\frac{1}{2}}(kr) A_n(\theta, \phi)$.

Now the equation of wave motions is unaffected if we multiply x, y, z and t by the same constant factor, i.e. if we multiply r and t by the same constant factor leaving θ and ϕ

unaltered; so that $A_n(\theta, \phi)$ may be taken to be independent of the arbitrary constant k which multiplies r and t.

Hence

$$\lim_{k \to 0} e^{ikct} r^{-\frac{1}{2}} k^{-n-\frac{1}{2}} J_{n+\frac{1}{2}}(kr) A_n(\theta, \phi)$$

is a solution of the equation of wave motions; and therefore $r^n A_n(\theta, \phi)$ is a solution (independent of t) of the equation of wave motions, and is consequently a solution of Laplace's equation; it is, accordingly, permissible to take $A_n(\theta, \phi)$ to be any surface harmonic of degree n; *and so we obtain the result that*

$$r^{-1/2} J_{n+\frac{1}{2}}(kr) P_n^m(\cos\theta) \begin{Bmatrix} \cos \\ \sin \end{Bmatrix} (m\phi) \begin{Bmatrix} \cos \\ \sin \end{Bmatrix} (ckt)$$

is a particular solution of the equation of wave motions.

18.611 Application of §18.61 to a physical problem

The solution just obtained for the equation of wave motions may be used in the following manner to determine the periods of free vibration of air contained in a rigid sphere.

The velocity potential V satisfies the equation of wave motions and the boundary condition is that $\dfrac{\partial V}{\partial r} = 0$ when $r = a$, where a is the radius of the sphere. Hence

$$V = r^{-1/2} J_{n+\frac{1}{2}}(kr) P_n^m(\cos\theta) \begin{Bmatrix} \cos \\ \sin \end{Bmatrix} (m\phi) \begin{Bmatrix} \cos \\ \sin \end{Bmatrix} (ckt)$$

gives a possible motion if k is so chosen that

$$\frac{d}{dr} \{r^{-1/2} J_{n+\frac{1}{2}}(kr)\}_{r=a} = 0.$$

This equation determines k; on using §17.24, we see that it may be written in the form

$$\tan ka = f_n(ka),$$

where $f_n(ka)$ is a rational function of ka.

In particular the radial vibrations, in which V is independent of θ and ϕ, are given by taking $n = 0$; then the equation to determine k becomes simply

$$\tan ka = ka;$$

and the pitches of the fundamental radial vibrations correspond to the roots of this equation.

18.7 Miscellaneous examples

Example 18.1 If V be a solution of Laplace's equation which is symmetrical with respect to OZ, and if $V = f\{z\}$ on OZ, shew that if $f(\zeta)$ be a function which is analytic in a domain of values (which contains the origin) of the complex variable ζ, then

$$V = \frac{1}{\pi} \int_0^\pi f\{z + i(x^2 + y^2)^{\frac{1}{2}} \cos\phi\} \, d\phi$$

at any point of a certain three-dimensional region.

Deduce that the potential of a uniform circular ring of radius c and of mass M lying in the plane XOY with its centre at the origin is

$$\frac{M}{\pi} \int_0^\pi [c^2 + \{z + i(x^2 + y^2)^{\frac{1}{2}} \cos \phi\}^2]^{-\frac{1}{2}} \, d\phi.$$

Example 18.2 (Dougall) If V be a solution of Laplace's equation, which is of the form $e^{mi\phi} F(\rho, z)$, where (ρ, ϕ, z) are cylindrical coordinates, and if this solution is approximately equal to $\rho^m e^{mi\phi} f(z)$ near the axis of z, where $f(\zeta)$ is of the character described in Example 18.1, shew that

$$V = \frac{m! \, \rho^m \, e^{mi\phi}}{\Gamma(m + \frac{1}{2})\Gamma(\frac{1}{2})} \int_0^\pi f(z + i\rho \cos t) \sin^{2m} t \, dt.$$

Example 18.3 (Forsyth [217]) If u be determined as a function of x, y and z by means of the equation

$$Ax + By + Cz = 1,$$

where A, B, C are functions of u such that

$$A^2 + B^2 + C^2 = 0,$$

shew that (subject to certain general conditions) any function of u is a solution of Laplace's equation.

Example 18.4 (Sylvester [616]) A, B are two points outside a sphere whose centre is C. A layer of attracting matter on the surface of the sphere is such that its surface density σ_P at P is given by the formula

$$\sigma_P \to (AP \cdot BP)^{-1}.$$

Shew that the total quantity of matter is unaffected by varying A and B so long as $CA \cdot CB$ and $A\hat{C}B$ are unaltered; and prove that this result is equivalent to the theorem that the surface integral of two harmonics of different degrees taken over the sphere is zero.

Example 18.5 (Appell [32]) Let $V(x, y, z)$ be the potential function defined analytically as due to particles of masses $\lambda + i\mu$, $\lambda - i\mu$ at the points $(a + ia', b + ib', c + ic')$ and $(a - ia', b - ib', c - ic')$ respectively. Shew that $V(x, y, z)$ is infinite at all points of a certain real circle, and if the point (x, y, z) describes a circuit intertwined once with this circle the initial and final values of $V(x, y, z)$ are numerically equal, but opposite in sign.

Example 18.6 Find the solution of Laplace's equation analytic in the region for which $a < r < A$, it being given that on the spheres $r = a$ and $r = A$ the solution reduces to

$$\sum_{n=0}^{\infty} c_n P_n(\cos \theta), \quad \sum_{n=0}^{\infty} C_n P_n(\cos \theta),$$

respectively.

Example 18.7 (Trinity, 1893) Let O' have coordinates $(0, 0, c)$, and let

$$P\hat{O}Z = \theta, \quad P\hat{O}'Z = \theta', \quad PO = r, \quad PO' = r'.$$

Shew that

$$\frac{P_n(\cos\theta')}{r'^{n+1}} = \begin{cases} \frac{P_n(\cos\theta)}{r^{n+1}} + (n+1)\frac{cP_{n+1}(\cos\theta)}{r^{n+2}} + \frac{(n+1)(n+2)}{2!}\frac{c^2 P_{n+2}(\cos\theta)}{r^{n+3}} + \cdots, & \text{if } r > c \\ (-1)^n \left\{ \frac{1}{c^{n+1}} + (n+1)\frac{r P_1(\cos\theta)}{c^{n+2}} + \frac{(n+1)(n+2)}{2!}\frac{r^2 P_2(\cos\theta)}{c^{n+3}} + \cdots \right\} & \text{if } r < c. \end{cases}$$

Obtain a similar expansion for $r'^m P'_n(\cos\theta)$.

Example 18.8 (St John's, 1899)   At a point (r, θ, ϕ) outside a uniform oblate spheroid whose semi-axes are a, b and whose density is ρ, shew that the potential is

$$4\pi\rho a^2 b \left[\frac{1}{3r} - \frac{m^2}{3\cdot 5}\frac{P_2(\cos\theta)}{r^3} + \frac{m^4}{5\cdot 7}\frac{P_4(\cos\theta)}{r^5} - \cdots \right],$$

where $m^2 = a^2 - b^2$ and $r > m$. Obtain the potential at points for which $r < m$,

Example 18.9 (Bauer [57])   Shew that

$$e^{ir\cos\theta} = \left(\tfrac{1}{2}\pi\right)^{\frac{1}{2}} \sum_{n=0}^{\infty} i^n (2n+1) r^{-\frac{1}{2}} P_n(\cos\theta) J_{n+\frac{1}{2}}(r).$$

Example 18.10 (Lamé)   Shew that if $x\pm iy = h\cos h(\xi\pm i\eta)$, the equation of two-dimensional wave motions in the coordinates ξ and η is

$$\frac{\partial^2 V}{\partial\xi^2} + \frac{\partial^2 V}{\partial\eta^2} = \frac{h^2}{c^2}(\cosh^2\xi - \cos^2\eta)\frac{\partial^2 V}{\partial t^2}.$$

Note   Examples 18.10, 18.11, 18.12 and 18.14 are most easily proved by using Lamé's result, [400], that if (λ, μ, ν) be orthogonal coordinates for which the line-element is given by the formula

$$(\delta x)^2 + (\delta y)^2 + (\delta z)^2 = (H_1\delta\lambda)^2 + (H_2\delta\mu)^2 + (H_3\delta\nu)^2,$$

Laplace's equation in these coordinates is

$$\frac{\partial}{\partial\lambda}\left(\frac{H_2 H_3}{H_1}\frac{\partial V}{\partial\lambda}\right) + \frac{\partial}{\partial\mu}\left(\frac{H_3 H_1}{H_2}\frac{\partial V}{\partial\mu}\right) + \frac{\partial}{\partial\nu}\left(\frac{H_1 H_2}{H_3}\frac{\partial V}{\partial\nu}\right) = 0.$$

A simple method (due to W. Thomson [625]) of proving this result, by means of arguments of a physical character, is reproduced by Lamb [398, §111]. Analytical proofs, based on Lamé's proof, are given by Bertrand [68, p. 181–187] and Goursat [256, p. 155–159]; and a most compact proof is due to Neville [491]. Another proof is given by Heine [287, p. 303–306].

Example 18.11 (Niven [503])   Let $x = (c + r\cos\theta)\cos\phi$, $y = (c + r\cos\theta)\sin\phi$, $z = r\sin\theta$; shew that the surfaces for which r, θ, ϕ respectively are constant form an orthogonal system; and shew that Laplace's equation in the coordinates r, θ, ϕ is

$$\frac{\partial}{\partial r}\left\{ r(c + r\cos\theta)\frac{\partial V}{\partial r} \right\} + \frac{1}{r}\frac{\partial}{\partial\theta}\left\{ (c + r\cos\theta)\frac{\partial V}{\partial\theta} \right\} + \frac{r}{c + r\cos\theta}\frac{\partial^2 V}{\partial\phi^2} = 0.$$

Example 18.12 (Hicks [303])   Let P have cartesian coordinates (x, y, z) and polar coordinates (r, θ, ϕ). Let the plane POZ meet the circle $x^2 + y^2 = k^2$, $z = 0$ in the points α, γ; and let

$$\alpha\hat{P}_\gamma = \omega, \quad \log(P\alpha/P\gamma) = \sigma.$$

Shew that Laplace's equation in the coordinates σ, ω, ϕ is

$$\frac{\partial}{\partial \sigma} \left\{ \frac{\sinh \sigma}{\cosh \sigma - \cos \omega} \frac{\partial V}{\partial \sigma} \right\} + \frac{\partial}{\partial \omega} \left\{ \frac{\sinh \sigma}{\cosh \sigma - \cos \omega} \frac{\partial V}{\partial \omega} \right\}$$

$$+ \frac{1}{\sinh \sigma (\cosh \sigma - \cos \omega)} \frac{\partial^2 V}{\partial \phi^2} = 0;$$

and shew that a solution is

$$V = (\cosh \sigma - \cos \omega)^{\frac{1}{2}} \cos n\omega \cos m\phi P^m_{n-\frac{1}{2}} (\cosh \sigma).$$

Example 18.13   Shew that

$$(R^2 + \rho^2 - 2R\rho \cos \phi + c^2)^{-\frac{1}{2}} = \sum_{m=0}^{\infty} \frac{e^{-\frac{1}{2}m\pi i}}{\pi} \int_0^{\infty} dk \int_{-\pi}^{\pi} e^{-ck} J_m(k\rho) e^{ikR\cos n} \cos mu\, du,$$

and deduce an expression for the potential of a particle in terms of Bessel functions.

Example 18.14 (Lamé)   Shew that if a, b, c are constants and λ, μ, ν are confocal coordinates, defined as the roots of the equation in ε

$$\frac{x^2}{a^2 + \varepsilon} + \frac{y^2}{b^2 + \varepsilon} + \frac{z^2}{c^2 + \varepsilon} = 1,$$

then Laplace's equation may be written

$$\Delta_\lambda (\mu - \nu) \frac{\partial}{\partial \lambda} \left\{ \Delta_\lambda \frac{\partial V}{\partial \lambda} \right\} + \Delta_\mu (\nu - \lambda) \frac{\partial}{\partial \mu} \left\{ \Delta_\mu \frac{\partial V}{\partial \mu} \right\} + \Delta_\nu (\lambda - \mu) \frac{\partial}{\partial \nu} \left\{ \Delta_\nu \frac{\partial V}{\partial \nu} \right\} = 0,$$

where $\Delta_\lambda = \sqrt{(a^2 + \lambda)(b^2 + \lambda)(c^2 + \lambda)}$.

Example 18.15 (Bateman [53])   Shew that a general solution of the equation of wave motions is

$$V = \int_{-\pi}^{\pi} F(x \cos \theta + y \sin \theta + iz, y + iz \sin \theta + ct \cos \theta, \theta)\, d\theta.$$

Example 18.16   If $U = f(x, y, z, t)$ be a solution of

$$\frac{1}{a^2} \frac{\partial U}{\partial t} = \frac{\partial^2 U}{\partial x^2} + \frac{\partial^2 U}{\partial y^2} + \frac{\partial^2 U}{\partial z^2},$$

prove that another solution of the equation is

$$U = t^{-\frac{3}{2}} f \left(\frac{x}{t}, \frac{y}{t}, \frac{z}{t}, -\frac{1}{t} \right) \exp \left(-\frac{x^2 + y^2 + z^2}{4a^2 t} \right).$$

Example 18.17 (Bateman [53])   Shew that a general solution of the equation of wave motions, when the motion is independent of ϕ, is

$$\int_{-\pi}^{\pi} f(z + i\rho \cos \theta, ct + \rho \sin \theta)\, d\theta$$

$$+ \int_0^b \int_{-\pi}^{\pi} \operatorname{arcsin} h \left(\frac{a + z + ct \cos \theta}{\rho \sin \theta} \right) F(a, \theta)\, d\theta\, da,$$

where ρ, ϕ, z are cylindrical coordinates and a, b are arbitrary constants.

Example 18.18 (Bateman [54])   If $V = f(x, y, z)$ is a solution of Laplace's equation, shew that

$$V = \frac{1}{(x - iy)^{\frac{1}{2}}} f\left(\frac{r^2 - a^2}{2(x - iy)}, \frac{r^2 + a^2}{2i(x - iy)}, \frac{az}{x - iy}\right)$$

is another solution.

Example 18.19 (Bateman [54])   If $U = f(x, y, z, t)$ is a solution of the equation of wave motions, shew that another solution is

$$U = \frac{1}{z - ct} f\left(\frac{x}{z - ct}, \frac{y}{z - ct}, \frac{r^2 - 1}{2(z - ct)}, \frac{r^2 + 1}{2c(z - ct)}\right).$$

Example 18.20 (Bateman [54])   If $l = x - iy$, $m = z + i\omega$, $n = x^2 + y^2 + z^2 + \omega^2$ and $\lambda = x + iy$, $\mu = z - i\omega$, $\nu = -1$, so that $l\lambda + m\mu + n\nu = 0$, shew that any homogeneous solution, of degree zero, of

$$\frac{\partial^2 U}{\partial x^2} + \frac{\partial^2 U}{\partial y^2} + \frac{\partial^2 U}{\partial z^2} + \frac{\partial^2 U}{\partial \omega^2} = 0$$

satisfies

$$\frac{\partial^2 U}{\partial l \partial \lambda} + \frac{\partial^2 U}{\partial m \partial \mu} + \frac{\partial^2 U}{\partial n \partial \nu} = 0;$$

and obtain a solution of this equation in the form

$$l^{-\alpha} \lambda^{-\alpha'} m^{-\beta} \mu^{-\beta'} n^{-\gamma} \nu^{-\gamma'} P \begin{Bmatrix} a, & b, & c \\ \alpha, & \beta, & \gamma, & \zeta \\ \alpha', & \beta', & \gamma' \end{Bmatrix},$$

where $l\lambda = (b - c)(\zeta - a)$, $m\mu = (c - a)(\zeta - b)$, $n\nu = (a - b)(\zeta - c)$.

Example 18.21 (Blades [75])   *Note.* The functions introduced in this example and the next are known as *internal* and *external spheroidal harmonics* respectively.

If (r, θ, ϕ) are spheroidal coordinates, defined by the equations

$$x = c(r^2 + 1)^{\frac{1}{2}} \sin\theta \cos\phi, \qquad y = c(r^2 + 1)^{\frac{1}{2}} \sin\theta \sin\phi, \qquad z = cr \cos\theta,$$

where x, y, z are rectangular coordinates and c is a constant, shew that, when n and m are integers,

$$\int_{-\pi}^{\pi} P_n\left(\frac{x \cos t + y \sin t + iz}{c}\right) \begin{Bmatrix} \cos \\ \sin \end{Bmatrix} (mt)\, dt$$

$$= 2\pi \frac{(n - m)!}{(n + m)!} P_n^m(ir) P_n^m(\cos\theta) \begin{Bmatrix} \cos \\ \sin \end{Bmatrix} (m\phi).$$

Example 18.22 (Jeffery [358])   With the notation of Example 18.21, shew that, if $z \neq 0$,

$$\int_{-\pi}^{\pi} Q_n\left(\frac{x \cos t + y \sin t + iz}{c}\right) \begin{Bmatrix} \cos \\ \sin \end{Bmatrix} (mt)\, dt = 2\pi \frac{(n - m)!}{(n + m)!} Q_n^m(ir) P_n^m(\cos\theta) \begin{Bmatrix} \cos \\ \sin \end{Bmatrix} (m\phi).$$

Example 18.23 (Donkin [186]; Hobson [322])   Prove that the most general solution of Laplace's equation which is of degree zero in x, y, z is expressible in the form

$$V = f\left(\frac{x + iy}{r + z}\right) + F\left(\frac{x - iy}{r + z}\right),$$

where f and F are arbitrary functions.

19

Mathieu Functions

19.1 The differential equation of Mathieu

The preceding five chapters have been occupied with the discussion of functions which belong to what may be generally described as the hypergeometric type, and many simple properties of these functions are now well known.

In the present chapter we enter upon a region of Analysis which lies beyond this, and which is, as yet, only very imperfectly explored.

The functions which occur in Mathematical Physics and which come next in order of complication to functions of hypergeometric type are called *Mathieu functions*; these functions are also known as *the functions associated with the elliptic cylinder*. They arise from the equation of two-dimensional wave motion, namely

$$\frac{\partial^2 V}{\partial x^2} + \frac{\partial^2 V}{\partial y^2} = \frac{1}{c^2}\frac{\partial^2 V}{\partial t^2}.$$

This partial differential equation occurs in the theory of the propagation of electromagnetic waves; if the electric vector in the wave-front is parallel to OZ and if E denotes the electric force, while $(H_x, H_y, 0)$ are the components of magnetic force, Maxwell's fundamental equations are

$$\frac{1}{c^2}\frac{\partial E}{\partial t} = \frac{\partial H_y}{\partial x} - \frac{\partial H_x}{\partial y}, \qquad \frac{\partial H_x}{\partial t} = -\frac{\partial E}{\partial y}, \qquad \frac{\partial H_y}{\partial t} = \frac{\partial E}{\partial x},$$

c denoting the velocity of light; and these equations give at once

$$\frac{1}{c^2}\frac{\partial^2 E}{\partial t^2} = \frac{\partial^2 E}{\partial x^2} + \frac{\partial^2 E}{\partial y^2}.$$

In the case of the scattering of waves, propagated parallel to OX, incident on an elliptic cylinder for which OX and OY are axes of a principal section, the boundary condition is that E should vanish at the surface of the cylinder.

The same partial differential equation occurs in connexion with the vibrations of a uniform plane membrane, the dependent variable being the displacement perpendicular to the membrane; if the membrane be in the shape of an ellipse with a rigid boundary, the boundary condition is the same as in the electromagnetic problem just discussed.

The differential equation was discussed by Mathieu [457, p. 137] in 1868 in connexion with the problem of vibrations of an elliptic membrane in the following manner: Suppose that the membrane, which is in the plane XOY when it is in equilibrium, is vibrating with

426

frequency p. Then, if we write

$$V = u(x, y)\cos(pt + \varepsilon),$$

the equation becomes

$$\frac{\partial^2 u}{\partial x^2} + \frac{\partial^2 u}{\partial y^2} + \frac{p^2}{c^2} u = 0.$$

Let the foci of the elliptic membrane be $(\pm h, 0, 0)$, and introduce new real variables. (The introduction of these variables is due to Lamé, who called ξ the *thermometric parameter*. They are more usually known as *confocal coordinates*. See Lamé [404, 1ere Leçon].) Let ξ, η, defined by the complex equation

$$x + iy = h\cosh(\xi + i\eta),$$

so that $x = h\cosh\xi\cos\eta$, $y = h\sinh\xi\sin\eta$. The curves, on which ξ or η is constant, are evidently ellipses or hyperbolas confocal with the boundary; if we take $\xi \geq 0$ and $-\pi < \eta \leq \pi$, to each point $(x, y, 0)$ of the plane corresponds one and only one[1] value of (ξ, η).

The differential equation for u transforms into[2]

$$\frac{\partial^2 u}{\partial \xi^2} + \frac{\partial^2 u}{\partial \eta^2} + \frac{h^2 p^2}{c^2}\left(\cosh^2\xi - \cos^2\eta\right) u = 0.$$

If we assume a solution of this equation of the form

$$u = F(\xi)\,G(\eta),$$

where the factors are functions of ξ only and of η only respectively, we see that

$$\left\{\frac{1}{F(\xi)}\frac{d^2 F(\xi)}{d\xi^2} + \frac{h^2 p^2}{c^2}\cosh^2\xi\right\} = -\left\{\frac{1}{G(\eta)}\frac{d^2 G(\eta)}{d\eta^2} - \frac{h^2 p^2}{c^2}\cos^2\eta\right\}.$$

Since the left-hand side contains ξ but not η, while the right-hand side contains η but not ξ, $F(\xi)$ and $G(\eta)$ must be such that each side is a constant, A, say, since ξ and η are independent variables. We thus arrive at the equations

$$\frac{d^2 F(\xi)}{d\xi^2} + \left(\frac{h^2 p^2}{c^2}\cosh^2\xi - A\right)F(\xi) = 0, \qquad \frac{d^2 G(\eta)}{d\eta^2} - \left(\frac{h^2 p^2}{c^2}\cos^2\eta - A\right)G(\eta) = 0.$$

By a slight change of independent variable in the former equation, we see that *both of these equations are linear differential equations, of the second order, of the form*

$$\frac{d^2 u}{dz^2} + (a + 16q\cos 2z)u = 0,$$

where a and q are constants. (Their actual values are $a = A - h^2 p^2/(2c^2)$, $q = h^2 p^2/(32c^2)$; the factor 16 is inserted to avoid powers of 2 in the solution.) It is obvious that every point (infinity excepted) is a regular point of this equation.

This is the equation which is known as *Mathieu's equation* and, in certain circumstances (§19.2), particular solutions of it are called *Mathieu functions*.

[1] This may be seen most easily by considering the ellipses obtained by giving ξ various positive values. If the ellipse be drawn through a definite point (ξ, η) of the plane, η is the eccentric angle of that point on the ellipse.

[2] A proof of this result, due to Lamé, is given in numerous textbooks; see the Note to Example 18.10, Chapter 18.

19.11 The form of the solution of Mathieu's equation

In the physical problems which suggested Mathieu's equation, the constant a is not given
a priori, and we have to consider how it is to be determined. It is obvious from physical
considerations in the problem of the membrane that $u(x, y)$ is a *one-valued* function of
position, and is consequently unaltered by increasing η by 2π; and the condition[3] $G(\eta+2\pi) =$
$G(\eta)$ is sufficient to determine a set of values of a in terms of q. And it will appear later
(§§19.4, sub:19.41) that, when a has not one of these values, the equation

$$G(\eta + 2\pi) = G(\eta)$$

is no longer true.

When a is thus determined, q (and thence p) is determined by the fact that $F(\xi) = 0$ on
the boundary; and so the periods of the free vibrations of the membrane are obtained.

Other problems of Mathematical Physics which involve Mathieu functions in their solution
are

  (i) Tidal waves in a cylindrical vessel with an elliptic boundary,
 (ii) Certain forms of steady vortex motion in an elliptic cylinder,
(iii) The decay of magnetic force in a metal cylinder (Maclaurin [450]).

The equation also occurs in a problem of Rigid Dynamics which is of general interest (Young
[685]).

19.12 Hill's equation

A differential equation, similar to Mathieu's but of a more general nature, arises in Hill's
[306] method of determining the motion of the Lunar Perigee[4], and in Adams' determination
of the motion of the Lunar Node [8]. Hill's equation is

$$\frac{d^2u}{dz^2} + \left(\theta_0 + 2\sum_{n=1}^{\infty} \theta_n \cos 2nz\right) u = 0.$$

The theory of Hill's equation is very similar to that of Mathieu's (in spite of the increase in
generality due to the presence of the infinite series), so the two equations will, to some extent,
be considered together. In the astronomical applications $\theta_0, \theta_1, \ldots$ are *known* constants, so
the problem of choosing them in such a way that the solution may be periodic does not arise.
The solution of Hill's equation in the Lunar Theory is, in fact, not periodic.

19.2 Periodic solutions of Mathieu's equation

We have seen that in physical (as distinguished from astronomical) problems the constant a
in Mathieu's equation has to be chosen to be such a function of q that the equation possesses
a periodic solution.

Let this solution be $G(z)$; then $G(z)$, in addition to being periodic, is an integral function
of z. Three possibilities arise as to the nature of $G(z)$:

---

[3] An elementary analogue of this result is that a solution of $\frac{d^2u}{dz^2} + au = 0$ has period 2π if, and only if, a is the
     square of an integer.
[4] Hill's memoir was originally published in 1877 at Cambridge, U.S.A.

(i) $G(z)$ may be an *even* function of z;

(ii) $G(z)$ may be an *odd* function of z;

(iii) $G(z)$ may be neither even nor odd.

In case (iii), $\frac{1}{2}\{G(z) + G(-z)\}$ is an *even* periodic solution and $\frac{1}{2}\{G(z) - G(-z)\}$ is an *odd* periodic solution of Mathieu's equation, these two solutions forming a fundamental system. It is therefore sufficient to confine our attention to periodic solutions of Mathieu's equation which are either even or odd. These solutions, *and these only*, will be called *Mathieu functions*.

It will be observed that, since the roots of the indicial equation at $z = 0$ are 0 and 1, two even (or two odd) periodic solutions of Mathieu's equation cannot form a fundamental system. But, so far, there seems to be no reason why Mathieu's equation, for special values of a and q, should not have one even and one odd periodic solution; for comparatively small values of $|q|$ it can be seen [Example 19.3.3, (ii) and (iii)] that Mathieu's equation has two periodic solutions only in the trivial case in which $q = 0$; the result that there are never pairs of periodic solutions for larger values of $|q|$ is a special case of a theorem due to Hille [309]. See also Ince [335].

19.21 An integral equation satisfied by even Mathieu functions

It will now be shewn that, if $G(\eta)$ is any even Mathieu function, *then* $G(\eta)$ *satisfies the homogeneous integral equation*

$$G(\eta) = \lambda \int_{-\pi}^{\pi} e^{k \cos \eta \cos \theta} G(\theta)\, d\theta,$$

where $k = \sqrt{32q}$. This result is suggested by the solution of Laplace's equation given in §18.3. This integral equation and the expansions of §19.3 were published by Whittaker [673] in 1912. The integral equation was known to him as early as 1904 [55, p. 193].

For, if $x + iy = h \cosh(\xi + i\eta)$ and if $F(\xi)$ and $G(\eta)$ are solutions of the differential equations

$$\frac{d^2 F(\xi)}{d\xi^2} - (A + m^2 h^2 \cosh^2 \xi)F(\xi) = 0, \qquad \frac{d^2 G(\eta)}{d\eta^2} + (A + m^2 h^2 \cos^2 \eta)G(\eta) = 0,$$

then, by §19.1, $F(\xi)G(\eta)e^{miz}$ is a particular solution of Laplace's equation. If this solution is a special case of the general solution

$$\int_{-\pi}^{\pi} f(h \cosh \xi \, \cos \eta \, \cos \theta + h \sinh \xi \, \sin \eta \, \sin \theta + iz, \theta)\, d\theta,$$

given in §18.3, it is natural to expect that

$$f(\upsilon, \theta) \equiv F(0)e^{m\upsilon} \phi(\theta),$$

where $\phi(\theta)$ is a function of θ to be determined. The constant $F(0)$ is inserted to simplify the algebra. Thus

$$F(\xi)G(\eta)e^{miz} = \int_{-\pi}^{\pi} F(0)\phi(\theta) \exp \{mh \cosh \xi \, \cos \eta \, \cos \theta + mh \sinh \xi \, \sin \eta \, \sin \theta + miz\}\, d\theta.$$

Since ξ and η are independent, we may put $\xi = 0$; and we are thus led to consider the possibility of Mathieu's equation possessing a solution of the form

$$G(\eta) = \int_{-\pi}^{\pi} e^{mh \cos \eta \cos \theta} \phi(\theta) \, d\theta.$$

19.22 *Proof that the even Mathieu functions satisfy the integral equation*

It is readily verified (§5.31) that, if $\phi(\theta)$ be analytic in the range $(-\pi, \pi)$ and if $G(\eta)$ be *defined* by the equation

$$G(\eta) = \int_{-\pi}^{\pi} e^{mh \cos \eta \cos \theta} \phi(\theta) \, d\theta,$$

then $G(\eta)$ is an even periodic integral function of η and

$$\frac{d^2 G(\eta)}{d\eta^2} + (A + m^2 h^2 \cos^2 \eta) G(\eta)$$

$$= \int_{-\pi}^{\pi} \left\{ m^2 h^2 (\sin^2 \eta \cos^2 \theta + \cos^2 \eta) - mh \cos \eta \cos \theta + A \right\} e^{mh \cos \eta \cos \theta} \phi(\theta) \, d\theta$$

$$= - \left[\{ mh \sin \theta \cos \eta \, \phi(\theta) + \phi'(\theta) \} e^{mh \cos \eta \cos \theta} \right]_{-\pi}^{\pi}$$

$$+ \int_{-\pi}^{\pi} \left\{ \phi''(\theta) + (A + m^2 h^2 \cos^2 \theta) \phi(\theta) \right\} e^{mh \cos \eta \cos \theta} \, d\theta,$$

on integrating by parts.

But if $\phi(\theta)$ be a periodic *function (with period 2π) such that*

$$\phi''(\theta) + (A + m^2 h^2 \cos^2 \theta) \phi(\theta) = 0,$$

both the integral and the integrated part vanish; that is to say, $G(\eta)$, defined by the integral, is a periodic solution of Mathieu's equation.

Consequently $G(\eta)$ is an even periodic solution of Mathieu's equation if $\phi(\theta)$ is a periodic solution of Mathieu's equation formed with the same constants; and therefore $\phi(\theta)$ is a constant multiple of $G(\theta)$; let it be $\lambda G(\theta)$. In the case when the Mathieu equation has two periodic solutions, if this case exist, we have $\phi(\theta) = \lambda G(\theta) + G_1(\theta)$ where $G_1(\theta)$ is an odd periodic function; but

$$\int_{-\pi}^{\pi} e^{mh \cos \eta \cos \theta} G_1(\theta) \, d\theta$$

vanishes, so the subsequent work is unaffected.

If we take a and q as the parameters of the Mathieu equation instead of A and mh, it is obvious that $mh = \sqrt{32q} = k$. We have thus proved that, if $G(\eta)$ be an even periodic solution of Mathieu's equation, then

$$G(\eta) = \lambda \int_{-\pi}^{\pi} e^{k \cos \eta \cos \theta} G(\theta) \, d\theta,$$

which is the result stated in §19.21.

From §11.23, it is known that this integral equation has a solution only when λ has one of the 'characteristic values'. It will be shewn in §19.3 that for such values of λ, the integral equation affords a simple means of constructing the even Mathieu functions.

Example 19.2.1   Shew that the odd Mathieu functions satisfy the integral equation

$$G(\eta) = \lambda \int_{-\pi}^{\pi} \sin(k \sin \eta \sin \theta) \, G(\theta) \, d\theta.$$

Example 19.2.2   Shew that both the even and the odd Mathieu functions satisfy the integral equation

$$G(\eta) = \lambda \int_{-\pi}^{\pi} e^{ik \sin \eta \sin \theta} G(\theta) \, d\theta.$$

Example 19.2.3   Shew that when the eccentricity of the fundamental ellipse tends to zero, the confluent form of the integral equation for the even Mathieu functions is

$$J_n(x) = \frac{1}{2\pi i^n} \int_{-\pi}^{\pi} e^{ix \cos \theta} \cos n\theta \, d\theta.$$

19.3 The construction of Mathieu functions

We shall now make use of the integral equation of §19.21 to construct Mathieu functions; the canonical form of Mathieu's equation will be taken as

$$\frac{d^2u}{dz^2} + (a + 16q \cos 2z)u = 0.$$

In the special case when q is zero, the periodic solutions are obtained by taking $a = n^2$, where n is any integer; the solutions are then

$$1, \cos z, \cos 2z, \ldots, \sin z, \sin 2z, \ldots.$$

The Mathieu functions, which reduce to these when $q \to 0$, will be called

$$\mathrm{ce}_0(z, q), \ \mathrm{ce}_1(z, q), \ \mathrm{ce}_2(z, q), \ldots, \mathrm{se}_1(z, q), \ \mathrm{se}_2(z, q), \ldots.$$

To make the functions precise, we take the coefficients of $\cos nz$ and $\sin nz$ in the respective Fourier series for $\mathrm{ce}_n(z, q)$ and $\mathrm{se}_n(z, q)$ to be unity. The functions $\mathrm{ce}_n(z, q)$, $\mathrm{se}_n(z, q)$ will be called *Mathieu functions of order n*.

Let us now construct $\mathrm{ce}_0(z, q)$. Since $\mathrm{ce}_0(z, 0) = 1$, we see that $\lambda \to (2\pi)^{-1}$ as $q \to 0$. Accordingly we suppose that, for general values of q, the characteristic value of λ which gives rise to $\mathrm{ce}_0(z, q)$ can be expanded in the form

$$(2\pi\lambda)^{-1} = 1 + \alpha_1 q + \alpha_2 q^2 + \cdots,$$

and that $\mathrm{ce}_0(z, q) = 1 + q\beta_1(z) + q^2\beta_2(z) + \cdots$, where $\alpha_1, \alpha_2, \ldots$ are numerical constants and $\beta_1(z), \beta_2(z), \ldots$ are periodic functions of z which are independent of q and which contain no constant term.

On substituting in the integral equation, we find that

$$(1 + \alpha_1 q + \alpha_2 q^2 + \cdots) \left(1 + q\beta_1(z) + q^2\beta_2(z) + \cdots\right)$$
$$= \frac{1}{2\pi} \int_{-\pi}^{\pi} \left(1 + \sqrt{32q} \cos z \cos \theta + 16q \cos^2 z \cos^2 \theta + \cdots\right)$$
$$\times \left(1 + q\beta_1(\theta) + q^2\beta_2(\theta) + \cdots\right) d\theta.$$

Equating coefficients of successive powers of q in this result and making use of the fact that $\beta_1(z), \beta_2(z), \ldots$ contain no constant term, we find in succession

$$\alpha_1 = 4, \qquad \beta_1(z) = 4\cos 2z,$$

$$\alpha_2 = 14, \qquad \beta_2(z) = 2\cos 4z,$$

$$\vdots \qquad\qquad \vdots$$

and we thus obtain the following expansion:

$$\mathrm{ce}_0(z,q) = 1 + \left(4q - 28q^3 + \tfrac{2^7 \cdot 29}{9}q^5 - \cdots\right)\cos 2z + \left(2q^2 - \tfrac{160}{9}q^4 + \cdots\right)\cos 4z$$
$$+ \left(\tfrac{4}{9}q^3 - \tfrac{13}{3}q^5 + \cdots\right)\cos 6z + \left(\tfrac{1}{18}q^4 - \cdots\right)\cos 8z + \left(\tfrac{1}{225}q^5 - \cdots\right)\cos 10z + \cdots,$$

the terms not written down being $O(q^6)$ as $q \to 0$.

The value of a is

$$-32q^2 + 224q^4 - \frac{2^{10}\cdot 29}{9}q^6 + O(q^8);$$

it will be observed that the coefficient of $\cos 2z$ in the series for $\mathrm{ce}_0(z,q)$ is $-a/(8q)$.

The Mathieu functions of higher order may be obtained in a similar manner from the same integral equation and from the integral equation of Example 19.2.1. The consideration of the convergence of the series thus obtained is postponed to §19.61.

Example 19.3.1 (Whittaker)   Obtain the following expansions:

(i) $\;\mathrm{ce}_0(z,q) = 1 + \displaystyle\sum_{r=1}^{\infty} \left\{ \frac{2^{r+1}q^r}{r!\,r!} - \frac{2^{r+3}r(3r+4)q^{r+2}}{(r+1)!(r+1)!} + O(q^{r+4}) \right\}\cos 2rz;$

(ii)

$$\mathrm{ce}_1(z,q) = \cos z + \sum_{r=1}^{\infty} \left\{ \frac{2^r q^r}{(r+1)!r!} - \frac{2^{r+1}rq^{r+1}}{(r+1)!(r+1)!} \right.$$
$$\left. + \frac{2^r q^{r+2}}{(r-1)!(r+2)!} + O(q^{r+3}) \right\}\cos(2r+1)z;$$

(iii)

$$\mathrm{se}_1(z,q) = \sin z + \sum_{r=1}^{\infty} \left\{ \frac{2^r q^r}{(r+1)!\,r!} + \frac{2^{r+1}rq^{r+1}}{(r+1)!(r+1)!} \right.$$
$$\left. + \frac{2^r q^{r+2}}{(r-1)!\,(r+2)!} + O(q^{r+3}) \right\}\sin(2r+1)z,$$

(iv)

$$\mathrm{ce}_2(z,q) = \left\{ -2q + \frac{40}{3}q^3 + O(q^5) \right\} + \cos 2z$$
$$+ \sum_{r=1}^{\infty} \left\{ \frac{2^{r+1}q^r}{r!\,(r+2)!} + \frac{2^{r+1}r(47r^2 + 222r + 247)\,q^{r+2}}{3^2 \cdot (r+2)!(r+3)!} + O(q^{r+4}) \right\}\cos(2r+2)z,$$

where, in each case, the constant implied in the symbol O depends on r but not on z. The leading terms of these series, as given in Example 19.4 at the end of the chapter, were obtained by Mathieu.

Example 19.3.2 (Mathieu) Shew that the values of a associated with (i) $\mathrm{ce}_0(z,q)$, (ii) $\mathrm{ce}_1(z,q)$, (iii) $\mathrm{se}_1(z,q)$, (iv) $\mathrm{ce}_2(z,q)$ are respectively:

(i) $-32q^2 + 224q^4 - \frac{2^{10}\cdot 29}{9}q^6 + O(q^8)$;

(ii) $1 - 8q - 8q^2 + 8q^3 - \frac{8}{3}q^4 + O(q^5)$;

(iii) $1 + 8q - 8q^2 - 8q^3 - \frac{8}{3}q^4 + O(q^5)$;

(iv) $4 + \frac{80}{3}q^2 - \frac{6104}{27}q^4 + O(q^6)$.

Example 19.3.3 Shew that, if n be an integer,

$$\mathrm{ce}_{2n+1}(z,q) = (-1)^n \, \mathrm{se}_{2n+1}(z + \tfrac{1}{2}\pi, -q).$$

19.31 The integral formulae for the Mathieu functions

Since all the Mathieu functions satisfy a homogeneous integral equation with a symmetrical nucleus (Example 19.2.3), it follows (§11.61) that

$$\int_{-\pi}^{\pi} \mathrm{ce}_m(z,q)\,\mathrm{ce}_n(z,q)\,dz = 0 \qquad (m \neq n),$$

$$\int_{-\pi}^{\pi} \mathrm{se}_m(z,q)\,\mathrm{se}_n(z,q)\,dz = 0 \qquad (m \neq n),$$

$$\int_{-\pi}^{\pi} \mathrm{ce}_m(z,q)\,\mathrm{se}_n(z,q)\,dz = 0.$$

Example 19.3.4 Obtain expansions of the form:

$$\exp(k \cos z \cos \theta) = \sum_{n=0}^{\infty} A_n \, \mathrm{ce}_n(z,q)\,\mathrm{ce}_n(\theta,q),$$

$$\cos(k \sin z \sin \theta) = \sum_{n=0}^{\infty} B_n \, \mathrm{ce}_n(z,q)\,\mathrm{ce}_n(\theta,q),$$

$$\sin(k \sin z \sin \theta) = \sum_{n=0}^{\infty} C_n \, \mathrm{se}_n(z,q)\,\mathrm{se}_n(\theta,q),$$

where $k = \sqrt{32q}$.

Example 19.3.5 Obtain the expansion

$$e^{iz \sin \phi} = \sum_{n=-\infty}^{\infty} J_n(z) e^{ni\phi}$$

as a confluent form of the last two expansions of Example 19.3.4.

19.4 The nature of the solution of Mathieu's general equation; Floquet's theory

We shall now discuss the nature of the solution of Mathieu's equation when the parameter *a* is no longer restricted so as to give rise to periodic solutions; this is the case which is of importance in astronomical problems, as distinguished from other physical applications of the theory.

The method is applicable to any linear equation with *periodic* coefficients which are one-valued functions of the independent variable; the nature of the general solution of particular equations of this type has long been perceived by astronomers, by inference from the circumstances in which the equations arise. These inferences have been confirmed by the following analytical investigation which was published in 1883 by Floquet [211, p. 47]. Floquet's analysis is a natural sequel to Picard's theory of differential equations with doubly-periodic coefficients (§20.1), and to the theory of the fundamental equation due to Fuchs and Hamburger.

Let $g(z)$, $h(z)$ be a fundamental system of solutions of Mathieu's equation (or, indeed, of any linear equation in which the coefficients have period 2π); then, if $F(z)$ be any other integral of such an equation, we must have

$$F(z) = Ag(z) + Bh(z),$$

where *A* and *B* are definite constants.

Since $g(z + 2\pi)$, $h(z + 2\pi)$ are obviously solutions of the equation[5], they can be expressed in terms of the continuations of $g(z)$ and $h(z)$ by equations of the type

$$g(z + 2\pi) = \alpha_1 g(z) + \alpha_2 h(z), \qquad h(z + 2\pi) = \beta_1 g(z) + \beta_2 h(z),$$

where α_1, α_2, β_1, β_2 are definite constants; and then

$$F(z + 2\pi) = (A\alpha_1 + B\beta_1)g(z) + (A\alpha_2 + B\beta_2)h(z).$$

Consequently $F(z + 2\pi) = kF(z)$, where k is a constant[6], if A and B are chosen so that

$$A\alpha_1 + B\beta_1 = kA, \qquad A\alpha_2 + B\beta_2 = kB.$$

These equations will have a solution, other than $A = B = 0$, if, and only if,

$$\begin{vmatrix} \alpha_1 - k & \beta_1 \\ \alpha_2 & \beta_2 - k \end{vmatrix} = 0;$$

and if *k* be taken to be either root of this equation, the function $F(z)$ can be constructed so as to be a solution of the differential equation such that

$$F(z + 2\pi) = kF(z).$$

Defining μ by the equation $k = e^{2\pi\mu}$ and writing $\phi(z)$ for $e^{-\mu z}F(z)$, we see that

$$\phi(z + 2\pi) = e^{-\mu(z+2\pi)}F(z + 2\pi) = \phi(z).$$

---

[5] These solutions may not be identical with $g(z)$, $h(z)$ respectively, as the solution of an equation with periodic coefficients is not necessarily periodic. To take a simple case, $u = e^z \sin z$ is a solution of $\frac{du}{dz} - (1 + \cot z)u = 0$.

[6] The symbol *k* is used in this particular sense only in this section. It must not be confused with the constant *k* of §19.21, which was associated with the parameter *q* of Mathieu's equation.

Hence the differential equation has a particular solution of the form $e^{\mu z}\phi(z)$, where $\phi(z)$ is a periodic function with period 2π.

We have seen that in physical problems, the parameters involved in the differential equation have to be so chosen that $k = 1$ is a root of the quadratic, and a solution is periodic. In general, however, in astronomical problems, in which the parameters are given, $k \neq 1$ and there is no periodic solution.

In the particular case of Mathieu's general equation or Hill's equation, a fundamental system of solutions[7] is then $e^{\mu z}\phi(z)$, $e^{-\mu z}\phi(-z)$, since the equation is unaltered by writing $-z$ for z; so that the complete solution of Mathieu's general equation is then

$$u = c_1 e^{\mu z}\phi(z) + c_2 e^{-\mu z}\phi(-z),$$

where c_1, c_2 are arbitrary constants, and μ is a definite function of a and q.

Example 19.4.1 Shew that the roots of the equation

$$\begin{vmatrix} \alpha_1 - k & \beta_1 \\ \alpha_2 & \beta_2 - k \end{vmatrix} = 0,$$

are independent of the particular pair of chosen solutions, $g(z)$ and $h(z)$.

19.41 Hill's method of solution

Now that the general functional character of the solution of equations with periodic coefficients has been found by Floquet's theory, it might be expected that the determination of an explicit expression for the solutions of Mathieu's and Hill's equations would be a comparatively easy matter; this however is not the case. For example, in the particular case of Mathieu's general equation, a solution has to be obtained in the form

$$y = e^{\mu z}\phi(z),$$

where $\phi(z)$ is periodic and μ is a function of the parameters a and q. The crux of the problem is to determine μ; when this is done, the determination of $\phi(z)$ presents comparatively little difficulty.

The first successful method of attacking the problem was published by Hill in the memoir cited in §19.12; since the method for Hill's equation is no more difficult than for the special case of Mathieu's general equation, we shall discuss the case of Hill's equation, viz.

$$\frac{d^2 u}{dz^2} + J(z)u = 0,$$

where $J(z)$ is an even function of z with period π. Two cases are of interest, the analysis being the same in each:

(I) The astronomical case when z is real and, for real values of z, $J(z)$ can be expanded in the form

$$J(z) = \theta_0 + 2\theta_1 \cos 2z + 2\theta_2 \cos 4z + 2\theta_3 \cos 6z + \cdots ;$$

the coefficients θ_n are known constants and $\sum_{n=0}^{\infty} \theta_n$ converges absolutely.

---

[7] The ratio of these solutions is not even periodic; still less is it a constant.

(II) The case when z is a complex variable and $J(z)$ is analytic in a strip of the plane (containing the real axis), whose sides are parallel to the real axis. The expansion of $J(z)$ in the Fourier series $\theta_0 + 2 \sum_{n=1}^{\infty} \theta_n \cos 2nz$ is then valid (§9.11) throughout the interior of the strip, and, as before, $\sum_{n=0}^{\infty} \theta_n$ converges absolutely.

Defining θ_{-n} to be equal to θ_n, we assume

$$u = e^{\mu z} \sum_{n=-\infty}^{\infty} b_n e^{2niz}$$

as a solution of Hill's equation.

Note In case (II) this is the solution analytic in the strip (§§10.2, 19.4); in case (I) it will have to be shewn ultimately (see the note at the end of §19.42) that the values of b_n which will be determined are such as to make $\sum_{n=-\infty}^{\infty} n^2 b_n$ absolutely convergent, in order to justify the processes which we shall now carry out.

On substitution in the equation, we find

$$\sum_{n=-\infty}^{\infty} (\mu + 2ni)^2 \, b_n e^{(\mu+2ni)z} + \left(\sum_{n=-\infty}^{\infty} \theta_n \, e^{2niz} \right)\left(\sum_{n=-\infty}^{\infty} b_n \, e^{(\mu+2ni)z} \right) = 0.$$

Multiplying out the absolutely convergent series and equating coefficients of powers of e^{2iz} to zero (§9.6, §9.632), we obtain the system of equations

$$(\mu + 2ni)^2 \, b_n + \sum_{n=-\infty}^{\infty} \theta_m \, b_{n-m} = 0 \qquad (n = \ldots, -2, -1, 0, 1, 2, \ldots).$$

If we eliminate the coefficients b_n determinantally (after dividing the typical equation by $\theta_0 - 4n^2$ to secure convergence) we obtain[8] Hill's determinantal equation:

$$\begin{vmatrix}
\vdots & \vdots & \vdots & \vdots & \vdots & \vdots & \vdots & \vdots & \vdots \\
\cdots & \frac{(i\mu+4)^2-\theta_0}{4^2-\theta_0} & \frac{-\theta_1}{4^2-\theta_0} & \frac{-\theta_2}{4^2-\theta_0} & \frac{-\theta_3}{4^2-\theta_0} & \frac{-\theta_4}{4^2-\theta_0} & \cdots \\
\cdots & \frac{-\theta_1}{2^2-\theta_0} & \frac{(i\mu+2)^2-\theta_0}{2^2-\theta_0} & \frac{-\theta_1}{2^2-\theta_0} & \frac{-\theta_2}{2^2-\theta_0} & \frac{-\theta_3}{2^2-\theta_0} & \cdots \\
\cdots & \frac{-\theta_2}{0^2-\theta_0} & \frac{-\theta_1}{0^2-\theta_0} & \frac{(i\mu)^2-\theta_0}{0^2-\theta_0} & \frac{-\theta_1}{0^2-\theta_0} & \frac{-\theta_2}{0^2-\theta_0} & \cdots \\
\cdots & \frac{-\theta_3}{2^2-\theta_0} & \frac{-\theta_2}{2^2-\theta_0} & \frac{-\theta_1}{2^2-\theta_0} & \frac{(i\mu-2)^2-\theta_0}{2^2-\theta_0} & \frac{-\theta_1}{2^2-\theta_0} & \cdots \\
\cdots & \frac{-\theta_4}{4^2-\theta_0} & \frac{-\theta_3}{4^2-\theta_0} & \frac{-\theta_2}{4^2-\theta_0} & \frac{-\theta_1}{4^2-\theta_0} & \frac{(i\mu-4)^2-\theta_0}{4^2-\theta_0} & \cdots \\
\vdots & \vdots & \vdots & \vdots & \vdots & \vdots & \vdots & \vdots & \vdots
\end{vmatrix} = 0.$$

We write $\Delta(i\mu)$ for the determinant, so the equation determining μ is

$$\Delta(i\mu) = 0.$$

---

[8] Since the coefficients b_n are not all zero, we may obtain the infinite determinant as the eliminant of the system of linear equations by multiplying these equations by suitably chosen cofactors and adding up.

19.42 The evaluation of Hill's determinant

We shall now obtain an extremely simple expression for Hill's determinant, namely

$$\Delta(i\mu) \equiv \Delta(0) - \sin^2\left(\frac{\pi i\mu}{2}\right)\operatorname{cosec}^2\left(\frac{\pi}{2}\sqrt{\theta_0}\right).$$

Adopting the notation of §2.8, we write

$$\Delta(i\mu) \equiv [A_{m,n}],$$

where

$$A_{m,n} = \frac{(i\mu - 2m)^2 - \theta_0}{4m^2 - \theta_0}, \qquad A_{m,n} = \frac{-\theta_{m-n}}{4m^2 - \theta_0} \qquad (m \neq n).$$

The determinant $[A_{m,n}]$ is only *conditionally* convergent, since the product of the principal diagonal elements does not converge absolutely (§§2.81, 2.7). We can, however, obtain an *absolutely* convergent determinant, $\Delta_1(i\mu)$, by dividing the linear equations of §19.41 by $\theta_0 - (i\mu - 2n)^2$ instead of dividing by $\theta_0 - 4n^2$. We write this determinant $\Delta_1(i\mu)$ in the form $[B_{m,n}]$, where

$$B_{m,m} = 1, \qquad B_{m,n} = \frac{-\theta_{m-n}}{(2m - i\mu)^2 - \theta_0} \qquad (m \neq n).$$

The absolute convergence of $\sum_{n=0}^{\infty} \theta_n$ secures the convergence of the determinant $[B_{m,n}]$, except when μ has such a value that the denominator of one of the expressions $B_{m,n}$ vanishes.

From the definition of an infinite determinant (§2.8) it follows that

$$\Delta(i\mu) = \Delta_1(i\mu) \lim_{p\to\infty} \prod_{n=-p}^{p} \left\{ \frac{\theta_0 - (i\mu - 2n)^2}{\theta_0 - 4n^2} \right\},$$

and so

$$\Delta(i\mu) = -\Delta_1(i\mu) \frac{\sin\frac{\pi}{2}(i\mu - \sqrt{\theta_0})\sin\frac{\pi}{2}(i\mu + \sqrt{\theta_0})}{\sin^2(\frac{\pi}{2}\sqrt{\theta_0})}.$$

Now, if the determinant $\Delta_1(i\mu)$ be written out in full, it is easy to see: (i) that $\Delta_1(i\mu)$ is an even periodic function of μ with period $2i$; (ii) that $\Delta_1(i\mu)$ is an analytic function (cf. §§2.81, 3.34, 5.3) of μ (except at its obvious simple poles), which tends to unity as the real part of μ tends to $\pm\infty$.

If now we choose the constant K so that the function $D(\mu)$, defined by the equation

$$D(\mu) \equiv \Delta_1(i\mu) - K \left\{ \cot\frac{\pi}{2}(i\mu + \sqrt{\theta_0}) - \cot\frac{\pi}{2}(i\mu - \sqrt{\theta_0}) \right\},$$

has no pole at the point $\mu = i\sqrt{\theta_0}$, then, since $D(\mu)$ is an even periodic function of μ, it follows that $D(\mu)$ has no pole at any of the points

$$2ni \pm i\sqrt{\theta_0},$$

where n is any integer.

The function $D(\mu)$ is therefore a periodic function of μ (with period $2i$) which has no poles, and which is obviously bounded as $\operatorname{Re}(\mu) \to \pm\infty$. The conditions postulated in Liouville's theorem (§5.63) are satisfied, and so $D(\mu)$ is a constant; making $\mu \to +\infty$, we see that this constant is unity.

Therefore

$$\Delta_1(i\mu) = 1 + K \left\{ \cot \tfrac{\pi}{2}(i\mu + \sqrt{\theta_0}) - \cot \tfrac{\pi}{2}(i\mu - \sqrt{\theta_0}) \right\},$$

and so

$$\Delta(i\mu) = - \frac{\sin \tfrac{\pi}{2}(i\mu - \sqrt{\theta_0}) \sin \tfrac{\pi}{2}(i\mu - \sqrt{\theta_0})}{\sin^2(\tfrac{\pi}{2}\sqrt{\theta_0})} + 2K \cot(\tfrac{\pi}{2}\sqrt{\theta_0}).$$

To determine K, put $\mu = 0$; then

$$\Delta(0) = 1 + 2K \cot(\tfrac{\pi}{2}\sqrt{\theta_0}).$$

Hence, on subtraction,

$$\Delta(i\mu) = \Delta(0) - \frac{\sin^2(\tfrac{\pi}{2}i\mu)}{\sin^2(\tfrac{\pi}{2}\sqrt{\theta_0})},$$

which is the result stated.

The roots of Hill's determinantal equation are therefore the roots of the equation

$$\sin^2 \left(\frac{\pi}{2}i\mu \right) = \Delta(0) \sin^2 \left(\frac{\pi}{2}\sqrt{\theta_0} \right).$$

When μ has thus been determined, the coefficients b_n can be determined in terms of b_0 and cofactors of $\Delta(i\mu)$; and the solution of Hill's differential equation is complete.

Note   In case (I) of §19.41, the convergence of $\sum |b_n|$ follows from the rearrangement theorem of §2.82; for $\sum n^2 |b_n|$ is equal to $|b_0| \sum\limits_{m=-\infty}^{\infty} |C_{m,0}| \div |C_{0,0}|$, where $C_{m,n}$ is the cofactor of $B_{m,n}$ in $\Delta(i\mu)$; and $\sum |C_{m,0}|$ is the determinant obtained by replacing the elements of the row through the origin by numbers whose moduli are bounded.

It was shewn by Hill that, for the purposes of his astronomical problem, a remarkably good approximation to the value of μ could be obtained by considering only the three central rows and columns of his determinant.

19.5 The Lindemann–Stieltjes theory of Mathieu's general equation

Up to the present, Mathieu's equation has been treated as a linear differential equation with periodic coefficients. Some extremely interesting properties of the equation have been obtained by Lindemann [437] by the substitution $\zeta = \cos^2 z$, which transforms the equation into an equation with rational coefficients, namely

$$4\zeta(1 - \zeta)\frac{d^2u}{d\zeta^2} + 2(1 - 2\zeta)\frac{du}{d\zeta} + (a - 16q + 32q\zeta)u = 0.$$

Note   This equation, though it somewhat resembles the hypergeometric equation, is of higher type than the equations dealt with in Chapters 14 and 16, inasmuch as it has two regular singularities at 0 and 1 and an irregular singularity at ∞; whereas the three singularities of the hypergeometric equation are all regular, while the equation for $W_{k,m}(z)$ has one irregular singularity and only one regular singularity.

We shall now give a short account of Lindemann's analysis, with some modifications due to Stieltjes [603]. The analysis is very similar to that employed by Hermite in his lectures at the École Polytechnique in 1872–1873 [297, p. 118–122] in connexion with Lamé's equation. See §23.7.

19.51 Lindemann's form of Floquet's theorem

Since Mathieu's equation (in Lindemann's form) has singularities at $\zeta = 0$ and $\zeta = 1$, the exponents at each being $0, \frac{1}{2}$, there exist solutions of the form

$$y_{00} = \sum_{n=0}^{\infty} a_n \zeta^n, \qquad y_{01} = \zeta^{1/2} \sum_{n=0}^{\infty} b_n \zeta^n,$$

$$y_{10} = \sum_{n=0}^{\infty} a'_n (1 - \zeta)^n, \qquad y_{11} = (1 - \zeta)^{1/2} \sum_{n=0}^{\infty} b'_n (1 - \zeta)^n;$$

the first two series converge when $|\zeta| < 1$, the last two when $|1 - \zeta| < 1$.

When the ζ-plane is cut along the real axis from 1 to $+\infty$ and from 0 to $-\infty$, the four functions defined by these series are one-valued in the cut plane; and so relations of the form

$$y_{10} = \alpha y_{00} + \beta y_{01}, \qquad y_{11} = \gamma y_{00} + \delta y_{01}$$

will exist throughout the cut plane.

Now suppose that ζ describes a closed circuit round the origin, so that the circuit crosses the cut from $-\infty$ to 0; the analytic continuation of y_{10} is $\alpha y_{00} - \beta y_{01}$ (since y_{00} is unaffected by the description of the circuit, but y_{01} changes sign) and the continuation of y_{11} is $\gamma y_{00} - \delta y_{01}$; and so $A y_{10}^2 + B y_{11}^2$ will be unaffected by the description of the circuit if

$$A(\alpha y_{00} + \beta y_{01})^2 + B(\gamma y_{00} + \delta y_{01})^2 \equiv A(\alpha y_{00} - \beta y_{01})^2 + B(\gamma y_{00} - \delta y_{01})^2;$$

i.e., if $A\alpha\beta + B\gamma\delta = 0$. Also $A y_{10}^2 + B y_{11}^2$ obviously has not a branch-point at $\zeta = 1$, and so, if $A\alpha\beta + B\gamma\delta = 0$, this function has no branch-points at 0 or 1, and, as it has no other possible singularities in the finite part of the plane, *it must be an integral function of ζ.*

The two expressions

$$A^{1/2} y_{10} + i B^{1/2} y_{11}, \qquad A^{1/2} y_{10} - i B^{1/2} y_{11}$$

are consequently two solutions of Mathieu's equation whose product is an integral function of ζ. This amounts to the fact (§19.4) that the product of $e^{\mu z} \phi(z)$ and $e^{-\mu z} \phi(-z)$ is a *periodic integral* function of z.

19.52 The determination of the integral function associated with Mathieu's equation

The integral function $F(z) \equiv A y_{10}{}^2 + B y_{11}{}^2$, just introduced, can be determined without difficulty; for, if y_{10} and y_{11} are any solutions of

$$\frac{d^2 u}{d\zeta^2} + P(\zeta) \frac{du}{d\zeta} + Q(\zeta) u = 0,$$

their squares (and consequently any linear combination of their squares) satisfy the equation (see Appell [31])

$$\frac{d^3 y}{d\zeta^3} + 3P(\zeta)\frac{d^2 y}{d\zeta^2} + \left[P'(\zeta) + 4Q(\zeta) + 2\{P(\zeta)\}^2 \right]\frac{dy}{d\zeta} + 2\left[Q'(\zeta) + 2P(\zeta)Q(\zeta)\right] y = 0;$$

in the case under consideration, this result reduces to

$$\zeta(1 - \zeta)\frac{d^3 F(\zeta)}{d\zeta^3} + \frac{3}{2}(1 - 2\zeta)\frac{d^2 F(\zeta)}{d\zeta^2} + (a - 1 - 16q + 32q\zeta)\frac{dF(\zeta)}{d\zeta} + 16qF(\zeta) = 0.$$

Let the Maclaurin series for $F(\zeta)$ be $\sum\limits_{n=0}^{\infty} c_n\zeta^n$; on substitution, we easily obtain the recurrence formula for the coefficients c_n, namely

$$v_{n+1}c_{n+2} = u_n c_{n+1} + c_n,$$

where

$$u_n = -\frac{(n+1)\{(n+1)^2 - a + 16q\}}{16q(2n+1)}, \qquad v_n = -\frac{n(n+1)(2n+1)}{32q(2n-1)}.$$

At first sight, it appears from the recurrence formula that c_0 and c_1 can be chosen arbitrarily, and the remaining coefficients $c_2, c_3, \ldots$ calculated in terms of them; but the third order equation has a singularity at $\zeta = 1$, and the series thus obtained would have only unit radius of convergence. It is necessary to choose the value of the ratio c_1/c_0 so that the series may converge for all values of ζ.

The recurrence formula, when written in the form

$$(c_n/c_{n+1}) = u_n + \frac{v_{n+1}}{(c_{n+1}/c_{n+2})},$$

suggests the consideration of the infinite continued fraction

$$u_n + \cfrac{v_{n+1}}{u_{n+1} + \cfrac{v_{n+2}}{u_{n+2} + \cdots}} = \lim_{m\to\infty}\left\{ u_n + \cfrac{v_{n+1}}{u_{n+1} + \cfrac{\ddots}{\quad\cfrac{v_{n+m}}{u_{n+m}}}} \right\}.$$

The continued fraction on the right can be written (see Sylvester [618, p. 446])

$$u_n = \frac{K(n, n+m)}{K(n+1, n+m)},$$

where

$$K(n, n+m) = \begin{vmatrix} 1 & v_{n+1}/u_n & 0 & \cdots & \cdots \\ -u_{n+1}^{-1} & 1 & v_{n+2}/u_{n+1} & \cdots & \cdots \\ 0 & -u_{n+2}^{-1} & 1 & \cdots & \cdots \\ \vdots & \vdots & \vdots & \vdots & \vdots \\ \cdots & \cdots & \cdots & -u_{n+m}^{-1} & 1 \end{vmatrix}$$

The limit of this, as $m \to \infty$, is a convergent determinant of von Koch's type (by Example 2.8.1); and since

$$\sum_{r=n}^{\infty} \left| \frac{v_{r+1}}{u_r u_{r+1}} \right| \to 0 \quad \text{as } n \to \infty,$$

it is easily seen that $K(n, \infty) \to 1$ as $n \to \infty$.

Therefore, if

$$\frac{c_n}{c_{n+1}} = \frac{u_n K(n, \infty)}{K(n+1, \infty)},$$

then c_n satisfies the recurrence formula and, since $c_{n+1}/c_n \to 0$ as $n \to \infty$, the resulting series for $F(\zeta)$ is an integral function. From the recurrence formula it is obvious that all the coefficients c_n are finite, since they are finite when n is sufficiently large. The construction of the integral function $F(\zeta)$ has therefore been effected.

19.53 *The solution of Mathieu's equation in terms of $F(\zeta)$*

If w_1 and w_2 be two particular solutions of

$$\frac{d^2 u}{d\zeta^2} + P(\zeta)\frac{du}{d\zeta} + Q(\zeta)u = 0,$$

then (Abel [3], primes denote differentiations with regard to ζ)

$$w_2 w_1' - w_1 w_2' = C \exp\left\{ -\int_0^\zeta P(\zeta)\, d\zeta \right\},$$

where C is a definite constant. Taking w_1 and w_2 to be those two solutions of Mathieu's general equation whose product is $F(\zeta)$, we have

$$\frac{w_1'}{w_1} - \frac{w_2'}{w_2} = \frac{C}{\zeta^{1/2}(1-\zeta)^{1/2}F(\zeta)}, \qquad \frac{w_1'}{w_1} - \frac{w_2'}{w_2} = \frac{F'(\zeta)}{F(\zeta)},$$

the latter following at once from the equation $w_1 w_2 = F(\zeta)$.

Solving these equations for w_1'/w_1 and w_2'/w_2, and then integrating, we at once get

$$w_1 = \gamma_1 \{F(\zeta)\}^{1/2} \exp\left\{ \frac{C}{2} \int_0^\zeta \frac{d\zeta}{\zeta^{1/2}(1-\zeta)^{1/2}F(\zeta)} \right\},$$

$$w_2 = \gamma_2 \{F(\zeta)\}^{1/2} \exp\left\{ -\frac{C}{2} \int_0^\zeta \frac{d\zeta}{\zeta^{1/2}(1-\zeta)^{1/2}F(\zeta)} \right\},$$

where γ_1, γ_2 are constants of integration; obviously no real generality is lost by taking $c_0 = \gamma_1 = \gamma_2 = 1$.

From the former result we have, for small values of $|\zeta|$,

$$w_1 = 1 + C\zeta^{1/2} + \tfrac{1}{2}(c_1 + C^2)\zeta + O(\zeta^{3/2}),$$

while, in the notation of §19.51, we have $a_1/a_0 = -a/2 + 8q$.

Hence $C^2 = 16q - a - c_1$. This equation determines C in terms of a, q and c_1, the value of c_1 being $\dfrac{K(1, \infty)}{u_0 K(0, \infty)}$.

Example 19.5.1   If the solutions of Mathieu's equation be $e^{\pm\mu z}\phi(\pm z)$, where $\phi(z)$ is periodic, shew that

$$\pi\mu = \pm C \int_0^\pi \frac{dz}{F(\cos^2 z)}.$$

Example 19.5.2 (Stieltjes)   Shew that the zeros of $F(\zeta)$ are all simple, unless $C = 0$. *Hint*. If $F(\zeta)$ could have a repeated zero, w_1 and w_2 would then have an essential singularity.

19.6 A second method of constructing the Mathieu function

So far, it has been assumed that all the various series of §19.3 involved in the expressions for $\mathrm{ce}_N(z,q)$ and $\mathrm{se}_N(z,q)$ are convergent. It will now be shewn that $\mathrm{ce}_N(z,q)$ and $\mathrm{se}_N(z,q)$ are integral functions of z and that the coefficients in their expansions q which converge absolutely when $|q|$ is sufficiently small. (The essential part of this theorem is the proof of the convergence of the series which occur in the coefficients; it is already known (§§10.2, 10.21) that solutions of Mathieu's equation are integral functions of z, and (in the case of *periodic* solutions) the existence of the Fourier expansion follows from §9.11.)

To obtain this result for the functions $\mathrm{ce}_N(z,q)$, we shall shew how to determine a particular integral of the equation

$$\frac{d^2u}{dz^2} + (a + 16q\cos 2z)u = \psi(a,q)\cos Nz$$

in the form of a Fourier series converging over the whole z-plane, where $\psi(a,q)$ is a function of the parameters a and q. The equation $\psi(a,q) = 0$ then determines a relation between a and q which gives rise to a Mathieu function. The reader who is acquainted with the method of Frobenius [227] as applied to the solution of linear differential equations in power series will recognise the resemblance of the following analysis to his work.

Write $a = N^2 + 8p$, where N is zero or a positive or negative integer. Mathieu's equation becomes

$$\frac{d^2u}{dz^2} + N^2u = -8(p + 2q\cos 2z)u.$$

If p and q are neglected, a solution of this equation is $u = \cos Nz = U_0(z)$, say. To obtain a closer approximation, write $-8(p + 2q\cos 2z)U_0(z)$ as a sum of cosines, i.e. in the form

$$-8\{q\cos(N-2)z + p\cos Nz + q\cos(N+2)z\} = V_1(z), \quad \text{say}.$$

Then, instead of solving $\dfrac{d^2u}{dz^2} + N^2u = V_1(z)$, suppress the terms[9] in $V_1(z)$ which involve $\cos Nz$; i.e. consider the function $W_1(z)$ where[10]

$$W_1(z) = V_1(z) + 8p\cos Nz.$$

---

[9] The reason for this suppression is that the particular integral of $\dfrac{d^2u}{dz^2} + N^2u = \cos Nz$ contains non-periodic terms.

[10] Unless $N = 1$, in which case $W_1(z) = V_1(z) + 8(p + q)\cos z$.

A particular integral of

$$\frac{d^2u}{dz^2} + N^2u = W_1(z)$$

is

$$u = 2\left\{\frac{q}{1(1-N)}\cos(N-2)z + \frac{q}{1(1+N)}\cos(N+2)z\right\} = U_1(z), \text{ say.}$$

Now express $-8(p + 2q\cos 2z)U_1(z)$ as a sum of cosines; calling this sum $V_2(z)$, choose α_2 to be such a function of p and q that $V_2(z) + \alpha_2\cos Nz$ contains no term in $\cos Nz$; and let $V_2(z) + \alpha_2\cos Nz = W_2(z)$.

Solve the equation $\dfrac{d^2u}{dz^2} + N^2u = W_2(z)$, and continue the process. Three sets of functions $U_m(z), V_m(z), W_m(z)$ are thus obtained, such that $U_m(z)$ and $W_m(z)$ contain no term in $\cos Nz$ when $m \neq 0$:

$$W_m(z) = V_m(z) + \alpha_m\cos Nz, \qquad V_m(z) = -8(p + 2q\cos 2z)U_{m-1}(z),$$

and

$$\frac{d^2U_m(z)}{dz^2} + N^2U_m(z) = W_m(z),$$

where α_m is a function of p and q but not of z.

It follows that

$$\left\{\frac{d^2}{dz^2} + N^2\right\}\sum_{m=0}^{n} U_m(z) = \sum_{m=1}^{n} W_m(z)$$

$$= \sum_{m=1}^{n} V_m(z) + \left(\sum_{m=1}^{n}\alpha_m\right)\cos Nz$$

$$= -8(p + 2q\cos 2z)\sum_{m=0}^{n-1} U_{m-1}(z) + \left(\sum_{m=1}^{n}\alpha_m\right)\cos Nz.$$

Therefore, if $U(z) = \sum\limits_{m=0}^{\infty} U_m(z)$ be a uniformly convergent series of analytic functions throughout a two-dimensional region in the z-plane, we have (§5.3)

$$\frac{d^2U(z)}{dz^2} + (a + 16q\cos 2z)U(z) = \psi(a,q)\cos Nz,$$

where

$$\psi(a,q) = \sum_{m=1}^{\infty}\alpha_m.$$

It is obvious that, if a be so chosen that $\psi(a,q) = 0$, then $U(z)$ reduces to $\mathrm{ce}_N(z)$.

A similar process can obviously be carried out for the functions $\mathrm{se}_N(z,q)$ by making use of sines of multiples of z.

19.61 The convergence of the series defining Mathieu functions

We shall now examine the expansion of §19.6 more closely, with a view to investigating the convergence of the series involved.

When $n \geq 1$, we may obviously write

$$U_n(z) = \sum_{r=1}^{*} \beta_{n,r} \cos(N - 2r)z + \sum_{r=1}^{n} \alpha_{n,r} \cos(N + 2r)z,$$

the asterisk denoting that the first summation ceases at the greatest value of r for which $r \leq \frac{1}{2}N$.

Since $\left\{ \dfrac{d^2}{dz^2} + N^2 \right\} U_{n+1}(z) = \alpha_{n+1} \cos Nz - 8(p + 2q \cos 2z)U_n(z)$, it follows on equating coefficients of $\cos(N \pm 2r)z$ on each side of the equation[11] that

$$\alpha_{n+1} = 8q(\alpha_{n,1} + \beta_{n,1})$$
$$r(r + N)\alpha_{n+1,r} = 2\{p\alpha_{n,r} + q(\alpha_{n,r-1} + \alpha_{n,r+1})\} \qquad (r = 1, 2, \ldots),$$
$$r(r - N)\beta_{n+1,r} = 2\{p\beta_{n,r} + q(\beta_{n,r-1} + \beta_{n,r+1})\} \qquad (r \leq \tfrac{1}{2}N).$$

These formulae hold universally with the following conventions[12]:

(i) $\alpha_{n,0} = \beta_{n,0} = 0$ for $n = 1, 2 \ldots$; $\alpha_{n,r} = \beta_{n,r} = 0, r > n$;

(ii) $\beta_{n,\frac{1}{2}N+1} = \beta_{n,\frac{1}{2}N-1}$ when N is even and $r = \frac{1}{2}N$;

(iii) $\beta_{n,\frac{1}{2}(N+1)} = \beta_{n,\frac{1}{2}(N-1)}$ when N is odd and $r = \frac{1}{2}(N - 1)$.

The reader will easily obtain the following special formulae:

(I) $\alpha_1 = 8p$, for $N \neq 1$; $\quad \alpha_1 = 8(p + q)$, for $N = 1$;

(II) $\alpha_{n,n} = \dfrac{(2q)^n N!}{n!(N+n)!}$, for $N \neq 0$; $\alpha_{n,n} = \dfrac{2^{n+1}q^n}{(n!)^2}$, for $N = 0$.

(III) $\alpha_{n,r}$ and $\beta_{n,r}$ are homogeneous polynomials of degree n in p and q.

If

$$\sum_{n=r}^{\infty} \alpha_{n,r} = A_r, \qquad \sum_{n=r}^{\infty} \beta_{n,r} = B_r,$$

we have

$$\psi(a, q) = 8p + 8q(A_1 + B_1) \qquad (N \neq 1),$$
$$r(r + N)A_r = 2\{pA_r + q(A_{r-1} + A_{r+1})\} \qquad \text{(A)}$$
$$r(r - N)B_r = 2\{pB_r + q(B_{r-1} + B_{r+1})\}, \qquad \text{(B)}$$

where $A_0 = B_0 = 1$ and B_r is subject to conventions due to (ii) and (iii) above.

Now write $w_r = -q\{r(r + N) - 2p\}^{-1}$, $w_r' = -q\{r(r - N) - 2p\}^{-1}$. The result of eliminating $A_1, A_2, \ldots, A_{r-1}, A_{r+1}, \ldots$ from the set of equations (A) is

$$A_r \Delta_0 = (-1)^r w_1 w_2 \cdots w_r \Delta_r,$$

---

[11] When $N = 0$ or 1 these equations must be modified by the suppression of all the coefficients $\beta_{n,r}$.

[12] The conventions (ii) and (iii) are due to the fact that $\cos z = \cos(-z)$, $\cos 2z = \cos(-2z)$.

where Δ_r is the infinite determinant of von Koch's type (§2.82)

$$\Delta_r = \begin{vmatrix} 1 & w_{r+1} & 0 & 0 & \cdots \\ w_{r+2} & 1 & w_{r+2} & 0 & \cdots \\ 0 & w_{r+3} & 1 & w_{r+3} & \cdots \\ \vdots & \vdots & \vdots & \vdots & \vdots \end{vmatrix}$$

The determinant converges absolutely (Example 2.8.1) if no denominator vanishes; and $\Delta_r \to 1$ as $r \to \infty$ (cf. §19.52). If p and q be given such values that $\Delta_0 \neq 0$, $2p \neq r(r+N)$, where $r = 1, 2, 3, \ldots$, the series

$$\sum_{r=1}^{\infty} (-1)^r w_1 w_2 \cdots w_r \Delta_r \Delta_0^{-1} \cos(N+2r)z$$

represents an integral function of z.

In like manner $B_r D_0 = (-1)^r w_1' w_2' \cdots w_r' D_r$, where D_r is the finite determinant

$$\begin{vmatrix} 1 & w_{r+1}' & 0 & \cdots \\ w_{r+2}' & 1 & w_{r+2}' & \cdots \\ \vdots & \vdots & \vdots & \vdots \end{vmatrix},$$

the last row being

$$0, 0, \ldots, 0, 2w'_{\frac{1}{2}N}, 1$$

or

$$0, 0, \ldots, 0, w'_{\frac{1}{2}(N-1)}, 1 + w'_{\frac{1}{2}(N-1)}$$

according as N is even or odd.

The series $\sum\limits_{n=0}^{\infty} U_n(z)$ is therefore

$$\cos Nz + \Delta_0^{-1} \sum_{r=1}^{\infty} (-1)^r w_1 w_2 \cdots w_r \Delta_r \cos(N+2r)z$$

$$+ D_0^{-1} \sum_{r=1}^{r < \frac{1}{2}N} (-1)^r w_1' w_2' \cdots w_r' D_r \cos(N-2r)z,$$

these series converging uniformly in any bounded domain of values of z, so that term-by-term differentiations are permissible.

Further, the condition $\psi(a, q) = 0$ is equivalent to $p = q \left(\frac{w_1 \Delta_1}{\Delta_0} + \frac{w_1' D_1}{D_0} \right)$, i.e.

$$p \Delta_0 D_0 - q(w_1 \Delta_1 D_0 + w_1' D_1 \Delta_0) = 0.$$

If we multiply by

$$\prod_{r=1}^{\infty} \left\{ 1 - \frac{2p}{r(r+N)} \right\} \prod_{r=1}^{r < \frac{1}{2}N} \left\{ 1 - \frac{2p}{r(r-N)} \right\},$$

the expression on the left becomes an integral function of both p and q, $\Psi(a, q)$, say; the terms of $\Psi(a, q)$, which are of lowest degrees in p and q, are respectively p and

$$q^2 \left\{ \frac{1}{N-1} - \frac{1}{N+1} \right\}.$$

Now expand

$$\frac{1}{2\pi i} \int \frac{p}{\Psi(N^2 + 8p, q)} \frac{\partial \Psi(N^2 + 8p, q)}{\partial p} \, dp$$

in ascending powers of q (cf. §7.31), the contour being a small circle in the p-plane, with centre at the origin, and $|q|$ being so small that $\Psi(N^2 + 8p, q)$ has only one zero inside the contour. Then it follows, just as in §7.31, that, for sufficiently small values of $|q|$, we may expand p as a power series in q commencing[13] with a term in q^2; and if $|q|$ be sufficiently small D_0 and Δ_0 will not vanish, since both are equal to 1 when $q = 0$.

On substituting for p in terms of q throughout the series for $U(z)$, we see that the series involved in $\mathrm{ce}_N(z, q)$ are absolutely convergent when $|q|$ is sufficiently small.

The series involved in $\mathrm{se}_N(z, q)$ may obviously be investigated in a similar manner.

19.7 The method of change of parameter

This appears in Whittaker [674]. The methods of Hill and of Lindemann–Stieltjes are effective in determining μ, but only after elaborate analysis. Such analysis is inevitable, as μ is by no means a simple function of q; this may be seen by giving q an assigned real value and making a vary from $-\infty$ to $+\infty$; then μ alternates between real and complex values, the changes taking place when, with the Hill–Mathieu notation, $\Delta(0) \sin^2(\frac{1}{2}\pi \sqrt{a})$ passes through the values 0 and 1; the complicated nature of this condition is due to the fact that $\Delta(0)$ is an elaborate expression involving both a and q.

It is, however, possible to express μ and a in terms of q and of a new parameter σ, and the results are very well adapted for purposes of numerical computation when $|q|$ is small. (They have been applied to Hill's problem by Ince [332].)

The introduction of the parameter σ is suggested by the series for $\mathrm{ce}_1(z, q)$ and $\mathrm{se}_1(z, q)$ given in Example 19.3.1; a consideration of these series leads us to investigate the potentialities of a solution of Mathieu's general equation in the form $y = e^{\mu z} \phi(z)$, where

$$\phi(z) = \sin(z - \sigma) + a_3 \cos(3z - \sigma) + b_3 \sin(3z - \sigma) + a_5 \cos(5z - \sigma) + b_5 \sin(5z - \sigma) + \cdots,$$

the parameter σ being rendered definite by the fact that no term in $\cos(z - \sigma)$ is to appear in $\phi(z)$; the special functions $\mathrm{se}_1(z, q)$, $\mathrm{ce}_1(z, q)$ are the cases of this solution in which σ is 0 or $\frac{1}{2}\pi$.

On substituting this expression in Mathieu's equation, the reader will have no difficulty in

---

[13] If $N = 1$ this result has to be modified, since there is an additional term q on the right and the term $q^2/(N-1)$ does not appear.

obtaining the following approximations, valid for[14] small values of q and real values of σ:

$$\mu = 4q \sin 2\sigma - 12q^3 \sin 2\sigma - 12q^4 \sin 4\sigma + O(q^5),$$

$$a = 1 + 8q \cos 2\sigma + (-16 + 8 \cos 4\sigma)q^2 - 8q^3 \cos 2\sigma$$
$$+ \left(\tfrac{256}{3} - 88 \cos 4\sigma\right) q^4 + O(q^5),$$

$$a_3 = 3q^2 \sin 2\sigma + 3q^3 \sin 4\sigma + \left(-\tfrac{274}{9} \sin 2\sigma + 9 \sin 6\sigma\right) q^4 + O(q^5),$$

$$b_3 = q + q^2 \cos 2\sigma + \left(-\tfrac{14}{3} + 5 \cos 4\sigma\right) q^3 + \left(-\tfrac{74}{9} \cos 2\sigma + 7 \cos 6\sigma\right) q^4 + O(q^5),$$

$$a_5 = \tfrac{14}{9}q^3 \sin 2\sigma + \tfrac{44}{27}q^4 \sin 4\sigma + O(q^5),$$

$$b_5 = \tfrac{1}{3}q^2 + \tfrac{4}{9}q^3 \cos 2\sigma + \left(-\tfrac{155}{54} + \tfrac{82}{27} \cos 4\sigma\right) q^4 + O(q^5),$$

$$a_7 = \tfrac{35}{108}q^4 \sin 2\sigma + O(q^5), \qquad b_7 = \tfrac{1}{18}q^3 + \tfrac{1}{12}q^4 \cos 2\sigma + O(q^5),$$

$$a_9 = O(q^5), \qquad b_9 = \tfrac{1}{180}q^4 + O(q^5),$$

the constants involved in the various functions $O(q^5)$ depending on σ.

The domains of values of q and σ for which these series converge have not yet been determined[15]. If the solution thus obtained be called $\Lambda(z, \sigma, q)$, then $\Lambda(z, \sigma, q)$ and $\Lambda(z, -\sigma, q)$ form a fundamental system of solutions of Mathieu's general equation if $\mu \neq 0$.

Example 19.7.1 Shew that, if $\sigma = 0.5 \times i$ and $q = 0.01$, then

$$a = 1.124, 841, 4 \cdots, \qquad \mu = i \times 0.046, 993, 5 \cdots;$$

shew also that, if $\sigma = i$ and $q = 0.01$, then

$$a = 1.321, 169, 3 \cdots, \qquad \mu = i \times 0.145, 027, 6 \cdots.$$

Example 19.7.2 Obtain the equations

$$\mu = 4q \sin 2\sigma - 4qa_3,$$
$$a = 1 + 8q \cos 2\sigma - \mu^2 - 8qb_3,$$

expressing μ and a in finite terms as functions of q, σ, a_3 and b_3.

Example 19.7.3 Obtain the recurrence formulae

$$\{-4n(n+1) + 8q \cos 2\sigma - 8qb_3 \pm 8qi(2n+1)(a_3 - \sin 2\sigma)\} z_{2n+1} + 8q(z_{2n-1} + z_{2n+3}) = 0,$$

where z_{2n+1} denotes $b_{2n+1} + ia_{2n+1}$ or $b_{2n+1} - ia_{2n+1}$, according as the upper or lower sign is taken.

19.8 The asymptotic solution of Mathieu's equation

If in Mathieu's equation

$$\frac{d^2u}{dz^2} + \left(a + \frac{1}{2}K^2 \cos 2z\right) u = 0$$

---

[14] The parameters q and σ are to be regarded as fundamental in this analysis, instead of a and q as hitherto.

[15] It seems highly probable that, if $|q|$ is sufficiently small, the series converge for all real values of σ, and also for complex values of σ for which $|\operatorname{Im} \sigma|$ is sufficiently small. It may be noticed that, when q is real, real and purely imaginary values of σ correspond respectively to real and purely imaginary values of μ.

we write $k \sin z = \xi$, we get

$$(\xi^2 - k^2)\frac{d^2u}{d\xi^2} + \xi\frac{du}{d\xi} + (\xi^2 - M^2)u = 0,$$

where $M^2 \equiv a + \frac{1}{2}k^2$.

This equation has an irregular singularity at infinity. From its resemblance to Bessel's equation, we are led to write $u = e^{i\xi}\xi^{-\frac{1}{2}}v$, and substitute

$$v = 1 + (a_1/\xi) + (a_2/\xi^2) + \cdots$$

in the resulting equation for v; we then find that

$$a_1 = -\frac{1}{2}i\left(\frac{1}{4} - M^2 + K^2\right), \qquad a_2 = t\frac{1}{8}\left(\frac{1}{4} - M^2 + k^2\right)\left(\frac{9}{4} - M^2 + k^2\right) + \frac{1}{4}k^2,$$

the general coefficient being given by the recurrence formula

$$2i(r+1)a_{r+1} = \left\{\frac{1}{4} - M^2 + k^2 + r(r+1)\right\} + (2r-1)ik^2a_{r-1} - \left(r^2 - 2r + \frac{3}{4}\right)k^2a_{r-2}.$$

The two series

$$e^{i\xi}\xi^{-\frac{1}{2}}\left(1 + \frac{a_1}{\xi} + \frac{a_2}{\xi^2} + \cdots\right), \qquad e^{-i\xi}\xi^{-\frac{1}{2}}\left(1 + \frac{a_1}{\xi} + \frac{a_2}{\xi^2} - \cdots\right)$$

are formal solutions of Mathieu's equation, reducing to the well-known asymptotic solutions of Bessel's equation (§17.5) when $k \to 0$. The complete formulae which connect them with the solutions $e^{\pm\mu z}\phi(\pm z)$ have not yet been published, though some steps towards obtaining them have been made by Dougall [188].

19.9 Miscellaneous examples

Example 19.1   Shew that, if $k = \sqrt{32q}$,

$$2\pi \operatorname{ce}_0(z, q) = \operatorname{ce}_0(0, q)\int_{-\pi}^{\pi}\cos(k \sin z \sin \theta)\operatorname{ce}_0(\theta, q)\,d\theta.$$

Example 19.2   Shew that the even Mathieu functions satisfy the integral equation

$$G(z) = \lambda\int_{-\pi}^{\pi}J_0\{ik(\cos z + \cos \theta)\}G(\theta)\,d\theta.$$

Example 19.3   Shew that the equation

$$(az^2 + c)\frac{d^2u}{dz^2} + 2az\frac{du}{dz} + (\lambda^2cz^2 + m)u = 0$$

(where a, c, λ, m are constants) is satisfied by

$$u = \int e^{\lambda z s}v(s)\,ds$$

taken round an appropriate contour, provided that $v(s)$ satisfies

$$(as^2 + c)\frac{d^2v(s)}{ds^2} + 2as\frac{dv(s)}{ds} + (\lambda^2cs^2 + m)v(s) = 0,$$

which is the same as the equation for u.

Derive the integral equations satisfied by the Mathieu functions as particular cases of this result.

Example 19.4 (Mathieu) Shew that, if powers of q above the fourth are neglected, then

$$\mathrm{ce}_1(z, q) = \cos z + q \cos 3z + q^2(\tfrac{1}{3} \cos 5z - \cos 3z)$$
$$+ q^3(\tfrac{1}{18} \cos 7z - \tfrac{4}{9} \cos 5z + \tfrac{1}{3} \cos 3z)$$
$$+ q^4(\tfrac{1}{180} \cos 9z - \tfrac{1}{12} \cos 7z + \tfrac{1}{6} \cos 5z + \tfrac{11}{9} \cos 3z),$$
$$\mathrm{se}_1(z, q) = \sin z + q \sin 3z + q^2 (\tfrac{1}{3} \sin 5z + \sin 3z)$$
$$+ q^3(\tfrac{1}{18} \sin 7z + \tfrac{4}{9} \sin 5z + \tfrac{1}{3} \sin 3z)$$
$$+ q^4(\tfrac{1}{180} \sin 9z + \tfrac{1}{12} \sin 7z + \tfrac{1}{6} \sin 5z - \tfrac{11}{9} \sin 3z),$$
$$\mathrm{ce}_2(z, p) = \cos 2z + q(\tfrac{2}{3} \cos 4z - 2) + \tfrac{1}{6}q^2 \cos 6z$$
$$+ q^3 (\tfrac{1}{45} \cos 8z + \tfrac{43}{47} \cos 4z + \tfrac{40}{3})$$
$$+ q^4(\tfrac{1}{540} \cos 10z + \tfrac{293}{540} \cos 6z).$$

Example 19.5 (Mathieu) Shew that

$$\mathrm{ce}_3 = \cos 3z + q(- \cos z + \tfrac{1}{2} \cos 5z) + q^2(\cos z + \tfrac{1}{10} \cos 7z)$$
$$+ q^3(-\tfrac{1}{2} \cos z + \tfrac{7}{40} \cos 5z + \tfrac{1}{90} \cos 9z) + O(q^4),$$

and that, in the case of this function

$$a = 9 + 4q^2 - 8q^3 + O(q^4).$$

Example 19.6 Shew that, if $y(z)$ be a Mathieu function, then a second solution of the corresponding differential equation is

$$y(z) \int^s \{y(t)\}^{-2} \, dt.$$

Shew that a second solution[16] of the equation for $\mathrm{ce}_0(z, q)$ is

$$z \, \mathrm{ce}_0(z, q) - 4q \sin 2z - 3q^2 \sin 4z - \cdots.$$

Example 19.7 If $y(z)$ be a solution of Mathieu's general equation, shew that

$$\frac{y(z + 2\pi) + y(z - 2\pi)}{y(z)}$$

is constant.

Example 19.8 Express the Mathieu functions as series of Bessel functions in which the coefficients are multiples of the coefficients in the Fourier series for the Mathieu functions. *Hint.* Substitute the Fourier series under the integral sign in the integral equations of §19.22.

Example 19.9 Shew that the confluent form of the equations for $\mathrm{ce}_n(z, q)$ and $\mathrm{se}_n(z, q)$, when the eccentricity of the fundamental ellipse tends to zero, is, in each case, the equation satisfied by $J_n(ik \cos z)$.

---

[16] This solution is called $\mathrm{in}_0(z, q)$; the second solutions of the equations satisfied by Mathieu functions have been investigated by Ince [333]. See also §19.2.

Example 19.10   Obtain the parabolic cylinder functions of Chapter 16 as confluent forms of the Mathieu functions, by making the eccentricity of the fundamental ellipse tend to unity.

Example 19.11   Shew that $ce_n(z, q)$ can be expanded in series of the form

$$\sum_{m=0}^{\infty} A_m \cos^{2m} z \quad \text{or} \quad \sum_{m=0}^{\infty} B_m \cos^{2m+1} z,$$

according as n is even or odd; and that these series converge when $|\cos z| < 1$.

Example 19.12   With the notation of Example 19.11, shew that, if

$$ce_n(z, q) = \lambda_n \int_{-\pi}^{\pi} e^{k \cos s \cos \theta} ce_n(\theta, q) \, d\theta,$$

then λ_n is given by one or other of the series

$$A_0 = 2\pi \lambda_n \sum_{m=0}^{\infty} \frac{2m}{2^{2m}(m!)^2} A_m, \qquad B_0 = 2\pi \lambda_n k \sum_{m=0}^{\infty} \frac{(2m+1)!}{2^{2m+1} m!(m+1)!} B_m,$$

provided that these series converge.

Example 19.13   Shew that the differential equation satisfied by the product of any two solutions of Bessel's equation for functions of order n is

$$\vartheta(\vartheta - 2n)(\vartheta + 2n)u + 4z^2(\vartheta + 1)u = 0,$$

where ϑ denotes $z\frac{d}{dz}$.

Shew that one solution of this equation is an integral function of z; and thence, by the methods of §§19.5–19.53, obtain the Bessel functions, discussing particularly the case in which n is an integer.

Example 19.14   Shew that an approximate solution of the equation

$$\frac{d^2u}{dz^2} + (A + k^2 \sinh^2 z)u = 0$$

is

$$u = C(\operatorname{cosech} z)^{1/2} \sin(k \cosh z + \varepsilon),$$

where C and ε are constants of integration; it is to be assumed that k is large, A is not very large and z is not small.

20

Elliptic Functions. General Theorems and the Weierstrassian Functions

20.1 Doubly-periodic functions

A most important property of the circular functions $\sin z$, $\cos z$, $\tan z, \ldots$ is that, if $f(z)$ denote any one of them,

$$f(z + 2\pi) = f(z),$$

and hence $f(z + 2n\pi) = f(z)$, for all integer values of n. It is on account of this property that the circular functions are frequently described as *periodic functions* with period 2π. To distinguish them from the functions which will be discussed in this and the two following chapters, they are called *singly-periodic functions*.

Let ω_1, ω_2 be any two numbers (real or complex) *whose ratio*[1] *is not purely real*. A function which satisfies the equations

$$f(z + 2\omega_1) = f(z), \qquad f(z + 2\omega_2) = f(z),$$

for all values of z for which $f(z)$ exists, is called a *doubly-periodic* function of z, with periods $2\omega_1$, $2\omega_2$. A doubly-periodic function which is analytic (except at poles), and which has no singularities other than poles in the finite part of the plane, is called an *elliptic function*.

Note What is now known as an *elliptic integral*[2] occurs in the researches of Jakob Bernoulli on the *Elastica*. Maclaurin, Fagnano, Legendre, and others considered such integrals in connexion with the problem of rectifying an arc of an ellipse; the idea of 'inverting' an elliptic integral (§21.7) to obtain an elliptic function is due to Abel, Jacobi and Gauss.

The periods $2\omega_1$, $2\omega_2$ play much the same part in the theory of elliptic functions as is played by the single period in the case of the circular functions.

Before actually constructing any elliptic functions, and, indeed, before establishing the existence of such functions, it is convenient to prove some general theorems (§20.11–§20.14) concerning properties common to all elliptic functions; this procedure, though not strictly logical, is convenient because a large number of the properties of particular elliptic functions can be obtained at once by an appeal to these theorems.

Example 20.1.1 The differential coefficient of an elliptic function is itself an elliptic function.

---

[1] If ω_2/ω_1 is real, the parallelograms defined in §20.11 collapse, and the function reduces to a singly-periodic function when ω_2/ω_1 is rational; and when ω_2/ω_1 is irrational, it has been shewn by Jacobi [350] that the function reduces to a constant.

[2] A brief discussion of elliptic integrals will be found in §22.7–§22.741.

20.11 Period-parallelograms

The study of elliptic functions is much facilitated by the geometrical representation afforded by the Argand diagram.

Suppose that in the plane of the variable z we mark the points $0, 2\omega_1, 2\omega_2, 2\omega_1 + 2\omega_2$, and, generally, all the points whose complex coordinates are of the form $2m\omega_1 + 2n\omega_2$, where m and n are integers.

Join in succession consecutive points of the set $0, 2\omega_1, 2\omega_1 + 2\omega_2, 2\omega_2, 0$, and we obtain a parallelogram. If there is no point ω inside or on the boundary of this parallelogram (the vertices excepted) such that

$$f(z + \omega) = f(z)$$

for all values of z, this parallelogram is called a *fundamental period-parallelogram* for an elliptic function with periods $2\omega_1, 2\omega_2$.

It is clear that the z-plane may be covered with a network of parallelograms equal to the fundamental period-parallelogram and similarly situated, each of the points $2m\omega_1 + 2n\omega_2$ being a vertex of four parallelograms.

These parallelograms are called *period-parallelograms*, or *meshes*; for all values of z, the points $z, z + 2\omega_1, \ldots, z + 2m\omega_1 + 2n\omega_2, \ldots$ manifestly occupy corresponding positions in the meshes; any pair of such points are said to be *congruent* to one another. The congruence of two points z, z' is expressed by the notation $z' \equiv z \mod 2\omega_1, 2\omega_2$.

From the fundamental property of elliptic functions, it follows that an elliptic function assumes the same value at every one of a set of congruent points; and so *its values in any mesh are a mere repetition of its values in any other mesh*.

For purposes of integration it is not convenient to deal with the actual meshes if they have singularities of the integrand on their boundaries; on account of the periodic properties of elliptic functions nothing is lost by taking as a contour, not an actual mesh, but a parallelogram obtained by translating a mesh (without rotation) in such a way that none of the poles of the integrands considered are on the sides of the parallelogram. Such a parallelogram is called a *cell*. Obviously the values assumed by an elliptic function in a cell are a mere repetition of its values in any mesh.

A set of poles (or zeros) of an elliptic function in any given cell is called an *irreducible* set; all other poles (or zeros) of the function are congruent to one or other of them.

20.12 Simple properties of elliptic functions

Theorem (I)   *The number of poles of an elliptic function in any cell is finite.*

For, if not, the poles would have a limit point, by the two-dimensional analogue of §2.21. This point is (§5.61) an essential singularity of the function; and so, by definition, the function is not an elliptic function.

Theorem (II)   *The number of zeros of an elliptic function in any cell is finite.*

For, if not, the reciprocal of the function would have an infinite number of poles in the cell, and would therefore have an essential singularity; and this point would be an essential singularity of the original function, which would therefore not be an elliptic function. This argument presupposes that the function is not identically zero.

Theorem (III)   *The sum of the residues of an elliptic function $f(z)$, at its poles in any cell is zero.*

Let C be the contour formed by the edges of the cell, and let the corners of the cell be t, $t + 2\omega_1, t + 2\omega_1 + 2\omega_2, t + 2\omega_2$.

Note   In future, the periods of an elliptic function will not be called $2\omega_1$, $2\omega_2$ indifferently; but that one will be called $2\omega_1$ which makes the ratio ω_2/ω_1 *have a positive imaginary part*; and then, if C be described in the sense indicated by the order of the corners given above, the description of C is *counter-clockwise*.

Throughout the chapter, we shall denote by the symbol C the contour formed by the edges of a cell.

The sum of the residues of $f(z)$ at its poles inside C is

$$\frac{1}{2\pi i} \int_C f(z)\,dz = \frac{1}{2\pi i} \left\{ \int_t^{t+2\omega_1} + \int_{t+2\omega_1}^{t+2\omega_1+2\omega_2} + \int_{t+2\omega_1+2\omega_2}^{t+2\omega_2} + \int_{t+2\omega_2}^{t} \right\} f(z)\,dz.$$

In the second and third integrals write $z+2\omega_1$, $z+2\omega_2$ respectively for z, and the right-hand side becomes

$$\frac{1}{2\pi i} \int_t^{t+2\omega_1} \{f(z) - f(z + 2\omega_2)\}\,dz - \frac{1}{2\pi i} \int_t^{t+2\omega_2} \{f(z) - f(z + 2\omega_1)\}\,dz,$$

and each of these integrals vanishes in virtue of the periodic properties of $f(z)$; and so $\int_C f(z)\,dz = 0$, and the theorem is established.

Theorem (IV: Liouville's theorem[3])   *An elliptic function, $f(z)$, with no poles in a cell is merely a constant.*

For if $f(z)$ has no poles inside the cell, it is analytic (and consequently bounded) inside and on the boundary of the cell (Corollary 3.6.2), that is to say, there is a number K such that $|f(z)| < K$ when z is inside or on the boundary of the cell. From the periodic properties of $f(z)$ it follows that $f(z)$ is analytic and $|f(z)| < K$ for all values of z; and so, by §5.63, $f(z)$ is a constant.

(This modification of the theorem of §5.63 is the result on which Liouville based his lectures on elliptic functions.) It will be seen later that a very large number of theorems concerning elliptic functions can be proved by the aid of this result.

20.13 The order of an elliptic function

It will now be shewn that, if $f(z)$ be an elliptic function and c be any constant, *the number of roots of the equation*

$$f(z) = c$$

which lie in any cell depends only on $f(z)$, and not on c; this number is called the *order* of the elliptic function, and is equal to the number of poles of $f(z)$ in the cell.

---

[3]   This modification of the theorem of §5.63 is the result on which Liouville based his lectures on elliptic functions.

By §6.31, the difference between the number of zeros and the number of poles of $f(z) - c$ which lie in the cell C is

$$\frac{1}{2\pi i} \int_C \frac{f'(z)}{f(z) - c}\, dz.$$

Since $f'(z + 2\omega_1) = f'(z + 2\omega_2) = f'(z)$, by dividing the contour into four parts, precisely, as in §20.12 Theorem (III), we find that this integral is zero.

Therefore the number of zeros of $f(z) - c$ is equal to the number of poles of $f(z) - c$; but any pole of $f(z) - c$ is obviously a pole of $f(z)$ and conversely; hence the number of zeros of $f(z) - c$ is equal to the number of poles of $f(z)$, which is independent of c; the required result is therefore established.

Note   In determining the order of an elliptic function by counting the number of its irreducible poles, it is obvious, from §6.31, that each pole has to be reckoned according to its multiplicity.

The order of an elliptic function is *never less than* 2; for an elliptic function of order 1 would have a single irreducible pole; and if this point actually were a pole (and not an ordinary point) the residue there would not be zero, which is contrary to the Theorem (III) above.

So far as singularities are concerned, the simplest elliptic functions are those of order 2. Such functions may be divided into two classes: (i) those which have a single irreducible double pole, at which the residue is zero in accordance with (III) above; (ii) those which have two simple poles at which, by (III), the residues are numerically equal but opposite in sign.

Functions belonging to these respective classes will be discussed in this chapter and in Chapter 22 under the names of Weierstrassian and Jacobian elliptic functions respectively; and it will be shewn that any elliptic function is expressible in terms of functions of either of these types.

20.14 Relation between the zeros and poles of an elliptic function

We shall now shew that *the sum of the affixes of a set of irreducible zeros of an elliptic function is congruent to the sum of the affixes of a set of irreducible poles.*

For, with the notation previously employed, it follows, from §6.3, that the difference between the sums in question is

$$\begin{aligned}
\frac{1}{2\pi i} \int_C \frac{z f'(z)}{f(z)}\, dz &= \frac{1}{2\pi i} \left\{ \int_t^{t+2\omega_1} + \int_{t+2\omega_1}^{t+2\omega_1+2\omega_2} + \int_{t+2\omega_1+2\omega_2}^{t+2\omega_2} + \int_{t+2\omega_2}^t \right\} \frac{z f'(z)}{f(z)}\, dz \\
&= \frac{1}{2\pi i} \int_t^{t+2\omega_1} \left\{ \frac{z f'(z)}{f(z)} - \frac{(z+2\omega_2)f'(z+2\omega_2)}{f(z+2\omega_2)} \right\} dz \\
&\quad - \frac{1}{2\pi i} \int_t^{t+2\omega_2} \left\{ \frac{z f'(z)}{f(z)} - \frac{(z+2\omega_1)f'(z+2\omega_1)}{f(z+2\omega_1)} \right\} dz \\
&= \frac{1}{2\pi i} \left\{ -2\omega_2 \int_t^{t+2\omega_1} \frac{f'(z)}{f(z)}\, dz + 2\omega_1 \int_t^{t+2\omega_2} \frac{f'(z)}{f(z)}\, dz \right\} \\
&= \frac{1}{2\pi i} \left\{ -2\omega_2 \left[\log f(z)\right]_t^{t+2\omega_1} + 2\omega_1 \left[\log f(z)\right]_t^{t+2\omega_2} \right\},
\end{aligned}$$

on making use of the substitutions used in §20.12 and of the periodic properties of $f(z)$ and $f'(z)$.

Now $f(z)$ has the same values at the points $t + 2\omega_1$, $t + 2\omega_2$ as at t, so the values of $\log f(z)$ at these points can only differ from the value of $\log f(z)$ at t by integer multiples of $2\pi i$, say $-2n\pi i$, $2m\pi i$; then we have

$$\frac{1}{2\pi i} \int_C \frac{z f'(z)}{f(z)} \, dz = 2m\omega_1 + 2n\omega_2,$$

and so the sum of the affixes of the zeros minus the sum of the affixes of the poles is a period; and this is the result which had to be established.

20.2 The construction of an elliptic function. Definition of $\wp(z)$

It was seen in §20.1 that elliptic functions may be expected to have some properties analogous to those of the circular functions. It is therefore natural to introduce elliptic functions into analysis by some definition analogous to one of the definitions which may be made the foundation of the theory of circular functions.

One mode of developing the theory of the circular functions is to start from the series $\sum_{m=-\infty}^{\infty} (z - m\pi)^{-2}$; calling this series $(\sin z)^{-2}$, it is possible to deduce all the known properties of $\sin z$; the method of doing so is briefly indicated in §20.222.

The analogous method of founding the theory of elliptic functions is to define the function $\wp(z)$ by the equation[4]

$$\wp(z) = \frac{1}{z^2} + {\sum_{m,n}}' \left\{ \frac{1}{(z - 2m\omega_1 - 2n\omega_2)^2} - \frac{1}{(2m\omega_1 + 2n\omega_2)^2} \right\},$$

where ω_1, ω_2 satisfy the conditions laid down in §20.1, and §20.12 Theorem (III); the summation extends over all integer values (positive, negative and zero) of m and n, simultaneous zero values of m and n excepted.

For brevity, we write $\Omega_{m,n}$ in place of $2m\omega_1 + 2n\omega_2$, so that

$$\wp(z) = z^{-2} + {\sum_{m,n}}' \left\{ (z - \Omega_{m,n})^{-2} - \Omega_{m,n}^{-2} \right\}.$$

When m and n are such that $|\Omega_{m,n}|$ is large, the general term of the series defining $\wp(z)$ is $O(|\Omega_{m,n}|^{-3})$, and so (§3.4) the series converges absolutely and uniformly (with regard to z) except near its poles, namely the points $\Omega_{m,n}$. Therefore (§5.3), $\wp(z)$ is analytic throughout the whole z-plane except at the points $\Omega_{m,n}$, where it has double poles.

The introduction of this function $\wp(z)$ is due to Weierstrass [663, p. 245–255]. The subject-matter of the greater part of this chapter is due to Weierstrass, and is contained in his lectures, of which an account has been published by Schwarz [589]. See also Cayley [132]

---

[4] Throughout the chapter $\sum_{m,n}$ will be written to denote a summation over all integer values of m and n, a prime being inserted ($\sum_{m,n}'$) when the term for which $m = n = 0$ has to be omitted from the summation. It is also customary to write $\wp'(z)$ for the derivative of $\wp(z)$. The use of the prime in two senses will not cause confusion.

and Eisenstein [193, 194]. We now proceed to discuss properties of $\wp(z)$, and in the course of the investigation it will appear that $\wp(z)$ is an elliptic function with periods $2\omega_1$, $2\omega_2$.

Note   For purposes of numerical computation the series for $\wp(z)$ is useless on account of the slowness of its convergence. Elliptic functions free from this defect will be obtained in Chapter 21.

Example 20.2.1   Prove that

$$\wp(z) = \left(\frac{\pi}{2\omega_1}\right)^2 \left[-\frac{1}{3} + \sum_{n=-\infty}^{\infty} \operatorname{cosec}^2\left(\frac{z - 2n\omega_2}{2\omega_1}\pi\right) - {\sum_{n=-\infty}^{\infty}}' \operatorname{cosec}^2\left(\frac{n\omega_2}{\omega_1}\pi\right) \right].$$

20.21 Periodicity and other properties of $\wp(z)$

Since the series for $\wp(z)$ is a uniformly convergent series of analytic functions, term-by-term differentiation is legitimate (§5.3), and so

$$\wp'(z) = \frac{d}{dz}\wp(z) = -2\sum_{m,n} \frac{1}{(z - \Omega_{m,n})^3}.$$

The function $\wp'(z)$ is an odd function of z; for, from the definition of $\wp'(z)$, we at once get

$$\wp'(-z) = 2\sum_{m,n} (z + \Omega_{m,n})^{-3}.$$

But the set of points $-\Omega_{m,n}$ is the same as the set $\Omega_{m,n}$ and so the terms of $\wp'(z)$ are just the same as those of $-\wp'(-z)$, but in a different order. But, the series for $\wp'(z)$ being absolutely convergent (§3.4), the derangement of the terms does not affect its sum, and therefore

$$\wp'(-z) = -\wp'(z).$$

In like manner, the terms of the absolutely convergent series

$$\sum_{m,n}' \left\{ (z + \Omega_{m,n})^{-2} - \Omega_{m,n}^{-2} \right\}$$

are the terms of the series

$$\sum_{m,n}' \left\{ (z - \Omega_{m,n})^{-2} - \Omega_{m,n}^{-2} \right\}$$

in a different order, and hence

$$\wp(-z) = \wp(z);$$

that is to say, $\wp(z)$ is an even function of z.
    Further,

$$\wp'(z + 2\omega_1) = -2\sum_{m,n} (z - \Omega_{m,n} + 2\omega_1)^{-3};$$

but the set of points $\Omega_{m,n} - 2\omega_1$ is the same as the set $\Omega_{m,n}$, so the series for $\wp'(z + 2\omega_1)$ is a derangement of the series for $\wp'(z)$. The series being absolutely convergent, we have

$$\wp'(z + 2\omega_1) = \wp'(z);$$

that is to say, $\wp'(z)$ *has the period* $2\omega_1$; in like manner it has the period $2\omega_2$.

Since $\wp'(z)$ is analytic except at its poles, it follows from this result that $\wp'(z)$ *is an elliptic function.*

If now we integrate the equation $\wp'(z + 2\omega_1) = \wp'(z)$, we get

$$\wp(z + 2\omega_1) = \wp(z) + A,$$

where A is constant. Putting $z = -\omega_1$ and using the fact that $\wp(z)$ is an even function, we get $A = 0$, so that

$$\wp(z + 2\omega_1) = \wp(z);$$

in like manner $\wp(z + 2\omega_2) = \wp(z)$.

Since $\wp(z)$ has no singularities but poles, it follows from these two results that $\wp(z)$ *is an elliptic function.*

There are other methods of introducing both the circular and elliptic functions into analysis; for the circular functions the following may be noticed:

(1) The geometrical definition in which $\sin z$ is the ratio of the side opposite the angle z to the hypotenuse in a right-angled triangle of which one angle is z. This is the definition given in elementary text-books on Trigonometry; from our point of view it has various disadvantages, some of which are stated in the Appendix.

(2) The definition by the power series

$$\sin z = z - \frac{z^3}{3!} + \frac{z^5}{5!} - \cdots$$

(3) The definition by the product

$$\sin z = z \left(1 - \frac{z^2}{\pi^2}\right) \left(1 - \frac{z^2}{2^2 \pi^2}\right) \left(1 - \frac{z^2}{3^2 \pi^2}\right) \cdots$$

(4) The definition by 'inversion' of an integral

$$z = \int_0^{\sin z} (1 - t^2)^{-\frac{1}{2}} dt.$$

The periodicity properties may be obtained easily from (4) by taking suitable paths of integration (cf. Forsyth [220, §104]), but it is extremely difficult to prove that $\sin z$ defined in this way is an analytic function.

The reader will see later (§§22.82, 22.1, 20.42, 20.22 and Example 20.5.4) that elliptic functions may be defined by definitions analogous to each of these, with corresponding disadvantages in the cases of the first and fourth.

Example 20.2.2 Deduce the periodicity of $\wp(z)$ directly from its definition as a double series. *Hint.* It is not difficult to justify the necessary derangement.

20.22 *The differential equation satisfied by* $\wp(z)$

We shall now obtain an equation satisfied by $\wp(z)$, which will prove to be of great importance in the theory of the function.

The function $\wp(z) - z^{-2}$, which is equal to $\displaystyle\sum_{m,n}{}' \left\{(z - \Omega_{m,n})^{-2} - \Omega_{m,n}^{-2}\right\}$, is analytic in a region of which the origin is an internal point, and it is an even function of z. Consequently, by Taylor's theorem, we have an expansion of the form

$$\wp(z) - z^{-2} = \tfrac{1}{20}g_2 z^2 + \tfrac{1}{28}g_3 z^4 + O(z^3)$$

valid for sufficiently small values of $|z|$. It is easy to see that

$$g_2 = 60 \sum_{m,n}{}' \Omega_{m,n}^{-4}, \qquad g_3 = 140 \sum_{m,n}{}' \Omega_{m,n}^{-6}.$$

Thus $\wp(z) = z^{-2} + \tfrac{1}{20}g_2 z^2 + \tfrac{1}{28}g_3 z^4 + O(z^6)$; differentiating this result, we have

$$\wp'(z) = -2z^{-3} + \tfrac{1}{10}g_2 z + \tfrac{1}{7}g_3 z^3 + O(z^5).$$

Cubing and squaring these respectively, we get

$$\wp^3(z) = z^{-6} + \tfrac{3}{20}g_2 z^{-2} + \tfrac{3}{28}g_3 + O(z^2),$$

$$(\wp')^2(z) = 4z^{-6} - \tfrac{2}{5}g_2 z^{-2} - \tfrac{4}{7}g_3 + O(z^2).$$

Hence $(\wp')^2(z) - 4\wp^3(z) = -g_2 z^{-2} - g_3 + O(z^2)$, and so

$$(\wp')^2(z) - 4\wp^3(z) + g_2\wp(z) + g_3 = O(z^2).$$

That is to say, the function $(\wp')^2(z) - 4\wp^3(z) + g_2\wp(z) + g_3$, which is obviously an elliptic function, is analytic at the origin, and consequently it is also analytic at all congruent points. But such points are the only possible singularities of the function, and *so it is an elliptic function with no singularities*; it is therefore a constant (by Liouville's theorem). On making $z \to 0$, we see that this constant is zero.

Thus, finally, the function $\wp(z)$ *satisfies the differential equation*

$$(\wp')^2(z) = 4\wp^3(z) - g_2\wp(z) - g_3,$$

where g_2 and g_3 (called the *invariants*) are given by the equations

$$g_2 = 60 \sum_{m,n}{}' \Omega_{m,n}^{-4}, \qquad g_3 = 140 \sum_{m,n}{}' \Omega_{m,n}^{-6}.$$

Conversely, given the equation

$$\left(\frac{dy}{dz}\right)^2 = 4y^3 - g_2 y - g_3,$$

if numbers ω_1, ω_2 *can be determined*[5] *such that*

$$g_2 = 60 \sum_{m,n}{}' \Omega_{m,n}^{-4}, \qquad g_3 = 140 \sum_{m,n}{}' \Omega_{m,n}^{-6},$$

---

[5] The difficult problem of establishing the existence of such numbers ω_1 and ω_2 when g_2 and g_3 are given is solved in §21.73.

then the general solution of the differential equation is

$$y = \wp(\pm z + \alpha),$$

where α is the constant of integration. This may be seen by taking a new dependent variable u defined by the equation[6] $y = \wp(u)$, when the differential equation reduces to $\left(\dfrac{du}{dz}\right)^2 = 1$.

Since $\wp(z)$ is an even function of z, we have $y = \wp(z \pm \alpha)$, and so the solution of the equation can be written in the form

$$y = \wp(z + \alpha)$$

without loss of generality.

Example 20.2.3 Deduce from the differential equation that, if

$$\wp(z) = z^{-2} + \sum_{n=1}^{\infty} c_{2n} z^{2n},$$

then

$$c_2 = \frac{g_2}{2^2 \cdot 5}, \qquad c_4 = \frac{g_3}{2^2 \cdot 7}, \qquad c_6 = \frac{g_2^2}{2^4 \cdot 3 \cdot 5^2},$$

and

$$c_8 = \frac{3 g_2 g_3}{2^4 \cdot 5 \cdot 7 \cdot 11}, \qquad c_{10} = \frac{g_2^3}{2^5 \cdot 3 \cdot 5^3 \cdot 13} + \frac{g_3^2}{2^4 \cdot 7^2 \cdot 13}, \qquad c_{12} = \frac{g_2^2 g_3}{2^5 \cdot 3 \cdot 5^2 \cdot 7 \cdot 11}.$$

20.221 *The integral formula for* $\wp(z)$

Consider the equation

$$z = \int_{\zeta}^{\infty} (4t^3 - g_2 t - g_3)^{-\frac{1}{2}} \, dt,$$

determining z in terms of ζ; the path of integration may be any curve which does not pass through a zero of $4t^3 - g_2 t - g_3$.

On differentiation, we get

$$\left(\frac{d\zeta}{dz}\right)^2 = 4\zeta^3 - g_2 \zeta - g_3,$$

and so $\zeta = \wp(z + \alpha)$, where α is a constant.

Make $\zeta \to \infty$; then $z \to 0$, since the integral converges, and so α is a pole of the function $\wp$; i.e., α is of the form $\Omega_{m,n}$, and so $\zeta = \wp(z + \Omega_{m,n}) = \wp(z)$.

The result that the equation $z = \displaystyle\int_{\zeta}^{\infty} (4t^3 - g_2 t - g_3)^{-\frac{1}{2}} \, dt$ is equivalent to the equation $\zeta = \wp(z)$ is sometimes written in the form

$$z = \int_{\wp(z)}^{\infty} (4t^3 - g_2 t - g_3)^{-\frac{1}{2}} \, dt.$$

---

[6] This equation in u always has solutions, by §20.13

20.222 *An illustration from the theory of the circular functions*

The theorems obtained in §20.2–§20.221 may be illustrated by the corresponding results in the theory of the circular functions. Thus we may deduce the properties of the function $\operatorname{cosec}^2 z$ from the series $\sum\limits_{m=-\infty}^{\infty} (z - m\pi)^{-2}$ in the following manner:

Denote the series by $f(z)$; the series converges absolutely and uniformly[7] (with regard to z) except near the points $m\pi$ at which it obviously has double poles. Except at these points, $f(z)$ is analytic. The effect of adding any multiple of π to z is to give a series whose terms are the same as those occurring in the original series; since the series converges absolutely, the sum of the series is unaffected, and so $f(z)$ is *a periodic function of z with period π.*

Now consider the behaviour of $f(z)$ in the strip for which $-\tfrac{1}{2}\pi \le \operatorname{Re}(z) \le \tfrac{1}{2}\pi$. From the periodicity of $f(z)$, the value of $f(z)$ at any point in the plane is equal to its value at the corresponding point of the strip. In the strip $f(z)$ has one singularity, namely $z = 0$; and $f(z)$ is bounded as $z \to \infty$ in the strip, because the terms of the series for $f(z)$ are small compared with the corresponding terms of the comparison series $\sum\limits_{m=-\infty}^{\infty} m^{-2}$.

In a domain including the point $z = 0$, $f(z) - z^{-2}$ is analytic, and is an even function; and consequently there is a Maclaurin expansion

$$f(z) - z^{-2} = \sum_{n=0}^{\infty} a_{2n} z^{2n},$$

valid when $|z| < \pi$. It is easily seen that

$$a_{2n} = \frac{2}{\pi^{2n+2}}(2n + 1) \sum_{m=1}^{\infty} m^{-2n-2},$$

hence $a_0 = \tfrac{1}{3}$ and $a_2 = 6\pi^{-4} \sum\limits_{m=1}^{\infty} m^{-4} = \tfrac{1}{15}$. Hence, for small values of $|z|$

$$f(z) = z^{-2} + \tfrac{1}{3} + \tfrac{1}{15}z^2 + O(z^4).$$

Differentiating this result twice, and also squaring it, we have

$$f''(z) = 6z^{-4} + \frac{2}{15} + O(z^2), \qquad f^2(z) = z^{-4} + \frac{2}{3}z^{-2} + \frac{11}{45} + O(z^2).$$

It follows that

$$f''(z) - 6f^2(z) + 4f(z) = O(z^2).$$

That is to say, the function $f''(z) - 6f^2(z) + 4f(z)$ is analytic at the origin and it is obviously periodic. Since its only possible singularities are at the points $m\pi$, it follows from the periodic property of the function that it is an integral function. Further, it is bounded as $z \to \infty$ in the strip $-\tfrac{1}{2}\pi \le \operatorname{Re}(z) \le \tfrac{1}{2}\pi$, since $f(z)$ is bounded and so is[8] $f''(z)$. Hence $f''(z) - 6f^2(z) + 4f(z)$ is bounded in the strip, and therefore from its periodicity it is bounded

---

[7] By comparison with the series $\sum\limits_{m=-\infty}^{\infty} m^{-2}$.

[8] The series for $f'(z)$ may be compared with $\sum\limits_{m=-\infty}^{\infty} m^{-4}$.

everywhere. By Liouville's theorem (§5.63) it is therefore a constant. By making $z \to 0$, we see that the constant is zero. Hence the function $\operatorname{cosec}^2(z)$ satisfies the equation

$$f''(z) = 6f^2(z) - 4f(z).$$

Multiplying by $2f'(z)$ and integrating, we get

$$f'^2(z) = 4f^2(z)(f(z) - 1) + c,$$

where c is a constant, which is easily seen to be zero on making use of the power series for $f'(z)$ and $f(z)$. We thence deduce that

$$2z = \int_{f(z)}^{\infty} t^{-1}(t-1)^{-\frac{1}{2}} \, dt,$$

when an appropriate path of integration is chosen.

Example 20.2.4 If $y = \wp(z)$ and primes denote differentiations with regard to z, shew that

$$\frac{3y''^2}{4y'^4} - \frac{y'''}{2y'^3} = \frac{3}{16} \left\{ (y - e_1)^{-2} + (y - e_2)^{-2} + (y - e_3)^{-2} \right\}$$

$$- \frac{3}{8} y(y - e_1)^{-1}(y - e_2)^{-1}(y - e_3)^{-1},$$

where e_1, e_2, e_3 are the roots of the equation $4t^3 - g_2 t - g_3 = 0$. *Hint.* We have $y'^2 = 4y^3 - g_2 y - g_3 = 4(y - e_1)(y - e_2)(y - e_3)$. Differentiating logarithmically and dividing by y', we have

$$\frac{2y''}{y'^2} = \sum_{r=1}^{3} (y - e_r)^{-1}.$$

Differentiating again, we have

$$\frac{2y'''}{y'^3} - \frac{4y''^2}{y'^4} = - \sum_{r=1}^{3} (y - e_r)^{-2}.$$

Adding this equation multiplied by $\frac{1}{4}$ to the square of the preceding equation, multiplied by $\frac{1}{16}$, we readily obtain the desired result.

It should be noted that the left-hand side of the equation is half the Schwarzian derivative (see Cayley [137]) of z with respect to y; and so z is the quotient of two solutions of the equation

$$\frac{d^2 v}{dy^2} + \left\{ \frac{3}{16} \sum_{r=1}^{3} (y - e_r)^{-2} - \frac{3}{8} y \prod_{r=1}^{3} (y - e_r)^{-1} \right\} v = 0.$$

Example 20.2.5 Obtain the *properties of homogeneity* of the function $\wp(z)$; namely that

$$\wp \left(\lambda z \Big|_{\omega_2}^{\omega_1} \right) = \lambda^{-2} \wp \left(z \Big|_{\omega_2}^{\omega_1} \right) \qquad \wp(\lambda z; \lambda^{-4} g_2, \lambda^{-6} g_3) = \lambda^{-2} \wp(z; g_2, g_3),$$

where $\wp \left(z \Big|_{\omega_2}^{\omega_1} \right)$ denotes the function formed with periods $2\omega_1$, $2\omega_2$ and $\wp(z; g_2, g_3)$ denotes the function formed with invariants g_2, g_3. *Hint.* The former is a direct consequence of the definition of $\wp(z)$ by a double series; the latter may then be derived from the double series defining the g invariants.

20.3 The addition-theorem for the function $\wp(z)$

The function $\wp(z)$ possesses what is known as an *addition-theorem*; that is to say, there exists a formula expressing $\wp(z + y)$ as an algebraic function of $\wp(z)$ and $\wp(y)$ for general values of z and y. It is, of course, unnecessary to consider the special cases when y, or z, or $y + z$ is a period.

Consider the equations

$$\wp'(z) = A\wp(z) + B, \qquad \wp'(y) = A\wp(y) + B,$$

which determine A and B in terms of z and y unless $\wp(z) = \wp(y)$, i.e. unless[9] $z \equiv \pm y$ mod $(2\omega_1, 2\omega_2)$.

Now consider $\wp'(\zeta) - A\wp(\zeta) - B$, *qua* function of ζ. It has a triple pole at $\zeta = 0$ and consequently it has three, and only three, irreducible zeros, by §20.13; the sum of these is a period, by §20.14, and as $\zeta = z$, $\zeta = y$ are two zeros, the third irreducible zero must be congruent to $-z - y$. Hence $-z - y$ is a zero of $\wp'(\zeta) - A\wp(\zeta) - B$, and so

$$\wp'(-z - y) = A\wp(-z - y) + B.$$

Eliminating A and B from this equation and the equations by which A and B were defined, we have

$$\begin{vmatrix} \wp(z) & \wp'(z) & 1 \\ \wp(y) & \wp'(y) & 1 \\ \wp(z + y) & -\wp'(z + y) & 1 \end{vmatrix} = 0.$$

Since the derived functions occurring in this result can be expressed algebraically in terms of $\wp(z)$, $\wp(y)$, $\wp(z+y)$ respectively (§20.22), this result really expresses $\wp(z+y)$ algebraically in terms of $\wp(z)$ and $\wp(y)$. It is therefore an *addition-theorem*.

Other methods of obtaining the addition-theorem are indicated in Examples 20.3.1 and 20.3.2, and §20.312.

A symmetrical form of the addition-theorem may be noticed, namely that, if $u + v + w = 0$, then

$$\begin{vmatrix} \wp(u) & \wp'(u) & 1 \\ \wp(v) & \wp'(v) & 1 \\ \wp(w) & \wp'(w) & 1 \end{vmatrix} = 0.$$

20.31 Another form of the addition-theorem

Retaining the notation of §20.3, we see that the values of ζ, which make $\wp'(\zeta) - A\wp(\zeta) - B$ vanish, are congruent to one of the points $z, y, -z - y$.

---

[9] The function $\wp(z) - \wp(y)$, *qua* function of z, has double poles at points congruent to $z = 0$, and no other singularities; it therefore (§20.13) has only two irreducible zeros; and the points congruent to $z = \pm y$ therefore give *all* the zeros of $\wp(z) - \wp(y)$.

Hence $(\wp')^2(\zeta)-\{A\wp(\zeta)+B\}^2$ vanishes when ζ is congruent to any of the points $z, y, -z-y$. And so

$$4\wp^3(\zeta) - A^2\wp^2(\zeta) - (2AB + g_2)\wp(\zeta) - (B^2 + g_3)$$

vanishes when $\wp(\zeta)$ is equal to any one of $\wp(z)$, $\wp(y)$, $\wp(z+y)$.

For general values of z and y, $\wp(z)$, $\wp(y)$ and $\wp(z+y)$ are unequal and so they are all the roots of the equation

$$4Z^3 - A^2 Z^2 - (2AB + g_2)Z - (B^2 + g_3) = 0.$$

Consequently, by the ordinary formula for the sum of the roots of a cubic equation,

$$\wp(z) + \wp(y) + \wp(z+y) = \tfrac{1}{4}A^2,$$

and so

$$\wp(z+y) = \frac{1}{4}\left\{ \frac{\wp'(z) - \wp'(y)}{\wp(z) - \wp(y)} \right\}^2 - \wp(z) - \wp(y),$$

on solving the equations by which A and B were defined.

This result expresses $\wp(z+y)$ explicitly in terms of functions of z and of y.

20.311 The duplication formula for $\wp(z)$

The forms of the addition-theorem which have been obtained are both nugatory when $y = z$. But the result of §20.31 is true, in the case of any given value of z, for general values of y. Taking the limiting form of the result when y approaches z, we have

$$\lim_{y \to z} \wp(z+y) = \frac{1}{4} \lim_{y \to z} \left\{ \frac{\wp'(z) - \wp'(y)}{\wp(z) - \wp(y)} \right\}^2 - \wp(z) - \lim_{y \to z} \wp(y).$$

From this equation, we see that, if $2z$ is not a period, we have

$$\wp(2z) = \frac{1}{4} \lim_{h \to 0} \left\{ \frac{\wp'(z) - \wp'(z+h)}{\wp(z) - \wp(z+h)} \right\}^2 - 2\wp(z)$$

$$= \frac{1}{4} \lim_{h \to 0} \left\{ \frac{-h\wp''(z) + O(h^2)}{-h\wp'(z) + O(h^2)} \right\}^2 - 2\wp(z),$$

on applying Taylor's theorem to $\wp(z+h)$, $\wp'(z+h)$; and so

$$\wp(2z) = \frac{1}{4} \left\{ \frac{\wp''(z)}{\wp'(z)} \right\}^2 - 2\wp(z),$$

unless $2z$ is a period. This result is called the *duplication formula*.

Example 20.3.1 Prove that

$$\frac{1}{4} \left\{ \frac{\wp'(z) - \wp'(y)}{\wp(z) - \wp(y)} \right\}^2 - \wp(z) - \wp(z+y),$$

qua function of z, has no singularities at points congruent with $z = 0$, $\pm y$; and, by making use of Liouville's theorem, deduce the addition-theorem.

Example 20.3.2 Apply the process indicated in Example 20.3.1 to the function

$$\begin{vmatrix} \wp(z) & \wp'(z) & 1 \\ \wp(y) & \wp'(y) & 1 \\ \wp(z+y) & -\wp'(z+y) & 1 \end{vmatrix}$$

and deduce the addition-theorem.

Example 20.3.3 Shew that

$$\wp(z+y) + \wp(z-y) = \{\wp(z) - \wp(y)\}^{-2} \left[\{2\wp(z)\wp(y) - \tfrac{1}{2}g_2\} \{\wp(z) + \wp(y)\} - g_3 \right].$$

Hint. By the addition-theorem we have

$$\wp(z+y) + \wp(z-y) = \frac{1}{4} \left\{ \frac{\wp'(z) - \wp'(y)}{\wp(z) - \wp(y)} \right\}^2 - \wp(z) - \wp(y)$$

$$+ \frac{1}{4} \left\{ \frac{\wp'(z) + \wp'(y)}{\wp(z) - \wp(y)} \right\}^2 - \wp(z) - \wp(y),$$

$$= \frac{1}{2} \frac{\wp'^2(z) + \wp'^2(y)}{\{\wp(z) - \wp(y)\}^2} - 2\{\wp(z) + \wp(y)\}.$$

Replacing $\wp'^2(z)$ and $\wp'^2(y)$ by $4\wp^3(z) - g_2\wp(z) - g_3$ and $4\wp^3(y) - g_2\wp(y) - g_3$ respectively, and reducing, we obtain the required result.

Example 20.3.4 (Trinity, 1905) Shew, by Liouville's theorem, that

$$\frac{d}{dz} \{\wp(z-a)\,\wp(z-b)\}$$

$$= \wp(a-b)\{\wp'(z-a) + \wp'(z-b)\} - \wp'(a-b)\{\wp(z-a) - \wp(z-b)\}.$$

20.312 Abel's method for proving the addition-theorem for $\wp(z)$.

The following outline of a method of establishing the addition-theorem for $\wp(z)$ is instructive, though a completely rigorous proof would be long and tedious. This method is due to Abel [2, 4].

Let the invariants of $\wp(z)$ be g_2, g_3; take rectangular axes OX, OY in a plane, and consider the intersections of the cubic curve

$$y^2 = 4x^3 - g_2 x - g_3$$

with a variable line $y = mx + n$. If any point (x_1, y_1) be taken on the cubic, the equation in z

$$\wp(z) - x_1 = 0$$

has two solutions $+z_1, -z_1$ (§20.13) and all other solutions are congruent to these two.

Since $(\wp')^2(z) = 4\wp^3(z) - g_2\wp(z) - g_3$, we have $(\wp')^2(z) = y_1^2$; choose z_1 to be the solution for which $\wp'(z_1) = +y_1$, not $-y_1$. A number z_1 thus chosen will be called the *parameter* of (x_1, y_1) on the cubic. Now the abscissae x_1, x_2, x_3 of the intersections of the cubic with the variable line are the roots of

$$\phi(x) \equiv 4x^3 - g_2 x - g_3 - (mx + n)^2 = 0,$$

and so $\phi(x) \equiv 4(x - x_1)(x - x_2)(x - x_3)$.

The variation δx_r in one of these abscissae due to the variation in position of the line consequent on small changes δm, δn in the coefficients m, n is given by the equation

$$\phi'(x_r)\delta x_r + \frac{\partial\phi}{\partial m}\delta m + \frac{\partial\phi}{\partial n}\delta n = 0,$$

and so $\phi'(x_r)\delta x_r = 2(mx_r + n)(x_r\delta m + \delta n)$, whence

$$\sum_{r=1}^{3} \frac{\delta x_r}{mx_r + n} = 2\sum_{r=1}^{3} \frac{x_r\delta m + \delta n}{\phi'(x_r)},$$

provided that x_1, x_2, x_3 are unequal, so that $\phi'(x_r) \neq 0$.

Now, if we put $x(x\delta m + \delta n)/\phi(x)$, *qua* function of x, into partial fractions, the result is

$$\sum_{r=1}^{3} \frac{A_r}{(x - x_r)},$$

where

$$A_r = \lim_{x \to x_r} x(x\delta m + \delta n)\frac{(x - x_r)}{\phi(x)} = x_r(x_r\delta m + \delta n) \lim_{x \to x_r} \frac{(x - x_r)}{\phi(x)}$$

$$= \frac{x_r(x_r\delta m + \delta n)}{\phi'(x_r)},$$

by Taylor's theorem.

Putting $x = 0$, we get $\sum_{r=1}^{3} \delta x_r/y_r = 0$, i.e., $\sum_{r=1}^{3} \delta z_r = 0$. That is to say, *the sum of the parameters of the points of intersection is a constant independent of the position of the line.*

Vary the line so that all the points of intersection move off to infinity (no two points coinciding during this process), and it is evident that $z_1 + z_2 + z_3$ is equal to the sum of the parameters when the line is the line at infinity; but when the line is at infinity, each parameter is a period of $\wp(z)$ and therefore $z_1 + z_2 + z_3$ is a period of $\wp(z)$. Hence the sum of the parameters of three collinear points on the cubic is congruent to zero. This result having been obtained, the determinantal form of the addition-theorem follows as in §20.3.

20.32 *The constants* e_1, e_2, e_3

It will now be shewn that $\wp(\omega_1)$, $\wp(\omega_2)$, $\wp(\omega_3)$ (where $\omega_3 = -\omega_1 - \omega_2$), are all unequal; and, if their values be e_1, e_2, e_3, then e_1, e_2, e_3 are the roots of the equation $4t^3 - g_2t - g_3 = 0$.

First consider $\wp'(\omega_1)$. Since $\wp'(z)$ is an odd periodic function, we have

$$\wp'(\omega_1) = -\wp'(-\omega_1) = -\wp'(2\omega_1 - \omega_1) = -\wp'(\omega_1),$$

and so $\wp'(\omega_1) = 0$. Similarly $\wp'(\omega_2) = \wp'(\omega_3) = 0$.

Since $\wp'(z)$ is an elliptic function whose only singularities are triple poles at points congruent to the origin, $\wp'(z)$ has three, and only three (§20.13), irreducible zeros. Therefore the only zeros of $\wp'(z)$ are points congruent to ω_1, ω_2, ω_3.

Next consider $\wp(z) - e_1$. This vanishes at ω_1, and, since $\wp'(\omega_1) = 0$, it has a double zero at ω_1. Since $\wp(z)$ has only two irreducible poles, it follows from (§20.13) that the only zeros of $\wp(z) - e_1$ are congruent to ω_1. In like manner, the only zeros of $\wp(z) - e_2$, $\wp(z) - e_3$ are double zeros at points congruent to ω_2, ω_3, respectively.

Hence $e_1 \neq e_2 \neq e_3$. For if $e_1 = e_2$, then $\wp(z) - e_1$ has a zero at ω_2, which is a point not congruent to ω_1.

Also, since $(\wp')^2(z) = 4\wp^3(z) - g_2\wp(z) - g_3$ and since $\wp'(z)$ vanishes at $\omega_1, \omega_2, \omega_3$, it follows that $4\wp^3(z) - g_2\wp(z) - g_3$ vanishes when $\wp(z) = e_1$, e_2 or e_3. That is to say, e_1, e_2, e_3 are the roots of the equation

$$4t^3 - g_2 t - g_3 = 0.$$

From the well-known formulae connecting roots of equations with their coefficients, it follows that

$$e_1 + e_2 + e_3 = 0,$$
$$e_2 e_3 + e_3 e_1 + e_1 e_2 = -\tfrac{1}{4} g_2,$$
$$e_1 e_2 e_3 = \tfrac{1}{4} g_3.$$

Example 20.3.5   When g_2 and g_3 are real and the discriminant $g_2{}^3 - 27g_3^2$ is positive, shew that e_1, e_2, e_3 are all real; choosing them so that $e_1 > e_2 > e_3$, shew that

$$\omega_1 = \int_{e_1}^{\infty} (4t^3 - g_2 t - g_3)^{-\frac{1}{2}}\, dt, \qquad \text{and} \qquad \omega_3 = -i \int_{-\infty}^{e_3} (g_3 + g_2 t - 4t^3)^{-\frac{1}{2}}\, dt,$$

so that ω_1 is real and ω_3 a pure imaginary.

Example 20.3.6   Shew that, in the circumstances of Example 20.3.5, $\wp(z)$ is real on the perimeter of the rectangle whose corners are 0, ω_3, $\omega_1 + \omega_3$, ω_1.

20.33 The addition of a half-period to the argument of $\wp(z)$

From the form of the addition-theorem given in §20.31, we have

$$\wp(z + \omega_1) + \wp(z) + \wp(\omega_1) = \frac{1}{4} \left\{ \frac{\wp'(z) - \wp'(\omega_1)}{\wp(z) - \wp(\omega_1)} \right\}^2,$$

and so, since

$$\wp'^2(z) = 4 \prod_{r=1}^{3} \{\wp(z) - e_r\},$$

we have

$$\wp(z + \omega_1) = \frac{\{\wp(z) - e_2\} \{\wp(z) - e_3\}}{\wp(z) - e_1} - \wp(z) - e_1$$

i.e.

$$\wp(z + \omega_1) = e_1 + \frac{(e_1 - e_2)(e_1 - e_3)}{\wp(z) - e_1},$$

on using the result $e_1 + e_2 + e_3 = 0$; this formula expresses $\wp(z + \omega_1)$ in terms of $\wp(z)$.

Example 20.3.7   Shew that

$$\wp\left(\tfrac{1}{2}\omega_1\right) = e_1 \pm \{(e_1 - e_2)(e_1 - e_3)\}^{1/2}.$$

Example 20.3.8 (Math. Trip. 1913)   From the formula for $\wp(z + \omega_2)$ combined with the result of Example 20.3.7, shew that

$$\wp\left(\tfrac{1}{2}\omega_1 + \omega_2\right) = e_1 \mp \{(e_1 - e_2)(e_1 - e_3)\}^{1/2}.$$

Example 20.3.9   Shew that the value of $\wp'(z)\wp'(z + \omega_1)\,\wp'(z + \omega_2)\,\wp'(z + \omega_3)$ is equal to the discriminant of the equation $4t^3 - g_2t - g_3 = 0$. *Hint*. Differentiating the result of §20.33, we have

$$\wp'(z + \omega_1) = -(e_1 - e_2)(e_1 - e_3)\,\wp'(z)\,\{\wp(z) - e_1\}^{-2};$$

from this and analogous results, we have

$$\wp'(z)\wp'(z + \omega_1)\,\wp'(z + \omega_2)\,\wp'(z + \omega_3)$$

$$= (e_1 - e_2)^2\,(e_2 - e_3)^2\,(e_3 - e_1)^2\,\wp'^4(z)\prod_{r=1}^{3}\{\wp(z) - e_r\}^{-2}$$

$$= 16\,(e_1 - e_2)^2\,(e_2 - e_3)^2\,(e_3 - e_1)^2,$$

which is the discriminant $g_2^3 - 27g_3^2$ in question.

Example 20.3.10 (Math. Trip. 1913)   Shew that, with appropriate interpretations of the radicals,

$$\wp'\left(\tfrac{1}{2}\omega_1\right) = -2\,\{(e_1 - e_2)(e_1 - e_3)\}^{\frac{1}{2}}\left\{(e_1 - e_2)^{\frac{1}{2}} + (e_1 - e_3)^{\frac{1}{2}}\right\}.$$

Example 20.3.11   Shew that, with appropriate interpretations of the radicals,

$$\{\wp(2z) - e_2\}^{\frac{1}{2}} + \{\wp(2z) - e_3\}^{\frac{1}{2}}\,\{\wp(2z) - e_3\}^{\frac{1}{2}}\,\{\wp(2z) - e_1\}^{\frac{1}{2}}$$

$$+ \{\wp(2z) - e_1\}^{\frac{1}{2}}\,\{\wp(2z) - e_2\}^{\frac{1}{2}} = \wp(z) - \wp(2z).$$

20.4 Quasi-periodic functions. The function $\zeta(z)$

We shall next introduce the function $\zeta(z)$ defined by the equation[10]

$$\frac{d\zeta(z)}{dz} = -\wp(z),$$

coupled with the condition $\lim_{z\to 0}\left\{\wp(z) - z^{-1}\right\} = 0$.

Since the series for $\wp(z) - z^{-2}$ converges uniformly throughout any domain from which the neighbourhoods of the points[11] $\Omega'_{m,n}$ are excluded, we may integrate term-by-term (§4.7) and get

$$\zeta(z) - z^{-1} = -\int_0^z \left\{\wp(z) - z^{-2}\right\}\,dz$$

$$= -\sum_{m,n}{}' \int_0^x \left\{(z - \Omega_{m,n})^{-2} - \Omega_{m,n}^{-2}\right\}\,dz,$$

[10] This function should not, of course, be confused with the zeta-function of Riemann, discussed in Chapter 13.
[11] The symbol $\Omega'_{m,n}$ is used to denote all the points $\Omega_{m,n}$ with the exception of the origin (cf. §20.2).

and so

$$\zeta(z) = \frac{1}{z} + \sum_{m,n}' \left\{ \frac{1}{z - \Omega_{m,n}} + \frac{1}{\Omega_{m,n}} + \frac{z}{\Omega_{m,n}^2} \right\}.$$

The reader will easily see that the general term of this series is

$$O\left(|\Omega_{m,n}|^{-3}\right) \qquad \text{as} \quad |\Omega_{m,n}| \to \infty;$$

and hence (cf. §20.2), $\zeta(z)$ is an analytic function of z over the whole z-plane except at simple poles (the residue at each pole being $+1$) at all the points of the set $\Omega_{m,n}$.

It is evident that

$$-\zeta(-z) = \frac{1}{z} + \sum_{m,n}' \left\{ \frac{1}{z + \Omega_{m,n}} - \frac{1}{\Omega_{m,n}} + \frac{z}{\Omega_{m,n}^2} \right\},$$

and, since this series consists of the terms of the series for $\zeta(z)$ deranged in the same way as in the corresponding series of §20.21, we have, by §2.52,

$$\zeta(-z) = -\zeta(z),$$

that is to say, $\zeta(z)$ *is an odd function of* z.

Following up the analogy of §20.222, we may compare $\zeta(z)$ with the function $\cot z$ defined by the series

$$z^{-1} + \sum_{m=-\infty}^{\infty} \left\{ (z - m\pi)^{-1} + (m\pi)^{-1} \right\}$$

the equation $\dfrac{d}{dz} \cot z = -\operatorname{cosec}^2 z$ corresponding to $\dfrac{d}{dz} \zeta(z) = -\wp(z)$.

20.41 The quasi-periodicity of the function $\zeta(z)$

The heading of §20.4 was an anticipation of the result, which will now be proved, that $\zeta(z)$ is not a doubly-periodic function of z; and the effect on $\zeta(z)$ of increasing z by $2\omega_1$, or by $2\omega_2$ will be considered. It is evident from (20.12) in §20.12 that $\zeta(z)$ cannot be an elliptic function, in view of the fact that the residue of $\zeta(z)$ at every pole is $+1$.

If now we integrate the equation $\wp(z + 2\omega_1) = \wp(z)$, we get

$$\zeta(z + 2\omega_1) = \zeta(z) + 2\eta_1,$$

where $2\eta_1$ is the constant introduced by integration; putting $z = -\omega_1$ and taking account of the fact that $\zeta(z)$ is an odd function, we have $\eta_1 = \zeta(\omega_1)$. In like manner, $\zeta(z + 2\omega_2) = \zeta(z) + 2\eta_2$, where $\eta_2 = \zeta(\omega_2)$.

Example 20.4.1 (Frobenius and Stickelberger, [229]). Prove by Liouville's theorem that, if $x + y + z = 0$, then

$$\{\zeta(x) + \zeta(y) + \zeta(z)\}^2 + \zeta'(x) + \zeta'(y) + \zeta'(z) = 0.$$

Note This result is a pseudo-addition-theorem. It is not a true addition-theorem since $\zeta'(x)$, $\zeta'(y)$, $\zeta'(z)$ are not algebraic functions of $\zeta(x)$, $\zeta(y)$, $\zeta(z)$.

Example 20.4.2 (Math. Trip. 1894)   Prove by Liouville's theorem that

$$\begin{vmatrix} 1 & \wp(x) & \wp^2(x) \\ 1 & \wp(y) & \wp^2(y) \\ 1 & \wp(z) & \wp^2(z) \end{vmatrix}^2 \Bigg/ \begin{vmatrix} 1 & \wp(x) & \wp'(x) \\ 1 & \wp(y) & \wp'(y) \\ 1 & \wp(z) & \wp'(z) \end{vmatrix} = \zeta(x+y+z) - \zeta(x) - \zeta(y) - \zeta(z).$$

Obtain a generalisation of this theorem involving n variables.

20.411 The relation between η_1 and η_2

We shall now shew that

$$\eta_1\omega_2 - \eta_2\omega_1 = \frac{\pi i}{2}.$$

To obtain this result consider $\displaystyle\int_C \zeta(z)\,dz$ taken round the boundary of a cell. There is one pole of $\zeta(z)$ inside the cell, the residue there being $+1$. Hence

$$\int_C \zeta(z)\,dz = 2\pi i.$$

Modifying the contour integral in the manner of §20.12, we get

$$2\pi i = \int_t^{t+2\omega_1} \{\zeta(z) - \zeta(z+2\omega_2)\}\,dz - \int_t^{t+2\omega_2} \{\zeta(z) - \zeta(z+2\omega_1)\}\,dz$$

$$= -2\eta_2 \int_t^{t+2\omega_1} dt + 2\eta_1 \int_t^{t+2\omega_2} dt,$$

and so $2\pi i = -4\eta_2\omega_1 + 4\eta_1\omega_2$, which is the required result.

20.42 The function $\sigma(z)$

We shall next introduce the function $\sigma(z)$, defined by the equation

$$\frac{d}{dz}\log\sigma(z) = \zeta(z)$$

coupled with the condition $\lim_{z\to 0}\sigma(z)/z = 1$.

On account of the uniformity of convergence of the series for $\zeta(z)$, except near the poles of $\zeta(z)$, we may integrate the series term-by-term. Doing so, and taking the exponential of each side of the resulting equation, we get

$$\sigma(z) = z\prod_{m,n}' \left\{ \left(1 - \frac{z}{\Omega_{m,n}}\right)\exp\left(\frac{z}{\Omega_{m,n}} + \frac{z^2}{2\Omega_{m,n}^2}\right)\right\};$$

the constant of integration has been adjusted in accordance with the condition stated.

By the methods employed in §§20.2, 20.12, 20.4, the reader will easily obtain the following results:

(I) The product for $\sigma(z)$ converges absolutely and uniformly in any bounded domain of values of z.

(II) The function $\sigma(z)$ is an odd integral function of z with simple zeros at all the points $\Omega_{m,n}$.

The function $\sigma(z)$ may be compared with the function $\sin z$ defined by the product

$$z \prod_{m=-\infty}^{\infty}{}' \left\{ \left(1 - \frac{z}{m\pi}\right) e^{z/(m\pi)} \right\},$$

the relation $\dfrac{d}{dz} \log \sin z = \cot z$ corresponding to $\dfrac{d}{dz} \log \sigma(z) = \zeta(z)$.

20.421 The quasi-periodicity of the function $\sigma(z)$

If we integrate the equation

$$\zeta(z + 2\omega_1) = \zeta(z) + 2\eta_1,$$

we get $\sigma(z + 2\omega_1) = ce^{2\eta_1 z}\sigma(z)$, where c is the constant of integration; to determine c, we put $z = -\omega_1$, and then

$$\sigma(\omega_1) = -ce^{-2\eta_1\omega_1}\sigma(\omega_1).$$

Consequently $c = -e^{2\eta_1\omega_1}$, and $\sigma(z + 2\omega_1) = -e^{2\eta_1(z+\omega_1)}\sigma(z)$. In like manner $\sigma(z + 2\omega_2) = e^{2\eta_2(z+\omega_2)}\sigma(z)$.

These results exhibit the behaviour of $\sigma(z)$ when z is increased by a period of $\wp(z)$. If, as in §20.32, we write $\omega_3 = \omega_1 - \omega_2$, then three other sigma-functions are defined by the equations

$$\sigma_r(z) = e^{-\eta_r z} \frac{\sigma(z + \omega_r)}{\sigma(\omega_r)} \qquad (r = 1, 2, 3).$$

The four sigma-functions are analogous to the four theta-functions discussed in Chapter 21 (see §21.9).

Example 20.4.3 Shew that, if m and n are any integers,

$$\sigma(z + 2m\omega_1 + 2m\omega_2)$$
$$= (-1)^{m+n}\sigma(z)\exp\left\{(2m\eta_1 + 2n\eta_2)z + 2m^2\eta_1\omega_1 + 4mn\eta_1\omega_2 + 2n^2\eta_2\omega_2\right\},$$

and deduce that $\eta_1\omega_2 - \eta_2\omega_1$ is an integer multiple of $\frac{1}{2}\pi i$.

Example 20.4.4 Shew that, if $q = \exp(\pi i\omega_2/\omega_1)$, so that $|q| < 1$, and if

$$F(z) = \exp\left(\frac{\eta_1 z^2}{2\omega_1}\right) \sin\left(\frac{\pi z}{2\omega_1}\right) \prod_{n=1}^{\infty} \left\{1 - 2q^{2n}\cos\frac{\pi z}{\omega_1} + q^{4n}\right\},$$

then $F(z)$ is an integral function with the same zeros as $\sigma(z)$ and also $F(z)/\sigma(z)$ is a doubly-periodic function of z with periods $2\omega_1$, $2\omega_2$.

Example 20.4.5 Deduce from Example 20.4.4, by using Liouville's theorem, that

$$\sigma(z) = \frac{2\omega_1}{\pi}\exp\left(\frac{\eta_1 z^2}{2\omega_1}\right)\sin\left(\frac{\pi z}{2\omega_1}\right)\prod_{n=1}^{\infty}\left\{\frac{1 - 2q^{2n}\cos(\pi z/\omega_1) + q^{4n}}{(1 - q^{2n})^2}\right\}.$$

Example 20.4.6 Obtain the result of Example 20.4.5 by expressing each factor on the right as a singly infinite product.

20.5 Formulae expressing any elliptic function in terms of Weierstrassian functions with the same periods

There are various formulae analogous to the expression of any rational fraction as

(I) a quotient of two sets of products of linear factors,
(II) a sum of partial fractions; of the first type there are two formulae involving sigma-functions and Weierstrassian elliptic functions respectively; of the second type there is a formula involving derivatives of zeta-functions.

These formulae will now be obtained.

20.51 The expression of any elliptic function in terms of $\wp(z)$ and $\wp'(z)$

Let $f(z)$ be any elliptic function, and let $\wp(z)$ be the Weierstrassian elliptic function formed with the same periods $2\omega_1, 2\omega_2$.

We first write

$$f(z) = \frac{1}{2}[f(z) + f(-z)] + \frac{1}{2}\left[\frac{f(z) - f(-z)}{\wp'(z)}\right]\wp'(z).$$

The functions

$$f(z) + f(-z), \qquad \{f(z) - f(-z)\}/\wp'(z)$$

are both *even* functions, and they are obviously elliptic functions when $f(z)$ is an elliptic function.

The solution of the problem before us is therefore effected *if we can express any even elliptic function $\phi(z)$, say, in terms of $\wp(z)$.*

Let a be a zero of $\phi(z)$ in any cell; then the point in the cell congruent to $-a$ will also be a zero. The irreducible zeros of $\phi(z)$ may therefore be arranged in two sets, say $a_1, a_2, \ldots, a_n$ and certain points congruent to $-a_1, -a_2, \ldots, -a_n$. In like manner, the irreducible poles may be arranged in two sets, say $b_1, b_2, \ldots, b_n$ and certain points congruent to $-b_1, -b_2, \ldots, -b_n$.

Consider now the function[12]

$$\frac{1}{\phi(z)}\prod_{r=1}^{n}\left\{\frac{\wp(z) - \wp(a_r)}{\wp(z) - \wp(b_r)}\right\}.$$

It is an elliptic function of z, and clearly it has no poles; for the zeros of $\phi(z)$ are zeros (of the same order of multiplicity) of the numerator of the product, and the zeros of the denominator of the product are poles (of the same order) of $\phi(z)$. Consequently by Liouville's theorem it is a constant, A_1, say.

Therefore

$$\phi(z) = A_1 \prod_{r=1}^{n}\left\{\frac{\wp(z) - \wp(a_r)}{\wp(z) - \wp(b_r)}\right\},$$

---

[12] If any one of the points a_r or b_r is congruent to the origin, we omit the corresponding factor $\wp(z) - \wp(a_r)$ or $\wp(z) - \wp(b_r)$. The zero (or pole) of the product and the zero (or pole) of $\phi(z)$ at the origin are then of the same order of multiplicity. In this product, and in that of §20.53, factors corresponding to multiple zeros and poles have to be repeated the appropriate number of times.

that is to say, $\phi(z)$ has been expressed as a rational function of $\wp(z)$.

Carrying out this process with each of the functions

$$f(z) + f(-z), \qquad \{f(z) - f(-z)\} / \wp'(z)$$

we obtain the theorem that *any elliptic function $f(z)$ can be expressed, in terms of the Weierstrassian elliptic functions $\wp(z)$ and $\wp'(z)$ with the same periods, the expression being rational in $\wp(z)$ and linear in $\wp'(z)$.*

20.52 *The expression of any elliptic function as a linear combination of zeta-functions and their derivatives*

Let $f(z)$ be any elliptic function with periods $2\omega_1, 2\omega_2$. Let a set of irreducible poles of $f(z)$ be $a_1, a_2, \ldots, a_n$, and let the principal part (§5.61) of $f(z)$ near the pole a_k be

$$\frac{c_{k,1}}{z - a_k} + \frac{c_{k,2}}{(z - a_k)^2} + \cdots + \frac{c_{k,r_k}}{(z - a_k)^{r_k}}.$$

Then we can shew that

$$f(z) = A_2 + \sum_{k=1}^{n} \left\{ c_{k,1}\zeta(z - a_k) - c_{k,2}\zeta'(z - a_k) + \cdots + \frac{(-1)^{r_k-1}c_{k,r_k}}{(r_k - 1)!}\zeta^{(r_k-1)}(z - a_k) \right\},$$

where A_2 is a constant, and $\zeta^{(s)}(z)$ denotes $\dfrac{d^s}{dz^s}\zeta(z)$. Denoting the summation on the right by $F(z)$, we see that

$$F(z + 2\omega_1) - F(z) = \sum_{k=1}^{n} 2\eta_1 c_{k,1},$$

by §20.41, since all the derivatives of the zeta-functions are periodic. But $\sum_{k=1}^{n} c_{k,1}$ is the sum of the residues of $f(z)$ at all of its poles in a cell, and is consequently (§20.12) zero. Therefore $F(z)$ has period $2\omega_1$, and similarly it has period $2\omega_2$; and so $f(z) - F(z)$ is an elliptic function.

Moreover $F(z)$ has been so constructed that $f(z) - F(z)$ has no poles at the points $a_1, a_2, \ldots, a_n$; and hence it has no poles in a certain cell. It is consequently a constant, A_2, by Liouville's theorem.

Thus the function $f(z)$ can be expanded in the form

$$A_2 + \sum_{k=1}^{n} \sum_{s=1}^{r_k} \frac{(-1)^{s-1}}{(s - 1)!} c_{k,s}\, \zeta^{(s-1)}(z - a_k).$$

This result is of importance in the problem of integrating an elliptic function $f(z)$ when the principal part of its expansion at each of its poles is known; for we obviously have

$$\int^z f(z)\, dz = A_2 z + \sum_{k=1}^{n} \left[c_{k,1} \log \sigma(z - a_k) \right.$$

$$\left. + \sum_{s=2}^{r_k} \frac{(-1)^{s-1}}{(s - 1)!} c_{k,s}\, \zeta^{(s-2)}(z - a_k) \right] + C,$$

where C is a constant of integration.

Example 20.5.1 Shew by the method of this article that

$$\wp^2(z) = \tfrac{1}{6}\wp''(z) + \tfrac{1}{12}g_2,$$

and deduce that

$$\int^z \wp^2(z)\,dz = \tfrac{1}{6}\wp'(z) + \tfrac{1}{12}g_2 z + C,$$

where C is a constant of integration.

20.53 The expression of any elliptic function as a quotient of sigma-functions

Let $f(z)$ be any elliptic function, with periods $2\omega_1$ and $2\omega_2$, and let a set of irreducible zeros of $f(z)$ be $a_1, a_2, \ldots, a_n$. Then (§20.14) we can choose a set of poles $b_1, b_2, \ldots, b_n$ such that all poles of $f(z)$ are congruent to one or other of them and

$$a_1 + a_2 + \cdots + a_n = b_1 + b_2 + \cdots + b_n.$$

Multiple zeros or poles are, of course, to be reckoned according to their degree of multiplicity; to determine $b_1, b_2, \ldots, b_n$, we choose $b_1, b_2, \ldots, b_{n-1}, b'_n$ to be the set of poles in the cell in which $a_1, a_2, \ldots, a_n$ lie, and then choose b_n, congruent to b'_n, in such a way that the required equation is satisfied.

Consider now the function

$$\prod_{r=1}^{n} \frac{\sigma(z - a_r)}{\sigma(z - b_r)}.$$

This product obviously has the same poles and zeros as $f(z)$; also the effect of increasing z by $2\omega_1$ is to multiply the function by

$$\prod_{r=1}^{n} \frac{\exp\{2\eta_1(z - a_r)\}}{\exp\{2\eta_1(z - b_r)\}} = 1.$$

The function therefore has period $2\omega_1$ (and in like manner it has period $2\omega_2$), and so the quotient

$$f(z) \Big/ \prod_{r=1}^{n} \frac{\sigma(z - a_r)}{\sigma(z - b_r)}$$

is an elliptic function with no zeros or poles. By Liouville's theorem, it must be a constant, A_3 say. Thus the function $f(z)$ can be expressed in the form

$$f(z) = A_3 \prod_{r=1}^{n} \frac{\sigma(z - a_r)}{\sigma(z - b_r)}.$$

An elliptic function is consequently determinate (save for a multiplicative constant) when its periods and a set of irreducible zeros and poles are known.

Example 20.5.2 Shew that

$$\wp(z) - \wp(y) = -\frac{\sigma(z + y)\sigma(z - y)}{\sigma^2(z)\sigma^2(y)}.$$

Example 20.5.3 Deduce by differentiation, from Example 20.5.1, that

$$\frac{1}{2}\frac{\wp'(z) - \wp'(y)}{\wp(z) - \wp(y)} = \zeta(z + y) - \zeta(z) - \zeta(y),$$

and by further differentiation obtain the addition-theorem for $\wp(z)$.

Example 20.5.4 If $\sum_{r=1}^{n} a_r = \sum_{r=1}^{n} b_r$, shew that

$$\sum_{r=1}^{n} \frac{\sigma(a_r - b_1)\sigma(a_r - b_2)\cdots\sigma(a_r - b_n)}{\sigma(a_r - a_1)\sigma(a_r - a_2)\cdots * \cdots\sigma(a_r - a_n)} = 0,$$

the $*$ denoting that the vanishing factor $\sigma(a_r - a_r)$ is to be omitted.

Example 20.5.5 Shew that

$$\wp(z) - e_r = \frac{\sigma_r^2(z)}{\sigma^2(z)} \qquad (r = 1, 2, 3).$$

[It is customary to define $\{\wp(z) - e_r\}^{1/2}$ to mean $\sigma_r(z)/\sigma(z)$, not $-\sigma_r(z)/\sigma(z)$.]

Example 20.5.6 Establish, by Example 20.5.1, the *three-term equation*, namely,

$$\sigma(z + a)\,\sigma(z - a)\,\sigma(b + c)\,\sigma(b - c) + \sigma(z + b)\,\sigma(z - b)\,\sigma(c + a)\,\sigma(c - a)$$
$$+ \sigma(z + c)\,\sigma(z - c)\sigma(a + b)\,\sigma(a - b) = 0.$$

This result is due to Weierstrass; see p. 47 of the edition of his lectures by Schwarz [589].

The equation is characteristic of the sigma-function; it has been proved by Halphen [268, Vol. I, p. 187] that no function essentially different from the sigma-function satisfies an equation of this type. See Example 20.38.

20.54 The connexion between any two elliptic functions with the same periods

We shall now prove the important result that *an algebraic relation exists between any two elliptic functions, $f(z)$ and $\phi(z)$, with the same periods*. For, by §20.51, we can express $f(z)$ and $\phi(z)$ as rational functions of the Weierstrassian functions $\wp(z)$ and $\wp'(z)$ with the same periods, so that

$$f(z) = R_1(\wp(z), \wp'(z)), \qquad \phi(z) = R_2(\wp(z), \wp'(z)),$$

where R_1 and R_2 denote rational functions of two variables.

Eliminating $\wp(z)$ and $\wp'(z)$ algebraically from these two equations and

$$(\wp')^2(z) = 4\wp^3(z) - g_2\wp(z) - g_3,$$

we obtain an algebraic relation connecting $f(z)$ and $\phi(z)$; and the theorem is proved.

A particular case of the proposition is that every elliptic function is connected with its derivate by an algebraic relation.

If now we take the orders of the elliptic functions $f(z)$ and $\phi(z)$ to be m and n respectively, then, corresponding to any given value of $f(z)$ there is (§20.13) a set of m irreducible values of z, and consequently there are m values (in general distinct) of $\phi(z)$. So, corresponding to each value of f, there are m values of ϕ and, similarly, to each value of ϕ correspond n values of f.

The relation between $f(z)$ and $\phi(z)$ is therefore (in general) of degree m in ϕ and n in f. The relation *may* be of lower degree. Thus, if $f(z) = \wp(z)$, of order 2, and $\phi(z) = \wp^2(z)$, of order 4, the relation is $f^2 = \phi$.

As an illustration of the general result take $f(z) = \wp(z)$, of order 2, and $\phi(z) = \wp'(z)$, of order 3. The relation should be of degree 2 in ϕ and of degree 3 in f; this is, in fact, the case, for the relation is $\phi^2 = 4f^3 - g_2 f - g_3$.

Example 20.5.7   If u, v, w are three elliptic functions of their argument of the second order with the same periods, shew that, in general, there exist two distinct relations which are linear in each of u, v, w namely

$$Auvw + Bvw + Cwu + Duv + Eu + Fv + Gw + H = 0,$$

$$A'uvw + B'vw + C'wu + D'uv + E'u + F'v + G'w + H' = 0,$$

where $A, B, \dots, H'$ are constants.

20.6 On the integration of $\left(a_0 x^4 + 4a_1 x^3 + 6a_2 x^2 + 4a_3 x + a_4\right)^{-1/2}$

It will now be shewn that certain problems of integration, which are insoluble by means of elementary functions only, can be solved by the introduction of the function $\wp(z)$.

Let $a_0 x^4 + 4a_1 x^3 + 6a_2 x^2 + 4a_3 x + a_4 \equiv f(x)$ be any quartic polynomial which has no repeated factors; and let its invariants (see Burnside and Panton [111]) be

$$g_2 \equiv a_0 a_4 - 4a_1 a_3 + 3a_2^2,$$

$$g_3 \equiv a_0 a_2 a_4 + 2a_1 a_2 a_3 - a_2^3 - a_0 a_3^2 - a_1^2 a_4.$$

Let $z = \displaystyle\int_{x_0}^{x} \{f(t)\}^{-\frac{1}{2}}\, dt$, where x_0 is any root of the equation $f(x) = 0$; then, if the function $\wp(z)$ be constructed (see §21.73) with the invariants g_2 and g_3, *it is possible to express x as a rational function of $\wp(z; g_2, g_3)$*.

Note   The reason for assuming that $f(x)$ has no repeated factors is that, when $f(x)$ has a repeated factor, the integration can be effected with the aid of circular or logarithmic functions only. For the same reason, the case in which $a_0 = a_1 = 0$ need not be considered.

By Taylor's theorem, we have

$$f(t) = 4A_3(t - x_0) + 6A_2(t - x_0)^2 + 4A_1(t - x_0)^3 + A_0(t - x_0)^4,$$

(since $f(x_0) = 0$), where

$$A_0 = a_0, \quad A_1 = a_0 x_0 + a_1, \quad A_2 = a_0 x_0^2 + 2a_1 x_0 + a_2,$$

$$A_3 = a_0 x_0^3 + 3a_1 x_0^2 + 3a_2 x_0 + a_3.$$

On writing $(t - x_0)^{-1} = \tau$, $(x - x_0)^{-1} = \xi$, we have

$$z = \int_{\xi}^{\infty} \left\{4A_3 \tau^3 + 6A_2 \tau^2 + 4A_1 \tau + A_0\right\}^{-\frac{1}{2}}\, d\tau.$$

To remove the second term in the cubic involved, write[13]

$$\tau = A_3^{-1}\left(\sigma - \tfrac{1}{2}A_2\right), \qquad \xi = A_3^{-1}\left(s - \tfrac{1}{2}A_2\right),$$

and we get

$$z = \int_3^{\infty} \left\{4\sigma^3 - (3A_2^2 - 4A_1A_3)\sigma - (2A_1A_2A_3 - A_2^3 - A_0A_3^2)\right\}^{-\frac{1}{2}} d\sigma.$$

The reader will verify, without difficulty, that $3A_2^2 - 4A_1A_3$ and $2A_1A_2A_3 - A_2^3 - A_0A_3^2$ are respectively equal to g_2 and g_3, the invariants of the original quartic, and so $s = \wp(z; g_2, g_3)$. Now $x = x_0 + A_3\left\{s - \tfrac{1}{2}A_2\right\}^{-1}$, and hence

$$x = x_0 + \tfrac{1}{4}f'(x_0)\left\{\wp(z; g_2, g_3) - \tfrac{1}{24}f''(x_0)\right\}^{-1},$$

so that x has been expressed as a rational function of $\wp(z; g_2, g_3)$.

This formula for x is to be regarded as the integral equivalent of the relation

$$z = \int_{x_0}^x \left\{f(t)\right\}^{-\frac{1}{2}} dt.$$

Example 20.6.1   With the notation of this article, shew that

$$\{f(x)\}^{\frac{1}{2}} = \frac{-f'(x_0)\wp'(z)}{4\left\{\wp(z) - \tfrac{1}{24}f''(x_0)\right\}^2}.$$

Example 20.6.2 (Weierstrass)   Shew that, if

$$z = \int_a^x \left\{f(t)\right\}^{-\frac{1}{2}} dt,$$

where a is *any* constant, not necessarily a zero of $f(x)$, and $f(x)$ is a quartic polynomial with no repeated factors, then

$$x = a + \frac{\{f(a)\}^{\frac{1}{2}}\,\wp'(z) + \tfrac{1}{2}f'(a)\left\{\wp(z) - \tfrac{1}{24}f''(a)\right\} + \tfrac{1}{24}f(a)f'''(a)}{2\left\{\wp(z) - \tfrac{1}{24}f''(a)\right\}^2 - \tfrac{1}{48}f(a)f^{iv}(a)}$$

the function $\wp(z)$ being formed with the invariants of the quartic $f(x)$.

Note   This result was first published in 1865, in an Inaugural-dissertation at Berlin by Biermann, who ascribed it to Weierstrass. An alternative result, due to Mordell [473], is that, if

$$z = \int_{a,b}^{x,y} \frac{y\,dx - x\,dy}{\sqrt{f(x,y)}},$$

where $f(x,y)$ is a homogeneous quartic whose Hessian is $h(x,y)$, then we may take

$$x = a\wp'(z)\sqrt{f} + \tfrac{1}{2}\wp(z)f_b + \tfrac{1}{2}h_b,$$
$$y = b\wp'(z)\sqrt{f} - \tfrac{1}{2}\wp(z)f_a - \tfrac{1}{2}h_a,$$

where f and h stand for $f(a,b)$ and $h(a,b)$, and suffixes denote partial differentiations.

[13] This substitution is legitimate since $A_3 \neq 0$; for the equation $A_3 = 0$ involves $f(x) = 0$ having $x = x_0$ as a repeated root.

Example 20.6.3   Shew that, with the notation of Example 20.6.1

$$\wp(z) = \frac{\{f(x)f(a)\}^{\frac{1}{2}} + f(a)}{2(x-a)^2} + \frac{f'(a)}{4(x-a)} + \frac{f''(a)}{24},$$

and

$$\wp'(z) = -\left\{\frac{f(x)}{(x-a)^3} - \frac{f'(x)}{4(x-a)^2}\right\}\{f(a)\}^{\frac{1}{2}} - \left\{\frac{f(a)}{(x-d)^3} + \frac{f'(a)}{4(x-a)^2}\right\}\{f(x)\}^{\frac{1}{2}}.$$

20.7 The uniformisation of curves of genus unity

The theorem of §20.6 may be stated somewhat differently thus:

If the variables x and y are connected by an equation of the form

$$y^2 = a_0 x^4 + 4a_1 x^3 + 6a_2 x^2 + 4a_3 x + a_4,$$

then they can be expressed as one-valued functions of a variable z by the equations

$$x = x_0 + \frac{1}{4}f'(x_0)\left\{\wp(z) - \tfrac{1}{24}f''(x_0)\right\}^{-1}$$

$$y = -\frac{1}{4}f'(x_0)\wp'(z)\left\{\wp(z) - \tfrac{1}{24}f''(x_0)\right\}^{-2}$$

where $f(x) = a_0 x^4 + 4a_1 x^3 + 6a_2 x^2 + 4a_3 x + a_4 x_0$ *is any zero of* $f(x)$, *and the function* $\wp(z)$ *is formed with the invariants of the quartic; and z is such that*

$$z = \int_{x_0}^x \{f(t)\}^{-\frac{1}{2}}\,dt.$$

This term employs the word *uniform* in the sense *one-valued*. To prevent confusion with the idea of uniformity as explained in Chapter 3, throughout the present work we have used the phrase *one-valued function* as being preferable to *uniform function*.

It is obvious that y is a two-valued function of x and x is a four-valued function of y; and the fact, that x and y can be expressed as *one-valued* functions of the variable z, makes this variable z of considerable importance in the theory of algebraic equations of the type considered; z is called the *uniformising variable* of the equation

$$y^2 = a_0 x^4 + 4a_1 x^3 + 6a_2 x^2 + 4a_3 x + a_4.$$

The reader who is acquainted with the theory of algebraic plane curves will be aware that they are classified according to their *deficiency* or *genus*, a number whose geometrical significance is that it is the difference between the number of double points possessed by the curve and the maximum number of double points which can be possessed by a curve of the same degree as the given curve.

Curves whose deficiency is zero are called *unicursal*. If $f(x, y) = 0$ is the equation of a unicursal curve, it is well known (see Salmon [571, Chapter 2]) that x and y can be expressed as *rational functions of a parameter*. Since rational functions are one-valued, this parameter is a *uniformising variable* for the curve in question.

Next consider curves of genus unity; let $f(x, y) = 0$ be such a curve; then it has been shewn by Clebsch [150] that x and y can be expressed as rational functions of ξ and η where

η_2 is a polynomial in ξ of degree three or four. (A proof of the result of Clebsch is given by Forsyth [220, §248]. See also Cayley [134].)

Hence, by §20.6, ξ and η can be expressed as rational functions of $\wp(z)$ and $\wp'(z)$ (these functions being formed with suitable invariants), and so x and y can be expressed as one-valued (elliptic) functions of z, which is therefore a uniformising variable for the equation under consideration.

When the genus of the algebraic curve $f(x, y) = 0$ is greater than unity, the uniformisation can be effected by means of what are known as *automorphic functions*. Two classes of such functions of genus greater than unity have been constructed, the first by Weber [658], the other by Whittaker [669]. The analogue of the period-parallelogram is known as the 'fundamental polygon'. In the case of Weber's functions this polygon is 'multiply-connected', i.e. it consists of a region containing islands which have to be regarded as not belonging to it; whereas in the case of the second class of functions, the polygon is 'simply-connected', i.e. it contains no such islands. The latter class of functions may therefore be regarded as a more immediate generalisation of elliptic functions. See Ford [213].

20.8 Miscellaneous examples

Example 20.1   Shew that

$$\wp(z + y) - \wp(z - y) = -\wp'(z)\wp'(y)\{\wp(z) - \wp(y)\}^{-2}.$$

Example 20.2 (Math. Trip. 1897)   Prove that

$$\wp(z) - \wp(z + y + w) = 2\frac{\partial}{\partial z}\frac{\sum \wp^2(z)\{\wp(y) - \wp(w)\}}{\sum \wp'(z)\{\wp(y) - \wp(w)\}},$$

where, on the right-hand side, the subject of differentiation is symmetrical in z, y, and w.

Example 20.3 (Trinity, 1898)   Shew that

$$\begin{vmatrix} \wp'''(z - y) & \wp'''(y - w) & \wp'''(w - z) \\ \wp''(z - y) & \wp''(y - w) & \wp''(w - z) \\ \wp(z - y) & \wp(y - w) & \wp(w - z) \end{vmatrix} = \frac{1}{2}g_2 \begin{vmatrix} \wp'''(z - y) & \wp'''(y - w) & \wp'''(w - z) \\ \wp(z - y) & \wp(y - w) & \wp(w - z) \\ 1 & 1 & 1 \end{vmatrix}.$$

Example 20.4 (Math. Trip. 1897)   If $y = \wp(z) - e_1$, $y' = \dfrac{dy}{dz}$; shew that y is one of the values of

$$\left\{ y'\left(y - \frac{1}{4}\frac{d^2}{dz^2}\log y'\right)^{\frac{1}{2}} + (e_1 - e_2)(e_1 - e_3)\right\}^{\frac{1}{2}}.$$

Example 20.5 (Math. Trip. 1896)   Prove that

$$\sum \{\wp(z) - e\}\{\wp(y) - \wp(w)\}^2\{\wp(y + w) - e\}^{\frac{1}{2}}\{\wp(y - w) - e\}^{\frac{1}{2}} = 0,$$

where the sign of summation refers to the three arguments z, y, w, and e is any one of the roots e_1, e_2, e_3.

Example 20.6 (Math. Trip. 1894)   Shew that

$$\frac{\wp'(z + w_1)}{\wp'(z)} = -\left\{\frac{\wp\left(\frac{1}{2}w_1\right) - \wp(w_1)}{\wp(z) - \wp(w_1)}\right\}^2.$$

Example 20.7 (Math. Trip. 1894)   Prove that

$$\wp(2z) - \wp(w_1) = \{\wp'(z)\}^{-2}\left\{\wp(z) - \wp\left(\frac{1}{2}w_1\right)\right\}^2\left\{\wp(z) - \wp\left(w_2 + \frac{1}{2}w_1\right)\right\}^2.$$

Example 20.8 (Trinity, 1908)   Shew that

$$\wp(u + v)\,\wp(u - v) = \frac{\left\{\wp(u)\wp(v) + \frac{1}{4}g_2\right\}^2 + g_3\left\{\wp(u) + \wp(v)\right\}}{\left\{\wp(u) - \wp(v)\right\}^2}.$$

Example 20.9 (Math. Trip. 1914)   If $\wp(u)$ have primitive periods $2\omega_1$, $2\omega_2$ and $f(u) = \{\wp(u) - \wp(\omega_2)\}^{1/2}$, while $\wp_1(u)$ and $f_1(u)$ are similarly constructed with periods $2\omega_1/n$ and $2\omega_2$, prove that

$$\wp_1(u) = \wp(u) + \sum_{m=1}^{n-1}\left\{\wp\left(u + 2m\omega_1/n\right) - \wp\left(2m\omega_1/n\right)\right\},$$

and

$$f_1(u) = \frac{\displaystyle\prod_{m=0}^{n-1} f\left(u + 2m\omega_1/n\right)}{\displaystyle\prod_{m=1}^{n-1} f\left(2m\omega_1/n\right)}.$$

The first of the formulae is due to Kiepert [371].

Example 20.10 (Burnside [110])   If $x = \wp(u + a)$, $y = \wp(u - a)$, where a is constant, shew that the curve on which (x, y) lies is

$$\left(xy + cx + cy + \frac{1}{4}g_2\right)^2 = 4(x + y + c)\left(cxy - \frac{1}{4}g_3\right),$$

where $c = \wp(2a)$.

Example 20.11 (Trinity, 1909)   Shew that

$$2(\wp'')^3(u) - 3g_2(\wp'')^2(u) + g_2^3 = 27\left\{(\wp')^2(u) + g_3\right\}^2.$$

Example 20.12 (Trinity, 1905)   If $z = \displaystyle\int_{-\infty}^{x} \left(x^4 + 6cx^2 + e^2\right)^{-\frac{1}{2}} dx$, verify that

$$x = \frac{\frac{1}{2}\wp'(z)}{\wp(z) + c},$$

the elliptic function being formed with the roots $-c$, $\frac{1}{2}(c + e)$, $\frac{1}{2}(c - e)$.

Example 20.13 (Math. Trip. 1897)   If m be any constant, prove that

$$\frac{1}{\wp'(y)} \int \frac{e^{m\{\wp(z)-\wp(y)\}}\wp'^2(z)\,dz}{\wp(z)-\wp(y)} + \wp'(z) \int \frac{e^{m\{\wp(z)-\wp(y)\}}\,dy}{\wp(z)-\wp(y)}$$

$$= -\frac{1}{2}\sum_r \int\int \frac{e^{m\{\wp(z)-\wp(y)\}}\wp'^2(z)\,dz\,dy}{\{\wp(z)-e_r\}\{\wp(y)-e_r\}},$$

where the summation refers to the values 1, 2, 3 of r; and the integrals are indefinite.

Example 20.14 (Hermite [295])   Let $R(x) = Ax^4 + Bx^3 + Cx^2 + Dx + E$, and let $\xi = \phi(x)$ be the function defined by the equation

$$x = \int^\xi \{R(\xi)\}^{-\frac{1}{2}}\,d\xi,$$

where the lower limit of the integral is arbitrary. Shew that

$$\frac{2\phi'(a)}{\phi(x+y)-\phi(a)} = \frac{\phi'(a+y)+\phi'(a)}{\phi(a+y)-\phi(a)} + \frac{\phi'(a-y)+\phi'(a)}{\phi(a-y)-\phi(a)}$$

$$-\frac{\phi'(a+y)-\phi'(x)}{\phi(a+y)-\phi(x)} - \frac{\phi'(a-y)-\phi'(x)}{\phi(a-y)-\phi(x)}.$$

This formula is an addition-formula which is satisfied by every elliptic function of order 2.

Example 20.15 (De Brun [104])   Shew that, when the change of variables

$$\xi' = \frac{\xi}{\eta}, \qquad \eta' = \frac{\xi^3}{\eta^2}$$

is applied to the equations

$$\eta^2 + \eta(1+p\xi) + \xi^3 = 0, \qquad du - \frac{d\xi}{2\eta+1+p\xi} = 0,$$

they transform into the similar equations

$$(\eta')^2 + \eta'(1+p\xi') + (\xi')^3 = 0, \qquad du - \frac{d\xi'}{2\eta'+1+p\xi'} = 0.$$

Shew that the result of performing this change of variables three times in succession is a return to the original variables ξ, η; and hence prove that, if ξ and η be denoted as functions of u by $E(u)$ and $F(u)$ respectively, then

$$E(u+A) = \frac{E(u)}{F(u)}, \qquad F(u+A) = \frac{E^3(u)}{F^2(u)},$$

where A is one-third of a period of the functions $E(u)$ and $F(u)$.
    Shew that

$$E(u) = \tfrac{1}{12}p^2 - \wp(u; g_2, g_3),$$

where $g_2 = 2p + \tfrac{1}{12}p^4$, $g_3 = -1 - \tfrac{1}{6}p^3 - \tfrac{1}{216}p^6$.

Example 20.16 (Math. Trip. 1913)   Shew that

$$\wp'(z) = \frac{2\sigma(z + \omega_1)\sigma(z + \omega_2)\sigma(z - \omega_1 - \omega_2)}{\sigma^3(z)\,\sigma(\omega_1)\sigma(\omega_2)\sigma(\omega_1 + \omega_2)},$$

and

$$\wp''(z) = \frac{6\sigma(z + a)\sigma(z - a)\sigma(z + c)\sigma(z - c)}{\sigma^4(z)\sigma^2(a)\sigma^2(c)},$$

where $\wp(a) = (\frac{1}{12}g_2)^{\frac{1}{2}}$, $\wp(c) = -(\frac{1}{12}g_2)^{\frac{1}{2}}$.

Example 20.17 (Math. Trip. 1895)   Prove that

$$\wp(z - a)\,\wp(z - b) = \wp(a - b)\,\{\wp(z - a) + \wp(z - b) - \wp(a) - \wp(b)\}$$
$$+ \wp'(a - b)\,\{\zeta(z - a) - \zeta(z - b) + \zeta(a) - \zeta(b)\} + \wp(a)\wp(b).$$

Example 20.18 (Math. Trip. 1910)   Shew that

$$\frac{1}{2}\left\{\frac{\wp'(u) + \wp'(w)}{\wp(u) - \wp(w)} - \frac{\wp'(v) + \wp'(w)}{\wp(v) - \wp(w)}\right\} = -\zeta(w - u) + \zeta(w - v) + \zeta(v) - \zeta(u).$$

Example 20.19 (Math. Trip. 1912)   Shew that

$$\zeta(u_1) + \zeta(u_2) + \zeta(u_3) - \zeta(u_1 + u_2 + u_3)$$
$$= \frac{2\{\wp(u_1) - \wp(u_2)\}\,\{\wp(u_2) - \wp(u_3)\}\,\{\wp(u_3) - \wp(u_1)\}}{\wp'(u_1)\{\wp(u_2) - \wp(u_3)\} + \wp'(u_2)\,\{\wp(u_3) - \wp(u_1)\} + \wp'(u_3)\,\{\wp(u_1) - \wp(u_2)\}}.$$

Example 20.20   Shew that

$$\frac{\sigma(x + y + z)\sigma(x - y)\sigma(y - z)\sigma(z - x)}{\sigma^3(x)\sigma^3(y)\sigma^3(z)} = \frac{1}{2}\begin{vmatrix} 1 & \wp(x) & \wp'(x) \\ 1 & \wp(y) & \wp'(y) \\ 1 & \wp(z) & \wp'(z) \end{vmatrix}$$

Obtain the addition-theorem for the function $\wp(z)$ from this result.

Example 20.21 (Frobenius and Stickelberger [228]. See also Kiepert [370]; Hermite [290].)
Shew by induction, or otherwise, that

$$\begin{vmatrix} 1 & \wp(z_0) & \wp'(z_0) & \cdots & \wp^{(n-1)}(z_0) \\ 1 & \wp(z_1) & \wp'(z_1) & \cdots & \wp^{(n-1)}(z_1) \\ \vdots & \vdots & \vdots & \vdots & \vdots \\ 1 & \wp(z_n) & \wp'(z_n) & \cdots & \wp^{(n-1)}(z_n) \end{vmatrix}$$
$$= (-1)^{\frac{1}{2}n(n-1)}1!\,2!\,\cdots\,n!\frac{\sigma(z_0 + z_1 + \cdots + z_n)\prod\sigma(z_\lambda - z_\mu)}{\sigma^{n+1}(z_0)\cdots\sigma^{n+1}(z_n)},$$

where the product is taken for pairs of all integral values of λ and μ from 0 to n, such that
$\lambda < \mu$.

Example 20.22 (Math. Trip. 1911)   Express

$$\begin{vmatrix} 1 & \wp(x) & \wp^2(x) & \wp'(x) \\ 1 & \wp(y) & \wp^2(y) & \wp'(y) \\ 1 & \wp(z) & \wp^2(z) & \wp'(z) \\ 1 & \wp(u) & \wp^2(u) & \wp'(u) \end{vmatrix}$$

as a fraction whose numerator and denominator arc products of sigma-functions. Deduce that if $\alpha = \wp(x)$, $\beta = \wp(y)$, $\gamma = \wp(z)$, $\delta = \wp(u)$, where $x + y + z + u = 0$, then

$$(e_2 - e_3)\{(\alpha - e_1)(\beta - e_1)(\gamma - e_1)(\delta - e_1)\}^{\frac{1}{2}}$$
$$+ (e_3 - e_1)\{(\alpha - e_2)(\beta - e_2)(\gamma - e_2)(\delta - e_2)\}^{\frac{1}{2}}$$
$$+ (e_1 - e_2)\{(\alpha - e_3)(\beta - e_3)(\gamma - e_3)(\delta - e_3)\}^{\frac{1}{2}}$$
$$= (e_2 - e_3)(e_3 - e_1)(e_1 - e_2).$$

Example 20.23 (Math. Trip. 1905)   Shew that

$$2\zeta(2u) - 4\zeta(u) = \frac{\wp''(u)}{\wp'(u)},$$

$$3\zeta(3u) - 9\zeta(u) = \frac{\wp'^3(u)}{\wp^4(u) - \frac{1}{2}g_2\wp^2(u) - g_3\wp(u) - \frac{1}{48}g_2^2}.$$

Example 20.24 (Math. Trip. 1912)   Shew that

$$\frac{\sigma(2u)}{\sigma^4(u)} = -\wp'(u), \qquad \frac{\sigma(3u)}{\sigma^9(u)} = 3\wp(u)\wp'^2(u) - \frac{1}{4}\wp''^2(u),$$

and prove that $\sigma(nu)/\{\sigma(u)\}^{n^2}$ is a doubly-periodic function of u.

Example 20.25 (Math. Trip. 1895)   Prove that

$$\zeta(z - a) - \zeta(z - b) - \zeta(a - b) + \zeta(2a - 2b) = \frac{\sigma(z - 2a + b)\sigma(z - 2b + a)}{\sigma(2b - 2a)\sigma(z - a)\sigma(z - b)}.$$

Example 20.26 (Math. Trip. 1897)   Shew that, if $z_1 + z_2 + z_3 + z_4 = 0$, then

$$\left\{\sum \zeta(z_r)\right\}^3 = 3 \left\{\sum \zeta(z_r)\right\}\left\{\sum \wp(z_r)\right\} + \sum \wp'(z_r),$$

the summations being taken for $r = 1, 2, 3, 4$.

Example 20.27 (Painlevé [514])   Shew that every elliptic function of order n can be expressed as the quotient of two expressions of the form

$$a_1\wp(z + b) + a_2\wp'(z + b) + \cdots + a_n\wp^{(n-1)}(z + b),$$

where $b, a_1, a_2, \ldots, a_n$ are constants.

Example 20.28 (Math. Trip. 1914)   Taking $e_1 > e_2 > e_3$, $\wp(\omega) = e_1$, $\wp(\omega') = e_3$, consider the values assumed by

$$\zeta(u) - u\zeta(\omega')/\omega'$$

as u passes along the perimeter of the rectangle whose corners are $-\omega, \omega, \omega + \omega', -\omega - \omega'$.

Example 20.29 (Math. Trip. 1912)   Obtain an integral of the equation

$$\frac{1}{w}\frac{d^2w}{dz^2} = 6\wp(z) + 3b$$

in the form

$$\frac{d}{dz}\left[\frac{\sigma(z + c)}{\sigma(z)\sigma(c)} \exp\left\{\frac{z\wp'(c)}{b - 2\wp(c)} - z\zeta(c)\right\}\right],$$

where c is defined by the equation

$$(b^2 - 3g_2)\,\wp(c) = 3(b^3 + g_3).$$

Also, obtain another integral in the form

$$\frac{\sigma(z + a_1)\sigma(z + a_2)}{\sigma^2(z)}\,\exp\{-z\zeta(a_1) - z\zeta(a_2)\},$$

where $\wp(a_1) + \wp(a_2) = b$, $\wp'(a_1) + \wp'(a_2) = 0$, and neither $a_1 + a_2$ nor $a_1 - a_2$ is congruent to a period.

Example 20.30 (Math. Trip. 1893)   Prove that

$$g(z) = \frac{\sigma(z + z_1)\sigma(z + z_2)\sigma(z + z_3)\sigma(z + z_4)}{\sigma\{2z + \frac{1}{2}(z_1 + z_2 + z_3 + z_4)\}}$$

is a doubly-periodic function of z, such that

$$g(z) + g(z + \omega_1) + g(z + \omega_2) + g(z + \omega_1 + \omega_2)$$
$$= -2\sigma\left\{\tfrac{1}{2}(z_2 + z_1 - z_1 - z_4)\right\}\sigma\left\{\tfrac{1}{2}(z_3 + z_1 - z_2 - z_4)\right\}\sigma\left\{\tfrac{1}{2}(z_1 + z_2 - z_3 - z_4)\right\}.$$

Example 20.31 (Math. Trip. 1894)   If $f(z)$ be a doubly-periodic function of the third order, with poles at $z = c_1$, $z = c_2$, $z = c_3$, and if $\phi(z)$ be a doubly-periodic function of the second order with the same periods and poles at $z = \alpha$, $z = \beta$, its value in the neighbourhood of $z = \alpha$ being

$$\phi(z) = \frac{\lambda}{z - \alpha} + \lambda_1(z - \alpha) + \lambda_2(z - \alpha)^2 + \cdots,$$

prove that

$$\frac{1}{2}\lambda^2\{f''(\alpha) - f''(\beta)\} - \lambda\{f(\alpha) + f'(\beta)\}\sum_1^3 \phi(c_1) +$$
$$\{f(\alpha) - f(\beta)\}\left\{3\lambda\lambda_1 + \sum_1^3 \phi(c_2)\phi(c_3)\right\} = 0.$$

Example 20.32 (Math. Trip. 1893)   If $\lambda(z)$ be an elliptic function with two poles a_1, a_2, and if $z_1, z_2, \ldots, z_{2n}$ be $2n$ constants subject only to the condition

$$z_1 + z_2 + \cdots + z_{2n} = n(a_1 + a_2),$$

shew that the determinant whose ith row is

$$1, \lambda(z_i), \lambda^2(z_i), \ldots, \lambda^n(z_i), \lambda_1(z_i), \lambda(z_i)\lambda_1(z_i), \lambda^2(z_i)\lambda_1(z_i), \ldots, \lambda^{n-2}(z_i)\lambda_1(z_i)$$

where $\lambda_1(z_i)$ denotes the result of writing z_i for z in the derivative of $\lambda(z)$, vanishes identically.

Example 20.33 (Kiepert [371])   Deduce from Example 20.21 by a limiting process, or otherwise prove, that

$$\begin{vmatrix} \wp'(z) & \wp''(z) & \cdots & \wp^{(n-1)}(z) \\ \wp''(z) & \wp'''(z) & \cdots & \wp^{(n)}(z) \\ \vdots & \vdots & \vdots & \vdots \\ \wp^{(n-1)}(z) & \wp^{(n)}(z) & \cdots & \wp^{(2n-3)}(z) \end{vmatrix} = (-1)^{n-1} \{1!\, 2!\, \cdots\, (n-1)!\}^2 \frac{\sigma(nu)}{\{\sigma(u)\}^{n^2}}.$$

Example 20.34 (Math. Trip. 1895)   Shew that, provided certain conditions of inequality are satisfied,

$$\frac{\sigma(z+y)}{\sigma(z)\,\sigma(y)} e^{-\eta_1 zy/\omega_1} = \frac{\pi}{2\omega_1}\left(\cot\frac{\pi z}{2\omega_1} + \cot\frac{\pi y}{2\omega_1}\right) + \frac{2\pi}{\omega_1}\sum q^{2mn}\sin\frac{\pi}{\omega_1}(mz+ny),$$

where the summation applies to all positive integer values of m and n, and $q = \exp(\pi i\omega_2/\omega_1)$.

Example 20.35 (Math. Trip. 1896)   Assuming the formula

$$\sigma(z) = e^{\frac{\eta_1 z^2}{2\omega_1}}\frac{2\omega_1}{\pi}\sin\frac{\pi z}{2\omega_1}\prod_{n=1}^{\infty}\frac{1 - 2q^{2n}\cos\frac{\pi z}{\omega_1} + q^{4n}}{(1-q^{2n})^2},$$

prove that

$$\wp(z) = -\frac{\eta_1}{\omega_1} + \left(\frac{\pi}{2\omega_1}\right)^2\operatorname{cosec}^2\frac{\pi z}{2\omega_1} - 2\left(\frac{\pi}{\omega_1}\right)^2\sum_{n=1}^{\infty}\frac{nq^{2n}}{1-\zeta^{2n}}\cos\frac{n\pi z}{\omega_1}$$

when z satisfies the inequalities

$$-2\operatorname{Re}\left(\frac{\omega_2}{i\omega_1}\right) < \operatorname{Re}\left(\frac{z}{i\omega_1}\right) < 2\operatorname{Re}\left(\frac{\omega_2}{i\omega_1}\right).$$

Example 20.36 (Trinity, 1898)   Shew that if $2\bar{\omega}$ be any expression of the form $2m\omega_1 + 2n\omega_2$ and if

$$x = \wp\left(\tfrac{2}{5}\bar{\omega}\right) + \wp\left(\tfrac{4}{5}\bar{\omega}\right),$$

then x is a root of the sextic

$$x^6 - 5g_2 x^4 - 40g_3 x^3 - 5g_2^2 x^2 - 8g_2 g_3 x - 5g_3^2 = 0,$$

and obtain all the roots of the sextic.

Example 20.37 (Dolbnia [185])   Shew that

$$\int \{(x^2-a)(x^2-b)\}^{-\frac{1}{4}}\,dx = -\frac{1}{2}\log\frac{\sigma(z-z_0)}{\sigma(z+z_0)} + \frac{i}{2}\log\frac{\sigma(z-iz_0)}{\sigma(z+iz_0)},$$

where

$$x^2 = a + \frac{1}{6(\wp^2(z)-\wp^2(z_0))}, \quad g_2 = \frac{2b}{3a(a-b)}, \quad g_3 = 0, \quad \wp^2(z_0) = \frac{1}{6(a-b)}.$$

Example 20.38 (Hermite [294])   Prove that every analytic function $f(z)$ which satisfies the three-term equation

$$\sum_{a,b,c} f(z+a)f(z-a)f(b+c)f(b-c) = 0,$$

for general values of a, b, c and z, is expressible as a finite combination of elementary functions, together with a sigma-function (including a circular function or an algebraic function as degenerate cases). *Hint.* Put $z = a = b = c = 0$, and then $f(0) = 0$; put $b = c$, and then $f(a - b) + f(b - a) = 0$, so that $f(z)$ is an odd function.

If $F(z)$ is the logarithmic derivative of $f(z)$, the result of differentiating the relation with respect to b, and then putting $b = c$, is

$$\frac{f(z + a)f(z - a)f(2b) \, f'(0)}{f(z + b)f(z - b)f(a + b)f(a - b)} = F(z + b) - F(z - b) + F(a - b) - F(a + b).$$

Differentiate with respect to b, and put $b = 0$; then

$$\frac{f(z + a)f(z - a)\{f'(0)\}}{\{f(z)f(a)\}^2} = F'(z) - F''(a).$$

If $f'(0)$ were zero, $F'(z)$ would be a constant and, by integration, $f(z)$ would be of the form $A \exp(Bz + Cz^2)$, and this is an odd function only in the trivial case when it is zero.

If $f'(0) \neq 0$, and we write $F'(z) = -\phi(z)$, it is found that the coefficient of a^4 in the expansion of

$$12f(z + a)f(z - a)/\{f(z)\}^2$$

is $6\{\phi(z)\}^2 - \phi''(z)$, and the coefficient of a^4 in $12\{f(a)\}^2\{\phi(a) - \phi(z)\}$ is a linear function of $\phi(z)$. Hence $\phi''(z)$ is a quadratic function of $\phi(z)$; and when we multiply this function by $\phi'(z)$ and integrate we find that

$$\{\phi'(z)\}^2 = 4\{\phi(z)\}^3 + 12A\{\phi(z)\}^2 + 12B\phi(z) + 4C,$$

where A, B, C are constants. If the cubic on the right has no repeated factors, then, by §20.6, $\phi(z) = \wp(z + a) + A$, where a is a constant, and on integration

$$f(z) = \sigma(z + a) \exp(-\tfrac{1}{2}Az^2 - Kz - L),$$

where K and L are constants; since $f(z)$ is an odd function $a = K = 0$, and

$$f(z) = \sigma(z) \exp\{-\tfrac{1}{2}Az^2 - L\}.$$

If the cubic has a repeated factor, the sigma-function is to be replaced (cf. §20.222) by the sine of a multiple of z, and if the cubic is a perfect cube the sigma-function is to be replaced by a multiple of z.

21

The Theta-Functions

21.1 The definition of a theta-function

When it is desired to obtain definite numerical results in problems involving elliptic functions, the calculations are most simply performed with the aid of certain auxiliary functions known as *theta-functions*. These functions are of considerable intrinsic interest, apart from their connexion with elliptic functions, and we shall now give an account of their fundamental properties.

The theta-functions were first systematically studied by Jacobi [349], who obtained their properties by purely algebraical methods; and his analysis was so complete that practically all the results contained in this chapter (with the exception of the discussion of the problem of inversion in §§21.7 *et seq.*) are to be found in his works. In accordance with the general scheme of this book, we shall not employ the methods of Jacobi, but the more powerful methods based on the use of Cauchy's theorem. These methods were first employed in the theory of elliptic and allied functions by Liouville in his lectures and have since been given in several treatises on elliptic functions, the earliest of these works being that by Briot and Bouquet [100].

Note The first function of the theta-function type to appear in Analysis was the *partition function*[1] $\prod_{n=1}^{\infty}(1 - x^n z)^{-1}$ of Euler [207, §304]; by means of the results given in §21.3, it is easy to express theta-functions in terms of partition functions. Euler also obtained properties of products of the type

$$\prod_{n=1}^{\infty}(1 \pm x^n), \quad \prod_{n=1}^{\infty}(1 \pm x^{2n}), \quad \prod_{n=1}^{\infty}(1 \pm x^{2n-1}).$$

The associated series $\sum_{n=0}^{\infty} m^{\frac{1}{2}n(n+3)}$, $\sum_{n=0}^{\infty} m^{\frac{1}{2}n(n+1)}$ and $\sum_{n=0}^{\infty} m^{n^2}$ had previously occurred in the posthumous work of Jakob Bernoulli [66, p. 55]. Theta-functions also occur in Fourier [223, p. 265].

The theory of theta-functions was developed from the theory of elliptic functions by Jacobi in his *Fundamenta Nova Theoriae Functionum Ellipticarum* (1829) [349], reprinted in his *Ges. Werke* [354, Vol. I, pp. 49–239]; the notation there employed is explained in §21.62.

---

[1] The partition function and associated functions have been studied by Gauss [238, vol. II, p. 16–21; vol. III, p. 433–480] and Cauchy [126]. For a discussion of properties of various functions involving what are known as *Basic numbers* (which are closely connected with partition functions) see Jackson [340, 341, 342, 343] and Watson [649]. A fundamental formula in the theory of Basic numbers was given by Heine [287, vol. I, p. 107].

In his subsequent lectures, he introduced the functions discussed in this chapter; an account of these lectures (1838) is given by Borchardt [84]. The most important results contained in them seem to have been discovered in 1835, cf. Kronecker [387].

Let τ be a (constant) complex number whose imaginary part is *positive*; and write $q = e^{\pi i \tau}$, so that $|q| < 1$. Consider the function $\vartheta(z, q)$, defined by the series

$$\vartheta(z, q) = \sum_{n=-\infty}^{\infty} (-1)^n q^{n^2} e^{2niz},$$

qua function of the variable z. If A be any positive constant, then, when $|z| \leq A$, we have

$$\left| q^{n^2} e^{\pm 2niz} \right| \leq |q|^{n^2} e^{2nA},$$

n being a positive integer. Now d'Alembert's ratio (§2.36) for the series $\sum\limits_{n=-\infty}^{\infty} |q|^{n^2} e^{2nA}$ is $|q|^{2n+1} e^{2A}$, which tends to zero as $n \to \infty$. The series for $\vartheta(z, q)$ is therefore a series of analytic functions, uniformly convergent (§3.34) in any bounded domain of values of z, and so it is an integral function (§5.3, §5.64).

It is evident that

$$\vartheta(z, q) = 1 + 2 \sum_{n=1}^{\infty} (-1)^n q^{n^2} \cos 2nz,$$

and that $\vartheta(z + \pi, q) = \vartheta(z, q)$; further

$$\vartheta(z + \pi\tau, q) = \sum_{n=-\infty}^{\infty} (-1)^n q^{n^2} q^{2n} e^{2niz}$$

$$= -q^{-1} e^{-2iz} \sum_{n=-\infty}^{\infty} (-1)^{n+1} q^{(n+1)^2} e^{2(n+1)iz},$$

and so $\vartheta(z + \pi\tau, q) = -q^{-1} e^{-2iz} \vartheta(z, q)$.

In consequence of these results, $\vartheta(z, q)$ is called a *quasi doubly-periodic function* of z. The effect of increasing z by π or $\pi\tau$ is the same as the effect of multiplying $\vartheta(z, q)$ by 1 or $-q^{-1} e^{-2iz}$, and accordingly 1 and $-q^{-1} e^{-2iz}$ are called the *multipliers* or *periodicity factors* associated with the *periods* π and $\pi\tau$ respectively.

21.11 The four types of theta-functions

It is customary to write $\vartheta_4(z, q)$ in place of $\vartheta(z, q)$; the other three types of theta-functions are then defined as follows: The function $\vartheta_3(z, q)$ is defined by the equation

$$\vartheta_3(z, q) = \vartheta_4 \left(z + \tfrac{1}{2}\pi, q \right) = 1 + 2 \sum_{n=1}^{\infty} q^{n^2} \cos 2nz.$$

Next, $\vartheta_1(z, q)$ is defined in terms of $\vartheta_4(z, q)$ by the equation

$$\vartheta_1(z, q) = -ie^{iz + \frac{1}{4}\pi i\tau} \vartheta_4 \left(z + \tfrac{1}{2}\pi\tau, q\right)$$

$$= -i \sum_{n=-\infty}^{\infty} (-1)^n q^{\left(n + \frac{1}{2}\right)^2} e^{(2n+1)iz},$$

and hence[2]

$$\vartheta_1(z, q) = 2 \sum_{n=0}^{\infty} (-1)^n q^{\left(n + \frac{1}{2}\right)^2} \sin(2n + 1)z.$$

Lastly, $\vartheta_2(z, q)$ is defined by the equation

$$\vartheta_2(z, q) = \vartheta_1 \left(z + \tfrac{1}{2}\pi, q\right) = 2 \sum_{n=0}^{\infty} q^{\left(n + \frac{1}{2}\right)^2} \cos(2n + 1)z.$$

Writing down the series at length, we have

$$\vartheta_1(z, q) = 2q^{\frac{1}{4}} \sin z - 2q^{\frac{9}{4}} \sin 3z + 2q^{\frac{25}{4}} \sin 5z - \cdots,$$

$$\vartheta_2(z, q) = 2q^{\frac{1}{4}} \cos z + 2q^{\frac{9}{4}} \cos 3z + 2q^{\frac{25}{4}} \cos 5z + \cdots,$$

$$\vartheta_3(z, q) = 1 + 2q \cos 2z + 2q^4 \cos 4z + 2q^9 \cos 6z + \cdots,$$

$$\vartheta_4(z, q) = 1 - 2q \cos 2z + 2q^4 \cos 4z - 2q^9 \cos 6z + \cdots.$$

It is obvious that $\vartheta_1(z, q)$ is an *odd* function of z and that the other theta-functions are *even* functions of z.

The notation which has now been introduced is a modified form of that employed in the treatise of Tannery and Molk [620]; the only difference between it and Jacobi's notation is that $\vartheta_4(z, q)$ is written where Jacobi would have written $\vartheta(z, q)$. There are, unfortunately, several notations in use; a scheme, giving the connexions between them, will be found in §21.9.

For brevity, the parameter q will usually not be specified, so that $\vartheta_1(z), \ldots$ will be written for $\vartheta_1(z, q), \ldots$. When it is desired to exhibit the dependence of a theta-function on the parameter τ, it will be written $\vartheta(z|\tau)$. Also $\vartheta_2(0), \vartheta_3(0), \vartheta_4(0)$, will be replaced by $\vartheta_2, \vartheta_3, \vartheta_4$ respectively; and ϑ_1' will denote the result of making z equal to zero in the derivate of $\vartheta_1(z)$.

Example 21.1.1   Shew that

$$\vartheta_3(z, q) = \vartheta_3(2z, q^4) + \vartheta_2(2z, q^4),$$

$$\vartheta_4(z, q) = \vartheta_3(2z, q^4) - \vartheta_2(2z, q^4).$$

Example 21.1.2   Obtain the results

$\vartheta_1(z) =$	$-\vartheta_2\left(z + \tfrac{1}{2}\pi\right) =$	$-iM\vartheta_3\left(z + \tfrac{1}{2}\pi + \tfrac{1}{2}\pi\tau\right) =$	$-iM\vartheta_4\left(z + \tfrac{1}{2}\pi\tau\right),$
$\vartheta_2(z) =$	$M\vartheta_3\left(z + \tfrac{1}{2}\pi\tau\right) =$	$M\vartheta_4\left(z + \tfrac{1}{2}\pi + \tfrac{1}{2}\pi\tau\right) =$	$\vartheta_1\left(z + \tfrac{1}{2}\pi\right),$
$\vartheta_3(z) =$	$\vartheta_4\left(z + \tfrac{1}{2}\pi\right) =$	$M\vartheta_1\left(z + \tfrac{1}{2}\pi + \tfrac{1}{2}\pi\tau\right) =$	$M\vartheta_2\left(z + \tfrac{1}{2}\pi\tau\right),$
$\vartheta_4(z) =$	$-iM\vartheta_1\left(z + \tfrac{1}{2}\pi\tau\right) =$	$-iM\vartheta_2\left(z + \tfrac{1}{2}\pi + \tfrac{1}{2}\pi\tau\right) =$	$\vartheta_3\left(z + \tfrac{1}{2}\pi\right),$

---

[2] Throughout the chapter, the many-valued function q^λ is to be interpreted to mean $\exp(\lambda\pi i\tau)$.

where $M = q^{\frac{1}{4}} e^{iz}$.

Example 21.1.3 Shew that the multipliers of the theta-functions associated with the periods $\pi, \pi\tau$ are given by the scheme

	$\vartheta_1(z)$	$\vartheta_2(z)$	$\vartheta_3(z)$	$\vartheta_4(z)$
π	-1	-1	1	1
$\pi\tau$	$-N$	N	N	$-N$

where $N = q^{-1} e^{-2iz}$.

Example 21.1.4 If $\vartheta(z)$ be any one of the four theta-functions and $\vartheta'(z)$ its derivative with respect to z, shew that

$$\frac{\vartheta'(z + \pi)}{\vartheta(z + \pi)} = \frac{\vartheta'(z)}{\vartheta(z)}, \quad \frac{\vartheta'(z + \pi\tau)}{\vartheta(z + \pi\tau)} = -2i + \frac{\vartheta'(z)}{\vartheta(z)}.$$

21.12 The zeros of the theta-functions

From the quasi-periodic properties of the theta-functions it is obvious that if $\vartheta(z)$ be any one of them, and if z_0 be any zero of $\vartheta(z)$, then

$$z_0 + m\pi + n\pi\tau$$

is also a zero of $\vartheta(z)$, for all integral values of m and n.

It will now be shewn that if C be a cell with corners $t, t + \pi, t + \pi + \pi\tau, t + \pi\tau$, then $\vartheta(z)$ has one and only one zero inside C.

Since $\vartheta(z)$ is analytic throughout the finite part of the z-plane, it follows, from §6.31, that the number of its zeros inside C is

$$\frac{1}{2\pi i} \int_C \frac{\vartheta'(z)}{\vartheta(z)} \, dz.$$

Treating the contour after the manner of §20.12, we see that

$$\frac{1}{2\pi i} \int_C \frac{\vartheta'(z)}{\vartheta(z)} \, dz = \frac{1}{2\pi i} \int_t^{t+\pi} \left\{ \frac{\vartheta'(z)}{\vartheta(z)} - \frac{\vartheta'(z + \pi\tau)}{\vartheta(z + \pi\tau)} \right\} dz$$

$$- \frac{1}{2\pi i} \int_t^{t+\pi\tau} \left\{ \frac{\vartheta'(z)}{\vartheta(z)} - \frac{\vartheta'(z + \pi)}{\vartheta(z + \pi)} \right\} dz$$

$$= \frac{1}{2\pi i} \int_t^{t+\pi} 2i \, dz,$$

by Example 21.1.4. Therefore

$$\frac{1}{2\pi i} \int_C \frac{\vartheta'(z)}{\vartheta(z)} \, dz = 1,$$

that is to say, $\vartheta(z)$ has one simple zero only inside C; this is the theorem stated.

Since one zero of $\vartheta_1(z)$ is obviously $z = 0$, it follows that the zeros of $\vartheta_1(z), \vartheta_2(z), \vartheta_3(z), \vartheta_4(z)$ are the points congruent respectively to $0, \frac{1}{2}\pi, \frac{1}{2}\pi + \frac{1}{2}\pi\tau, \frac{1}{2}\pi\tau$. The reader will observe that these four points form the corners of a parallelogram described counter-clockwise.

21.2 The relations between the squares of the theta-functions

It is evident that, if the theta-functions be regarded as functions of a single variable z, this variable can be eliminated from the equations defining any pair of theta-functions, the result being a relation[3] between the functions which might be expected, on general grounds, to be non-algebraic; there are, however, extremely simple relations connecting any *three* of the theta-functions; these relations will now be obtained.

Each of the four functions $\vartheta_1^2(z)$, $\vartheta_2^2(z)$, $\vartheta_3^2(z)$, $\vartheta_4^2(z)$ is analytic for all values of z and has periodicity factors 1, $q^{-2} e^{-4iz}$ associated with the periods π, $\pi\tau$; and each has a double zero (and no other zeros) in any cell.

From these considerations it is obvious that, if a, b, a' and b' are suitably chosen constants, each of the functions

$$\frac{a\vartheta_1^2(z) + b\vartheta_4^2(z)}{\vartheta_2^2(z)}, \quad \frac{a'\vartheta_1^2(z) + b'\vartheta_4^2(z)}{\vartheta_3^2(z)}$$

is a *doubly-periodic function* (with periods $\pi, \pi\tau$) having at most only a *simple* pole in each cell. By §20.13, such a function is merely a constant; and obviously we can adjust a, b, a', b' so as to make the constants, in each of the cases under consideration, equal to unity.

There exist, therefore, relations of the form

$$\vartheta_2^2(z) = a\vartheta_1^2(z) + b\vartheta_4^2(z), \quad \vartheta_3^2(z) = a'\vartheta_1^2(z) + b'\vartheta_4^2(z).$$

To determine a, b, a', b', give z the special values $\tfrac{1}{2}\pi\tau$ and 0; since

$$\vartheta_2\left(\tfrac{1}{2}\pi\tau\right) = q^{-\frac{1}{4}}\vartheta_3, \quad \vartheta_4\left(\tfrac{1}{2}\pi\tau\right) = 0, \quad \vartheta_1\left(\tfrac{1}{2}\pi\tau\right) = iq^{-\frac{1}{4}}\vartheta_4,$$

we have $\vartheta_3^2 = -a\vartheta_4^2$, $\vartheta_2^2 = b\vartheta_4^2$; $\vartheta_2^2 = -a'\vartheta_4^2$, $\vartheta_3^2 = b'\vartheta_4^2$. Consequently, we have obtained the relations

$$\vartheta_2^2(z)\vartheta_4^2 = \vartheta_4^2(z)\vartheta_2^2 - \vartheta_1^2(z)\vartheta_3^2,$$
$$\vartheta_3^2(z)\vartheta_4^2 = \vartheta_4^2(z)\vartheta_3^2 - \vartheta_1^2(z)\vartheta_2^2.$$

If we write $z + \tfrac{1}{2}\pi$ for z, we get the additional relations

$$\vartheta_1^2(z)\vartheta_4^2 = \vartheta_3^2(z)\vartheta_2^2 - \vartheta_2^2(z)\vartheta_3^2,$$
$$\vartheta_4^2(z)\vartheta_4^2 = \vartheta_3^2(z)\vartheta_3^2 - \vartheta_2^2(z)\vartheta_2^2.$$

By means of these results it is possible to express any theta-function in terms of any other pair of theta-functions.

Corollary 21.2.1   *Writing $z = 0$ in the last relation, we have*

$$\vartheta_2^4 + \vartheta_4^4 = \vartheta_3^4,$$

that is to say

$$16q(1 + q^{1\cdot2} + q^{2\cdot3} + q^{3\cdot4} + \cdots)^4 + (1 - 2q + 2q^4 - 2q^9 + \cdots)^4$$
$$= (1 + 2q + 2q^4 + 2q^9 + \cdots)^4.$$

---

[3] The analogous relation for the functions $\sin z$ and $\cos z$ is, of course, $\sin^2 z + \cos^2 z = 1$.

21.21 The addition-formulae for the theta-functions

The results just obtained are particular cases of formulae containing two variables; these formulae are not addition-theorems in the strict sense, as they do not express theta-functions of $z + y$ algebraically in terms of theta-functions of z and y, but all involve theta-functions of $z - y$ as well as of $z + y$, z and y.

To obtain one of these formulae, consider $\vartheta_3(z + y)\vartheta_3(z - y)$ *qua* function of z. The periodicity factors of this function associated with the periods π and $\pi\tau$ are 1 and $q^{-1}e^{-2i(z+y)}$. $q^{-1}e^{-2i(z-y)} = q^{-2}e^{-4iz}$.

But the function $a\vartheta_3^2(z) + b\vartheta_1^2(z)$ has the same periodicity factors, and we can obviously choose the ratio $a : b$ so that *the doubly-periodic function*

$$\frac{a\vartheta_3^2(z) + b\vartheta_1^2(z)}{\vartheta_3(z + y)\vartheta_3(z - y)}$$

has no poles at the zeros of $\vartheta_3(z - y)$; it then has, at most, a single simple pole in any cell, namely the zero of $\vartheta_3(z + y)$ in that cell, and consequently (§20.13) it is a constant, i.e. independent of z; and, as only the ratio $a : b$ is so far fixed, we may choose a and b so that the constant is unity.

We then have to determine a and b from the identity in z,

$$a\vartheta_3^2(z) + b\vartheta_1^2(z) \equiv \vartheta_3(z + y)\vartheta_3(z - y).$$

To do this, put z in turn equal to 0 and $\frac{1}{2}\pi + \frac{1}{2}\pi\tau$, and we get

$$a\vartheta_3^2 = \vartheta_3^2(y),$$
$$b\vartheta_1^2\left(\tfrac{1}{2}\pi + \tfrac{1}{2}\pi\tau\right) = \vartheta_3\left(\tfrac{1}{2}\pi + \tfrac{1}{2}\pi\tau + y\right)\vartheta_3\left(\tfrac{1}{2}\pi + \tfrac{1}{2}\pi\tau - y\right);$$

and so $a = \vartheta_3^2(y)/\vartheta_3^2$ and $b = \vartheta_1^2(y)/\vartheta_3^2$.

We have therefore obtained an addition-formula, namely

$$\vartheta_3(z + y)\vartheta_3(z - y)\vartheta_3^2 = \vartheta_3^2(y)\vartheta_3^2(z) + \vartheta_1^2(y)\vartheta_1^2(z).$$

The set of formulae, of which this is typical, will be found in Examples 21.1 and 21.2 at the end of this chapter.

21.22 Jacobi's fundamental formulae

The addition-formulae just obtained are particular cases of a set of identities first given by Jacobi [354, vol. I, p. 505] who obtained them by purely algebraical methods; each identity involves as many as four independent variables, w, x, y, z.

Let w', x', y', z' be defined in terms of w, x, y, z by the set of equations

$$2w' = -w + x + y + z,$$
$$2x' = w - x + y + z,$$
$$2y' = w + x - y + z,$$
$$2z' = w + x + y - z.$$

Readers will easily verify that the connexion between w, x, y, z and w', x', y', z' is a reciprocal one[4]. For brevity[5], write $[r]$ for $\vartheta_r(w)\vartheta_r(x)\vartheta_r(y)\vartheta_r(z)$ and $[r]'$ for $\vartheta_r(w')\vartheta_r(x')\vartheta_r(y')\vartheta_r(z')$.

Consider $[3]$, $[1]'$, $[2]'$, $[3]'$, $[4]'$ *qua* functions of z. The effect of increasing z by π or $\pi\tau$ is to transform the functions in the first row of the following table into those in the second or third row respectively.

	$[3]$	$[1]'$	$[2]'$	$[3]'$	$[4]'$
(π)	$[3]$	$-[2]'$	$-[1]'$	$[4]'$	$[3]'$
$(\pi\tau)$	$N[3]$	$-N[4]'$	$N[3]'$	$N[2]'$	$-N[1]'$

For brevity, N has been written in place of $q^{-1}e^{-2is}$. Hence both $-[1]' + [2]' + [3]' + [4]'$ and $[3]$ have periodicity factors 1 and N, and so their quotient is a doubly-periodic function with, at most, a single simple pole in any cell, namely the zero of $\vartheta_3(z)$ in that cell.

By §20.13, this quotient is merely a constant, i.e. independent of z and considerations of symmetry shew that it is also independent of w, x and y.

We have thus obtained the result

$$A[3] = -[1]' + [2]' + [3]' + [4]',$$

where A is independent of w, x, y, z; to determine A put $w = x = y = z = 0$, and we get

$$A\vartheta_3{}^4 = \vartheta_2^4 + \vartheta_3^4 + \vartheta_4^4;$$

and so, by Corollary 21.2.1, we see that $A = 2$. Therefore

$$2[3] = -[1]' + [2]' + [3]' + [4]'. \tag{i}$$

This is one of Jacobi's formulae; to obtain another, increase w, x, y, z (and therefore also w', x', y', z') by $\frac{1}{2}\pi$; and we get

$$2[4] = [1]' - [2]' + [3]' + [4]'. \tag{ii}$$

Increasing all the variables in (i) and (ii) by $\frac{1}{2}\pi\tau$, we obtain the further results

$$2[2] = [1]' + [2]' + [3]' - [4]', \tag{iii}$$

$$2[1] = [1]' + [2]' - [3]' + [4]'. \tag{iv}$$

Note   There are 256 expressions of the form $\vartheta_p(w)\vartheta_q(x)\vartheta_r(y)\vartheta_s(z)$ which can be obtained from $\vartheta_3(w)\vartheta_3(x)\vartheta_3(y)\vartheta_3(z)$ by increasing w, x, y, z by suitable half-periods, but only those in which the suffixes p, q, r, s are either equal in pairs or all different give rise to formulae not containing quarter-periods on the right-hand side.

Example 21.2.1   Shew that

$$[1] + [2] = [1]' + [2]', \qquad [2] + [3] = [2]' + [3]',$$
$$[1] + [4] = [1]' + [4]', \qquad [3] + [4] = [3]' + [4]',$$
$$[1] + [3] = [2]' + [4]', \qquad [2] + [4] = [1]' + [3]'.$$

---

[4] In Jacobi's work the signs of w, x', y', z' are changed throughout so that the complete symmetry of the relations is destroyed; the symmetrical forms just given are due to H. J. S. Smith [596].

[5] The idea of this abridged notation is to be traced in H. J. S. Smith's memoir. It seems, however, not to have been used before Kronecker [385].

Example 21.2.2   By writing $w + \frac{1}{2}\pi$, $x + \frac{1}{2}\pi$ for w, x (and consequently $y' + \frac{1}{2}\pi$, $z' + \frac{1}{2}\pi$ for y', z'), shew that

$$[3344] + [2211] = [4433]' + [1122]',$$

where $[3344]$ means $\vartheta_3(w)\vartheta_3(x)\vartheta_4(y)\vartheta_4(z)$, etc.

Example 21.2.3   Shew that

$$2[1234] = [3412]' + [2143]' - [1234]' + [4321]'.$$

Example 21.2.4   Shew that

$$\vartheta_1^4(z) + \vartheta_3^4(z) = \vartheta_2^4(z) + \vartheta_4^4(z).$$

21.3 Jacobi's expressions for the theta-functions as infinite products

We shall now establish the result [349, p. 145]

$$\vartheta_4(z) = G \prod_{n=1}^{\infty} (1 - 2q^{m-1} \cos 2z + q^{4n-2}),$$

(where G is independent of z), and three similar formulae.

Let

$$f(z) = \prod_{n=1}^{\infty} (1 - q^{2n-1} e^{2iz}) \prod_{n=1}^{\infty} (1 - q^{2n-1} e^{-2iz});$$

each of the two products converges absolutely and uniformly in any bounded domain of values of z, by §3.341, on account of the absolute convergence of $\sum_{n=1}^{\infty} q^{2n-1}$; hence $f(z)$ is analytic throughout the finite part of the z-plane, and so it is an integral function.

The zeros of $f(z)$ are simple zeros at the points where

$$e^{2iz} = e^{(2n+1)\pi i \tau}, \qquad (n = \ldots, -2, -1, 0, 1, 2, \ldots)$$

i.e. where $2iz = (2n+1)\pi i \tau + 2m\pi i$; so that $f(z)$ and $\vartheta_4(z)$ have the same zeros; consequently the quotient $\vartheta_4(z)/f(z)$ has neither zeros nor poles in the finite part of the plane.

Now, obviously $f(z + \pi) = f(z)$; and

$$f(z + \pi\tau) = \prod_{n=1}^{\infty} (1 - q^{2n+1} e^{2iz}) \prod_{n=1}^{\infty} (1 - q^{2n-3} e^{-2iz})$$

$$= f(z)(1 - q^{-1} e^{-2iz})/(1 - qe^{2iz})$$

$$= -q^{-1} e^{-2iz} f(z).$$

That is to say $f(z)$ and $\vartheta_4(z)$ have the same periodicity factors (Example 21.1.3). Therefore $\vartheta_4(z)/f(z)$ is a doubly-periodic function with no zeros or poles, and so (§20.12) it is a constant G, say; consequently

$$\vartheta_4(z) = G \prod_{n=1}^{\infty} (1 - 2q^{2n-1} \cos 2z + q^{4n-2}).$$

(It will appear in §21.42 that $G = \prod_{n=1}^{\infty}(1 - q^{2n})$.)

Write $z + \frac{1}{2}\pi$ for z in this result, and we get

$$\vartheta_3(z) = G\prod_{n=1}^{\infty}(1 + 2q^{2n-1}\cos 2z + q^{4n-2}).$$

Also

$$\vartheta_1(z) = -iq^{\frac{1}{4}}e^{iz}\vartheta_4\left(z + \tfrac{1}{2}\pi\tau\right)$$

$$= -iq^{\frac{1}{4}}e^{iz}G\prod_{n=1}^{\infty}(1 - q^{2n}e^{2iz})\prod_{n=1}^{\infty}(1 - q^{2n-2}e^{-2iz})$$

$$= 2Gq^{\frac{1}{4}}\sin z\prod_{n=1}^{\infty}(1 - q^{2n}e^{2iz})\prod_{n=1}^{\infty}(1 - q^{2n}e^{-2iz}),$$

and so

$$\vartheta_1(z) = 2Gq^{\frac{1}{4}}\sin z\prod_{n=1}^{\infty}(1 - 2q^{2n}\cos 2z + q^{4n}),$$

while

$$\vartheta_2(z) = \vartheta_1\left(z + \tfrac{1}{2}\pi\right)$$

$$= 2Gq^{\frac{1}{4}}\cos z\prod_{n=1}^{\infty}(1 + 2q^{2n}\cos 2z + q^{4n}).$$

Example 21.3.1 (Jacobi)   Shew that

$$\left\{\prod_{n=1}^{\infty}(1 - q^{2n-1})\right\}^8 + 16q\left\{\prod_{n=1}^{\infty}(1 + q^{2n})\right\}^8 = \left\{\prod_{n=1}^{\infty}(1 + q^{2n-1})\right\}^8.$$

Jacobi [349, p. 90] describes this result as 'aequatio identica satis abstrusa'.

21.4 The differential equation satisfied by the theta-functions

We may regard $\vartheta_3(z|\tau)$ as a function of two independent variables z and τ; and it is permissible to differentiate the series for $\vartheta_3(z|\tau)$ any number of times with regard to z or τ, on account of the uniformity of convergence of the resulting series (Corollary 4.7.1); in particular

$$\frac{\partial^2\vartheta_3(z|\tau)}{\partial z^2} = -4\sum_{n=-\infty}^{\infty}n^2\exp(n^2\pi i\tau + 2niz)$$

$$= -\frac{4}{\pi i}\frac{\partial\vartheta_3(z|\tau)}{\partial\tau}.$$

Consequently, the function $\vartheta_3(z|\tau)$ satisfies the partial differential equation

$$\frac{\pi i}{4}\frac{\partial^2 y}{\partial z^2} + \frac{\partial y}{\partial\tau} = 0.$$

The reader will readily prove that the other three theta-functions also satisfy this equation.

21.41 A relation between theta-functions of zero argument

The remarkable result that

$$\vartheta_1'(0) = \vartheta_2(0)\vartheta_3(0)\vartheta_4(0)$$

will now be established. Several proofs of this important proposition have been given, but none are simple. Jacobi's original proof [354, vol. I, pp. 515–517], though somewhat more difficult than the proof given here, is well worth study. For a different method of proof of the preliminary formula given in the text, see Example 21.21. It is first necessary to obtain some formulae for differential coefficients of all the theta-functions.

Since the resulting series converge uniformly, except near the zeros of the respective theta-functions, we may differentiate the formulae for the logarithms of theta-functions, obtainable from §21.3, as many times as we please.

Denoting differentiations with regard to z by primes, we thus get

$$\vartheta_3'(z) = \vartheta_3(z) \left[\sum_{n=1}^{\infty} \frac{2iq^{2n-1} e^{2iz}}{1 + q^{2n-1} e^{2iz}} - \sum_{n=1}^{\infty} \frac{2iq^{2n-1} e^{-2iz}}{1 + q^{2n-1} e^{-2iz}} \right],$$

$$\vartheta_3''(z) = \vartheta_3'(z) \left[\sum_{n=1}^{\infty} \frac{2iq^{2n-1} e^{2iz}}{1 + q^{2n-1} e^{2iz}} - \sum_{n=1}^{\infty} \frac{2iq^{2n-1} e^{-2iz}}{1 + q^{2n-1} e^{-2iz}} \right]$$

$$+ \vartheta_3(z) \left[\sum_{n=1}^{\infty} \frac{(2i)^2 q^{2n-1} e^{2iz}}{(1 + q^{2n-1} e^{2iz})^2} + \sum_{n=1}^{\infty} \frac{(2i)^2 q^{2n-1} e^{-2iz}}{(1 + q^{2n-1} e^{-2iz})^2} \right].$$

Making $z \to 0$, we get

$$\vartheta_3'(0) = 0, \qquad \vartheta_3''(0) = -8\vartheta_3(0) \sum_{n=1}^{\infty} \frac{q^{2n-1}}{(1 + q^{2n-1})^2}.$$

In like manner,

$$\vartheta_4'(0) = 0, \qquad \vartheta_4''(0) = 8\vartheta_4(0) \sum_{n=1}^{\infty} \frac{q^{2n-1}}{(1 - q^{2n-1})^2},$$

$$\vartheta_2'(0) = 0, \qquad \vartheta_2''(0) = \vartheta_2(0) \left[-1 - 8 \sum_{n=1}^{\infty} \frac{q^{2n}}{(1 + q^{2n})^2} \right];$$

and, if we write $\vartheta_1(z) = \sin z \times \phi(z)$, we get

$$\phi'(0) = 0, \qquad \phi''(0) = 8\phi(0) \sum_{n=1}^{\infty} \frac{q^{2n}}{(1 - q^{2n})^2}.$$

If, however, we differentiate the equation $\vartheta_1(z) = \sin z \times \phi(z)$ three times, we get

$$\vartheta_1'(0) = \phi(0), \qquad \vartheta_1'''(0) = 3\phi''(0) - \phi(0).$$

Therefore

$$\frac{\vartheta_1'''(0)}{\vartheta_1'(0)} = 24 \sum_{n=1}^{\infty} \frac{q^{2n}}{(1 - q^{2n})^2} - 1;$$

and

$$1 + \frac{\vartheta_2''(0)}{\vartheta_2(0)} + \frac{\vartheta_3''(0)}{\vartheta_3(0)} + \frac{\vartheta_4''(0)}{\vartheta_4(0)}$$

$$= 8\left[-\sum_{n=1}^{\infty}\frac{q^{2n}}{(1+q^{2n})^2} - \sum_{n=1}^{\infty}\frac{q^{2n-1}}{(1+q^{2n-1})^2} + \sum_{n=1}^{\infty}\frac{q^{2n-1}}{(1-q^{2n-1})^2}\right]$$

$$= 8\left[-\sum_{n=1}^{\infty}\frac{q^{n}}{(1+q^{n})^2} + \sum_{n=1}^{\infty}\frac{q^{n}}{(1-q^{n})^2} - \sum_{n=1}^{\infty}\frac{q^{2n}}{(1-q^{2n})^2}\right],$$

on combining the first two series and writing the third as the difference of two series. If we add corresponding terms of the first two series in the last line, we get at once

$$1 + \frac{\vartheta_2''(0)}{\vartheta_2(0)} + \frac{\vartheta_3''(0)}{\vartheta_3(0)} + \frac{\vartheta_4''(0)}{\vartheta_4(0)} = 24\sum_{n=1}^{\infty}\frac{q^{2n}}{(1-q^{2n})^2} = 1 + \frac{\vartheta_1'''(0)}{\vartheta_1'(0)}.$$

Utilising the differential equations of §21.4, this may be written

$$\frac{1}{\vartheta_1'(0|\tau)}\frac{d\vartheta_1'(0|\tau)}{d\tau}$$

$$= \frac{1}{\vartheta_2(0|\tau)}\frac{d\vartheta_2(0|\tau)}{d\tau} + \frac{1}{\vartheta_3(0|\tau)}\frac{d\vartheta_3(0|\tau)}{d\tau} + \frac{1}{\vartheta_4(0|\tau)}\frac{d\vartheta_4(0\tau)}{d\tau}.$$

Integrating with regard to τ, we get

$$\vartheta_1'(0,q) = C\vartheta_2(0,q)\vartheta_3(0,q)\vartheta_4(0,q),$$

where C is a constant (independent of q). To determine C, make $q \to 0$; since

$$\lim_{q\to 0} q^{-\frac{1}{4}}\vartheta_1' = 2, \quad \lim_{q\to 0} q^{-\frac{1}{4}}\vartheta_2 = 2, \quad \lim_{q\to 0}\vartheta_3 = 1, \quad \lim_{q\to 0}\vartheta_4 = 1,$$

we see that $C = 1$; and so

$$\vartheta_1' = \vartheta_2\vartheta_3\vartheta_4,$$

which is the result stated.

21.42 The value of the constant G

From the result just obtained, we can at once deduce the value of the constant G which was introduced in §21.3.

For, by the formulae of that section,

$$\vartheta_1' = \phi(0) = 2q^{\frac{1}{4}}G\prod_{n=1}^{\infty}(1-q^{2n})^2, \qquad \vartheta_2 = 2q^{\frac{1}{4}}G\prod_{n=1}^{\infty}(1-q^{2n})^2,$$

$$\vartheta_3' = G\prod_{n=1}^{\infty}(1+q^{2n-1})^2, \qquad \vartheta_4 = G\prod_{n=1}^{\infty}(1+q^{2n-1})^2,$$

and so, by §21.41, we have

$$\prod_{n=1}^{\infty}(1 - q^{2n})^2 = G^2 \prod_{n=1}^{\infty}(1 + q^{2n})^2 \prod_{n=1}^{\infty}(1 + q^{2n-1})^2 \prod_{n=1}^{\infty}(1 - q^{2n-1})^2.$$

Now all the products converge absolutely, since $|q| < 1$, and so the following rearrangements are permissible:

$$\left\{\prod_{n=1}^{\infty}(1 - q^{2n-1}) \prod_{n=1}^{\infty}(1 - q^{2n})\right\} \times \left\{\prod_{n=1}^{\infty}(1 + q^{2n-1}) \prod_{n=1}^{\infty}(1 + q^{2n})\right\}$$

$$= \prod_{n=1}^{\infty}(1 - q^n) \prod_{n=1}^{\infty}(1 + q^n)$$

$$= \prod_{n=1}^{\infty}(1 - q^{2n}),$$

the first step following from the consideration that all positive integers are comprised under the forms $2n - 1$ and $2n$.

Hence the equation determining G is

$$\prod_{n=1}^{\infty}(1 - q^{2n})^2 = G^2,$$

and so $G = \pm \prod_{n=1}^{\infty}(1 - q^{2n})$.

To determine the ambiguity in sign, we observe that G is an analytic function of q (and consequently one-valued) throughout the domain $|q| < 1$; and from the product for $\vartheta_3(z)$, we see that $G \to 1$ as $q \to 0$. Hence the plus sign must always be taken; and so we have established the result

$$G = \prod_{n=1}^{\infty}(1 - q^{2n}).$$

Example 21.4.1 Shew that

$$\vartheta_1' = 2q^{\frac{1}{4}} G^3.$$

Example 21.4.2 Shew that

$$\vartheta_4 = \prod_{n=1}^{\infty}(1 - q^{2n-1})(1 - q^n).$$

Example 21.4.3 Shew that

$$1 + 2\sum_{n=1}^{\infty} q^{n^2} = \prod_{n=1}^{\infty}(1 - q^{2n})(1 + q^{2n-1})^2.$$

21.43 Connexion of the sigma-function with the theta-functions

It has been seen (Example 20.4.6) that the function $\sigma(z|\omega_1,\omega_2)$, formed with the periods $2\omega_1$, $2\omega_2$, is expressible in the form

$$\sigma(z) = \frac{2\omega_1}{\pi} \exp\left(\frac{\eta_1 z^2}{2\omega_1}\right) \sin\left(\frac{\pi z}{2\omega_1}\right) \prod_{n=1}^{\infty} \left\{\left(1 - 2q^{2n}\cos\frac{\pi z}{\omega_1} + q^{4n}\right)(1 - q^{2n})^{-2}\right\},$$

where $q = \exp(\pi i \omega_2/\omega_1)$.

If we compare this result with the product of §21.4 for $\vartheta_1(z|\tau)$, we see at once that

$$\sigma(z) = \frac{2\omega_1}{\pi} \exp\left(\frac{\eta_1 z^2}{2\omega_1}\right) \cdot \frac{1}{2}q^{-1/4} \prod_{n=1}^{\infty} (1 - q^{2n})^{-3} \vartheta_1\left(\frac{\pi z}{2\omega_1}\bigg|\frac{\omega_2}{\omega_1}\right).$$

To express η_1 in terms of theta-functions, take logarithms and differentiate twice, so that

$$-\wp(z) = \frac{\eta_1}{\omega_1} - \left(\frac{\pi}{2\omega_1}\right)^2 \mathrm{cosec}^2\left(\frac{\pi z}{2\omega_1}\right) + \left(\frac{\pi}{2\omega_1}\right)^2 \left[\frac{\phi''(v)}{\phi(v)} - \left\{\frac{\phi'(v)}{\phi(v)}\right\}^2\right],$$

where $v = \dfrac{\pi z}{2\omega_1}$ and the function ϕ is that defined in §21.41.

Expanding in ascending powers of z and equating the terms independent of z in this result, we get

$$0 = \frac{\eta_1}{\omega_1} - \frac{1}{3}\left(\frac{\pi}{2\omega_1}\right)^2 + \left(\frac{\pi}{2\omega_1}\right)^2 \frac{\phi''(0)}{\phi(0)},$$

and so

$$\eta_1 = -\frac{\pi^2}{12\omega_1} \frac{\vartheta_1'''}{\vartheta_1'}.$$

Consequently $\sigma(z|\omega_1,\omega_2)$ can be expressed in terms of theta-functions by the formula

$$\sigma(z|\omega_1,\omega_2) = \frac{2\omega_1}{\pi\vartheta_1'} \exp\left(-\frac{v^2 \vartheta_1'''}{6\vartheta_1'}\right) \vartheta_1\left(v\bigg|\frac{\omega_2}{\omega_1}\right),$$

where $v = \dfrac{\pi z}{2\omega_1}$.

Example 21.4.4   Prove that

$$\eta_2 = -\left(\frac{\pi^2 \omega_2 \vartheta_1''}{12\omega_1^2 \vartheta_1'} + \frac{\pi i}{2\omega_1}\right).$$

21.5 The expression of elliptic functions by means of theta-functions

It has just been seen that theta-functions are substantially equivalent to sigma-functions, and so, corresponding to the formulae of §§20.5–20.53, there will exist expressions for elliptic functions in terms of theta-functions. From the theoretical point of view, the formulae of §§20.5–20.53 are the more important on account of their symmetry in the periods, but in practice the theta-function formulae have two advantages: (i) that theta-functions are more readily computed than sigma-functions; (ii) that theta-functions have a specially simple behaviour with respect to the real period, which is generally the significant period in applications of elliptic functions in Applied Mathematics.

Let $f(z)$ be an elliptic function with periods $2\omega_1$, $2\omega_2$; let a fundamental set of zeros $(\alpha_1, \alpha_2, \ldots, \alpha_n)$ and poles $(\beta_1, \beta_2, \ldots, \beta_n)$ be chosen, so that

$$\sum_{r=1}^{n} (\alpha_r - \beta_r) = 0,$$

as in §20.53.

Then, by the methods of §20.53, the reader will at once verify that

$$f(z) = A_3 \prod_{r=1}^{n} \left\{ \vartheta_1 \left(\frac{\pi z - \pi \alpha_r}{2\omega_1} \middle| \frac{\omega_2}{\omega_1} \right) \middle/ \vartheta_1 \left(\frac{\pi z - \pi \beta_r}{2\omega_1} \middle| \frac{\omega_2}{\omega_1} \right) \right\},$$

where A_3 is a constant; and if

$$\sum_{m=1}^{m_r} A_{r,m} (z - \beta_r)^{-m}$$

be the principal part of $f(z)$ at its pole β_r, then, by the methods of §20.52,

$$f(z) = A_2 + \sum_{r=1}^{m} \left\{ \sum_{r=1}^{m_r} \frac{(-1)^{m-1} A_{r,m}}{(m-1)!} \frac{d^m}{dz^m} \log \vartheta_1 \left(\frac{\pi z - \pi \beta_r}{2\omega_1} \middle| \frac{\omega_2}{\omega_1} \right) \right\},$$

where A_2 is a constant.

This formula is important in connexion with the integration of elliptic functions. An example of an application of the formula to a dynamical problem will be found in §22.741.

Example 21.5.1 Shew that

$$\frac{\vartheta_3^2(z)}{\vartheta_1^2(z)} = -\frac{\vartheta_3^2}{\vartheta_1'^2} \frac{d}{dz} \frac{\vartheta_1'(z)}{\vartheta_1(z)} + \frac{\vartheta_3 \vartheta_3''}{\vartheta_1'^3},$$

and deduce that

$$\int_z^{\pi/2} \frac{\vartheta_3^2(z)}{\vartheta_1^2(z)} \, dz = \frac{\vartheta_3^2}{\vartheta_1'^2} \frac{\vartheta_1'(z)}{\vartheta_1(z)} + \left(\frac{\pi}{2} - z \right) \frac{\vartheta_3 \vartheta_3''}{\vartheta_1'^3}.$$

21.51 Jacobi's imaginary transformation

If an elliptic function be constructed with periods $2\omega_1$, $2\omega_2$, such that

$$\mathrm{Im}\,(\omega_2/\omega_1) > 0,$$

it might be convenient to regard the periods as being $2\omega_2$, $-2\omega_1$; for these numbers are periods and, if $\mathrm{Im}(\omega_2/\omega_1) > 0$, then also $\mathrm{Im}\,(-\omega_1/\omega_2) > 0$. In the case of the elliptic functions which have been considered up to this point, the periods have appeared in a symmetrical manner and nothing is gained by this point of view. But in the case of the theta-functions, which are only quasi-periodic, the behaviour of the function with respect to the real period π is quite different from its behaviour with respect to the complex period $\pi\tau$. Consequently, in view of the result of §21.43, we may expect to obtain transformations of theta-functions in which the period-ratios of the two theta-functions involved are respectively τ and $-1/\tau$.

The transformations of the four theta-functions were first obtained by Jacobi [346], who obtained them from the theory of elliptic functions; but[6] Poisson [532, p. 592] had previously obtained a formula identical with one of the transformations and the other three transformations can be obtained from this one by elementary algebra. A direct proof of the transformations is due to Landsberg [407], who used the methods of contour integration. (This method is indicated in Example 6.17 of Chapter 6.) The investigation of Jacobi's formulae, which we shall now give, is based on Liouville's theorem; the precise formula which we shall establish is

$$\vartheta_3(z|\tau) = (-i\tau)^{-\frac{1}{2}} \exp\left(\frac{z^2}{\pi i \tau}\right) \vartheta_3\left(\frac{z}{\tau} \middle| -\frac{1}{\tau}\right),$$

where $(-i\tau)^{-\frac{1}{2}}$ is to be interpreted by the convention $|\arg(-i\tau)| < \frac{1}{2}\pi$.

For brevity, we shall write $-1/\tau \equiv \tau'$, $q' = \exp(\pi i \tau')$.

The only zeros of $\vartheta_3(z|\tau)$ and $\vartheta_3(\tau' z|\tau')$ are simple zeros at the points at which

$$z = m\pi + n\pi\tau + \tfrac{1}{2}\pi + \tfrac{1}{2}\pi\tau, \qquad \tau' z = m'\pi + n'\pi\tau' + \tfrac{1}{2}\pi + \tfrac{1}{2}\pi\tau'$$

respectively, where m, n, m', n' take all integer values; taking $m' = -n - 1$, $n' = m$, we see that the quotient

$$\psi(z) \equiv \exp\left(\frac{z^2}{\pi i \tau}\right) \vartheta_3\left(\frac{z}{\tau} \middle| -\frac{1}{\tau}\right) \middle/ \vartheta_3(z|\tau)$$

is an integral function with no zeros.

Also

$$\psi(z + \pi\tau)/\psi(z) = \exp\left(\frac{2z\pi\tau + \pi^2\tau^2}{\pi i \tau}\right) \middle/ q^{-1} e^{-2iz} = 1,$$

while

$$\psi(z - \pi)/\psi(z) = \exp\left(\frac{-2z\pi + \pi^2}{\pi i \tau}\right) \times q'^{-1} e^{-2iz/r} = 1.$$

Consequently $\psi(z)$ is a doubly-periodic function with no zeros or poles; and so (§20.12) $\psi(z)$ must be a constant, A (independent of z). Thus,

$$A\vartheta_3(z|\tau) = \exp(i\tau' z^2/\pi)\vartheta_3(z\tau'|\tau')$$

and writing $z + \tfrac{1}{2}\pi$, $z + \tfrac{1}{2}\pi\tau$, $z + \tfrac{1}{2}\pi + \tfrac{1}{2}\pi\tau$ in turn for z, we easily get

$$A\vartheta_4(z|\tau) = \exp(i\tau' z^2/\pi)\vartheta_2(z\tau'|\tau'),$$
$$A\vartheta_2(z|\tau) = \exp(i\tau' z^2/\pi)\vartheta_4(z\tau'|\tau'),$$
$$A\vartheta_1(z|\tau) = -i \exp(i\tau' z^2/\pi)\vartheta_1(z\tau'|\tau').$$

We still have to prove that $A = (-i\tau)^{\frac{1}{2}}$; to do so, differentiate the last equation and then put $z = 0$; we get

$$A\vartheta_1'(0|\tau) = -i\tau'\vartheta_1'(0|\tau').$$

---

[6] The special case of the formula in which $z = 0$ had been given earlier by Poisson, [531, p. 420].

But

$$\vartheta_1'(0|\tau) = \vartheta_2(0|\tau)\vartheta_3(0|\tau)\vartheta_4(0|\tau)$$

and

$$\vartheta_1'(0|\tau') = \vartheta_2(0|\tau')\vartheta_3(0|\tau')\vartheta_4(0|\tau');$$

on dividing these results and substituting, we at once get $A^{-2} = -i\tau'$, and so $A = \pm(-i\tau)^{\frac{1}{2}}$. To determine the ambiguity in sign, we observe that

$$A\vartheta_3(0|\tau) = \vartheta_3(0|\tau'),$$

both the theta-functions being analytic functions of τ when Im $(\tau) > 0$; thus A is analytic and single-valued in the upper half τ-plane. Since the theta-functions are both positive when τ is a pure imaginary, the *plus* sign must then be taken. Hence, by the theory of analytic continuation, we *always* have

$$A = +(-i\tau)^{\frac{1}{2}};$$

this gives the transformation stated.

It has thus been shewn that

$$\sum_{n=-\infty}^{\infty} e^{n^2 \pi i \tau + 2niz} = \frac{1}{\sqrt{-i\tau}} \sum_{n=-\infty}^{\infty} e^{(z-n\pi)^2/(\pi i \tau)}.$$

Example 21.5.2 Shew that

$$\frac{\vartheta_4(0|\tau)}{\vartheta_3(0|\tau)} = \frac{\vartheta_2(0|\tau')}{\vartheta_3(0|\tau')}$$

when $\tau\tau' = -1$.

Example 21.5.3 Shew that

$$\frac{\vartheta_2(0|\tau + 1)}{\vartheta_3(0|\tau + 1)} = e^{\frac{1}{4}\pi i} \frac{\vartheta_2(0|\tau)}{\vartheta_4(0|\tau)}.$$

Example 21.5.4 Shew that

$$\prod_{n=1}^{\infty} \left(\frac{1 - q^{2n-1}}{1 + q^{2n-1}} \right) = \pm 2^{\frac{1}{2}} q'^{\frac{1}{8}} \prod_{n=1}^{\infty} \left(\frac{1 + q'^{2n}}{1 + q'^{2n-1}} \right);$$

and shew that the plus sign should be taken.

21.52 Landen's type of transformation

A transformation of elliptic integrals (§22.7), which is of historical interest, is due to Landen (§22.42); this transformation follows at once from a transformation connecting theta-functions with parameters τ and 2τ, namely

$$\frac{\vartheta_3(z|\tau)\vartheta_4(z|\tau)}{\vartheta_4(2z|2\tau)} = \frac{\vartheta_3(0|\tau)\vartheta_4(0|\tau)}{\vartheta_4(0|2\tau)},$$

which we shall now prove.

The zeros of $\vartheta_3(z|\tau)\vartheta_4(z|\tau)$ are simple zeros at the points where $z = \left(m + \frac{1}{2}\right)\pi + \left(n + \frac{1}{2}\right)\pi\tau$

and $z = m\pi + \left(n + \frac{1}{2}\right)\pi\tau$, with m and n taking all integral values; these are the points where $2z = m\pi + \left(n + \frac{1}{2}\right)\pi \cdot 2\tau$, which are the zeros of $\vartheta_4(2z|2\tau)$. Hence the quotient

$$\frac{\vartheta_3(z|\tau)\vartheta_4(z|\tau)}{\vartheta_4(2z|2\tau)}$$

has no zeros or poles. Moreover, associated with the periods π and $\pi\tau$, it has multipliers 1 and $(q^{-1}e^{-2iz})(-q^{-1}e^{-2iz})/(-q^{-2}e^{-4iz}) = 1$; it is therefore a doubly-periodic function, and is consequently (§20.12) a constant. The value of this constant may be obtained by putting $z = 0$ and we then have the result stated.

If we write $z + \frac{1}{2}\pi\tau$ for z, we get a corresponding result for the other theta-functions, namely

$$\frac{\vartheta_2(z|\tau)\vartheta_1(z|\tau)}{\vartheta_1(2z|2\tau)} = \frac{\vartheta_3(0|\tau)\vartheta_4(0|\tau)}{\vartheta_4(0|2\tau)}.$$

21.6 The differential equations satisfied by quotients of theta-functions

From Example 21.1.3 it is obvious that the function $\vartheta_1(z)/\vartheta_4(z)$ has periodicity factors -1, $+1$ associated with the periods π, $\pi\tau$ respectively; and consequently its derivative

$$\frac{\vartheta_1'(z)\vartheta_4(z) - \vartheta_4'(z)\vartheta_1(z)}{\vartheta_4^2(z)}$$

has the same periodicity factors.

But it is easy to verify that $\vartheta_2(z)\vartheta_3(z)/\vartheta_4^2(z)$ has periodicity factors -1, $+1$; and consequently, if $\phi(z)$ be defined as the quotient

$$\frac{\vartheta_1'(z)\vartheta_4(z) - \vartheta_4'(z)\vartheta_1(z)}{\vartheta_2(z)\vartheta_3(z)},$$

then $\phi(z)$ is doubly-periodic with periods π and $\pi\tau$; and the only possible poles of $\phi(z)$ are simple poles at points congruent to $\frac{1}{2}\pi$ and $\frac{1}{2}\pi + \frac{1}{2}\pi\tau$. Now consider $\phi\left(z + \frac{1}{2}\pi\tau\right)$; from the relations of §21.11, namely

$$\vartheta_1\left(z + \tfrac{1}{2}\pi\tau\right) = iq^{-\frac{1}{4}}e^{-iz}\vartheta_4(z), \quad \vartheta_4\left(z + \tfrac{1}{2}\pi\tau\right) = iq^{-\frac{1}{4}}e^{-iz}\vartheta_1(z),$$

$$\vartheta_2\left(z + \tfrac{1}{2}\pi\tau\right) = q^{-\frac{1}{4}}e^{-iz}\vartheta_3(z), \quad \vartheta_3\left(z + \tfrac{1}{2}\pi\tau\right) = q^{-\frac{1}{4}}e^{-iz}\vartheta_2(z),$$

we easily see that

$$\phi\left(z + \tfrac{1}{2}\pi\tau\right) = \frac{-\vartheta_4'(z)\vartheta_1(z) + \vartheta_1'(z)\vartheta_4(z)}{\vartheta_3(z)\vartheta_2(z)}.$$

Hence $\phi(z)$ is doubly-periodic with periods π and $\frac{1}{2}\pi\tau$; *and, relative to these periods, the only possible poles of $\phi(z)$ are simple poles at points congruent to $\frac{1}{2}\pi$.* Therefore (§20.12), $\phi(z)$ is a constant; and making $z \to 0$, we see that the value of this constant is $\{\vartheta_1'\vartheta_4\}/\{\vartheta_2\vartheta_3\} = \vartheta_4^2$.

We have therefore established the important result that

$$\frac{d}{dz}\left\{\frac{\vartheta_1(z)}{\vartheta_4(z)}\right\} = \vartheta_4^2 \frac{\vartheta_2(z)}{\vartheta_4(z)} \cdot \frac{\vartheta_3(z)}{\vartheta_4(z)};$$

writing $\xi \equiv \vartheta_1(z)/\vartheta_4(z)$ and making use of the results of §21.2, we see that

$$\left(\frac{d\xi}{dz}\right)^2 = (\vartheta_2^2 - \xi^2\vartheta_3^2)(\vartheta_3^2 - \xi^2\vartheta_2^2).$$

This differential equation possesses the solution $\vartheta_1(z)/\vartheta_4(z)$. It is not difficult to see that the general solution is $\pm\vartheta_1(z+\alpha)/\vartheta_4(z+\alpha)$ where α is the constant of integration; since this quotient changes sign when α is increased by π, the negative sign may be suppressed without affecting the generality of the solution.

Example 21.6.1　Shew that

$$\frac{d}{dz}\left\{\frac{\vartheta_2(z)}{\vartheta_4(z)}\right\} = -\vartheta_3^2 \frac{\vartheta_1(z)}{\vartheta_4(z)}\frac{\vartheta_3(z)}{\vartheta_4(z)}.$$

Example 21.6.2　Shew that

$$\frac{d}{dz}\left\{\frac{\vartheta_3(z)}{\vartheta_4(z)}\right\} = -\vartheta_2^2 \frac{\vartheta_1(z)}{\vartheta_4(z)}\frac{\vartheta_2(z)}{\vartheta_4(z)}.$$

21.61 The genesis of the Jacobian elliptic function sn *u*

The differential equation

$$\left(\frac{d\xi}{dz}\right)^2 = (\vartheta_2^2 - \xi^2\vartheta_3^2)(\vartheta_3^2 - \xi^2\vartheta_2^2),$$

which was obtained in §21.6, may be brought to a canonical form by a slight change of variable.

Write[7] $\xi\vartheta_3/\vartheta_2 = y$, $z\vartheta_3^2 = u$; then, if $k^{\frac{1}{2}}$ be written in place of ϑ_2/ϑ_3, the equation determining y in terms of u is

$$\left(\frac{dy}{du}\right)^2 = (1 - y^2)(1 - k^2 y^2).$$

This differential equation has the particular solution

$$y = \frac{\vartheta_3}{\vartheta_2}\frac{\vartheta_1\left(u\vartheta_3^{-2}\right)}{\vartheta_4\left(u\vartheta_3^{-2}\right)}.$$

The function of u on the right has multipliers -1, $+1$ associated with the periods $\pi\vartheta_3^2$, $\pi\tau\vartheta_3{}^2$; it is therefore a doubly-periodic function with periods $2\pi\vartheta_3^2$, $\pi\tau\vartheta_3^2$. In any cell, it has two simple poles at the points congruent to $\frac{1}{2}\pi\tau\vartheta_3{}^2$ and $\pi\vartheta_3^2 + \frac{1}{2}\pi\tau\vartheta_3^2$; and, on account of the nature of the quasi-periodicity of y, the residues at these points are equal and opposite in sign; the zeros of the function are the points congruent to 0 and $\pi\vartheta_3^2$.

It is customary to regard y as depending on k rather than on q; and to exhibit y as a function of u and k, we write[8]

$$y = \mathrm{sn}(u, k), \quad \text{or simply} \quad y = \mathrm{sn}\, u.$$

---

[7]　Notice, from the formulae of §21.3, that $\vartheta_2 \neq 0$, $\vartheta_3 \neq 0$ when $|q| < 1$, except when $q = 0$, in which case the theta-functions degenerate; the substitutions are therefore legitimate.

[8]　Jacobi and other early writers used the notation sinam in place of sn.

It is now evident that $\text{sn}(u,k)$ is an elliptic function of the second of the types described in §20.13; when $q \to 0$ (so that $k \to 0$), it is easy to see that $\text{sn}(u,k) \to \sin u$.

The constant k is called the *modulus*; if $k'^{\frac{1}{2}} = \vartheta_4/\vartheta_3$, so that $k^2 + k'^2 = 1$, k' is called the *complementary modulus*. The quasi-periods $\pi\vartheta_3^2, \pi\tau\vartheta_3^2$ are usually written $2K, 2iK'$, so that $\text{sn}(u,k)$ has periods $4K, 2iK'$. From §21.51, we see that $2K' = \pi\vartheta_3^2(0|\tau')$, so that K' is the same function of τ' as K is of τ, when $\tau\tau' = -1$.

Example 21.6.3   Shew that

$$\frac{d}{dz}\frac{\vartheta_2(z)}{\vartheta_4(z)} = -\vartheta_3^2\frac{\vartheta_1(z)}{\vartheta_4(z)}\frac{\vartheta_3(z)}{\vartheta_4(z)};$$

and deduce that, if

$$y = \frac{\vartheta_4}{\vartheta_2}\frac{\vartheta_2(z)}{\vartheta_4(z)}, \quad \text{and} \quad u = z\vartheta_3^2,$$

then

$$\left(\frac{dy}{du}\right)^2 = (1-u^2)(k'^2 + k^2u^2).$$

Example 21.6.4   Shew that

$$\frac{d}{dz}\frac{\vartheta_3(z)}{\vartheta_4(z)} = -\vartheta_2^2\frac{\vartheta_1(z)}{\vartheta_4(z)}\frac{\vartheta_2(z)}{\vartheta_4(z)};$$

and deduce that, if $y = \dfrac{\vartheta_4}{\vartheta_3}\dfrac{\vartheta_3(z)}{\vartheta_4(z)}$, and $u = z\vartheta_3^2$, then

$$\left(\frac{dy}{du}\right)^2 = (1-u^2)(u^2 - k'^2).$$

Example 21.6.5   Obtain the following results:

$$\left(\frac{2kK}{\pi}\right)^{\frac{1}{2}} = \vartheta_2 = 2q^{\frac{1}{4}}(1 + q^2 + q^6 + q^{12} + q^{20} + \cdots),$$

$$\left(\frac{2K}{\pi}\right)^{\frac{1}{2}} = \vartheta_3 = 1 + 2q + 2q^4 + 2q^9 + \cdots,$$

$$\left(\frac{2k'K}{\pi}\right)^{\frac{1}{2}} = \vartheta_4 = 1 - 2q + 2q^4 + 2q^9 - \cdots,$$

$$K' = K\pi^{-1}\log(1/q).$$

These results are convenient for calculating k, k', K, K' when q is given.

21.62 Jacobi's earlier notation. The theta-function $\Theta(u)$ and the eta-function $\text{H}(u)$

The presence of the factors ϑ_3^{-2} in the expression for $\text{sn}(u,k)$ renders it sometimes desirable to use the notation which Jacobi employed in the *Fundamenta Nova* [349], and subsequently discarded. The function which is of primary importance with this notation is $\Theta(u)$, defined by the equation

$$\Theta(u) = \vartheta_4(u\vartheta_3^{-2}|\tau),$$

so that the periods associated with $\Theta(u)$ are $2K$ and $2iK'$.

The function $\Theta(u + K)$ then replaces $\vartheta_3(z)$; and in place of $\vartheta_1(z)$ we have the function $H(u)$ defined by the equation

$$H(u) = -iq^{-\frac{1}{4}}e^{i\pi u/(2K)}\Theta(u + iK') = \vartheta_1(u\vartheta_3^{-2}|\tau),$$

and $\vartheta_2(z)$ is replaced by $H(u + K)$.

The reader will have no difficulty in translating the analysis of this chapter into Jacobi's earlier notation.

Example 21.6.6   If $\Theta'(u) = \dfrac{d\Theta(u)}{du}$, shew that the singularities of $\dfrac{\Theta'(u)}{\Theta(u)}$ are simple poles at the points congruent to $iK' \pmod{2K, 2iK'}$; and the residue at each singularity is 1.

Example 21.6.7   Shew that

$$H'(0) = \frac{\pi}{2}K^{-1}H(K)\Theta(0)\Theta(K).$$

21.7 The problem of inversion

Up to the present, the Jacobian elliptic function $\mathrm{sn}(u, k)$ has been implicitly regarded as depending on the parameter q rather than on the modulus k; and it has been shewn that it satisfies the differential equation

$$\left(\frac{d\,\mathrm{sn}\,u}{du}\right)^2 = (1 - \mathrm{sn}^2 u)(1 - k^2\,\mathrm{sn}^2 u), \tag{21.1}$$

where

$$k^2 = \frac{\vartheta_2^4(0, q)}{\vartheta_3^4(0, q)}. \tag{21.2}$$

But, in those problems of Applied Mathematics in which elliptic functions occur, we have to deal with the solution of the differential equation

$$\left(\frac{dy}{du}\right)^2 = (1 - y^2)(1 - k^2y^2)$$

in which the *modulus k* is given, and we have no *a priori* knowledge of the value of q; and, to prove the existence of an analytic function $\mathrm{sn}(u, k)$ which satisfies this equation, we have to shew that a number τ exists[9] such that

$$k^2 = \frac{\vartheta_2^4(0|\tau)}{\vartheta_3^4(0|\tau)}. \tag{21.3}$$

When this number τ has been shewn to exist, the function $\mathrm{sn}(u, k)$ can be constructed as a

---

[9]   The existence of a number τ, for which $\operatorname{Im}\tau > 0$, involves the existence of a number q such that $|q| < 1$. An alternative procedure would be to discuss the differential equation directly, after the manner of Chapter 10.

quotient of theta-functions, satisfying the differential equation and possessing the properties of being doubly-periodic and analytic except at simple poles; and also

$$\lim_{u \to 0} \frac{\text{sn}(u, k)}{u} = 1.$$

That is to say, we can *invert* the integral

$$u = \int_0^y \frac{dt}{(1 - t^2)^{\frac{1}{2}}(1 - k^2 t^2)^{\frac{1}{2}}},$$

so as to obtain the equation $y = \text{sn}(u, k)$.

The difficulty, of course, arises in shewing that the equation

$$c = \frac{\vartheta_2^4(0|\tau)}{\vartheta_3^4(0|\tau)}, \tag{21.4}$$

(where c has been written for k^2), has a solution.

When $0 < c < 1$, it is easy to shew that a solution exists. (This is the case which is of practical importance.) From the identity given in Corollary 21.2.1, it is evident that it is sufficient to prove the existence of a solution of the equation $1 - c = \vartheta_4^4(0|\tau)/\vartheta_3^4(0|\tau)$, which may be written

$$1 - c = \prod_{n=1}^{\infty} \left(\frac{1 - q^{2n-1}}{1 + q^{2n-1}} \right)^8. \tag{21.5}$$

Now, as q increases from 0 to 1, the product on the right is continuous and steadily decreases from 1 to 0; and so (§3.63) it passes through the value $1 - c$ once and only once. Consequently a solution of the equation in τ exists and the problem of inversion may be regarded as solved.

21.71 The problem of inversion for complex values of c. The modular functions $f(\tau)$, $g(\tau)$, $h(\tau)$

The problem of inversion may be regarded as a problem of Integral Calculus, and it may be proved, by somewhat lengthy algebraical investigations involving a discussion of the behaviour of

$$\int_0^y (1 - t^2)^{-\frac{1}{2}}(1 - k^2 t^2)^{-\frac{1}{2}} \, dt,$$

when y lies on a 'Riemann surface', that the problem of inversion possesses a solution. For an exhaustive discussion of this aspect of the problem, the reader is referred to Hancock [271].

It is, however, more in accordance with the spirit of this work to prove by Cauchy's method (§6.31) that the equation $c = \vartheta_2^4(0|\tau)/\vartheta_3^4(0|\tau)$ has one root lying in a certain domain of the τ-plane and that (subject to certain limitations) this root is an analytic function of c, when c is regarded as variable. It has been seen that the existence of this root yields the solution of the inversion problem, so that the existence of the Jacobian elliptic function with given modulus k will have been demonstrated.

The method just indicated has the advantage of exhibiting the potentialities of what are known as *modular functions*. The general theory of these functions (which are of great importance in connexion with the *Theories of Transformation of Elliptic Functions*) has been considered in a treatise by Klein and Fricke [378].

Let

$$f(\tau) = 16e^{\pi i\tau} \prod_{n=1}^{\infty} \left\{ \frac{1 + e^{2n\pi i\tau}}{1 + e^{(2n-1)\pi i\tau}} \right\}^8 = \frac{\vartheta_2^4(0|\tau)}{\vartheta_3^4(0|\tau)},$$

$$g(\tau) = \prod_{n=1}^{\infty} \left\{ \frac{1 - e^{(2n-1)\pi i\tau}}{1 + e^{(2n-1)\pi i\tau}} \right\}^8 = \frac{\vartheta_4^4(0|\tau)}{\vartheta_3^4(0|\tau)},$$

$$h(\tau) = -f(\tau)/g(\tau).$$

Then, if $\tau\tau' = -1$, the functions just introduced possess the following properties:

$$f(\tau + 2) = f(\tau), \quad g(\tau + 2) = g(\tau), \quad f(\tau) + g(\tau) = 1,$$
$$f(\tau + 1) = h(\tau), \quad f(\tau') = g(\tau), \quad g(\tau') = f(\tau),$$

by Corollary 21.2.1 and Example 21.5.2.

It is easy to see that as $\mathrm{Im}\,(\tau) \to +\infty$, the functions $\frac{1}{16}e^{-\pi i\tau} f(\tau) = f_1(\tau)$ and $g(\tau)$ tend to unity, uniformly with respect to $\mathrm{Re}\,(\tau)$, when $-1 \le \mathrm{Re}\,(\tau) \le 1$; and the derivatives of these two functions (with regard to τ) tend uniformly to zero in the same circumstances. This follows from the expressions for the theta-functions as power series in q, it being observed that $|q| \to 0$ as $\mathrm{Im}\,(\tau) \to +\infty$.

21.711 The principal solution of $f(\tau) - c = 0$

It has been seen in §6.31 that, if $f(\tau)$ is analytic inside and on any contour, $2\pi i$ times the number of roots of the equation $f(\tau) - c = 0$ inside the contour is equal to

$$\int \frac{1}{f(\tau) - c} \frac{df(\tau)}{d\tau} \, d\tau,$$

taken round the contour in question.

Take the contour $ABCDEFE'D'C'B'A$ shewn in the figure, it being supposed temporarily that $f(\tau) - c$ has no zero actually on the contour. The values of $f(\tau)$ at points on the contour are discussed in §21.712.

The contour is constructed in the following manner: FE is drawn parallel to the real axis, at a large distance from it. AB is the inverse of FE with respect to the circle $|\tau| = 1$. BC is the inverse of ED with respect to $|\tau| = 1$, D being chosen so that $D_1 = A0$. By elementary geometry, it follows that, since C and D are inverse points and 1 is its own inverse, the circle on $D1$ as diameter passes through C; and so the arc CD of this circle is the reflexion of the arc AB in the line $\mathrm{Re}\,(\tau) = \frac{1}{2}$. The left-hand half of the figure is the reflexion of the right-hand half in the line $\mathrm{Re}\,(\tau) = 0$.

It will now be shewn that, unless[10] $c \ge 1$ or $c \le 0$, the equation $f(\tau) - c = 0$ has one, and only one, root inside the contour, provided that FE is sufficiently distant from the real axis. This root will be called the *principal root* of the equation.

To establish the existence of this root, consider $\displaystyle\int \frac{1}{f(\tau) - c} \frac{df(\tau)}{d\tau} \, d\tau$ taken along the

---

[10] It is shewn in §21.712 that, if $c \ge 1$ or $c \le 0$, then $f(\tau) - c$ has a zero on the contour.

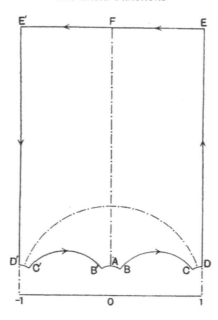

various portions of the contour. Since $f(\tau + 2) = f(\tau)$, we have

$$\left\{\int_{DE} + \int_{E'D'}\right\} \frac{1}{f(\tau) - c} \frac{df(\tau)}{d\tau} d\tau = 0.$$

Also, as τ describes BC and $B'C'$, $\tau'(= -1/\tau)$ describes $E'D'$ and ED respectively; and so

$$\left\{\int_{BC} + \int_{C'B'}\right\} \frac{1}{f(\tau) - c} \frac{df(\tau)}{d\tau} d\tau = \left\{\int_{BC} + \int_{C'B'}\right\} \frac{1}{g(\tau') - c} \frac{dg(\tau')}{d\tau} d\tau$$

$$= \left\{\int_{E'D'} + \int_{DE}\right\} \frac{1}{g(\tau') - c} \frac{dg(\tau')}{d\tau'} d\tau'$$

$$= 0,$$

because $g(\tau' + 2) = g(\tau')$, and consequently corresponding elements of the integrals cancel.

Since $f(\tau \pm 1) = h(\tau)$, we have

$$\left\{\int_{D'C'} + \int_{CD}\right\} \frac{1}{f(\tau) - c} \frac{df(\tau)}{d\tau} d\tau = \int_{B'AB} \frac{1}{h(\tau) - c} \frac{dh(\tau)}{d\tau} d\tau;$$

but, as τ' describes $B'AB$, τ describes EE', and so the integral round the complete contour reduces to

$$\int_{EE'} \left\{\frac{1}{f(\tau) - c} \frac{df(\tau)}{d\tau} + \frac{1}{h(\tau') - c} \frac{dh(\tau')}{d\tau} + \frac{1}{f(\tau') - c} \frac{df(\tau')}{d\tau}\right\} d\tau$$

$$= \int_{EE'} \left\{\frac{1}{f(\tau) - c} \frac{df(\tau)}{d\tau} - \frac{1}{h(\tau)\{1 - ch(\tau)\}} \frac{dh(\tau)}{d\tau} + \frac{1}{g(\tau) - c} \frac{dg(\tau)}{d\tau}\right\} d\tau.$$

Now as EE' moves off to infinity[11], $f(\tau) - c \to -c \neq 0$, $g(\tau) - c \to 1 - c \neq 0$, and so the

---

[11] It has been supposed temporarily that $c \neq 0$ and $c \neq 1$.

limit of the integral is

$$- \lim \int_{EE'} \frac{1}{1 - c \cdot h(\tau)} \frac{d}{d\tau} \{\log h(\tau)\} \, d\tau$$

$$= \lim \int_{E'E} \frac{1}{1 - c \cdot h(\tau)} \left\{ \pi i + \frac{d \log f_1(\tau)}{d\tau} - \frac{d \log g(\tau)}{d\tau} \right\} \, d\tau.$$

But $1 - c \cdot h(\tau) \to 1$, $f_1(\tau) \to 1$, $g_1(\tau) \to 1$, $\dfrac{df_1(\tau)}{d\tau} \to 0$, $\dfrac{dg(\tau)}{d\tau} \to 0$, and so the limit of the integral is

$$\int_{E'E} \pi i \, d\tau = 2\pi i.$$

Now, if we choose EE' to be initially so far from the real axis that $f(\tau) - c$, $1 - ch(\tau)$, $g(\tau) - c$ have no zeros when τ is above EE', then the contour will pass over no zeros of $f(\tau) - c$ as EE' moves off to infinity and the radii of the arcs CD, $D'C'$, $B'AB$ diminish to zero; and then the integral will not change as the contour is modified, and so the original contour integral will be $2\pi i$, and the number of zeros of $f(\tau) - c$ inside the original contour will be precisely 1.

21.712 The values of the modular function $f(\tau)$ on the contour considered

We now have to discuss the point mentioned at the beginning of §21.711, concerning the zeros of $f(\tau) - c$ on the lines[12] joining ± 1 to $\pm 1 + i\infty$ and on the semicircles of $0BC1$, $(-1)C'B'0$.

As τ goes from 1 to $1 + i\infty$ or from -1 to $-1 + i\infty$, $f(\tau)$ goes from $-\infty$ to 0 through real negative values. So, if c is negative, we make an indentation in DE and a corresponding indentation in $D'E'$; and the integrals along the indentations cancel in virtue of the relation $f(\tau + 2) + f(\tau)$.

As τ describes the semicircle $0BG1$, τ' goes from $-1 + i\infty$ to -1, and $f(\tau) = g(\tau') = 1 - f(\tau')$, and goes from 1 to $+\infty$ through real values; it would be possible to make indentations in BC and $B'C'$ to avoid this difficulty, but we do not do so for the following reason: the effect of changing the sign of the imaginary part of the number c is to change the sign of the real part of τ. Now, if $0 < \mathrm{Re}\,(c) < 1$ and $\mathrm{Im}\,(c)$ be small, this merely makes τ cross $0F$ by a short path; if $\mathrm{Re}\,(c) < 0$, τ goes from DE to $D'E'$ (or *vice versa*) and the value of q alters only slightly; but if $\mathrm{Re}\,(c) > 1$, τ goes from BC to $B'C'$, and so q is not a one-valued function of c so far as circuits round $c = +1$ are concerned; to make q a one-valued function of c, we cut the c-plane from $+1$ to $+\infty$; and then for values of c in the cut plane, q is determined as a one-valued analytic function of c, say $q(c)$, by the formula $q(c) = e^{\pi i \tau(c)}$ where

$$\tau(c) = \frac{1}{2\pi i} \int \frac{\tau}{f(\tau) - c} \frac{df(\tau)}{d\tau} \, d\tau,$$

as may be seen from §6.3, by using the method of §5.22.

If c describes a circuit not surrounding the point $c = 1$, $q(c)$ is one-valued, but $\tau(c)$ is one-valued only if, in addition, the circuit does not surround the point $c = 0$.

---

[12] We have seen that EE' can be so chosen that $f(\tau) - c$ has no zeros either on EE' or on the small circular arcs.

21.72 The periods, regarded as functions of the modulus

Since $K = \frac{1}{2}\pi\vartheta_3^2(0, q)$ we see from §21.712 that K is a one-valued analytic function of $c(= k^2)$ when a cut from 1 to $+\infty$ is made in the c-plane; but since $K' = -i\tau K$, we see that K' is not a one-valued function of c unless an additional cut is made from 0 to $-\infty$; it will appear later (§22.32) that the cut from 1 to $+\infty$ which was necessary so far as K is concerned is *not* necessary as regards K'.

21.73 The inversion-problem associated with Weierstrassian elliptic functions

It will now be shewn that, when invariants g_2 and g_3 are given, such that $g_2^3 \neq 27g_3^2$, it is possible to construct the Weierstrassian elliptic function with these invariants; that is to say, we shall shew that *it is possible to construct periods $2\omega_1$, $2\omega_2$ such that the function $\wp(z|\omega_1, \omega_2)$ has invariants g_2 and g_3*. On the actual calculation of the periods, see R. T. A. Innes [337].

The problem is solved if we can obtain a solution of the differential equation

$$\left(\frac{dy}{dz}\right)^2 = 4y^3 - g_2 y - g_3$$

of the form $y = \wp(z|\omega_1, \omega_2)$. We proceed to effect the solution of the equation with the aid of theta-functions.

Let $v = Az$, where A is a constant to be determined presently. By the methods of §21.6, it is easily seen that

$$\vartheta_2'(v)\vartheta_1(v) - \vartheta_1'(v)\vartheta_2(v) = -\vartheta_3(v)\vartheta_4(v)\vartheta_2^2,$$

and hence, using the results of §21.2, we have

$$\left\{\frac{d}{dz}\frac{\vartheta_2^2(v)}{\vartheta_1^2(v)}\vartheta_3^2\vartheta_4^2\right\}^2 = 4A^2\left(\frac{\vartheta_2^2(v)}{\vartheta_1^2(v)}\vartheta_3^2\vartheta_4^2\right)\left(\frac{\vartheta_2^2(v)}{\vartheta_1^2(v)}\vartheta_3^2\vartheta_4^2 + \vartheta_4^4\right)\left(\frac{\vartheta_2^2(v)}{\vartheta_1^2(v)}\vartheta_3^2\vartheta_4^2 + \vartheta_3^4\right).$$

Now let e_1, e_2, e_3 be the roots of the equation $4y^3 - g_2 y - g_3 = 0$, chosen in such an order that $(e_1 - e_2)/(e_1 - e_3)$ is not[13] a real number greater than unity or negative.

In these circumstances the equation

$$\frac{e_1 - e_2}{e_1 - e_3} = \frac{\vartheta_4^4(0|\tau)}{\vartheta_3^4(0|\tau)}$$

possesses a solution (§21.712) such that $\mathrm{Im}\,(r) > 0$; this equation determines the parameter τ of the theta-functions, which has, up till now, been at our disposal.

Choosing τ in this manner, let A be next chosen so that[14] $A^2\vartheta_4^4 = e_1 - e_2$. Then the function

$$y = A^2\frac{\vartheta_2^2(v|\tau)}{\vartheta_1^2(v|\tau)}\vartheta_3^2(0|\tau)\vartheta_4^2(0|\tau) + e_1$$

---

[13] If $\dfrac{e_i - e_j}{e_i - e_k} > 1$, then $0 < \dfrac{e_i - e_k}{e_i - e_j} < 1$; and if $\dfrac{e_i - e_j}{e_i - e_k} < 0$, then $1 - \dfrac{e_i - e_j}{e_i - e_k} > 1$, and

$\dfrac{e_j - e_i}{e_j - e_k} = \left\{1 - \dfrac{e_i - e_k}{e_i - e_j}\right\}^{-1} < 1$. The values $0, 1, \infty$ of $(e_1 - e_2)/(e_1 - e_3)$ are excluded since $g_2^3 \neq 27g_3^2$.

[14] The sign attached to A is a matter of indifference, since we deal exclusively with *even* functions of v and z.

satisfies the equation

$$\left(\frac{dy}{dz}\right)^2 = 4(y - e_1)(y - e_2)(y - e_3).$$

The periods of y, *qua* function of z, are πA, $\pi\tau/A$; calling these $2\omega_1$, $2\omega_2$ we have $\text{Im}\,(\omega_2/\omega_1) > 0$. The function $\wp(z|\omega_1, \omega_2)$ may be constructed with these periods, and it is easily seen that

$$\wp(z) - A^2 \frac{\vartheta_2^2(v|\tau)}{\vartheta_1^2(v|\tau)} \vartheta_3^2(0|\tau)\vartheta_4^2(0|\tau) - e_1$$

is an elliptic function with no pole at the origin[15]; it is therefore a constant, C, say.

If G_2, G_3 be the invariants of $\wp(z|\omega_1, \omega_2)$, we have

$$4\wp^3(z) - G_2\wp(z) - G_3 = \wp'^2(z) = 4\{\wp(z) - C - e_1\}\{\wp(z) - C - e_2\}\{\wp(z) - C - e_3\},$$

and so, comparing coefficients of powers of $\wp(z)$, we have

$$0 = 12C, \qquad G_2 = g_2 - 12C^2, \qquad G_3 = g_3 - g_2C + 4C^3.$$

Hence $C = 0$, $G_2 = g_2$, $G_3 = g_3$; and so the function $\wp(z|\omega_1, \omega_2)$ with the required invariants has been constructed.

21.8 The numerical computation of elliptic functions

The series proceeding in ascending powers of q are convenient for calculating theta-functions generally, even when $|q|$ is as large as 0.9. But it usually happens in practice that the modulus k is given and the calculation of K, K' and q is necessary. It will be seen later (§§22.301, 22.32) that K, K' are expressible in terms of hypergeometric functions, by the equations

$$K = \frac{\pi}{2} F\left(\frac{1}{2}, \frac{1}{2}; 1; k^2\right), \qquad K' = \frac{\pi}{2} F\left(\frac{1}{2}, \frac{1}{2}; 1; k'^2\right);$$

but these series converge slowly except when $|k|$ and $|k'|$ respectively are quite small; so that the series are never simultaneously suitable for numerical calculations.

To obtain more convenient series for numerical work, we first calculate q as a root of the equation $k = \vartheta_2^2(0, q)/\vartheta_3^2(0, q)$, and then obtain K from the formula $K = \frac{1}{2}\pi\vartheta_3^2(0, q)$ and K' from the formula

$$K' = \pi^{-1} K \log_e(1/q).$$

The equation $k = \vartheta_2^2(0, q)/\vartheta_3^2(0, q)$ is equivalent to[16] $\sqrt{k'} = \vartheta_4(0, q)/\vartheta_3(0, q)$.

Writing $2\varepsilon = \dfrac{1 - \sqrt{k'}}{1 + \sqrt{k'}}$ (so that $0 < \varepsilon < \frac{1}{2}$ when $0 < k < 1$), we get

$$2\varepsilon = \frac{\vartheta_3(0, q) - \vartheta_4(0, q)}{\vartheta_3(0, q) + \vartheta_4(0, q)} = \frac{\vartheta_2(0, q^4)}{\vartheta_3(0, q^4)}.$$

We have seen (§§21.71–21.712) that this equation in q^4 possesses a solution which is an

---

[15] The terms in z^{-2} cancel, and there is no term in z^{-1} because the function is even.

[16] In numerical work $0 < k < 1$, and so q is positive and $0 < \sqrt{k'} < 1$.

analytic function of ε^4 when $|\varepsilon| < \frac{1}{2}$; and so q will be expansible in a Maclaurin series in powers of ε in this domain. The theta-functions do not vanish when $|q| < 1$ except at $q = 0$, so this gives the only possible branch point.

It remains to determine the coefficients in this expansion from the equation

$$\varepsilon = \frac{q + q^9 + q^{25} + \cdots}{1 + 2q^4 + 2q^{16} + \cdots},$$

which may be written $q = \varepsilon + 2q^4\varepsilon - q^9 + 2q^{16}\varepsilon - q^{25} + \cdots$; the reader will easily verify by continually substituting $\varepsilon + 2q^4\varepsilon - q^0 + \cdots$ for q wherever q occurs on the right that the first two terms[17] are given by

$$q = \varepsilon + 2\varepsilon^5 + 15\varepsilon^9 + 150\varepsilon^{13} + O(\varepsilon^{17}).$$

It has just been seen that this series converges when $|\varepsilon| < \frac{1}{2}$.

Note   The first two terms of this expansion usually suffice; thus, even if k be as large as $\sqrt{0.8704} = 0.933 \cdots$, $\varepsilon = \frac{1}{8}$, $2\varepsilon^5 = 0.0000609$, $15\varepsilon^9 = 0.0000002$.

Example 21.8.1   Given $k = k' = 1/\sqrt{2}$, calculate q, K, K' by means of the expansion just obtained, and also by observing that $\tau = i$, so that $q = e^{-\pi}$.

$$q = 0.0432139, \qquad K = K' = 1.854075.$$

21.9 The notations employed for the theta-functions

The following scheme indicates the principal systems of notation which have been employed by various writers; the symbols in any one column all denote the same function.

$\vartheta_1(\pi z)$	$\vartheta_2(\pi z)$	$\vartheta_3(\pi z)$	$\vartheta(\pi z)$	Jacobi
$\vartheta_1(z)$	$\vartheta_2(z)$	$\vartheta_3(z)$	$\vartheta_4(z)$	Tannery and Molk
$\theta_1(\omega z)$	$\theta_2(\omega z)$	$\theta_3(\omega z)$	$\theta(\omega z)$	Briot and Bouquet
$\theta_1(z)$	$\theta_2(z)$	$\theta_3(z)$	$\theta_0(z)$	Weierstrass, Halphen, Hancock
$\theta(z)$	$\theta_1(z)$	$\theta_3(z)$	$\theta_2(z)$	Jordan, Harkness and Morley

The notation employed by Hermite, H. J. S. Smith and some other mathematicians is expressed by the equation

$$\theta_{\mu,\nu}(x) = \sum_{n=-\infty}^{\infty} (-1)^{n\nu} q^{\frac{1}{4}(2n+\mu)^2} e^{i\pi(2n+\mu)x/a}; \qquad (\mu = 0, 1; \; \nu = 0, 1)$$

with this notation the results of Example 21.1.3 take the very concise form

$$\theta_{\mu,\nu}(x + a) = (-1)^{\mu}\theta_{\mu,\nu}(x), \qquad \theta_{\mu,\nu}(x + a\tau) = (-1)^{\nu}q^{-1}e^{-2i\pi x/a}\theta_{\mu,\nu}(x).$$

Cayley employs Jacobi's earlier notation (§21.62). The advantage of the Weierstrassian notation is that unity (instead of π) is the real period of $\theta_3(z)$ and $\theta_0(z)$. Jordan's notation exhibits the analogy between the theta-functions and the three sigma-functions defined in §20.421. The reader will easily obtain relations, similar to that of §21.43, connecting $\theta_r(z)$ with $\sigma_r(2\omega_1 z)$ when $r = 1, 2, 3$.

---

[17] This expansion was given by Weierstrass [663, p. 276].

21.10 Miscellaneous examples

Example 21.1 (Jacobi)   Obtain the addition-formulae

$$\vartheta_1(y+z)\vartheta_1(y-z)\vartheta_4^2 = \vartheta_3^2(y)\vartheta_2^2(z) - \vartheta_2^2(y)\vartheta_3^2(z)$$
$$= \vartheta_1^2(y)\vartheta_4^2(z) - \vartheta_4^2(y)\vartheta_1^2(z),$$
$$\vartheta_2(y+z)\vartheta_2(y-z)\vartheta_4^2 = \vartheta_4^2(y)\vartheta_2^2(z) - \vartheta_1^2(y)\vartheta_3^2(z)$$
$$= \vartheta_2^2(y)\vartheta_4^2(z) - \vartheta_3^2(y)\vartheta_1^2(z),$$
$$\vartheta_3(y+z)\vartheta_3(y-z)\vartheta_4^2 = \vartheta_4^2(y)\vartheta_3{}^2(z) - \vartheta_1^2(y)\vartheta_2^2(z)$$
$$= \vartheta_3^2(y)\vartheta_4^2(z) - \vartheta_2^2(y)\vartheta_1^2(z),$$
$$\vartheta_4(y+z)\vartheta_4(y-z)\vartheta_4^2 = \vartheta_3^2(y)\vartheta_3^2(z) - \vartheta_2^2(y)\vartheta_2^2(z)$$
$$= \vartheta_4^2(y)\vartheta_4^2(z) - \vartheta_1^2(y)\vartheta_1^2(z).$$

Example 21.2 (Jacobi)   Obtain the addition-formulae

$$\vartheta_4(y+z)\vartheta_4(y-z)\vartheta_2^2 = \vartheta_4^2(y)\vartheta_2^2(z) + \vartheta_3^2(y)\vartheta_1^2(z)$$
$$= \vartheta_2^2(y)\vartheta_4^2(z) + \vartheta_1^2(y)\vartheta_3^2(z),$$
$$\vartheta_4(y+z)\vartheta_4(y-z)\vartheta_3^2 = \vartheta_4^2(y)\vartheta_3^2(z) + \vartheta_2^2(y)\vartheta_1^2(z)$$
$$= \vartheta_3^2(y)\vartheta_4^2(z) + \vartheta_1^2(y)\vartheta_2^2(z);$$

and, by increasing y by half periods, obtain the corresponding formulae for

$$\vartheta_r(y+z)\vartheta_r(y-z)\vartheta_2^2 \quad \text{and} \quad \vartheta_r(y+z)\vartheta_r(y-z)\vartheta_3^2,$$

where $r = 1, 2, 3$.

Example 21.3 (Jacobi)   Obtain the formulae

$$\vartheta_1(y \pm z)\vartheta_2(y \mp z)\vartheta_3\vartheta_4 = \vartheta_1(y)\vartheta_2(y)\vartheta_3(z)\vartheta_4(z)$$
$$\pm \vartheta_3(y)\vartheta_4(y)\vartheta_1(z)\vartheta_2(z),$$
$$\vartheta_1(y \pm z)\vartheta_3(y \mp z)\vartheta_2\vartheta_4 = \vartheta_1(y)\vartheta_3(y)\vartheta_2(z)\vartheta_4(z)$$
$$\pm \vartheta_2(y)\vartheta_4(y)\vartheta_1(z)\vartheta_3(z),$$
$$\vartheta_1(y \pm z)\vartheta_4(y \mp z)\vartheta_2\vartheta_3 = \vartheta_1(y)\vartheta_4(y)\vartheta_2(z)\vartheta_3(z)$$
$$\pm \vartheta_2(y)\vartheta_3(y)\vartheta_1(z)\vartheta_4(z),$$
$$\vartheta_2(y \pm z)\vartheta_3(y \mp z)\vartheta_2\vartheta_3 = \vartheta_2(y)\vartheta_3(y)\vartheta_2(z)\vartheta_3(z)$$
$$\mp \vartheta_1(y)\vartheta_4(y)\vartheta_1(z)\vartheta_4(z),$$
$$\vartheta_2(y \pm z)\vartheta_4(y \mp z)\vartheta_2\vartheta_4 = \vartheta_2(y)\vartheta_4(y)\vartheta_2(z)\vartheta_4(z)$$
$$\mp \vartheta_1(y)\vartheta_3(y)\vartheta_1(z)\vartheta_3(z),$$
$$\vartheta_3(y \pm z)\vartheta_4(y \mp z)\vartheta_3\vartheta_4 = \vartheta_3(y)\vartheta_4(y)\vartheta_3(z)\vartheta_4(z)$$
$$\mp \vartheta_1(y)\vartheta_2(y)\vartheta_1(z)\vartheta_2(z).$$

Example 21.4 (Jacobi)   Obtain the duplication-formulae

$$\vartheta_2(2y)\vartheta_2\vartheta_4^2 = \vartheta_2^2(y)\vartheta_4^2(y) - \vartheta_1^2(y)\vartheta_3^2(y),$$
$$\vartheta_3(2y)\vartheta_3\vartheta_4^2 = \vartheta_3^2(y)\vartheta_4^2(y) - \vartheta_1^2(y)\vartheta_2^2(y),$$
$$\vartheta_4(2y)\vartheta_4^3 = \vartheta_3^4(y) - \vartheta_2^4(y) = \vartheta_4^4(y) - \vartheta_1^4(y).$$

Example 21.5 (Jacobi)   Obtain the duplication-formulae

$$\vartheta_1(2y)\vartheta_2\vartheta_3\vartheta_4 = 2\vartheta_1(y)\vartheta_2(y)\vartheta_3(y)\vartheta_4(y).$$

Example 21.6   Obtain duplication-formulae from the results indicated in Example 21.2.

Example 21.7   Shew that, with the notation of §21.22,

$$[1] - [2] = [4]' - [3]', \quad [1] - [3] = [1]' - [3]', \quad [1] - [4] = [2]' - [3]',$$
$$[2] - [3] = [1]' - [4]', \quad [2] - [4] = [2]' - [4]', \quad [3] - [4] = [2]' - [1]'.$$

Example 21.8 (Jacobi)   Shew that

$$2[1122] = [1122]' + [2211]' - [4433]' + [3344]',$$
$$2[1133] = [1133]' + [3311]' - [4422]' + [2244]',$$
$$2[1144] = [1144]' + [4411]' - [3322]' + [2233]',$$
$$2[2233] = [2233]' + [3322]' - [4411]' + [1144]',$$
$$2[2244] = [2244]' + [4422]' - [3311]' + [1133]',$$
$$2[3344] = [3344]' + [4433]' - [2211]' + [1122]'.$$

Example 21.9   Obtain the formulae

$$2\pi^{-1}Kk^{\frac{1}{2}} = 2q^{\frac{1}{4}} \prod_{n=1}^{\infty}\{(1 - q^{2n})^2(1 - q^{2n-1})^{-2}\},$$

$$k^{\frac{1}{2}}k'^{-\frac{1}{2}} = 2q^{\frac{1}{4}} \prod_{n=1}^{\infty}\{(1 + q^{2n})^2(1 - q^{2n-1})^{-2}\}.$$

Example 21.10 (Jacobi)   Deduce the following results from Example 21.9:

$$\prod_{n=1}^{\infty}(1 - q^{2n-1})^6 = 2q^{\frac{1}{4}}k'k^{-\frac{1}{2}}, \qquad \prod_{n=1}^{\infty}(1 + q^{2n-1})^6 = 2q^{\frac{1}{4}}(kk')^{-\frac{1}{2}},$$

$$\prod_{n=1}^{\infty}(1 - q^{2n})^6 = 2\pi^{-3}q^{-\frac{1}{2}}kk'K^3, \qquad \prod_{n=1}^{\infty}(1 + q^{2n})^6 = \frac{1}{4}q^{-\frac{1}{2}}kk'^{-\frac{1}{2}},$$

$$\prod_{n=1}^{\infty}(1 - q^n)^6 = 4\pi^{-3}q^{-\frac{1}{4}}k^{\frac{1}{2}}k'^2K^3, \qquad \prod_{n=1}^{\infty}(1 + q^n)^6 = \frac{1}{2}q^{-\frac{1}{4}}k^{\frac{1}{2}}k'^{-1}.$$

Example 21.11   By considering $\displaystyle\int \frac{\vartheta_4'(z)}{\vartheta_4(z)} e^{2niz} \, dz$ taken along the contour formed by the parallelogram whose corners are $-\frac{1}{2}\pi, \frac{1}{2}\pi, \frac{1}{2}\pi + \pi\tau, -\frac{1}{2}\pi + \pi\tau$, shew that, when n is a positive integer,

$$(1 - q^{2n}) \int_{-\pi/2}^{\pi/2} \frac{\vartheta_4'(z)}{\vartheta_4(z)} e^{2niz} \, dz = 2\pi i q^n,$$

and deduce that, when $|\,\mathrm{Im}\,(z)| < \frac{1}{2}\,\mathrm{Im}\,(\pi\tau)$,

$$\frac{\vartheta_4'(z)}{\vartheta_4(z)} = 4 \sum_{n=1}^{\infty} \frac{q^n \sin 2nz}{1 - q^{2n}}.$$

Example 21.12 (Jacobi)   Obtain the following expansions:

$$\frac{\vartheta_1'(z)}{\vartheta_1(z)} = \cot z + 4 \sum_{n=1}^{\infty} \frac{q^{2n} \sin 2nz}{1 - q^{2n}},$$

$$\frac{\vartheta_2'(z)}{\vartheta_2(z)} = -\tan z + 4 \sum_{n=1}^{\infty} \frac{(-1)^n q^{2n} \sin 2nz}{1 - q^{2n}},$$

$$\frac{\vartheta_3'(z)}{\vartheta_3(z)} = 4 \sum_{n=1}^{\infty} \frac{(-1)^n q^n \sin 2nz}{1 - q^{2n}},$$

each expansion being valid in the strip of the z-plane in which the series involved is absolutely convergent.

Example 21.13 (Math. Trip. 1908)   Shew that, if $|\operatorname{Im} y| < \operatorname{Im} \pi\tau$ and $|\operatorname{Im} z| < \operatorname{Im} \pi\tau$, then

$$\frac{\vartheta_1(y+z)\vartheta_1'}{\vartheta_1(y)\vartheta_1(z)} = \cot y + \cot z + 4 \sum_{m=1}^{\infty} \sum_{n=1}^{\infty} q^{2mn} \sin(2my + 2nz).$$

Example 21.14 (Math. Trip. 1903)   Shew that, if $|\operatorname{Im}(z)| < \frac{1}{2} \operatorname{Im}(\pi\tau)$, then

$$\frac{Kk^{\frac{1}{2}}}{\pi} \frac{\vartheta_4}{\vartheta_4(z)} = \frac{1}{2}\alpha_0 + \sum_{n=1}^{\infty} \alpha_n \cos 2nz,$$

where $\alpha_n = 2 \sum_{m=0}^{\infty} (-1)^m q^{(m+\frac{1}{2})(2n+m+\frac{1}{2})}$. *Hint.* Obtain a reduction formula for a_n by considering $\int \{\vartheta_4(z)\}^{-1} e^{2niz} \, dz$ taken round the contour of Example 21.11.

Example 21.15   Shew that

$$\frac{\vartheta_1'(z)}{\vartheta_1(z)} - \left[\cot z + 4 \sum_{n=1}^{\infty} \frac{q^{2n} \sin 2z}{1 - 2q^{2n} \cos 2z + q^{4n}} \right]$$

is a doubly-periodic function of z with no singularities, and deduce that it is zero. Prove similarly that

$$\frac{\vartheta_2'(z)}{\vartheta_2(z)} = -\tan z - 4 \sum_{n=1}^{\infty} \frac{q^{2n} \sin 2z}{1 + 2q^{2n} \cos 2z + q^{4n}},$$

$$\frac{\vartheta_3'(z)}{\vartheta_3(z)} = -4 \sum_{n=1}^{\infty} \frac{q^{2n-1} \sin 2z}{1 + 2q^{2n-1} \cos 2z + q^{4n-2}},$$

$$\frac{\vartheta_4'(z)}{\vartheta_4(z)} = 4 \sum_{n=1}^{\infty} \frac{q^{2n-1} \sin 2z}{1 - 2q^{2n-1} \cos 2z + q^{4n-2}}.$$

Example 21.16   Obtain the values of k, k', K, K' correct to six places of decimals when $q = \frac{1}{10}$. *Answer.* $k = 0.895769$, $k' = 0.444518$, $K = 2.262700$, $K' = 1.658414$.

Example 21.17   Shew that, if $w + x + y + z = 0$, then, with the notation of §21.22,

$$[3] + [1] = [2] + [4],$$
$$[1234] + [3412] + [2143] + [4321] = 0.$$

Example 21.18   Shew that

$$\frac{\vartheta_4'(y)}{\vartheta_4(y)} + \frac{\vartheta_4'(z)}{\vartheta_4(z)} - \frac{\vartheta_4'(y+z)}{\vartheta_4(y+z)} = \vartheta_2\vartheta_3\frac{\vartheta_1(y)\vartheta_1(z)\vartheta_1(y+z)}{\vartheta_4(y)\vartheta_4(z)\vartheta_4(y+z)}.$$

Example 21.19   By putting $x = y = z, w = 3x$ in Jacobi's fundamental formulae:

$$\vartheta_1^3(x)\vartheta_1(3x) + \vartheta_4^3(x)\vartheta_4(3x) = \vartheta_4^3(2x)\vartheta_4,$$

$$\vartheta_3^3(x)\vartheta_3(3x) - \vartheta_4^3(x)\vartheta_4(3x) = \vartheta_2^3(2x)\vartheta_2,$$

$$\vartheta_2^3(x)\vartheta_2(3x) + \vartheta_4^3(x)\vartheta_4(3x) = \vartheta_3^3(2x)\vartheta_3.$$

Example 21.20 (Trinity, 1882)   Deduce from Example 21.19 that

$$\{\vartheta_1^3(x)\vartheta_1(3x)\vartheta_4^2 + \vartheta_4^3(x)\vartheta_4(3x)\vartheta_4^2\}^{\frac{2}{3}}$$

$$+ \{\vartheta_3^3(x)\vartheta_3(3x)\vartheta_2^2 - \vartheta_4^3(x)\vartheta_4(3x)\vartheta_2^2\}^{\frac{2}{3}}$$

$$= \{\vartheta_2^3(x)\vartheta_2(3x)\vartheta_3^2 + \vartheta_4^3(x)\vartheta_4(3x)\vartheta_3^2\}^{\frac{2}{3}}.$$

Example 21.21   Deduce from Liouville's theorem that

$$\frac{2\vartheta_1(z)\vartheta_2(z)\vartheta_3(z)\vartheta_4(z)}{\vartheta_1(2z)\vartheta_2(0)\vartheta_3(0)\vartheta_4(0)}$$

is constant, and, by making $z \to 0$, that it is equal to 1. Hence, by comparing coefficients of z^2 in the expansions of

$$\log\frac{\vartheta_1(2z)}{2\vartheta_1(z)} \quad \text{and} \quad \log\frac{\vartheta_2(z)}{\vartheta_2(0)} + \log\frac{\vartheta_3(z)}{\vartheta_3(0)} + \log\frac{\vartheta_4(z)}{\vartheta_4(0)}$$

by Maclaurin's theorem, deduce that

$$\frac{\vartheta_1'''(0)}{\vartheta_1'(0)} = \frac{\vartheta_2''(0)}{\vartheta_2(0)} + \frac{\vartheta_3''(0)}{\vartheta_3(0)} + \frac{\vartheta_4''(0)}{\vartheta_4(0)}.$$

Hence, after the manner of §21.41, deduce that

$$\vartheta_1'(0) = \vartheta_2(0)\vartheta_3(0)\vartheta_4(0).$$

This method of obtaining the preliminary formula of §21.41 was suggested to the authors by Mr. C. A. Stewart.

22

The Jacobian Elliptic Functions

22.1 Elliptic functions with two simple poles

In the course of proving general theorems concerning elliptic functions at the beginning of Chapter 20, it was shewn that two classes of elliptic functions were simpler than any others so far as their singularities were concerned, namely the elliptic functions of order 2. The first class consists of those with a single double pole (with zero residue) in each cell, the second consists of those with two simple poles in each cell, the sum of the residues at these poles being zero.

An example of the first class, namely $\wp(z)$, was discussed at length in Chapter 20; in the present chapter we shall discuss various examples of the second class, known as *Jacobian elliptic functions*. These functions were introduced by Jacobi, but many of their properties were obtained independently by Abel, who used a different notation. See §22.7.

It will be seen (§22.122) that, in certain circumstances, the Jacobian functions degenerate into the ordinary circular functions; accordingly, a notation (invented by Jacobi and modified by Gudermann and Glaisher) will be employed which emphasises an analogy between the Jacobian functions and the circular functions.

From the theoretical aspect, it is most simple to regard the Jacobian functions as quotients of theta-functions (§21.61). But as many of their fundamental properties can be obtained by quite elementary methods, without appealing to the theory of theta-functions, we shall discuss the functions without making use of Chapter 21 except when it is desirable to do so for the sake of brevity or simplicity.

22.11 The Jacobian elliptic functions, sn u, cn u, dn u

It was shewn in §21.61 that if

$$y = \frac{\vartheta_3}{\vartheta_2} \frac{\vartheta_1(u/\vartheta_3^2)}{\vartheta_4(u/\vartheta_3^2)};$$

the theta-functions being formed with parameter τ, then

$$\left(\frac{dy}{du}\right)^2 = (1 - y^2)(1 - k^2 y^2),$$

where $k^{1/2} = \vartheta_2(0|\tau)/\vartheta_3(0|\tau)$. Conversely, if the constant k (called the *modulus*[1]) be given, then, unless $k^2 \geq 1$ or $k^2 \leq 0$, a value of τ can be found (§§21.7–21.712) for which

---

[1] If $0 < k < 1$, and θ is the acute angle such that $\sin \theta = k$, then θ is called the *modular angle*.

$\vartheta_2^4(0|\tau)/\vartheta_3^4(0|\tau) = k^2$, so that the solution of the differential equation

$$\left(\frac{dy}{du}\right)^2 = (1 - y^2)(1 - k^2 y^2)$$

subject to the condition $\left(\dfrac{dy}{du}\right)_{u=y=0} = 1$ is

$$y = \frac{\vartheta_3}{\vartheta_2} \frac{\vartheta_1(u/\vartheta_3^2)}{\vartheta_4(u/\vartheta_3^2)}$$

the theta-functions being formed with the parameter τ which has been determined.

The differential equation may be written

$$u = \int_0^y (1 - t^2)^{-\frac{1}{2}} (1 - k^2 t^2)^{-\frac{1}{2}} \, dt,$$

and, by the methods of §21.73, it may be shewn that, if y and u are connected by this integral formula, y may be expressed in terms of u as the quotient of two theta-functions, in the form already given.

Thus, if

$$u = \int_0^y (1 - t^2)^{-1/2} (1 - k^2 t^2)^{-1/2} \, dt,$$

y may be regarded as the function of u defined by the quotient of the theta-functions, so that y is an analytic function of u except at its singularities, which are all simple poles; to denote this functional dependence, we write

$$y = \operatorname{sn}(u, k),$$

or simply $y = \operatorname{sn} u$, when it is unnecessary to emphasise the modulus. The modulus will always be inserted when it is not k.

The function $\operatorname{sn} u$ is known as a *Jacobian elliptic function* of u, and

$$\operatorname{sn} u = \frac{\vartheta_3}{\vartheta_2} \frac{\vartheta_1(u/\vartheta_3{}^2)}{\vartheta_4(u/\vartheta_3{}^2)}. \tag{A}$$

Note   Unless the theory of the theta-functions is assumed, it is exceedingly difficult to shew that the integral formula defines y as a function of u which is analytic except at simple poles. See Hancock [271].

Now write

$$\operatorname{cn}(u, k) = \frac{\vartheta_4}{\vartheta_2} \frac{\vartheta_2(u/\vartheta_3{}^2)}{\vartheta_4(u/\vartheta_3{}^2)} \tag{B}$$

$$\operatorname{dn}(u, k) = \frac{\vartheta_4}{\vartheta_3} \frac{\vartheta_3 \, (u/\vartheta_3{}^2)}{\vartheta_4 \, (u/\vartheta_3{}^2)}. \tag{C}$$

Then, from the relation of §21.6, we have

$$\frac{d}{du} \operatorname{sn} u = \operatorname{cn} u \, \operatorname{dn} u \tag{I}$$

and from the relations of §21.2 we have

$$\operatorname{sn}^2 u + \operatorname{cn}^2 u = 1 \tag{II}$$

$$k^2 \operatorname{sn}^2 u + \operatorname{dn}^2 u = 1 \tag{III}$$

and, obviously,

$$\operatorname{cn} 0 = \operatorname{dn} 0 = 1. \tag{IV}$$

Note We shall now discuss the properties of the functions sn u, cn u, dn u as defined by the equations (A), (B), (C) by using the four relations (I), (II), (III), (IV); these four relations are sufficient to make sn u, cn u, dn u determinate functions of u. It will be assumed, when necessary, that sn u, cn u, dn u are one-valued functions of u, analytic except at their poles; it will also be assumed that they are one-valued analytic functions of k^2 when cuts are made in the plane of the complex variable k^2 from 1 to $+\infty$ and from 0 to $-\infty$.

22.12 Simple properties of sn u, cn u, dn u

From the integral

$$u = \int_0^y (1 - t^2)^{-1/2} (1 - k^2 t^2)^{-1/2} \, dt, \tag{22.1}$$

it is evident, on writing $-t$ for t, that, if the sign of y be changed, the sign of u is also changed. *Hence* sn u *is an odd function of* u.

Since $\operatorname{sn}(-u) = -\operatorname{sn}(u)$, it follows from (II) that $\operatorname{cn}(-u) = \pm \operatorname{cn} u$; on account of the one-valuedness of cn u, by the theory of analytic continuation it follows that either the upper sign, or else the lower sign, must always be taken. In the special case $u = 0$, the upper sign has to be taken, and so it has to be taken always; hence $\operatorname{cn}(-u) = \operatorname{cn}(u)$, *and* cn u *is an even function of* u. In like manner, dn u is an even function of u.

These results are also obvious from the definitions (A), (B) and (C) of §22.11.

Next, let us differentiate the equation $\operatorname{sn}^2 u + \operatorname{cn}^2 u = 1$; on using equation (I), we get

$$\frac{d \operatorname{cn} u}{du} = -\operatorname{sn} u \, \operatorname{dn} u;$$

in like manner, from equations (III) and (I) we have

$$\frac{d \operatorname{dn} u}{du} = -k^2 \operatorname{sn} u \, \operatorname{cn} u.$$

22.121 The complementary modulus

If $k^2 + k'^2 = 1$ and $k' \to +1$ as $k \to 0$, k' is known as the *complementary modulus*. On account of the cut in the k^2-plane from 1 to $+\infty$, k' is a one-valued function of k.

Note With the aid of the theta-functions, we can make $k'^{1/2}$ one-valued, by defining it to be

$$\vartheta_4(0|\tau)/\vartheta_3(0|\tau).$$

Example 22.1.1 Shew that, if

$$u = \int_y^1 (1 - t^2)^{-\frac{1}{2}} (k'^2 + k^2 t^2)^{-\frac{1}{2}} \, dt$$

then $y = \mathrm{cn}(u, k)$. Also, shew that, if

$$u = \int_y^1 (1 - t^2)^{-\frac{1}{2}} (t^2 - k'^2)^{-\frac{1}{2}} \, dt,$$

then $y = \mathrm{dn}(u, k)$. These results are sometimes written in the form

$$u = \int_{\mathrm{cn}\,u}^1 (1 - t^2)^{-\frac{1}{2}} (k'^2 + k^2 t^2)^{-\frac{1}{2}} \, dt = \int_{\mathrm{dn}\,u}^1 (1 - t^2)^{-\frac{1}{2}}(t^2 - k'^2)^{-\frac{1}{2}} \, dt.$$

22.122 Glaisher's notation for quotients

A short and convenient notation has been invented by Glaisher [252] to express reciprocals and quotients of the Jacobian elliptic functions; the reciprocals are denoted by reversing the order of the letters which express the function, thus

$$\mathrm{ns}\,u = 1/\mathrm{sn}\,u, \quad \mathrm{nc}\,u = 1/\mathrm{cn}\,u, \quad \mathrm{nd}\,u = 1/\mathrm{dn}\,u;$$

while quotients are denoted by writing in order the first letters of the numerator and denominator functions, thus

$$\mathrm{sc}\,u = \mathrm{sn}\,u/\mathrm{cn}\,u, \quad \mathrm{sd}\,u = \mathrm{sn}\,u/\mathrm{dn}\,u, \quad \mathrm{cd}\,u = \mathrm{cn}\,u/\mathrm{dn}\,u,$$

$$\mathrm{cs}\,u = \mathrm{cn}\,u/\mathrm{sn}\,u, \quad \mathrm{ds}\,u = \mathrm{dn}\,u/\mathrm{sn}\,u, \quad \mathrm{dc}\,u = \mathrm{dn}\,u/\mathrm{cn}\,u.$$

Note   Jacobi's notation for the functions $\mathrm{sn}\,u$, $\mathrm{cn}\,u$, $\mathrm{dn}\,u$ was $\mathrm{sinam}\,u$, $\mathrm{cosam}\,u$, $\Delta\,\mathrm{am}\,u$, the abbreviations now in use being due to Gudermann [262], who also wrote $\mathrm{tn}\,u$, as an abbreviation for $\mathrm{tanam}\,u$, in place of what is now written $\mathrm{sc}\,u$.

The reason for Jacobi's notation was that he regarded the inverse of the integral

$$u = \int_0^\phi (1 - k^2 \sin^2 \theta)^{-\frac{1}{2}} \, d\theta$$

as fundamental, and wrote[2] $\phi = \mathrm{am}\,u$; he also wrote $\Delta\phi = (1 - k^2 \sin^2 \phi)^{\frac{1}{2}}$ for $\dfrac{d\phi}{du}$.

Example 22.1.2   Obtain the following results:

$$u = \int_0^{\mathrm{sc}\,u} (1 + t^2)^{-\frac{1}{2}} (1 + k'^2 t^2)^{-\frac{1}{2}} \, dt = \int_{\mathrm{cs}\,u}^\infty (t^2 + 1)^{-\frac{1}{2}} (t^2 - k'^2)^{-\frac{1}{2}} \, dt$$

$$= \int_0^{\mathrm{sd}\,u} (1 - k'^2 t^2)^{-\frac{1}{2}} (1 + k^2 t^2)^{-\frac{1}{2}} \, dt = \int_{\mathrm{ds}\,u}^\infty (t^2 - k'^2)^{-\frac{1}{2}} (t^2 + k^2)^{-\frac{1}{2}} \, dt$$

$$= \int_{\mathrm{cd}\,u}^1 (1 - t^2)^{-\frac{1}{2}} (1 - k^2 t^2)^{-\frac{1}{2}} \, dt = \int_{\mathrm{dc}\,u}^\infty (t^2 - 1)^{-\frac{1}{2}} (t^2 - k^2)^{-\frac{1}{2}} \, dt$$

$$= \int_{\mathrm{ns}\,u}^\infty (t^2 - 1)^{-\frac{1}{2}} (t^2 - k^2)^{-\frac{1}{2}} \, dt = \int_1^{\mathrm{nc}\,u} (t^2 - 1)^{-\frac{1}{2}} (K'^2 t^2 + k^2)^{-\frac{1}{2}} \, dt$$

$$= \int_1^{\mathrm{nd}\,u} (t^2 - 1)^{-\frac{1}{2}} (1 - K'^2 t^2)^{-\frac{1}{2}} \, dt.$$

[2] Jacobi [349, p. 30]. As $k \to 0$, $\mathrm{am}\,u \to u$.

22.2 The addition-theorem for the function sn *u*

We shall now shew how to express $\mathrm{sn}(u+v)$ in terms of the Jacobian elliptic functions of u and v; the result will be the addition-theorem for the function sn u; it will be an addition-theorem in the strict sense, as it can be written in the form of an algebraic relation connecting sn u, sn v, $\mathrm{sn}(u+v)$.

Note   There are numerous methods of establishing the result; the one given is essentially due to Euler [199][3], who was the first to obtain (in 1756, 1757) the integral of

$$\frac{dx}{\sqrt{X}} + \frac{dy}{\sqrt{Y}} = 0.$$

in the form of an algebraic relation between x and y, when X denotes a quartic function of x and Y is the same quartic function of y. Three other methods are given as examples, at the end of this section. Another method is given by Legendre [422, vol. I, p. 20], and an interesting geometrical proof was given by Jacobi [344].

Suppose that u and v vary while $u + v$ remains constant and equal to α, say, so that

$$\frac{dv}{du} = -1.$$

Now introduce, as new variables, s_1 and s_2 defined by the equations

$$s_1 = \mathrm{sn}\,u, \qquad s_2 = \mathrm{sn}\,v,$$

so that[4]

$$\dot{s_1}^2 = (1 - s_1^2)\,(1 - k^2 s_1^2), \quad \text{and} \quad \dot{s_2}^2 = (1 - s_2^2)\,(1 - k^2 s_2^2)$$

since $\dot{v}^2 = 1$.

Differentiating with regard to u and dividing by $2\dot{s_1}$ and $2\dot{s_2}$ respectively, we find that, for general values of u and v (i.e. those values for which cn u dn u and cn v dn v do not vanish)

$$\ddot{s_1} = -(1 + k^2)s_1 + 2k^2 s_1^3, \qquad \ddot{s_2} = -(1 + k^2)s_2 + 2k^2 s_2^3.$$

Hence, by some easy algebra,

$$\frac{\ddot{s_1}s_2 - \ddot{s_2}s_1}{\dot{s_1}^2 s_2^2 - \dot{s_2}^2 s_1^2} = \frac{2k^2 s_1 s_2(s_1^2 - s_2^2)}{(s_2^2 - s_1^2)(1 - k^2 s_1^2 s_2^2)},$$

and so

$$(\dot{s_1}s_2 - \dot{s_2}s_1)^{-1}\frac{d}{du}(\dot{s_1}s_2 - \dot{s_2}s_1) = (1 - k^2 s_1^2 s_2^2)^{-1}\frac{d}{du}(1 - k^2 s_1^2 s_2^2);$$

on integrating this equation we have

$$\frac{\dot{s_1}s_2 - \dot{s_2}s_1}{1 - k^2 s_1^2 s_2^2} = C,$$

---

[3] Euler had obtained some special cases of this result a few years earlier.
[4] For brevity, we shall denote differential coefficients with regard to u by dots, thus

$$\dot{v} \equiv \frac{dv}{du}, \qquad \ddot{v} \equiv \frac{d^2 v}{du^2}.$$

where C is the constant of integration.

Replacing the expressions on the left by their values in terms of u and v we get

$$\frac{\text{cn}\, u \,\text{dn}\, u \,\text{sn}\, v + \text{cn}\, v \,\text{dn}\, v \,\text{sn}\, u}{1 - k^2 \,\text{sn}^2\, u \,\text{sn}^2\, v} = C.$$

That is to say, we have two integrals of the equation $du + dv = 0$, namely

(i) $u + v = \alpha$, and

(ii) $\dfrac{\text{sn}\, u \,\text{cn}\, v \,\text{dn}\, v + \text{sn}\, v \,\text{cn}\, u \,\text{dn}\, u}{1 - k^2 \,\text{sn}^2\, u \,\text{sn}^2\, v} = C,$

each integral involving an arbitrary constant. By the general theory of differential equations of the first order, these integrals cannot be functionally independent, and so

$$\frac{\text{sn}\, u \,\text{cn}\, v \,\text{dn}\, v + \text{sn}\, v \,\text{cn}\, u \,\text{dn}\, u}{1 - k^2 \,\text{sn}^2\, u \,\text{sn}^2\, v}$$

is expressible as a function of $u + v$; call this function $f(u + v)$.

On putting $v = 0$, we see that $f(u) = \text{sn}\, u$; and so the function f is the sn function. *We have thus demonstrated the result that*

$$\text{sn}(u + v) = \frac{\text{sn}\, u \,\text{cn}\, v \,\text{dn}\, v + \text{sn}\, v \,\text{cn}\, u \,\text{dn}\, u}{1 - k^2 \,\text{sn}^2\, u \,\text{sn}^2\, v},$$

which is the addition-theorem. Using an obvious notation (due to Glaisher [250]), we may write

$$\text{sn}(u + v) = \frac{s_1 c_2 d_2 + s_2 c_1 d_1}{1 - k^2 s_1^2 s_2^2}.$$

Example 22.2.1   Obtain the addition-theorem for $\sin u$ by using the results

$$\left(\frac{d \sin u}{du}\right)^2 = 1 - \sin^2 u, \qquad \left(\frac{d \sin v}{dv}\right)^2 = 1 - \sin^2 v.$$

Example 22.2.2 (Abel)   Prove from first principles that

$$\left(\frac{\partial}{\partial v} - \frac{\partial}{\partial u}\right) \frac{s_1 c_2 d_2 + s_2 c_1 d_1}{1 - k^2 s_1^2 s_2^2} = 0,$$

and deduce the addition-theorem for $\text{sn}\, u$.

Example 22.2.3 (Cayley)   Shew that

$$\text{sn}(u + v) = \frac{s_1^2 - s_2^2}{s_1 c_2 d_2 - s_2 c_1 d_1} = \frac{s_1 c_1 d_2 + s_2 c_2 d_1}{c_1 c_2 + s_1 d_1 s_2 d_2} = \frac{s_1 d_1 c_2 + s_2 d_2 c_1}{d_1 d_2 + k^2 s_1 s_2 c_1 c_2}.$$

Example 22.2.4 (Jacobi)   Obtain the addition-theorem for $\text{sn}\, u$ from the results

$$\vartheta_1(y + z)\vartheta_4(y - z)\vartheta_2\vartheta_3 = \vartheta_1(y)\vartheta_4(y)\vartheta_2(z)\vartheta_3(z) + \vartheta_2(y)\vartheta_3(y)\vartheta_1(z)\vartheta_4(z),$$
$$\vartheta_4(y + z)\,\vartheta_4(y - z)\vartheta_4^2 = \vartheta_4^2(y)\vartheta_4^2(z) - \vartheta_1^2(y)\vartheta_1^2(z),$$

given in Chapter 21, Examples 21.1 and 21.3.

Example 22.2.5   Assuming that the coordinates of any point on the curve

$$y^2 = (1 - x^2)(1 - k^2 x^2)$$

can be expressed in the form (sn u, cn u dn u), obtain the addition-theorem for sn u by Abel's method (§20.312).

Note   Consider the intersections of the given curve with the variable curve $y = 1 + mx + nx^2$; one is $(0, 1)$; let the others have parameters u_1, u_2, u_3, of which u_1, u_2 may be chosen arbitrarily by suitable choice of m and n. Shew that $u_1 + u_2 + u_3$ is constant, by the method of §20.312, and deduce that this constant is zero by taking

$$m = 0, \quad n = -\tfrac{1}{2}(1 + k^2).$$

Observe also that, by reason of the relations

$$(k^2 - n^2) x_1 x_2 x_3 = 2m, \qquad (k^2 - n^2)(x_1 + x_2 + x_3) = 2mn,$$

we have

$$
\begin{aligned}
x_3(1 - k^2 x_1^2 x_2^2) &= x_3 - \left(1 + \frac{n^2}{k^2 - n^2}\right) 2m x_1 x_2 \\
&= x_3 - 2m x_1 x_2 - n x_1 x_2 (x_1 + x_2 + x_3) \\
&= (x_1 + x_2 + x_3 - n x_1 x_2 x_3) - (x_1 + x_2) - 2m x_1 x_2 - n x_1 x_2 (x_1 + x_2) \\
&= -x_1 y_2 - x_2 y_1.
\end{aligned}
$$

22.21 *The addition-theorems for* cn *u and* dn *u*

We shall now establish the results

$$cn(u + v) = \frac{cn\,u\,cn\,v - sn\,u\,sn\,v\,dn\,u\,dn\,v}{1 - k^2\,sn^2\,u\,sn^2\,v},$$

$$dn(u + v) = \frac{dn\,u\,dn\,v - k^2\,sn\,u\,sn\,v\,cn\,u\,cn\,v}{1 - k^2\,sn^2\,u\,sn^2\,v};$$

the most simple method of obtaining them is from the formula for sn$(u + v)$.

Using the notation introduced at the end of §22.2, we have

$$
\begin{aligned}
(1 - k^2 s_1^2 s_2^2)^2\,cn^2(u + v) &= (1 - k^2 s_1^2 s_2^2)^2 \{1 - sn^2(u + v)\} \\
&= (1 - k^2 s_1^2 s_2^2)^2 - (s_1 c_2 d_2 + s_2 c_1 d_1)^2 \\
&= 1 - 2k^2 s_1^2 s_2^2 + k^4 s_1^4 s_2^4 - 2 s_1 s_2 c_1 c_2 d_1 d_2 \\
&\quad - s_1^2(1 - s_2^2)(1 - k^2 s_2^2) - s_2^2(1 - s_1^2)(1 - k^2 s_1^2) \\
&= (1 - s_1^2)(1 - s_2^2) + s_1^2 s_2^2 (1 - k^2 s_1^2)(1 - k^2 s_2^2) - 2 s_1 s_2 c_1 c_2 d_1 d_2 \\
&= (c_1 c_2 - s_1 s_2 d_1 d_2)^2
\end{aligned}
$$

and so

$$cn(u + v) = \pm \frac{c_1 c_2 + s_1 s_2 d_1 d_2}{1 - k^2 s_1^2 s_2^2}.$$

But both of these expressions are one-valued functions of u, analytic except at isolated poles and zeros, and it is inconsistent with the theory of analytic continuation that their ratio should be $+1$ for some values of u, and -1 for other values, so the ambiguous sign is really definite; putting $u = 0$, we see that the plus sign has to be taken. The first formula is consequently proved.

The formula for $\mathrm{dn}(u + v)$ follows in like manner from the identity

$$(1 - k^2 s_1^2 s_2^2)^2 - k^2 (s_1 c_2 d_2 + s_2 c_1 d_1)^2$$
$$\equiv (1 - k^2 s_1^2)(1 - k^2 s_2^2) + k^4 s_1^2 s_2^2 (1 - s_1^2)(1 - s_2^2) - 2k^2 s_1 s_2 c_1 c_2 d_1 d_2,$$

the proof of which is left to the reader.

Example 22.2.6 (Jacobi)   Shew that

$$\mathrm{dn}(u + v)\,\mathrm{dn}(u - v) = \frac{d_2{}^2 - k^2 s_1{}^2 c_2{}^2}{1 - k^2 s_1{}^2 s_2{}^2}.$$

Note   A set of 33 formulae of this nature connecting functions of $u + v$ and of $u - v$ is given in Jacobi [349, p. 32–34].

Example 22.2.7 (Cayley)   Shew that

$$\frac{\partial}{\partial u} \frac{\mathrm{cn}\,u + \mathrm{cn}\,v}{\mathrm{sn}\,u\,\mathrm{dn}\,v + \mathrm{sn}\,v\,\mathrm{dn}\,u} = \frac{\partial}{\partial v} \frac{\mathrm{cn}\,u + \mathrm{cn}\,v}{\mathrm{sn}\,u\,\mathrm{dn}\,v + \mathrm{sn}\,v\,\mathrm{dn}\,u},$$

so that $(\mathrm{cn}\,u + \mathrm{cn}\,v)/(\mathrm{sn}\,u\,\mathrm{dn}\,v + \mathrm{sn}\,v\,\mathrm{dn}\,u)$ is a function of $u + v$ only; and deduce that it is equal to $\{1 + \mathrm{cn}(u + v)\}/\mathrm{sn}(u + v)$. Obtain a corresponding result for the function $(s_1 c_2 + s_2 c_1)/(d_1 + d_2)$.

Example 22.2.8 (Jacobi)   Shew that

$$1 - k^2 \mathrm{sn}^2(u + v)\,\mathrm{sn}^2(u - v) = (1 - k^2 \mathrm{sn}^4 u)(1 - k^2 \mathrm{sn}^4 v)(1 - k^2 \mathrm{sn}^2 u\,\mathrm{sn}^2 v)^{-2},$$
$$k'^2 + k^2 \mathrm{cn}^2(u + v)\,\mathrm{cn}^2(u - v) = (k'^2 + k^2 \mathrm{cn}^4 u)(k'^2 + k^2 \mathrm{cn}^4 v)(1 - k^2 \mathrm{sn}^2 u\,\mathrm{sn}^2 v)^{-2}.$$

Example 22.2.9   Obtain the addition-theorems for $\mathrm{cn}(u + v)$, $\mathrm{dn}(u + v)$ by the method of Example 22.2.1.

Example 22.2.10   Using Glaisher's abridged notation [251], namely

$$s, c, d = \mathrm{sn}\,u,\ \mathrm{cn}\,u,\ \mathrm{dn}\,u, \quad \text{and} \quad S, C, D = \mathrm{sn}\,2u,\ \mathrm{cn}\,2u,\ \mathrm{dn}\,2u,$$

prove that

$$S = \frac{2scd}{1 - k^2 s^4}, \quad C = \frac{1 - 2s^2 + k^2 s^4}{1 - k^2 s^4}, \quad D = \frac{1 - 2k^2 s^2 + k^2 s^4}{1 - k^2 s^4},$$
$$s = \frac{(1 + S)^{1/2} - (1 - S)^{1/2}}{(1 + kS)^{1/2} + (1 - kS)^{1/2}}.$$

Example 22.2.11 (Glaisher)   With the notation of Example 22.2.10, shew that

$$s^2 = \frac{1 - C}{1 + D} = \frac{1 - D}{k^2(1 + C)} = \frac{D - k^2 C - k'^2}{k^2(D - C)} = \frac{D - C}{k'^2 + D - k^2 C},$$
$$c^2 = \frac{D + C}{1 + D} = \frac{D + k^2 C - k'^2}{k^2(1 + C)} = \frac{k'^2(1 - D)}{k^2(D - C)} = \frac{k'^2(1 + C)}{k'^2 + D - k^2 C},$$
$$d^2 = \frac{k'^2 + D + k^2 C}{1 + D} = \frac{D + C}{1 + C} = \frac{k'^2(1 - C)}{D - C} = \frac{k'^2(1 + D)}{k'^2 + D - k^2 C}.$$

22.3 The constant K

We have seen that, if

$$u = \int_0^y (1 - t^2)^{-1/2}(1 - k^2 t^2)^{-1/2}\, dt,$$

then $y = \operatorname{sn}(u, k)$.

If we take the upper limit to be unity (the path of integration being a straight line), it is customary to denote the value of the integral by the symbol K, so that $\operatorname{sn}(K, k) = 1$.

Note It will be seen in §22.302 that this definition of K is equivalent to the definition as $\frac{1}{2}\pi\vartheta_3^2$ in §22.61.

It is obvious that $\operatorname{cn} K = 0$ and $\operatorname{dn} K = \pm k'$; to fix the ambiguity in sign, suppose $0 < k < 1$, and trace the change in $(1 - k^2 t^2)^{\frac{1}{2}}$ as t increases from 0 to 1; since this expression is initially unity and as neither of its branch points (at $t = \pm k^{-1}$) is encountered, the final value of the expression is positive, and so it is $+k'$; and therefore, since $\operatorname{dn} K$ is a continuous function of k, its value is always $+k'$.

The elliptic functions of K are thus given by the formulae

$$\operatorname{sn} K = 1, \qquad \operatorname{cn} K = 0, \qquad \operatorname{dn} K = k'.$$

22.301 The expression of K in terms of k

In the integral defining K, write $t = \sin\phi$, and we have at once

$$K = \int_0^{\pi/2} (1 - k^2 \sin^2 \phi)^{-1/2}\, d\phi.$$

When $|k| < 1$, the integrand may be expanded in a series of powers of k, the series converging uniformly with regard to ϕ (by §3.34, since $\sin^{2n}\phi \le 1$); integrating term-by-term (§4.7), we at once get

$$K = \frac{\pi}{2} F\left(\tfrac{1}{2}, \tfrac{1}{2}; 1; k^2\right) = \frac{\pi}{2} F\left(\tfrac{1}{2}, \tfrac{1}{2}; 1; c\right),$$

where $c = k^2$. By the theory of analytic continuation, this result holds for all values of c when a cut is made from 1 to $+\infty$ in the c-plane, since both the integrand and the hypergeometric function are one-valued and analytic in the cut plane.

Example 22.3.1 (Legendre [422]) Shew that

$$\frac{d}{dk}\left(kk'^2 \frac{dK}{dk}\right) = kK.$$

22.302 The equivalence of the definitions of K

Taking $u = \frac{1}{2}\pi\vartheta_3^2$ in §21.61, we see at once that $\operatorname{sn}(\frac{1}{2}\pi\vartheta_3^2) = 1$ and so $\operatorname{cn}(\frac{1}{2}\pi\vartheta_3^2) = 0$. Consequently, $1 - \operatorname{sn} u$ has a *double* zero at $\frac{1}{2}\pi\vartheta_3^2$. Therefore, since the number of poles of $\operatorname{sn} u$ in the cell with corners 0, $2\pi\vartheta_3^2$, $\pi(\tau + 1)\,\vartheta_3^2$, $\pi(\tau - 1)\,\vartheta_3^2$ is two, it follows from §20.13 that the *only* zeros of $1 - \operatorname{sn} u$ are at the points $u = \frac{1}{2}\pi(4m + 1 + 2n\tau)\vartheta_3^2$, where m and n are integers. Therefore, with the definition of §22.3,

$$K = \frac{\pi}{2}(4m + 1 + 2n\tau)\vartheta_3^2.$$

Now take τ to be a pure imaginary, so that $0 < k < 1$, and K is real; and we have $n = 0$, so that

$$\frac{\pi}{2}(4m + 1)\vartheta_3^2 = \int_0^{\pi/2} (1 - k^2 \sin^2 \phi)^{-\frac{1}{2}} \, d\phi,$$

where m is a positive integer or zero; it is obviously not a negative integer.

If m is a positive integer, since $\int_0^a (1 - k^2 \sin^2 \phi)^{-\frac{1}{2}} \, d\phi$ is a continuous function of a and so passes through all values between 0 and K as a increases from 0 to $\frac{1}{2}\pi$, we can find a value of a *less* than $\frac{1}{2}\pi$, such that

$$\frac{K}{(4m + 1)} = \frac{\pi}{2}\vartheta_3^2 = \int_0^a (1 - k^2 \sin^2 \phi)^{-\frac{1}{2}} \, d\phi;$$

and so $\mathrm{sn}(\frac{1}{2}\pi\vartheta_3^2) = \sin a < 1$, which is untrue, since $\mathrm{sn}(\frac{1}{2}\pi\vartheta_3^2) = 1$. Therefore m *must be zero*, that is to say we have

$$K = \tfrac{1}{2}\pi\vartheta_3^2.$$

But both K and $\frac{\pi}{2}\vartheta_3^2$ are analytic functions of k when the c-plane is cut from 1 to $+\infty$, and so, by the theory of analytic continuation, this result, proved when $0 < k < 1$, persists throughout the cut plane.

The equivalence of the definitions of K has therefore been established.

Example 22.3.2   By considering the integral

$$\int_0^{(1+)} (1 - t^2)^{-\frac{1}{2}}(1 - k^2 t^2)^{-\frac{1}{2}} \, dt,$$

shew that $\mathrm{sn}\, 2K = 0$.

Example 22.3.3   Prove that

$$\mathrm{sn}\, \tfrac{1}{2}K = (1 + k')^{-\frac{1}{2}}, \qquad \mathrm{cn}\, \tfrac{1}{2}K = k'^{\frac{1}{2}}(1 + k')^{-\frac{1}{2}}, \qquad \mathrm{dn}\, \tfrac{1}{2}K = k'^{\frac{1}{2}}.$$

Note   Notice that when $u = \tfrac{1}{2}K$, $\mathrm{cn}\, 2u = 0$. The simplest way of determining the signs to be attached to the various radicals is to make $k \to 0$, $k' \to 1$, and then $\mathrm{sn}\, u$, $\mathrm{cn}\, u$, $\mathrm{dn}\, u$ degenerate into $\sin u$, $\cos u$, 1.

Example 22.3.4   Prove, by means of the theory of theta-functions, that

$$\mathrm{cs}\, \tfrac{1}{2}K = \mathrm{dn}\, \tfrac{1}{2}K = k'^{\frac{1}{2}}.$$

22.31 The periodic properties (associated with K) of the Jacobian elliptic functions

The intimate connexion of K with periodic properties of the functions $\mathrm{sn}\, u$, $\mathrm{cn}\, u$, $\mathrm{dn}\, u$, which may be anticipated from the periodic properties of theta-functions associated with $\frac{1}{2}\pi$, will now be demonstrated directly from the addition-theorem.

By §22.2, we have

$$\mathrm{sn}(u + K) = \frac{\mathrm{sn}\, u \; \mathrm{cn}\, K \; \mathrm{dn}\, K - \mathrm{sn}\, K \; \mathrm{cn}\, u \; \mathrm{dn}\, u}{1 - k^2 \, \mathrm{sn}^2 u \; \mathrm{sn}^2 K} = \mathrm{cd}\, u.$$

In like manner, from §22.21,

$$cn(u + K) = -k' \, sd \, u, \qquad dn(u + K) = k' \, nd \, u.$$

Hence

$$sn(u + 2K) = \frac{cn(u + K)}{dn(u + K)} = -\frac{k' \, sd \, u}{k' \, nd \, u} = - \, sn \, u,$$

and, similarly,

$$cn(u + 2K) = - \, cn \, u, \qquad dn(u + 2K) = dn \, u.$$

Finally,

$$sn(u + 4K) = - \, sn(u + 2K) = sn \, u, \qquad cn(u + 4K) = cn \, u.$$

Thus 4K is a period of each of the functions sn *u,* cn *u, while* dn *u has the smaller period* 2K.

Example 22.3.5 Obtain the results

$$sn(u + K) = cd \, u, \qquad cn(u + K) = -k' \, sd \, u, \qquad dn(u + K) = k' \, nd \, u,$$

directly from the definitions of sn *u,* cn *u,* dn *u* as quotients of theta-functions.

Example 22.3.6 Shew that cs *u* cs(K − u) = k'.

22.32 *The constant K'*

We shall denote the integral

$$\int_0^1 (1 - t^2)^{-1/2} \, (1 - k'^2 t^2)^{-1/2} \, dt \tag{22.2}$$

by the symbol K', so that K' is the same function of $k'^2 \, (= c')$ as K is of $K^2 \, (= c)$; and so

$$K' = \frac{\pi}{2} \, F\left(\frac{1}{2}, \frac{1}{2}; 1; k'^2\right),$$

when the c'-plane is cut from 1 to $+\infty$, i.e. when the c-plane is cut from 0 to $-\infty$.

Note To shew that this definition of K' is equivalent to the definition of §21.61, we observe that if $\tau\tau' = -1$, then K is the *one-valued* function of k^2, in the cut plane, defined by the equations

$$K = \frac{\pi}{2}\vartheta_3^2(0|\tau), \qquad k^2 = \vartheta_2^4(0|\tau)/\vartheta_3^4(0|\tau),$$

while, with the definition of §21.51,

$$K' = \frac{1}{2}\pi\vartheta_3^2(0|\tau'), \qquad k'^2 = \vartheta_2^4(0|\tau')/\vartheta_3^4(0|\tau'),$$

so that K' must be the same function of $(k')^2$ as K is of k^2; and this is consistent with the integral definition of K' as

$$\int_0^1 (1 - t^2)^{-1/2} \, (1 - k'^2 t^2)^{-1/2} \, dt. \tag{22.3}$$

It will now be shewn that, if the c-plane be cut from 0 to $-\infty$ and from 1 to $+\infty$, then, in the cut plane, K' may be defined by the equation

$$K' = \int_1^{1/k} (s^2 - 1)^{-\frac{1}{2}} (1 - k^2 s^2)^{-\frac{1}{2}} \, ds.$$

First suppose that $0 < k < 1$, so that $0 < k' < 1$, and then the integrals concerned are real. In the integral

$$\int_0^1 (1 - t^2)^{-\frac{1}{2}} (1 - k'^2 t^2)^{-\frac{1}{2}} \, dt$$

make the substitution

$$s = (1 - k'^2 t^2)^{-\frac{1}{2}},$$

which gives

$$(s^2 - 1)^{\frac{1}{2}} = k' t (1 - k'^2 t^2)^{-\frac{1}{2}},$$
$$(1 - k^2 s^2)^{\frac{1}{2}} = k' (1 - t^2)^{\frac{1}{2}} (1 - k'^2 t^2)^{-\frac{1}{2}},$$
$$\frac{ds}{dt} = \frac{k'^2 t}{(1 - k'^2 t^2)^{\frac{3}{2}}},$$

it being understood that the positive value of each radical is to be taken. On substitution, we at once get the result stated, namely that

$$K' = \int_1^{1/k} (s^2 - 1)^{-\frac{1}{2}} (1 - k^2 s^2)^{-\frac{1}{2}} \, ds,$$

provided that $0 < k < 1$; the result has next to be extended to complex values of k.

Note   Consider $\int_0^{1/k} (1 - t^2)^{-\frac{1}{2}} (1 - k^2 t^2)^{-\frac{1}{2}} \, dt$, the path of integration passing above the point 1, and not crossing the imaginary axis[5]. The path may be taken to be the straight lines joining 0 to $1 - \delta$ and $1 + \delta$ to k^{-1} together with a semicircle of (small) radius δ above the real axis. If $(1 - t^2)^{\frac{1}{2}}$ and $(1 - k^2 t^2)^{\frac{1}{2}}$ reduce to $+1$ at $t = 0$, the value of the former at $1 + \delta$ is

$$e^{-\frac{1}{2}\pi i} \delta^{\frac{1}{2}} (2 + \delta)^{\frac{1}{2}} = -i(t^2 - 1)^{\frac{1}{2}},$$

where each radical is positive; while the value of the latter at $t = 1$ is $+k'$ when k is real, and hence by the theory of analytic continuation it is always $+k'$.

Make $\delta \to 0$, and the integral round the semicircle tends to zero like $\delta^{\frac{1}{2}}$; and so

$$\int_0^{1/k} (1 - t^2)^{-\frac{1}{2}} (1 - k^2 t^2)^{-\frac{1}{2}} \, dt = K + i \int_1^{1/k} (t^2 - 1)^{-\frac{1}{2}} (1 - k^2 t^2)^{-\frac{1}{2}} \, dt.$$

Now

$$\int_0^{1/k} (1 - t^2)^{-\frac{1}{2}} (1 - k^2 t^2)^{-\frac{1}{2}} \, dt = \int_0^1 (k^2 - u^2)^{-\frac{1}{2}} (1 - u^2)^{-\frac{1}{2}} \, du,$$

which[6] is analytic throughout the cut plane, while K is analytic throughout the cut plane.

---

[5] $\mathrm{Re}\, k > 0$ because $|\arg c| < \pi$.
[6] The path of integration passes above the point $u = k$.

Hence $\int_1^{1/k} (t^2 - 1)^{-\frac{1}{2}} (1 - k^2 t^2)^{-\frac{1}{2}} \, dt$ is analytic throughout the cut plane, and as it is equal to the analytic function K' when $0 < k < 1$, the equality persists throughout the cut plane; that is to say

$$K' = \int_1^{1/k} (t^2 - 1)^{-\frac{1}{2}} (1 - k^2 t^2)^{-\frac{1}{2}} \, dt,$$

when the c-plane is cut from 0 to $-\infty$ and from 1 to $+\infty$.

Since

$$K + iK' = \int_0^{1/k} (1 - t^2)^{-\frac{1}{2}} (1 - k^2 t^2)^{-\frac{1}{2}} \, dt,$$

we have $\operatorname{sn}(K + iK') = 1/k$, $\operatorname{dn}(K + iK') = 0$; while the value of $\operatorname{cn}(K + iK')$ is the value of $(1 - t^2)^{\frac{1}{2}}$ when t has followed the prescribed path to the point $1/k$, and so its value is $-ik'/k$, not $+ik'/k$.

Example 22.3.7 Shew that

$$\frac{1}{2} \int_0^1 \{t(1 - t)(1 - k^2 t)\}^{-\frac{1}{2}} \, dt = \frac{1}{2} \int_{1/k^2}^{\infty} \{t(t - 1)(k^2 t - 1)\}^{-\frac{1}{2}} \, dt = K,$$

$$\frac{1}{2} \int_{-\infty}^0 \{-t(1 - t)(1 - k^2 t)\}^{-\frac{1}{2}} \, dt = \frac{1}{2} \int_1^{1/k^2} \{t(t - 1)(1 - k^2 t)\}^{-\frac{1}{2}} \, dt = K'.$$

Example 22.3.8 Shew that K' satisfies the same linear differential equation as K given in Example 22.3.1.

22.33 The periodic properties (associated with $K + iK'$) of the Jacobian elliptic functions

If we make use of the three equations

$$\operatorname{sn}(K + iK') = k^{-1}, \quad \operatorname{cn}(K + iK') = -ik'/k, \quad \operatorname{dn}(K + iK') = 0,$$

we get at once, from the addition-theorems for $\operatorname{sn} u, \operatorname{cn} u, \operatorname{dn} u$, the following results:

$$\operatorname{sn}(u + K + iK') = \frac{\operatorname{sn} u \, \operatorname{cn}(K + iK') \, \operatorname{dn}(K + iK') + \operatorname{sn}(K + iK') \, \operatorname{cn} u \, \operatorname{dn} u}{1 - k^2 \, \operatorname{sn}^2 u \, \operatorname{sn}^2(K + iK')}$$

$$= k^{-1} \operatorname{dc} u,$$

and similarly

$$\operatorname{cn}(u + K + iK') = -ik'k^{-1} \operatorname{nc} u,$$

$$\operatorname{dn}(u + K + iK') = ik' \operatorname{sc} u.$$

By repeated applications of these formulae we have

$$\left\{ \begin{array}{lll} \operatorname{sn}(u + 2K + 2iK'') &=& -\operatorname{sn} u, \\ \operatorname{cn}(u + 2K + 2iK') &=& \operatorname{cn} u, \\ \operatorname{dn}(u + 2K + 2iK') &=& -\operatorname{dn} u, \end{array} \right. \qquad \left\{ \begin{array}{lll} \operatorname{sn}(u + 4K + 4iK') &=& \operatorname{sn} u, \\ \operatorname{cn}(u + 4K + 4iK') &=& \operatorname{cn} u, \\ \operatorname{dn}(u + 4K + 4iK') &=& \operatorname{dn} u. \end{array} \right.$$

Hence the functions sn *u and* dn *u have period* $4K + 4iK'$, *while* cn *u has the smaller period* $2K + 2iK'$. The double periodicity of sn *u* may be inferred from dynamical considerations. See Whittaker [678, §44].

22.34 The periodic properties (associated with iK') of the Jacobian elliptic functions

By the addition-theorem we have

$$\operatorname{sn}(u + iK') = \operatorname{sn}(u - K + K + iK')$$
$$= k^{-1} \operatorname{dc}(u - K)$$
$$= k^{-1} \operatorname{ns} u.$$

Similarly we find the equations

$$\operatorname{cn}(u + iK') = -ik^{-1} \operatorname{ds} u,$$
$$\operatorname{dn}(u + iK') = -i \operatorname{cs} u.$$

By repeated applications of these formulae we have

$$\left\{ \begin{array}{lll} \operatorname{sn}(u + 2iK') &=& \operatorname{sn} u, \\ \operatorname{cn}(u + 2iK') &=& -\operatorname{cn} u, \\ \operatorname{dn}(u + 2iK') &=& -\operatorname{dn} u, \end{array} \right. \qquad \left\{ \begin{array}{lll} \operatorname{sn}(u + 4iK') &=& \operatorname{sn} u, \\ \operatorname{cn}(u + 4iK') &=& \operatorname{cn} u, \\ \operatorname{dn}(u + 4iK') &=& \operatorname{dn} u. \end{array} \right.$$

Hence the functions cn *u and* dn *u have period* $4iK'$, *while* sn *u has the smaller period* $2iK'$.

Example 22.3.9   Obtain the formulae

$$\operatorname{sn}(u + 2mK + 2niK') = (-1)^m \operatorname{sn} u,$$
$$\operatorname{cn}(u + 2mK + 2niK') = (-1)^{m+n} \operatorname{cn} u,$$
$$\operatorname{dn}(u + 2mK + 2niK') = (-1)^n \operatorname{dn} u.$$

22.341 The behaviour of the Jacobian elliptic functions near the origin and near iK'

We have

$$\frac{d}{du} \operatorname{sn} u = \operatorname{cn} u \, \operatorname{dn} u,$$

$$\frac{d^3}{du^3} \operatorname{sn} u = 4k^2 \operatorname{sn}^2 u \, \operatorname{cn} u \, \operatorname{dn} u - \operatorname{cn} u \, \operatorname{dn} u \, (\operatorname{dn}^2 u + k^2 \operatorname{cn}^2 u).$$

Hence, by Maclaurin's theorem, we have, for small values of $|u|$,

$$\operatorname{sn} u = u - \frac{1}{6}(1 + k^2)u^3 + O(u^5),$$

on using the fact that sn *u* is an *odd* function. In like manner

$$\operatorname{cn} u = 1 - \frac{1}{2}u^2 + O(u^4),$$

$$\operatorname{dn} u = 1 - \frac{1}{2}k^2 u^2 + O(u^4).$$

It follows that

$$sn(u + iK') = k^{-1} \, ns \, u$$

$$= \frac{1}{ku} \left\{ 1 - \frac{1}{6}(1 + k^2) u^2 + O(u^4) \right\}^{-1}$$

$$= \frac{1}{ku} + \frac{1 + k^2}{6k} u + O(u^3);$$

and similarly

$$cn(u + iK') = \frac{-i}{ku} + \frac{2k^2 - 1}{6k} iu + O(u^3),$$

$$dn(u + iK') = -\frac{i}{u} + \frac{2 - k^2}{6} iu + O(u^3).$$

It follows that at the point iK' the functions sn v, cn v, dn v *have simple poles with residues* k^{-1}, $-ik^{-1}$, $-i$ *respectively.*

Example 22.3.10 Obtain the residues of sn u, cn u, dn u at iK' by the theory of theta-functions.

22.35 *General description of the functions* sn u, cn u, dn u

The foregoing investigations of the functions sn u, cn u and dn u may be summarised in the following terms:

(I) The function sn u is a doubly-periodic function of u with periods $4K, 2iK'$. It is analytic except at the points congruent to iK' or to $2K + iK'$ (mod $4K$, $2iK'$); these points are simple poles, the residues at the first set all being k^{-1}; and the residues at the second set all being $-k^{-1}$ and the function has a simple zero at all points congruent to 0 (mod $2K$, $2iK'$).

It may be observed that sn u is the only function of u satisfying this description; for if $\phi(u)$ were another such function, sn $u - \phi(u)$ would have no singularities and would be a doubly-periodic function; hence (§20.12) it would be a constant, and this constant vanishes, as may be seen by putting $u = 0$; so that $\phi(u) \equiv$ sn u.

When $0 < k^2 < 1$, it is obvious that K and K' are real, and sn u is real for real values of u and is a pure imaginary when u is a pure imaginary.

(II) The function cn u is a doubly-periodic function of u with periods $4K$ and $2K + 2iK'$. It is analytic except at points congruent to iK' or to $2K + ik'$ mod $4K$, $2K + 2iK'$; these points are simple poles, the residues at the first set being $-ik^{-1}$, and the residues at the second set being ik^{-1}; and the function has a simple zero at all points congruent to K mod $2K$, $2iK'$.

(III) The function dn u is a doubly-periodic function of u with periods $2K$ and $4iK'$. It is analytic except at points congruent to iK' or to $3iK'$ (mod $2K$, $4iK'$); these points are simple poles, the residues at the first set being $-i$, and the residues at the second set being i; and the function has a simple zero at all points congruent to $K + iK'$ mod $2K$, $2iK'$.

Note   To see that the functions have no zeros or poles other than those just specified, recourse must be had to their definitions in terms of theta-functions.

22.351 The connexion between Weierstrassian and Jacobian elliptic functions

If e_1, e_2, e_3 be any three distinct numbers whose sum is zero, and if we write

$$y = e_3 + \frac{e_1 - e_3}{\mathrm{sn}^2(\lambda u, k)},$$

we have

$$\left(\frac{dy}{du}\right)^2 = 4(e_1 - e_3)^2 \lambda^2 \, \mathrm{ns}^2 \, \lambda u \, \mathrm{cs}^2 \, \lambda u \, \mathrm{ds}^2 \, \lambda u$$

$$= 4(e_1 - e_3)^2 \lambda^2 \, \mathrm{ns}^2 \, \lambda u (\mathrm{ns}^2 \, \lambda u - 1)(\mathrm{ns}^2 \, \lambda u - k^2)$$

$$= 4\lambda^2 (e_1 - e_3)^{-1}(y - e_3)(y - e_1)\left\{y - k^2(e_1 - e_3) - e_3\right\}.$$

Hence, if $\lambda^2 = e_1 - e_3$ and $k^2 = (e_2 - e_3)/(e_1 - e_3)$, then y satisfies the equation[7]

$$\left(\frac{dy}{du}\right)^2 = 4y^3 - g_2 y - g_3,$$

and so

$$e_3 + (e_1 - e_3) \, \mathrm{ns}^2 \left\{u(e_1 - e_3)^{\frac{1}{2}}, \sqrt{\frac{e_2 - e_3}{e_1 - e_3}}\right\} = \wp(u + a; g_2, g_3),$$

where a is a constant. Making $u \to 0$, we see that a is a period, and so

$$\wp(u; g_2, g_3) = e_3 + (e_1 - e_3) \, \mathrm{ns}^2 \left\{u(e_1 - e_3)^{\frac{1}{2}}\right\},$$

the Jacobian elliptic function having its modulus given by the equation

$$k^2 = \frac{e_2 - e_3}{e_1 - e_3}.$$

22.4 Jacobi's imaginary transformation

The result of §21.51, which gave a transformation from theta-functions with parameter τ to theta-functions with parameter $\tau' = -1/\tau$, naturally produces a transformation of Jacobian elliptic functions; this transformation is expressed by the equations

$$\mathrm{sn}(iu, k) = i \, \mathrm{sc}(u, k'), \qquad \mathrm{cn}(iu, k) = \mathrm{nc}(u, k'), \qquad \mathrm{dn}(iu, k) = \mathrm{dc}(u, k')$$

(Jacobi [349, p. 34–35]). Abel [2, p. 104] derives the double periodicity of elliptic functions from this result. (See a letter of Jan. 12, 1828, from Jacobi to Legendre [354, p. 402].)

Suppose, for simplicity, that $0 < c < 1$ and $y > 0$; let

$$\int_0^{iy} (1 - t^2)^{-\frac{1}{2}} (1 - k^2 t^2)^{-\frac{1}{2}} \, dt = iu,$$

---

[7] The values of g_2 and g_3 are, as usual, $-\frac{1}{4}\sum e_2 e_3$ and $\frac{1}{4} e_1 e_2 e_3$.

so that $iy = \text{sn}(iu, k)$; take the path of integration to be a straight line, and we have

$$\text{cn}(iu, k) = (1 + y^2)^{\frac{1}{2}}, \qquad \text{dn}(iu, k) = (1 + k^2 y^2)^{\frac{1}{2}}.$$

Now put $y = \eta/(1 - \eta^2)^{\frac{1}{2}}$, where $0 < \eta < 1$, so that the range of values of t is from 0 to $i\eta/(1 - \eta^2)^{\frac{1}{2}}$ and hence, if $t = it_1/(1 - t_1^2)^{\frac{1}{2}}$, the range of values of t_1 is from 0 to η. Then $dt = i(1 - t_1^2)^{-\frac{3}{2}} dt_1$, $(1 - t^2)^{\frac{1}{2}} = (1 - t_1^2)^{-\frac{1}{2}}$,

$$1 - k^2 t^2 = (1 - k'^2 t_1^2)^{-\frac{1}{2}}(1 - t_1^2)^{-\frac{1}{2}},$$

and we have

$$iu = \int_0^{\eta} (1 - t_1^2)^{-\frac{1}{2}}(1 - k'^2 t_1^2)^{-\frac{1}{2}} i \, dt_1,$$

so that $\eta = \text{sn}(u, k')$ and therefore $y = \text{sc}(u, k')$. We have thus obtained the result that

$$\text{sn}(iu, k) = i \, \text{sc}(u, k').$$

Also $\text{cn}(iu, k) = (1 + y^2)^{\frac{1}{2}} = (1 - \eta^2)^{-\frac{1}{2}} = \text{nc}(u, k')$, and

$$\text{dn}(iu, k) = (1 - k^2 y^2)^{\frac{1}{2}} = (1 - k'^2 \eta^2)^{\frac{1}{2}}(1 - \eta^2)^{-\frac{1}{2}} = \text{dc}(u, k').$$

Now $\text{sn}(iu, k)$ and $i \, \text{sc}(u, k')$ are one-valued functions of u and k (in the cut c-plane) with isolated poles. Hence by the theory of analytic continuation the results proved for real values of u and k hold for general complex values of u and k.

22.41 Proof of Jacobi's imaginary transformation by the aid of theta-functions

The results just obtained may be proved very simply by the aid of theta-functions. Thus, from §21.61,

$$\text{sn}(iu, k) = \frac{\vartheta_3(0|\tau)}{\vartheta_2(0|\tau)} \frac{\vartheta_1(iz|\tau)}{\vartheta_4(iz|\tau)},$$

where $z = u/\vartheta_3^2(0|\tau)$, and so, by §21.51,

$$\text{sn}(iu, k) = \frac{\vartheta_3(0|\tau')}{\vartheta_4(0|\tau')} \frac{-i\vartheta_1(iz\tau'|\tau')}{\vartheta_2(iz\tau'|\tau')}$$
$$= -i \, \text{sc}(v, k'),$$

where $v = iz\tau'\vartheta_3^2(0|\tau') = iz\tau'(-i\tau)\vartheta_3^2(0|\tau) = -u$, so that, finally, $\text{sn}(iu, k) = i \, \text{sc}(u, k')$.

Example 22.4.1 Prove that $\text{cn}(iu, k) = \text{nc}(u, k')$, $\text{dn}(iu, k) = \text{dc}(u, k')$ by the aid of theta-functions.

Example 22.4.2 Shew that

$$\text{sn}\left(\tfrac{1}{2}iK', k\right) = i \, \text{sc}\left(\tfrac{1}{2}K', k'\right) = ik^{-\frac{1}{2}},$$
$$\text{cn}\left(\tfrac{1}{2}iK', k\right) = (1 + k)^{\frac{1}{2}} k^{-\frac{1}{2}},$$
$$\text{dn}\left(\tfrac{1}{2}iK', k\right) = (1 + k)^{\frac{1}{2}}.$$

Note There is great difficulty in determining the signs of $\text{sn}\,\tfrac{1}{2} iK'$, $\text{cn}\,\tfrac{1}{2} iK'$, $\text{dn}\,\tfrac{1}{2} iK'$, if any method other than Jacobi's transformation is used.

Example 22.4.3   Shew that

$$\operatorname{sn} \tfrac{1}{2}(K + iK') = \frac{(1 + k)^{\frac{1}{2}} + i(1 - k)^{\frac{1}{2}}}{\sqrt{2k}},$$

$$\operatorname{cn} \tfrac{1}{2}(K + iK') = \frac{(1 - i)\sqrt{k'}}{\sqrt{2k}},$$

$$\operatorname{dn} \tfrac{1}{2}(K + iK') = \frac{k'^{\frac{1}{2}}\{(1 + k')^{\frac{1}{2}} - i(1 - k)^{\frac{1}{2}}\}}{\sqrt{2}}.$$

Example 22.4.4 (Glaisher)   If $0 < k < 1$ and if θ be the modular angle, shew that

$$\operatorname{sn} \tfrac{1}{2}(K + iK') = e^{\frac{1}{4}\pi i - \frac{1}{2}i\theta} \sqrt{\operatorname{cosec} \theta},$$

$$\operatorname{cn} \tfrac{1}{2}(K + iK') = e^{-\frac{1}{4}\pi i} \sqrt{\cot \theta},$$

$$\operatorname{dn} \tfrac{1}{2}(K + iK') = e^{-\frac{1}{2}i\theta} \sqrt{\cos \theta}.$$

22.42 Landen's transformation

This appears in [406]. We shall now obtain the formula

$$\int_0^{\phi_1} (1 - k_1^2 \sin^2 \theta_1)^{-\frac{1}{2}} \, d\theta_1 = (1 + k') \int_0^{\phi} (1 - k^2 \sin^2 \theta)^{-\frac{1}{2}} \, d\theta,$$

where

$$\sin \phi_1 = (1 + k') \sin \phi \, \cos \phi \, (1 - k^2 \sin^2 \phi)^{-\frac{1}{2}}$$

and

$$k_1 = (1 - k')/(1 + k').$$

This formula, of which Landen was the discoverer, may be expressed by means of Jacobian elliptic functions in the form

$$\operatorname{sn} \{(1 + k')u, \, k_1\} = (1 + k') \operatorname{sn}(u, k) \operatorname{cd}(u, k),$$

on writing $\phi = \operatorname{am} u$, $\phi_1 = \operatorname{am} u_1$.

To obtain this result, we make use of the equations of §21.52, namely

$$\frac{\vartheta_3(z|\tau) \, \vartheta_4(z|\tau)}{\vartheta_4(2z|2\tau)} = \frac{\vartheta_2(z|\tau) \, \vartheta_1(z|\tau)}{\vartheta_1(2z|2\tau)} = \frac{\vartheta_3(0|\tau)\vartheta_4(0|\tau)}{\vartheta_4(0|2\tau)}.$$

Write[8] $\tau_1 = 2\tau$, and let k_1, Λ, Λ' be the modulus and quarter-periods formed with parameter τ_1; then the equation

$$\frac{\vartheta_1(z|\tau) \, \vartheta_2(z|\tau)}{\vartheta_3(z|\tau) \, \vartheta_4(z|\tau)} = \frac{\vartheta_1(2z|\tau_1)}{\vartheta_4(2z|\tau_1)}$$

may obviously be written

$$k \, \operatorname{sn}(2Kz/\pi, k) \operatorname{cd}(2Kz/\pi, k) = k_1^{\frac{1}{2}} \operatorname{sn}(4\Delta z/\pi, k_1). \qquad (A)$$

---

[8] It will be supposed that $|\operatorname{Re} \tau| < \tfrac{1}{2}$, to avoid difficulties of sign which arise if $\operatorname{Re}(\tau_1)$ does not lie between ± 1. This condition is satisfied when $0 < k < 1$, for τ is then a pure imaginary.

To determine k_1 in terms of k, put $z = \pi/4$, and we immediately get

$$\frac{k}{1 + k'} = k_1^{1/2},$$

which gives, on squaring, $k_1 = (1 - k')/(1 + k')$, as stated above.

To determine Λ, divide equation (A) by z, and then make $z \to 0$; and we get $2Kk = 4k_1^{1/2}\Lambda$, so that $\Lambda = \frac{1}{2}(1 + k')K$. Hence, writing u in place of $2Kz/\pi$, we at once get from (A)

$$(1 + k')\,\mathrm{sn}(u, k)\,\mathrm{cd}(u, k) = \mathrm{sn}\left\{(1 + k')u, k_1\right\},$$

since $4\Lambda z/\pi = 2\Lambda u/K = (1 + k')\,u$; so that Landen's result has been completely proved.

Example 22.4.5 Shew that $\Lambda'/\Lambda = 2K'/K$, and thence that $\Lambda' = (1 + k')K'$.

Example 22.4.6 Shew that

$$\mathrm{cn}\left\{(1 + k')u, k_1\right\} = \left\{1 - (1 + k')\,\mathrm{sn}^2(u, k)\right\}\mathrm{nd}(u, k),$$
$$\mathrm{dn}\left\{(1 + k')u, k_1\right\} = \left\{k' + (1 - k')\,\mathrm{cn}^2(u, k)\right\}\mathrm{nd}(u, k).$$

Example 22.4.7 Shew that

$$\mathrm{dn}(u, k) = (1 - k')\,\mathrm{cn}\left\{(1 + k')u, k_1\right\} + (1 + k')\,\mathrm{dn}\left\{(1 + k')u, k_1\right\},$$

where $k = 2k_1^{\frac{1}{2}}/(1 + k_1)$.

22.421 Transformations of elliptic functions

The formula of Landen is a particular case of what is known as a transformation of elliptic functions; a transformation consists in the expression of elliptic functions with parameter τ in terms of those with parameter $(a + b\tau)/(c + d\tau)$, where a, b, c, d are integers. We have had another transformation in which $a = -1$, $b = 0$, $c = 0$, $d = 1$, namely Jacobi's imaginary transformation. For the general theory of transformations, which is outside the range of this book, the reader is referred to Jacobi [349], to Klein and Fricke [378], and to Cayley [136].

Example 22.4.8 By considering the transformation $\tau_2 = \tau \pm 1$, shew, by the method of §22.42, that

$$\mathrm{sn}(k'u, k_2) = k'\,\mathrm{sd}(u, k),$$

where $k_2 = \pm ik/k'$, and the upper or lower sign is taken according as $\mathrm{Re}\,\tau < 0$ or $\mathrm{Re}\,\tau > 0$; and obtain formulae for $\mathrm{cn}(k'u, k_2)$ and $\mathrm{dn}(k'u, k_2)$.

22.5 Infinite products for the Jacobian elliptic functions

The products for the theta-functions, obtained in §21.3, at once yield products for the Jacobian elliptic functions [349, p. 84–115]; writing $u = 2Kx/\pi$, we obviously have, from §22.11, formulae (A), (B) and (C),

$$\operatorname{sn} u = 2q^{\frac{1}{4}} k^{-\frac{1}{2}} \sin x \prod_{n=1}^{\infty} \left\{ \frac{1 - 2q^{2n} \cos 2x + q^{4n}}{1 - 2q^{2n-1} \cos 2x + q^{4n-2}} \right\},$$

$$\operatorname{cn} u = 2q^{\frac{1}{4}} k'^{\frac{1}{2}} k^{-\frac{1}{2}} \cos x \prod_{n=1}^{\infty} \left\{ \frac{1 + 2q^{2n} \cos 2x + q^{4n}}{1 - 2q^{2n-1} \cos 2x + q^{4n-2}} \right\},$$

$$\operatorname{dn} u = k'^{\frac{1}{2}} \prod_{n=1}^{\infty} \left\{ \frac{1 + 2q^{2n-1} \cos 2x + q^{4n-2}}{1 - 2q^{2n-1} \cos 2x + q^{4n-2}} \right\}.$$

From these results the products for the nine reciprocals and quotients can be written down.

There are twenty-four other formulae which may be obtained in the following manner: From the duplication-formulae in Example 22.2.10 we have

$$\frac{1 - \operatorname{cn} u}{\operatorname{sn} u} = \operatorname{sn}\left(\frac{u}{2}\right) \operatorname{dc}\left(\frac{u}{2}\right), \quad \frac{1 + \operatorname{dn} u}{\operatorname{sn} u} = \operatorname{ds}\left(\frac{u}{2}\right) \operatorname{nc}\left(\frac{u}{2}\right), \quad \frac{\operatorname{dn} u + \operatorname{cn} u}{\operatorname{sn} u} = \operatorname{cn}\left(\frac{u}{2}\right) \operatorname{ds}\left(\frac{u}{2}\right).$$

Take the first of these, and use the products for $\operatorname{sn} \frac{1}{2}u$, $\operatorname{cn} \frac{1}{2}u$, $\operatorname{dn} \frac{1}{2}u$; we get

$$\frac{1 - \operatorname{cn} u}{\operatorname{sn} u} = \frac{1 - \cos x}{\sin x} \prod_{n=1}^{\infty} \left\{ \frac{1 - 2(-q)^n \cos x + q^{2n}}{1 + 2(-q)^n \cos x + q^{2n}} \right\},$$

on combining the various products. Write $u + K$ for u, $x + \frac{1}{2}\pi$ for x, and we have

$$\frac{\operatorname{dn} u + k' \operatorname{sn} u}{\operatorname{cn} u} = \frac{1 + \sin x}{\cos x} \prod_{n=1}^{\infty} \left\{ \frac{1 + 2(-q)^n \sin x + q^{2n}}{1 - 2(-q)^n \sin x + q^{2n}} \right\}.$$

Writing $u + iK'$ for u in these formulae we have

$$k \operatorname{sn} u + i \operatorname{dn} u = i \prod_{n=1}^{\infty} \left\{ \frac{1 + 2i(-1)^n q^{n-\frac{1}{2}} \sin x - q^{2n-1}}{1 - 2i(-1)^n q^{n-\frac{1}{2}} \sin x - q^{2n-1}} \right\},$$

and the expression for $k \operatorname{cd} u + ik' \operatorname{nd} u$ is obtained by writing $\cos x$ for $\sin x$ in this product.

From the identities $(1 - \operatorname{cn} u)(1 + \operatorname{cn} u) \equiv \operatorname{sn}^2 u$, $(k \operatorname{sn} u + i \operatorname{dn} u)(k \operatorname{sn} u - i \operatorname{dn} u) \equiv 1$, etc., we at once get four other formulae, making eight in all; the other sixteen follow in the same way from the expressions for $\operatorname{ds} \frac{1}{2}u \operatorname{nc} \frac{1}{2}u$ and $\operatorname{cn} \frac{1}{2}u \operatorname{ds} \frac{1}{2}u$. The reader may obtain these as an example, noting specially the following:

$$\operatorname{sn} u + i \operatorname{cn} u = ie^{-ix} \prod_{n=1}^{\infty} \left\{ \frac{(1 - q^{4n-3} e^{2ix})(1 - q^{4n-1} e^{-2ix})}{(1 - q^{4n-1} e^{2ix})(1 - q^{4n-3} e^{-2ix})} \right\}.$$

Example 22.5.1   Shew that

$$\operatorname{dn}\left(\frac{K + iK'}{2}\right) = k'^{\frac{1}{2}} \prod_{n=1}^{\infty} \left\{ \frac{(1 + iq^{2n-\frac{1}{2}})(1 - iq^{2n-\frac{3}{2}})}{(1 - iq^{2n-\frac{1}{2}})(1 + iq^{2n-\frac{3}{2}})} \right\}$$

$$= k'^{\frac{1}{2}} \prod_{n=0}^{\infty} \left\{ \frac{1 - (-1)^n iq^{n+\frac{1}{2}}}{1 + (-1)^n iq^{n+\frac{1}{2}}} \right\}.$$

Example 22.5.2 (Jacobi [349])   Deduce from Example 22.5.1 and from Example 22.4.4 that, if θ be the modular angle, then

$$e^{-\frac{1}{2}i\theta} = \prod_{n=0}^{\infty} \left\{ \frac{1 - (-1)^n \, iq^{n+\frac{1}{2}}}{1 + (-1)^n \, iq^{n+\frac{1}{3}}} \right\},$$

and thence, by taking logarithms, obtain Jacobi's result

$$\frac{1}{4}\theta = \sum_{n=0}^{\infty} (-1)^n \arctan q^{n+\frac{1}{2}} = \arctan \sqrt{q} - \arctan \sqrt{q^3} + \arctan \sqrt{q^5} - \cdots,$$

'quae inter formulas elegantissimas censeri debet'.

Example 22.5.3 (Jacobi [349])   By expanding each term in the equation

$$\log \operatorname{sn} u = \log(2q^{\frac{1}{4}}) - \frac{\log k}{2} + \log \sin x + \sum_{n=1}^{\infty} \{ \log(1 - q^{2n} e^{2ix})$$

$$+ \log(1 - q^{2n} e^{-2ix}) - \log(1 - q^{2n-1} e^{2ix}) - \log(1 - q^{2n-1} e^{-2ix}) \}$$

in powers of $e^{\pm 2ix}$, and rearranging the resulting double series, shew that

$$\log \operatorname{sn} u = \log(2q^{\frac{1}{4}}) - \frac{\log k}{2} + \log \sin x + \sum_{m=1}^{\infty} \frac{2q^m \cos 2mx}{m(1 + q^m)},$$

when $|\operatorname{Im} z| < \frac{\pi}{2} \operatorname{Im} \tau$. Obtain similar series for $\log \operatorname{cn} u$, $\log \operatorname{dn} u$.

Example 22.5.4 (Glaisher [248])   Deduce from Example 22.5.3 that

$$\int_0^K \log \operatorname{sn} u \, du = -\frac{1}{4}\pi K' - \frac{1}{2}K \log k.$$

22.6 Fourier series for the Jacobian elliptic functions

If $u \equiv 2Kx/\pi$, $\operatorname{sn} u$ is an odd periodic function of x (with period 2π), which obviously satisfies Dirichlet's conditions (§9.2) for real values of x; and therefore (§9.22) we may expand $\operatorname{sn} u$ as a Fourier sine-series in sines of multiples of x, thus

$$\operatorname{sn} u = \sum_{n=1}^{\infty} b_n \sin nx,$$

the expansion being valid for all real values of x. It is easily seen that the coefficients b_n are given by the formula

$$\pi i b_n = \int_{-\pi}^{\pi} \operatorname{sn} u \exp(nix) \, dx.$$

(These results are substantially due to Jacobi [349, p. 101].) To evaluate this integral, consider $\int \operatorname{sn} u \exp(nix) \, dx$ taken round the parallelogram whose corners are $-\pi$, π, $\pi\tau$, $-2\pi + \pi\tau$. From the periodic properties of $\operatorname{sn} u$ and $\exp(nix)$, we see that $\int_{\pi}^{\pi\tau}$ cancels $\int_{-2\pi+\pi\tau}^{-\pi}$; and so, since $-\pi + \frac{1}{2}\pi\tau$ and $\frac{1}{2}\pi\tau$ are the only poles of the integrand (*qua* function of x) inside the contour, with residues[9]

$$-k^{-1}\left(\tfrac{1}{2}\pi K\right)\exp\left(-ni\pi + \tfrac{1}{2}n\pi i\tau\right) \quad\text{and}\quad k^{-1}\left(\tfrac{1}{2}\pi K\right)\exp\left(\tfrac{1}{2}n\pi i\tau\right)$$

respectively, we have

$$\left\{\int_{-\pi}^{\pi} - \int_{-2\pi+\pi\tau}^{\pi\tau}\right\}\operatorname{sn} u\exp(nix)\,dx = \frac{\pi^2 i}{Kk}q^{\frac{1}{2}n}\{1 - (-1)^n\}.$$

Writing $x - \pi + \pi\tau$ for x in the second integral, we get

$$\{1 + (-1)^n q^n\}\int_{-\pi}^{\pi}\operatorname{sn} u\exp(nix)\,dx = \frac{\pi^2 i}{Kk}q^{\frac{1}{2}n}\{1 - (-1)^n\}.$$

Hence,

$$b_n = \begin{cases} 0 & \text{when } n \text{ is even,} \\ \dfrac{2\pi}{Kk}\dfrac{q^{\frac{1}{2}n}}{1 - q^n} & \text{when } n \text{ is odd.} \end{cases}$$

Consequently

$$\operatorname{sn} u = \frac{2\pi}{Kk}\left\{\frac{q^{\frac{1}{2}}\sin x}{1 - q} + \frac{q^{\frac{3}{2}}\sin 3x}{1 - q^3} + \frac{q^{\frac{5}{2}}\sin 5x}{1 - q^5} + \cdots\right\},$$

when x is real; but the right-hand side of this equation is analytic when $q^{\frac{1}{2}n}\exp(nix)$ and $q^{\frac{1}{2}n}\exp(-nix)$ both tend to zero as $n \to \infty$, and the left-hand side is analytic except at the poles of $\operatorname{sn} u$.

Hence both sides are analytic in the strip (in the plane of the complex variable x) which is defined by the inequality $|\operatorname{Im} x| < \frac{1}{2}\pi\operatorname{Im}\tau$. And so, by the theory of analytic continuation, we have the result

$$\operatorname{sn} u = \frac{2\pi}{Kk}\sum_{n=0}^{\infty}\frac{q^{n+\frac{1}{2}}\sin(2n + 1)x}{1 - q^{2n+1}},$$

(where $u = 2Kx/\pi$), valid throughout the strip $|\operatorname{Im} x| < \frac{1}{2}\pi\operatorname{Im}\tau$.

Example 22.6.1   Shew that, if $u = 2Kx/\pi$, then

$$\operatorname{cn} u = \frac{2\pi}{Kk}\sum_{n=0}^{\infty}\frac{q^{n+\frac{1}{2}}\cos(2n + 1)x}{1 + q^{2n+1}},$$

$$\operatorname{dn} u = \frac{\pi}{2K} + \frac{2\pi}{K}\sum_{n=1}^{\infty}\frac{q^n\cos 2nx}{1 + q^{2n}},$$

$$\operatorname{am} u = \int_0^u \operatorname{dn} t\,dt = x + \sum_{n=1}^{\infty}\frac{2q^n\sin 2nx}{n(1 + q^{2n})},$$

these results being valid when $|\operatorname{Im} x| < \frac{1}{2}\pi\operatorname{Im}\tau$.

[9] The factor $\frac{1}{2}\pi/K$ has to be inserted because we are dealing with $\operatorname{sn}(2Kx/\pi)$.

Example 22.6.2 By writing $x + \frac{1}{2}\pi$ for x in results already obtained, shew that, if

$$u = 2Kx/\pi \quad \text{and} \quad |\operatorname{Im} x| < \frac{1}{2}\pi \operatorname{Im} \tau,$$

then

$$\operatorname{cd} u = \frac{2\pi}{Kk} \sum_{n=0}^{\infty} \frac{(-1)^n q^{n+\frac{1}{2}} \cos(2n+1)x}{1 - q^{2n+1}},$$

$$\operatorname{sd} u = \frac{2\pi}{Kkk'} \sum_{n=0}^{\infty} \frac{(-1)^n q^{n+\frac{1}{2}} \sin(2n+1)x}{1 + q^{2n+1}},$$

$$\operatorname{nd} u = \frac{\pi}{2Kk'} + \frac{2\pi}{Kk'} \sum_{n=1}^{\infty} \frac{(-1)^n q^n \cos 2nx}{1 + q^{2n}}.$$

22.61 Fourier series for reciprocals of Jacobian elliptic functions

In the result of §22.6, write $u + iK'$ for u and consequently $x + \frac{1}{2}\pi\tau$ for x; then we see that, if $0 > \operatorname{Im} x > -\pi \operatorname{Im} \tau$,

$$\operatorname{sn}(u + iK') = \frac{2\pi}{Kk} \sum_{n=0}^{\infty} \frac{q^{n+\frac{1}{2}} \sin(2n+1)(x + \frac{1}{2}\pi\tau)}{1 - q^{2n+1}},$$

and so (§22.34)

$$\operatorname{ns} u = -\frac{i\pi}{K} \sum_{n=0}^{\infty} q^{n+\frac{1}{2}} \left\{ q^{n+\frac{1}{2}} e^{(2n+1)ix} - q^{-n-\frac{1}{2}} e^{-(2n+1)ix} \right\} / (1 - q^{2n+1}),$$

$$= -\frac{i\pi}{K} \sum_{n=0}^{\infty} \left\{ 2iq^{2n+1} \sin(2n+1)x + (1 - q^{-2n-1}) e^{-(2n+1)ix} \right\} / (1 - q^{2n+1}),$$

$$= \frac{2\pi}{K} \sum_{n=0}^{\infty} \frac{q^{2n+1} \sin(2n+1)x}{1 - q^{2n+1}} - \frac{i\pi}{K} \sum_{n=0}^{\infty} e^{-(2n+1)ix}.$$

That is to say

$$\operatorname{ns} u = \frac{\pi}{2K} \operatorname{cosec} x + \frac{2\pi}{K} \sum_{n=0}^{\infty} \frac{q^{2n+1} \sin(2n+1)x}{1 - q^{2n+1}}.$$

But, apart from isolated poles at the points $x = n\pi$, each side of this equation is an analytic function of x in the strip in which

$$\pi \operatorname{Im} \tau > \operatorname{Im} x > -\pi \operatorname{Im} \tau$$

a strip double the width of that in which the equation has been proved to be true; and so, by the theory of analytic continuation, this expansion for $\operatorname{ns} u$ is valid throughout the wider strip, except at the points $x = n\pi$.

Example 22.6.3 Obtain the following expansions, valid throughout the strip $|\operatorname{Im} x| <$

$\pi \operatorname{Im} \tau$ except at the poles of the first term on the right-hand sides of the respective expansions:

$$\operatorname{ds} u = \frac{\pi}{2K} \operatorname{cosec} x - \frac{2\pi}{K} \sum_{n=0}^{\infty} \frac{q^{2n+1} \sin(2n+1)x}{1 + q^{2n+1}},$$

$$\operatorname{cs} u = \frac{\pi}{2K} \cot x - \frac{2\pi}{K} \sum_{n=1}^{\infty} \frac{q^{2n} \sin(2nx)}{1 + q^{2n}},$$

$$\operatorname{dc} u = \frac{\pi}{2K} \sec x - \frac{2\pi}{K} \sum_{n=0}^{\infty} \frac{(-1)^n q^{2n+1} \cos(2n+1)x}{1 - q^{2n+1}},$$

$$\operatorname{nc} u = \frac{\pi}{2Kk'} \sec x - \frac{2\pi}{Kk'} \sum_{n=0}^{\infty} \frac{(-1)^n q^{2n+1} \cos(2n+1)x}{1 + q^{2n+1}},$$

$$\operatorname{sc} u = \frac{\pi}{2Kk'} \tan x - \frac{2\pi}{Kk'} \sum_{n=1}^{\infty} \frac{(-1)^n q^{2n} \sin 2nx}{1 + q^{2n}}.$$

22.7 Elliptic integrals

An integral of the form $\int R(w, x)\, dx$, where R denotes a rational function of w and x, and w^2 *is a quartic, or cubic function of x* (without repeated factors), is called an *elliptic integral*. Strictly speaking, it is only called an elliptic integral when it cannot be integrated by means of the elementary functions, and consequently involves one of the three kinds of elliptic integrals introduced in §22.72.

Note   Elliptic integrals are of considerable historical importance, owing to the fact that a very large number of important properties of such integrals were discovered by Euler and Legendre before it was realised that the *inverses* of certain standard types of such integrals, rather than the integrals themselves, should be regarded as fundamental functions of analysis.

The first mathematician to deal with elliptic *functions* as opposed to elliptic *integrals* was Gauss (§22.8), but the first results published were by Abel [2] and Jacobi. Jacobi announced his discovery in two letters (dated June 13, 1827 and August 2, 1827) to Schumacher, who published extracts from them in [345] in September 1827 – the month in which Abel's memoir appeared.

The results obtained by Abel were brought to the notice of Legendre by Jacobi immediately after the publication by Legendre of [422]. In the supplement (tome III. (1828), p. 1), Legendre comments on their discoveries in the following terms: "À peine mon ouvrage avait-il vu le jour, à peine son titre pouvait-il être connu des savans étrangers, que j'appris, avec autant d'étonnement que de satisfaction, que deux jeunes géomètres, MM. *Jacobi* (C.-G.-J.) de Koenigsberg et *Abel* de Christiania, avaient réussi, par leurs travaux particuliers, á perfectionner considérablement la théorie des fonctions elliptiques dans ses points les plus élevés."

An interesting correspondence between Legendre and Jacobi was printed in [423]; in one of the letters Legendre refers to the claim of Gauss to have made in 1809 many of the discoveries published by Jacobi and Abel. The validity of this claim was established by Schering (see Gauss [238, vol. III, pp. 493–494]), though the researches of Gauss [238, vol. III, pp. 404–460] remained unpublished until after his death.

We shall now give a brief outline of the important theorem that every elliptic integral can be evaluated by the aid of theta-functions, combined with the elementary functions of analysis; it has already been seen (§20.6) that this process can be carried out in the special case of $\int w^{-1}\,dx$, since the Weierstrassian elliptic functions can easily be expressed in terms of theta-functions and their derivatives (§21.73).

Note The most important case practically is that in which R is a real function of x and w, which are themselves real on the path of integration; it will be shewn how, in such circumstances, the integral may be expressed in a real form.

Since $R(w, x)$ is a *rational* function of w and x we may write

$$R(w, x) \equiv P(w, x)/Q(w, x),$$

where P and Q denote polynomials in w and x; then we have

$$R(w, x) \equiv \frac{w P(w, x)\, Q(-w, x)}{w Q(w, x)\, Q(-w, x)}.$$

Now $Q(w, x)Q(-w, x)$ is a *rational function of w^2 and x*, since it is unaffected by changing the sign of w; it is therefore expressible as a rational function of x.

If now we multiply out $w P(w, x)\, Q(-w, x)$ and substitute for w^2 in terms of x wherever it occurs in the expression, we ultimately reduce it to a polynomial in x and w, the polynomial being *linear in w*. We thus have an identity of the form

$$R(w, x) \equiv \frac{R_1(x) + w R_2(x)}{w},$$

by reason of the expression for w^2 as a quartic in x; where R_1 and R_2 denote rational functions of x.

Now $\int R_2(x)\,dx$ can be evaluated by means of elementary functions only. The integration of *rational* functions of one variable is discussed in textbooks on Integral Calculus; so the problem is reduced to that of evaluating $\int w^{-1} R_1(x)\,dx$. To carry out this process it is necessary to obtain a canonical expression for w^2, which we now proceed to do.

22.71 The expression of a quartic as the product of sums of squares

It will now be shewn that *any quartic (or cubic[10]) in x (with no repeated factors) can be expressed in the form*

$$\left\{ A_1(x - \alpha)^2 + B_1(x - \beta)^2 \right\} \left\{ A_2(x - \alpha)^2 + B_2(x - \beta)^2 \right\},$$

where, if the coefficients in the quartic are real A_1, B_1, A_2, B_2, α, β are all real.

[10] In the following analysis, a cubic may be regarded as a quartic in which the coefficient of x^4 vanishes.

To obtain this result, we observe that any quartic can be expressed in the form $S_1 S_2$ where S_1, S_2 are quadratic in x, say[11]

$$S_1 \equiv a_1 x^2 + 2b_1 x + c_1, \qquad S_2 \equiv a_2 x^2 + 2b_2 x + c_2.$$

Now, λ being a constant, $S_1 - \lambda S_2$ will be a perfect square in x if

$$(a_1 - \lambda a_2)(c_1 - \lambda c_2) - (b_1 - \lambda b_2)^2 = 0.$$

Let the roots of this equation be λ_1, λ_2; then, by hypothesis, numbers α, β exist such that

$$S_1 - \lambda_1 S_2 \equiv (a_1 - \lambda_1 a_2)(x - \alpha)^2, \qquad S_1 - \lambda_2 S_2 \equiv (a_1 - \lambda_2 a_2)(x - \beta)^2;$$

on solving these as equations in S_1, S_2, we obviously get results of the form

$$S_1 \equiv A_1(x - \alpha)^2 + B_1(x - \beta)^2, \qquad S_2 \equiv A_2(x - \alpha)^2 + B_2(x - \beta)^2,$$

and the required reduction of the quartic has been effected.

Note   If the quartic is real and has two or four complex factors, let S_1 have complex factors; then λ_1 and λ_2 are real and distinct since

$$(a_1 - \lambda a_2)(c_1 - \lambda c_2) - (b_1 - \lambda b_2)^2$$

is positive when $\lambda = 0$ and negative[12] when $\lambda = a_1/a_2$.

When S_1 and S_2 have real factors, say $(x - \xi_1)(x - \xi_1')$, $(x - \xi_2)(x - \xi_2')$, the condition that λ_1 and λ_2 should be real is easily found to be

$$(\xi_1 - \xi_2)(\xi_1' - \xi_2)(\xi_1 - \xi_2')(\xi_1' - \xi_2') > 0,$$

a condition which is satisfied when the zeros of S_1 and those of S_2 do not interlace; this was, of course, the reason for choosing the factors S_1 and S_2 of the quartic in such a way that their zeros do not interlace.

22.72 The three kinds of elliptic integrals

Let α, β be determined by the rule just obtained in §22.71, and, in the integral $\int w^{-1} R_1(x)\, dx$, take a new variable t defined by the equation[13] $t = (x - \alpha)/(x - \beta)$; we then have

$$\frac{dx}{w} = \pm \frac{(\alpha - \beta)^{-1}\, dt}{\{(A_1 t^2 + B_1)(A_2 t^2 + B_2)\}^{1/2}}.$$

---

[11] If the coefficients in the quartic are real, the factorisation can be carried out so that the coefficients in S_1 and S_2 are real. In the special case of the quartic having four real linear factors, these factors should be associated in pairs (to give S_1 and S_2) in such a way that the roots of one pair do not interlace the roots of the other pair; the reason for this will seen in the note at the end of the section.

[12] Unless $a_1 : a_2 = b_1 : b_2$, in which case $S_1 \equiv a_1(x - a)^2 + B_1$, and $S_2 \equiv a_2(x - a)^2 + B_2$.

[13] It is rather remarkable that Jacobi did not realise the existence of this homographic substitution; in his reduction he employed a quadratic substitution, equivalent to the result of applying a Landen transformation to the elliptic functions which we shall introduce.

If we write $R_1(x)$ in the form $\pm(\alpha - \beta)R_s(t)$, where R_s is rational, we get

$$\int \frac{R_1(x)\,dx}{w} = \int \frac{R_s(t)\,dt}{\{(A_1t^2 + B_1)(A_2t^2 + B_2)\}^{1/2}}.$$

Now $R_3(t) + R_3(-t) = 2R_4(t^2)$, $R_3(t) + R_3(-t) = 2tR_5(t^2)$, where R_4 and R_5 are rational functions of t^2, and so

$$R_3(t) = R_4(t^2) + tR_5(t^2).$$

But $\int \{(A_1t^2 + B_1)(A_2t^2 + B_2)\}^{-1/2}\, tR_5(t^2)\, dt$ can be evaluated in terms of elementary functions by taking t^2 as a new variable (see, e.g., Hardy [280]); so that, if we put $R_4(t^2)$ into partial fractions, the problem of integrating $\int R(w, x)\, dx$ has been reduced to the integration of integrals of the following types:

$$\int t^{2m} \left\{(A_1t^2 + B_1)(A_2t^2 + B_2)\right\}^{-\frac{1}{2}}\, dt,$$

$$\int (1 + Nt^2)^{-m} \left\{(A_1t^2 + B_1)(A_2t^2 + B_2)\right\}^{-\frac{1}{2}}\, dt;$$

in the former of these m is an integer, in the latter m is a positive integer and $N \neq 0$.

By differentiating expressions of form

$$t^{2m-1} \left\{(A_1t^2 + B_1)(A_2t^2 + B_2)\right\}^{\frac{1}{2}}, \quad t(1 + Nt^2)^{1-m} \left\{(A_1t^2 + B_1)(A_2t^2 + B_2)\right\}^{\frac{1}{2}},$$

it is easy to obtain reduction formulae by means of which the above integrals can be expressed in terms of one of the three canonical forms:

(i) $\int \left\{(A_1t^2 + B_1)(A_2t^2 + B_2)\right\}^{-\frac{1}{2}}\, dt;$

(ii) $\int t^2 \left\{(A_1t^2 + B_1)(A_2t^2 + B_2)\right\}^{-\frac{1}{2}}\, dt,$

(iii) $\int (1 + Nt^2)^{-1} \left\{(A_1t^2 + B_1)(A_2t^2 + B_2)\right\}^{-\frac{1}{2}}\, dt.$

These integrals were called by Legendre [421, vol. I, p. 19] *elliptic integrals of the first, second and third kinds*, respectively.

The elliptic integral of the first kind presents no difficulty, as it can be integrated at once by a substitution based on the integral formulae of §§22.121, 22.122; thus, if A_1, B_1, A_2, B_2 are all positive and $A_2B_1 > A_1B_2$, we write

$$A_1^{\frac{1}{2}}t = B_1^{\frac{1}{2}} \operatorname{cs}(u, k) \qquad [k'^2 = A_1B_2/A_2B_1].$$

Example 22.7.1 Verify that, in the case of *real* integrals, the following scheme gives all possible essentially different arrangements of sign, and determine the appropriate substitutions necessary to evaluate the corresponding integrals.

A_1	+	+	−	+	+	−
B_1	+	−	+	−	−	+
A_2	+	+	+	+	−	−
B_2	+	+	+	−	+	+

Example 22.7.2 (Glaisher)   Shew that

$$\int \operatorname{sn} u \, du = \frac{1}{2k} \log \frac{1 - k \operatorname{cd} u}{1 + k \operatorname{cd} u}, \qquad \int \operatorname{cn} u \, du = k^{-1} \arctan(k \operatorname{sd} u)$$

$$\int \operatorname{dn} u \, du = \operatorname{am} u, \qquad \int \operatorname{sc} u \, du = \frac{1}{2k'} \log \frac{\operatorname{dn} u + k'}{\operatorname{dn} u - k'},$$

$$\int \operatorname{ds} u \, du = \frac{1}{2} \log \frac{1 - \operatorname{cn} u}{1 + \operatorname{cn} u}, \qquad \int \operatorname{dc} u \, du = \frac{1}{2} \log \frac{1 + \operatorname{sn} u}{1 - \operatorname{sn} u},$$

and obtain six similar formulae by writing $u + K$ for u.

Example 22.7.3   Prove, by differentiation, the equivalence of the following twelve expressions:

$$u - k^2 \int \operatorname{sn}^2 u \, du, \qquad k'^2 u + k^2 \int \operatorname{cn}^2 u \, du,$$

$$\int \operatorname{dn}^2 u \, du, \qquad u - \operatorname{dn} u \operatorname{cs} u - \int \operatorname{ns}^2 u \, du,$$

$$k'^2 u + \operatorname{dn} u \operatorname{sc} u - k'^2 \int \operatorname{nc}^2 u \, du, \qquad k^2 \operatorname{sn} u \operatorname{cd} u + k'^2 \int \operatorname{nd}^2 u \, du,$$

$$\operatorname{dn} u \operatorname{sc} u - k'^2 \int \operatorname{sc}^2 u \, du, \qquad k'^2 u + k^2 \operatorname{sn} u \operatorname{cd} u + k^2 k'^2 \int \operatorname{sd}^2 u \, du,$$

$$u + k^2 \operatorname{sn} u \operatorname{cd} u - k^2 \int \operatorname{cd}^2 u \, du, \qquad -\operatorname{dn} u \operatorname{cs} u - \int \operatorname{cs}^2 u \, du,$$

$$k'^2 u - \operatorname{dn} u \operatorname{cs} u - \int \operatorname{ds}^2 u \, du, \qquad u + \operatorname{dn} u \operatorname{sc} u - \int \operatorname{dc}^2 u \, du.$$

Example 22.7.4 (Jacobi; Glaisher [252])   Shew that

$$\frac{d^2 \operatorname{sn}^n u}{du^2} = n(n-1) \operatorname{sn}^{n-2} u - n^2(1 + k^2) \operatorname{sn}^n u + n(n+1)k^2 \operatorname{sn}^{n+2} u,$$

and obtain eleven similar formulae for the second differential coefficients of $\operatorname{cn}^n u, \operatorname{dn}^n u, \ldots,$ $\operatorname{nd}^n u$. What is the connexion between these formulae and the reduction formula for

$$\int t^n \{(A_1 t^2 + B_1)(A_2 t^2 + B_2)\}^{-1/2} \, dt?$$

Example 22.7.5   By means of §20.6 shew that, if a and b are positive,

$$\int_{-a}^{a} \{(a^2 - x^2)(x^2 + b^2)\}^{-\frac{1}{2}} dx = \int_{e_1}^{\infty} (4s^3 - g_2 s - g_3)^{-\frac{1}{2}} ds,$$

where e_1 is the real root of the cubic and

$$g_2 = \tfrac{1}{12}(a^2 - b^2)^2 - a^2 b^2, \qquad g_3 = -\tfrac{1}{216}(a^2 - b^2)\{(a^2 - b^2)^2 + 36a^2 b^2\};$$

and prove that, if $g_2 = 0$, then a and b are given by the equations

$$a^2 - b^2 = -3(2g_3)^{\frac{1}{3}}, \qquad a^2 + b^2 = 2\sqrt{3}|2g_3|^{\frac{1}{3}}.$$

Example 22.7.6 Deduce from Example 22.7.5, combined with the integral formula for
$\operatorname{cn} u$, that, if g_3 is positive,

$$\int_{e_1}^{\infty} (4s^3 - g_3)^{-\frac{1}{2}} \, ds = 2(a^2 + b^2)^{-\frac{1}{2}} K, \qquad \int_{e_1}^{\infty} (4s^3 - g_3)^{-\frac{1}{2}} \, ds = 2(a^2 + b^2)^{-\frac{1}{2}} K',$$

where $a^2 = (\sqrt{3} - \frac{3}{2})(2g_3)^{\frac{1}{3}}$, $b^2 = (\sqrt{3} + \frac{3}{2})(2g_3)^{\frac{1}{3}}$ and the modulus is $a(a^2 + b^2)^{-\frac{1}{2}}$.

22.73 The elliptic integral of the second kind. The function $E(u)$

To reduce an integral of the type

$$\int t^2 \{(A_1 t^2 + B_1)(A_2 t^2 + B_2)\}^{-\frac{1}{2}} \, dt,$$

we employ the same elliptic function substitution as in the case of that elliptic integral of
the first kind which has the same expression under the radical. We are thus led to one of the
twelve integrals

$$\int \operatorname{sn}^2 u \, du, \qquad \int \operatorname{cn}^2 u \, du, \qquad \dots, \qquad \int \operatorname{nd}^2 u \, du.$$

By Example 22.7.3, these are all expressible in terms of u, elliptic functions of u and
$\int \operatorname{dn}^2 u \, du$; it is convenient to regard

$$E(u) \equiv \int_0^u \operatorname{dn}^2 u \, du$$

as the fundamental elliptic integral of the second kind, in terms of which all others can be
expressed; when the modulus has to be emphasised, we write $E(u, k)$ in place of $E(u)$. This
notation was introduced by Jacobi [347, p. 373]. In the *Fundamenta Nova* [349], he wrote
$E(\operatorname{am} u)$ where we write $E(u)$.

We observe that

$$\frac{dE(u)}{du} = \operatorname{dn}^2 u, \qquad E(0) = 0.$$

Further, since $\operatorname{dn}^2 u$ is an even function with double poles at the points $2mK + (2n + 1)iK$,
the residue at each pole being zero, it is easy to see that $E(u)$ is an odd one-valued[14] function
of u with simple poles at the poles of $\operatorname{dn} u$.

It will now be shewn that $E(u)$ may be expressed in terms of theta-functions; the most
convenient type to employ is the function $\Theta(u)$.

Consider

$$\frac{d}{du} \left\{ \frac{\Theta'(u)}{\Theta(u)} \right\};$$

it is a doubly-periodic function of u with double poles at the zeros of $\Theta(u)$, i.e. at the poles
of $\operatorname{dn} u$, and so, if A be a suitably chosen constant,

$$\operatorname{dn}^2 u - A \frac{d}{du} \left\{ \frac{\Theta'(u)}{\Theta(u)} \right\}$$

---

[14] Since the residues of $\operatorname{dn}^2 u$ are zero, the integral defining $E(u)$ is independent of the path chosen (§6.1).

is a doubly-periodic function of u, with periods $2K$, $2iK'$, with only a single simple pole in any cell. It is therefore a constant; this constant is usually written in the form E/K. To determine the constant A, we observe that the principal part of $\mathrm{dn}^2 u$ at iK' is $-(u - iK')^{-2}$, by §22.341; and the residue of $\Theta'(u)/\Theta(u)$ at this pole is unity, so the principal part of

$$\frac{d}{du}\left\{\frac{\Theta'(u)}{\Theta(u)}\right\}$$

is $-(u - iK')^{-2}$. Hence $A = 1$, so

$$\mathrm{dn}^2 u = \frac{d}{du}\left\{\frac{\Theta'(u)}{\Theta(u)}\right\} + \frac{E}{K}.$$

Integrating and observing that $\Theta'(0) = 0$, we get

$$E(u) = \Theta'(u)/\Theta(u) + uE/K.$$

Since $\Theta'(K) = 0$, we have $E(K) = E$; hence

$$E = \int_0^k \mathrm{dn}^2 u\, du = \int_0^{\pi/2} (1 - k^2 \sin^2 \phi)^{\frac{1}{2}}\, d\phi = \frac{\pi}{2} F\left(-\tfrac{1}{2}, \tfrac{1}{2}; 1; k^2\right).$$

It is usual (cf. §22.3) to call K and E *the complete* elliptic integrals of the first and second kinds. Tables of them *qua* functions of the modular angle are given by Legendre [422, vol. II].

Example 22.7.7   Shew that $E(u + 2nK) = E(u) + 2nE$, where n is any integer.

Example 22.7.8   By expressing $\Theta(u)$ in terms of the function $\vartheta_4(\frac{1}{2}\pi u/K)$, and expanding about the point $u = iK'$, shew that

$$E = \tfrac{1}{3}\{2 - k^2 - \vartheta_1'''/(\vartheta_3^4 \vartheta_1')\}K.$$

22.731 The zeta-function Z(u)

The function $E(u)$ is not periodic in either $2K$ or in $2iK'$, but, associated with these periods, it has additive constants $2E$, $\{2iK'E - \pi i\}/K$; it is convenient to have a function of the same general type as $E(u)$ which is singly-periodic, and such a function is

$$Z(u) \equiv \Theta'(u)/\Theta(u);$$

from this definition, we have[15]

$$Z(u) = E(u) - uE/K, \qquad \Theta(u) = \Theta(0) \exp\left\{\int_0^u Z(t)\, dt\right\}.$$

---

[15] The integral in the expression for $\Theta(u)$ is not one-valued as $Z(t)$ has residue 1 at its poles; but the difference of the integrals taken along any two paths with the same end points is $2n\pi i$ where n is the number of poles enclosed, and the exponential of the integral is therefore one-valued, as it should be, since $\Theta(u)$ is one-valued.

22.732 *The addition-formulae for E(u) and Z(u)*

Consider the expression

$$\frac{\Theta'(u+v)}{\Theta(u+v)} - \frac{\Theta'(u)}{\Theta(u)} - \frac{\Theta'(v)}{\Theta(v)} + k^2 \, \text{sn} \, u \, \text{sn} \, v \, \text{sn}(u+v)$$

qua function of u. It is doubly-periodic[16] (periods $2K$ and $2iK'$) with simple poles congruent to iK' and to $iK' - v$; the residue of the first two terms at iK' is -1, and the residue of $\text{sn} \, u \, \text{sn}(u+v)$ is $k^{-1} \, \text{sn} \, v \, \text{sn}(iK'+v) = k^{-2}$. Hence the function is doubly-periodic and has no poles at points congruent to iK' or (similarly) at points congruent to $iK' - v$. By Liouville's theorem, it is therefore a constant, and, putting $u = 0$, we see that the constant is zero.

Hence we have the addition-formulae

$$Z(u) + Z(v) - Z(u+v) = k^2 \, \text{sn} \, u \, \text{sn} \, u \, \text{sn}(u+v),$$
$$E(u) + E(v) - E(u+v) = k^2 \, \text{sn} \, u \, \text{sn} \, u \, \text{sn}(u+v).$$

Note Since $Z(u)$ and $E(u)$ are not doubly-periodic, it is possible to prove that no *algebraic* relation can exist connecting them with sn u, cn u and dn u, so these are not *addition-theorems* in the strict sense. A theorem due to Weierstrass states that an analytic function, $f(z)$, possessing an addition-theorem in the strict sense must be either (i) an algebraic function of z, or (ii) an algebraic function of $\exp(\pi i z/\omega)$, or (iii) an algebraic function of $\wp(z|\omega_1,\omega_2)$; where ω, ω_1, ω_2, are suitably chosen constants. See Forsyth [220, Chapter 13].

22.733 *Jacobi's imaginary transformation of Z(u)*

From §21.51 it is fairly evident that there must be a transformation of Jacobi's type for the function $Z(u)$. To obtain it [349, p. 161], we translate the formula

$$\vartheta_2(ix|\tau) = (-i\tau)^{1/2} \exp(-i\tau'x^2/\pi)\vartheta_4(ix\tau'|\tau')$$

into Jacobi's earlier notation, when it becomes

$$H(iu + K, k) = (-i\tau)^{\frac{1}{2}} \exp\left(\frac{\pi u^2}{4KK'}\right) \Theta(u, k'),$$

and hence

$$\text{cn}(iu, k) = (-i\tau)^{\frac{1}{2}} \exp\left(\frac{\pi u^2}{4KK'}\right) \frac{\vartheta_4(0|\tau)}{\vartheta_2(0|\tau)} \frac{\Theta(u, k')}{\Theta(iu, k)}.$$

Taking the logarithmic differential of each side, we get, on making use of §22.4,

$$Z(iu, k) = i \, \text{dn}(u, k') \, \text{sc}(u, k') - iZ(u, k') - \frac{\pi i u}{2KK'}.$$

---

[16] $2iK'$ is a period since the additive constants for the first two terms cancel.

22.734 Jacobi's imaginary transformation of $E(u)$

It is convenient to obtain the transformation of $E(u)$ directly from the integral definition; we have

$$E(iu, k) = \int_0^{iu} \mathrm{dn}^2(t, k)\, dt = \int_0^u \mathrm{dn}^2(it', k) i\, dt'$$

$$= i \int_0^u \mathrm{dc}^2(t', k')\, dt',$$

on writing $t = it'$ and making use of §22.4.

Hence, from Example 22.7.3, we have

$$E(iu, k) = i \left\{ u + \mathrm{dn}(u, k')\,\mathrm{sc}(u, k') - \int_0^u \mathrm{dn}^2(t', k')\, dt' \right\},$$

and so

$$E(iu, k) = iu + i\,\mathrm{dn}(u, k')\,\mathrm{sc}(u, k') - iE(u, k').$$

This is the transformation stated.

It is convenient to write E' to denote the same function of k' as E is of k, i.e. $E' = E(K', k')$, so that

$$E(2iK', k) = 2i(K' - E').$$

22.735 Legendre's relation

From the transformations of $E(u)$ and $Z(u)$ just obtained, it is possible to derive a remarkable relation connecting the two kinds of complete elliptic integrals, namely

$$EK' + E'K - KK' = \frac{\pi}{2}.$$

For we have, by the transformations of §§22.733, 22.734,

$$E(iu, k) - Z(iu, k) = iu - i\left\{E(u, k') - Z(u, k')\right\} + \frac{\pi iu}{2KK'},$$

and on making use of the connexion between the functions $E(u, k)$ and $Z(u, k)$, this gives

$$iuE/K = iu - i\left\{uE'/K'\right\} + \pi iu/(2KK').$$

Since we may take $u \neq 0$, the result stated follows at once from this equation; it is the analogue of the relation $\eta_1\omega_2 - \eta_2\omega_2 = \pi i/2$ which arose in the Weierstrassian theory (§20.411).

Example 22.7.9   Shew that

$$E(u + K) - E(u) = E - k^2 \,\mathrm{sn}\,u\,\mathrm{cd}\,u.$$

Example 22.7.10   Shew that

$$E(2u + 2iK') = E(2u) + 2i(K' - E').$$

Example 22.7.11   Deduce from Example 22.7.10 that

$$E(u + iK') = \frac{1}{2}E(2u + 2iK') + \frac{1}{2}k^2 \,\mathrm{sn}^2(u + iK')\,\mathrm{sn}(2u + 2iK')$$

$$= E(u) + \mathrm{cn}\,u\,\mathrm{ds}\,u + i(K' - E').$$

Example 22.7.12 Shew that

$$E(u + K + iK') = E(u) - \operatorname{sn} u \operatorname{dc} u + E + i(K' - E').$$

Example 22.7.13 (Jacobi) Obtain the expansions, valid when $| \operatorname{Im} x | < \frac{1}{2}\pi \operatorname{Im} \tau$,

$$(kK)^2 \operatorname{sn}^2 u = K^2 - KE - 2\pi^2 \sum_{n=1}^{\infty} \frac{nq^n \cos 2nx}{1 - q^{2n}}, \qquad KZ(u) = 2\pi \sum_{n=1}^{\infty} \frac{q^n \sin 2nx}{1 - q^{2n}}.$$

22.736 *Properties of the complete elliptic integrals regarded as functions of the modulus*

If, in the formulae $E = \displaystyle\int_0^{\pi/2} (1 - k^2 \sin^2 \phi)^{\frac{1}{2}} \, d\phi$, we differentiate under the sign of integration (§4.2), we have

$$\frac{dE}{dk} = -\int_0^{\pi/2} k \sin^2 \phi (1 - k^2 \sin^2 \phi)^{-\frac{1}{2}} \, d\phi = \frac{E - K}{k}.$$

Treating the formula for K in the same manner, we have

$$\frac{dK}{dk} = \int_0^{\pi/2} k \sin^2 \phi (1 - k^2 \sin^2 \phi)^{-\frac{1}{2}} \, d\phi = k \int_0^K \operatorname{sd}^2 u \, du$$

$$= \frac{1}{kk'^2} \left\{ \int_0^K \operatorname{dn}^2 u \, du - \left[k'^2 u \right]_0^K \right\},$$

by Example 22.7.3 so that

$$\frac{dK}{dk} = \frac{E}{kk'^2} - \frac{K}{k}.$$

If we write $k^2 = c, k'^2 = c'$, these results assume the forms

$$2\frac{dE}{dc} = \frac{E - k}{c}, \qquad 2\frac{dK}{dc} = \frac{E - Kc'}{cc'}.$$

Example 22.7.14 Shew that

$$2\frac{dE'}{dc} = \frac{K' - E'}{c'}, \qquad 2\frac{dK'}{dc} = \frac{cK' - E'}{cc'}.$$

Example 22.7.15 Shew, by differentiation with regard to c, that $EK' + E'K - KK'$ is constant.

Example 22.7.16 (Legendre) Shew that K and K' are solutions of

$$\frac{d}{dk} \left\{ kk'^2 \frac{du}{dk} \right\} = ku,$$

and that E and $E' - K'$ are solutions of

$$k'^2 \frac{d}{dk} \left(k \frac{du}{dk} \right) + ku = 0.$$

22.737 The values of the complete elliptic integrals for small values of k

From the integral definitions of E and K it is easy to see, by expanding in powers of k, that

$$\lim_{k \to 0} K = \lim_{k \to 0} E = \frac{\pi}{2}$$

and

$$\lim_{k \to 0} \frac{K - E}{k^2} = \frac{\pi}{4}.$$

In like manner,

$$\lim_{k \to 0} E' = \int_0^{\pi/2} \cos \phi \, d\phi = 1.$$

It is not possible to determine $\lim_{k \to 0} K'$ in the same way because $(1 - k'^2 \sin^2 \phi)^{-1/2}$ is discontinuous at $\phi = 0$, $k = 0$; but it follows from Example 14.21 on page 312 that, when $|\arg k| < \pi$,

$$\lim_{k \to 0} \left\{ K' - \log \frac{4}{k} \right\} = 0.$$

This result is also deducible from the formulae $2iK' = \pi k \vartheta_3^2$, $k = \vartheta_2^2 / \vartheta_3^2$, by making $q \to 0$; or it may be proved for real values of k by the following elementary method. By §22.32,

$$K' = \int_k^1 (t^2 - k^2)^{-\frac{1}{2}} (1 - t^2)^{-\frac{1}{2}} \, dt;$$

now, when $k < t < \sqrt{k}$, $(1 - t^2)$ lies between 1 and $1 - k$; and, when $\sqrt{k} < t < 1$, $(t^2 - k^2)/t^2$ lies between 1 and $1 - k$. Therefore K' lies between

$$\int_k^{\sqrt{k}} (t^2 - k^2)^{-\frac{1}{2}} \, dt + \int_{\sqrt{k}}^1 t^{-1} (1 - t^2)^{-\frac{1}{2}} \, dt$$

and

$$(1 - k)^{-\frac{1}{2}} \left\{ \int_k^{\sqrt{k}} (t^2 - k^2)^{-\frac{1}{2}} \, dt + \int_{\sqrt{k}}^1 t^{-1} (1 - t^2)^{\frac{1}{2}} \, dt \right\};$$

and therefore

$$K' = (1 - \theta k)^{-\frac{1}{2}} \left\{ \log \frac{\sqrt{k} + \sqrt{k - k^2}}{k} - \log \frac{\sqrt{k}}{1 + \sqrt{1 - k}} \right\}$$

$$= (1 - \theta k)^{-\frac{1}{2}} \left[2 \log \left\{ 1 + \sqrt{1 - k} \right\} - \log k \right],$$

where $0 \le \theta \le 1$. Now

$$\lim_{k \to 0} \left[2(1 - \theta k)^{-\frac{1}{2}} \log \left\{ 1 + \sqrt{1 - k} \right\} - \log 4 \right] = 0, \quad \lim_{k \to 0} \left\{ 1 - (1 - \theta k)^{-\frac{1}{2}} \right\} \log k = 0,$$

and therefore $\lim_{k \to 0} \{ K' - \log(4/k) \} = 0$, which is the required result.

Example 22.7.17 Deduce Legendre's relation from Example 22.7.15 by making $k \to 0$.

22.74 The elliptic integral of the third kind

To evaluate an integral of the type[17]

$$\int (1 + Nt^2)^{-1} \left\{ (A_1 t^2 + B_1)(A_2 t^2 + B_2) \right\}^{-\frac{1}{2}} dt$$

in terms of known functions, we make the substitution made in the corresponding integrals of the first and second kinds (§§22.72, 22.73). The integral is thereby reduced to

$$\int \frac{\alpha + \beta \operatorname{sn}^2 u}{1 + v \operatorname{sn}^2 u} \, du = \alpha u + (\beta - \alpha v) \int \frac{\operatorname{sn}^2 u}{1 + v \operatorname{sn}^2 u} \, du,$$

where α, β, v are constants; if $v = 0, -1, \infty$ or $-k^2$ the integral can be expressed in terms of integrals of the first and second kinds; for other values of v we determine the *parameter a* by the equation $v = -k^2 \operatorname{sn}^2 a$, and then it is evidently permissible to take as the fundamental integral of the third kind

$$\Pi(u, a) = \int_0^u \frac{k^2 \operatorname{sn} a \operatorname{cn} a \operatorname{dn} a \operatorname{sn}^2 u}{1 - k^2 \operatorname{sn}^2 a \operatorname{sn}^2 u} \, du.$$

To express this in terms of theta-functions, we observe that the integrand may be written in the form

$$k^2 \operatorname{sn} u \operatorname{sn} a \left\{ \operatorname{sn}(u + a) + \operatorname{sn}(u - a) \right\} = Z(u - a) - Z(u + a) + 2Z(a),$$

by the addition-theorem for the zeta-function; making use of the formula $Z(u) = \Theta'(u)/\Theta(u)$, we at once get

$$\Pi(u, a) = \frac{1}{2} \log \frac{\Theta(u - a)}{\Theta(u + a)} + uZ(a),$$

a result which shews that $\Pi(u, a)$ is a many-valued function of u with logarithmic singularities at the zeros of $\Theta(u \pm a)$.

Example 22.7.18 (Legendre)   Obtain the addition-formula[18]

$$\Pi(u, a) + \Pi(v, a) - \Pi(u + v, a) = \frac{1}{2} \log \frac{\Theta(u + v + a) \, \Theta(u - a) \, \Theta(v - a)}{\Theta(u + v - a) \, \Theta(u + a) \, \Theta(v + a)}$$

$$= \frac{1}{2} \log \frac{1 - k^2 \operatorname{sn} a \operatorname{sn} u \operatorname{sn} v \operatorname{sn}(u + v - a)}{1 + k^2 \operatorname{sn} a \operatorname{sn} u \operatorname{sn} v \operatorname{sn}(u + v + a)}.$$

Hint. Take $x : y : z : w = u : v : \pm a : u + v \pm a$ in Jacobi's fundamental formula

$$[4] + [1] = [4]' + [1]'.$$

Example 22.7.19 (Legendre; Jacobi)   Shew that

$$\Pi(u, a) - \Pi(a, u) = uZ(a) - aZ(u).$$

This is known as the formula for interchange of argument and parameter.

---

[17] Legendre [421, p. 17], [422, vol. I, pp. 14–18, 74, 75]; Jacobi [349, pp. 137–172]; we employ Jacobi's notation, not Legendre's.

[18] No fewer than 96 forms have been obtained for the expression on the right. See Glaisher [249].

Example 22.7.20 (Jacobi)   Shew that

$$\Pi(u+a) + \Pi(u,b) - \Pi(u, a+b)$$
$$= \frac{1}{2} \log \frac{1 - k^2 \operatorname{sn} a \operatorname{sn} b \operatorname{sn} u \operatorname{sn}(a+b-u)}{1 + k^2 \operatorname{sn} a \operatorname{sn} b \operatorname{sn} u \operatorname{sn}(a+b+u)} + uk^2 \operatorname{sn} a \operatorname{sn} b \operatorname{sn}(a+b).$$

This is known as the formula for addition of parameters.

Example 22.7.21 (Jacobi)   Shew that

$$\Pi(iu, ia + K, k) = \Pi(u, a + K', k').$$

Example 22.7.22 (Jacobi)   Shew that

$$\Pi(u+v, a+b) + \Pi(u-v, a-b) - 2\Pi(u,a) - 2\Pi(v,b)$$
$$= -k^2 \operatorname{sn} a \operatorname{sn} b \{(u+v) \operatorname{sn}(a+b) - (u-v) \operatorname{sn}(a-b)\}$$
$$+ \frac{1}{2} \log \frac{1 - k^2 \operatorname{sn}^2(u-a) \operatorname{sn}^2(v-b)}{1 + k^2 \operatorname{sn}^2(u+a) \operatorname{sn}^2(v+b)},$$

and obtain special forms of this result by putting v or b equal to zero.

22.741 *A dynamical application of the elliptic integral of the third kind*

It is evident from the expression for $\Pi(u, a)$ in terms of theta-functions that if u, a, k are real, the *average* rate of increase of $\Pi(u,a)$ as u increases is $Z(a)$, since $\Theta(u \pm a)$ is periodic with respect to the real period $2K$.

This result determines the mean precession about the invariable line in the motion of a rigid body relative to its centre of gravity under forces whose resultant passes through its centre of gravity. It is evident that, for purposes of computation, a result of this nature is preferable to the corresponding result in terms of sigma-functions and Weierstrassian zeta-functions, for the reasons that the theta-functions have a specially simple behaviour with respect to their real period – the period which is of importance in Applied Mathematics – and that the q-series are much better adapted for computation than the product by which the sigma-function is most simply defined.

22.8 The lemniscate functions

The integral $\displaystyle\int_0^x (1 - t^4)^{-1/2} \, dt$ occurs in the problem of rectifying the arc of the lemniscate. The equation of the lemniscate being $r^2 = a^2 \cos 2\theta$, it is easy to derive the equation $\left(\dfrac{ds}{dr}\right)^2 = \dfrac{a^4}{a^4 - r^4}$ from the formula $\left(\dfrac{ds}{dr}\right)^2 = 1 + \left(\dfrac{r\,d\theta}{dr}\right)^2$. If the integral be denoted by ϕ, we shall express the relation between ϕ and x by writing $x = \operatorname{sinlemn} \phi$. (Gauss [238, vol. III, p. 404] wrote sl and cl for sinlemn and coslemn.) In like manner, if

$$\phi_1 = \int_x^1 (1 - t^4)^{-1/2} \, dt, \qquad \frac{1}{2}\tilde{\omega} = \int_0^1 (1 - t^4)^{-1/2} \, dt,$$

we write $x = \operatorname{coslemn} \phi_1$, and we have the relation

$$\operatorname{sinlemn} \phi = \operatorname{coslemn}\left(\tfrac{1}{2}\tilde{\omega} - \phi\right).$$

These *lemniscate functions*, which were the first functions[19] defined by the inversion of an integral, can easily be expressed in terms of elliptic functions with modulus $1/\sqrt{2}$; for, from the formula (Example 22.1.2)

$$u = \int_0^{\text{sd } u} \left\{(1 - k'^2 y^2)(1 + k^2 y^2)\right\}^{-\frac{1}{2}} dy,$$

it is easy to see (on writing $y = t\sqrt{2}$) that

$$\text{sinlemn } \phi = 2^{-\frac{1}{2}} \text{ sd}(\phi\sqrt{2}, 1/\sqrt{2});$$

similarly, $\text{coslemn } \phi = \text{cn}(\phi\sqrt{2}, 1/\sqrt{2})$. Further, $\frac{1}{2}\tilde{\omega}$ is the smallest positive value of ϕ for which

$$\text{cn}(\phi\sqrt{2}, 1/\sqrt{2}) = 0,$$

so that $\tilde{\omega} = \sqrt{2}K_0$, the suffix attached to the complete elliptic integral denoting that it is formed with the particular modulus $1/\sqrt{2}$. This result renders it possible to express K_0 in terms of Gamma-functions, thus

$$K_0 = 2^{\frac{1}{2}} \int_0^1 (1 - t^4)^{-\frac{1}{2}} dt = 2^{-\frac{3}{2}} \int_0^1 u^{-\frac{3}{4}} (1 - u)^{-\frac{1}{2}} du = 2^{-\frac{3}{2}} \Gamma\left(\tfrac{1}{4}\right) \Gamma\left(\tfrac{1}{2}\right) / \Gamma\left(\tfrac{3}{4}\right)$$

$$= \tfrac{1}{4}\pi^{-\frac{1}{2}} \left\{\Gamma\left(\tfrac{1}{4}\right)\right\}^2,$$

a result first obtained by Legendre [421, vol. I, p. 209]. The value of K_0 is $1.85407468\cdots$, where $\tilde{\omega} = 2.62205756\cdots$. Since $k = k'$ when $k = 1/\sqrt{2}$, it follows that $K_0 = K_0'$, and so $q_0 = e^{-\pi}$.

Example 22.8.1 Express K_0 in terms of Gamma-functions by using Kummer's formula (see Chapter 14 Example 14.12).

Example 22.8.2 By writing $t = (1 - u^2)^{\frac{1}{2}}$ in the formula

$$E_0 = \int_0^1 (1 - \tfrac{1}{2}t^2)^{\frac{1}{2}} (1 - t^2)^{-\frac{1}{2}} dt,$$

shew that

$$2^{\frac{1}{2}} E_0 = \int_0^1 (1 - u^4)^{-\frac{1}{2}} du + \int_0^1 u^2 (1 - u^4)^{-\frac{1}{2}} du,$$

and deduce that $2E_0 - K_0 = 2\pi^{\frac{3}{2}} \left\{\Gamma\left(\tfrac{1}{4}\right)\right\}^{-2}$.

Example 22.8.3 Deduce Legendre's relation (§22.735) from Example 22.8.2 combined with Example 22.7.15.

Example 22.8.4 Shew that

$$\text{sinlemn}^2 \phi = \frac{1 - \text{coslemn}^2 \phi}{1 + \text{coslemn}^2 \phi}.$$

[19] Gauss [234, p. 404]. The idea of investigating the functions occurred to Gauss on January 8, 1797.

22.81 The values of K and K' for special values of k

It has been seen that, when $k = 1/\sqrt{2}$, K can be evaluated in terms of Gamma-functions, and $K = K'$; this is a special case of a general theorem (see Abel [4, p. 184]) that, whenever

$$\frac{K'}{K} = \frac{a + b\sqrt{n}}{c + d\sqrt{n}},$$

where a, b, c, d, n are integers, k is a root of an algebraic equation with integral coefficients.

This theorem is based on the theory of the transformation of elliptic functions and is beyond the scope of this book; but there are three distinct cases in which k, K, K' all have fairly simple values, namely:

(I) $k = \sqrt{2} - 1$, $K' = K\sqrt{2}$;
(II) $k = \sin\frac{\pi}{12}$, $K' = K\sqrt{3}$;
(III) $k = \tan^2\frac{\pi}{8}$, $K' = 2K$.

Of these we shall give a brief investigation. For some similar formulae of a less simple nature, see Kronecker [383, 384].

(I) *The quarter-periods with the modulus $\sqrt{2} - 1$.*

Landen's transformation gives a relation between elliptic functions with any modulus k and those with modulus $k_1 = (1 - k')/(1 + k')$; and the quarter-periods Λ, Λ' associated with the modulus k_1 satisfy the relation $\Lambda'/\Lambda = 2K'/K$.

If we choose k so that $k_1 = k'$, then $\Lambda = K'$ and $k_1' = k$ so that $\Lambda' = K$; and the relation $\Lambda'/\Lambda = 2K'/K$ gives $\Lambda'^2 = 2\Lambda^2$. Therefore the quarter-periods Λ, Λ' associated with the modulus k_1 given by the equation $k_1 = (1 - k_1)/(1 + k_1)$ are such that $\Lambda' = \pm\Lambda\sqrt{2}$; i.e. if $k_1 = \sqrt{2} - 1$, then $\Lambda' = \Lambda\sqrt{2}$ (since Λ, Λ' obviously are both positive).

(II) *The quarter-periods associated with the modulus $\sin\frac{\pi}{12}$.*

The case of $k = \sin\frac{\pi}{12}$ was discussed by Legendre [421, vol. I, pp. 59, 210], [422, vol. I, pp. 59–60]; he obtained the remarkable result that, with this value of k,

$$K' = K\sqrt{3}.$$

This result follows from the relation between definite integrals

$$\int_{-\infty}^{1} (1 - x^3)^{-\frac{1}{2}}\, dx = \sqrt{3} \int_{1}^{\infty} (x^3 - 1)^{-\frac{1}{2}}\, dx.$$

To obtain this relation, consider $\int (1 - z^3)^{-\frac{1}{2}}\, dz$ taken round the contour formed by the part of the real axis (indented at $z = 1$ by an arc of radius R^{-1}) joining the points 0 and R, the line joining $Re^{\frac{1}{2}\pi i}$ to 0 and the arc of radius R joining the points R and $Re^{\frac{1}{2}\pi i}$; as $R \to \infty$, the integral round the arc tends to zero, as does the integral round the indentation, and so, by Cauchy's theorem,

$$\int_{0}^{1} (1 - x^3)^{-\frac{1}{2}}\, dx + i \int_{1}^{\infty} (x^3 - 1)^{-\frac{1}{2}}\, dx + e^{\frac{1}{8}\pi i} \int_{\infty}^{0} (1 + x^3)^{-\frac{1}{2}}\, dx = 0,$$

on writing x and $xe^{\pi i/8}$ respectively for z on the two straight lines. Writing

$$I_1 = \int_0^1 (1-x^3)^{-\frac{1}{2}}\,dx = I_2 = \int_1^\infty (x^3-1)^{-\frac{1}{2}}\,dx,$$

$$I_3 = \int_0^\infty (1+x^3)^{-\frac{1}{2}}\,dx = \int_{-\infty}^0 (1-x^3)^{-\frac{1}{2}}\,dx,$$

we have $I_1 + iI_2 = \frac{1}{2}(1+i\sqrt{3})I_3$; so, equating real and imaginary parts, $I_1 = \frac{1}{2}I_3, I_2 = \frac{1}{2}I_3\sqrt{3}$, and therefore $I_1 + I_3 - I_2\sqrt{3} = \frac{1}{2}I_3 + I_3 - \frac{3}{2}I_3 = 0$, which is the relation stated[20].

Now, by Example 22.7.6,

$$I_2 = 4(\alpha^2+\beta^2)^{-\frac{1}{2}}K, \qquad I_1 + I_3 = 4(\alpha^2+\beta^2)^{-\frac{1}{2}}K',$$

where the modulus is $\alpha(\alpha^2+\beta^2)^{-\frac{1}{2}}$ and

$$\alpha^2 = 2\sqrt{3}-3, \qquad \beta^2 = 2\sqrt{3}+3,$$

so that $k^2 = \frac{1}{4}(2-\sqrt{3}) = \sin^2\frac{\pi}{12}$. We therefore have

$$3^{-\frac{1}{4}}\cdot 2K = 3^{-\frac{3}{4}}\cdot 2K' = I_2 = 3^{\frac{1}{2}}I_1 = 3^{-\frac{1}{2}}\int_0^1 t^{-\frac{2}{3}}(1-t)^{-\frac{1}{2}}\,dt = \frac{1}{3}\pi^{\frac{1}{2}}\Gamma\left(\tfrac{1}{6}\right)/\Gamma\left(\tfrac{2}{3}\right),$$

when the modulus k is $\sin\frac{\pi}{12}$.

(III) *The quarter-periods with the modulus* $\tan^2\frac{\pi}{8}$.

If, in Landen's transformation (§22.42), we take $k = 1/\sqrt{2}$, we have $\Lambda'/\Lambda = 2K'/K = 2$; now this value of k gives

$$k_1 = \frac{\sqrt{2}-1}{\sqrt{2}+1} = \tan^2\frac{\pi}{8};$$

and the corresponding quarter-periods Λ, Λ' are $\frac{1}{2}(1+1/\sqrt{2})K_0$ and $(1+1/\sqrt{2})K_0$.

Example 22.8.5 Discuss the quarter-periods when k has the values

$$(2\sqrt{2}-2)^{\frac{1}{2}}, \quad \sin(5\pi/12), \quad \text{and} \quad 2^{\frac{5}{4}}(\sqrt{2}-1).$$

Example 22.8.6 (Glaisher [244]) Shew that

$$2^{\frac{1}{4}}e^{-\frac{1}{24}\pi} = \prod_{n=0}^\infty (1+e^{-(2n+1)\pi}); \qquad 3^{\frac{1}{4}}e^{-\frac{1}{18}\pi}\sqrt{3} = \prod_{n=1}^\infty (1-e^{-2n\pi/\sqrt{3}}).$$

Example 22.8.7 Express the coordinates of any point on the curve $y^2 = x^3 - 1$ in the form

$$x = 1 + \frac{3^{\frac{1}{2}}(1-\operatorname{cn}u)}{1+\operatorname{cn}u}, \qquad y = \frac{2\cdot 3^{\frac{3}{4}}\operatorname{sn}u\,\operatorname{dn}u}{(1+\operatorname{cn}u)^2},$$

where the modulus of the elliptic functions is $\sin\frac{\pi}{12}$, and shew that $\frac{dx}{du} = 3^{-\frac{1}{4}}y$. By considering $\int_1^\infty y^{-1}\,dx = 3^{-\frac{1}{4}}\int_0^{2K}du$, evaluate K in terms of Gamma-functions when $k = \sin\frac{\pi}{12}$.

[20] Another method of obtaining the relation is to express I_1, I_2, I_3 in terms of Gamma-functions by writing $t^{\frac{1}{3}}$, $t^{-\frac{1}{3}}, (t^{-1}-1)^{\frac{1}{3}}$ respectively for x in the integrals by which I_1, I_2, I_3 are defined.

Example 22.8.8   Shew that, when $y^2 = x^3 - 1$,

$$\int_1^\infty y^{-3}(x-1)^2 dx = \left[-\frac{2}{3} y^{-1}(1-x^{-1})^2 \right]_1^\infty + \frac{4}{3} \int_1^\infty (x^{-2}y^{-1} - x^{-3}y^{-1}) \, dx;$$

and thence, by using Example 22.2.2 and expressing the last integral in terms of Gamma-functions by the substitution $x = t^{-\frac{1}{3}}$, obtain the formula of Legendre [421, p. 60] connecting the first and second complete elliptic integrals with modulus $\sin \frac{\pi}{12}$:

$$\frac{\pi}{4\sqrt{3}} = K \left\{ E - \frac{\sqrt{3}+1}{2\sqrt{3}} K \right\}.$$

Example 22.8.9   By expressing the coordinates of any point on the curve $y^2 = 1 - x^3$ in the form

$$x = 1 - \frac{3^{\frac{1}{2}}(1 - \operatorname{cn} v)}{1 + \operatorname{cn} v}, \qquad y = \frac{2 \cdot 3^{\frac{3}{4}} \operatorname{sn} v \operatorname{dn} v}{(1 + \operatorname{cn} v)^2},$$

in which the modulus of the elliptic functions is $\sin \frac{5\pi}{12}$, and evaluating

$$\left\{ \int_{-\infty}^0 + \int_0^1 \right\} y^{-3}(1-x)^2 \, dx$$

in terms of Gamma-functions, obtain Legendre's result that when $k = \sin \frac{\pi}{12}$,

$$\frac{\pi\sqrt{3}}{4} = K' \left\{ E' - \frac{\sqrt{3}-1}{2\sqrt{3}} K' \right\}.$$

It is interesting to observe that, when Legendre had proved by differentiation that $EK' + E'K - KK'$ is constant, he used the results of Examples 22.2.3 and 22.2.4 to determine the constant, before using the methods of Examples 22.8.3 and 22.7.17.

22.82 *A geometrical illustration of the functions* sn *u*, cn *u*, dn *u*

A geometrical representation of Jacobian elliptic functions with $k = 1/\sqrt{2}$ is afforded by the arc of the lemniscate, as has been seen in §22.8; to represent the Jacobian functions with any modulus k, $(0 < k < 1)$, we may make use of a *curve described on a sphere*, known as *Seiffert's spherical spiral*; Seiffert [591]. Take a sphere of radius unity with centre at the origin, and let the cylindrical polar coordinates of any point on it be (ρ, ϕ, z), so that the arc of a curve traced on the sphere is given by the formula[21]

$$(ds)^2 = \rho^2 (d\phi)^2 + (1 - \rho^2)^{-1} (d\rho)^2.$$

Seiffert's spiral is defined by the equation

$$\phi = ks,$$

where s is the arc measured from the pole of the sphere (i.e. the point where the axis of z meets the sphere) and k is a positive constant, less than unity[22].

---

[21] This is an obvious transformation of the formula $(d\delta)^2 = (ds)^2 + \rho^2 (d\phi)^2 + (dz)^2$ when ρ and z are connected by the relation $\rho^2 + z^2 = 1$.

[22] If $k > 1$, the curve is imaginary.

For this curve we have

$$(ds)^2(1 - k^2\rho^2) = (1 - \rho^2)^{-1}(d\rho)^2,$$

and so, since s and ρ vanish together,

$$\rho = \mathrm{sn}(s, k).$$

The cylindrical polar coordinates of any point on the curve expressed in terms of the arc measured from the pole are therefore

$$(\rho, \phi, z) = (\mathrm{sn}\, s, k s, \mathrm{cn}\, s);$$

and dn s is easily seen to be the cosine of the angle at which the curve cuts the meridian. Hence it may be seen that, if K be the arc of the curve from the pole to the equator, then sn s and cn s have period $4K$, while dn s has period $2K$.

22.9 Miscellaneous examples

Example 22.1 (Math. Trip. 1904)   Shew that one of the values of

$$\left\{ \left(\frac{\mathrm{dn}\, u + \mathrm{cn}\, u}{1 + \mathrm{cn}\, u} \right)^{1/2} + \left(\frac{\mathrm{dn}\, u - \mathrm{cn}\, u}{1 - \mathrm{cn}\, u} \right)^{1/2} \right\} \left\{ \left(\frac{1 - \mathrm{sn}\, u}{\mathrm{dn}\, u - k'\, \mathrm{sn}\, u} \right)^{1/2} + \left(\frac{1 + \mathrm{sn}\, u}{\mathrm{dn}\, u + k'\, \mathrm{sn}\, u} \right)^{1/2} \right\}$$

is $2(1 + k')$.

Example 22.2 (Math. Trip. 1911)   If $x + iy = \mathrm{sn}^2(u + iv)$ and $x - iy = \mathrm{sn}^2(u - iv)$, shew that

$$\left\{ (x - 1)^2 + y^2 \right\}^{\frac{1}{2}} = (x^2 + y^2)^{\frac{1}{2}} \, \mathrm{dn}(2u) + \mathrm{cn}(2u).$$

Example 22.3   Shew that

$$\{1 \pm \mathrm{cn}(u + v)\} \{1 \pm \mathrm{cn}(u - v)\} = \frac{(\mathrm{cn}\, u \pm \mathrm{cn}\, v)^2}{1 - k^2\, \mathrm{sn}^2\, u\, \mathrm{sn}^2\, v}.$$

Example 22.4 (Jacobi)   Shew that

$$1 + \mathrm{cn}(u + v)\, \mathrm{cn}(u - v) = \frac{\mathrm{cn}^2\, u + \mathrm{cn}^2\, v}{1 - k^2\, \mathrm{sn}^2\, u\, \mathrm{sn}^2\, v}.$$

Example 22.5 (Math. Trip. 1909)   Express $\dfrac{1 + \mathrm{cn}(u + v)\, \mathrm{cn}(u - v)}{1 + \mathrm{dn}(u + v)\, \mathrm{dn}(u - v)}$ as a function of $\mathrm{sn}^2\, u + \mathrm{sn}^2\, v$.

Example 22.6 (Jacobi)   Shew that

$$\mathrm{sn}(u - v)\, \mathrm{dn}(u + v) = \frac{\mathrm{sn}\, u\, \mathrm{dn}\, u\, \mathrm{cn}\, v - \mathrm{sn}\, v\, \mathrm{dn}\, v\, \mathrm{cn}\, u}{1 - k^2\, \mathrm{sn}^2\, u\, \mathrm{sn}^2\, v}.$$

Example 22.7 (Math. Trip. 1914)   Shew that

$$\{1 - (1 + k')\, \mathrm{sn}\, u\, \mathrm{sn}(u + k)\} \{1 - (1 - k')\, \mathrm{sn}\, u\, \mathrm{sn}(u + K)\} = \{\mathrm{sn}(u + K) - \mathrm{sn}\, u\}^2.$$

Example 22.8   Shew that

$$\operatorname{sn}\left(u + \tfrac{1}{2}K\right) = (1 + k')^{-\frac{1}{2}}\frac{k' \operatorname{sn} u + \operatorname{cn} u \operatorname{dn} u}{1 - (1 - k') \operatorname{sn}^2 u},$$

$$\operatorname{sn}\left(u + \tfrac{1}{2}iK'\right) = k^{-\frac{1}{2}}\frac{(1 + k) \operatorname{sn} u + i \operatorname{cn} u \operatorname{dn} u}{1 + k \operatorname{sn}^2 u}.$$

Example 22.9 (Jacobi)   Shew that

$$\sin\{\operatorname{am}(u + v) + \operatorname{am}(u - v)\} = \frac{2 \operatorname{sn} u \operatorname{cn} u \operatorname{dn} v}{1 - k^2 \operatorname{sn}^2 u \operatorname{sn}^2 v},$$

$$\cos\{\operatorname{am}(u + v) - \operatorname{am}(u - v)\} = \frac{\operatorname{cn}^2 v - \operatorname{sn}^2 v \operatorname{dn}^2 u}{1 - k^2 \operatorname{sn}^2 u \operatorname{sn}^2 v}.$$

Example 22.10 (Trinity, 1903)   Shew that

$$\operatorname{dn}(u + v) \operatorname{dn}(u - v) = \frac{\operatorname{ds}^2 u \operatorname{ds}^2 v + k^2 k'^2}{\operatorname{ns}^2 u \operatorname{ns}^2 v - k^2},$$

and hence express

$$\left[\frac{\wp(u + v) - e_2}{\wp(u + v) - e_3} \cdot \frac{\wp(u - v) - e_2}{\wp(u - v) - e_3}\right]^{1/2}$$

as a rational function of $\wp(u)$ and $\wp(v)$.

Example 22.11 (Trinity, 1906)   From the formulae for $\operatorname{cn}(2K - u)$ and $\operatorname{dn}(2K - u)$ combined with the formulae for $1 + \operatorname{cn} 2u$ and $1 + \operatorname{dn} 2u$, shew that

$$\left(1 - \operatorname{cn}\frac{2K}{3}\right)\left(1 + \operatorname{dn}\frac{2K}{3}\right) = 1.$$

Example 22.12 (Trinity, 1906)   With notation similar to that of §22.2, shew that

$$\frac{c_1 d_2 - c_2 d_1}{s_1 - s_2} = \frac{\operatorname{cn}(u_1 + u_2) - \operatorname{dn}(u_1 + u_2)}{\operatorname{sn}(u_1 + u_2)};$$

and deduce that, if $u_1 + u_2 + u_3 + u_4 = 2K$, then

$$(c_1 d_2 - c_2 d_1)(c_3 d_4 - c_4 d_3) = k'^2(s_1 - s_2)(s_3 - s_4).$$

Example 22.13 (Math. Trip. 1907)   Shew that, if $u + v + w = 0$, then

$$1 - \operatorname{dn}^2 u - \operatorname{dn}^2 v - \operatorname{dn}^2 w + 2 \operatorname{dn} u \operatorname{dn} v \operatorname{dn} w = k^4 \operatorname{sn}^2 u \operatorname{sn}^2 v \operatorname{sn}^2 w.$$

Example 22.14 (Math. Trip. 1910)   By Liouville's theorem or otherwise, shew that

$$\operatorname{dn} u \operatorname{dn}(u + w) - \operatorname{dn} v \operatorname{dn}(v + w) =$$
$$k^2 \{\operatorname{sn} v \operatorname{cn} u \operatorname{sn}(v + w) \operatorname{cn}(u + w) - \operatorname{sn} u \operatorname{cn} v \operatorname{sn}(u + w) \operatorname{cn}(v + w)\}.$$

Example 22.15 (Math. Trip. 1894)   Shew that

$$\sum \operatorname{cn} u_2 \operatorname{cn} u_3 \operatorname{sn}(u_2 - u_3) \operatorname{dn} u_1 + \operatorname{sn}(u_2 - u_3) \operatorname{sn}(u_3 - u_1) \operatorname{sn}(u_1 - u_2) \operatorname{dn} u_1 \operatorname{dn} u_2 \operatorname{dn} u_3 = 0,$$

the summation applying to the suffices 1, 2, 3.

Example 22.16   Obtain the formulae

$$\operatorname{sn} 3u = A/D, \qquad \operatorname{cn} 3u = B/D, \qquad \operatorname{dn} 3u = C/D,$$

where

$$A = 3s - 4(1 + k^2)s^3 + 6k^2 s^5 - k^4 s^9,$$
$$B = c\left\{1 - 4s^2 + 6k^2 s^4 - 4k^4 s^6 + k^4 s^8\right\},$$
$$C = d\left\{1 - 4k^2 s^2 + 6k^2 s^4 - 4k^4 s^6 + k^4 s^8\right\},$$
$$D = 1 - 6k^2 s^4 + 4k^2(1 + k^2)s^6 - 3k^4 s^8,$$

and $s = \operatorname{sn} u$, $c = \operatorname{cn} u$, $d = \operatorname{dn} u$.

Example 22.17 (Trinity, 1912)   Shew that

$$\frac{1 - \operatorname{dn} 3u}{1 + \operatorname{dn} 3u} = \left(\frac{1 - \operatorname{dn} u}{1 + \operatorname{dn} u}\right)\left(\frac{1 + a_1 \operatorname{dn} u + a_2 \operatorname{dn}^2 u + a_3 \operatorname{dn}^3 u + a_4 \operatorname{dn}^4 u}{1 - a_1 \operatorname{dn} u + a_2 \operatorname{dn}^2 u - a_3 \operatorname{dn}^3 u - a_4 \operatorname{dn}^4 u}\right)^2,$$

where a_1, a_2, a_3, a_4 are constants to be determined.

Example 22.18 (Math. Trip. 1908)   If $P(u) = \left(\dfrac{1 + \operatorname{dn} 3u}{1 + \operatorname{dn} u}\right)^{\frac{1}{2}}$, shew that

$$\frac{P(u) + P(u + 2iK')}{P(u) - P(u + 2iK')} = -\frac{\operatorname{sn} 2u \operatorname{cn} u}{\operatorname{cn} 2u \operatorname{sn} u}.$$

Determine the poles and zeros of $P(u)$ and the first term in the expansion of the function about each pole and zero.

Example 22.19 (Glaisher [245])   Shew that

$$\operatorname{sn}(u_1 + u_2 + u_3) = \frac{A}{D}, \qquad \operatorname{cn}(u_1 + u_2 + u_3) = \frac{B}{D}, \qquad \operatorname{dn}(u_1 + u_2 + u_3) = \frac{C}{D},$$

where

$$A = s_1 s_2 s_3 \left\{-1 - k^2 + 2k^2 \sum s_1^2 - (k^2 + k^4) \sum s_2^2 s_3^2 + 2k^4 s_1^2 s_2^2 s_3^2\right\}$$
$$\quad + \sum \left\{s_1 c_2 c_3 d_2 d_3 \left(1 + 2k^2 s_2^2 s_3^2 - k^2 \sum s_2^2 s_3^2\right)\right\},$$
$$B = c_1 c_2 c_3 \left\{1 - k^2 \sum s_2^2 s_3^2 + 2k^4 s_1^2 s_2^2 s_3^2\right\}$$
$$\quad + \sum \left\{c_1 s_2 s_3 d_2 d_3 \left(-1 + 2k^2 s_2^2 s_3^2 + 2k^2 s_1^2 - k^2 \sum s_2^2 s_3^2\right)\right\},$$
$$C = d_1 d_2 d_3 \left\{1 - k^2 \sum s_2^2 s_3^2 + 2k^2 s_1^2 s_2^2 s_3^2\right\}$$
$$\quad + k^2 \sum \left\{d_1 s_2 s_3 c_2 c_3 \left(-1 + 2k^2 s_2^2 s_3^2 + 2s_1^2 - k^2 \sum s_2^2 s_3^2\right)\right\},$$
$$D = 1 - 2k^2 \sum s_2^2 s_3^2 + 4(k^2 + k^4)s_1^2 s_2^2 s_3^2 - 2k^4 s_1^2 s_2^2 s_3^2 \sum s_1^2 + k^4 \sum s_2^4 s_3^4,$$

and the summations refer to the suffices 1, 2, 3.

Example 22.20 (Cayley [130])   Shew that

$$\operatorname{sn}(u_1 + u_2 + u_3) = \frac{A'}{D'}, \qquad \operatorname{cn}(u_1 + u_2 + u_3) = \frac{B'}{D'}, \qquad \operatorname{dn}(u_1 + u_2 + u_3) = \frac{C'}{D'},$$

where

$$A' = \sum s_1 c_2 c_3 d_2 d_3 - s_1 s_2 s_3 \left(1 + k^2 - k^2 \sum s_1^2 + k^4 s_1^2 s_2^2 s_3^2 \right),$$

$$B' = c_1 c_2 c_3 \left(1 - k^4 s_1^2 s_2^2 s_3^2 \right) - d_1 d_2 d_3 \sum s_2 s_3 c_1 d_1,$$

$$C' = d_1 d_2 d_3 \left(1 - k^2 s_1^2 s_2^2 s_3^2 \right) - k^2 c_1 c_2 c_3 \sum s_2 s_3 c_1 d_1,$$

$$D' = 1 - k^2 \sum s_2^2 s_3^2 + (k^2 + k^4) s_1^2 s_2^2 s_3^2 - k^2 s_1 s_2 s_3 \sum s_1 c_2 c_3 d_2 d_3.$$

Example 22.21 (Cayley [131])   By applying Abel's method (§20.312) to the intersections of the twisted curve $x^2 + y^2 = 1$, $z^2 + k^2 x^2 = 1$ with the variable plane $lx + my + nz = 1$, shew that, if

$$u_1 + u_2 + u_3 + u_4 = 0,$$

then

$$\begin{vmatrix} s_1 & c_1 & d_1 & 1 \\ s_2 & c_2 & d_2 & 1 \\ s_3 & c_3 & d_3 & 1 \\ s_4 & c_4 & d_4 & 1 \end{vmatrix} = 0.$$

Obtain this result also from the equation

$$(s_2 - s_1)(c_3 d_4 - c_4 d_3) + (s_4 - s_3)(c_1 d_2 - c_2 d_1) = 0,$$

which may be proved by the method of Example 22.12.

Example 22.22 (Forsyth [215])   Shew that

$$(s_4^2 - s_3^2)(c_1^2 d_2^2 - c_2^2 d_1^2) = (s_2^2 - s_1^2)(c_3^2 d_4^2 - c_4^2 d_3^2),$$

by expressing each side in terms of s_1, s_2, s_3, s_4; and deduce from Example 22.21 that, if $u_1 + u_2 + u_3 + u_4 = 0$, then

$$s_4 c_1 d_2 + s_3 c_2 d_1 + s_2 c_3 d_4 + s_1 c_4 d_3 = 0,$$
$$s_4 c_2 d_1 + s_3 c_1 d_2 + s_2 c_4 d_3 + s_1 c_3 d_4 = 0.$$

Example 22.23 (Gudermann [262])   Deduce from Jacobi's fundamental theta-function formulae that, if $u_1 + u_2 + u_3 + u_4 = 0$, then

$$k'^2 - k^2 k'^2 s_1 s_2 s_3 s_4 + k^2 c_1 c_2 c_3 c_4 - d_1 d_2 d_3 d_4 = 0. \tag{22.4}$$

Example 22.24 (H. J. S. Smith [595])   Deduce from Jacobi's fundamental theta-function formulae that, if $u_1 + u_2 + u_3 + u_4 = 0$, then

$$k'^2(s_1 s_2 c_3 c_4 - c_1 c_2 s_3 s_4) - d_1 d_2 + d_3 d_4 = 0,$$
$$k'^2(s_1 s_2 - s_3 s_4) + d_1 d_2 c_3 c_4 - c_1 c_2 d_3 d_4 = 0,$$
$$s_1 s_2 d_3 d_4 - d_1 d_2 s_3 s_4 + c_3 c_4 - c_1 c_2 = 0.$$

Example 22.25 (Math. Trip. 1905)   If $u_1 + u_2 + u_3 + u_4 = 0$, shew that the cross-ratio of $\operatorname{sn} u_1$, $\operatorname{sn} u_2$, $\operatorname{sn} u_3$, $\operatorname{sn} u_4$ is equal to the cross-ratio of $\operatorname{sn}(u_1 + K)$, $\operatorname{sn}(u_2 + K)$, $\operatorname{sn}(u_3 + K)$, $\operatorname{sn}(u_4 + K)$.

Example 22.26 (Math. Trip. 1913)   Shew that

$$\begin{vmatrix} \operatorname{sn}^2(u+v) & \operatorname{sn}(u+v)\operatorname{sn}(u-v) & \operatorname{sn}^2(u-v) \\ \operatorname{cn}^2(u+v) & \operatorname{cn}(u+v)\operatorname{cn}(u-v) & \operatorname{cn}^2(u-v) \\ \operatorname{dn}^2(u+v) & \operatorname{dn}(u+v)\operatorname{dn}(u-v) & \operatorname{dn}^2(u-v) \end{vmatrix} = -\frac{8k'^2 s_1 s_2^3 c_1 c_2 d_1 d_2}{(1 - k^2 s_1^2 s_2^2)^3}.$$

Example 22.27 (Math. Trip. 1901)   Find all systems of values of u and v for which $\operatorname{sn}^2(u+iv)$ is real when u and v are real and $0 < k^2 < 1$.

Example 22.28 (Math. Trip. 1902)   If $k' = \frac{1}{4}(a^{-1} - a)^2$, where $0 < a < 1$, shew that

$$\operatorname{sn}^2\left(\frac{K}{4}\right) = \frac{4a^3}{(1 + a^2)(1 + 2a - a^2)},$$

and that $\operatorname{sn}^2\left(\frac{3K}{4}\right)$ is obtained by writing $-a^{-1}$ for a in this expression.

Example 22.29 (Math. Trip. 1899)   If the values of $\operatorname{cn} z$, which are such that $\operatorname{cn}(3z) = a$, are $c_1, c_2, \ldots, c_9$, shew that

$$3k^4 \prod_{r=1}^{9} c_r + k'^4 \sum_{r=1}^{9} c_r = 0.$$

Example 22.30 (King's, 1900)   If

$$\frac{a + \operatorname{sn}(u+v)}{a + \operatorname{sn}(u-v)} = \frac{b + \operatorname{cn}(u+v)}{b + \operatorname{cn}(u-v)} = \frac{c + \operatorname{dn}(u+v)}{c + \operatorname{dn}(u-v)},$$

and if none of $\operatorname{sn} v$, $\operatorname{cn} u$, $\operatorname{dn} u$, $1 - k^2 \operatorname{sn}^2 u \operatorname{sn}^2 v$ vanishes, shew that u is given by the equation

$$k^2(k'^2 a^2 + b^2 - c^2)\operatorname{sn}^2 u = k'^2 + k^2 b^2 - c^2.$$

Example 22.31 (Math. Trip. 1912)   Shew that

$$1 - \operatorname{sn}\left(\frac{2Kx}{\pi}\right) = (1 - \sin x)\prod_{n=1}^{\infty}\left\{\frac{(1 - q^{2n-1})^2}{(1 + q^{2n})^2}\frac{(1 - 2q^n \sin x + q^{2n})^2}{(1 - 2q^{2n-1}\cos 2x + q^{4n-2})}\right\}.$$

Example 22.32 (Math. Trip. 1904)   Shew that

$$\frac{1 - \operatorname{sn}\left(\frac{2Kx}{\pi}\right)}{\left\{\operatorname{dn}\left(\frac{2Kx}{\pi}\right) - k'\operatorname{sn}\left(\frac{2Kx}{\pi}\right)\right\}^{\frac{1}{2}}} = \prod_{n=1}^{\infty}\left\{\frac{1 - 2q^{2n-1}\sin x + q^{4n-2}}{1 + 2q^{2n-1}\sin x + q^{4n-2}}\right\}.$$

Example 22.33 (Trinity, 1904)   Shew that if k be so small that k^4 may be neglected, then

$$\operatorname{sn} u = \sin u - \tfrac{1}{4}k^2 \cos u \cdot (u - \sin u \cos u),$$

for small values of u.

Example 22.34 (Math. Trip. 1907)   Shew that, if $|\operatorname{Im} x| < \frac{\pi}{2} \operatorname{Im} \tau$, then

$$\log \operatorname{cn}\left(\frac{2Kx}{\pi}\right) = \log \cos x - \sum_{n=1}^{\infty} \frac{4q^n \sin^2 nx}{n\{1 + (-q)^n\}}.$$

Hint. Integrate the Fourier series for $\operatorname{sn}\left(\frac{2Kx}{\pi}\right) \operatorname{dc}\left(\frac{2Kx}{\pi}\right)$.

Example 22.35 (Math. Trip. 1906)   Shew that

$$\int_0^{K/4} \frac{1 - k^2 \operatorname{sn}^4 u}{\operatorname{cn}^2 u \, \operatorname{dn}^2 u} \, du = \left\{(1 + k')^{\frac{1}{2}} - 1\right\} / k'^{\frac{3}{4}}.$$

Hint. Express the integrand in terms of functions of $2u$.

Example 22.36 (Math. Trip. 1912)   Shew that

$$\int \frac{\operatorname{cn} v \, du}{\operatorname{sn} v - \operatorname{sn} u} = \log \frac{\vartheta_1(\frac{1}{2}x + \frac{1}{2}y - \frac{1}{2}\pi)\vartheta_1(\frac{1}{2}x + \frac{1}{2}y - \frac{1}{2}\pi - \frac{1}{2}\pi\tau)}{\vartheta_1(\frac{1}{2}x - \frac{1}{2}y)\vartheta_1(\frac{1}{2}x - \frac{1}{2}y - \frac{1}{2}\pi\tau)} - x\frac{\vartheta_1'(y + \frac{1}{2}\pi\tau)}{\vartheta_1(y + \frac{1}{2}\pi\tau)},$$

where $2Kx = \pi u$, $2Ky = \pi v$.

Example 22.37 (Math. Trip. 1903)   Shew that

$$(1 + k')k'^2 \int_0^K \frac{\operatorname{sn}^3 u \, du}{(1 + \operatorname{cu} u) \, \operatorname{dn}^2 u} = 1.$$

Example 22.38 (St John's, 1914)   Shew that

$$k \int_{\alpha-\beta}^{\alpha+\beta} \operatorname{sn} u \, du = \log \frac{1 + k \operatorname{sn} \alpha \operatorname{sn} \beta}{1 - k \operatorname{sn} \alpha \operatorname{sn} \beta}.$$

Example 22.39 (Math. Trip. 1902)   By integrating $\int e^{2iz} \operatorname{dn} u \operatorname{cs} u \, dz$ round a rectangle whose corners are $\pm\frac{1}{2}\pi$, $\pm\frac{1}{2}\pi + \infty i$ (where $2Kz = \pi u$) and then integrating by parts, shew that, if $0 < k^2 < 1$, then

$$\int_0^K \cos\left(\frac{\pi u}{K}\right) \log \operatorname{sn} u \, du = \frac{K}{2} \tanh\left(\frac{1}{2}\pi i\tau\right).$$

Example 22.40   Shew that K and K' satisfy the equation

$$c(1 - c)\frac{d^2u}{dc^2} + (1 - 2c)\frac{du}{dc} - \frac{1}{4}u = 0,$$

where $c = k^2$; and deduce that they satisfy Legendre's equation for functions of degree $-\frac{1}{2}$ with argument $1 - 2k^2$.

Example 22.41   Express the coordinates of any point on the curve $x^3 + y^3 = 1$ in the form

$$x = \frac{2 \cdot 3^{\frac{1}{4}} \operatorname{sn} u \operatorname{dn} u - (1 - \operatorname{cn} u)^2}{2 \cdot 3^{\frac{1}{4}} \operatorname{sn} u \operatorname{dn} u + (1 - \operatorname{cn} u)^2}, \qquad y = \frac{2^{\frac{11}{6}} \cos \frac{\pi}{12}(1 - \operatorname{cn} u)\{1 + \tan \frac{\pi}{12} \operatorname{cn} u\}}{2 \cdot 3^{\frac{1}{4}} \operatorname{sn} u \operatorname{dn} u + (1 - \operatorname{cn} u)^2},$$

the modulus of the elliptic functions being $\sin \frac{\pi}{12}$; and shew that

$$\int_x^1 (1 - x^3)^{-\frac{2}{3}} \, dx = \int_0^y (1 - y^3)^{-\frac{2}{3}} \, dy = 2^{-\frac{8}{3}} \cdot 3^{\frac{1}{4}} u.$$

Shew further that the sum of the parameters of three collinear points on the cubic is a period. See Richelot [554] and Cayley [138]. A uniformising variable for the general cubic in the canonical form

$$X^3 + Y^3 + Z^3 + 6mXYZ = 0$$

has been obtained by Bobek [77, p. 251]. Dixon [181] has developed the theory of elliptic functions by taking the equivalent curve $x^3 + y^3 - 3axy = 1$ as fundamental, instead of the curve $y^2 = (1 - x^2)(1 - k^2x^2)$.

Example 22.42 (Math. Trip. 1911) Express $\int_0^2 \{(2x - x^2)(4x^2 + 9)\}^{-\frac{1}{2}} dx$ in terms of a complete elliptic integral of the first kind with a real modulus.

Example 22.43 (Math. Trip. 1899) If $u = \int_x^\infty \{(t + 1)(t^2 + t + 1)\}^{-\frac{1}{2}} dt$, express x in terms of Jacobian elliptic functions of u with a real modulus.

Example 22.44 (Math. Trip. 1914) If $u = \int_0^\infty (1 + t^2 - 2t^4)^{-\frac{1}{2}} dt$, express x in terms of u by means of either Jacobian or Weierstrassian elliptic functions.

Example 22.45 (Trinity, 1881) Shew that

$$e^{-\pi} + e^{-9\pi} + e^{-25\pi} + \cdots = \frac{(2^{\frac{1}{4}} - 1)\Gamma(\frac{1}{4})}{2^{\frac{11}{4}}\pi^{\frac{3}{4}}}.$$

Example 22.46 (Gauss [238]; Math. Trip. 1895) When $a > x > \beta > \gamma$, reduce the integrals

$$\int_x^a \{(a - t)(t - \beta)(t - \gamma)\}^{-\frac{1}{2}} dt, \qquad \int_\beta^x \{(a - t)(t - \beta)(t - \gamma)\}^{-\frac{1}{2}} dt$$

by the substitutions

$$x - \gamma = (a - \gamma) \operatorname{dn}^2 u, \qquad x - \gamma = (\beta - \gamma) \operatorname{nd}^2 v$$

respectively, where $k^2 = (\alpha - \beta)/(a - \gamma)$. Deduce that, if $u + v = K$, then

$$1 - \operatorname{sn}^2 u - \operatorname{sn}^2 v + k^2 \operatorname{sn}^2 u \operatorname{sn}^2 v = 0.$$

By the substitution $y = (a - t)(t - \beta)/(t - \gamma)$ applied to the above integral taken between the limits β and a, obtain the Gaussian form of Landen's transformation,

$$\int_0^{\pi/2} (a_1^2 \cos^2 \theta + b_1^2 \sin^2 \theta)^{-\frac{1}{2}} d\theta = \int_0^{\pi/2} (a^2 \cos^2 \theta + b^2 \sin^2 \theta)^{-\frac{1}{2}} d\theta,$$

where a_1, b_1 are the arithmetic and geometric means between a and b.

Example 22.47 (Math. Trip. 1903) Shew that

$$\operatorname{sc} u = -k'^{-1}\{\zeta(u - K) - \zeta(u - K - 2iK') - \zeta(2iK')\},$$

where the zeta-functions are formed with periods $2\omega_1, 2\omega_2 = 2K, 4iK'$.

Example 22.48 (Math. Trip. 1911)   Shew that $E - k'^2 K$ satisfies the equation

$$4cc' \frac{d^2 u}{dc^2} = u,$$

where $c = k^2$, and obtain the primitive of this equation.

Example 22.49 (Trinity, 1906)   Shew that

$$n \int_0^1 k^n K' \, dk = (n-1) \int_0^1 k^{n-2} E' \, dk,$$

$$(n+2) \int_0^1 k^n E' \, dk = (n+1) \int_0^1 k^n K' \, dk.$$

Example 22.50 (Trinity, 1896)   If $u = \frac{1}{2} \int_0^x \{t(1-t)(1-ct)\}^{-\frac{1}{2}} \, dt$, shew that

$$c(c-1) \frac{d^2 u}{dc^2} + (2c-1) \frac{du}{dc} + \frac{1}{4} u = \frac{1}{4} \left\{ \frac{x(1-x)}{(1-cx)^3} \right\}^{\frac{1}{2}}.$$

Example 22.51 (Math. Trip. 1906)   Shew that the primitive of

$$\frac{du}{dk} + \frac{u^2}{k} + \frac{k}{1-k^2} = 0$$

is $u = \dfrac{A(E-K) + A'E'}{AE + A'(E'-K')}$, where A, A' are constants.

Example 22.52 (Math. Trip. 1910)   Deduce from the addition-formula for $E(u)$ that, if

$$u_1 + u_2 + u_3 + u_4 = 0,$$

then $(\operatorname{sn} u_1 \operatorname{sn} u_2 - \operatorname{sn} u_3 \operatorname{sn} u_4) \operatorname{sn}(u_1 + u_2)$ is unaltered by any permutation of suffices.

Example 22.53 (Math. Trip. 1913)   Shew that

$$E(3u) - 3E(u) = \frac{-8k^2 s^3 c^3 d^3}{1 - 6k^2 s^4 + 4(k^2 + k^4)s^6 - 3k^4 s^8}.$$

Example 22.54 (Math. Trip. 1904)   Shew that

$$3k^4 \int_0^{2k} u \operatorname{cd}^4 u \, du = 2K \left\{ (2+k^2)K - 2(1+k^2) E \right\}.$$

Hint. Write $u = K + v$.

Example 22.55 (Math. Trip. 1908)   By considering the curves $y^2 = x(1-x)(1-k^2 x)$, $y = l + mx + nx^2$, shew that, if $u_1 + u_2 + u_3 + u_4 = 0$, then

$$E(u_1) + E(u_2) + E(u_3) + E(u_4) = k \left\{ \sum_{r=1}^4 sr^2 + 2c_1 c_2 c_3 c_4 - 2s_1 s_2 s_3 s_4 - 2 \right\}^{\frac{1}{2}}.$$

Example 22.56 (Forsyth [216])   By the method of Example 22.21, obtain the following seven expressions for $E(u_1) + E(u_2) + E(u_3) + E(u_4)$ when $u_1 + u_2 + u_3 + u_4 = 0$:

$$\frac{-k^2 s_1 s_2 s_3 s_4}{1 + k^2 s_1 s_2 s_3 s_4} \sum_{r=1}^{4} \frac{c_r d_r}{s_r}, \qquad \frac{k^2 d_1 d_2 d_3 d_4}{k^2 + d_1 d_2 d_3 d_4} \sum_{r=1}^{4} \frac{s_r c_r}{d_r},$$

$$\frac{k^2 c_1 c_2 c_3 c_4}{k^2 c_1 c_2 c_3 c_4 - k'^2} \sum_{r=1}^{4} \frac{s_r d_r}{c_r}, \qquad \frac{k^2 s_1 s_2 s_3 s_4 d_1 d_2 d_3 d_4}{k^2 k'^2 s_1 s_2 s_3 s_4 - d_1 d_2 d_3 d_4} \sum_{r=1}^{4} \frac{c_r}{s_r d_r},$$

$$\frac{-k^2 c_1 c_2 c_3 c_4 d_1 d_2 d_3 d_4}{d_1 d_2 d_3 d_4 + k^2 c_1 c_2 c_3 c_4} \sum_{r=1}^{4} \frac{s_r}{c_r d_r}, \qquad \frac{k^2 s_1 s_2 s_3 s_4 + c_1 c_2 c_3 c_4}{c_1 c_2 c_3 c_4 + k'^2 s_1 s_2 s_3 s_4} \sum_{r=1}^{4} \frac{d_r}{s_r c_r},$$

$$- k^2 \left\{ (s_1 s_2 s_3 s_4)^{-1} + (c_1 c_2 c_3 c_4)^{-1} + k^4 (d_1 d_2 d_3 d_4)^{-1} \right\}^{-1} \sum_{r=1}^{4} \frac{1}{s_r c_r d_r}.$$

Example 22.57 (Jacobi)   Shew that

$$\left(\frac{2k}{\pi} \right)^2 \mathrm{ns}^2 \left(\frac{2Kx}{\pi} \right) = \mathrm{cosec}^2 x + \frac{4K(K - E)}{\pi^2} - 8 \sum_{n=1}^{\infty} \frac{n q^{2n} \cos 2nx}{1 - q^{2n}},$$

when $| \operatorname{Im} x | < \pi \operatorname{Im} \tau$; and, by differentiation, deduce that

$$6 \left(\frac{2K}{\pi} \right)^4 \mathrm{ns}^4 \left(\frac{2Kx}{\pi} \right) = 6 \, \mathrm{cosec}^4 x + 4 \left[(1 + k^2) \left(\frac{2K}{\pi} \right)^2 - 1 \right] \mathrm{cosec}^2 x$$

$$+ 64(1 + k^2) \frac{K^3 (K - E)}{\pi^4} - 2k^2 \left(\frac{2K}{\pi} \right)^4$$

$$- 32 \sum_{n=1}^{\infty} n \left[(1 + k^2) \left(\frac{2K}{\pi} \right)^2 - n^2 \right] \frac{q^{2n} \cos 2nx}{1 - q^{2n}}.$$

Shew also that, when $| \operatorname{Im} x | < \frac{\pi}{2} \operatorname{Im} \tau$,

$$\mathrm{sn}^3 \left(\frac{2Kx}{\pi} \right) = \sum_{n=0}^{\infty} \left\{ \frac{1 + k^2}{2k^3} - \frac{(2n + 1)^2}{2k^3} \left(\frac{\pi}{2K} \right)^2 \right\} \frac{2\pi q^{n + \frac{1}{2}} \sin(2n + 1)x}{k(1 - q^{2n+1})}.$$

Example 22.58 (Ramanujan [548])   Shew that, if a be the semi-major axis of an ellipse whose eccentricity is $\sin \frac{\pi}{12}$, the perimeter of the ellipse is

$$a \left(\frac{\pi}{\sqrt{3}} \right)^{\frac{1}{2}} \left\{ \left(1 + \frac{1}{\sqrt{3}} \right) \frac{\Gamma(\frac{1}{3})}{\Gamma(\frac{5}{6})} + \frac{2\Gamma(\frac{5}{6})}{\Gamma(\frac{1}{3})} \right\}.$$

Example 22.59 (Trinity, 1882)   Deduce from Example 21.19 of Chapter 21 that

$$k^2 \mathrm{cn}^3 (2u) = \frac{-k'^2 + \mathrm{dn}^3 u \, \mathrm{dn}(3u)}{1 + k^2 \mathrm{sn}^3 u \, \mathrm{sn}(3u)}, \qquad \mathrm{dn}^3 (2u) = \frac{k'^2 + k^2 \mathrm{cn}^3 u \, \mathrm{cn}(3u)}{1 + k^2 \mathrm{sn}^3 u \, \mathrm{sn}(3u)}.$$

Example 22.60   From the formula $\mathrm{sd}(iu, k) = i \, \mathrm{sd}(u, k')$ deduce that

$$\frac{1}{K} \sum_{n=0}^{\infty} \frac{(-1)^n q^{n + \frac{1}{2}}}{1 + q^{2n+1}} \sinh \left((n + \tfrac{1}{2}) \frac{\pi u}{K} \right) = \frac{1}{K'} \sum_{n=0}^{\infty} \frac{(-1)^n q_1^{n + \frac{1}{2}}}{1 + q_1^{2n+1}} \sin \left((n + \tfrac{1}{2}) \frac{\pi u}{K'} \right),$$

where $q = \exp(-\pi K'/K)$, $q_1 = \exp(-\pi K/K')$, and u lies inside the parallelogram whose vertices are $\pm iK \pm K'$. By integrating from u to K', from 0 to u and again from u to K', prove that

$$\frac{\pi^3}{64}(K'^2 - K^2 - u^2) + K^2 \sum_{n=0}^{\infty} \frac{(-1)^n q^{n+\frac{1}{2}}}{(2n+1)^3(1+q^{2n+1})} \cosh\left((n+\tfrac{1}{2})\frac{\pi u}{K}\right)$$

$$= K'2 \sum_{n=0}^{\infty} \frac{(-)^n q_1^{n+\frac{1}{2}}}{(2n+1)^3(1+q_1^{2n+1})} \cos\left((n+\tfrac{1}{2})\frac{\pi u}{K'}\right).$$

A formula which may be derived from this by writing $u = \xi + i\eta$, where ξ and η are real, and equating imaginary parts on either side of the equation was obtained by Thompson and Tait [626, p. 249], but they failed to observe that their formula was nothing but a consequence of Jacobi's imaginary transformation. The formula was suggested to Thompson and Tait by the solution of a problem in the theory of Elasticity.

23

Ellipsoidal Harmonics and Lamé's Equation

23.1 The definition of ellipsoidal harmonics

It has been seen earlier in this work (§18.4) that solutions of Laplace's equation, which are analytic near the origin and which are appropriate for the discussion of physical problems connected with a sphere, may be conveniently expressed as linear combinations of functions of the type

$$r^n P_n(\cos\theta), \qquad r^n P_n^m(\cos\theta) \, \begin{matrix} \cos \\ \sin \end{matrix} \, m\phi,$$

where n and m are positive integers (zero included).

When $P_n(\cos\theta)$ is resolved into a product of factors which are linear in $\cos^2\theta$ (multiplied by $\cos\theta$ when n is odd), we see that, if $\cos\theta$ is replaced by z/r, then the zonal harmonic $r^n P_n(\cos\theta)$ is expressible as a product of factors which are linear in x^2, y^2 and z^2, the whole being multiplied by z when n is odd. The tesseral harmonics are similarly resoluble into factors which are linear in x^2, y^2 and z^2 multiplied by one of the eight products $1, x, y, z, yz, zx, xy, xyz$.

The surfaces on which any given zonal or tesseral harmonic vanishes are surfaces on which either θ or ϕ has some constant value, so that they are circular cones or planes, the coordinate planes being included in certain cases.

When we deal with physical problems connected with ellipsoids, the structure of spheres, cones and planes associated with polar coordinates is replaced by a structure of confocal quadrics. The property of spherical harmonics which has just been explained suggests the construction of a set of harmonics which shall vanish on certain members of the confocal system.

Such harmonics are known as *ellipsoidal* harmonics; they were studied by Lamé [402, 403] in the early part of the nineteenth century by means of confocal coordinates. The expressions for ellipsoidal harmonics in terms of Cartesian coordinates were obtained many years later by W. D. Niven [504], and the following account of their construction is based on his researches.

The fundamental ellipsoid is taken to be

$$\frac{x^2}{a^2} + \frac{y^2}{b^2} + \frac{z^2}{c^2} = 1,$$

and any confocal quadric is

$$\frac{x^2}{a^2 + \theta} + \frac{y^2}{b^2 + \theta} + \frac{z^2}{c^2 + \theta} = 1,$$

where θ is a constant. It will be necessary to consider sets of such quadrics, and it conduces

567

to brevity to write

$$\frac{x^2}{a^2 + \theta_p} + \frac{y^2}{b^2 + \theta_p} + \frac{z^2}{c^2 + \theta_p} - 1 \equiv \Theta_p, \qquad \frac{x^2}{a^2 + \theta_p} + \frac{y^2}{b^2 + \theta_p} + \frac{z^2}{c^2 + \theta_p} \equiv K_p.$$

The equation of any member of the set is then $\Theta_p = 0$. The analysis is made more definite by taking the x-axis as the longest axis of the fundamental ellipsoid and the z-axis as the shortest, so that $a > b > c$.

23.2 The four species of ellipsoidal harmonics

A consideration of the expressions for spherical harmonics in factors indicates that there are four possible species of ellipsoidal harmonics to be investigated. These are included in the scheme

$$\left\{ 1, \quad \begin{matrix} x, & yz, \\ y, & zx, & xyz \\ z, & xy, \end{matrix} \right\} \Theta_1 \Theta_2 \cdots \Theta_m,$$

where one or other of the expressions in { } is to multiply the product $\Theta_1 \Theta_2 \cdots \Theta_m$.

If we write for brevity

$$\Theta_1 \Theta_2 \cdots \Theta_m = \Pi(\Theta),$$

any harmonic of the form $\Pi(\Theta)$ will be called *an ellipsoidal harmonic of the first species*. A harmonic of any of the three forms[1] $x\Pi(\Theta)$, $y\Pi(\Theta)$, $z\Pi(\Theta)$ will be called *an ellipsoidal harmonic of the second species*. A harmonic of any of the three forms $yz\Pi(\Theta)$, $zx\Pi(\Theta)$, $xy\Pi(\Theta)$ will be called *an ellipsoidal harmonic of the third species*. And a harmonic of the form $xyz\Pi(\Theta)$ will be called *an ellipsoidal harmonic of the fourth species*.

The terms of highest degree in these species of harmonics are of degrees $2m$, $2m+1$, $2m+2$, $2m + 3$ respectively. It will appear subsequently (§23.26) that $2n + 1$ linearly independent harmonics of degree n can be constructed, and hence that the terms of degree n in these harmonics form a fundamental system (§18.3) of harmonics of degree n.

We now proceed to explain in detail how to construct harmonics of the first species and to give a general account of the construction of harmonics of the other three species. The reader should have no difficulty in filling up the lacunae in this account with the aid of the corresponding analysis given in the case of functions of the first species.

23.21 The construction of ellipsoidal harmonics of the first species

As a simple case let us first consider the harmonics of the first species which are of the second degree. Such a harmonic must be simply of the form Θ_1.

Now the effect of applying Laplace's operator, namely

$$\frac{\partial^2}{\partial x^2} + \frac{\partial^2}{\partial y^2} + \frac{\partial^2}{\partial z^2}$$

---

[1] The three forms will be distinguished by being described as different *types* of the species.

to

$$\frac{x^2}{a^2 + \theta_1} + \frac{y^2}{b^2 + \theta_1} + \frac{z^2}{c^2 + \theta_1} - 1$$

is

$$\frac{2}{a^2 + \theta_1} + \frac{2}{b^2 + \theta_1} + \frac{2}{c^2 + \theta_1},$$

and so Θ_1 is a harmonic if θ_1 is a root of the quadratic equation

$$(\theta + b^2)(\theta + c^2) + (\theta + c^2)(\theta + a^2) + (\theta + a^2)(\theta + b^2) = 0.$$

This quadratic has one root between $-c^2$ and $-b^2$ and another between $-b^2$ and $-a^2$. Its roots are therefore unequal, and, by giving θ_1 the value of each root in turn, we obtain two[2] ellipsoidal harmonics of the first species of the second degree.

Next consider the general product $\Theta_1 \Theta_2 \cdots \Theta_m$; this product will be denoted by $\Pi(\Theta)$ and it will be supposed that it has no repeated factors – a supposition which will be justified later (§23.43). If we temporarily regard $\Theta_1, \Theta_2, \ldots, \Theta_m$ as a set of auxiliary variables, the ordinary formula of partial differentiation gives

$$\frac{\partial \Pi(\Theta)}{\partial x} = \sum_{p=1}^{m} \frac{\partial \Pi(\Theta)}{\partial \Theta_p} \frac{\partial \Theta_p}{\partial x} = \sum_{p=1}^{m} \frac{\partial \Pi(\Theta)}{\partial \Theta_p} \cdot \frac{2x}{a^2 + \theta_p},$$

and, if we differentiate again,

$$\frac{\partial^2 \Pi(\Theta)}{\partial x^2} = \sum_{p=1}^{m} \frac{\partial \Pi(\Theta)}{\partial \Theta_p} \cdot \frac{2}{a^2 + \theta_p} + \sum_{p \neq q} \frac{\partial^2 \Pi(\Theta)}{\partial \Theta_p \partial \Theta_q} \cdot \frac{8x^2}{(a^2 + \theta_p)(a^2 + \theta_q)},$$

where the last summation extends over all *unequal* pairs of the integers $1, 2, \ldots, m$. The terms for which $p = q$ may be omitted because none of the expressions $\Theta_1, \Theta_2, \ldots, \Theta_m$ enters into $\Pi(\Theta)$ to a degree higher than the first.

It follows that the result of applying Laplace's operator to $\Pi(\Theta)$ is

$$\sum_{p=1}^{m} \frac{\partial \Pi(\Theta)}{\partial \Theta_p} \left\{ \frac{2}{a^2 + \theta_p} + \frac{2}{b^2 + \theta_p} + \frac{2}{c^2 + \theta_p} \right\}$$

$$+ \sum_{p \neq q} \frac{\partial^2 \Pi(\Theta)}{\partial \Theta_p \partial \Theta_q} \left\{ \frac{8x^2}{(a^2 + \theta_p)(a^2 + \theta_q)} + \frac{8y^2}{(b^2 + \theta_p)(b^2 + \theta_q)} + \frac{8z^2}{(c^2 + \theta_p)(c^2 + \theta_q)} \right\}.$$

Now

$$\sum_{\left(\substack{x, \, y, \, z \\ a, \, b, \, c}\right)} \frac{x^2}{(a^2 + \theta_p)(a^2 + \theta_q)} = \frac{\Theta_p - \Theta_q}{\theta_q - \theta_p}$$

and $\partial \Pi(\Theta)/\partial \Theta_p$ consists of the product $\Pi(\Theta)$ with the factor Θ_p omitted, while $\partial^2 \Pi(\Theta)/\partial \Theta_p \partial \Theta_q$ consists of the product $\Pi(\Theta)$ with the factors Θ_p and Θ_q omitted. That is to say

$$\Theta_p \frac{\partial^2 \Pi(\Theta)}{\partial \Theta_p \partial \Theta_q} = \frac{\partial \Pi(\Theta)}{\partial \Theta_q}, \qquad \Theta_q \frac{\partial^2 \Pi(\Theta)}{\partial \Theta_p \partial \Theta_q} = \frac{\partial \Pi(\Theta)}{\partial \Theta}.$$

[2] The complete set of 5 ellipsoidal harmonics of the second degree is composed of these two together with the three harmonics yz, zx, xy, which are of the third species.

If we make these substitutions, we see that

$$\left[\frac{\partial^2}{\partial x^2} + \frac{\partial^2}{\partial y^2} + \frac{\partial^2}{\partial z^2} \right] \Pi(\Theta)$$

may be written in the form

$$\sum_{p=1}^{m} \frac{\partial \Pi(\Theta)}{\partial \Theta_p} \left\{ \frac{2}{a^2 + \theta_p} + \frac{2}{b^2 + \theta_p} + \frac{2}{c^2 + \theta_p} + \sum_{q=1}^{m}{}' \frac{8}{\theta_p - \theta_q} \right\},$$

the prime indicating that the term for which $q = p$ has to be omitted from the summation.

If $\Pi(\Theta)$ is to be a harmonic it is annihilated by Laplace's operator; and it will certainly be so annihilated if it is possible to choose $\theta_1, \theta_2, \ldots, \theta_m$ so that each of the equations

$$\frac{1}{a^2 + \theta_p} + \frac{1}{b^2 + \theta_p} + \frac{1}{c^2 + \theta_p} + \sum{}' \frac{4}{\theta_p - \theta_q} = 0$$

is satisfied, where p takes the values $1, 2, \ldots, m$.

Now let θ be a variable and let $\Lambda_1(\theta)$ denote the polynomial of degree m in θ

$$\prod_{q=1}^{m} (\theta - \theta_q).$$

If $\Lambda_1'(\theta)$ denotes $d\Lambda_1(\theta)/d\theta$; then, by direct differentiation, it is seen that $\Lambda_1'(\theta)$ is equal to the sum of all products of $\theta - \theta_1, \theta - \theta_2, \ldots, \theta - \theta_m, m - 1$ at a time, and $\Lambda_1''(\theta)$ is twice the sum of all products of the same expressions, $m - 2$ at a time. Hence, if θ be given the special value θ_p, the quotient $\Lambda_1''(\theta_p)/\Lambda_1'(\theta_p)$ becomes equal to twice the sum of the reciprocals of $\theta_p - \theta_1, \theta_p - \theta_2, \ldots, \theta_p - \theta_m$ (the expression $\theta_p - \theta_p$ being omitted).

Consequently the set of equations derived from the hypothesis that $\prod_{p=1}^{m} \Theta_p$ is a harmonic shews that the expression

$$\frac{1}{a^2 + \theta} + \frac{1}{b^2 + \theta} + \frac{1}{c^2 + \theta} + \frac{2\Lambda_1''(\theta)}{\Lambda_1'(\theta)}$$

vanishes whenever θ has any of the special values $\theta_1, \theta_2, \ldots, \theta_m$.

Hence the expression

$$(a^2 + \theta)(b^2 + \theta)(c^2 + \theta)\Lambda_1''(\theta) + \frac{1}{2} \left\{ \sum_{a,b,c} (b^2 + \theta)(c^2 + \theta) \right\} \Lambda_1'(\theta)$$

is a polynomial in θ which vanishes when θ has any of the values $\theta_1, \theta_2, \ldots, \theta_m$, and so it has $\theta - \theta_1, \theta - \theta_2, \ldots, \theta - \theta_m$ as factors. Now this polynomial is of degree $m + 1$ in θ and the coefficient of θ^{m+1} is $m(m + \frac{1}{2})$. Since m of the factors are known, the remaining factor must be of the form

$$m \left(m + \tfrac{1}{2} \right) \theta + \tfrac{1}{4}C,$$

where C is a constant which will be determined subsequently.

We have therefore shewn that

$$(a^2 + \theta)(b^2 + \theta)(c^2 + \theta)\Lambda_1''(\theta) + \frac{1}{2}\left\{\sum_{a,b,c}(b^2 + \theta)(c^2 + \theta)\right\}\Lambda_1'(\theta)$$

$$= \left\{m\left(m + \tfrac{1}{2}\right)\theta + \tfrac{1}{4}C\right\}\Lambda_1(\theta).$$

That is to say, any ellipsoidal harmonic of the first species of (even) degree n is expressible in the form

$$\prod_{p=1}^{n/2}\left\{\frac{x^2}{a^2 + \theta_p} + \frac{y^2}{b^2 + \theta_p} + \frac{z^2}{c^2 + \theta_p} - 1\right\}$$

where $\theta_1, \theta_2, \ldots, \theta_{\frac{1}{2}n}$ are the zeros of a polynomial $\Lambda_1(\theta)$ of degree $\frac{1}{2}n$; and this polynomial must be a solution of a differential equation of the type

$$4\sqrt{(a^2 + \theta)(b^2 + \theta)(c^2 + \theta)}\frac{d}{d\theta}\left[\sqrt{(a^2 + \theta)(b^2 + \theta)(c^2 + \theta)}\frac{d\Lambda_1(\theta)}{d\theta}\right]$$

$$= \{n(n + 1)\theta + C\}\Lambda_1(\theta).$$

This equation is known as *Lamé's differential equation*. It will be investigated in considerable detail in §23.4, and in the course of the investigation it will be shewn that: (I) there are precisely $\frac{1}{2}n + 1$ different real values of C for which the equation has a solution which is a polynomial in θ of degree $\frac{1}{2}n$; and (II) these polynomials have no repeated factors.

The analysis of this section may then be reversed step by step to establish the existence of $\frac{1}{2}n + 1$ ellipsoidal harmonics of the first species of (even) degree n, and the elementary theory of the harmonics of the first species will then be complete. The corresponding results for harmonics of the second, third and fourth species will now be indicated briefly, the notation already introduced being adhered to so far as possible.

23.22 *Ellipsoidal harmonics of the second species*

We take $x\prod_{p=1}^{m}\Theta_p$ as a typical harmonic of the second species of degree $2m + 1$. The result of applying Laplace's operator to it is

$$x\left[\sum_{p=1}^{m}\frac{\partial\Pi(\Theta)}{\partial\Theta_p}\left\{\frac{6}{a^2 + \theta_p} + \frac{2}{b^2 + \theta_p} + \frac{2}{c^2 + \theta_p}\right\}\right.$$

$$\left. + \sum_{p\ne q}\frac{\partial^2\Pi(\Theta)}{\partial\Theta_p\partial\Theta_q}\left\{\frac{8x^2}{(a^2 + \theta_p)(a^2 + \theta_q)} + \frac{8y^2}{(b^2 + \theta_p)(b^2 + \theta_q)} + \frac{8z^2}{(c^2 + \theta_p)(c^2 + \theta_q)}\right\}\right],$$

and this has to vanish. Consequently, if

$$\Lambda_2(\theta) \equiv \prod_{q=1}^{m}(\theta - \theta_q),$$

we find, by the reasoning of §23.21, that $\Lambda_2(\theta)$ is a solution of the differential equation

$$(a^2 + \theta)(b^2 + \theta)(c^2 + \theta)\Lambda_2''(\theta)$$
$$+ \tfrac{1}{2}\{3(b^2 + \theta)(c^2 + \theta) + (c^2 + \theta)(a^2 + \theta) + (a^2 + \theta)(b^2 + \theta)\}\Lambda_2'(\theta)$$
$$= \left\{ m\left(m + \frac{3}{2}\right)\theta + \frac{1}{4}C_2 \right\}\Lambda_2(\theta),$$

where C_2 is a constant to be determined.

If now we write $\Lambda_2(\theta) \equiv \Lambda(\theta)/\sqrt{a^2 + \theta}$, we find that $\Lambda(\theta)$ is a solution of the differential equation

$$4\sqrt{(a^2 + \theta)(b^2 + \theta)(c^2 + \theta)}\frac{d}{d\theta}\left[\sqrt{(a^2 + \theta)(b^2 + \theta)(c^2 + \theta)}\frac{d\Lambda(\theta)}{d\theta} \right]$$
$$= \{(2m + 1)(2m + 2)\theta + C\}\Lambda(\theta),$$

where $C = C_2 + b^2 + c^2$. It will be observed that the last differential equation is of the same type as the equation derived in §23.21, the constant n being still equal to the degree of the harmonic, which, in the case now under consideration, is $2m + 1$.

Hence the discussion of harmonics of the second species is reduced to the discussion of solutions of Lamé's differential equation. In the case of harmonics of the first type the solutions are required to be polynomials in θ multiplied by $\sqrt{a^2 + \theta}$; the corresponding factors for harmonics of the second and third types are $\sqrt{b^2 + \theta}$ and $\sqrt{c^2 + \theta}$ respectively. It will be shewn subsequently that precisely $m + 1$ values of C can be associated with each of the three types, so that, in all, $3m + 3$ harmonics of the second species of degree $2m + 1$ are obtained.

23.23 Ellipsoidal harmonics of the third species

We take $yz\prod\limits_{p=1}^{m}(\Theta_p)$ as a typical harmonic of the third species of degree $2m + 2$. The result of applying Laplace's operator to it is

$$yz\left[\sum_{p=1}^{m}\frac{\partial\Pi(\Theta)}{\partial\Theta_p}\left\{ \frac{2}{a^2 + \theta_p} + \frac{6}{b^2 + \theta_p} + \frac{6}{c^2 + \theta_p} \right\} \right.$$
$$\left. + \sum_{p\neq q}\frac{\partial^2\Pi(\Theta)}{\partial\Theta_p\partial\Theta_q}\left\{ \frac{8x^2}{(a^2 + \theta_p)(a^2 + \theta_q)} + \frac{8y^2}{(b^2 + \theta_p)(b^2 + \theta_q)} + \frac{8z^2}{(c^2 + \theta_p)(c^2 + \theta_q)} \right\} \right],$$

and this has to vanish. Consequently, if

$$\Lambda_3(\theta) \equiv \prod_{q=1}^{m}(\theta - \theta_q),$$

we find, by the reasoning of §23.21, that $\Lambda_3(\theta)$ is a solution of the differential equation

$$(a^2 + \theta)(b^2 + \theta)(c^2 + \theta)\Lambda_3''(\theta)$$
$$+ \tfrac{1}{2}\{(b^2 + \theta)(c^2 + \theta) + 3(c^2 + \theta)(a^2 + \theta) + 3(a^2 + \theta)(b^2 + \theta)\}\Lambda_3'(\theta)$$
$$= \left\{ m\left(m + \frac{5}{2}\right)\theta + \frac{1}{4}C_3 \right\}\Lambda_3(\theta),$$

where C_3 is a constant to be determined.

If now we write $\Lambda_3(\theta) \equiv \Lambda(\theta)/\sqrt{(b^2 + \theta)(c^2 + \theta)}$, we find that $\Lambda(\theta)$ is a solution of the differential equation

$$4\sqrt{(a^2 + \theta)(b^2 + \theta)(c^2 + \theta)}\,\frac{d}{d\theta}\left[\sqrt{(a^2 + \theta)(b^2 + \theta)(c^2 + \theta)}\,\frac{d\Lambda(\theta)}{d\theta}\right]$$

$$= \{(2m + 2)(2m + 3)\theta + C\}\Lambda(\theta),$$

where $C = C_3 + 4a^2 + b^2 + c^2$.

It will be observed that the last equation is of the same type as the equation derived in §23.21, the constant n being still equal to the degree of the harmonic, which, in the case now under consideration, is $2m + 2$.

Hence the discussion of harmonics of the third species is reduced to the discussion of solutions of Lamé's differential equation. In the case of harmonics of the first type, the solutions are required to be polynomials in θ multiplied by $\sqrt{(b^2 + \theta)(c^2 + \theta)}$; the corresponding factors for harmonics of the second and third types are $\sqrt{(c^2 + \theta)(a^2 + \theta)}$ and $\sqrt{(a^2 + \theta)(b^2 + \theta)}$ respectively. It will be shewn subsequently that precisely $m + 1$ values of C can be associated with each of the three types, so that, in all, $3m + 3$ harmonics of the third species of degree $2m + 2$ are obtained.

23.24 Ellipsoidal harmonics of the fourth species

The harmonic of the fourth species of degree $2m + 3$ is expressible in the form $xyz \prod_{p=1}^{m} \Theta_p$.

The result of applying Laplace's operator to it is

$$xyz\left[\sum_{p=1}^{m}\frac{\partial\Pi(\Theta)}{\partial\Theta_p}\left\{\frac{6}{a^2 + \theta_p} + \frac{6}{b^2 + \theta_p} + \frac{6}{c^2 + \theta_p}\right\}\right.$$
$$\left. + \sum_{p\neq q}\frac{\partial^2\Pi(\Theta)}{\partial\Theta_p\partial\Theta_q}\left\{\frac{8x^2}{(a^2 + \theta_p)(a^2 + \theta_q)} + \frac{8y^2}{(b^2 + \theta_p)(b^2 + \theta_q)} + \frac{8z^2}{(c^2 + \theta_p)(c^2 + \theta_q)}\right\}\right],$$

and this has to vanish. Consequently, if $\Lambda_4(\theta) \equiv \prod_{q=1}^{m}(\theta - \theta_q)$, we find by the reasoning of §23.21 that $\Lambda_4(\theta)$ is a solution of the equation

$$(a^2 + \theta)(b^2 + \theta)(c^2 + \theta)\Lambda_4''(\theta) + \frac{3}{2}\left\{\sum_{a,b,c}(b^2 + \theta)(c^2 + \theta)\right\}\Lambda_4'(\theta)$$

$$= \left\{m\left(m + \frac{7}{2}\right)\theta + \frac{1}{4}C_4\right\}\Lambda_4(\theta),$$

where C_4 is a constant to be determined.

If now we write

$$\Lambda_4(\theta) \equiv \Lambda(\theta)/\sqrt{(a^2 + \theta)(b^2 + \theta)(c^2 + \theta)},$$

we find that $\Lambda(\theta)$ is a solution of the differential equation

$$4\sqrt{(a^2+\theta)(b^2+\theta)(c^2+\theta)}\,\frac{d}{d\theta}\left[\sqrt{(a^2+\theta)(b^2+\theta)(c^2+\theta)}\,\frac{d\Lambda(\theta)}{d\theta}\right]$$
$$= \{(2m+3)(2m+4)\theta + C\}\Lambda(\theta),$$

where $C = C_4 + 4(a^2+b^2+c^2)$.

It will be observed that the last equation is of the same type as the equation derived in §23.21, the constant n being still equal to the degree of the harmonic which, in the case now under consideration, is $2m+3$.

Hence the discussion of harmonics of the fourth species is reduced to the discussion of solutions of Lamé's differential equation. The solutions are required to be polynomials in θ multiplied by $\sqrt{(a^2+\theta)(b^2+\theta)(c^2+\theta)}$. It will be shewn subsequently that precisely $m+1$ values of C can be associated with solutions of this type, so that $m+1$ harmonics of the fourth species of degree $2m+3$ are obtained.

23.25 *Niven's expressions for ellipsoidal harmonics in terms of homogeneous harmonics*

If $G_n(x,y,z)$ denotes any of the harmonics of degree n which have just been tentatively constructed, then $G_n(x,y,z)$ consists of a finite number of terms of degrees $n, n-2, n-4, \ldots$ in x, y, z. If $H_n(x,y,z)$ denotes the aggregate of terms of degree n, it follows from the homogeneity of Laplace's operator that $H_n(x,y,z)$ is itself a solution of Laplace's equation, and it may obviously be obtained from $G_n(x,y,z)$ by replacing the factors Θ_p, which occur in the expression of $G_n(x,y,z)$ as a product, by the factors K_p.

It has been shewn by Niven [504, pp. 243–245] that $G_n(x,y,z)$ may be derived from $H_n(x,y,z)$ by applying to the latter function the differential operator

$$1 - \frac{D^2}{2(2n-1)} + \frac{D^4}{2\cdot4\cdot(2n-1)(2n-3)} - \frac{D^6}{2\cdot4\cdot6\cdot(2n-1)(2n-3)(2n-5)} + \cdots,$$

where D^2 stands for

$$a^2\frac{\partial^2}{\partial x^2} + b^2\frac{\partial^2}{\partial y^2} + c^2\frac{\partial^2}{\partial z^2};$$

and terms containing powers of D higher than the nth may be omitted from the operator.

We shall now give a proof of this result for any harmonic of the first species. The proofs for harmonics of the other three species are left to the reader as examples. A proof applicable to functions of all four species has been given by Hobson [312]. In constructing the proof given in the text, several modifications have been made in Niven's proof.

For such harmonics the degree is even and we write

$$G_n(x,y,z) = \prod_{p=1}^{n/2}\Theta_p = \prod_{p=1}^{n/2}(K_p-1) = S_n - S_{n-2} + S_{n-4} - \cdots,$$

where $S_n, S_{n-2}, S_{n-4}, \ldots$ are homogeneous functions of degrees $n, n-2, n-4, \ldots$ respectively,

and

$$S_n = H_n(x, y, z) = \prod_{p=1}^{n/2} K_p.$$

The function S_{n-2r} is evidently the sum of the products of $K_1, K_2, \ldots, K_{\frac{1}{2}n}$ taken $\frac{1}{2}n - r$ at a time.

If $K_1, K_2, \ldots, K_{\frac{1}{2}n}$ be regarded as an auxiliary system of variables, then, by the ordinary formula of partial differentiation

$$\frac{\partial S_{n-2r}}{\partial x} = \sum_{p=1}^{n/2} \frac{\partial S_{n-2r}}{\partial K_p} \frac{\partial K_p}{\partial x}$$

$$= \sum_{p=1}^{n/2} \frac{\partial S_{n-2r}}{\partial K_p} \cdot \frac{2x}{a^2 + \theta_p};$$

and, if we differentiate again,

$$\frac{\partial^2 S_{n-2r}}{\partial x^2} = \sum_{p=1}^{n/2} \frac{\partial S_{n-2r}}{\partial K_p} \frac{2}{a^2 + \theta_p} + \sum_{p \neq 1} \frac{\partial^2 S_{n-2r}}{\partial K_p \partial K_q} \frac{8x^2}{(a^2 + \theta_p)(a^2 + \theta_q)}.$$

The terms in $\partial^2 S_{n-2r}/\partial K_p{}^2$ can be omitted because each of the functions K_p does not occur in S_{n-2r} to a degree higher than the first. It follows that

$$D^2 S_{n-2r} = \sum_{p=1}^{n/2} \frac{\partial S_{n-2r}}{\partial K_p} \left\{ \frac{2a^2}{a^2 + \theta_p} + \frac{2b^2}{b^2 + \theta_p} + \frac{2c^2}{c^2 + \theta_p} \right\}$$

$$+ \sum_{p \neq q} \frac{\partial^2 S_{n-2r}}{\partial K_p \partial K_q} \left\{ \frac{8a^2 x^2}{(a^2 + \theta_p)(a^2 + \theta_q)} + \frac{8b^2 y^2}{(b^2 + \theta_p)(b^2 + \theta_q)} + \frac{8c^2 z^2}{(c^2 + \theta_p)(c^2 + \theta_q)} \right\}.$$

It will now be shewn that the expression on the right is a constant multiple of S_{n-2r-2}. We first observe that

$$\sum_{\left(\substack{x, \ y, \ z \\ a, \ b, \ c} \right)} \frac{a^2 x^2}{(a^2 + \theta_p)(a^2 + \theta_q)} = \frac{\theta_p K_p - \theta_q K_q}{\theta_p - \theta_q}$$

and that, by the differential equation of §23.21,

$$\sum_{a,b,c} \frac{a^2}{a^2 + \theta_p} = 3 - \theta_p \sum_{a,b,c} \frac{1}{a^2 + \theta_p} = 3 + \theta_p \sum_{q=1}^{n/2}{}' \frac{4}{\theta_p - \theta_q},$$

so that

$$D^2 S_{n-2r} = 6 \sum_{p=1}^{n/2} \frac{\partial S_{n-2r}}{\partial K_p} + 8 \sum_{p=1}^{n/2} \theta_p \frac{\partial S_{n-2r}}{\partial K_p} \left\{ \sum_{q=1}^{n/2}{}' \frac{1}{\theta_p - \theta_q} \right\} + 8 \sum_{p \neq q} \frac{\partial^2 S_{n-2r}}{\partial K_p \partial K_q} \frac{\theta_p K_p - \theta_q K_q}{\theta_p - \theta_q}.$$

Now $\partial S_{n-2r}/\partial K_p$ is the sum of the products of the expressions $K_1, K_2, \ldots, K_{\frac{1}{2}n}$ (K_p being omitted) taken $\frac{1}{2}n - r - 1$ at a time; and $K_q \partial^2 S_{n-2r} / \partial K_p \partial K_q$ consists of those terms of this sum which contain K_q as a factor. Hence

$$\frac{\partial S_{n-2r}}{\partial K_p} - K_q \frac{\partial^2 S_{n-2r}}{\partial K_p \partial K_q}$$

is equal to the sum of the products of the expressions $K_1, K_2, \ldots, K_{n/2}$ (both K_p and K_q being omitted) taken $\frac{n}{2} - r - 1$ at a time; and therefore, by symmetry, we have

$$\frac{\partial S_{n-2r}}{\partial K_p} - K_q \frac{\partial^2 S_{n-2r}}{\partial K_p \partial K_q} = \frac{\partial S_{n-2r}}{\partial K_p} - K_q \frac{\partial^2 S_{n-2r}}{\partial K_p \partial K_q},$$

so that

$$\frac{\partial^2 S_{n-2r}}{\partial K_p \partial K_q} = \left\{ \frac{\partial S_{n-2r}}{\partial K_p} - \frac{\partial S_{n-2r}}{\partial K_q} \right\} \bigg/ (K_q - K_p).$$

On substituting by this formula for the second differential coefficients, it is found that

$$D^2 S_{n-2r} = \sum_{p=1}^{n/2} \frac{\partial S_{n-2r}}{\partial K_p} \left[6 + 8\theta_p \sum_{q=1}^{n/2}{}' \frac{1}{\theta_p - \theta_q} - 8 \sum_{q=1}^{n/2}{}' \frac{\theta_p K_p - \theta_q K_q}{(\theta_p - \theta_q)(K_p - K_q)} \right]$$

$$= \sum_{p=1}^{n/2} \frac{\partial S_{n-2r}}{\partial K_p} \left[6 - 8 \sum_{q=1}^{n/2}{}' \frac{K_q}{K_p - K_q} \right]$$

$$= (4n - 2) \sum_{p=1}^{n/2} \frac{\partial S_{n-2r}}{\partial K_p} - 8 \sum_{p \neq q} \left\{ K_p \frac{\partial S_{n-2r}}{\partial K_p} - K_q \frac{\partial S_{n-2r}}{\partial K_q} \right\} \bigg/ (K_p - K_q).$$

Now we may write S_{n-2r} in the form

$$\overline{S}_{n-2r} + K_p \overline{S}_{n-2r-2} + K_q \overline{S}_{n-2r-2} + K_p K_q \overline{S}_{n-2r-4},$$

where $\overline{S}_{2m}$ denotes the sum of the products of the expressions $K_1, K_2, \ldots, K_{n/2}$ (with K_p and K_q both being omitted) taken m at a time; and we then see that

$$K_p \frac{\partial S_{n-2r}}{\partial K_p} - K_q \frac{\partial S_{n-2r}}{\partial K_q} = (K_p - K_q)\overline{S}_{n-2r-2}.$$

Hence

$$D^2 S_{n-2r} = (4n - 2) \sum_{p=1}^{n/2} \frac{\partial S_{n-2r}}{\partial K_p} - 8 \sum_{p \neq 1} \overline{S}_{n-2r-2}.$$

Now it is clear that the expression on the right is a homogeneous symmetric function of $K_1, K_2, \ldots, K_{n/2}$ of degree $n/2 - r - 1$, and it contains no power of any of the expressions $K_1, K_2, \ldots, K_{n/2}$ to a degree higher than the first. It is therefore a multiple of S_{n-2r-2}. To determine the multiple we observe that when S_{n-2r-2} is written out at length it contains $\binom{n/2}{r-1}$ terms while the number of terms in

$$(4n - 2) \sum_{p=1}^{n/2} \frac{\partial S_{n-2r}}{\partial K_p} - 8 \sum_{p \neq q} \overline{S}_{n-2r-2}$$

is

$$\frac{1}{2}n(4n-2)\binom{n/2-1}{r} - 8\binom{n/2}{2}\cdot\binom{n/2-2}{r-1}.$$

The multiple is therefore

$$\left[\frac{1}{2}n(4n-2)\binom{n/2-1}{r} - 8\binom{n/2}{2}\cdot\binom{n/2-2}{r-1}\right] \Big/ \binom{n/2}{r+1}$$

and this is equal to $(2r+2)(2n-2r-1)$. It has consequently been proved that

$$D^2 S_{n-2r} = (2r+2)(2n-2r-1)S_{n-2r-2}.$$

It follows at once by induction that

$$S_{n-2r} = \frac{D^{2r} S_n}{2\cdot 4\cdots 2r\cdot(2n-1)(2n-3)\cdots(2n-2r+1)},$$

and the formula

$$G_n(x,y,z) = \left[\sum_{r=0}^{n/2} \frac{(-1)^r D^{2r}}{2\cdot 4\cdots 2r\cdot(2n-1)(2n-3)\cdots(2n-2r+1)}\right] H_n(x,y,z)$$

is now obvious when $G_n(x,y,z)$ is an ellipsoidal harmonic of the first species.

Example 23.2.1 Prove Niven's formula when $G_n(x,y,z)$ is an ellipsoidal harmonic of the second, third or fourth species.

Example 23.2.2 Obtain the symbolic formula

$$G_n(x,y,z) = \Gamma\left(\tfrac{1}{2}-n\right)\cdot\left(\tfrac{1}{2}D\right)^{n+\frac{1}{2}} I_{-n-\frac{1}{2}}(D)\cdot H_n(x,y,z).$$

23.26 Ellipsoidal harmonics of degree n

The results obtained and stated in §§23.21–23.24 shew that when n is even, there are $n/2+1$ harmonics of the first species and $\frac{3}{2}n$ harmonics of the third species; when n is odd there are $\frac{3}{2}(n+1)$ harmonics of the second species and $\frac{1}{2}(n-1)$ harmonics of the fourth species, so that, in either case, there are $2n+1$ harmonics in all. It follows from §18.3 that, if the terms of degree n in these harmonics are linearly independent, they form a fundamental system of harmonics of degree n; and any homogeneous harmonic of degree n is expressible as a linear combination of the homogeneous harmonics which are obtained by selecting the terms of degree n from the $2n+1$ ellipsoidal harmonics.

In order to prove the results concerning the number of harmonics of degree n and to establish their linear independence, it is necessary to make an intensive study of Lamé's equation; but before we pursue this investigation we shall study the construction of ellipsoidal harmonics in terms of confocal coordinates.

Note These expressions for ellipsoidal harmonics are of historical importance in view of Lamé's investigations, but the expressions which have just been obtained by Niven's method are, in some respects, more suitable for physical applications.

For applications of ellipsoidal harmonics to the investigation of the Figure of the Earth, and for the reduction of the harmonics to forms adapted for numerical computation, the reader is referred to the memoir by G. H. Darwin [165].

23.3 Confocal coordinates

If (X, Y, Z) denote current coordinates in three-dimensional space, and if a, b, c are positive $(a > b > c)$, the equation

$$\frac{X^2}{a^2} + \frac{Y^2}{b^2} + \frac{Z^2}{c^2} = 1$$

represents an ellipsoid; the equation of any confocal quadric is

$$\frac{X^2}{a^2 + \theta} + \frac{Y^2}{b^2 + \theta} + \frac{Z^2}{c^2 + \theta} = 1,$$

and θ is called the *parameter* of this quadric.

The quadric passes through a particular point (x, y, z) if θ is chosen so that

$$\frac{x^2}{a^2 + \theta} + \frac{y^2}{b^2 + \theta} + \frac{z^2}{c^2 + \theta} = 1.$$

Whether θ satisfies this equation or not, it is convenient to write

$$1 - \frac{x^2}{a^2 + \theta} - \frac{y^2}{b^2 + \theta} - \frac{z^2}{c^2 + \theta} \equiv \frac{f(\theta)}{(a^2 + \theta)(b^2 + \theta)(c^2 + \theta)},$$

and, since $f(\theta)$ is a cubic function of θ, it is clear that, in general, three quadrics of the confocal system pass through any particular point (x, y, z).

To determine the species of these three quadrics, we construct the following table.

θ	$f(\theta)$
$-\infty$	$-\infty$
$-a^2$	$-x^2(a^2 - b^2)(a^2 - c^2)$
$-b^2$	$y^2(a^2 - b^2)(b^2 - c^2)$
$-c^2$	$-z^2(a^2 - c^2)(b^2 - c^2)$
$+\infty$	$+\infty$

It is evident from this table that the equation $f(\theta) = 0$ has three real roots λ, μ, ν, and if they are arranged so that $\lambda > \mu > \nu$, then

$$\lambda > -c^2 > \mu > -b^2 > \nu > -a^2;$$

and also $f(\theta) \equiv (\theta - \lambda)(\theta - \mu)(\theta - \nu)$.

From the values of λ, μ, ν it is clear that the surfaces, on which θ has the respective values λ, μ, ν, are an ellipsoid, an hyperboloid of one sheet and an hyperboloid of two sheets.

Now take the identity in θ,

$$1 - \frac{x^2}{a^2 + \theta} - \frac{y^2}{b^2 + \theta} - \frac{z^2}{c^2 + \theta} \equiv \frac{(\theta - \lambda)(\theta - \mu)(\theta - \nu)}{(a^2 + \theta)(b^2 + \theta)(c^2 + \theta)},$$

and multiply it, in turn, by $a^2 + \theta$, $b^2 + \theta$, $c^2 + \theta$; and after so doing, replace θ by $-a^2$, $-b^2$, $-c^2$ respectively. It is thus found that

$$x^2 = \frac{(a^2 + \lambda)(a^2 + \mu)(a^2 + \nu)}{(a^2 - b^2)(a^2 - c^2)},$$

$$y^2 = -\frac{(b^2 + \lambda)(b^2 + \mu)(b^2 + \nu)}{(a^2 - b^2)(b^2 - c^2)},$$

$$z^2 = \frac{(c^2 + \lambda)(c^2 + \mu)(c^2 + \nu)}{(a^2 - c^2)(b^2 - c^2)}.$$

From these equations it is clear that, if (x, y, z) be any point of space and if λ, μ, ν denote the parameters of the quadrics confocal with

$$\frac{X^2}{a^2} + \frac{Y^2}{b^2} + \frac{Z^2}{c^2} = 1$$

which pass through the point, then (x^2, y^2, z^2) are uniquely determinate in terms of (λ, μ, ν) and vice versa.

The parameters (λ, μ, ν) are called the *confocal coordinates* of the point (x, y, z) relative to the fundamental ellipsoid

$$\frac{X^2}{a^2} + \frac{Y^2}{b^2} + \frac{Z^2}{c^2} = 1.$$

It is easy to shew that confocal coordinates form an orthogonal system; for consider the direction cosines of the tangent to the curve of intersection of the surfaces (μ) and (ν); these direction cosines are proportional to

$$\left(\frac{\partial x}{\partial \lambda}, \frac{\partial y}{\partial \lambda}, \frac{\partial z}{\partial \lambda} \right),$$

and since

$$\frac{\partial x}{\partial \lambda}\frac{\partial x}{\partial \mu} + \frac{\partial y}{\partial \lambda}\frac{\partial y}{\partial \mu} + \frac{\partial z}{\partial \lambda}\frac{\partial z}{\partial \mu} = \frac{1}{4} \sum_{a,b,c} \frac{a^2 + \nu}{(a^2 - b^2)(a^2 - c^2)} = 0,$$

it is evident that the directions

$$\left(\frac{\partial x}{\partial \lambda}, \frac{\partial y}{\partial \lambda}, \frac{\partial z}{\partial \lambda} \right), \left(\frac{\partial x}{\partial \mu}, \frac{\partial y}{\partial \mu}, \frac{\partial z}{\partial \mu} \right)$$

are perpendicular; and, similarly, each of these directions is perpendicular to

$$\left(\frac{\partial x}{\partial \nu}, \frac{\partial y}{\partial \nu}, \frac{\partial z}{\partial \nu} \right).$$

It has therefore been shewn that the three systems of surfaces, on which λ, μ, ν respectively are constant, form a triply orthogonal system.

Hence the square of the line-element, namely $(\delta x)^2 + (\delta y)^2 + (\delta z)^2$, is expressible in the form $(H_1\delta\lambda)^2 + (H_2\delta\mu)^2 + (H_3\delta\nu)^2$, where

$$H_1^2 = \left(\frac{\partial x}{\delta\lambda} \right)^2 + \left(\frac{\partial y}{\delta\lambda} \right)^2 + \left(\frac{\partial z}{\delta\lambda} \right)^2,$$

with similar expressions in μ and ν for H_2^2 and H_3^2.

To evaluate H_1^2 in terms of (λ, μ, ν), observe that

$$H_1^2 = \frac{1}{4x^2}\left(\frac{\partial x^2}{\delta\lambda}\right)^2 + \frac{1}{4y^2}\left(\frac{\partial y^2}{\delta\lambda}\right)^2 + \frac{1}{4z^2}\left(\frac{\partial z^2}{\delta\lambda}\right)^2$$

$$= \frac{1}{4}\sum_{a,b,c} \frac{(a^2 + \mu)(a^2 + \nu)}{(a^2 + \lambda)(a^2 - b^2)(a^2 - c^2)}.$$

But, if we express

$$\frac{(\lambda - \mu)(\lambda - \nu)}{(a^2 + \lambda)(b^2 + \lambda)(c^2 + \lambda)},$$

qua function of λ, as a sum of partial fractions, we see that it is precisely equal to

$$\sum_{a,b,c} \frac{(a^2 + \mu)(a^2 + \nu)}{(a^2 + \lambda)(a^2 - b^2)(a^2 - c^2)},$$

and consequently

$$H_1^2 = \frac{(\lambda - \mu)(\lambda - \nu)}{4(a^2 + \lambda)(b^2 + \lambda)(c^2 + \lambda)}.$$

The values of H_2^2 and H_3^2 are obtained from this expression by cyclical interchanges of (λ, μ, ν).

Note   Formulae equivalent to those of this section were obtained by Lamé [401].

Example 23.3.1   With the notation of this section, shew that

$$x^2 + y^2 + z^2 = a^2 + b^2 + c^2 + \lambda + \mu + \nu.$$

Example 23.3.2   Shew that

$$4H_1^2 = \frac{x^2}{(a^2 + \lambda)^2} + \frac{y^2}{(b^2 + \lambda)^2} + \frac{z^2}{(c^2 + \lambda)^2}.$$

23.31 *Uniformising variables associated with confocal coordinates*

It has been seen in §23.3 that when the Cartesian coordinates (x, y, z) are expressed in terms of the confocal coordinates (λ, μ, ν), the expressions so obtained are not one-valued functions of (λ, μ, ν). To avoid the inconvenience thereby produced, we express (λ, μ, ν) in terms of three new variables (u, v, w) respectively by writing

$$\wp(u) = \lambda + \tfrac{1}{3}(a^2 + b^2 + c^2),$$
$$\wp(v) = \mu + \tfrac{1}{3}(a^2 + b^2 + c^2),$$
$$\wp(w) = \nu + \tfrac{1}{3}(a^2 + b^2 + c^2),$$

the invariants g_2 and g_3 of the Weierstrassian elliptic functions being defined by the identity

$$4(a^2 + \lambda)(b^2 + \lambda)(c^2 + \lambda) \equiv 4\wp^3(u) - g_2\wp(u) - g_3.$$

The discriminant associated with the elliptic functions (cf. Example 20.3.9) is

$$16 \, (a^2 - b^2)^2 (b^2 - c^2)^2 (c^2 - a^2)^2,$$

and so it is positive; and, therefore[3], of the periods $2\omega_1$, $2\omega_2$ and $2\omega_3$, $2\omega_1$ is positive while $2\omega_3$ is a pure imaginary; and $2\omega_2$ has its real part negative, since $\omega_1 + \omega_2 + \omega_3 = 0$; the imaginary part of ω_2 is positive since $\operatorname{Im} \omega_2 / \omega_1 > 0$.

In these circumstances $e_1 > e_2 > e_3$, and so we have

$$3e_1 = a^2 + b^2 - 2c^2, \qquad 3e_2 = c^2 + a^2 - 2b^2, \qquad 3e_3 = b^2 + c^2 - 2a^2.$$

Next we express (x, y, z) in terms of (u, v, w); we have

$$
\begin{aligned}
x^2 &= \frac{(a^2 + \lambda)\,(a^2 + \mu)\,(a^2 + \nu)}{(a^2 - b^2)\,(a^2 - c^2)} \\
&= \frac{\{\wp(u) - e_3\}\,\{\wp(v) - e_3\}\,\{\wp(w) - e_3\}}{(e_1 - e_3)\,(e_2 - e_3)} \\
&= \frac{\sigma_3{}^2(u)\,\sigma_3{}^2(v)\,\sigma_3{}^2(w)}{\sigma^2(u)\,\sigma^2(v)\,\sigma^2(w)} \cdot \frac{\sigma^2(\omega_1)\,\sigma^2(\omega_2)}{\sigma_3{}^2(\omega_1)\,\sigma_3{}^2(\omega_2)},
\end{aligned}
$$

by Example 20.5.4. Therefore, by §20.421, we have

$$x = \pm e^{-\eta_3 \omega_3} \sigma^2(\omega_3) \frac{\sigma_3(u)\sigma_3(v)\sigma_3(w)}{\sigma(u)\sigma(v)\sigma(w)}$$

and similarly

$$y = \pm e^{-\eta_2 \omega_2} \sigma^2(\omega_2) \frac{\sigma_2(u)\sigma_2(v)\sigma_2(w)}{\sigma(u)\sigma(v)\sigma(w)}$$

$$z = \pm e^{-\eta_1 \omega_1} \sigma^2(\omega_1) \frac{\sigma_1(u)\sigma_1(v)\sigma_1(w)}{\sigma(u)\sigma(v)\sigma(w)}.$$

The effect of increasing each of u, v, w by $2\omega_3$ is to change the sign of the expression given for x while the expressions for y and z remain unaltered; and similar statements hold for increases by $2\omega_2$ and $2\omega_1$; and again each of the three expressions is changed in sign by changing the signs of u, v, w.

Hence, if the upper signs be taken in the ambiguities, there is a unique correspondence between all sets of values of (x, y, z), real or complex, and all the sets of values of (u, v, w) whose three representative points lie in any given cell.

The uniformisation is consequently effected by taking

$$
\left\{
\begin{aligned}
x &= e^{-\eta_3 \omega_3} \sigma^2(\omega_3) \frac{\sigma_3(u)\sigma_3(v)\sigma_3(w)}{\sigma(u)\sigma(v)\sigma(w)}, \\
y &= e^{-\eta_2 \omega_2} \sigma^2(\omega_3) \frac{\sigma_2(u)\sigma_2(v)\sigma_2(w)}{\sigma(u)\sigma(v)\sigma(w)}, \\
z &= e^{-\eta_1 \omega_1} \sigma^2(\omega_1) \frac{\sigma_1(u)\sigma_1(v)\sigma_1(w)}{\sigma(u)\sigma(v)\sigma(w)}.
\end{aligned}
\right.
$$

Formulae which differ from these only by the interchange of the suffixes 1 and 3 were given by Halphen [269, p. 459].

---

[3] See §20.32, Example 20.3.5.

23.32 Laplace's equation referred to confocal coordinates

It has been shewn by Lamé and by W. Thomson that Laplace's equation when referred to *any* system of orthogonal coordinates (λ, μ, ν) assumes the form

$$\frac{\partial}{\partial \lambda}\left\{\frac{H_2 H_3}{H_1} \cdot \frac{\partial V}{\partial \lambda}\right\} + \frac{\partial}{\partial \mu}\left\{\frac{H_3 H_1}{H_2} \cdot \frac{\partial V}{\partial \mu}\right\} + \frac{\partial}{\partial \nu}\left\{\frac{H_1 H_2}{H_3} \cdot \frac{\partial V}{\partial \nu}\right\} = 0,$$

where (H_1, H_2, H_3) are to be determined from the consideration that

$$(H_1 \delta\lambda)^2 + (H_2 \delta\mu)^2 + (H_3 \delta\nu)^2$$

is to be the square of the line-element. Although W. Thomson's proof of this result, based on arguments of a physical character, is extremely simple, all the analytical proofs are either very long or else severely compressed.

It has, however, been shewn by Lamé [403, pp. 133–136] that, in the special case in which (λ, μ, ν) represent confocal coordinates, Laplace's equation assumes a simple form obtainable without elaborate analysis; when the uniformising variables (u, v, w) of §23.31 are adopted as coordinates, the form of Laplace's equation becomes still simpler.

By straightforward differentiation it may be proved that, when *any* three independent functions (λ, μ, ν) of (x, y, z) are taken as independent variables, then

$$\frac{\partial^2 V}{\partial x^2} + \frac{\partial^2 V}{\partial y^2} + \frac{\partial^2 V}{\partial z^2}$$

transforms into

$$\sum_{\lambda,\mu,\nu} \left[\left(\frac{\partial\lambda}{\partial x}\right)^2 + \left(\frac{\partial\lambda}{\partial y}\right)^2 + \left(\frac{\partial\lambda}{\partial z}\right)^2\right]\frac{\partial^2 V}{\partial\lambda^2}$$

$$+ 2\sum_{\lambda,\mu,\nu}\left[\frac{\partial\mu}{\partial x}\frac{\partial\nu}{\partial x} + \frac{\partial\mu}{\partial y}\frac{\partial\nu}{\partial y} + \frac{\partial\mu}{\partial z}\frac{\partial\nu}{\partial z}\right]\frac{\partial^2 V}{\partial\mu\partial\nu}$$

$$+ \sum_{\lambda,\mu,\nu}\left[\frac{\partial^2\lambda}{\partial x^2} + \frac{\partial^2\lambda}{\partial y^2} + \frac{\partial^2\lambda}{\partial z^2}\right]\frac{\partial V}{\partial\lambda}.$$

In order to reduce this expression, we observe that λ satisfies the equation

$$\frac{x^2}{a^2 + \lambda} + \frac{y^2}{b^2 + \lambda} + \frac{z^2}{c^2 + \lambda} = 1$$

and so, by differentiation with x, y, z as independent variables,

$$\frac{2x}{a^2 + \lambda} - \left\{\frac{x^2}{(a^2 + \lambda)^2} + \frac{y^2}{(b^2 + \lambda)^2} + \frac{z^2}{(c^2 + \lambda)^2}\right\}\frac{\partial\lambda}{\partial x} = 0,$$

$$\frac{2}{a^2 + \lambda} - \frac{4x}{(a^2 + \lambda)^2}\frac{\partial\lambda}{\partial x} + 2\left\{\frac{x^2}{(a^2 + \lambda)^3} + \frac{y^2}{(b^2 + \lambda)^3} + \frac{z^2}{(c^2 + \lambda)^3}\right\}\left(\frac{\partial\lambda}{\partial x}\right)^2$$

$$- \left\{\frac{x^2}{(a^2 + \lambda)^2} + \frac{y^2}{(b^2 + \lambda)^2} + \frac{z^2}{(c^2 + \lambda)^2}\right\}\frac{\partial^2\lambda}{\partial x^2} = 0.$$

Hence

$$\frac{2x}{a^2 + \lambda} = 4H_1^3 \frac{\partial \lambda}{\partial x},$$

$$\frac{2}{a^2 + \lambda} - \frac{2x^2}{(a^2 + \lambda)^3 H_1^2} + \frac{x^2}{2H_1^4(a^2 + \lambda)^2} \sum_{\left(\begin{smallmatrix} x, \, y, \, z \\ a, \, b, \, c \end{smallmatrix}\right)} \frac{x^2}{(a^2 + \lambda)^3} = 4H_1^2 \frac{\partial^2 \lambda}{\partial x^2},$$

with similar equations in μ, ν and y, z.

From equations of the first type it is seen that the coefficient of $\dfrac{\partial^2 V}{\partial \lambda^2}$ is $\dfrac{1}{H_1^2}$ and the coefficient of $\dfrac{\partial^2 V}{\partial \mu \partial \nu}$ is zero; and if we add up equations of the second type obtained by interchanging x, y, z cyclically, it is found that

$$4H_1^2 \left\{ \frac{\partial^2 \lambda}{\partial x^2} + \frac{\partial^2 \lambda}{\partial y^2} + \frac{\partial^2 \lambda}{\partial z^2} \right\} = \sum_{a,b,c} \frac{2}{a^2 + \lambda},$$

with similar equations in μ and ν.

If, for brevity, we write

$$\sqrt{(a^2 + \lambda)(b^2 + \lambda)(c^2 + \lambda)} \equiv \Delta_\lambda,$$

with similar meanings for Δ_μ and Δ_ν, we see that

$$\frac{\partial^2 \lambda}{\partial x^2} + \frac{\partial^2 \lambda}{\partial y^2} + \frac{\partial^2 \lambda}{\partial z^2} = \frac{\Delta_\lambda^2}{(\lambda - \mu)(\lambda - \nu)} \left\{ \frac{2}{a^2 + \lambda} + \frac{2}{b^2 + \lambda} + \frac{2}{c^2 + \lambda} \right\}$$

$$= \frac{4\Delta_\lambda}{(\lambda - \mu)(\lambda - \nu)} \frac{d\Delta_\lambda}{d\lambda},$$

and so Laplace's equation assumes the form

$$\sum_{\lambda, \mu, \nu} \frac{4}{(\lambda - \mu)(\lambda - \nu)} \left[\Delta_\lambda^2 \frac{\partial^2 V}{\partial \lambda^2} + \Delta_\lambda \frac{d\Delta_\lambda}{d\lambda} \frac{\partial V}{\partial \lambda} \right] = 0,$$

that is to say

$$(\mu - \nu)\Delta_\lambda \frac{\partial}{\partial \lambda} \left\{ \Delta_\lambda \frac{\partial V}{\partial \lambda} \right\} + (\nu - \lambda)\Delta_\mu \frac{\partial}{\partial \mu} \left\{ \Delta_\mu \frac{\partial V}{\partial \mu} \right\} + (\lambda - \mu)\Delta_\nu \frac{\partial}{\partial \nu} \left\{ \Delta_\nu \frac{\partial V}{\partial \nu} \right\} = 0.$$

The equivalent equation with (u, v, w) as independent variables is simply

$$\{\wp(v) - \wp(w)\} \frac{\partial^2 V}{\partial u^2} + \{\wp(w) - \wp(u)\} \frac{\partial^2 V}{\partial v^2} + \{\wp(u) - \wp(v)\} \frac{\partial^2 V}{\partial w^2} = 0,$$

or, more briefly,

$$(\mu - \nu)\frac{\partial^2 V}{\partial u^2} + (\nu - \lambda)\frac{\partial^2 V}{\partial v^2} + (\lambda - \mu)\frac{\partial^2 V}{\partial w^2} = 0.$$

The last three equations will be regarded as canonical forms of Laplace's equation in the subsequent analysis.

23.33 Ellipsoidal harmonics referred to confocal coordinates

When Niven's function Θ_p, defined as

$$\frac{x^2}{a^2 + \theta_p} + \frac{y^2}{b^2 + \theta_p} + \frac{z^2}{c^2 + \theta_p} - 1,$$

is expressed in terms of the confocal coordinates (λ, μ, ν) of the point (x, y, z), it assumes the form

$$-\frac{(\lambda - \theta_p)(\mu - \theta_p)(\nu - \theta_p)}{(a^2 + \theta_p)(b^2 + \theta_p)(c^2 + \theta_p)},$$

and consequently, when constant factors of the form

$$-(a^2 + \theta_p)(b^2 + \theta_p)(c^2 + \theta_p)$$

are omitted, ellipsoidal harmonics assume the form

$$\left\{ \begin{matrix} & x, & yz & \\ 1, & y, & zx, & xyz \\ & z, & xy, & \end{matrix} \right\} \prod_{p=1}^{m}(\lambda - \theta_p) \prod_{p=1}^{m}(\mu - \theta_p) \prod_{p=1}^{m}(\nu - \theta_p).$$

If now we replace x, y, z by their values in terms of λ, μ, ν, we see that *any ellipsoidal harmonic is expressible in the form of a constant multiple of* ΛMN, *where* Λ *is a function of* λ *only, and M and N are the same functions of* μ *and* ν *respectively as* Λ *is of* λ. Further Λ is a polynomial of degree m in λ multiplied, in the case of harmonics of the second, third or fourth species, by one, two or three of the expressions $\sqrt{a^2 + \lambda}$, $\sqrt{b^2 + \lambda}$, $\sqrt{c^2 + \lambda}$.

Since the polynomial involved in Λ is $\prod_{p=1}^{m}(\lambda - \theta_p)$, it follows from a consideration of §§23.21–23.24 that Λ is a solution of Lamé's differential equation

$$4\sqrt{(a^2 + \lambda)(b^2 + \lambda)(c^2 + \lambda)}\,\frac{d}{d\lambda}\left[\sqrt{(a^2 + \lambda)(b^2 + \lambda)(c^2 + \lambda)}\,\frac{d\Lambda}{d\lambda}\right]$$
$$= \{n(n + 1)\lambda + C\}\,\Lambda,$$

where n is the degree of the harmonic in (x, y, z).

This result may also be attained from a consideration of solutions of Laplace's equation which are of the type[4]

$$V = \Lambda MN,$$

where Λ, M, N are functions only of λ, μ, ν respectively.

For if we substitute this expression in Laplace's equation, as transformed in §23.32, on division by V, we find that

$$\frac{\wp(v) - \wp(w)}{\Lambda}\frac{d^2\Lambda}{du^2} + \frac{\wp(w) - \wp(u)}{M}\frac{d^2M}{dv^2} + \frac{\wp(u) - \wp(v)}{N}\frac{d^2N}{dw^2} = 0.$$

---

[4] A harmonic which is the product of three functions, each of which depends on one coordinate only, is sometimes called a *normal solution* of Laplace's equation. Thus normal solutions with polar coordinates are (§18.31)

$$r^n P_n^m(\cos \theta)_{\sin}^{\cos} m\phi.$$

The last two terms, *qua* functions of *u*, are linear functions of $\wp(u)$, and so $\dfrac{1}{\Lambda}\dfrac{d^2\Lambda}{du^2}$ must be a linear function of $\wp(u)$; since it is independent of the coordinates *v* and *w*, we have

$$\frac{1}{\Lambda}\frac{d^2\Lambda}{du^2} = \{K\wp(u) + B\},$$

where *K* and *B* are constants.

If we make this substitution in the differential equation, we get a linear function of $\wp(u)$ equated (identically) to zero, and so the coefficients in this linear function must vanish; that is to say

$$K\{\wp(v) - \wp(w)\} - \frac{1}{M}\frac{d^2M}{dv^2} + \frac{1}{N}\frac{d^2N}{dw^2} = 0,$$

$$B\{\wp(v) - \wp(w)\} + \frac{\wp(w)}{M}\frac{d^2M}{dv^2} - \frac{\wp(v)}{N}\frac{d^2N}{dw^2} = 0,$$

and on solving these with the observation that $\wp(v) - \wp(w)$ is not identically zero, we obtain the three equations

$$\frac{d^2\Lambda}{du^2} = \{K\wp(u) + B\}\Lambda,$$

$$\frac{d^2M}{dv^2} = \{K\wp(v) + B\}\,M,$$

$$\frac{d^2N}{dw^2} = \{K\wp(w) + B\}\,N.$$

When λ is taken as independent variable, the first equation becomes

$$4\Delta_\lambda\frac{d}{d\lambda}\left\{\Delta_\lambda\frac{d\Lambda}{d\lambda}\right\} = \{K\lambda + B + \tfrac{1}{3}K(a^2 + b^2 + c^2)\}\Lambda,$$

and this is the equation already obtained for Λ, the degree *n* of the harmonic being given by the formula

$$n(n + 1) = K.$$

We have now progressed so far with the study of ellipsoidal harmonics as is convenient without making use of properties of Lamé's equation.

We now proceed to the detailed consideration of this equation.

23.4 Various forms of Lamé's differential equation

We have already encountered two forms of Lamé's equation, namely

$$4\Delta_\lambda\frac{d}{d\lambda}\left\{\Delta_\lambda\frac{d\Lambda}{d\lambda}\right\} = \{n(n + 1)\lambda + C\}\Lambda,$$

and this may also be written

$$\frac{d^2\Lambda}{d\lambda^2} + \left\{\frac{\frac{1}{2}}{a^2 + \lambda} + \frac{\frac{1}{2}}{b^2 + \lambda} + \frac{\frac{1}{2}}{c^2 + \lambda}\right\}\frac{d\Lambda}{d\lambda} = \frac{\{n(n + 1)\lambda + C\}\Lambda}{4(a^2 + \lambda)(b^2 + \lambda)(c^2 + \lambda)},$$

which may be termed the algebraic form; and

$$\frac{d^2\Lambda}{du^2} = \{n(n+1)\wp(u) + B\}\Lambda,$$

which, since it contains the Weierstrassian elliptic function $\wp(u)$, may be termed the Weierstrassian form; the constants B and C are connected by the relation

$$B + \tfrac{1}{3}n(n+1)(a^2 + b^2 + c^2) = C.$$

If we take $\wp(u)$ as a new variable, which will be called ξ, we obtain the slightly modified algebraic form (cf. §10.6)

$$\frac{d^2\Lambda}{d\xi^2} + \left\{ \frac{\tfrac{1}{2}}{\xi - e_1} + \frac{\tfrac{1}{2}}{\xi - e_2} + \frac{\tfrac{1}{2}}{\xi - e_3} \right\} \frac{d\Lambda}{d\xi} = \frac{\{n(n+1)\xi + B\}\Lambda}{4(\xi - e_1)(\xi - e_2)(\xi - e_3)}.$$

This differential equation has singularities at e_1, e_2, e_3 at which the exponents are 0, $\tfrac{1}{2}$ in each case; and a singularity at infinity, at which the exponents are $-\tfrac{1}{2}n$, $\tfrac{1}{2}(n+1)$.

Note The Weierstrassian form of the equation has been studied by Halphen [268, II, pp. 457–531]. The algebraic forms have been studied by Stieltjes [604], Klein [374], and Bôcher [78].

The more general differential equation with four arbitrary singularities at which the exponents are arbitrary (save that the sum of all the exponents at all the singularities is 2) has been discussed by Heun [300]; the gain in generality by taking the singularities arbitrary is only apparent, because by a homographic change of the independent variable one of them can be transferred to the point at infinity, and then a change of origin is sufficient to make the sum of the complex coordinates of the three finite singularities equal to zero.

Another important form of Lamé's equation is obtained by using the notation of Jacobian elliptic functions; if we write

$$z_1 = u\sqrt{e_1 - e_3},$$

the Weierstrassian form becomes

$$\frac{d^2\Lambda}{dz_1^2} = \left[n(n+1) \left\{ \frac{e_3}{e_1 - e_3} + \mathrm{ns}^2 z_1 \right\} + \frac{B}{e_1 - e_3} \right] \Lambda,$$

and putting $z_1 = \alpha - iK'$, where $2iK'$ is the imaginary period of $\mathrm{sn}\, z_1$, we obtain the simple form

$$\frac{d^2\Lambda}{d\alpha^2} = \{n(n+1)k^2 \,\mathrm{sn}^2\,\alpha + A\}\Lambda,$$

where A is a constant connected with B by the relation $B + e_3 n(n+1) = A(e_1 - e_3)$.

Note The Jacobian form has been studied by Hermite [291], published separately, Paris, 1885.

In studying the properties of Lamé's equation, it is best not to use one form only, but to take the form best fitted for the purpose in hand. For practical applications the Jacobian form, leading to the theta-functions, is the most suitable. For obtaining the properties of the solutions of the equation, the best form to use is, in general, the second algebraic form, though in some problems analysis is simpler with the Weierstrassian form.

23.41 Solutions in series of Lamé's equation

Let us now assume a solution of Lamé's equation, which may be written

$$4(\xi - e_1)(\xi - e_2)(\xi - e_3)\frac{d^2\Lambda}{d\xi^2} + (6\xi^2 - \tfrac{1}{2}g_2)\frac{d\Lambda}{d\xi} - \{n(n+1)\xi + B\}\Lambda = 0,$$

in the form

$$\Lambda = \sum_{r=0}^{\infty} b_r(\xi - e_2)^{n/2-r}.$$

The series on the right, if it is a solution, will converge (§10.31) for sufficiently small values of $|\xi - e_2|$; but our object will be not the discussion of the convergence but the choice of B in such a way that the series may terminate, so that considerations of convergence will be superfluous.

The result of substituting this series for Λ on the left-hand side of the differential equation and arranging the result in powers of $\xi - e_2$ is minus the series

$$4\sum_{r=0}^{\infty}(\xi - e_2)^{n/2-r+1}[r(n-r+\tfrac{1}{2})b_r - \{3e_2(n/2 - r + 1)^2 - \tfrac{1}{4}n(n+1)e_2 - \tfrac{1}{4}B\}b_{r-1}$$

$$+ (e_1 - e_2)(e_2 - e_3)(\tfrac{1}{2}n - r + 2)(\tfrac{1}{2}n - r + \tfrac{3}{2})b_{r-2}],$$

in which the coefficients b_r with negative suffixes are to be taken to be zero.

Hence, if the series is to be a solution, the relation connecting successive coefficients is

$$r(n-r+\tfrac{1}{2})b_r = \{3e_2(\tfrac{1}{2}n - r + 1)^2 - \tfrac{1}{4}n(n+1)e_2 - \tfrac{1}{4}B\}b_{r-1}$$

$$- (e_1 - e_2)(e_2 - e_3)(\tfrac{1}{2}n - r + 2)(\tfrac{1}{2}n - r + \tfrac{3}{2})b_{r-2},$$

and

$$\left(n - \frac{1}{2}\right)b_1 = \left\{\frac{3}{4}n^2 e_2 - \frac{1}{4}n(n+1)e_2 - \frac{1}{4}B\right\}b_0.$$

If we take $b_0 = 1$, as we may do without loss of generality, the coefficients b_r are seen to be functions of B with the following properties:

(i) b_r is a polynomial in B of degree r;
(ii) the sign of the coefficient of B^r in b_r is that of $(-1)^r$, provided that $r \le n$: the actual coefficient of B^r is

$$\frac{(-1)^r}{2 \cdot 4 \cdots 2r(2n-1)(2n-3) \cdots (2n-2r+1)};$$

(iii) if e_1, e_2, e_3, and B are real and $e_1 > e_2 > e_3$, then, if $b_{r-1} = 0$, the values of b_r and b_{r-2} are opposite in sign, provided that $r < \tfrac{1}{2}(n+3)$ and $r < n$.

Now suppose that n is even and that we choose B in such a way that

$$b_{n/2+1} = 0.$$

If this choice is made, the recurrence formula shews that

$$b_{n/2+2} = 0,$$

by putting $r = \frac{1}{2}n + 2$ in the formula in question; and if both $b_{n/2+1}$ and $b_{n/2+2}$ are zero *the subsequent recurrence formulae are satisfied by taking*

$$b_{n/2+3} = b_{n/2+4} = \cdots = 0.$$

Hence the condition that Lamé's equation should have a solution which is a polynomial in ξ is that B should be a root of a certain algebraic equation of degree $n/2 + 1$, when n is even.

When n is odd, we take $b_{\frac{1}{2}(n+1)}$ to vanish and then $b_{\frac{1}{2}(n+3)}$ also vanishes, and so do the subsequent coefficients; so that the condition, when n is odd, is that B should be a root of a certain algebraic equation of degree $\frac{1}{2}(n + 1)$.

It is easy to shew that, when $e_1 > e_2 > e_3$, these algebraic equations have all their roots real. For the properties (ii) and (iii) shew that, *qua* functions of B, the expressions $b_0, b_1, b_2, \ldots, b_r$ form a set of Sturm's functions [615] when $r < \frac{1}{2}(n + 3)$, and so the equation

$$b_{\frac{1}{2}n+1} = 0 \quad \text{or} \quad b_{\frac{1}{2}(n+1)} = 0$$

has all its roots real and unequal. This procedure is due to Liouville [439].

Hence, when the constants e_1, e_2, e_3 are real (which is the case of practical importance, as was seen in §23.31), there are $\frac{1}{2}n + 1$ real and distinct values of B for which Lamé's equation has a solution of the type

$$\sum_{r=0}^{n/2} b_r(\xi - e_2)^{n/2-r}$$

when n is even; and there are $\frac{1}{2}(n + 1)$ real and distinct values of B for which Lamé's equation has a solution of the type

$$\sum_{r=0}^{\frac{1}{2}(n-1)} b_r(\xi - e_2)^{n/2-r}$$

when n is odd.

Note When the constants e_1, e_2, e_3 are *not* all real, it is possible for the equation satisfied by B to have equal roots; the solutions of Lamé's equation in such cases have been discussed by Cohn [152].

Example 23.4.1 Discuss solutions of Lamé's equation of the types

(i) $(\xi - e_1)^{\frac{1}{2}} \displaystyle\sum_{r=0}^{\infty} b'_r(\xi - e_2)^{\frac{1}{2}n-r-\frac{1}{2}}$,

(ii) $(\xi - e_3)^{\frac{1}{2}} \displaystyle\sum_{r=0}^{\infty} b''_r(\xi - e_2)^{\frac{1}{2}n-r-\frac{1}{2}}$,

(iii) $(\xi - e_1)^{\frac{1}{2}}(\xi - e_3)^{\frac{1}{2}} \displaystyle\sum_{r=0}^{\infty} b'''_r(\xi - e_2)^{\frac{1}{2}n-r-1}$,

obtaining the recurrence relations

(i) $r\left(n - r + \frac{1}{2}\right) b'_r = \left\{ 3e_2 \left(\frac{1}{2}n - r + \frac{1}{2}\right)^2 + (e_2 - e_3)\left(\frac{1}{2}n - r + \frac{3}{4}\right) - \frac{1}{4}n(n + 1) \right.$

$$\left. e_2 - \frac{1}{4}B \right\} b'_{r-1} - (e_1 - e_2)(e_2 - e_3)\left(\frac{1}{2}n - r + \frac{3}{2}\right)\left(\frac{1}{2}n - r + 1\right) b'_{r-2},$$

(ii) $r(n - r + \frac{1}{2})b''_r = \left\{ 3e_2 \left(\frac{1}{2}n - r + \frac{1}{2} \right)^2 - (e_1 - e_3)(\frac{1}{2}n - r + \frac{3}{4}) - \frac{1}{4}n(n + 1) \right.$

$\left. e_2 - \frac{1}{4}B \right\} b''_{r-1} - (e_1 - e_2)(e_2 - e_3) \left(\frac{1}{2}n - r + \frac{3}{2} \right) \left(\frac{1}{2}n - r + 1 \right) b''_{r-2},$

(iii) $r(n - r + \frac{1}{2})b'''_r = \left\{ 3e_2 \left(\frac{1}{2}n - r + \frac{1}{2} \right)^2 - \frac{1}{4}e_2(n^2 + n + 1) - \frac{1}{4}B \right\} b'''_{r-1}$

$- (e_1 - e_2)(e_2 - e_3)(\frac{1}{2}n - r + 1)(\frac{1}{2}n - r + \frac{1}{2})b'''_{r-2}.$

Example 23.4.2 With the notation of Example 23.4.1 shew that the numbers of real distinct values of B for which Lamé's equation is satisfied by terminating series of the several species are

(i) $\frac{1}{2}(n - 1)$ or $\frac{1}{2}(n - 2)$;
(ii) $\frac{1}{2}(n - 1)$ or $\frac{1}{2}(n - 2)$;
(iii) $\frac{1}{2}(n - 2)$ or $\frac{1}{2}(n - 3)$.

23.42 The definition of Lamé functions

When we collect the results which have been obtained in §23.41, it is clear that, given the equation

$$\frac{d^2 \Lambda}{du^2} = [n(n + 1)\wp(u) + B] \Lambda,$$

n being a positive integer, there are $2n + 1$ values of B for which the equation has a solution of one or other of the four species described in §§23.21–23.24.

If, when such a solution is expanded in descending powers of ξ, the coefficient of the leading term $\xi^{n/2}$ is taken to be unity, as was done in §23.41, the function so obtained is called a *Lamé function of degree n, of the first kind*, of the first (second, third or fourth) species. The $2n + 1$ functions so obtained are denoted by the symbol

$$E_n^m(\xi); \qquad (m = 1, 2, \ldots, 2n + 1),$$

and, when we have to deal with only one such function, it may be denoted by the symbol

$$E_n(\xi).$$

Note Tables of the expressions representing Lamé functions for $n = 1, 2, \ldots, 10$ have been compiled by Guerritore [263].

Example 23.4.3 Obtain the five Lamé functions of degree 2, namely

$$\lambda + \frac{1}{3} \sum a^2 \pm \frac{1}{3} \sqrt{\sum a^4 - \sum b^2 c^2},$$

$$\sqrt{\lambda + b^2}\sqrt{\lambda + c^2}, \qquad \sqrt{\lambda + c^2}\sqrt{\lambda + a^2}, \qquad \sqrt{\lambda + a^2}\sqrt{\lambda + b^2}.$$

Example 23.4.4 Obtain the seven Lamé functions of degree 3, namely

$$\sqrt{(\lambda + a^2)(\lambda + b^2)(\lambda + c^2)},$$

and six functions obtained by interchanges of a, b, c in the expressions

$$\sqrt{\lambda + a^2} \left[\lambda + \frac{1}{5}(a^2 + 2b^2 + 2c^2) \pm \frac{1}{5}\sqrt{a^4 + 4b^4 + 4c^4 - 7b^2c^2 - c^2a^2 - a^2b^2} \right].$$

23.43 The non-repetition of factors in Lamé functions

It will now be shewn that all the rational linear factors of $E_n^m(\xi)$ are *unequal*. This result follows most simply from the differential equation which $E_n^m(\xi)$ satisfies; for, if $\xi - \xi_1$ be any factor of $E_n^m(\xi)$, where ξ_1 is not one of the numbers e_1, e_2 or e_3, then ξ_1 is a regular point of the equation (§10.3), and any solution of the equation which, when expanded in powers of $\xi - \xi_1$, does not begin with a term in $(\xi - \xi_1)^0$ or $(\xi - \xi_1)^1$ must be identically zero.

Again, if ξ_1 were one of the numbers e_1, e_2 or e_3, the indicial equation appropriate to ξ_1 would have the roots 0 and $\frac{1}{2}$, and so the expansion of $E_n^m(\xi)$ in ascending powers of ξ_1 would begin with a term in $(\xi - \xi_1)^0$ or $(\xi - \xi_1)^{\frac{1}{2}}$.

Hence, in no circumstances has $E_n^m(\xi)$, *qua* function of ξ, a repeated factor.

The determination of the numbers $\theta_1, \theta_2, \ldots, \theta_m$ introduced in §§23.21–23.24 may now be regarded as complete; for it has been seen that solutions of Lamé's equation can be constructed with non-repeated factors, and the values of $\theta_1, \theta_2, \ldots$ which correspond to the roots of $E_n^m(\xi) = 0$ satisfy the equations which are requisite to ensure that Niven's products are solutions of Laplace's equation.

It still remains to be shewn that the $2n + 1$ ellipsoidal harmonics constructed in this way form a fundamental system of solutions of degree n of Laplace's equation.

23.44 The linear independence of Lamé functions

It will now be shewn that the $2n + 1$ Lamé functions $E_n^m(\xi)$ which are of degree n are linearly independent; that is to say, that no linear relation can exist which connects them identically for general values of ξ.

In the first place, if such a linear relation existed in which functions of different species were involved, it is obvious that by suitable changes of signs of the radicals $\sqrt{\xi - e_1}$, $\sqrt{\xi - e_2}$, $\sqrt{\xi - e_3}$ we could obtain other relations which, on being combined by addition or subtraction with the original relation, would give rise to two (or more) linear relations each of which involved functions restricted not merely to be of the same species but also of the same type.

Let one of these latter relations, if it exists, be

$$\sum a_m E_n^m(\xi) \equiv 0 \qquad (a_m \neq 0)$$

and let this relation involve r of the functions.

Operate on this identity $r - 1$ times with the operator

$$\frac{d^2}{du^2} - n(n + 1)\xi.$$

The results of the successive operations are

$$\sum a_m (B_n^m)^s E_n^m(\xi) \equiv 0 \qquad (s = 1, 2, \ldots, r - 1),$$

where B_n^m is the particular value of B which is associated with $E_n^m(\xi)$.

Eliminate $a_1, a_2, \ldots, a_r$ from the r equations now obtained; and it is found that

$$
\begin{vmatrix}
1 & 1 & 1 & \cdots & 1 \\
B_n^1 & B_n^2 & B_n^3 & \cdots & B_n^r \\
\vdots & \vdots & \vdots & \vdots & \vdots \\
(B_n^1)^{r-1} & (B_n^2)^{r-1} & \cdots & \cdots & (B_n^r)^{r-1}
\end{vmatrix} = 0.
$$

Now the only factors of the determinant on the left are differences of the numbers B_n^m, and these differences cannot vanish, by §23.41. Hence the determinant cannot vanish and so the postulated relation does not exist.

The linear independence of the $2n + 1$ Lamé functions of degree n is therefore established.

23.45 The linear independence of ellipsoidal harmonics

Let $G_n^m(x, y, z)$ be the ellipsoidal harmonic of degree n associated with $E_n^m(\xi)$, and let $H_n^m(x, y, z)$ be the corresponding homogeneous harmonic.

It is now easy to shew that not only are the $2n + 1$ harmonics of the type $G_n^m(x, y, z)$ linearly independent, but also the $2n + 1$ harmonics of the type $H_n^m(x, y, z)$ are linearly independent.

In the first place, if a linear relation existed between harmonics of the type $G_n^m(x, y, z)$, then, when we expressed these harmonics in terms of confocal coordinates (λ, μ, ν), we should obtain a linear relation between Lamé functions of the type $E_n^m(\xi)$ where $\xi = \lambda + \frac{1}{3}(a^2 + b^2 + c^2)$, and it has been seen that no such relation exists.

Again, if a linear relation existed between homogeneous harmonics of the type $H_n^m(x, y, z)$, by operating on the relation with Niven's operator (§23.25),

$$
1 - \frac{D^2}{2(2n - 1)} + \frac{D^4}{2 \cdot 4(2n - 1)(2n - 3)} - \cdots,
$$

we should obtain a linear relation connecting functions of the type $G_n^m(x, y, z)$, and since it has just been seen that no such relation exists, it follows that the homogeneous harmonics of degree n are linearly independent.

23.46 Stieltjes' theorem on the zeros of Lamé functions

It has been seen that any Lamé function of degree n is expressible in the form

$$
(\theta + a^2)^{\kappa_1} (\theta + b^2)^{\kappa_2} (\theta + c^2)^{\kappa_3} \cdot \prod_{p=1}^{m} (\theta - \theta_p),
$$

where $\kappa_1, \kappa_2, \kappa_3$ are equal to 0 or $\frac{1}{2}$ and the numbers $\theta_1, \theta_2, \ldots, \theta_m$ are real and unequal both to each other and to $-a^2, -b^2, -c^2$; and $\frac{1}{2}n = m + \kappa_1 + \kappa_2 + \kappa_3$. When $\kappa_1, \kappa_2, \kappa_3$ are given the number of Lamé functions of this degree and type is $m + 1$.

The remarkable result has been proved by Stieltjes [604] that these $m + 1$ functions can be arranged in order in such a way that the rth function of the set has $r - 1$ of its zeros[5] between

---

[5] The zeros $-a^2, -b^2, -c^2$ are to be omitted from this enumeration, $\theta_1, \theta_2, \ldots, \theta_m$ only being taken into account.

$-a^2$ and $-b^2$ and the remaining $m - r + 1$ of its zeros between $-b^2$ and $-c^2$, and, incidentally, that, for *all* the $m + 1$ functions, $\theta_1, \theta_2, \ldots, \theta_m$ lie between $-a^2$ and $-c^2$.

To prove this result, let $\phi_1, \phi_2, \ldots, \phi_m$ be any real variables such that

$$\begin{cases} -a^2 \le \phi_p \le -b^2, & (p = 1, 2, \ldots, r - 1) \\ -b^2 \le \phi_p \le -c^2, & (p = r, r + 1 \ldots, m) \end{cases}$$

and consider the product

$$\Pi = \prod_{p=1}^{m} \left[|(\phi_p + a^2)|^{\kappa_1 + \frac{1}{4}} \cdot |(\phi_p + b^2)|^{\kappa_2 + \frac{1}{4}} \cdot |(\phi_p + c^2)|^{\kappa_3 + \frac{1}{4}} \right] \prod_{p \ne q} |(\phi_p - \phi_q)|.$$

This product is zero when all the variables ϕ_p have their least values and also when all have their greatest values; when the variables ϕ_p are unequal both to each other and to $-a^2$, $-b^2$, $-c^2$, then Π is positive and it is obviously a continuous bounded function of the variables.

Hence there is a set of values of the variables for which Π attains its upper bound, which is positive and not zero (cf. §3.62).

For this set of values of the variables the conditions for a maximum give

$$\frac{\partial \log \Pi}{\partial \phi_1} = \frac{\partial \log \Pi}{\partial \phi_2} = \cdots = 0,$$

that is to say

$$\frac{\kappa_1 + \frac{1}{4}}{\phi_p + a^2} + \frac{\kappa_2 + \frac{1}{4}}{\phi_p + b^2} + \frac{\kappa_3 + \frac{1}{4}}{\phi_p + c^2} + \sum_{q=1}^{m}{}' \frac{1}{\phi_p - \phi_q} = 0,$$

where p assumes in turn the values $1, 2, \ldots, m$.

Now this system of equations is precisely the system by which $\theta_1, \theta_2, \ldots, \theta_p$ are determined (cf. §§23.21–23.24); and *so the system of equations determining* $\theta_1, \theta_2, \ldots, \theta_m$ *has a solution for which*

$$\begin{cases} -a^2 < \theta_p < -b^2, & (p = 1, 2, \ldots, r - 1) \\ -b^2 < \theta_p < -c^2, & (p = r, r + 1 \ldots, m). \end{cases}$$

Hence, if r has any of the values $1, 2, \ldots, m + 1$, a Lamé function exists with $r - 1$ of its zeros between $-a^2$ and $-b^2$ and the remaining $m - r + 1$ zeros between $-b^2$ and $-c^2$.

Since there are $m + 1$ Lamé functions of the specified type, they are all obtained when r is given in turn the values $1, 2, \ldots, m + 1$; and this is the theorem due to Stieltjes.

An interesting statical interpretation of the theorem was given by Stieltjes, namely that if $m + 3$ particles which attract one another according to the law of the inverse distance are placed on a line, and three of these particles, whose masses are $\kappa_1 + \frac{1}{4}$, $\kappa_2 + \frac{1}{4}$, $\kappa_3 + \frac{1}{4}$, are fixed at points with coordinates $-a^2$, $-b^2$, $-c^2$, the remainder being of unit mass and free to move on the line, then $\log \Pi$ is the gravitational potential of the system; and the positions of equilibrium of the system are those in which the coordinates of the moveable particles are $\theta_1, \theta_2, \ldots, \theta_m$, i.e. the values of θ for which a certain one of the Lamé functions of degree $2(m + \kappa_1 + \kappa_2 + \kappa_3)$ vanishes.

Example 23.4.5 (Stieltjes) Discuss the positions of the zeros of polynomials which satisfy an equation of the type

$$\frac{d^2\Lambda}{d\theta^2} + \sum_{s=1}^{r} \frac{1-\alpha_s}{\theta - \alpha_s} \frac{d\Lambda}{d\theta} + \frac{\phi_{r-2}(\theta)}{r \prod_{s=1}^{r}(\theta - \alpha_s)} \Lambda = 0,$$

where $\phi_{r-2}(\theta)$ is a polynomial of degree $r-2$ in θ in which the coefficient of θ^{r-2} is

$$-m \left\{ m + r - 1 - \sum_{x=1}^{r} \alpha_s \right\},$$

m being a positive integer, and the remaining coefficients in $\phi_{r-2}(\theta)$ are determined from the consideration that the equation has a polynomial solution.

23.47 Lamé functions of the second kind

The functions $E_n^m(\xi)$, hitherto discussed, are known as Lamé functions of the *first kind*. It is easy to verify that an independent solution of Lamé's equation

$$\frac{d^2\Lambda}{du^2} = \{n(n+1)\xi + B_n^m\} \Lambda$$

is the function $F_n^m(\xi)$ defined by the equation

$$F_n^m(\xi) = (2n+1)E_n^m(\xi) \int_0^u \frac{du}{\left\{E_n^m(\xi)\right\}^2},$$

and $F_n^m(\xi)$ is termed a Lamé function of the *second kind*. This definition of the function $F_n^m(\xi)$ is due to Heine [285].

From this formula it is clear that, near $u = 0$,

$$F_n^m(\xi) = (2n+1)u^{-n} \{1 + O(u)\} \int_0^u u^{2n} \{1 + O(u)\} \, du = u^{n+1} \{1 + O(u)\},$$

and we obviously have

$$E_n^m(\xi) = u^{-n} \{1 + O(u)\} .$$

It is clear from these results that $F_n^m(\xi)$ can never be a Lamé function of the first kind, *and so there is no value of B_n^m for which Lamé's equation is satisfied by two Lamé functions of the first kind of different species or types.*

It is possible to obtain an expression for $F_n^m(\xi)$ which is free from quadratures, analogous to Christoffel's formula for $Q_n(z)$, given on Chapter 15, Example 15.29. We shall give the analysis in the case when $E_n^m(\xi)$ is of the first species. The only irreducible poles of $1/\left\{E_n^m(\xi)\right\}^2$, *qua* function of u, are at a set of points $u_1, u_2, \ldots, u_n$ which are none of them periods or half periods.

Near any one of these points we have an expansion of the form

$$E_n^m(\xi) = k_1(u - u_r) + k_2(u - u_r)^2 + k_3(u - u_r)^3 + \cdots,$$

and, by substitution of this series in the differential equation, it is found that k_2 is zero.

Hence the principal part of $1/\{E_n^m(\xi)\}^2$ near u_r is

$$\frac{1}{k_1^2(u-u_r)^2},$$

and the residue is zero.

Hence we can find constants A_r such that

$$\{E_n^m(\xi)\}^{-2} - \sum_{r=1}^n A_r \wp(u-u_r)$$

has no poles at any points congruent to any of the points u_r; it is therefore a constant A, by Liouville's theorem, since it is a doubly periodic function of u.

Hence

$$\int_0^u \frac{du}{\{E_n^m(\xi)\}^2} = Au - \sum_{r=1}^n A_r \{\zeta(u-u_r) + \zeta(u_r)\}.$$

Now the points u_r can be grouped in pairs whose sum is zero, since $E_n^m(\xi)$ is an even function of u.

If we take $u_{n-r} = -u_{r+1}$, we have

$$\int_0^u \frac{du}{\{E_n^m(\xi)\}^2} = Au - \sum_{r=1}^{n/2} A_r \{\zeta(u-u_r) + \zeta(u+u_r)\}$$

$$= Au - 2\zeta(u) \sum_{r=1}^{n/2} A_r - \sum_{r=1}^{n/2} \frac{A_r \wp'(u)}{\wp(u) - \wp(u_r)},$$

and therefore

$$F_n^m(\xi) = (2n+1) \left\{ Au - 2\zeta(u) \sum_{r=1}^{n/2} A_r \right\} E_n^m(\xi) + \wp'(u) w_{n/2-1}(\xi),$$

where $w_{n/2-1}(\xi)$ is a polynomial in ξ of degree $n/2 - 1$.

Example 23.4.6 Obtain formulae analogous to this expression for $F_n^m(\xi)$ when $E_n^m(\xi)$ is of the second, third or fourth species.

23.5 Lamé's equation in association with Jacobian elliptic functions

All the results which have so far been obtained in connexion with Lamé functions of course have their analogues in the notation of Jacobian elliptic functions, and, in the hands of Hermite (see §23.71), the use of Jacobian elliptic functions in the discussion of generalisations of Lamé's equation has produced extremely interesting results.

Unfortunately it is not possible to use Jacobian elliptic functions in which all the variables involved are real, without a loss of symmetry.

The symmetrical formulae may be obtained by taking new variables α, β, γ defined by the equations

$$\begin{cases} \alpha = iK' + u\sqrt{e_1 - e_3}, \\ \beta = iK' + v\sqrt{e_1 - e_3}, \\ \gamma = iK' + w\sqrt{e_1 - e_3}, \end{cases}$$

and then the formulae of §23.31 are equivalent to

$$\begin{cases} x = & k^2\sqrt{a^2 - c^2}\, \text{sn}\,\alpha\, \text{sn}\,\beta\, \text{sn}\,\gamma, \\ y = & -(k^2/k')\sqrt{a^2 - c^2}\, \text{cn}\,\alpha\, \text{cn}\,\beta\, \text{cn}\,\gamma, \\ z = & (i/k')\sqrt{a^2 - c^2}\, \text{dn}\,\alpha\, \text{dn}\,\beta\, \text{dn}\,\gamma, \end{cases}$$

the modulus of the elliptic functions being

$$\sqrt{\frac{a^2 - b^2}{a^2 - c^2}}.$$

The equation of the quadric of the confocal system on which α is constant is

$$\frac{X^2}{(a^2 - b^2)\,\text{sn}^2\,\alpha} - \frac{Y^2}{(a^2 - b^2)\,\text{cn}^2\,\alpha} - \frac{Z^2}{(a^2 - c^2)\,\text{dn}^2\,\alpha} = 1.$$

This is an ellipsoid if α lies between iK' and $K + iK'$; the quadric on which β is constant is an hyperboloid of one sheet if β lies between $K + iK'$ and K; and the quadric on which γ is constant is an hyperboloid of two sheets if γ lies between 0 and K; and with this determination of (α, β, γ) the point (x, y, z) lies in the positive octant.

It has already been seen (§23.4) that, with this notation, Lamé's equation assumes the form

$$\frac{d^2\Lambda}{d\alpha^2} = \{n(n + 1)k^2\,\text{sn}^2\,\alpha + A\}\,\Lambda,$$

and the solutions expressible as periodic functions of α will be called[6] $E_n^m(\alpha)$. The first species of Lamé's function is then a polynomial in $\text{sn}^2\,\alpha$, and generally the species may be defined by a scheme analogous to that of §23.2

$$\begin{cases} & \text{sn}\,\alpha, & \text{cn}\,\alpha\,\text{dn}\,\alpha, \\ 1, & \text{cn}\,\alpha, & \text{dn}\,\alpha\,\text{sn}\,\alpha, & \text{sn}\,\alpha\,\text{cn}\,\alpha\,\text{dn}\,\alpha \\ & \text{dn}\,\alpha, & \text{sn}\,\alpha\,\text{cn}\,\alpha, \end{cases} \prod_p (\text{sn}^2\,\alpha - \text{sn}^2\,\alpha_p).$$

23.6 The integral equation satisfied by Lamé functions of the first and second species

We shall now shew that, if $E_n^m(\alpha)$ is any Lamé function of the first species (n being even) or of the second species (n being odd) with $\text{sn}\,\alpha$ as a factor, then $E_n^m(\alpha)$ is a solution of the integral equation[7]

$$E_n^m(\alpha) = \lambda \int_{-2K}^{2K} P_n(k\,\text{sn}\,\alpha\,\text{sn}\,\theta)E_n^m(\theta)\,d\theta;$$

---

[6] There is no risk of confusing these with the corresponding functions $E_n^m(\xi)$.

[7] This integral equation and the corresponding formulae of (§23.62) associated with ellipsoidal harmonics were given by Whittaker [675]. Proofs of the formulae involving functions of the third and fourth species have not been previously published.

where λ is one of the 'characteristic numbers' (§11.23).

To establish this result we need the lemma that $P_n(k \operatorname{sn} \alpha \operatorname{sn} \theta)$ is annihilated by the partial differential operator

$$\frac{\partial^2}{\partial \alpha^2} - \frac{\partial^2}{\partial \theta^2} - n(n+1)k^2(\operatorname{sn}^2 \alpha - \operatorname{sn}^2 \theta).$$

To prove the lemma, observe that, when μ is written for brevity in place of $k \operatorname{sn} \alpha \operatorname{sn} \theta$, we have

$$\left\{ \frac{\partial^2}{\partial \alpha^2} - \frac{\partial^2}{\partial \theta^2} \right\} P_n(k \operatorname{sn} \alpha \operatorname{sn} \theta) = k^2 \left\{ \operatorname{cn}^2 \alpha \operatorname{dn}^2 \alpha \operatorname{sn}^2 \theta - \operatorname{cn}^2 \theta \operatorname{dn}^2 \theta \operatorname{sn}^2 \alpha \right\} P_n''(\mu)$$

$$+ 2k^3 \operatorname{sn} \alpha \operatorname{sn} \theta (\operatorname{sn}^2 \alpha - \operatorname{sn}^2 \theta) P_n'(\mu)$$
$$= k^2 (\operatorname{sn}^2 \alpha - \operatorname{sn}^2 \theta) \left[(\mu^2 - 1) P_n''(\mu) + 2\mu P_n'(\mu) \right]$$
$$= k^2 (\operatorname{sn}^2 \alpha - \operatorname{sn}^2 \theta) n(n+1) P_n(\mu),$$

when we use Legendre's differential equation (§15.13). And the lemma is established.

The result of applying the operator

$$\frac{\partial^2}{\partial \alpha^2} - n(n+1)k^2 \operatorname{sn}^2 \alpha - A_n^m$$

to the integral

$$\int_{-2K}^{2K} P_n(k \operatorname{sn} \alpha \operatorname{sn} \theta) E_n^m(\theta) \, d\theta$$

is now seen to be

$$\int_{-2K}^{2K} \left\{ \frac{\partial^2}{\partial \alpha^2} - n(n+1)k^2 \operatorname{sn}^2 \alpha - A_n^m \right\} P_n(k \operatorname{sn} \alpha \operatorname{sn} \theta) E_n^m(\theta) \, d\theta$$

$$= \int_{-2K}^{2K} \left[\left\{ \frac{\partial^2}{\partial \theta^2} - n(n+1)k^2 \operatorname{sn}^2 \theta - A_n^m \right\} P_n(k \operatorname{sn} \alpha \operatorname{sn} \theta) \right] E_n^m(\theta) \, d\theta,$$

and when we integrate twice by parts this becomes

$$\left[\frac{\partial P_n(k \operatorname{sn} \alpha \operatorname{sn} \theta)}{\partial \theta} E_n^m(\theta) - P_n(k \operatorname{sn} \alpha \operatorname{sn} \theta) \frac{dE_n^m(\theta)}{d\theta} \right]_{-2K}^{2K}$$

$$+ \int_{-2K}^{2K} P_n(k \operatorname{sn} \alpha \operatorname{sn} \theta) \left\{ \frac{d^2}{d\theta^2} - n(n+1)k^2 \operatorname{sn}^2 \theta - A_n^m \right\} E_n^m(\theta) \, d\theta = 0.$$

Hence it follows that the integral

$$\int_{-2K}^{2K} P_n(k \operatorname{sn} \alpha \operatorname{sn} \theta) E_n^m(\theta) \, d\theta$$

is annihilated by the operator

$$\frac{d^2}{d\alpha^2} - n(n+1)k^2 \operatorname{sn} \alpha - A_n^m,$$

and it is evidently a polynomial of degree n in $\operatorname{sn}^2 \alpha$. Since Lamé's equation has only one integral of this type[8], it follows that the integral is a multiple of $E_n^m(\alpha)$ if it is not zero; and the result is established.

Note It appears that *every* characteristic number associated with the equation

$$f(\alpha) = \lambda \int_{-2K}^{2K} P_n(k \operatorname{sn} \alpha \operatorname{sn} \theta) f(\theta)$$

yields a solution of Lamé's equation; cf. Ince [334].

Example 23.6.1 Shew that the nucleus of an integral equation satisfied by Lamé functions of the first species (n being even) or of the second species (n being odd) with cn α as a factor, may be taken to be

$$P_n \left(\frac{ik}{k'} \operatorname{cn} \alpha \operatorname{cn} \theta \right).$$

Example 23.6.2 Shew that the nucleus of an integral equation satisfied by Lamé functions of the first species (n being even) or of the second species (n being odd) with dn α as a factor, may be taken to be

$$P_n \left(\frac{1}{k'} \operatorname{dn} \alpha \operatorname{dn} \theta \right).$$

23.61 The integral equation satisfied by Lamé functions of the third and fourth species

The theorem analogous to that of §23.6, in the case of Lamé functions of the third and fourth species, is that any Lamé function of the fourth species (n being odd) or of the third species (n being even) with cn α dn α as a factor, satisfies the integral equation

$$E_n^m(\alpha) = \lambda \int_{-2K}^{2K} \operatorname{cn} \alpha \operatorname{dn} \alpha \operatorname{cn} \theta \operatorname{dn} \theta P_n''(k \operatorname{sn} \alpha \operatorname{sn} \theta) E_n^m(\theta) \, d\theta.$$

The preliminary lemma is that the nucleus

$$\operatorname{cn} \alpha \operatorname{dn} \alpha \operatorname{cn} \theta \operatorname{dn} \theta P_n''(k \operatorname{sn} \alpha \operatorname{sn} \theta),$$

like the nucleus of §23.6, is annihilated by the operator

$$\frac{\partial^2}{\partial \alpha^2} - \frac{\partial^2}{\partial \theta^2} - n(n+1)k^2(\operatorname{sn}^2 \alpha - \operatorname{sn}^2 \theta).$$

To verify the lemma observe that

$$\frac{\partial^2}{\partial \alpha^2} \left\{ \operatorname{cn} \alpha \operatorname{dn} \alpha P_n''(k \operatorname{sn} \alpha \operatorname{sn} \theta) \right\} = k^2 \operatorname{cn}^3 \alpha \operatorname{dn}^3 \alpha \operatorname{sn}^2 \theta P_n^{iv}(\mu)$$
$$- 3k \operatorname{sn} \alpha \operatorname{cn} \alpha \operatorname{dn} \alpha \operatorname{sn} \theta (dn^2 \alpha + k^2 \operatorname{cn}^2 \alpha) P_n'''(\mu)$$
$$- \operatorname{cn} \alpha \operatorname{dn} \alpha (dn^2 \alpha + k^2 \operatorname{cn}^2 \alpha - 4k^2 \operatorname{sn}^2 \alpha) P_n''(\mu),$$

---

[8] The other solution when expanded in descending powers of sn α begins with a term in $(\operatorname{sn} \alpha)^{-n-1}$.

and so

$$\left\{\frac{\partial^2}{\partial\alpha^2} - \frac{\partial^2}{\partial\theta^2}\right\} \cdot \left\{\operatorname{cn}\alpha\,\operatorname{dn}\alpha\,\operatorname{cn}\theta\,\operatorname{dn}\theta P_n''(k\operatorname{sn}\alpha\operatorname{sn}\theta)\right\}$$

$$= k\operatorname{cn}\alpha\,\operatorname{dn}\alpha\,\operatorname{cn}\theta\,\operatorname{dn}\theta(\operatorname{sn}^2\alpha - \operatorname{sn}^2\theta)\left\{(\mu^2-1)P_n^{\mathrm{iv}}(\mu)\right.$$
$$\left.+6\mu P_n'''(\mu) + 6P_n''(\mu)\right\}$$

$$= k^2\operatorname{cn}\alpha\,\operatorname{dn}\alpha\,\operatorname{cn}\theta\,\operatorname{dn}\theta(\operatorname{sn}^2\alpha - \operatorname{sn}^2\theta)\frac{d^3}{d\mu^3}\left\{(\mu^2-1)P_n'(\mu)\right\}$$

$$= k^2 n(n+1)\operatorname{cn}\alpha\,\operatorname{dn}\alpha\,\operatorname{cn}\theta\,\operatorname{dn}\theta(\operatorname{sn}^2\alpha - \operatorname{sn}^2\theta)P_n''(\mu),$$

and the lemma is established. The proof that $E_n^m(\alpha)$ satisfies the integral equation now follows precisely as in the case of the integral equation of §23.6.

Example 23.6.3　Shew that the nucleus of an integral equation which is satisfied by Lamé functions of the fourth species (n being odd) or of the third species (n being even) with $\operatorname{sn}a$ $\operatorname{dn}a$ as a factor, may be taken to be

$$\operatorname{sn}a\,\operatorname{dn}a\,\operatorname{sn}\theta\,\operatorname{dn}\theta P_n''\left(\frac{ik}{k'}\operatorname{cn}a\operatorname{cn}\theta\right).$$

Example 23.6.4　Shew that the nucleus of an integral equation which is satisfied by Lamé functions of the fourth species (n being odd) or of the third species (n being even) with $\operatorname{sn}a\operatorname{cn}a$ as a factor, may be taken to be

$$\operatorname{sn}a\,\operatorname{cn}a\,\operatorname{sn}\theta\,\operatorname{cn}\theta P_n''\left(\frac{1}{k'}\operatorname{dn}a\operatorname{dn}\theta\right).$$

Example 23.6.5　Obtain the following three integral equations satisfied by Lamé functions of the fourth species (n being odd) and of the third species (n being even):

(i)　$k^2\operatorname{sn}^2 aE_n^m(a) = \lambda\operatorname{cn}a\operatorname{dn}a\,ds\int_{-2K}^{2K}P_n(k\operatorname{sn}a\operatorname{sn}\theta)\frac{d}{d\theta}\left\{\frac{1}{\operatorname{cn}\theta\operatorname{dn}\theta}\frac{dE_n^m(\theta)}{d\theta}\right\}d\theta,$

(ii)　$-k^2\operatorname{cn}^2 aE_n^m(a) = \lambda k'^2\operatorname{sn}a\operatorname{dn}a\int_{-2K}^{2K}P_n\left(\frac{ik}{k'}\operatorname{cn}a\operatorname{cn}\theta\right)\frac{d}{d\theta}\left\{\frac{1}{\operatorname{sn}\theta\operatorname{dn}\theta}\frac{dE_n^m(\theta)}{d\theta}\right\}d\theta,$

(iii)　$k^2\operatorname{dn}^2 aE_n^m(a) = \lambda k'^2\operatorname{sn}a\operatorname{cn}a\int_{-2K}^{2K}P_n\left(\frac{1}{k'}\operatorname{dn}a\operatorname{dn}\theta\right)\frac{d}{d\theta}\left\{\frac{1}{\operatorname{sn}\theta\operatorname{cn}\theta}\frac{dE_n^m(\theta)}{d\theta}\right\}d\theta;$

in the case of functions of even order, the functions of the different types each satisfy one of these equations only.

23.62 Integral formulae for ellipsoidal harmonics

The integral equations just considered make it possible to obtain elegant representations of the ellipsoidal harmonic $G_n^m(x,y,z)$ and of the corresponding homogeneous harmonic $H_n^m(x,y,z)$ in terms of definite integrals.

From the general equation formula of §18.3, it is evident that $H_n^m(x,y,z)$ is expressible in the form

$$H_n^m(x,y,z) = \int_{-\pi}^{\pi}(x\cos t + y\sin t + iz)^n f(t)\,dt,$$

where $f(t)$ is a periodic function to be determined.

Now the result of applying Niven's operator D^2 to $(x \cos t + y \sin t + iz)^n$ is

$$n(n-1)(a^2 \cos^2 t + b^2 \sin^2 t - c^2)(x \cos t + y \sin t + iz)^{n-2},$$

and so, by Niven's formula (§23.25) we find that $G_n^m(x, y, z)$ is expressible in the form

$$G_n^m(x, y, z) =$$

$$\int_{-\pi}^{\pi} \left\{ \mathfrak{A}^n - \frac{n(n-1)}{2(2n-1)} \mathfrak{A}^{n-2}\mathfrak{B}^2 + \frac{n(n-1)(n-2)(n-3)}{2 \cdot 4(2n-1)(2n-3)} \mathfrak{A}^{n-4}\mathfrak{B}^4 - \cdots \right\} f(t)\, dt,$$

where $\mathfrak{A} \equiv x \cos t + y \sin t + iz$, $\mathfrak{B} \equiv \sqrt{(a^2 - c^2)\cos^2 t + (b^2 - c^2)\sin^2 t}$, so that

$$G_n^m(x, y, z) = \frac{2^n (n!)^2}{(2n)!} \int_{-\pi}^{\pi} \mathfrak{B}^n P_n \left(\frac{x \cos t + y \sin t + iz}{\sqrt{(a^2 - c^2)\cos^2 t + (b^2 - c^2)\sin^2 t}} \right) f(t)\, dt.$$

Now write $\sin t \equiv \operatorname{cd}\theta$, the modulus of the elliptic functions being, as usual, given by the equation

$$k^2 = \frac{a^2 - b^2}{a^2 - c^2}.$$

The new limits of integration are $-3K$ and K, but they may be replaced by $-2K$ and $2K$ on account of the periodicity of the integrand.

It is thus found that

$$G_n^m(x, y, z) = \int_{-2K}^{2K} P_n \left(\frac{k' x \operatorname{sn}\theta + y \operatorname{cn}\theta + iz \operatorname{dn}\theta}{\sqrt{b^2 - c^2}} \right) \phi(\theta)\, d\theta,$$

where $\phi(\theta)$ is a periodic function of θ, independent of x, y, z, which is, as yet, to be determined.

If we express the ellipsoidal harmonic as the product of three Lamé functions, with the aid of the formulae of §23.5 we find that

$$E_n^m(\alpha)E_n^m(\beta)E_n^m(\gamma) = C \int_{-2K}^{2K} P_n(\mu)\phi(\theta)\, d\theta,$$

where C is a known constant and

$$\mu \equiv k^2 \operatorname{sn}\alpha \operatorname{sn}\beta \operatorname{sn}\gamma \operatorname{sn}\theta - (k^2/k'^2) \operatorname{cn}\alpha \operatorname{cn}\beta \operatorname{cn}\gamma \operatorname{cn}\theta$$
$$- (1/k'^2) \operatorname{dn}\alpha \operatorname{dn}\beta \operatorname{dn}\gamma \operatorname{dn}\theta.$$

If the ellipsoidal harmonic is of the first species or of the second species and first type, we now give β and γ the special values $\beta = K$, $\gamma = K + iK'$, and we see that

$$C \int_{-2K}^{2K} P_n(k \operatorname{sn}\alpha \operatorname{sn}\theta)\, \phi(\theta)\, d\theta$$

is a solution of Lamé's equation, and so, by §23.6, $\phi(\theta)$ is a solution of Lamé's equation which can be no other[9] than a multiple of $E_n^m(\theta)$.

---

[9] If $\phi(\theta)$ involved the second solution, the integral would not converge.

Hence it follows that

$$G_n^m(x, y, z) = \lambda \int_{-2K}^{2K} P_n \left(\frac{k'x \operatorname{sn} \theta + y \operatorname{cn} \theta + iz \operatorname{dn} \theta}{\sqrt{b^2 - c^2}} \right) E_n^m(\theta) \, d\theta,$$

where λ is a constant.

If $G_n^m(x, y, z)$ be of the second species and of the second or third type we put $\beta = 0$, $\gamma = K + iK'$, or $\beta = 0$, $\gamma = K$ respectively, and we obtain anew the same formula.

It thus follows that if $G_n^m(x, y, z)$ be any ellipsoidal harmonic of the first or second species, then

$$G_n^m(x, y, z) = \lambda \int_{-2K}^{2K} P_n(\mu) E_n^m(\theta) \, d\theta,$$

$$H_n^m(x, y, z) = \lambda \frac{(2n)!}{2^n (n!)^2 (b^2 - c^2)^{n/2}} \int_{-\pi}^{\pi} (k'x \operatorname{sn} \theta + y \operatorname{cn} \theta + iz \operatorname{dn} \theta)^n E_n^m(\theta) \, d\theta,$$

where $\mu \equiv (k'x \operatorname{sn} \theta + y \operatorname{cn} \theta + iz \operatorname{dn} \theta)/\sqrt{b^2 - c^2}$.

23.63 Integral formulae for ellipsoidal harmonics of the third and fourth species

In order to obtain integral expressions for harmonics of the third and fourth species, we turn to the equation of §23.62, namely

$$E_n^m(\alpha) E_n^m(\beta) E_n^m(\gamma) = C \int_{-2K}^{2K} P_n(\mu) \phi(\theta) \, d\theta,$$

where

$$\mu \equiv k^2 \operatorname{sn} \alpha \operatorname{sn} \beta \operatorname{sn} \gamma \operatorname{sn} \theta - (k^2/k'^2) \operatorname{cn} \alpha \operatorname{cn} \beta \operatorname{cn} \gamma \operatorname{cn} \theta - (1/k'^2) \operatorname{dn} \alpha \operatorname{dn} \beta \operatorname{dn} \gamma \operatorname{dn} \theta;$$

this equation is satisfied by harmonics of *any* species.

Suppose now that $E_n^m(\alpha)$ is of the fourth species or of the first type of the third species so that it has $\operatorname{cn} \alpha \operatorname{dn} \alpha$ as a factor.

We next differentiate the equation with respect to β and γ, and then put $\beta = K, \gamma = K + iK'$. It is thus found that

$$E_n^m(\alpha) \left[\frac{d}{d\beta} E_n^m(\beta) \right]_{\beta=K} \left[\frac{d}{d\gamma} E_n^m(\gamma) \right]_{\gamma=K+iK'} = C \int_{-2K}^{2K} \left[\frac{\partial^2 P_n(\mu)}{\partial \beta \partial \gamma} \right]_{(\beta=K, \gamma=K+iK')} \phi(\theta) \, d\theta.$$

Now

$$\left[\frac{\partial P_n(\mu)}{\partial \gamma} \right]_{\gamma=K+iK'} = -(i/k') \operatorname{dn} \alpha \operatorname{dn} \beta \operatorname{dn} \theta P_n'(\mu),$$

so that

$$\left[\frac{\partial^2 P_n(\mu)}{\partial \beta \partial \gamma} \right]_{(\beta=K, \gamma=K+iK')} = -k \operatorname{cn} \alpha \operatorname{dn} \alpha \operatorname{dn} \theta P_n''(k \operatorname{sn} \alpha \operatorname{sn} \theta).$$

Hence

$$\int_{-2K}^{2K} \operatorname{cn} \alpha \operatorname{dn} \alpha \operatorname{cn} \theta \operatorname{dn} \theta P_n''(k \operatorname{sn} \alpha \operatorname{sn} \theta) \phi(\theta) \, d\theta$$

is a solution of Lamé's equation with cn α dn α as a factor; and so, by §23.61, $\phi(\theta)$ *can be none other than a constant multiple of* $E_n^m(\alpha)$.

We have thus found that the equation

$$G_n^m(x, y, z) = \lambda \int_{-2K}^{2K} P_n(\mu) E_n^m(\theta)\, d\theta$$

is satisfied by any ellipsoidal harmonic which has cn α dn α as a factor; the corresponding formula for the homogeneous harmonic is

$$H_n^m(x, y, z) = \lambda \frac{(2n)!}{2^n (n!)^2 (b^2 - c^2)^{n/2}} \int_{-2K}^{2K} (k' x \operatorname{sn} \theta + y \operatorname{cn} \theta + iz \operatorname{dn} \theta)^n E_n^m(\theta)\, d\theta.$$

Example 23.6.6 Shew that the equation of this section is satisfied by the ellipsoidal harmonics which have sn a dn a or sn a cn a as a factor.

23.7 Generalisations of Lamé's equation

Two obvious generalisations of Lamé's equation at once suggest themselves. In the first, the constant B has not one of the characteristic values B_n^m, for which a solution is expressible as an algebraic function of $\wp(u)$; and in the second, the degree n is no longer supposed to be an integer. The first generalisation has been fully dealt with by Hermite [291] and Halphen [269, pp. 494–502] but the only case of the second which has received any attention is that in which n is half of an odd integer; this has been discussed by Brioschi [99], Halphen [269, pp. 471–473] and Crawford [157].

We shall now examine the solution of the equation

$$\frac{d^2 \Lambda}{du^2} = \{n(n+1)\wp(u) + B\}\Lambda,$$

where B is arbitrary and n is a positive integer, by the method of Lindemann–Stieltjes already explained in connexion with Mathieu's equation (§§19.5, 19.52).

The product of any pair of solutions of this equation is a solution of

$$\frac{d^3 X}{du^3} - 4\{n(n+1)\wp(u) + B\}\frac{dX}{du} - 2n(n+1)\wp'(u)X = 0,$$

by §19.52. The algebraic form of this equation is

$$4(\xi - e_1)(\xi - e_2)(\xi - e_3)\frac{d^3 X}{d\xi^3} + 3(6\xi^2 - \tfrac{1}{2}g_2)\frac{d^2 X}{d\xi^2}$$

$$- 4\{(n^2 + n - 3)\xi + B\}\frac{dX}{d\xi} - 2n(n+1)X = 0.$$

If a solution of this in descending powers of $\xi - e_2$ be taken to be

$$X = \sum_{r=0}^{\infty} c_r (\xi - e_2)^{n-r}, \qquad (c_0 = 1)$$

the recurrence formula for the coefficients c_r is

$$4r(n - r + \tfrac{1}{2})(2n - r + 1)c_r$$
$$= (n - r + 1)\{12e_2(n - r)(n - r + 2) - 4e_2(n^2 + n - 3) - 4B\}c_{r-1}$$
$$- 2(n - r + 1)(n - r + 2)(e_1 - e_2)(e_2 - e_3)(2n - 2r + 3)c_{r-2}.$$

Write $r = n + 1$, and it is seen that $c_{n+1} = 0$; then write $r = n + 2$ and $c_{n+2} = 0$; and the recurrence formulae with $r > n + 2$ are all satisfied by taking

$$c_{n+3} = c_{n+4} = \cdots = 0.$$

Hence Lamé's generalised equation always has two solutions whose product is of the form

$$\sum_{r=0}^{n} c_r (\xi - e_2)^{n-r}.$$

This polynomial may be written in the form

$$\prod_{r=1}^{n} \{\wp(u) - \wp(a_r)\},$$

where $a_1, a_2, \ldots, a_n$ are, as yet, undetermined as to their signs; and the two solutions of Lamé's equation will be called Λ_1, Λ_2.

Two cases arise: (I) when Λ_1/Λ_2 is constant; (II) when Λ_1/Λ_2 is not constant.

(I)   The first case is easily disposed of; for unless the polynomial

$$\prod_{r=1}^{n} \{\xi - \wp(a_r)\}$$

is a perfect square in ξ, multiplied possibly by expressions of the type $\xi - e_1, \xi - e_2, \xi - e_3$, then the algebraic form of Lamé's equation has an indicial equation, one of whose roots is $\tfrac{1}{2}$, at one or more of the points $\xi = \wp(a_r)$; and this is not the case (§23.43).

Hence the polynomial must be a square multiplied possibly by one or more of $\xi - e_1$, $\xi - e_2, \xi - e_3$, and then Λ_1 is a Lamé function, so that B has one of the characteristic values B_n^m; and this is the case which has been discussed at length in §§23.1–23.47.

(II)   In the second case we have (§19.53)

$$\Lambda_1 \frac{d\Lambda_2}{du} - \Lambda_2 \frac{d\Lambda_1}{du} = 2\mathfrak{C},$$

where $\mathfrak{C}$ is a constant which is not zero. Then

$$\begin{cases} \dfrac{d\log\Lambda_2}{du} - \dfrac{d\log\Lambda_1}{du} = \dfrac{2\mathfrak{C}}{X}, \\[2mm] \dfrac{d\log\Lambda_2}{du} - \dfrac{d\log\Lambda_1}{du} = \dfrac{1}{X}\dfrac{dX}{du}, \end{cases}$$

so that

$$\frac{d\log\Lambda_1}{du} = \frac{1}{2X}\frac{dX}{du} - \frac{\mathfrak{C}}{X}, \qquad \frac{d\log\Lambda_2}{du} = \frac{1}{2X}\frac{dX}{du} + \frac{\mathfrak{C}}{X}.$$

On integration, we see that we may take

$$\Lambda_1 = \sqrt{X} \exp\left\{-\mathfrak{C}\int \frac{du}{X}\right\}.$$

Again, if we differentiate the equation

$$\frac{1}{\Lambda_1}\frac{d\Lambda_1}{du} = \frac{1}{2X}\frac{dX}{du} - \frac{\mathfrak{C}}{X},$$

we find that

$$\frac{1}{\Lambda_1}\frac{d^2\Lambda_1}{du^2} - \left\{\frac{1}{\Lambda_1}\frac{d\Lambda_1}{du}\right\}^2 \frac{1}{2X}\frac{d^2X}{du^2} - \frac{1}{2X^2}\left(\frac{dX}{du}\right)^2 + \frac{\mathfrak{C}}{X^2}\frac{dX}{du},$$

and hence, with the aid of Lamé's equation, we obtain the interesting formula

$$n(n+1)\wp(u) + B = \frac{1}{2X}\frac{d^2X}{du^2} - \left(\frac{1}{2X}\frac{dX}{du}\right)^2 + \frac{\mathfrak{C}^2}{X^2}.$$

If now $\xi_r \equiv \wp(a_r)$, we find from this formula (when multiplied by X^2), that, if u be given the special value a_r, then

$$\left(\frac{dX}{d\xi}\right)^2_{\xi=\xi_r} = \frac{4\mathfrak{C}^2}{\wp'^2(a_r)}.$$

We now fix the signs of $a_1, a_2, \ldots, a_n$ by taking

$$\left(\frac{dX}{d\xi}\right)_{\xi=\xi_r} = \frac{2\mathfrak{C}}{+\wp'(a_r)}.$$

And then, if we put $2\mathfrak{C}/X$, *qua* function ξ, into partial fractions, it is seen that

$$\frac{2\mathfrak{C}}{X} = \sum_{r=1}^{n}\frac{\wp'(a_r)}{\xi - \wp(a_r)} = \sum_{r=1}^{n}\{\zeta(u-a_r) - \zeta(u+a_r) + 2\zeta(a_r)\},$$

and therefore

$$\Lambda_1 = \left[\prod_{r=1}^{n}\{\wp(u) - \wp(a_r)\}\right]^{\frac{1}{2}}$$

$$\times \exp\left[\frac{1}{2}\sum_{r=1}^{n}\{\log\sigma(a_r + u) - \log\sigma(a_r - u) - 2u\zeta(a_r)\}\right],$$

whence it follows that (§20.53, Example 20.5.2)

$$\Lambda_1 = \prod_{r=1}^{n}\left\{\frac{\sigma(a_r + u)}{\sigma(u)\sigma(a_r)}\right\}\exp\left\{-u\sum_{r=1}^{n}\zeta(a_r)\right\},$$

and

$$\Lambda_2 = \prod_{r=1}^{n}\left\{\frac{\sigma(a_r - u)}{\sigma(u)\sigma(a_r)}\right\}\exp\left\{u\sum_{r=1}^{n}\zeta(a_r)\right\}.$$

The complete solution has therefore been obtained for arbitrary values of the constant B.

23.71 *The Jacobian form of the generalised Lamé equation*

We shall now construct the solution of the equation

$$\frac{d^2\Lambda}{d\alpha^2} = \left\{n(n+1)k^2 \operatorname{sn}^2\alpha + A\right\}\Lambda,$$

for general values of A, in a form resembling that of §23.6.

The solution which corresponds to that of §23.6 is seen to be

$$\Lambda = \prod_{r=1}^{n}\left\{\frac{\mathrm{H}(\alpha+\alpha_r)}{\Theta(\alpha)}\right\}e^{\rho\alpha},$$

where $\rho, \alpha_1, \alpha_2, \ldots, \alpha_n$ are constants to be determined. This solution was published in 1872 in Hermite's lithographed notes of his lectures delivered at the École polytechnique.

On differentiating this equation it is seen that

$$\frac{1}{\Lambda}\frac{d\Lambda}{d\alpha} = \sum_{r=1}^{n}\left\{\frac{\mathrm{H}'(\alpha+\alpha_r)}{\mathrm{H}(\alpha+\alpha_r)} - \frac{\Theta'(\alpha)}{\Theta(\alpha)}\right\} + \rho$$

$$= \sum_{r=1}^{n}\{Z(\alpha+\alpha_r+iK') - Z(\alpha)\} + \rho + \tfrac{1}{2}n\pi i/K,$$

so that

$$\frac{1}{\Lambda}\frac{d^2\Lambda}{d\alpha^2} - \left\{\frac{1}{\Lambda}\frac{d\Lambda}{d\alpha}\right\}^2 = \sum_{r=1}^{n}\left\{\operatorname{dn}^2(\alpha+\alpha_r+iK') - \operatorname{dn}^2\alpha\right\},$$

and therefore, since Λ is a solution of Lamé's equation, the constants $\rho, \alpha_1, \alpha_2, \ldots, \alpha_n$ are to be determined from the consideration that the equation

$$n(n+1)k^2\operatorname{sn}^2\alpha + A = \sum_{r=1}^{n}\left\{\operatorname{dn}^2(\alpha+\alpha_r+iK') - \operatorname{dn}^2\alpha\right\}$$

$$+ \left[\sum_{r=1}^{n}\{Z(\alpha+\alpha_r+iK') - Z(\alpha)\} + \rho + \tfrac{1}{2}n\pi i/K\right]^2$$

is to be an identity; that is to say

$$n^2k^2\operatorname{sn}^2\alpha + n + A + \sum_{r=1}^{n}\operatorname{cs}^2(\alpha+\alpha_r)$$

$$\equiv \left[\sum_{r=1}^{n}\{Z(\alpha+\alpha_r+iK') - Z(\alpha)\} + \rho + \tfrac{1}{2}n\pi i/K\right]^2.$$

Now both sides of the proposed identity are doubly periodic functions of α with periods $2K$, $2iK'$, and their singularities are double poles at points congruent to $-iK', -\alpha_1, -\alpha_2, \ldots, -\alpha_n$; the dominant terms near $-iK'$ and $-\alpha_r$ are respectively

$$\frac{n^2}{(\alpha+iK')^2}, \qquad -\frac{1}{(\alpha+\alpha_r)^2}$$

in the case of each of the expressions under consideration.

The residues of the expression on the left are all zero and so, if we choose $\rho, \alpha_1, \alpha_2, \ldots, \alpha_n$ so that the residues of the expression on the right are zero, it will follow from Liouville's theorem that the two expressions differ by a constant which can be made to vanish by proper choice of A.

We thus obtain $n + 2$ equations connecting $\rho, \alpha_1, \alpha_2, \ldots, \alpha_n$ with A, but these equations are not all independent.

It is easy to prove that, near $-\alpha_r$,

$$\sum_{r=1}^{n} \{Z(\alpha + \alpha_r + iK') - Z(\alpha)\} + \rho + \tfrac{1}{2}n\pi i/K$$

$$= \frac{1}{\alpha + \alpha_r} + \sum_{p=1}^{n}{}' Z(\alpha_p - \alpha_r + iK') + nZ(\alpha_r) + \rho + \tfrac{1}{2}(n-1)\pi i/K + O(\alpha + \alpha_r),$$

where the prime denotes that the term for which $p = r$ is omitted; and, near $-iK'$,

$$\sum_{r=1}^{n} \{Z(\alpha + \alpha_r + iK') - Z(\alpha)\} + \rho + \tfrac{1}{2}n\pi i/K$$

$$= -\frac{n}{\alpha + iK} + \sum_{r=1}^{n} Z(\alpha_r) + \rho + O(\alpha + iK').$$

Hence the residues of

$$\left[\sum_{r=1}^{n} \{Z(\alpha + \alpha_r + iK') - Z(\alpha)\} + \rho + \tfrac{1}{2}n\pi i/K\right]^2$$

will all vanish if $\rho, \alpha_1, \alpha_2, \ldots, \alpha_n$ are chosen so that the equations

$$\begin{cases} \displaystyle\sum_{p=1}^{n}{}' Z(\alpha_p - \alpha_r + iK') + nZ(\alpha_r) + \rho + \tfrac{1}{2}(n-1)\pi i/K = 0, \\[2mm] \displaystyle\sum_{r=1}^{n} Z(\alpha_r) + \rho = 0 \end{cases}$$

are all satisfied.

The last equation merely gives the value of ρ, namely

$$-\sum_{r=1}^{n} Z(\alpha_r),$$

and, when we substitute this value in the first system, we find that

$$\sum_{p=1}^{n}{}' [Z(\alpha_p + \alpha_r + iK') + Z(\alpha_r) - Z(\alpha_p) + \tfrac{1}{2}\pi i/K] = 0,$$

where $r = 1, 2, \ldots, n$. By §22.735, Example 22.7.10, the sum of the left-hand sides of these equations is zero, so they are equivalent to $n - 1$ equations at most; and, when $\alpha_1, \alpha_2, \ldots, \alpha_n$

have any values which satisfy them, the difference

$$\left[n^2 k^2 \operatorname{sn}^2 \alpha + n + A + \sum_{r=1}^{n} \operatorname{cs}^2(\alpha + \alpha_r) \right.$$

$$\left. - \left[\sum_{r=1}^{n} \left\{ Z(\alpha + \alpha_r + iK') + Z(\alpha) - Z(\alpha_r) + \frac{1}{2}\pi i / K \right\} \right] \right]^2$$

is constant. By taking $\alpha = 0$, it is seen that the constant is zero if

$$n + A + \sum_{r=1}^{n} \operatorname{cs}^2 \alpha_r = \left[\sum_{r=1}^{n} \left\{ Z(\alpha_r + iK') - Z(\alpha_r) + \frac{1}{2}\pi i / K \right\} \right]^2,$$

i.e. if

$$\left\{ \sum_{r=1}^{n} \operatorname{cn} \alpha_r \operatorname{ds} \alpha_r \right\}^2 - \sum_{r=1}^{n} \operatorname{ns}^2 \alpha_r = A.$$

We now reduce the system of n equations; with the notation of §22.2, if functions of a_p, a_r be denoted by the suffixes 1 and 2, it is easy to see that

$$Z(a_p - a_r + iK') + Z(a_r) - Z(a_p) + \tfrac{1}{2}\pi i / K$$

$$= Z(a_p - a_r + iK') + Z(a_r) - Z(a_p + iK') + c_1 d_1 / s_1$$

$$= k^2 \operatorname{sn}(a_p + iK') \operatorname{sn} a_r \operatorname{sn}(a_p + iK' - a_r) + c_1 d_1 / s_1$$

$$= \frac{s_2}{s_1 \operatorname{sn}(a_p - a_r)} + \frac{c_1 d_1}{s_1}$$

$$= \frac{s_2(s_1 c_2 d_2 + s_2 c_1 d_1) + c_1 d_1 (s_1^2 - s_2^2)}{s_1(s_1^2 - s_2^2)}$$

$$= \frac{s_1 c_1 d_1 + s_2 c_2 d_2}{s_1^2 - s_2^2}.$$

Consequently a solution of Lamé's equation

$$\frac{d^2 \Lambda}{d\alpha^2} = \left\{ n(n+1)k^2 \operatorname{sn}^2 \alpha + A \right\} \Lambda$$

is

$$\Lambda = \prod_{r=1}^{n} \left[\frac{H(\alpha + \alpha_r)}{\Theta(\alpha)} \exp\left\{ -\alpha Z(\alpha_r) \right\} \right],$$

provided that $\alpha_1, \alpha_2, \ldots, \alpha_n$ be chosen to satisfy the n independent equations comprised in the system

$$\sum_{p=1}^{n}{}' \frac{\operatorname{sn} \alpha_p \operatorname{cn} \alpha_p \operatorname{dn} \alpha_p + \operatorname{sn} \alpha_r \operatorname{cn} \alpha_r \operatorname{dn} \alpha_r}{\operatorname{sn}^2 \alpha_p - \operatorname{sn}^2 \alpha_r} = 0,$$

$$\left[\sum_{r=1}^{n} \operatorname{cn} \alpha_r \operatorname{ds} \alpha_r \right]^2 - \sum_{r=1}^{n} \operatorname{ns}^2 \alpha_r = A;$$

and if this solution of Lamé's equation is not doubly periodic, a second solution is

$$\prod_{r=1}^{n} \left[\frac{H(\alpha - \alpha_r)}{\Theta(\alpha)} \exp\{\alpha Z(\alpha_r)\} \right] = 0.$$

The existence of a solution of the system of $n + 1$ equations follows from §23.7.

23.8 Miscellaneous examples

Example 23.1 (Niven [504])   Obtain the formula

$$G_n(x, y, z) = \frac{2^n n}{(2n)!} \int_0^\infty D^{2n} P_n \left(\frac{u}{D}\right) e^{-u} \, du \cdot H_n(x, y, z).$$

Example 23.2 (Hobson [312])   Shew that

$$H_n \left(\frac{\partial}{\partial x}, \frac{\partial}{\partial y}, \frac{\partial}{\partial z}\right) \frac{1}{\sqrt{x^2 + y^2 + z^2}} = \frac{(-1)^n (2n)!}{2^n n!} \frac{H_n(x, y, z)}{(x^2 + y^2 + z^2)^{n+\frac{1}{2}}}.$$

Example 23.3 (Niven [505] and Hobson [312])   Shew that the 'external ellipsoidal harmonic' $F_n^m(\xi) E_n^m(\eta) E_n^m(\zeta)$ is a constant multiple of

$$H_n \left(\frac{\partial}{\partial x}, \frac{\partial}{\partial y}, \frac{\partial}{\partial z}\right) \left(1 + \frac{D^2}{2 \cdot (2n + 3)} + \frac{D^4}{2 \cdot 4(2n + 3)(2n + 5)} + \cdots \right) \frac{1}{\sqrt{x^2 + y^2 + z^2}}.$$

Example 23.4 (Haentzschel [266])   Discuss the confluent form of Lamé's equation when the invariants g_2 and g_3 of the Weierstrassian elliptic function are made to tend to zero; express the solution in terms of Bessel functions.

Example 23.5 (Hermite)   If v denotes $\frac{H(a+\mu)}{\Theta(a)} \exp[\{\lambda - Z(\mu)\} \alpha]$, where λ and μ, are constants, shew that Lamé's equation has a solution which is expressible as a linear combination of

$$\frac{d^{n-1} v}{da^{n-1}}, \frac{d^{n-3} v}{da^{n-3}}, \frac{d^{n-5} v}{da^{n-5}}, \dots,$$

where λ^2 and $\operatorname{sn}^2 \mu$ are algebraic functions of the constant A.

Example 23.6 (Stenberg [602])   Obtain solutions of

$$\frac{1}{w} \frac{d^2 w}{dz^2} = 12k^2 \operatorname{sn}^2 z - 4(1 + k^2) \pm 5\sqrt{1 - k^2 + k^4}.$$

Example 23.7 (Heun [301])   Discuss the solution of the equation

$$z(z - 1)(z - a)\frac{d^2 y}{dz^2} + \left[(a + \beta + 1)z^2 - \{a + \beta - \delta + 1 + (\gamma + \delta)a\} z + a\gamma\right] \frac{dy}{dz} + a\beta(z - q)y$$
$$= 0$$

in the form of the series

$$1 + \alpha\beta \sum_{n=1}^{\infty} \frac{G_n(q)(z/\alpha)^n}{n! \, \gamma(\gamma + 1)\cdots(\gamma + n)},$$

where

$$G_1(q) = q,$$
$$G_2(q) = \alpha\beta q^2 + \{(\alpha + \beta - \delta + 1) + (\gamma + \delta)\alpha\}\, q - \alpha\gamma,$$
$$G_{n+1}(q) = [n\,\{(\alpha + \beta - \delta + n) + (\gamma + \delta + n - 1)\alpha\} + \alpha\beta q]\, G_n(q)$$
$$- (\alpha + n - 1)(\beta + n - 1)(\gamma + n - 1)n\alpha G_{n-1}(q).$$

Example 23.8 (Heun [300]) Shew that the exponents at the singularities $0, 1, \alpha, \infty$ of Heun's equation are

$$(0, 1 - \gamma), \quad (0, 1 - \delta), \quad (0, 1 - \varepsilon), \quad (\alpha, \beta),$$

where $\gamma + \delta + \varepsilon = \alpha + \beta + 1$.

Example 23.9 (Heun, [300]) Obtain the following group of variables for Heun's equation, corresponding to the group

$$z, \quad 1 - z, \quad \frac{1}{z}, \quad \frac{1}{1 - z}, \quad \frac{z}{z - 1}, \quad \frac{z - 1}{z},$$

for the hypergeometric equation:

$$z, \quad 1 - z, \quad \frac{1}{z}, \quad \frac{1}{1 - z}, \quad \frac{z}{z - 1}, \quad \frac{z - 1}{z},$$
$$\frac{z}{\alpha}, \quad \frac{\alpha - z}{\alpha}, \quad \frac{\alpha}{z}, \quad \frac{\alpha}{\alpha - z}, \quad \frac{z}{z - \alpha}, \quad \frac{z - \alpha}{z},$$
$$\frac{z - \alpha}{1 - \alpha}, \quad \frac{z - 1}{\alpha - 1}, \quad \frac{1 - \alpha}{z - \alpha}, \quad \frac{\alpha - 1}{z - 1}, \quad \frac{z - \alpha}{z - 1}, \quad \frac{z - 1}{z - \alpha},$$
$$\frac{z - \alpha}{\alpha(z - 1)}, \quad \frac{(\alpha - 1)z}{\alpha(z - 1)}, \quad \frac{\alpha(z - 1)}{z - \alpha}, \quad \frac{\alpha(z - 1)}{(\alpha - 1)z}, \quad \frac{z - \alpha}{(1 - \alpha)z}, \quad \frac{(1 - \alpha)z}{z - \alpha}.$$

Example 23.10 If the series of Example 23.7 be called

$$F(\alpha, q; \alpha, \beta, \gamma, \delta; z),$$

obtain 192 solutions of the differential equation in the form of powers of z, $z - 1$ and $z - \alpha$ multiplied by functions of the type F. Heun gives 48 of these solutions.

Example 23.11 If $u = 2v$, shew that Lamé's equation

$$\frac{d^2\Lambda}{du^2} = \{n(n + 1)\wp(u) + B\}\,\Lambda$$

may be transformed into

$$\frac{d^2L}{dv^2} - 2n\frac{\wp''(v)}{\wp'(v)}\frac{dL}{dv} + 4\,\{n(2n - 1)\wp(v) - B\}\,L = 0,$$

by the substitution

$$\Lambda = \{\wp'(v)\}^{-n}\,L.$$

Example 23.12 (Brioschi [99]; Halphen)   If $\zeta = \wp(v)$, shew that a formal solution of the equation of Example 23.11 is

$$L = \sum_{r=0}^{\infty} b_r (\zeta - e_2)^{a-r},$$

provided that $(a - 2n)(a - n + \frac{1}{2}) = 0$ and that

$$4(a - r - 2n)\left(a - r - n + \tfrac{1}{2}\right)b_r$$
$$+ \left[12e_2(a - r + 1)(a - r - 2n + 1) + 4e_2 n(2n - 1) - 4B\right]b_{r-1}$$
$$- 4(e_1 - e_2)(e_2 - e_3)(a - r + 2)\left(a - r - n + \tfrac{3}{2}\right)b_{r-2} = 0.$$

Example 23.13 (Brioschi and Halphen)   Shew that, if n is half of an odd positive integer, a solution of the equation of Example 23.11 expressible in finite form is

$$L = \sum_{r=0}^{n-\frac{1}{2}} b_r (\zeta - e_2)^{2n-r},$$

provided that

$$4r\left(n - r + \tfrac{1}{2}\right)b_r + [12e_2(2n - r + 1)(r - 1) - 4e_2 n(2n - 1) + 4B]b_{r-1}$$
$$+ 4(e_1 - e_2)(e_2 - e_3)(2n - r + 2)\left(n - r + \tfrac{3}{2}\right)b_{r-2} = 0,$$

and B is so determined that $b_{n+\frac{1}{2}} = 0$.

Example 23.14 (Crawford)   Shew that, if n is half of an odd integer, a solution of the equation of Example 23.11 expressible in finite form is

$$L' = \sum_{p=0}^{n-\frac{1}{2}} b'_p (\zeta - e_2)^{n-p-\frac{1}{2}},$$

provided that

$$4p\left(n + p + \tfrac{1}{2}\right)b'_p - \left[12e_2\left(n - p + \tfrac{1}{2}\right)\left(n + p - \tfrac{1}{2}\right) - 4e_2 n(2n - 1)\right.$$
$$\left. + 4B\right]b'_{p-1} + 4(e_1 - e_2)(e_2 - e_3)\left(n - p + \tfrac{3}{2}\right)(p - 1)b'_{p-2} = 0$$

and $b'_{n+\frac{1}{2}} = 0$ is the equation which determines B.

Example 23.15 (Crawford)   With the notation of Examples 23.13 and 23.14 shew that, if

$$b'_p = (-1)^p (e_1 - e_2)^p (e_2 - e_3)^p c_{n-p-\frac{1}{2}},$$

the equations that determine $c_0, c_1, \ldots, c_{n-\frac{1}{2}}$ and those that determine $b_0, b_1, \ldots, b_{n-\frac{1}{2}}$ are identical; and deduce that, if one of the solutions of Lamé's equation (in which n is half of an odd integer) is expressible as an algebraic function of $\wp(v)$, so also is the other.

Example 23.16   Prove that the values of B determined in Example 23.13 are real when e_1, e_2 and e_3 are real.

Example 23.17 (Halphen [267])   Shew that the complete solution of

$$\frac{1}{\Lambda}\frac{d^2\Lambda}{du^2} = \frac{3}{4}\wp(u)$$

is

$$\Lambda = \left\{\wp'\left(\frac{1}{2}u\right)\right\}^{-\frac{1}{2}}\left\{A\wp\left(\frac{1}{2}u\right) + B\right\},$$

where A and B are arbitrary constants.

Example 23.18 (Jamet [357])   Shew that the complete solution of

$$\frac{1}{\Lambda}\frac{d^2\Lambda}{da^2} = \frac{3}{4}k^2 \operatorname{sn}^2 a - \frac{1}{4}(1 + k^2)$$

is

$$\Lambda = \left\{\operatorname{sn}\frac{1}{2}(C-a)\operatorname{cn}\frac{1}{2}(C-a)\operatorname{dn}\frac{1}{2}(C-a)\right\}^{-\frac{1}{2}}\left\{A + B\operatorname{sn}^2\frac{1}{2}(C-a)\right\},$$

where A and B are arbitrary constants and $C = 2K + iK'$.

Appendix

The Elementary Transcendental Functions

A.1 On certain results assumed in Chapters 1 to 4

It was convenient, in the first four chapters of this work, to assume some of the properties of the elementary transcendental functions, namely the exponential, logarithmic and circular functions; it was also convenient to make use of a number of results which the reader would be prepared to accept intuitively by reason of his familiarity with the geometrical representation of complex numbers by means of points in a plane.

To take two instances, (i) it was assumed (§2.7) that $\lim(\exp z) = \exp(\lim z)$, and (ii) the geometrical concept of an angle in the Argand diagram made it appear plausible that the argument of a complex number was a many-valued function, possessing the property that any two of its values differed by an integer multiple of 2π.

The assumption of results of the first type was clearly illogical; it was also illogical to base arithmetical results on geometrical reasoning. For, in order to put the foundations of geometry on a satisfactory basis, it is not only desirable to employ the axioms of arithmetic, but it is also necessary to utilise a further set of axioms of a more definitely geometrical character, concerning properties of points, straight lines and planes[1]. And, further, the arithmetical theory of the logarithm of a complex number appears to be a necessary preliminary to the development of a logical theory of angles.

Apart from this, it seems unsatisfactory to the aesthetic taste of the mathematician to employ one branch of mathematics as an essential constituent in the structure of another; particularly when the former has, to some extent, a material basis whereas the latter is of a purely abstract nature[2].

The reasons for pursuing the somewhat illogical and unaesthetic procedure, adopted in the earlier part of this work, were, firstly, that the properties of the elementary transcendental functions were required gradually in the course of Chapter 2, and it seemed undesirable that the course of a general development of the various infinite processes should be frequently interrupted in order to prove theorems (with which the reader was, in all probability, already familiar) concerning a single particular function; and, secondly, that (in connexion with the

---

[1] It is not our object to give any account of the foundations of geometry in this work. They are investigated by various writers, such as Whitehead [667] and Mathews [456]. A perusal of Chapters ɪ, xx, xxɪɪ and xxv of the latter work will convince the reader that it is even more laborious to develop geometry in a logical manner, from the minimum number of axioms, than it is to evolve the theory of the circular functions by purely analytical methods. A complete account of the elements both of arithmetic and of geometry has been given by Whitehead and Russell [668].

[2] Cf. Merz [467, p. 631 Note 2 and p. 707 Note 1], where a letter from Weierstrass to Schwarz is quoted. See also Sylvester [616], [618, p. 50].

assumption of results based on geometrical considerations) a purely arithmetical mode of development of Chapters 1 to 4, deriving no help or illustrations from geometrical processes, would have very greatly increased the difficulties of the reader unacquainted with the methods and the spirit of the analyst.

A.11 Summary of the Appendix

The general course of the Appendix is as follows:

In §§A.2–A.22, the exponential function is defined by a power series. From this definition, combined with results contained in Chapter 2, are derived the elementary properties (apart from the periodic properties) of this function. It is then easy to deduce corresponding properties of logarithms of positive numbers (§§A.3–A.33).

Next, the sine and cosine are defined by power series from which follows the connexion of these functions with the exponential function. A brief sketch of the manner in which the formulae of elementary trigonometry may be derived is then given (§§A.4–A.42).

The results thus obtained render it possible to discuss the periodicity of the exponential and circular functions by *purely arithmetical methods* (§§A.5, A.51).

In §§A.52–A.522, we consider, substantially, the continuity of the inverse circular functions. When these functions have been investigated, the theory of logarithms of complex numbers (§A.6) presents no further difficulty.

Finally, in §A.7, it is shewn that an angle, defined in a purely analytical manner, possesses properties which are consistent with the ordinary concept of an angle, based on our experience of the material world.

It will be obvious to the reader that we do not profess to give a complete account of the elementary transcendental functions, but we have confined ourselves to a brief sketch of the logical foundations of the theory[3]. The developments have been given by writers of various treatises, such as Hobson [321]; Hardy [277]; and Bromwich [102].

A.12 A logical order of development of the elements of analysis

The reader will find it instructive to read Chapters 1 to 4 and the Appendix a second time in the following order:

Chapter 1 (omitting[4] all of §1.5 except the first two paragraphs).

Chapter 2 to the end of §2.61 (omitting the examples in §§2.31–2.61).

Chapter 3 to the end of §3.34 and §§3.5–3.73.

The Appendix, §§A.2–A.6 (omitting §§A.32, A.33).

Chapter 2, the examples of §§2.31–2.61.

Chapter 3, §§3.341–3.4.

Chapter 4, inserting §§A.32, A.33, A.7 after §4.13.

Chapter 2, §§2.7–2.82.

He should try thus to convince himself that (in that order) it is possible to elaborate a

---

[3] In writing the Appendix, frequent reference has been made to the article on Algebraic Analysis in the *Encyklopä die der Math. Wissenschaften* by Pringsheim and Faber, to the same article translated and revised by Molk for the *Encyclopédie des Sciences Math.*, and to Tannery, *Introduction à la Théorie des Fonctions d'une Variable* (Paris, 1904).

[4] The properties of the argument (or phase) of a complex number are not required in the text before Chapter 5.

purely arithmetical development of the subject, in which the graphic and familiar language of geometry[5] is to be regarded as merely conventional.

A.2 The exponential function $\exp z$

The exponential function, of a complex variable z, is defined by the series[6]

$$\exp z = 1 + \frac{z}{1!} + \frac{z^2}{2!} + \frac{z^3}{3!} + \cdots = 1 + \sum_{n=1}^{\infty} \frac{z^n}{n!}.$$

This series converges absolutely for all values of z (real and complex) by D'Alembert's ratio test (§2.36) since $\lim_{n \to \infty} |z/n| = 0 < 1$; so the definition is valid for all values of z.

Further, the series converges uniformly throughout any bounded domain of values of z; for, if the domain be such that $|z| \le R$ when z is in the domain, then

$$|z^n/n!| \le R^n/n!,$$

and the uniformity of the convergence is a consequence of the test of Weierstrass (§3.34), by reason of the convergence of the series $1 + \sum_{n=1}^{\infty} \frac{R^n}{n!}$, in which the terms are independent of z.

Moreover, since, for any fixed value of n, $z^n/n!$ is a continuous function of z, it follows from §3.32 that the exponential function is continuous for all values of z; and hence (cf. §3.2), if z be a variable which tends to the limit ζ, we have

$$\lim_{z \to \zeta} \exp z = \exp \zeta.$$

A.21 The addition-theorem for the exponential function, and its consequences

From Cauchy's theorem on multiplication of absolutely convergent series (§2.53), it follows that[7]

$$(\exp z_1)(\exp z_2) = \left(1 + \frac{z_1}{1!} + \frac{z_1^2}{2!} + \cdots\right)\left(1 + \frac{z_2}{1!} + \frac{z_2^2}{2!} + \cdots\right)$$

$$= 1 + \frac{z_1 + z_2}{1!} + \frac{z_1^2 + 2z_1 z_2 + z_2^2}{2!} + \cdots$$

$$= \exp(z_1 + z_2),$$

---

[5] For example 'a point' for 'an ordered number-pair', 'the circle of unit radius with centre at the origin' for 'the set of ordered number-pairs (x, y) which satisfy the condition $x^2 + y^2 = 1$', 'the points of a straight line' for 'the set of ordered number-pairs (x, y) which satisfy a relation of the type $Ax + By + C = 0$', and so on.

[6] It was formerly customary to define $\exp z$ as $\lim_{n \to \infty} \left(1 + \frac{z}{n}\right)^n$, cf. Cauchy [120, pp. 167, 168, 309]. Cauchy also derived the properties of the function from the series, but his investigation when z is not rational is incomplete. See also Schlömilch [587, pp. 29, 178, 246]. Hardy [276] has pointed out that the limit definition has many disadvantages.

[7] The reader will at once verify that the general term in the product series is

$$\left(z_1^n + \binom{n}{1}z_1^{n-1}z_2 + \binom{n}{2}z_1^{n-2}z_2^2 + \cdots + z_2^n\right)/n! = (z_1 + z_2)^n/n!.$$

so that $\exp(z_1 + z_2)$ can be expressed in terms of exponential functions of z_1 and of z_2 by the formula

$$\exp(z_1 + z_2) = (\exp z_1)(\exp z_2).$$

This result is known as the *addition-theorem* for the exponential function. From it, we see by induction that

$$(\exp z_1)(\exp z_2) \cdots (\exp z_n) = \exp(z_1 + z_2 + \cdots + z_n),$$

and, in particular,

$$\{\exp z\} \{\exp(-z)\} = \exp 0 = 1.$$

From the last equation, it is apparent that there is no value of z for which $\exp z = 0$; for, if there were such a value of z, since $\exp(-z)$ would exist for this value of z, we should have $0 = 1$.

It also follows that, when x is real, $\exp x > 0$; for, from the series definition, $\exp x \geq 1$ when $x \geq 0$; and, when $x \leq 0$, $\exp x = 1/\exp(-x) > 0$.

Further, $\exp x$ is an *increasing* function of the real variable x; for, if $k > 0$,

$$\exp(x + k) - \exp x = \exp x \cdot \{\exp k - 1\} > 0,$$

because $\exp x > 0$ and $\exp k > 1$.

Also, since $\{\exp h - 1\}/h = 1 + (h/2!) + (h^2/3!) + \cdots$, and the series on the right is seen (by the methods of §A.2) to be continuous for all values of h, we have

$$\lim_{h \to 0} \frac{\exp h - 1}{h} = 1,$$

and so $\dfrac{d \exp z}{dz} = \lim_{h \to 0} \dfrac{\exp(z + h) - \exp z}{h} = \exp z.$

A.22 Various properties of the exponential function

Returning to the formula $(\exp z_1)(\exp z_2) \cdots (\exp z_n) = \exp(z_1 + z_2 + \cdots + z_n)$, we see that, when n is a positive integer,

$$(\exp z)^n = \exp(nz),$$

and

$$(\exp z)^{-n} = 1/(\exp z)^n = 1/\exp(nz) = \exp(-nz).$$

In particular, taking $z = 1$ and writing e in place of $\exp 1 = 2.71828 \cdots$, we see that, when m is an integer, positive or negative,

$$e^m = \exp m = 1 + (m/1!) + (m^2/2!) + \cdots$$

Also, if μ be any rational number (say p/q, where p and q are integers, q being positive)

$$(\exp \mu)^q = \exp \mu q = \exp p = e^p,$$

so that the qth power of $\exp \mu$ is e^p; that is to say, $\exp \mu$ is a value of $e^{p/q} = e^\mu$, and it is obviously (§A.21) the real positive value.

If x be an irrational-real number (defined by a section in which a_1 and a_2 are typical members of the L-class and the R-class respectively), the *irrational* power e^x is most simply *defined* as $\exp x$; we thus have, for all real values of x, rational and irrational,

$$e^x = 1 + \frac{x}{1!} + \frac{x^2}{2!} + \cdots,$$

an equation first given by Newton[8].

It is, therefore, legitimate to write e^x for $\exp x$ when x is real, and it is customary to write e^z for $\exp z$ when z is complex. The function e^z (which, of course, must not be regarded as being a power of e), thus defined, is subject to the ordinary laws of indices, viz.

$$Fe^z \cdot e^\zeta = e^{z+\zeta}, \quad e^{-z} = 1/e^{-z}.$$

Note Tannery [619] practically defines e^x, when x is irrational, as the *only* number X such that $e^{a_1} \leq X \leq e^{a_2}$, for every a_1 and a_2. From the definition we have given it is easily seen that such a *unique* number exists. For $\exp x(= X)$ satisfies the inequality, and if $X'(\neq X)$ also did so, then

$$\exp a_2 - \exp a_1 = e^{a_2} - e^{a_1} \geq |X' - X|,$$

so that, since the exponential function is continuous, $a_2 - a_1$ cannot be chosen arbitrarily small, and so (a_2, a_1) does not define a section.

A.3 Logarithms of positive numbers

It has been seen (§§A.2, A.21) that, when x is real, $\exp x$ is a positive continuous increasing function of x, and obviously $\exp x \to +\infty$ as $x \to +\infty$, while

$$\exp x = 1/\exp(-x) \to 0 \quad \text{as} \quad x \to -\infty.$$

If, then, a be any positive number, it follows from §3.63 that the equation in x,

$$\exp x = a,$$

has one real root and only one. This root (which is, of course, a function of a) will be written[9] $\text{Log}_e a$ or simply $\text{Log}\, a$; it is called the *Logarithm of the positive number a*. [10]

Since a one-one correspondence has been established between x and a, and since a is an increasing function of x, it must be that x be increasing function of a; that is to say, the Logarithm is an increasing function.

Example A.3.1 Deduce from §A.21 that $\text{Log}\, a + \text{Log}\, b = \text{Log}\, ab$.

---

[8] Newton [494] (written before 1669, but not published till 1711); it was also given both by Newton and by Leibniz in letters to Oldenburg in 1676; it was first published by Wallis in 1685 in his *Treatise on Algebra*, p. 343. The equation when x is irrational was explicitly stated by Schlömilch [587, p. 182].

[9] This is in agreement with the notation of most textbooks, in which Log denotes the principal value (see §A.6) of the logarithm of a complex number.

[10] Many mathematicians define the Logarithm by the integral formula given in §A.32. The reader should consult a memoir by Hurwitz [330] on the foundations of the theory of the logarithm.

A.31 The continuity of the Logarithm

It will now be shewn that, when a is positive, $\mathrm{Log}\,a$ is a continuous function of a.
  Let

$$\mathrm{Log}\,a = x, \qquad \mathrm{Log}(a + h) = x + k,$$

so that

$$e^x = a, \qquad e^{x+k} = a + h, \qquad 1 + \frac{h}{a} = e^k.$$

First suppose that $h > 0$, so that $k > 0$, and then

$$1 + h/a = 1 + k + \tfrac{1}{2}k^2 + \cdots > 1 + k,$$

and so $0 < k < h/a$; that is to say, $0 < \mathrm{Log}(a + h) - \mathrm{Log}\,a < h/a$.
  Hence, h being positive, $\mathrm{Log}(a + h) - \mathrm{Log}\,a$ can be made arbitrarily small by taking h sufficiently small.
  Next, suppose that $h < 0$, so that $k < 0$, and then $a/(a + h) = e^{-k}$. Hence (taking $0 < -h < \tfrac{1}{2}a$, as is obviously permissible) we get

$$a/(a + h) = 1 + (-k) + \tfrac{1}{2}k^2 + \cdots > 1 - k,$$

and so $-k < -1 + a/(a + h) = -h/(a + h) < -2h/a$. Therefore, whether h be positive or negative, if ε be an arbitrary positive number and if $|h|$ be taken less than both $\tfrac{1}{2}a$ and $\tfrac{1}{2}a\varepsilon$, we have

$$|\,\mathrm{Log}(a + h) - \mathrm{Log}\,a\,| < \varepsilon,$$

and so the condition for continuity (§3.2) is satisfied.

A.32 Differentiation of the Logarithm

Retaining the notation of §A.31, we see, from results there proved, that, if $h \to 0$ (a being fixed), then also $k \to 0$. Therefore, when $a > 0$,

$$\frac{d\,\mathrm{Log}\,a}{da} = \lim_{k \to 0} \frac{k}{e^{x+k} - e^x} = \frac{1}{e^x} = \frac{1}{a}.$$

Since $\mathrm{Log}\,1 = 0$, we have, by §4.13, Example 4.1.8,

$$\mathrm{Log}\,a = \int_1^a t^{-1}\,dt.$$

A.33 The expansion of $\mathrm{Log}(1 + a)$ in powers of a

From §A.32 we have

$$\begin{aligned}
\mathrm{Log}(1 + a) &= \int_0^a (1 + t)^{-1}\,dt \\
&= \int_0^a \left\{ 1 - t + t^2 - \cdots + (-1)^{n-1}t^{n-1} + (-1)^n t^n (1 + t)^{-1} \right\} dt \\
&= a - \tfrac{1}{2}a^2 + \tfrac{1}{3}a^3 - \cdots + (-1)^{n-1}\frac{1}{n}a^n + R_n,
\end{aligned}$$

where $R_n = (-1)^n \int_0^a t^n (1+t)^{-1} \, dt$.

Now, if $-1 < a < 1$, we have

$$|R_n| \leq \int_0^{|a|} t^n (1 - |a|)^{-1} \, dt$$

$$= |a|^{n+1} \{(n+1)(1 - |a|)\}^{-1}$$

$$\to 0 \quad \text{as} \quad n \to \infty.$$

Hence, when $-1 < a < 1$, $\text{Log}(1+a)$ can be expanded into the convergent series[11]

$$\text{Log}(1+a) = a - \tfrac{1}{2}a^2 + \tfrac{1}{3}a^3 - \cdots = \sum_{n=1}^{\infty} (-1)^{n-1} a^n / n.$$

If $a = +1$,

$$|R_n| = \int_0^1 t^n (1+t)^{-1} \, dt < \int_0^1 t^n \, dt = (n+1)^{-1} \to 0 \quad \text{as} \quad n \to \infty,$$

so the expansion is valid when $a = +1$; it is not valid when $a = -1$.

Example A.3.2 Shew that

$$\lim_{n \to \infty} \left(1 + \frac{1}{n}\right)^n = e.$$

Hint. We have

$$\lim_{n \to \infty} n \log \left(1 + \frac{1}{n}\right) = \lim_{n \to \infty} \left(1 - \frac{1}{2n} + \frac{1}{3n^2} - \cdots\right) = 1,$$

and the result required follows from the result of §A.2 that $\lim_{z \to \zeta} e^z = e^\zeta$.

A.4 The definition of the sine and cosine

The functions[12] $\sin z$ and $\cos z$ are defined analytically by means of power series, thus

$$\sin z = z - \frac{z^3}{3!} + \frac{z^5}{5!} - \cdots = \sum_{n=0}^{\infty} \frac{(-1)^n z^{2n+1}}{(2n+1)!},$$

$$\cos z = 1 - \frac{z^2}{2!} + \frac{z^4}{4!} - \cdots = 1 + \sum_{n=1}^{\infty} \frac{(-1)^n z^{2n}}{(2n)!};$$

these series converge absolutely for all values of z (real and complex) by §2.36, and so the definitions are valid for all values of z.

On comparing these series with the exponential series, it is apparent that the sine and

---

[11] This method of obtaining the logarithmic expansion is, in effect, due to Wallis [644].

[12] These series were given by Newton [494], see §A.22 footnote. The other trigonometrical functions are defined in the manner with which the reader is familiar, as quotients and reciprocals of sines and cosines.

cosine are not essentially new functions, but they can be expressed in terms of exponential functions by the equations[13]

$$2i \sin z = \exp(iz) - \exp(-iz), \qquad 2 \cos z = \exp(iz) + \exp(-iz).$$

It is obvious that $\sin z$ and $\cos z$ are odd and even functions of z respectively; that is to say

$$\sin(-z) = -\sin z, \qquad \cos(-z) = \cos z.$$

A.41 *The fundamental properties of* $\sin z$ *and* $\cos z$

It may be proved, just as in the case of the exponential function (§A.2), that the series for $\sin z$ and $\cos z$ converge uniformly in any bounded domain of values of z, and consequently that $\sin z$ and $\cos z$ are continuous functions of z for all values of z.

Further, it may be proved in a similar manner that the series

$$1 - \frac{z^2}{3!} + \frac{z^4}{5!} - \cdots$$

defines a continuous function of z for all values of z, and, in particular, this function is continuous at $z = 0$, and so it follows that

$$\lim_{z \to 0} \frac{\sin z}{z} = 1.$$

A.42 *The addition-theorems for* $\sin z$ *and* $\cos z$

By using Euler's equations (§A.4), it is easy to prove from properties of the exponential function that

$$\sin(z_1 + z_2) = \sin z_1 \cos z_2 + \cos z_1 \sin z_2$$

and

$$\cos(z_1 + z_2) = \cos z_1 \cos z_2 + \sin z_1 \sin z_2;$$

these results are known as *the addition-theorems* for $\sin z$ and $\cos z$.

It may also be proved, by using Euler's equations, that

$$\sin^2 z + \cos^2 z = 1.$$

By means of this result, $\sin(z_1 + z_2)$ can be expressed as an algebraic function of $\sin z_1$ and $\sin z_2$, while $\cos(z_1 + z_2)$ can similarly be expressed as an algebraic function of $\cos z_1$ and $\cos z_2$; so the addition-formulae may be regarded as addition-theorems in the strict sense (cf. §20.3 and the Note on page 547).

By differentiating Euler's equations, it is obvious that

$$\frac{d \sin z}{dz} = \cos z, \qquad \frac{d \cos z}{dz} = -\sin z.$$

---

[13] These equations were derived by Euler [they were given in a letter to Johann Bernoulli in 1740 and published in [198, p. 279] from the geometrical definitions of the sine and cosine, upon which the theory of the circular functions was then universally based.

Example A.4.1 Shew that

$$\sin 2z = 2 \sin z \cos z, \qquad \cos 2z = 2 \cos^2 z - 1;$$

these results are known as the duplication-formulae.

A.5 The periodicity of the exponential function

If z_1 and z_2 are such that $\exp z_1 = \exp z_2$, then, multiplying both sides of the equation by $\exp(-z_2)$, we get $\exp(z_1 - z_2) = 1$; and writing γ for $z_1 - z_2$, we see that, for all values of z and all integral values of n,

$$\exp(z + n\gamma) = \exp z \cdot (\exp \gamma)^n = \exp z.$$

The exponential function is then said *to have period* γ, since the effect of increasing z by γ, or by an integral multiple thereof, does not affect the value of the function.

It will now be shewn that such numbers γ (other than zero) actually exist, and that *all* the numbers γ, possessing the property just described, are comprised in the expression

$$2n\pi i, \qquad (n = \pm 1, \pm 2, \pm 3, \dots)$$

where π is a certain positive number[14] which happens to be greater than $2\sqrt{2}$ and less than 4.

A.51 The solution of the equation $\exp \gamma = 1$

Let $\gamma = \alpha + i\beta$, where α and β are real; then the problem of solving the equation $\exp \gamma = 1$ is identical with that of solving the equation

$$\exp \alpha \cdot \exp i\beta = 1.$$

Comparing the real and imaginary parts of each side of this equation, we have

$$\exp \alpha \cdot \cos \beta = 1, \qquad \exp \alpha \cdot \sin \beta = 0.$$

Squaring and adding these equations, and using the identity $\cos^2 \beta + \sin^2 \beta \equiv 1$, we get

$$\exp 2\alpha = 1.$$

Now if α were positive, $\exp 2\alpha$ would be greater than 1, and if α were negative, $\exp 2\alpha$ would be less than 1; *and so the only possible value for α is zero*. It follows that $\cos \beta = 1$, $\sin \beta = 0$. Now the equation $\sin \beta = 0$ is a necessary consequence of the equation $\cos \beta = 1$, on account of the identity $\cos^2 \beta + \sin^2 \beta \equiv 1$. It is therefore sufficient to consider solutions (if such solutions exist) of the equation $\cos \beta = 1$.

Instead, however, of considering the equation $\cos \beta = 1$, it is more convenient to consider the equation[15] $\cos x = 0$.

It will now be shewn that the equation $\cos x = 0$ has one root, and only one, lying between

---

[14] The fact that π is an irrational number, whose value is $3.14159 \cdots$, is irrelevant to the present investigation. For an account of attempts at determining the value of π, concluding with a proof of the theorem that π satisfies no algebraic equation with rational coefficients, see Hobson's monograph [320].

[15] If $\cos x = 0$, it is an immediate consequence of the duplication-formulae that $\cos 2x = -1$ and thence that $\cos 4x = 1$, so, if x is a solution of $\cos x = 0$, it follows that $4x$ is a solution of $\cos \beta = 1$.

0 and 2, and that this root exceeds $\sqrt{2}$; to prove these statements, we make use of the following considerations:

(I) The function $\cos x$ is certainly continuous in the range $0 \le x \le 2$.

(II) When $0 \le x \le \sqrt{2}$, we have[16]

$$1 - \frac{x^2}{2!} \ge 0, \qquad \frac{x^4}{4!} - \frac{x^6}{6!} \ge 0, \qquad \frac{x^8}{8!} - \frac{x^{10}}{10!} \ge 0, \dots,$$

and so, when $0 \le x \le \sqrt{2}$, $\cos x > 0$.

(III) The value of $\cos 2$ is

$$1 - 2 + \frac{2}{3} - \frac{2^6}{720}\left(1 - \frac{4}{7 \cdot 8}\right) - \frac{2^{10}}{10!}\left(1 - \frac{4}{11 \cdot 12}\right) - \dots = -\frac{1}{3} - \dots < 0.$$

(IV) When $0 < x \le 2$,

$$\frac{\sin x}{x} = \left(1 - \frac{x^2}{6}\right) + \frac{x^4}{120}\left(1 - \frac{x^2}{6 \cdot 7}\right) + \dots > 1 - \frac{x^2}{6} \ge \frac{1}{3},$$

and so, when $0 \le x \le 2$, $\sin x \ge \frac{1}{3}x$.

It follows from (II) and (III) combined with the results of (I) and of §3.63 that the equation $\cos x = 0$ has *at least* one root in the range $\sqrt{2} < x < 2$, and it has no root in the range $0 \le x \le \sqrt{2}$.

Further, there is *not more than* one root in the range $\sqrt{2} < x < 2$; for, suppose that there were two, x_1 and x_2 (with $x_2 > x_1$); then $0 < x_2 - x_1 < 2 - \sqrt{2} < 1$, and

$$\sin(x_2 - x_1) = \sin x_2 \cos x_1 - \sin x_1 \cos x_2 = 0,$$

and this is incompatible with (IV) which shews that $\sin(x_2 - x_1) \ge \frac{1}{3}(x_2 - x_1)$.

The equation $\cos x = 0$ *therefore has one and only one root lying between* 0 *and* 2. This root lies between $\sqrt{2}$ and 2, and it is called $\frac{1}{2}\pi$; and, as stated in the footnote to §A.5, its actual value happens to be $1.57079\cdots$.

From the addition-formulae, it may be proved at once by induction that

$$\cos n\pi = (-1)^n, \qquad \sin n\pi = 0,$$

where n is any integer. In particular, $\cos 2n\pi = 1$, where n is any integer.

Moreover, there is no value of β, other than those values which are of the form $2n\pi$, for which $\cos \beta = 1$; for if there were such a value, it must be real[17], and so we can choose the integer m so that

$$-\pi \le 2m\pi - \beta < \pi.$$

We then have

$$\sin|m\pi - \tfrac{1}{2}\beta| = \pm\sin(m\pi - \tfrac{1}{2}\beta) = \pm\sin\tfrac{1}{2}\beta = \pm2^{-\frac{1}{2}}(1 - \cos\beta)^{\frac{1}{2}} = 0,$$

and this is inconsistent[18] with $\sin|m\pi - \tfrac{1}{2}\beta| \ge \frac{1}{3}|m\pi - \tfrac{1}{2}\beta|$ unless $\beta = 2m\pi$.

[16] The symbol $\ge$ may be replaced by $>$ except when $x = \sqrt{2}$ in the first place where it occurs, and except when $x = 0$ in the other places.

[17] The equation $\cos\beta = 1$ implies that $\exp i\beta = 1$, and we have seen that this equation has no complex roots.

[18] The inequality is true by (IV) since $0 \le |m\pi - \tfrac{1}{2}\beta| \le \tfrac{1}{2}\pi < 2$.

Consequently the numbers $2n\pi$, $(n = 0, \pm 1, \pm 2, \ldots)$, and no others, have their cosines equal to unity.

It follows that a positive number π exists such that $\exp z$ *has period $2\pi i$ and that $\exp z$ has no period fundamentally distinct from $2\pi i$.*

The formulae of elementary trigonometry concerning the periodicity of the circular functions, with which the reader is already acquainted, can now be proved by analytical methods without any difficulty.

Example A.5.1   Shew that $\sin \pi/2$ is equal to 1, not to -1.

Example A.5.2   Shew that $\tan x > x$ when $0 < x < \frac{1}{2}\pi$. *Hint.* For $\cos x > 0$ and

$$\sin x - x \cos x = \sum_{n=1}^{\infty} \frac{x^{4n-1}}{(4n-1)!} \left\{ 4n - 2 - \frac{x^2}{4n+1} \right\},$$

and every term in the series is positive.

Example A.5.3   Shew that $1 - \dfrac{x^2}{2} + \dfrac{x^4}{24} - \dfrac{x^6}{720}$ is positive when $x = \frac{25}{16}$, and that $1 - \dfrac{x^2}{2} + \dfrac{x^4}{24}$ vanishes when $x = (6 - 2\sqrt{3})^{\frac{1}{2}} = 1.5924 \cdots$; and deduce that[19]

$$3.125 < \pi < 3.185.$$

A.52 The solution of a pair of trigonometrical equations

Let λ, μ be a pair of real numbers such that $\lambda^2 + \mu^2 = 1$. Then, if $\lambda \neq -1$, the equations

$$\cos x = \lambda, \qquad \sin x = \mu$$

have an infinity of solutions of which one and only one lies between[20] $-\pi$ and π.

First, let λ and μ be not negative; then (§3.63) the equation $\cos x = \lambda$ has at least one solution x_1 such that $0 \leq x_1 \leq \frac{1}{2}\pi$, since $\cos 0 = 1$, $\cos \frac{1}{2}\pi = 0$. The equation has not two solutions in this range, for if x_1 and x_2 were distinct solutions we could prove (cf. §A.51) that $\sin(x_1 - x_2) = 0$, and this would contradict §A.51 (IV), since

$$0 < |x_2 - x_1| \leq \tfrac{1}{2}\pi < 2.$$

Further, $\sin x_1 = +\sqrt{1 - \cos^2 x_1} = +\sqrt{1 - \lambda^2} = \mu$, so x_1 is a solution of *both* equations. The equations have no solutions in the ranges $(-\pi, 0)$ and $(\frac{1}{2}\pi, \pi)$ since, in these ranges, either $\sin x$ or $\cos x$ is negative. Thus the equations have one solution, and only one, in the range $(-\pi, \pi)$.

If λ or μ (or both) is negative, we may investigate the equations in a similar manner; the details are left to the reader.

It is obvious that, if x_1 is a solution of the equations, so also is $x_1 + 2n\pi$, where n is any integer, and therefore the equations have an infinity of real solutions.

---

[19]   See De Morgan [167, p. 316], for reasons for proving that $\pi > 3\frac{1}{8}$.
[20]   If $\lambda = -1$, $\pm\pi$ are solutions and there are no others in the range $(-\pi, \pi)$.

A.521 The principal solution of the trigonometrical equations

The unique solution of the equations $\cos x = \lambda$, $\sin x = \mu$ (where $\lambda^2 + \mu^2 = 1$) which lies between $-\pi$ and π is called the *principal solution*[21], and any other solution differs from it by an integer multiple of 2π.

The *principal value*[22] of the argument of a complex number $z(\neq 0)$ can now be defined analytically as the principal solution of the equations

$$|z| \cos \phi = \operatorname{Re} z, \qquad |z| \sin \phi = \operatorname{Im} z,$$

and then, if $z = |z|(\cos \theta + i \sin \theta)$, we must have $\theta = \phi + 2n\pi$, and θ is called *a value of the argument* of z, and is written $\arg z$ (cf. §1.5).

A.522 The continuity of the argument of a complex variable

It will now be shewn that it is possible to choose such a value of the argument $\theta(z)$, of a complex variable z, that it is a continuous function of z, provided that z does not pass through the value zero.

Let z_0 be a given value of z and let θ_0 be any value of its argument; then, to prove that $\theta(z)$ is continuous at z_0, it is sufficient to shew that a number θ_1 exists such that $\theta_1 = \arg z_1$ and that $|\theta_1 - \theta_0|$ can be made less than an arbitrary positive number ε by giving $|z_1 - z_0|$ any value less than some positive number η.

Let $z_0 = x_0 + i y_0$, $z_1 = x_1 + i y_1$. Also let $|z_1 - z_0|$ be chosen to be so small that the following inequalities are satisfied[23]:

(I) $|x_1 - x_0| < \frac{1}{2}|x_0|$, provided that $x_0 \neq 0$,
(II) $|y_1 - y_0| < \frac{1}{2}|y_0|$, provided that $y_0 \neq 0$,
(III) $|x_1 - x_0| < \frac{1}{4}\varepsilon|z_0|$, $|y_1 - y_0| < \frac{1}{4}\varepsilon|z_0|$.

From (I) and (II) it follows that $x_0 x_1$ and $y_0 y_1$ are not negative, and

$$x_0 x_1 \geq \frac{1}{2}x_0^2, \quad y_0 y_1 \geq \frac{1}{2}y_0^2,$$

so that $x_0 x_1 + y_0 y_1 \geq \frac{1}{2}|z_0|^2$.

Now let that value of θ_1 be taken which differs from θ_0 by less than π; then, since x_0 and x_1 have not opposite signs and y_0 and y_1 have not opposite signs[24], it follows from the solution of the equations of §A.52 that θ_1 and θ_0 *differ by less than* $\frac{1}{2}\pi$.

Now $\tan(\theta_1 - \theta_0) = \dfrac{x_0 y_1 - x_1 y_0}{x_0 x_1 + y_0 y_1}$, and so (§A.51 example A.5.2),

$$
\begin{aligned}
|\theta_1 - \theta_0| &\leq \frac{|x_0 y_1 - x_1 y_0|}{x_0 x_1 + y_0 y_1} \\
&= \frac{|x_0(y_1 - y_0) - y_0(x_1 - x_0)|}{x_0 x_1 + y_0 y_1} \\
&\leq 2|z_0|^{-2} \left\{ |x_0| \cdot |y_1 - y_0| + |y_0| \cdot |x_1 - x_0| \right\}.
\end{aligned}
$$

[21] If $\lambda = -1$, we take $+\pi$ as the principal solution.
[22] The term *principal value* was introduced in 1845 by Björling; see [74].
[23] (I) or (II) respectively is simply to be suppressed in the case (i) when $x_0 = 0$, or (ii) when $y_0 = 0$.
[24] The geometrical interpretation of these conditions is merely that z_0 and z_1 are not in different quadrants of the plane.

But $|x_0| \leq |z_0|$ and also $|y_0| \leq |z_0|$; therefore

$$|\theta_1 - \theta_0| \leq 2|z_0|^{-1}\{|y_1 - y_0| + |x_1 - x_0|\} < \varepsilon.$$

Further, if we take $|z_1 - z_0|$ less than $\frac{1}{2}|x_0|$, (if $x_0 \neq 0$) and $\frac{1}{2}|y_0|$, (if $y_0 \neq 0$) and $\frac{1}{4}\varepsilon|z_0|$, the inequalities (I), (II), (III) above are satisfied; so that, if η be the smallest of the three numbers[25] $\frac{1}{2}|x_0|, \frac{1}{2}|y_0|, \frac{1}{4}\varepsilon|z_0|$, by taking $|z_1 - z_0| < \eta$, we have $|\theta_1 - \theta_0| < \varepsilon$; and this is the condition that $\theta(z)$ should be a continuous function of the complex variable z.

A.6 Logarithms of complex numbers

The number ζ is said to be a *logarithm* of z if $z = e^\zeta$.

To solve this equation in ζ, write $\zeta = \xi + i\eta$, where ξ and η are real; and then we have

$$z = e^\xi(\cos\eta + i\sin\eta).$$

Taking the modulus of each side, we see that $|z| = e^\xi$, so that (§A.3), $\xi = \mathrm{Log}\,|z|$; and then

$$z = |z| \cdot (\cos\eta + i\sin\eta),$$

so that η must be a value of arg z.

The logarithm of a complex number is consequently a many-valued function, and it can be expressed in terms of more elementary functions by the equation

$$\log z = \log|z| + i\arg z.$$

The continuity of $\log z$ (when $z \neq 0$) follows from §§A.31 and A.522, since $|z|$ is a continuous function of z.

The differential coefficient of any particular branch of $\log z$ (§5.7) may be determined as in §A.32; and the expansion of §A.33 may be established for $\log(1 + a)$ when $|a| < 1$.

Corollary A.6.1 *If a^z be defined to mean $e^{z\log a}$, a^z is a continuous function of z and of a when $a \neq 0$.*

A.7 The analytical definition of an angle

Let z_1, z_2, z_3 be three complex numbers represented by the points P_1, P_2, P_3 in the Argand diagram. Then the angle between the lines (§A.12, footnote) P_1P_2 and P_1P_3 is defined to be any value of $\arg(z_3 - z_1) - \arg(z_2 - z_1)$.

It will now be shewn[26] that the area (defined as an integral), which is bounded by two radii of a given circle and the arc of the circle terminated by the radii, is proportional to one of the values of the angle between the radii, so that an angle (in the analytical sense) possesses the property which is given at the beginning of all textbooks on Trigonometry[27].

---

[25] If any of these numbers is zero, it is to be omitted.

[26] The proof here given applies only to acute angles; the reader should have no difficulty in extending the result to angles greater than $\frac{1}{2}\pi$, and to the case when OX is not one of the bounding radii.

[27] Euclid's definition of an angle does not, in itself, afford a *measure* of an angle; it is shewn in treatises on Trigonometry (cf. Hobson [321], ch. 1]) that an angle is measured by twice the area of the sector which the angle cuts off from a unit circle whose centre is at the vertex of the angle.

Let (x_1, y_1) be any point (both of whose coordinates are positive) of the circle $x^2 + y^2 = a^2$, $(a > 0)$. Let θ be the principal value of $\arg(x_1 + iy_1)$, so that $0 < \theta < \frac{1}{2}\pi$. Then the area bounded by OX and the line joining $(0,0)$ to (x_1, y_1) and the arc of the circle joining (x_1, y_1) to $(a, 0)$ is $\int_0^a f(x)\,dx$, where[28]

$$f(x) = x \tan \theta \quad (0 \leq x \leq a \cos \theta),$$
$$f(x) = (a^2 - x^2)^{\frac{1}{2}} \quad (a \cos \theta \leq x \leq a),$$

if an area be defined as meaning a suitably chosen integral.

It remains to be proved that $\int_0^a f(x)\,dx$ is proportional to θ.

Now

$$\int_0^a f(x)dx = \int_0^{a \cos \theta} x \tan \theta \, dx + \int_{a \cos \theta}^a (a^2 - x^2)^{\frac{1}{2}} \, dx$$

$$= \tfrac{1}{2}a^2 \sin \theta \cos \theta + \tfrac{1}{2} \int_{a \cos \theta}^a \left\{ a^2 (a^2 - x^2)^{-\frac{1}{2}} + \frac{d}{dx} x(a^2 - x^2)^{\frac{1}{2}} \right\} dx$$

$$= \tfrac{1}{2}a^2 \int_{a \cos \theta}^a (a^2 - x^2)^{-\frac{1}{2}} \, dx$$

$$= \tfrac{1}{2}a^2 \left\{ \int_0^1 (1 - t^2)^{-\frac{1}{2}} \, dt - \int_0^{\cos \theta} (1 - t^2)^{-\frac{1}{2}} \, dt \right\}$$

$$= \tfrac{1}{2}a^2 \left\{ \tfrac{1}{2}\pi - (\tfrac{1}{2}\pi - \theta) \right\}$$

$$= \tfrac{1}{2}a^2 \theta,$$

on writing $x = at$ and using the example worked out on Chapter 4.

That is to say, the area of the sector is proportional to the angle of the sector. To this extent, we have shewn that the popular conception of an angle is consistent with the analytical definition.

---

[28] The reader will easily see the geometrical interpretation of the integral by drawing a figure.

References

[1] N. ABEL, *Untersuchungen über die Reihe:* $1 + (m/1)x + m \cdot (m - 1)/(1 \cdot 2) \cdot x^2 + m \cdot (m - 1) \cdot (m - 2)/(1 \cdot 2 \cdot 3) \cdot x^3 + \cdots$ *o.s.w.*, J. reine angew. Math., 1 (1826), pp. 311–339.

[2] ——, *Recherches sur les fonctions elliptiques*, J. reine angew. Math., 2 (1827), pp. 101–181.

[3] ——, *Ueber einige bestimmte Integrale*, J. reine angew. Math., 2 (1827), pp. 22–30.

[4] ——, *Recherches sur les fonctions elliptiques. (Suite du mémoire Nr. 12. tom II.cah. 2 de ce journal)*, J. reine angew. Math., 3 (1828), pp. 160–190.

[5] ——, *Oeuvres Completes*, Grondhal and son, Christiania, 1881.

[6] ——, *Solution de quelques problèmes à l'aide d'intégrales définies*. In *Oeuvres completes de Niels Henkik Abel*, vol. 1, Christiana Imprimerie De Grondhal & Son, 1881.

[7] W. ABIKOFF, *The uniformization theorem*, Amer. Math. Monthly, 88 (1981), pp. 574–592.

[8] J. C. ADAMS, *On the Motion of the Moon's Node in the case when the Orbits of the Sun and Moon are supposed to have no Eccentricities, and when their mutual Inclination is supposed to be indefinitely small*, Monthly Notices R. A. S., 38 (1877), pp. 43–53.

[9] ——, *On the expression of the product of any two Legendre's coefficients by means of a series of Legendre's coefficients*, Proc. Roy. Soc. London, 27 (1878), pp. 63–71.

[10] ——, *On the value of Euler's constant*, Proc. Roy. Soc. London, 27 (1878), pp. 88–94.

[11] ——, *Table of the first sixty-two numbers of Bernoulli*, J. reine angew. Math., 85 (1878), pp. 269–272.

[12] L. V. AHLFORS, *Conformal Invariants. Topics in Geometric Function Theory*, McGraw-Hill Book Company, first ed., 1973.

[13] ——, *Complex Analysis. An Introduction to the Theory of Analytic Functions of One Complex Variable*, McGraw-Hill Book Company, third ed., 1979.

[14] J. R.. AIREY, *Tables of Neumann functions $G_n(x)$ and $Y_n(x)$*, Phil. Mag. Series 6, 21 (1911), pp. 658–663.

[15] W. S. ALDIS, *Tables for the solution of the equation $d^2y/dx^2 + 1/x \cdot dy/dx - (1 + n^2/x^2)y = 0$*, Proc. Royal Soc., 64 (1899), pp. 202–223.

[16] ——, *On the numerical computation of the functions $G_0(x)$, $G_1(x)$, and $J_n(x\sqrt{i})$*, Proc. Royal Soc., 66 (1900), pp. 32–43.

[17] E. AMIGUES, *Application du calcul des résidues*, Nouv. Ann. Math., 12 (1893), pp. 142–148.

[18] E. ANDING, *Sechsstellige Tafeln der Bessel'schen Funktionen imaginären Argumentes*, Leipzig, 1911.

[19] G. E. ANDREWS, *A simple proof of Jacobi's triple product identity*, Proc. Amer. Math. Soc., 16 (1965), pp. 333–334.

[20] G. E. ANDREWS, R. ASKEY, AND R. ROY, *Special Functions*, vol. 71 of Encyclopedia of Mathematics and its Applications, Cambridge University Press, 1999.

[21] G. E. ANDREWS AND B. C. BERNDT, *Ramanujan's Lost Notebook. Part I*, Springer, New York, 2005.

[22] ——, *Ramanujan's Lost Notebook. Part II*, Springer, New York, 2009.

[23] ——, *Ramanujan's Lost Notebook. Part III*, Springer, New York, 2012.

[24] ——, *Ramanujan's Lost Notebook. Part IV*, Springer, New York, 2013.

[25] ——, *Ramanujan's Lost Notebook. Part V*, Springer, New York, 2018.

[26] M. Y. ANTIMOROV, A. A. KOLYSHKIN, AND R. VAILLANCOURT, *Complex Variables*, Academic Press, New York, 1998.

[27] R. APÉRY, *Irrationalite de $\zeta(2)$ et $\zeta(3)$*, Astérisque, 61 (1979), pp. 11–13.

[28] P. Appell, *Évaluation d'une intégrale définie*, C. R. Acad. Sci. Paris, 87 (1878), pp. 874–876.

[29] ———, *Sur une classe d'èquations différentielles linéaires*, C. R. Acad. Sci. Paris, 91 (1878), pp. 684–717.

[30] ———, *Sur des fonctions de deux variables à trois ou quatre paires de périods*, C. R. Acad. Sci. Paris, 90 (1880), pp. 174–176.

[31] ———, *Sur la transformation des équations differentielles linéaires*, C. R. Acad. Sci. Paris, 90 (1880), pp. 211–214.

[32] ———, *Quelques remarques sur la théorie des potentiels multiformes*, Math. Ann., 30 (1887), pp. 155–156.

[33] J. R. Argand, *Essai sur une Manière de Représenter des Quantités Imaginaires dans les Constructions Géométriques*, Duminil-Lesueur, Paris, 1806.

[34] F. M. Arscott, *Periodic Differential Equations. An Introduction to Mathieu, Lamé and Allied Functions*, The MacMillan Company, 1964.

[35] P. Bachmann, *Niedere Zahlentheorie*, Leipzig–Berlin, 1902.

[36] J. Baik, P. Deift, and K. Johansson, *On the distribution of the length of the longest increasing subsequence of random permutations*, J. Amer. Math. Soc., 12 (1999), pp. 1119–1179.

[37] J. Baik, P. Deift, and T. Suidan, *Combinatorics and Random Matrix Theory*, vol. 172 of Graduate Studies in Mathematics, Amer. Math. Soc., 1st ed., 2016.

[38] D. H. Bailey, D. Borwein, and J. M. Borwein, *On Eulerian log-gamma integrals and Tornheim–Witten zeta functions*, The Ramanujan Journal, 36 (2015), pp. 43–68.

[39] H. F. Baker, *Abelian Functions*, Cambridge University Press, 1897. Reissued 1995.

[40] ———, *Abel's Theorem and the Allied Theory, including the Theory of the Theta Functions*, Cambridge University Press, 1897.

[41] ———, *On functions of several variables*, Proc. Royal Soc. London, 2 (1904), pp. 14–36.

[42] E. W. Barnes, *The theory of the G-function*, Quart. J. Math., 31 (1899), pp. 264–314.

[43] ———, *The genesis of the double gamma function*, Proc. London Math. Soc., 31 (1900), pp. 358–381.

[44] ———, *The theory of the gamma function*, Mess. Math., 29 (1900), pp. 64–128.

[45] ———, *The theory of the double gamma function*, Philos. Trans. Roy. Soc. A, 196 (1901), pp. 265–388.

[46] ———, *On the theory of the multiple gamma function*, Trans. Camb. Philos. Soc., 19 (1904), pp. 374–425.

[47] ———, *The MacLaurin sum-formula*, Proc. London Math. Soc., 2(3) (1905), pp. 253–272.

[48] ———, *A new development of the theory of the hypergeometric functions*, Proc. London Math. Soc., 6 (1908), pp. 141–177.

[49] ———, *On functions defined by simple types of hypergeometric series*, Trans. Camb. Phil. Soc., 206 (1908), pp. 253–279.

[50] ———, *A transformation of generalized hypergeometric series*, Quart. J. Math., 41 (1910), pp. 136–140.

[51] A. B. Basset, *An Elementary Treatise on Hydrodynamics and Sound*, Cambridge Deighton Bell and Co., London, George Bell and Sons, 1st ed., 1888.

[52] ———, *On the Potentials of the surfaces formed by the revolution of Limacons and Cardiods about their axes*, Proc. Camb. Phil. Soc., 6 (1889), pp. 2–18.

[53] H. Bateman, *The solution of partial differential equations by means of definite integrals*, Proc. London Math. Soc., 1 (1904), pp. 451–458.

[54] ———, *The conformal transformations of a space of four dimensions and their applications to geometrical optics*, Proc. London Math. Soc., 7 (1909), pp. 70–89.

[55] ———, *The solution of linear differential equations by means of definite integrals*, Trans. Camb. Phil. Soc., 21 (1912), pp. 171–196.

[56] ———, *An extension of Lagrange's expansion*, Trans. Amer. Math. Soc., 28 (1926), pp. 346–356.

[57] G. Bauer, *Von den Coefficienten der Reihen von Kugelfunctionen einer Variablen*, J. reine angew. Math., 56 (1859), pp. 101–121.

[58] ———, *Von der Coefficienten der Reihen von Kugelfunctionen einen Variablen*, J. reine angew. Math., 56 (1859), pp. 101–121.

[59] R. BEALS AND R. WONG, *Special Functions. A Graduate Text*, vol. 126 of Cambridge Studies in Advanced Mathematics, Cambridge University Press, 2010.

[60] B. BERNDT, *Ramanujan's Notebooks, Part I*, Springer-Verlag, New York, 1985.

[61] ———, *Ramanujan's Notebooks, Part II*, Springer-Verlag, New York, 1989.

[62] ———, *Ramanujan's Notebooks, Part III*, Springer-Verlag, New York, 1991.

[63] ———, *Ramanujan's Notebooks, Part IV*, Springer-Verlag, New York, 1994.

[64] ———, *Ramanujan's Notebooks, Part V*, Springer-Verlag, New York, 1998.

[65] D. BERNOULLI, *Theoremata de oscillationibus corporum filo flexili connexorum et catenao verticaliter suspensae*, Commentarii Academiae Scientiarum Petropolitanae, 6 (1738), pp. 108–122.

[66] J. BERNOULLI, *Ars Conjectandi*, Impression Anastaltique Culture er Civilisation [Reprinted in 1968], 1713.

[67] J. BERTRAND, *Règles sur la convergence des séries*, Journal de Mathématiques Pures et Appliquées, 7 (1842), pp. 35–54.

[68] ———, *Traitè de Calcul Différentielle et de calcul intégral*, Gauthier-Villars, Paris, 1864.

[69] F. W. BESSEL, *Einige Resultate aus Bradleys Beobachtungen*, Kónigsberger Archiv für Naturwissenschaft, 1 (1812), pp. 369–405.

[70] ———, *Untersuchung des Theils der planetarischen Störungen welcher aus der Bewegung der Sonne entsteht*, Abh. Ber. Acad. Wiss., (1824), pp. 1–52.

[71] D. BESSO, *Sull' integral seno e l'integral coseno*, Giornale di matematiche, 6 (1868), pp. 313–43.

[72] F. BEUKERS, *A note on the irrationality of $\zeta(2)$ and $\zeta(3)$*, Bull. London Math. Soc., 11 (1979), pp. 268–272.

[73] J. BINET, *Sur les intégrales définies eulériennes et sur leur application à la théorie des suites ainsi q''a l'évaluation des fonctions des grandes nombres*, Journal de l'Ecole Polytechnique, Paris, 16 (1839), pp. 123–343.

[74] E. G. BJÖRLING, *Quid in Analysis mathematica valeant signa illa x, log b(x), sin x, cos x, arcsin x. arccos x, disquisitio*, Archiv der Math. und Phys., 9 (1845), pp. 383–408.

[75] E. BLADES, *On spheroidal harmonics*, Proc. Edinburgh Math. Soc., 33 (1914), pp. 65–68.

[76] P. BOALCH, *From Klein to Painlevé via Fourier, Laplace and Jimbo*, Proc. London Math. Soc., 90 (2005), pp. 167–208.

[77] K. BOBEK, *Einleitung in die Theorie der Elliptischen Funktionen*, Leipzig, 1884.

[78] M. BÔCHER, *Ueber die Reihenentwickelungen der Potentialtheorie*, B. G. Teubner, 1894.

[79] ———, *On certain methods of Sturm and their application to the roots of Bessel's functions*, Bull. Amer. Math. Soc., 3 (1897), pp. 205–213.

[80] ———, *An Introduction to the Study of Integral Equations*, vol. 10 of Cambridge Tracts in Mathematics and Mathematical Physics, Cambridge University Press, 1909.

[81] B. BOLZANO, *Rein analytischer Beweis des Lehrsatzes*, Abh. der k. böhmischen Ges. der Wiss., 5 (1817).

[82] O. BONNET, *Mémoire sur la théorie générale des séries*, Mémoires des Savants étrangers of the Belgian Academy, 23 (1848), pp. 1–116.

[83] ———, *Remarques sur quelques intégrales définies*, Journal de Mathématiques Pures et Appliquées, 14 (1849), pp. 249–256.

[84] C. W. BORCHARDT, *Lecons sur les fonctions doublement périodiques faites en 1847 par M. J. Liouville*, J. reine angew. Math., 89 (1880), pp. 277–310.

[85] E. BOREL, *Leçons sur les Séries Divergentes*, Gauthier-Villars, 1901.

[86] ———, *Définition et domaine d'existence des fonctions monogénes uniformes*. In *Proceedings of the 1912 International Congress of Mathematicians, Cambridge, United Kingdom*, E. W. Hobson and A. E. H. Love, eds., Cambridge University Press, 1913, pp. 133–144.

[87] E. BOREL AND G. JULIA, *Leçons sur les Fonctions Monogènes Uniformes d'une Variable Vomplexe*, Gauthier-Villars, 1917.

[88] J. M. BORWEIN AND D. H. BAILEY, *Mathematics by Experiment: Plausible reasoning in the 21st century*, A. K. Peters, 1st ed., 2003.

[89] J. M. BORWEIN, D. H. BAILEY, AND R. GIRGENSOHN, *Experimentation in Mathematics: Computational Paths to Discovery*, A. K. Peters, 1st ed., 2004.

[90] J. M. Borwein and P. B. Borwein, *Pi and the AGM – A Study in Analytic Number Theory and Computational Complexity*, Wiley, New York, 1st ed., 1987.

[91] ——, *Ramanujan, modular equations, and approximations to pi or how to compute one billion digits of pi*, Amer. Math. Monthly, 96 (1989), pp. 201–219.

[92] J. M. Borwein and R. C. Corless, *Gamma and factorial in the Monthly*, Amer. Math. Monthly, 125 (2018), pp. 400–424.

[93] P. Borwein, S. Choi, B. Rooney, and A. Weirathmueller, *The Riemann hypothesis. A resource for the afficionado and virtuoso alike*, Canadian Mathematical Society, 1st ed., 2008.

[94] L. Bourget, *Mémoire sur les nombres de Cauchy et leur application à divers problémes de mécanique céleste*, Journal de Math., 6 (1861), pp. 33–54.

[95] ——, *Mémoire sur le mouvement vibratoire des membranes circulaires*, Ann. Sci. de l'École Norm. Sup., 3 (1866), pp. 55–95.

[96] ——, *Note sur les intégrales eulériennes: Extrait d'une lettre adressée à M. Ch. Hermite*, Acta Math., 1 (1882), pp. 295–296.

[97] ——, *Sur quelques intégrales définies: Extrait d'une lettre adressée à M. Ch. Hermite*, Acta Math., 1 (1882), pp. 363–367.

[98] P. W. Bridgeman, *On a certain development in Bessel's functions*, Phil. Mag., (6)XVI (1908), pp. 947–948.

[99] F. Brioschi, *Sur l'equation de Lamé*, C. R. Acad. Sci. Paris, 86 (1878), pp. 313–315.

[100] C. Briot and C. Bouquet, *Théorie des Fonctions Elliptiques*, Gauthier-Villars, Paris, 2nd ed., 1875.

[101] T. J. Bromwich, *The inversion of a repeated infinite integral*, Proc. London Math. Soc., 1 (1904), pp. 176–201.

[102] ——, *An Introduction to the Theory of Infinite Series*, Macmillan & Co., New York, NY, 2nd ed., 1926.

[103] L. V. Brouncker, *The squaring of the hyperbola, by an infinite series of rational numbers, together with its demonstration*, Phil. Trans., 3 (1668), pp. 645–649.

[104] F. de Brun, *Einige neue Formeln der Theorie der elliptischen Functionen, II*, Öfversigt af K. Vet. Akad., Stockholm, 54 (1897), pp. 523–532.

[105] Y. A. Brychkov, *Handbook of Special Functions. Derivatives, Integrals, Series and Other Formulas*, Taylor and Francis, Boca Raton, Florida, 2008.

[106] E. B. Burger and R. Tubbs, *Making Transcendence Transparent*, Springer-Verlag, New York, 2004.

[107] J. Burgess, *On the definite integral $\frac{2}{\sqrt{\pi}} \int_0^t e^{-2} dt$, with extended tables of values*, Trans. Roy. Soc. Edin., 39 (1900), pp. 257–322.

[108] H. Burkhardt, *Zur Theorie der trigonometrischen Reihen und der Entwicklungen nach Kugelfunktionen*, Sitzungsberichte der mathematisch-physikalischen Klasse der K. B. Akademie der Wissenschaften zu München, 39 (1909), pp. 1–23.

[109] H. H. Bürmann, *Rapport sur deux mémoires d'analyse du professeur Bürmann, by Lagrange and Adrien-Marie Legendre*, Mémoires de l'Institut National des Sciences et Arts, 2 (1799), pp. 13–17.

[110] W. S. Burnside, *Two notes on Weierstrass's $\wp(u)$*, Mess. Math., 21 (1891), pp. 84–87.

[111] W. S. Burnside and A. W. Panton, *The Theory of Equations*, Dublin Hodges, Figgia, 1886.

[112] O. Callandreau, *Sur le calcul des polynômes $X_n(\cos\theta)$ de Legendre pour les grandes valeurs de n*, Bull. des Sci. Math., 15 (1891), pp. 121–124.

[113] G. Cantor, *Beweis, dass eine für jeden reellen Werth von x durch eine trigonometrische Reihe gegebene Function $f(x)$ sich nur auf eine einzige Weise in dieser Form darstellen lässt*, J. reine angew. Math., 72 (1870), pp. 139–142.

[114] ——, *Ueber einen die trigonometrischen Reihen betreffenden Lehrsatz*, J. reine angew. Math., 72 (1870), pp. 130–138.

[115] ——, *Ueber trigonometrische Reihen*, Math. Ann., 4 (1871), pp. 139–143.

[116] ——, *Ueber die Ausdehnung eines Satzes aus der Theorie der trigonometrischen Reihen*, Math. Ann., 5 (1872), pp. 123–132.

[117] K. Carda, *Zur Darstellung der Bernoullischen Zahlen durch bestimmte Integrale*, Monatshefte für Math. und Phys., 5 (1894), pp. 321–324.

[118] L. CARLESON, *On convergence and growth of partial sums of Fourier series*, Acta Math., 116 (1966), pp. 135–157.

[119] H. S. CARSLAW, *Introduction to the Theory of Fourier's Series and Integrals*, Macmillan and co., London, 1921.

[120] A. L. CAUCHY, *Analyse Algebrique. Cours d'Analyse de l'Ecole Royale Polytechnique*, vol. I, L'Imprimerie Royale, Debure frères, Libraires du Roi et de la Bibliothèque du Roi; (reissued by Cambridge University Press, 2009), 1st ed., 1821.

[121] ———, *Memoire sur les Integrales Definies, Prises entre des Limites Imaginaires*, Gauthier-Villars, Paris, 1825.

[122] ———, *Sur les d'eveloppements des fonctions en séries périodiques*, Mém. de l'Acad. R. des Sci., 6 (1826), pp. 603–612.

[123] ———, *Exercises de Mathematiques*. Reprinted by Kessinger Publishing, LLC, September, 2010, 1828.

[124] ———, *Leçons sur le Calcul Différentiel*, Chez De Bure freres, libraires du roi et de la bibliothéque du roi, Paris, 1829.

[125] ———, *Résumés Analytiques*, L'Imprimerie Royale, 1833.

[126] ———, *Sur les fonctions alternées et sur diverse formules d'analyse*, C. R. Acad. Sci. Paris, Ser. I., 10 (1840), pp. 178–181.

[127] ———, *Mémoire sur quelques propositions fondamentales du calcul des residues*, C. R. Acad. Sci. Paris, Ser. I., 19 (1844), pp. 1337–1338.

[128] ———, *Sur la transformation de variables qui déterminant les mouvement d'une planète ou méme d'une comète en fonction explcite du temps, et sur le développment de ces fonctions en séries convergentes*, C. R. Acad. Sci., Paris, 38 (1854), pp. 990–993.

[129] ———, *Oeuvres Complètes d'Augustin Cauchy*, Gauthier-Villars, Paris, 1882–1919.

[130] A. CAYLEY, *Note sur l'addition des fonction elliptiques*, J. reine angew. Math., 41 (1852), pp. 57–65.

[131] ———, *On a formula in elliptic functions*, Mess. Math., 14 (1885), pp. 21–22.

[132] ———, *Mémoire sur les fonctions doublement périodiques*, Jour. de Math., 10 (1845), pp. 385–420.

[133] ———, *On a theorem relating to hypergeometric series*, Phil. Mag., 16 (1858), pp. 356–357.

[134] ———, *On bicursal curves*, Proc. London Math. Soc., 4 (1873), pp. 347–352.

[135] ———, *On Wroński's theorem*, Quart. J. Math. (Oxford), 12 (1873), pp. 221–228.

[136] ———, *An Elementary Treatise on Elliptic Functions*, Cambridge: Deighton, Bell; London: Bell, 1876.

[137] ———, *On the Schwarzian derivative, and the polyhedral functions*, Trans. Camb. Phil. Soc., 13 (1883), pp. 5–68.

[138] ———, *On the elliptic-function solution of the equation $x^3 + y^3 - 1 = 0$*, Proc. Camb. Phil. Soc., 4 (1883), pp. 106–109.

[139] ———, *On linear differential equations*, Quart. J. Math. (Oxford), 21 (1886), pp. 326–335.

[140] C. CELLÉRIER, *Note sur les principes fondamentaux de l'analyse*, Bull. des Sci. Math., 14 (1890), pp. 142–160.

[141] E. CESÀRO, *Sur la multiplication des séries*, Bull. des Sci. Math., 14 (1890), pp. 114–120.

[142] S. CHAPMAN, *On the general theory of summability, with application to Fourier's and other series*, Quart. Jour., 43 (1912), pp. 1–53.

[143] J. CHARTIER, *Règles de convergence des séries et des intégrales définies qui contiennent un facteur périodique; par le P*, Journal de Mathématiques Pures et Appliquées, 18 (1853), pp. 201–212.

[144] C. CHREE, *On the coefficients in certain series of Bessel's functions*, Phil. Mag., (6)XVII (1909), pp. 329–331.

[145] CHRISTOFFEL, *Über die Gaussische Quadratur und eine Verallgemeinerung derselben*, J. reine angew. Math., 55 (1858), pp. 61–82.

[146] G. CHRYSTAL, *Algebra. Part II*, A. and C. Black, London, 2nd ed., 1922.

[147] A. C. CLAIRAULT, *Mémoire sur l'Orbite Apparente du Soleil Autour de la Terre, en Ayant égard aux Perturbations Produites par les Actions de la Lune at des Planètes Principales*, Mémoires de Mathématique et de Physique de l'Académie Royales des Sciences, 9 (1754), pp. 801–870.

[148] T. CLAUSEN, *Demonstratio duarum celeberrimi Gaussii propositionum*, J. reine angew. Math., 3 (1828), p. 311.

[149] A. CLEBSCH, *Ueber die Reflexion an einer Kugelfläche*, J. reine angew. Math., 61 (1863), pp. 195–262.

[150] ———, *Ueber diejenigen Curven, deren Coordinaten sich als elliptische Functionen eines Parameters darstellen lassen*, J. reine angew. Math., 64 (1865), pp. 210–270.

[151] E. A. CODDINGTON AND N. LEVINSON, *Theory of Ordinary Differential Equations*, McGraw-Hill Book Co., 1955.

[152] F. COHN, *Uber Lamésche Funktionen mit komplexen Parametern*, Thesis, Königsberg University, 1888.

[153] J. B. CONREY, *More than two fifths of the zeros of the Riemann zeta function are on the critical line*, J. reine angew. Math., 399 (1989), pp. 1–16.

[154] ———, *The Riemann hypothesis*, Notices of the AMS, 50 (2003), pp. 341–353.

[155] E. T. COPSON, *An Introduction to the Theory of Functions of a Complex Variable*, Oxford University Press, Oxford, 1960.

[156] S. A. COREY, *The development of functions*, Ann. Math., 1 (1900), pp. 77–80.

[157] L. CRAWFORD, *On the solutions of Lamé's equation $\frac{d^2U}{du^2} = U\{n(n+1)\wp u + B\}$ in finite terms when $2n$ is an odd integer*, Quarterly J. Pure and Appl. Math., 27 (1895), pp. 93–98.

[158] E. CUNNINGHAM, *The ω-functions, a class of normal functions occurring in statistics*, Proc. Roy. Soc. 81, (1908), pp. 310–331.

[159] D'ALEMBERT, *Opuscules Mathématiques, ou Mémoirs sur Différens Sujets de Géometrie, de . . .*, David, 1768.

[160] G. DARBOUX, *Mémoire sur les fonctions discontinues*, Ann. de l'École Norm. Sup. Ser. 2, (1875), pp. 57–112.

[161] ———, *Mémoire sur l'approximation de fonctions de très-grands nombres et sur une classe étendue de développements en série*, C. R. Acad. Sci. Paris, 82 (1876), pp. 365–368; 404–406.

[162] ———, *Sur les dévelopment en série des fonctions d'une seule variable*, Journal de Mathématiques Pures et Appliquées, 2 (1876), pp. 291–312.

[163] ———, *Mémoire sur l'approximation des fonctions de très-grands nombres, et sur une classe ètendue de développements en série*, Journal de Mathématiques Pures et Appliquées, 4 (1878), pp. 5–56.

[164] ———, *Sur l'approximation des fonctions de très-grands nombres, et sur une classe ètendue de développements en série*, Journal de Math., 3 (1878), pp. 377–416.

[165] G. H. DARWIN, *Ellipsoidal harmonic analysis*, Phil. Trans., 197A (1901), pp. 461–557.

[166] P. J. DAVIS, *Leonhard Euler's integral: a historical profile of the Gamma function. in memorian: Milton Abramowitz*, Amer. Math. Monthly, 66 (1959), pp. 849–869.

[167] A. DE MORGAN, *A Budget of Paradoxes*, London, 1872.

[168] P. DEBYE, *Näherungsformeln für die Zylinderfunktionen für grosse Werte des Arguments und unbeschränkt veränderliche Werte des Index*, Math. Ann., 67 (1909), pp. 535–558.

[169] R. DEDEKIND, *Stetigkeit und Irrationale Zahlen*, F. Vieweg und sohn, 1872.

[170] U. DINI, *Serie di Fourier e altre rappresentazioni analitiche delle funzioni di una variabile reale*, T. Nistri, Pisa, 1880.

[171] A. DINNIK, *Tafeln der Besseischen Funktionen $J + i/3$ und $J_\pm$*, Archiv der Math. und Phys, 18 (1911), pp. 337–338.

[172] P. DIRICHLET, *Sur la convergence des séries trigonométriques qui servant à representer une fonction arbitraire entre des limits données*, J. reine angew. Math., 4 (1829), pp. 157–169.

[173] ———, *Beweis des Satzes, dass jede unbegrenzte arithmetische Progression*, Berliner Abh., (1837), pp. 45–81.

[174] ———, *Sur les séries dont le terme général dépend de deux angles, et qui servent à exprimer des fonctions arbitraires entre des limites donnée*, J. reine angew. Math., 17 (1837), pp. 35–56.

[175] ———, *Sur l'usage des intégrales définies dans la sommation des séries finies ou infinies*, J. reine angew. Math., 17 (1837), pp. 57–67.

[176] ———, *Démonstration de cette proposition: Toute progression arithmétique dont le premier terme et la raison sont des entiers sans diviseur commun contient une infinité de nombres premiers*, Journal de Mathématiques Pures et Appliquées, 4 (1839), pp. 393–422.

[177] ———, *Démonstration d'un théoreme d'Abel*, Journal de Math., 7 (1863), pp. 253–255.

[178] ———, *Sur les intégrales Euleriennes*. In *Werke*, L. Kronecker, ed., vol. 1, Berlin: Reimer, 1889, pp. 271–279.

[179] ———, *Vorlesungen über die Lehre von den einfachen und mehrfachen bestimmten Integralen*, Braunschweig: Vieweg, 1904.

[180] E. DIRKSEN, *Ueber die Convergenz einer nach den Sinussen und Cosinussen der Vielfachen eines Winkels fortschreitenden Reihe*, J. reine angew. Math., 4 (1829), pp. 170–178.

[181] A. C. DIXON, *On the double periodic functions arising out of the curve* $x^3 + y^3 - 3\alpha x y = 1$, Quart. Jour., 24, (1890), pp. 167–233.

[182] ———, *On Burmann's theorem*, Proc. London Math. Soc., 34 (1901), pp. 151–153.

[183] ———, *Summation of a certain series*, Proc. London Math. Soc., 35 (1902), pp. 284–291.

[184] ———, *On a certain double integral*, Proc. London Math. Soc., 2 (1905), pp. 8–15.

[185] M. J. DOLBNIA, *Sur l'intégrale* $\int \frac{dx}{\sqrt[4]{x^4+px^2+q}}$, Bulletin des sciences mathématiques Ser. 2, 19 (1895), pp. 67–84.

[186] W. F. DONKIN, *On the equation of Laplace's functions, &c*, Phil. Trans., 147, (1857), pp. 43–57.

[187] J. DOUGALL, *On Vandermonde's theorem, and some more general expansions*, Proc. Edinburgh Math. Soc., 25 (1906), pp. 114–132.

[188] ———, *The solution of Mathieu's differential equation*, Proc. Edinburgh Math. Soc., 34 (1915), pp. 176–196.

[189] P. DU BOIS-REYMOND, *Ueber den Gültigkeitsbereich der Taylorschen Reihenentwickelung*, Münchener Sitzungsberichte, Mathematisch-physicalische Classe, 6 (1876), pp. 225–237.

[190] B. DUBROVIN AND A. KAPAEV, *A Riemann–Hilbert approach to the Heun equation*, SIGMA, 14 (2018), 24 pages.

[191] H. M. EDWARDS, *Riemann's Zeta Function*, Academic Press, New York, 1974.

[192] M. EICHLER AND D. ZAGIER, *On the zeros of the Weierstrass ℘-function*, Math. Ann., 258 (1982), pp. 399–407.

[193] G. EISENSTEIN, *Beitrage zu Theorie der elliptischen Funktionen*, J. reine angew. Math., 35 (1847), pp. 137–184.

[194] ———, *Beitrage zu Theorie der elliptischen Funktionen. (Fortsetzung)*, J. reine angew. Math., 35 (1847), pp. 185–274.

[195] J. F. ENCKE, *Über die Methode der kleinsten Quadrate*, Berliner astronomisches. Jahrbuch, (1834), pp. 249–312.

[196] O. ESPINOSA AND V. MOLL, *On some definite integrals involving the Hurwitz zeta function. Part 2*, The Ramanujan Journal, 6 (2002), pp. 449–468.

[197] L. EULER, *Variae observationes circa series infinitas*, Commentationes Acad. Sci. Imp. Petropolitanae, 9 (1737), pp. 160–188.

[198] ———, *Recherches sur les racines imaginaires de equations*, Histoire de l'Académie Royale des Sciences et des Belles Lettres de Berlin, 5 (1749), pp. 212–288.

[199] ———, *De integratione aequationis differentialis*

$$\frac{m\,dx}{\sqrt{1 - x^4}} = \frac{n\,dy}{\sqrt{1 - y^4}}.$$

In *Opera Omnia*, vol. 20 of I, Teubner, Berlin, 1738-1914, pp. 58–79.

[200] ———, *Evolution formulae integralis* $\int x^{f-1} dx\, (\ell x)^{m/n}$ *integratione a valore* $x = 0$ *ad* $x = 1$ *extensa*. In *Opera Omnia*, vol. 17 of I, Teubner, Berlin, 1738-1914, pp. 316–357.

[201] ———, *Institutiones calculi differentialis*. In *Opera Omnia*, vol. 10 of I, Teubner, Berlin, 1738-1914.

[202] ———, *Institutiones Calculi Integralis*. In *Opera Omnia*, vol. 11 of I, Teubner, Berlin, 1738-1914.

[203] ———, *Methodus facilis inveniendi series per sinus*. In *Opera Omnia*, vol. 16 of I, Teubner, Berlin, 1738-1914, pp. 94–113.

[204] ——, *Methodus generalis summandi progressiones*. In *Opera Omnia*, vol. 14 of I, Teubner, Berlin, 1738-1914, pp. 42–72.

[205] ——, *De motu vibratorio tympanorum*, Novi commentarii academiae scientiarum Petropolitanae 10, (1764), 1766, pp. 243–260.

[206] ——, *De motu vibratorio tympanorum*. In *Opera Omnia*, vol. 10 of II, Teubner, Berlin, 1766, pp. 344–358.

[207] ——, *Introductio in Analysis Infinitorum*. English translation Introduction to Analysis of the Infinite, by J. D. Blantan, Editor, Springer-Verlag, New York, 1st ed., 1988.

[208] H. M. FARKAS AND I. KRA, *Theta Constants, Riemann Surfaces and the Modular Group*, vol. 37 of Graduate Studies in Mathematics, Springer-Verlag, New York, 2001.

[209] J. FAY, *Theta Functions on Riemann Surfaces*, vol. 352 of Lecture Notes in Mathematics, Springer-Verlag, New York, 1973.

[210] L. FEJÉR, *Untersuchungen über Fourier Reihen*, Math. Ann., 58 (1904), pp. 51–69.

[211] G. FLOQUET, *Sur les èquations différentielles linéaires à coefficients périodiques*, Annales de l'Ecole Normale Supérieure, 12 (1883), pp. 47–88.

[212] A. FOKAS, A. ITS, A. KAPAEV, AND V. NOVOKSHENOV, *Painlevé Transcendents: The Riemann–Hilbert Approach*, vol. 128 of Mathematical Surveys and Monographs, American Mathematical Society, Providence, 2006.

[213] L. R. FORD, *Introduction to the Theory of Automorphic Functions*, G. Bell and Sons, London, 1915.

[214] P. J. FORRESTER AND S. O. WARNAAR, *The importance of the Selberg integral*, Bull. Amer. Math. Soc., 45 (2008), pp. 489–534.

[215] A. R. FORSYTH, *Note on Professor Cayley's "Formula in elliptic functions"*, Messenger Math., 14 (1885), pp. 23–25.

[216] ——, *The addition-theorem for the second and third elliptic integral*, Messenger Math., 15 (1886), pp. 49–57.

[217] ——, *New solutions of some of the partial differential equations in mathematical physics*, Messeng. Math., 27 (1898), pp. 99–118,190.

[218] ——, *Theory of Differential Equations*, vol. 4, Cambridge University Press, 1902.

[219] ——, *Theory of Differential Equations*, vol. 6, Cambridge University Press, 1906.

[220] ——, *Theory of Functions of a Complex Variable*, Cambridge University Press, 1918.

[221] ——, *A Treatise on Differential Equations*, Courier Corporation, London, 1929.

[222] ——, *A treatise on differential equations*, Dover, 1996.

[223] J. B. FOURIER, *Théorie Analytique de la Chaleur*, F. Didot, Paris, 1822.

[224] I. FREDHOLM, *On a new method for solving the Dirichlet problem*, Stock. Öfv., 57 (1900), pp. 39–46.

[225] ——, *Sur une classe d'équations fonctionnelles*, Acta Math., 27 (1903), pp. 365–390.

[226] G. FROBENIUS, *Ueber die Entwickelung analytischer Functionen in Reihen, die nach gegebenen Functionen fortschreiten*, J. reine angew. Math., 73 (1871), pp. 1–30.

[227] ——, *Ueber die Integration der linearen Differentialgleichungen durch Reichen*, J. reine angew. Math., 76 (1873), pp. 214–235.

[228] G. FROBENIUS AND G. STICKELBERGER, *Theorie der elliptischen Functionen*, J. reine angew. Math., 83 (1877), pp. 175–179.

[229] ——, *Ueber die Addition und Multiplication der elliptischer Functionen*, J. reine angew. Math., 89 (1880), pp. 146–184.

[230] E. FÜRSTENAU, *Darstellung der reellen Wurzeln algebraischer Gleichungen durch Determinanten der Coefficienten*, Elwert, 1860.

[231] P. H. FUSS, *Correspondance Mathématique et Physique de quelques célèbres géomètres du XVII-éme siècle*, vol. I, St.-Pétersbourg, 1843.

[232] D. GAMBIOLI, *Sur le développement, suivant les puissances croissantes de x, de la fraction algébrique* $1/\sum_0^n a_r x^r$, Bologna, Memoire, (1892), pp. 39–44.

[233] C. F. GAUSS, *Integralium valores per approximationem inveniendi*, Commentationes Societatis Regiae Scientiarum Gottingensis (1814), pp. 39–76. Reprinted in *Werke*, 3 (1876), pp. 165–196.

[234] ——, *Arithmetisch Geometrisches Mittel*. In *Werke*, 3 (1876), pp. 361–403.

[235] ——, *Demonstration nova theorematis omnen functionem algebraicam rationalem integram unius variabilis in factores reales primi vel secundi gradus resolve posse*. In *Werke*, 3 (1876), pp. 3–31.

[236] ——, *Disquisitiones generales circa seriem infinitam*. In *Werke*, 3 (1876), pp. 123–162.

[237] ——, *Theorie residuorum biquadraticorum*. In *Werke*, 2 (1863), pp. 95–148.

[238] ——, *Werke*, Königlichen Gesellschaft der Wissenschaften zu Göttingen, 1863–1903.

[239] L. Gegenbauer, Über die functioner X_n^m. Wiener Sitzungsberichte, 68 (1874), pp. 357–367.

[240] ——, *Über einige bestimmte Integrale*, Wiener Sitzungsberichte, 70 (1874), pp. 433–443.

[241] L. v. Gegenbauer, *Bemerkung über die Bessel'schen Functionen*, Monatsh. f. Math. Phys., 102 (1893), p. 942.

[242] ——, *Bemerkung über die Bessel'schen Functionen*, Monatsh. f. Math. Phys., 8 (1897), pp. 383–384.

[243] J. W. Gibbs, *Fourier's series*, Nature, 59 (1899), p. 606.

[244] J. W. L. Glaisher, *On* sin ∞ *and* cos ∞, Mess. Math., 5 (1871), pp. 232–244.

[245] ——, *Formulae for the* sn, cn, dn, *of* $u + v + w$, Mess. Math., 11 (1882), pp. 45–77.

[246] ——, *Tables of the first 250 Bernoulli's numbers (to nine figures) and their logarithms (to ten figures)*, Trans. of Camb. Phil. Soc., (I) 12 (1873), pp. 384–391.

[247] ——, *Notes on Laplace's coefficients*, Proc. London Math. Soc., 6 (1874), pp. 126–136.

[248] ——, *On definite integrals involving elliptic functions*, Proc. London Math. Soc., 29 (1879), pp. 331–351.

[249] ——, *On the addition equation for the third elliptic integral*, Mess. Math., 10 (1881), pp. 124–135.

[250] ——, *On some elliptic functions and trigonometrical theorems*, Mess. Math., 10 (1881), pp. 92–97.

[251] ——, *Systems of formulae in elliptic functions*, Mess. Math., 10 (1881), pp. 104–111.

[252] ——, *On elliptic functions, I*, Mess. Math., 11 (1882), pp. 81–95.

[253] ——, *Factor Tables*, London, 1883.

[254] E. Goursat, *Sur la définition générale des fonctions analytiques, d'après Cauchy*, Trans. Amer. Math. Soc., 1 (1900), pp. 14–16.

[255] ——, *Cours d'Analyse Mathématique*, vol. I, Gauthier-Villars, Paris, 1902.

[256] ——, *Cours d'Analyse Mathématique*, vol. II, Gauthier-Villars, Paris, 1910.

[257] J. H. Grace, *On the zeros of a polynomial*, Proc. Camb. Phil. Soc., 11 (1902), pp. 352–357.

[258] I. S. Gradshteyn and I. M. Ryzhik, *Table of Integrals, Series, and Products*, Edited by D. Zwillinger and V. Moll. Academic Press, New York, 8th ed., 2015.

[259] A. Gray and G. B. Mathews, *A Treatise on Bessel Functions and their Applications to Physics*, Macmillan, London, 1895.

[260] R. Greene and S. Krantz, *Function Theory of One Complex Variable*, vol. 40 of Graduate Studies in Mathematics, American Mathematical Society, 2002.

[261] V. I. Gromak, I. Laine, and S. Shimomura, *Painlevé Differential Equations in the Complex Plane*, Walter de Gruyter, 2002.

[262] C. Gudermann, *Theorie der Modular-Functionen und der Modular-Integrale*, J. reine angew. Math., 18 (1838), pp. 1–54.

[263] G. Guerritore, *Calcolo delle funzioni di Lamé fio a quelle di grado 10*, Giorn. Math. Napoli, 47 (1909), pp. 164–172.

[264] A. Gutzmer, *Ein Satz über Potenzreihen*, Math. Ann., 32 (1888), pp. 596–600.

[265] J. Hadamard, *Résolution d'une question relative au déterminants*, Bull. Sci. Math., 17 (1893), pp. 240–246.

[266] E. Haentzschel, *Ueber den functionentheoretischen Zusammenhang zwischen den Lamé'schen, Laplace'schen und Bessel'schen Functionen*, Zeitschrift für Math. und Phys., 31 (1886), pp. 25–33.

[267] G. H. Halphen, *Mémoire sur la réduction des équations différentielles linéaires aux formes intégrables*, Mém. présentées par divers savants à l'Académie de Sciences, 28 (1880), pp. 1–260.

[268] ——, *Traité des Fonctions lliptiques et de Leurs Applications*, vol. 1, Gauthier-Villars, Paris, 1886.

[269] ——, *Traité des Fonctions Elliptiques et de Leurs Applications*, vol. 2, Gauthier-Villars, Paris, 1888.

[270] W. R. Hamilton, *On fluctuating functions*, Trans. Royal Irish Academy, 19 (1843), pp. 264–321.

[271] H. Hancock, *Lectures on the Theory of Elliptic Functions*, J. Wiley and Sons, New York, 1910.

[272] H. Hankel, *Die Euler'schen Integrale bei unbeschränkter Variabilität des Argumentes*, Zeitschrift für Math. und Phys., 9 (1864), pp. 1–21.

[273] ———, *Die Cylinderfunctionen erster und zweiter Art*, Math. Ann., 1 (1869), pp. 467–501.

[274] G. H. HARDY, *The elementary theory of Cauchy's principal values*, Proc. London Math. Soc., 34 (1901), pp. 16–40.

[275] ———, *Notes on some points in the Integral Calculus*. Reprinted in *Collected Papers of G. H. Hardy vol. V*, Oxford at the Clarendon Press, 1972, pp. 177–183 (258).

[276] ———, *Some theorems connected with Abel's theorem on the continuity of power series*, Proc. London Math. Soc., 2(4) (1907), pp. 247–265.

[277] ———, *A Course of Pure Mathematics*, Cambridge University Press, 1908.

[278] ———, *Theorems relating to the summability and convergence of slowly oscillating series*, Proc. London Math. Soc., 8 (1910), pp. 301–320.

[279] ———, *Sur les zéros de la fonction ζ(s) de Riemann*, C. R. Acad. Sci. Paris, 158 (1914), pp. 1012–1014.

[280] ———, *The Integration of Functions of a Single Variable*, vol. 2 of Cambridge University Tracts in Mathematical Physics, Cambridge University Press, 2nd ed., 1958.

[281] G. H. HARDY AND J. E. LITTLEWOOD, *Contributions to the theory of the Riemann zeta-function and the theory of the distribution of primes*, Proc. London Math. Soc., 19 (1920), pp. 119–196.

[282] C. J. HARGREAVE, *On the solution of linear differential equations*, Phil. Trans. R. Soc. Lond., 138 (1848), pp. 31–54.

[283] A. HARNACK, *Ueber Cauchy's zweiten Beweis für die Convergenz der Fourier'schen Reihen und eine damit verwandte ältere Method von Poisson*, Math. Ann., 32 (1888), pp. 175–202.

[284] L. HEFFTER, *Über eine Veranschanlichung von Funktionen einer komplexen Variabelen*, Zeit. für Math. und Phys., 44 (1899), pp. 235–236.

[285] E. HEINE, *Beitrag zur Theorie der Anziehung und der Wärme*, J. reine angew. Math., 29 (1845), pp. 185–208.

[286] ———, *Theorie der Anziehung eines Ellipsoids*, J. reine angew. Math., 42 (1851), pp. 70–82.

[287] ———, *Handbuch der Kugelfunktionen*, George Reimer, 1861.

[288] ———, *Ueber trigonometrische Reihen*, J. reine angew. Math., 71 (1870), pp. 353–365.

[289] C. HERMITE, *Sur un nouveau développement en série des fonctions*, C. R. Acad. Sci. Paris, 58 (1864), pp. 266–273.

[290] ———, *Extrait d'une lettre de M. Ch. Hermite adressée á M. L. Fuchs.*, J. reine angew. Math., 82 (1877), pp. 343–347.

[291] ———, *Sur quelques applications des fonctions elliptiques*, C. R. Acad. Sci. Paris, 85 (1877), pp. 689–695; 728–732; 821–826. Reprinted in *Oeuvres de Charles Hermite*, Tome 2, pp. 266–274.

[292] ———, *Sur une application du théorème de M. Mittag-Leffler, dans la théorie des fonctions. (Extrait d'une lettre adresseé à M. Mittag-Leffler de Stockholm par M. Ch. Hermite à Paris)*, J. reine angew. Math., 92 (1882), pp. 145–155.

[293] ———, *Sur les polynômes de Legendre*, Jornal de sciencias mathematicas e astronomicas, 6 (1885), pp. 81–84.

[294] ———, *Sur Quelques Applications des Fonctions Elliptiques*, Gauthier-Villars, 1885.

[295] ———, *Sur quelques propositions fondamentales de la théorie des fonctions elliptiques*, Proc. Math. Congress, American Mathematical Society, Vol. 1, (1896), pp. 105–115.

[296] ———, *Extraits de quelques lettres adressées à M. S. Pincherle*, Annili di Matematica, 3 (1901), pp. 57–72.

[297] ———, *Oeuvres de Charles Hermite*, vol. 3, Gauthier-Villars, 1912.

[298] C. HERMITE AND T. J. STIELTJES, *Correspondance d'Hermite et de Stieltjes*, vol. 2, Les Soins de B. Baillaud et H. Bourget, avec une prèface de Èmile Picard, Paris, Gauthier-Villars, 1905.

[299] K. HEUN, *Die Kugelfunctionen und Laméschen Functionen als Determinanten*, Gött. Nach., (1881).

[300] ———, *Zur Theorie der Riemann'schen Functionen zweiter Ordnung mit vier Verzweigungspunkten*, Math. Ann., 33 (1888), pp. 161–179.

[301] ———, *Beitrag zur Theorie der Lamé'schen Functionen*, Math. Ann., 33 (1888), pp. 180–196.

[302] W. HEYMANN, *Über hypergeometrische Funktionen, deren letztes Element speziell ist.*, Zeitschrift für Math. und Phys., 44 (1899), pp. 280–88.

[303] W. M. HICKS, *On toroidal functions*, Phil. Trans., 172 (1881), pp. 609–652.

[304] O. HIJAB, *Introduction to Calculus and Classical Analysis*, Springer-Verlag, New York, 1st ed., 1997.

[305] D. HILBERT, *Grundzüge einer allgemeinen Theorie der linearen Integralgleichungen*, Göttinger Nach., 1904 (1904), pp. 41–91.

[306] G. W. HILL, *On the part of the motion of lunar perigee which is a function of the mean motions of the sun and moon*, Acta Math., 8 (1886), pp. 1–36.

[307] M. J. M. HILL, *The continuation of certain fundamental power series*, Proc. London Math. Soc., 35 (1902), pp. 388–416.

[308] ———, *On a formula for the sum of a finite number of terms of the hypergeometric series when the fourth element is equal to unity*, Proc. London Math. Soc., 5 (1907), pp. 335–341.

[309] E. HILLE, *On the zeros of Mathieu functions*, Proc. London Math. Soc., (2)23 (1925), pp. 185–237.

[310] ———, *Ordinary Differential Equations in the Complex Domain*, Dover Publications Inc., 1976.

[311] M. HIRSCHHORN, *The power of q: A personal journey*, vol. 49 of Developments in Mathematics, Springer, 2017.

[312] E. W. HOBSON, *On a theorem in differentiation, and its application to spherical harmonics*, Proc. London Math. Soc., 24 (1892), pp. 55–67.

[313] ———, *On Bessel's functions, and relations connecting them with hyperspherical and spherical harmonics*, Proc. London Math. Soc., 25 (1893), pp. 49–75.

[314] ———, *On a type of spherical harmonics of unrestricted degree, order, and argument*, Phil. Trans. R. Soc. London, 187A (1896), pp. 443–531.

[315] ———, *Functions of a Real Variable*, Cambridge University Press, 1907.

[316] ———, *The Theory of Functions of a Real Variable and the Theory of Fourier Series*, Cambridge University Press, 1907.

[317] ———, *On a general convergence theorem, and the theory of the representation of a function by series of normal functions*, Proc. London Math. Soc., 6 (1908), pp. 349–395.

[318] ———, *On the representation of a function by series of Bessel's functions*, Proc. London Math. Soc., 7 (1908), pp. 359–388.

[319] ———, *On the representation of a function by a series of Legendre's functions*, Proc. London Math. Soc., 7 (1909), pp. 24–39.

[320] ———, *Squaring the Circle*, Cambridge University Press, 1913.

[321] ———, *A Treatise on Plane Trigonometry*, Cambridge University Press, 2nd ed., 1918.

[322] ———, *On the summability of generalized Fourier's series*, Proc. London Math. Soc., 22 (1924), pp. 420–424.

[323] ———, *A Treatise on Plane and Advanced Trigonometry*, Dover Publications, Inc., New York, 7th ed., 1928.

[324] M. HOCEVAR, *Über die unvollständige Gammafunction*, Zeitschrift für Math. und Phys., 21 (1876), pp. 449–450.

[325] O. HÖLDER, *Ueber die Eigenschaft der Gammafunction keiner algebraischen Differentialgleichungen zu genügen*, Math. Ann., 28 (1887), pp. 1–13.

[326] ———, *Über einen Mittelwertsatz*, Gott. Nachr., 2 (1889), pp. 38–47.

[327] A. HURWITZ, *Einige Eigenschaften der Dirichlet'schen Funktionen* $F(s) = \sum \left(\frac{D}{n} \right) \cdot \frac{1}{n^s}$, *die bei der Bestimmung der Klassenanzahlen Binärer quadratischer Formen auftreten*, Z. für Math. und Physik, 27 (1882), pp. 86–101.

[328] ———, *Uber die Fourierschen Konstanten integrierbarer Funktionen*, Math. Ann., 57 (1903), pp. 425–446.

[329] ———, *Note on certain iterated and multiple integrals*, Ann. Math., 9 (1908), pp. 183–192.

[330] ———, *Über die Einführung der elementaren transzendenten Funktionen in der algebraischen Analysis*, Math. Ann., 70 (1911), pp. 33–47.

[331] D. HUSEMÖLLER, *Elliptic Curves*, vol. 111 of Graduate Texts in Mathematics, Springer Verlag, 2nd ed., 2004.

[332] E. L. INCE, *On a general solution of Hill's equation*, Roy. Astronomical Soc., 55 (1915), pp. 436–448.

[333] ———, *Continued fractions associated with the hypergeometric equation*, Proc. Edinburgh Math. Soc., 34 (1916), pp. 2–15.

References

[334] ——, *On the connexion between linear differential systems and integral equations*, Proc. Royal Soc. Edin., 42 (1922), pp. 43–53.

[335] ——, *A proof of the impossibility of the coexistence of two Mathieu functions*, Proc. Camb. Phil. Soc., 21 (1922), pp. 117–120.

[336] ——, *Ordinary Differential Equations*, Dover, New York, 1956.

[337] R. T. A. INNES, *On the periods of the elliptic functions of Weierstrass*, Proc. Edin. Math. Soc., 27 (1907), pp. 357–368.

[338] J. G. ISHERWOOD, *Tables of the Bessel functions for pure imaginary values of the argument*, Proc. Manchester Lit. and Phil. Soc., 48 (XIX) (1904), pp. 1–3.

[339] K. IWASAKI, H. KIMURA, S. SHIMOMURA, AND M. YOSHIDA, *From Gauss to Painlevé: A Modern Theory of Special Functions*, Braunschweig: Fried. Vieweg and Sohn, 1991.

[340] F. H. JACKSON, *An Extension of the Theorem* $\frac{\prod(\gamma-\alpha-\beta-1)(\prod(\gamma-1)}{\prod(\gamma-\alpha-1)\prod(\gamma-\beta-1)}$, Proc. London Math. Soc., Series 1, 28 (1897), pp. 475–486.

[341] ——, *Series Connected with the Enumeration of Partitions*, Proc. London Math. Soc., Series 2, 1, (1904), pp. 63–88.

[342] ——, *The Application of Basic Numbers to Bessel's and Legendre's Functions*, Proc. London Math. Soc., Series 2, 2(1), (1905), pp. 192–220.

[343] ——, *A generalisation of the functions* $\Gamma(n)$ *and* x^n, Proc. Roy. Soc., 74 (1905), pp. 64–72.

[344] C. G. J. JACOBI, *Lieber die Anwendung der elliptischen Transcendenten auf ein bekanntes Problem, der Elementaregeometrie:"Die Relation zwisvchen der Dstanz der Mittelpiancte und den Radien zweir Kreise zu finden, von denen der einem unregelmäfsigen Polygon eingeschriebem, der andere demselben umgesschrieben ist*, J. reine angew. Math., 3 (1828), pp. 376–389.

[345] ——, *Demonstratio theorematis ad theoriam functionum ellipticarum spectantis*, Astr. Nach., 6 (10) (1828), pp. 133–141; and *Extraits de deux lettres de Mr. Jacobi de l'Université de Königsberg á l'éditeur*, 6 (3) (1828), pp. 33–37.

[346] ——, *Suite des notices sur les fonctions elliptiques*, J. reine angew. Math., 3 (1828), pp. 403–404.

[347] ——, *De functionibus ellipticis commentatio*, J. reine angew. Math., 4 (1829), pp. 371–390.

[348] ——, *Disquisitiones analyticae de fractionibus simplicibus*, in Gesammelte Werke, 1884, vol. III, Chelsea Publishing Company, 1829, pp. 1–44.

[349] ——, *Fundamenta nova theoriae functionum ellipticarum*, in Gesammelte Werke, 1884, vol. I, Chelsea Publishing Company, 1829, pp. 49–239.

[350] ——, *De functionibus duarum variabilium quadrupliciter periodicis, quibus theoria transcendentium Abelianarum innitur*, J. reine angew. Math., 13 (1835), pp. 55–78.

[351] ——, *Ueber die Entwicklung des Ausdrucks*, J. reine angew. Math., 26 (1843), pp. 81–87.

[352] ——, *Uber die Differentialgleichung, welcher die Reihen* $1 \pm 2q + 2q^4 \pm 2q^9 + etc..$ *etc. Genüge leisten*, J. reine angew. Math., 36 (1848), pp. 97–112.

[353] ——, *Versuch einer Berechnung der grossen Ungleichheit des Saturns nach einer strengen Entwickelung*, Astr. Nach., 28 (1849), pp. 81–94.

[354] ——, *Gesammelte Werke*, (2nd. ed.), New York: Chelsea Publ. Co., 1881.

[355] W. JACOBSTHAL, *Asymptotische Darstellung von Lösungen linearer Differentialgleichungen*, Math. Ann., 56 (1903), pp. 129–154.

[356] E. JAHNKE AND F. EMDE, *Funktionentafeln mit Formeln und Kurven*, Leipzig, Teubner, 1909.

[357] V. JAMET, *Sur un cas particulier de l'équation de Lamé*, C. R. Acad. Sci. Paris, 101 (1890), pp. 638–639.

[358] G. B. JEFFERY, *On spheroidal harmonics and allied functions*, Proc. Edinburgh Math. Soc., 33 (1914), pp. 118–121.

[359] J. L. W. V. JENSEN, *Remarques relatives aux réponses de MM. Franel et Kluyver (1895, pp. 153–157)*, L'Intermédiaire des Math., (1895), pp. 346–347.

[360] C. JORDAN, *Sur la série de Fourier*, C. R. Acad. Sci. Paris, 92 (1881), pp. 228–230.

[361] ——, *Cours d'Analyse de l'Ecole Polytechnique*, vol. I, Gauthier-Villars et fils, Paris, 1893.

[362] ——, *Cours d'Analyse de l'Ecole Polytechnique*, vol. II, Gauthier-Villars et fils, Paris, 1894.

[363] ——, *Cours d'Analyse de l'Ecole Polytechnique*, vol. III, Gauthier-Villars et fils, Paris, 1894.

[364] A. KALÄHNE, *Über die Wurzeln einiger Zylinderfunktionen und gewisser aus ihnen gebildeter Gleichungen*, Zeitschrift für Math. und Phys., 54 (1907), pp. 55–86.

[365] W. KAPTEYN, *Recherches sur les fonctions de Fourier–Bessel*, Ann. Sci. de l'Ecole Norm. Sup., (3)X (1893), pp. 91–120.

[366] ———, *Sur deus séries qui représentent la même fonction dans une partie du plan*, Niuew Archief, (2)3 (1898), pp. 225–229.

[367] N. M. KATZ AND P. SARNAK, *Zeros of zeta functions and symmetry*, Bull. Amer. Math. Soc., 36 (1999), pp. 1–26.

[368] Y. KATZNELSON, *An Introduction to Harmonic Analysis*, Cambridge University Press, 2004.

[369] J. P. KEATING AND N. C. SNAITH, *Random matrix theory and $\zeta(1/2 + it)$*, Comm. Math. Physics, 214 (2000), pp. 57–89.

[370] L. KIEPERT, *Wirkliche Ausführung der ganzzahligen Multiplication der elliptischen Functionen*, J. reine angew. Math., 76 (1873), pp. 21–33.

[371] ———, *Auflösung der Transformationsgleichungen und Division der elliptischen Functionen*, J. reine angew. Math., 76 (1873), pp. 34–44.

[372] F. KLEIN, *Ueber die hypergeometrische Function. Vorlesung, gehalten im Wintersemester 1893/94*, Göttingen, 1894.

[373] ———, *Vorlesung über Lamé'schen Funktionen (lithographed)*, Göttingen, 1894.

[374] ———, *Vorlesungen über lineare Differentielgleichungen (lithographed)*, Göttingen, 1894.

[375] ———, *Gauss wissenschaftliches Tagebuch 1796 – 1814*, Math. Ann., 57 (1903), pp. 1–34.

[376] ———, *Ueber Lineare Differentialgleichungen der Zweiten Ordnung: Vorlesung Gehalten im Sommersemester 1894*, Leipzig, Teubner, 1906.

[377] ———, *Developments of Mathematics in the 19th century*, Trans. Math. Sci. Press, Editor R. Hermann, Brookline, MA, 1928.

[378] F. KLEIN AND R. FRICKE, *Vorlesungen über die Theorie der Elliptischen Modulfunctionen*, B. G. Teubner, Leipzig, 1890.

[379] J. C. KLUYVER, *Sur les séries de factorielles*, C. R. Acad. Sci. Paris, 133 (1901), pp. 587–588.

[380] A. KNESER, *Ein Beitrag zur Theorie der Integralgleichungen*, Rend. del Circolo Matematico di Palermo, 22 (1906), pp. 233–240.

[381] A. KOLMOGOROV, *Une série de Fourier–Lebesgue divergent presque partout*, Fundamenta Mathematica, 4 (1923), pp. 324–328.

[382] ———, *Une série de Fourier–Lebesgue divergent partout*, C. R. Acad. Paris, 186 (1926), pp. 1327–1328.

[383] L. KRONECKER, *Eine Mathematische Mittheilung*, Monatsberichte der Königlichen Preussische Akademie des Wissenschaften zu Berlin, (1857), pp. 455–460.

[384] ———, *Über eine neue Eigenschaft der quadratischen Formen von negativer Determinante*, Monatsberichte der Königlichen Preussische Akademie des Wissenschaften zu Berlin, (1862), pp. 302–311.

[385] ———, *Bemerkungen über die Jacobischen Thetaformeln*, J. reine angew. Math., 102 (1888), pp. 260–272.

[386] ———, *Bemerkungen über die Darstellung von Reihen durch Integrale*, J. reine angew. Math., 105 (1889), pp. 345–354.

[387] ———, *Über die Zeit und die Art der Entstehung der jacob'schen Thetaformeln*, Sitzungsberichte der Königlich Preussischen Akademie der Wissenschaften zu Berlin, (1891), pp. 653–659.

[388] E. KUMMER, *Über die hypergeometrische Reihe*, J. reine angew. Math., 15 (1836), pp. 39–83.

[389] ———, *Über die hypergeometrische Reihe. (Fortzetzung)*, J. reine angew. Math., 15 (1836), pp. 127–172.

[390] E. E. KUMMER, *Beitrag zur theorie der Function $\Gamma(x)$*, J. reine angew. Math., 35 (1847), pp. 1–4.

[391] C. LAGRANGE, *Démonstration élémentaire de la loi supréme de Wronski*, Mémoires Couronnés et Mémoires des Savants Étrangers, 47 (2) (1886), pp. 3–8.

[392] J. L. LAGRANGE, *Memoire sur l'utilité de la méthode de prendre le milieu entre les résultas de plusiers observations*, Miscellanea Taurinenia, 1770–1773 (1770), pp. 167–232.

[393] ———, *Sur le problème de Kepler*, Hist. de l'Acad. R. des Sciences de Berlin, 25 (1770–1771), pp. 204–233.

[394] ———, In *Oeuvres*, vol. 1, Gauthier-Villars, Paris, 1869, p. 520.

[395] ———, *Leçons sur le Calcul des fonctions*, in Oeuvres, vol. 2, Gauthier-Villars, Paris, 1869, p. 25.

[396] ———, *Oeuvres*, Edited by J. A. Serret, Gauthier-Villars, Paris, 1870.

[397] E. LAGUERRE, *Sur l'integrale* $\int_x^\infty \frac{e^{-x}\,dx}{x}$, Bull. de la Soc. Math. France, 7 (1879), pp. 72–81.

[398] H. LAMB, *Hydrodynamics*, Cambridge University Press, 1895.

[399] ———, *An Elementary Course of Infinitesimal Calculus*, Cambridge University Press, 1897.

[400] G. LAMÉ, *Mémoire sur les lois de l'équilibre du fluide éthéré*, Journal de l'École Polyt., 23 (1834), pp. 191–288.

[401] ———, *Sur les surfaces isothermes dans les corps solides homogènes en équilibre de température*, Journal de Math. Pure et Appl., 2 (1837), pp. 147–183.

[402] ———, *Mémoire sur les axes des surfaces isothermes du second degré, considérés comme de fonctions de la température*, Journal de Math. Pure et Appl., 4 (1839), pp. 100–125.

[403] ———, *Mémoire sur l'équilibre des températures dans un ellipsoide á trois axes inégaux*, Journal de Math. Pure et Appl., 4 (1839), pp. 126–163.

[404] ———, *Leçns sur les Fonctions Inverses des Transcendantes et les Surfaces Isothermes*, Mallet-Bachelier, 1857.

[405] E. LANDAU, *Handbuch der Lehre von der Verteilung der Primzahlen*, Leipzig: B. G. Teubner, 1909.

[406] J. LANDEN, *An investigation of a general theorem for finding the length of any arc of any conic hyperbola, by means of two elliptic arcs, with some other new and useful theorems deduced therefrom*, Philos. Trans. Royal Soc. London, 65 (1775), pp. 283–289.

[407] G. LANDSBERG, *Zur Theorie der Gaussschen Summen und der linearen Transformation der Thetafunctionen*, J. reine angew. Math., 111 (1893), pp. 234–253.

[408] P. S. LAPLACE, *Mémoire sur la théorie de l'anneau de Saturne*, Histoire de l'Académie royale des sciences, avec les mémoires de mathématique et de physique, (1787), pp. 249–267.

[409] ———, *Traité de Mècanique Cèleste*, Paris, Duprat, 1799.

[410] ———, *Théorie Analytique des Probabilités*, 1812. Reproduced by Nabu Press, September, 2011.

[411] H. LAURENT, *Mémoire sur les fonctions de Legendre*, Journal de Mathématiques Pures et Appliquées, 1 (1875), pp. 373–398.

[412] ———, *Sur les nombres premiers*, Nouv. Ann. Math., 18 (1899), pp. 234–241.

[413] P. A. LAURENT, *Extension du théorème de M. Cauchy, relatif á la convergence du développement d'une fonction suivant les puissances ascendantes de la variable*, Comptes Rendus de la Academie des Sciences de Paris, 17 (1843), pp. 348–349.

[414] A. LAURINCIKAS AND R. GARUNKSTIS, *The Lerch Zeta Function*, Springer-Verlag, 1st. ed., 2003.

[415] P. D. LAX, *Functional Analysis*, J. Wiley and Sons, New York, 2002.

[416] J. G. LEATHEM, *Elements of the Mathematical Theory of Limits*, London, G. Bell and sons, Ltd., 1925.

[417] H. LEBESGUE, *Intégrale, longeur, aire*, Annali di Matematica Pura ed Applicata, 7 (1902), pp. 231–359.

[418] ———, *Leçons sur l'Intégration*, Gauthier-Villars, Paris, 1904.

[419] ———, *Leçons sur les Séries Trigonométriques*, Gauthier-Villars, Paris, 1906.

[420] A. M. LEGENDRE, *Recherches sur diverses sortes d'intégrales définies*, Mémoires par divers Savants à l'Académie des Sciences de l'Insitute de France, 10 (1785), pp. 416–509.

[421] ———, *Exercises de Calcul Intégral sur Diverses Ordres de Transcendentes et sur les Quadratures*, Paris, 1811.

[422] ———, *Traite des Fonctions Elliptiques et des Integrales Euleriennes*; three volumes, Firmin Didot Freres, Paris, 1825, 1826, 1828-1832.

[423] A. M. LEGENDRE AND C. JACOBI, *Correspondance mathématique entre Legendre et Jacobi*, J. reine angew. Math., 80 (1875), pp. 205–279.

[424] G. LEIBNITZ, *Letter from J. Bernoulli to Leibnitz, Ges. Werke*, Dritte Folge, III (Halle, 1855), p. 75.

[425] M. LERCH, *Contributions á la théorie des fonctions*, Sitz. Böhm. Akad., (1886), pp. 571–583.

[426] ———, *Note sur la fonction* $\Re(w,x,s) = \sum_{k=0}^\infty \frac{e^{2k\pi i x}}{(w+k)^3}$, Acta Math., 11 (1887), pp. 19–24.

[427] ——, *Ueber die Nichtdifferentiirbarkeit gewisser Functionen*, J. reine angew. Math., 103 (1888), pp. 126–138.

[428] ——, *O vlastnostech nekonečné řady* $\varphi(x,a) = \sum_{n=1}^{\infty} \frac{x^n}{n-a}$, Casopis, 21 (1892), pp. 65–68.

[429] ——, *Sur une intégrale définie*, Giornale di Matematiche, 31, (1893), pp. 171–172.

[430] ——, *Sur la différentiation d'une classe de séries trigonométriques*, Ann. de l'École norm. sup., 12 (1895), pp. 351–361.

[431] ——, *Uber eine Formel aus der Theorie der Gammafunction*, Monatshefte für Mathematik und Physik, 8 (1897), pp. 187–192.

[432] T. LEVI-CIVITA, *Sullo sviluppo delle funzioni implicite*, Rend. dei Lincei, 16 (1907), pp. 3–12.

[433] R. P. LEWIS, *A combinatorial proof of the triple product identity*, Amer. Math. Monthly, 91 (1984), pp. 420–423.

[434] A. M. LIAPOUNOFF, *Sur une série relative à la théorie de équations différentielles linéaires à coefficients périodiques*, Comptes Rendus de la Academie des Sciences de Paris, 123 (1896), pp. 1248–1252. See also *On a series encountered in the theory of linear differential equations of the second order with periodic coefficients* (in Russian), Zap. Akad. Nauk po Fiz.-matem. otd., 13 (2) (1902), pp. 1–70.

[435] E. LINDELÖF, *Sur les systèmes complets et le calcul des invariants differentiels des groupes continus finis*, Diss. et Acta Soc. Scient. Fennicae, 20 (1893), p. 1.

[436] ——, *Calcul de Résidues et ses Applications a la Théorie des Fonctions*, Gauthier-Villars, 1905.

[437] F. LINDEMANN, *Ueber die Differentialgleiching der Functionen des elliptischen Cylinders*, Math. Ann., 22 (1883), pp. 117–123.

[438] J. LIOUVILLE, *Premier Mémoire sur la Théorie des Équations différentielles linéaires et sur le développement des Fonctions en séries*, Journal de Mathématiques Pures et Appliquées, 3 (1838), pp. 561–614.

[439] ——, *Lettres sur diverses questions d'analyse et de physique mathématique concernant l'ellipsoïde, adressées à M. P.-H. Blanchet*, Journal de Mathématiques Pures et Appliquées, 11 (1846), pp. 217–236.

[440] R. LIPSCHITZ, *De explicatione per series trigonometricas instituenda functionum unius variabilis arbitrariarum, et praecipue earum, quae per variabilis spatium finitum valorum maximorum et minimorum numerum habent infinitum disquisitio*, J. reine angew. Math., 63 (1864), pp. 296–308.

[441] E. LOMMEL, *Studien über die Bessel'schen Functionen*, Leipzig, 1868.

[442] F. LONDON, *Ueber Doppelfolgen und Doppelreihen*, Math. Ann., 53 (1900), pp. 322–370.

[443] D. W. LOZIER, *The NIST Digital Library of Mathematical Functions Project*, Ann. Math. Art. Intel., 38 (2003), pp. 105–119.

[444] H. M. MACDONALD, *Note on Bessel functions*, Proc. London Math. Soc., 29 (1897), pp. 110–115.

[445] ——, *Zeros of the spherical harmonic $P_n^m(\mu)$ considered as a function of n*, Proc. London Math. Soc., 31 (1899), pp. 264–281.

[446] ——, *The addition-theorem for the Bessel functions*, Proc. London Math. Soc., 32 (1900), pp. 152–157.

[447] ——, *Note on the zeros of the spherical harmonic $P_n^{-m}(\mu)$*, Proc. London Math. Soc., 34 (1901), pp. 52–55.

[448] ——, *Note on the evaluation of a certain integral containing Bessel's functions*, Proc. London Math. Soc., (2)7 (1909), pp. 142–149.

[449] C. MACLAURIN, *A Treatise of Fluxions in Two Books*, Ruddimans, Edinburgh, 1742.

[450] R. C. MACLAURIN, *On the solutions of the equation $(V^2 + \kappa^2)\psi = 0$ in elliptic coordinates and their physical applications*, Trans. Camb. Phil. Soc., 17 (1898), pp. 41–108.

[451] W. MAGNUS AND W. WINKLER, *Hill's Equation*, Interscience Wiley, New York, 1st ed., 1966.

[452] R. S. MAIER, *The 192 solutions of the Heun equation*, Math. Comp., 76 (2007), pp. 811–843.

[453] S. MANGEOT, *Sur un mode de développement en série des fonctions algébriques explicites*, Ann. de l'École Norm. Sup., (3) 14 (1897), pp. 247–250.

[454] P. MANSION, *Continuité au sens analytique et continuité au sens vulgaire*, Mathesis, 19 (1899), pp. 129–131.

[455] A. I. MARKUSHEVICH, *Theory of Functions of a Complex Variable*, vol. 2, Prentice-Hall, Englewood Cliffs, NJ, 1965.

[456] G. B. MATHEWS, *Projective Geometry*, London, 1914.

[457] E. MATHIEU, *Memoire sur le Mouvement Vibratoire d'une Membrane de Forme Elliptique*, Jour. Math. Pures Appl., 13 (1868), pp. 137–203.

[458] B. MAZUR AND W. STEIN, *Prime Numbers and the Riemann Hypothesis*, Cambridge University Press, 2016.

[459] H. P. MCKEAN, *Integrable systems and algebraic curves*. In *Global Analysis (Proc. Biennial Sem. Canad. Math. Congr., Univ. Calgary, Calgary, Alta., 1978)*, vol. 755 of Lecture Notes in Math., Springer, Berlin, 1979, pp. 83–200.

[460] ———, *Fredholm determinants*, Cent. Eur. J. Math., 9 (2011), pp. 205–243.

[461] H. P. MCKEAN AND V. MOLL, *Elliptic Curves: Function Theory, Geometry, Arithmetic*, Cambridge University Press, 1997.

[462] H. P. MCKEAN AND E. TRUBOWITZ, *Hill's operator and hyperelliptic function theory in the presence of infinitely many branch points*, Comm. Pure Appl. Math., 29 (1976), pp. 143–226.

[463] H. P. MCKEAN AND P. VAN MOERBEKE, *The spectrum of Hill's equation*, Invent. Math., 30 (1975), pp. 217–274.

[464] F. G. MEHLER, *Über die Vertheilung der statischen Elektricität in einem von zwei Kugelkalotten begrenzten Körper*, J. reine angew. Math., 68 (1868), pp. 134–150.

[465] ———, *Notiz über die Dirichlet'schen Integralausdrücke für die Kugelfunction $P^n(\cos \vartheta)$ und über eine analoge Integralform für dir Cylinderfunction $J(x)$*, Math. Annalen, 5 (1872), pp. 141–144.

[466] E. MEISSEL, *Tafel der Bessel'sehen Functionen I_k^0 und I_k^1 von $k = 0$ bis $k = 15.5$ berechnet*, Mathematische Abhandlungen der Königlichen Akademie der Wissenschaften zu Berlin (1888), pp. 1–23.

[467] J. T. MERZ, *History of European Thought in the Nineteenth Century*, vol. 2, Blackwood and Sons, Edinburgh and London, 1903.

[468] P. D. MILLER, *Applied Asymptotic Analysis*, vol. 75 of Graduate Studies in Mathematics, American Mathematical Society, 2006.

[469] J. H. MICHELL, *The wave resistance of a ship*, Phil. Mag., 45 (1898), pp. 106–123.

[470] G. MITTAG-LEFFLER, *Funktions teoretiska studier, I. en ny serientveckling für funktionner af rational karakter*, Acta Soc. Scient. Fennicae, 11 (1880), pp. 273–293.

[471] G. MITTAG-LEFFLER, *Démonstration nouvelle du théoréme de Laurent*, Acta Math., 4 (1884), pp. 80–88.

[472] E. H. MOORE, *Concerning transcendentally transcendental functions*, Math. Ann., 48 (1897), pp. 49–74.

[473] L. J. MORDELL, *The inversion of the integral $u = \int \frac{y\,dx - x\,dy}{\sqrt{(a,b,c,d,e)(x,y)^4}}$*, Mess. Math., 44 (1915), pp. 138–141.

[474] G. MORERA, *Intorno all'integrale di Cauchy*, Rend.del Ist. Lombardo, 22 (1889), pp. 191–200.

[475] F. MORLEY, *On the series $1 + \left(\frac{p}{1}\right)^3 + \left(\frac{p(p+1)}{1\cdot 2}\right)^3 + \cdots$*, Proc. London Math. Soc., 34 (1901), pp. 397–402.

[476] P. M. MORSE AND H. FESHBACH, *Methods of Theoretical Physics*, New York, 1953.

[477] D. MUMFORD, *Curves and their Jacobians*, The University of Michigan Press, Ann Arbor, Michigan, 1975. Reprinted in *The Red Book of Varieties and Schemes*, Second edition, Springer, 2009.

[478] ———, *Tata Lectures on Theta, I*, vol. 28 of Progr. Math., Birkhäuser, Boston, 1983.

[479] ———, *Tata Lectures on Theta, II*, vol. 43 of Progr. Math., Birkhäuser, Boston, 1984.

[480] ———, *Tata Lectures on Theta, III*, Progr. Math., Birkhäuser, Boston, 1984.

[481] R. MURPHY, *On the inverse method of definite integrals, with physical applications*, Camb. Phil. Trans., 4 (1833), pp. 353–408.

[482] ———, *Elementary Principles of the Theories of Electricty, Heat, and Molecular Actions*, Cambridge: Printed at the Pitt Press by J. Smith for J. & J. Deighton, 1833.

[483] ———, *Second memoir on the inverse method of definite integrals*, Camb. Phil. Trans., 5 (1835), pp. 113–148; *Third memoir on the inverse method of definite integrals*, pp. 315–394.

[484] E. Netto, *Beitrage sur Integralrechnung*, Zeitschrift für Math. und Phys., 40 (1895), pp. 180–185.

[485] C. Neumann, *Über die Entwicklung einer Funktion nach den Kugelfunktionen*, Halle, 1862.

[486] ———, *Theorie der Bessel'schen Functionen: Ein analogon zur Theorie der Kugelfunctionen*, B. G. Teubner, Leipzig, 1867.

[487] ———, *Ueber die Entwickelung beliebig gegebener Functionen nach den Besselschen Functionen*, J. reine angew. Math., 67 (1867), pp. 310–314.

[488] ———, *Untersuchungen über das Logarithmische und Newton'sche Potential*, B. G. Teubner, Leipzig, 1877.

[489] ———, *Über die Kugelfunctionen P_n und Q_n, insbesondere über Entwicklung der Ausdrücke $P_n(zz_i + \sqrt{1-z^2}\sqrt{1-z_i^2}\cos\Phi)$ und $Q_n(zz_i + \sqrt{1-z^2}\sqrt{1-z_i^2}\cos\Phi)$ nach den Cosinus der Vielfachen von* Φ, Abhandlungen der Mathematisch-Physischen Classe de Königlich Sächsischen Gesellschaft der Wissenschaften, 13 (1887), pp. 401–475.

[490] J. Neumann, *Entwickelung der in elliptischen Coordinaten ausgedrückten reciproken Entfernung zweier Puncte in Reihen, welche nach den Laplace'schen $Y^{(n)}$ fortschreiten; und Anwendung dieser Reihen zur Bestimmung des magnetischen Zustandes eines Rotations-Ellipsoïds, welche durch vertheilende Kräfter erregt ist*, J. reine angew. Math., 37 (1848), pp. 21–50.

[491] E. H. Neville, *The genesis of Lamé's equation*, Quarterly Journal of Pure and Applied Mathematics, 49 (1923), pp. 338–352.

[492] D. J. Newman, *Simple analytic proof of the Prime Number Theorem*, Amer. Math. Monthly, 87 (1980), pp. 693–696.

[493] F. W. Newman, *On Γa, especially when a is negative*, Cambridge and Dublin Math. Journal, 3 (1848), pp. 57–60.

[494] I. Newton, *De Analysi per aequationes numero terminorum infinitas* (1669). In *The Mathematical Papers of Isaac Newton*, D. Whiteside, ed., vol. 2, Cambridge University Press, 1969.

[495] J. W. Nicholson, *The asymptotic expansion of Bessel functions of high order*, Phil. Mag., 14 (1907), pp. 697–707.

[496] ———, *On Bessel functions of equal argument and order*, Phil. Mag., 16, (1908), pp. 271–279.

[497] ———, *On the relation of Airy's integral to the Bessel functions*, Phil. Mag., 18, (1909), pp. 6–17.

[498] F. Nicole, *Sur le calcul des différences finies, et des sommes des suites*, Mém. de l'Acad. des Sci., Paris, (1717), pp. 38–47.

[499] N. Nielsen, *Sur les séries de factorielles*, C. R. Acad. Sci. Paris, 134 (1902), pp. 1273–1275.

[500] ———, *Handbuch der Theorie der Cylinderfunktionen*, B. G. Teubner, Leipzig, 1904.

[501] ———, *Sur la représentation asymptotique d'une série de factorielles*, Annales scientifiques de l'École Normale Supérieure, 21 (1904), pp. 449–458.

[502] ———, *Recherches sur le développement d'une fonction analytique en série de fonctions hypergéometriques*, Annales scientifiques de l'École Normale Supérieure, 30 (1913), pp. 121–171.

[503] W. D. Niven, *On a special form of Laplace's equation*, Mess. Math., 10 (1880), pp. 114–117.

[504] W. D. Niven, *On ellipsoidal harmonics*, Phil. Trans., 182a (1892), pp. 231–278.

[505] ———, *On the harmonics of a ring*, Proc. London Math. Soc., 24 (1892), pp. 373–377.

[506] N. E. Nörlund, *Sur les séries de facultés*, Acta Math., 37 (1914), pp. 327–387.

[507] ———, *Mémoire sur les polynomes de Bernoulli*, Acta Math., 43 (1920), pp. 121–196.

[508] F. W. J. Olver, *Asymptotics and Special Functions*, Academic Press. New York, 1974.

[509] F. W. J. Olver, D. W. Lozier, R. F. Boisvert, and C. W. Clark, eds., *NIST Handbook of Mathematical Functions*, Cambridge University Press, 2010.

[510] W. M. Orr, *Theorems relating to the product of two hypergeometric series*, Trans. Camb. Phil. Soc., 17 (1899), pp. 1–15.

[511] W. F. Osgood, *Some points in the elements of the theory of functions*, Bull. Amer. Math. Soc., 2 (1896), pp. 296–302.

[512] ———, *Problems in infinite series and definite integrals; with a statement of certain sufficient conditions which are fundamental in the theory of definite integrals*, Ann. Math., 3 (1901), pp. 129–146.

[513] ———, *Lehrbuch der Funktionentheorie*, B. G. Teubner, Leipzig und Berlin, 1907.

[514] P. Painlevé, *Mémoire sur les équations différentielles dont l'integrale générale est uniforme*, Bulletin de la Soc. Mat. France, 27 (1900), pp. 201–261.

[515] E. Papperitz, *Ueber verwanndtes-Functionen*, Math. Ann., 25 (1885), pp. 212–221.

[516] M. A. Parseval, *Méthode générale pour sommer, par le moyen des intégrales définies, la suite donnée par le théorème de m. Lagrange, au moyen de laquelle it trouve une valeur qui satisfait á une équation algébrique ou transcendente*, Mém. par divers savants, I (1805), pp. 639–648.

[517] K. Pearson, *A Mathematical Theory of Random Migration*, London, Dulau & Co.

[518] M. Petkovsek, H. Wilf, and D. Zeilberger, *A=B*, A. K. Peters, 1st. ed., 1996.

[519] J. Pierpoint, *The Theory of Functions of Real Variables*, Ginn and Company, 1908.

[520] S. Pincherle, *Una nuova estensione delle funzioni sferiche*, Memorie della Reale Accademia delle Scienze dell'Istituto di Bologna, Classe di Scienze Fisiche Series 5, 1 (1890), pp. 337–339.

[521] ———, *I sistemi ricorrenti di prim'ordine e di secondo grado*, Atti della R. Accad. dei Lincei, Rendiconti Ser. 4, 5 (1889), pp. 8–12.

[522] ———, *Un sistema d'integrali ellitici considerati come funzioni dell'invariante assoluto*, Atti della R. Accad. dei Lincei, Rendiconti Ser. 4, 7 (1891), pp. 74–87.

[523] ———, *Della validitá effetiva di alcuni sviluppi in series di funzioni*, Atti della R. Accad. dei Lincei, Rendiconti Ser. 5, 5 (1896), pp. 27–33.

[524] ———, *Sulle serie di fattoriali*, Atti della R. Accad. dei Lincei, Rendiconti Ser. 5, 11 (1902), pp. 139–144 and 417–426.

[525] ———, *Sulla serie di fattoriali generalizzate*, Rendiconti del Circolo Matematico di Palermo, 37 (1914), pp. 379–390.

[526] G. A. A. Plana, *Sur une nouvelle expression analytique des nombres Bernoulliens, propre à exprimer en terms finis la formula g'en'erale pour la sommation des suites*, Mem. Acad. Sci. Torino, 25 (1820), pp. 403–418.

[527] L. Pochhammer, *Zur Theorie der Euler'schen Integrale*, Math. Ann., 35 (1890), pp. 495–526.

[528] H. Poincaré, *Sur les déterminants d'ordre infini*, Bull. Soc. Math. France, 14 (1886), pp. 77–90.

[529] ———, *Sur les intégrales irrégulièrs: Des équations linéires*, Acta Math., 8 (1886), pp. 295–344.

[530] S. D. Poisson, *Second mémoire sur la distribution de la chaleur dans les corps solides*, Journal de l'École Polytechnique, Paris, 12 (1823), pp. 249–403.

[531] ———, *Suite du mémoire sur les intégrales définies et sur la sommation des séries*, Journal de l'École Polytechnique, Paris, 12 (1823), pp. 404–509.

[532] ———, *Mémoire sur le calcul numérique des intégrales définies*, Mém. de l'Acad. des Sci., Paris, 6 (1827), pp. 571–602.

[533] ———, *Mémoire sur l'équilibre et le mouvement des corps élastiques*, Mémoires de l'Académie Royale des sciences de l'Institut de France, 8 (1829), pp. 357–570.

[534] E. Poole, *Introduction to the Theory of Linear Differential Equations*, Oxford Univ. Press, Oxford, 1936.

[535] M. B. Porter, *Note on the roots of Bessel's functions*, Bull. Amer. Math. Soc., 4 (1898), pp. 274–275.

[536] J. Pöschel and E. Trubowitz, *Inverse Spectral Theory*, vol. 130 of Pure and Applied Mathematics, Academic Press, New York, 1987.

[537] A. Pringsheim, *Ueber das Verhalten gewisser Potenzreihen auf dem Convergenzkreise*, Math. Ann., 25 (1885), pp. 419–426.

[538] ———, *Zur Theorie der Gamma-Functionen*, Math. Ann., 31 (1888), pp. 455–481.

[539] ———, *Ueber die Convergenz unendlicher Producte*, Math. Ann., 33 (1889), pp. 119–154.

[540] ———, *Zur Theorie der Taylor'schen Reihe und der analytischen Functionen mit beschränktem Existenzbereich*, Math. Ann., 42 (1893), pp. 153–184.

[541] ———, *Uber Konvergenz und funktionentheretischen Charakter gewisser limitär-periodischer Kettenbrüche*, Münchner Sitzungsberichte, 27 (1897), pp. 101–152.

[542] ———, *Zur Theorie der zweifach unendlichen Zahlenfolgen*, Math. Ann., 53 (1900), pp. 289–321.

[543] A. Pringsheim and J. Molk, *Algorithmes illimités*. In *Encyclopédie des Sciences Mathématiques Pures at Appliquées*, J. Monk, ed., vol. I4, Gauthier-Villars, Paris, 1904.

[544] A. P. Prudnikov, Y. A. Brychkov, and O. I. Marichev, *Integrals and Series*. Five volumes, Gordon and Breach Science Publishers, 1992.

[545] F. E. Prym, *Zur Theorie der Gammafunction*, J. reine angew. Math., 82 (1877), pp. 165–172.

[546] J. L. Raabe, *Angenäherte Bestimmung der Factorenfolge* $1 \cdot 2 \cdot 3 \cdot 4 \cdot 5 \cdots \cdot n = \Gamma(1+n) = \int x^n e^{-x} dx$, *wenn n eine sehr grosse Zahl ist.*, J. reine angew. Math., 25 (1843), pp. 146–59.

[547] ——, *Zurückführung einiger Summen und bestimmten Integrale auf die Jacob–Bernoullische Function*, J. reine angew. Math., 42 (1851), pp. 348–367.

[548] S. Ramanujan, *Modular equations and approximations to π*, Quart. J. Math., 45 (1914), pp. 350–372.

[549] L. Ravut, *Extension du théoréme de Cauchy aux fonctions d'une variable complexe de la forme* $\rho e^{i e^{i h} \alpha}$, Nouv. Annales de Math., 16 (1897), pp. 365–367.

[550] J. W. S. Rayleigh, *The Theory of Sound*, London, Macmillan and Co., 1877.

[551] R. Reiff, *Geschichte der Unendlichen Reichen*, Tübingen, H. Laupp, 1889.

[552] R. Remmert, *Theory of Complex Functions*, Readings in Mathematics, Springer-Verlag, New York, 1991.

[553] E. Reyssat, *Quelques Aspects des Surfaces de Riemann*, vol. 77 of Progress in Mathematics, Birkhäuser, Boston, 1989.

[554] F. J. Richelot, *Note sur le théorème relatif à une certaine fonction transcendante démontré dans No. 22. cah. 3. du présent volume*, J. reine angew. Math., 9 (1832), pp. 407–408.

[555] G. F. B. Riemann, *Zur Theorie der Nobili'schen Farbenringe*, Ann. der Phys. und Chemie, 95 (1855), pp. 130–139.

[556] ——, *Beiträge zur Theorie der durch die Gauss'sche Reihe $F(\alpha, \beta, \gamma, x)$ darstellbaren Functionen*, Abh. Kön. Ges. Wiss. Göttingen Math., 7 (1857).

[557] ——, *Ueber die Anzahl der Primzahlen unter einer gegebenen Grösse*, Monatberichte der Berliner Akademie, 7 (1859), pp. 671–680.

[558] ——, *Gesammelte Mathematische Werke und Wissenschaftlicher Nachlass*, B. G. Teubner, 1876.

[559] M. Riesz, *Sur la sommation des séries de Dirichlet*, C. R. Acad. Sci. Paris, 149 (1909), pp. 18–21.

[560] J. F. Ritt, *On the differentiability of asymptotic series*, Bull. Amer. Math. Soc., 24 (1918), pp. 225–227.

[561] M. Rodrigues, *Mémoire sur l'attraction des sphéroïdes*, Corresp. sur l'École polytechnique, 3 (1816), pp. 361–385.

[562] D. Romik, *The Surprising Mathematics of Longest Increasing Subsequences*, Cambridge University Press, 2015.

[563] A. Ronveaux, *Heun's Differential Equations*, Oxford Science Publications, Clarendon Press, 1995.

[564] E. J. Routh, *On an expansion of the potential functions i/r^{n-1} in Legendre's functions*, Proc. London Math. Soc., 26 (1894), pp. 481–491.

[565] H. L. Royden, *Real Analysis*, Prentice-Hall, Englewood Cliffs, N.J. 07632, third ed., 1988.

[566] S. N. M. Ruijsenaars, *On Barnes' multiple zeta and gamma function*, Adv. Math., 156 (2000), pp. 107–132.

[567] B. Russell, *Introduction to Mathematical Philosophy*, The Macmillan Company, New York, 1st ed., 1919.

[568] L. Saalschütz, *Bemerkungen über die Gammafunction mit negativen Argumenten*, Zeitschrift fúr Math. und Phys., 32 (1887), pp. 246–250.

[569] L. Saalschütz, *Weitere bemerkungen über die Gammafunction mit negativen Argumenten*, Zeitschrift fúr Math. und Phys., 33 (1888), pp. 363–374.

[570] ——, *Eine summations formel*, Zeitschrift fúr Math. und Phys., 35 (1887), pp. 186–189.

[571] G. Salmon, *A Treatise on the Higher Plane Curves*, Hodges, Foster, and Figgs. Dublin, 1879.

[572] H. G. Savidge, *Tables of the ber and bei and ker and kei functions, with further formulae for their computation*, Phil. Mag., 19 (1910), pp. 49–58.

[573] P. Schafheitlin, *Ueber die Darstellung der hypergeometrischen Reihe durch ein bestimmtes Integrals*, Math. Ann., 30 (1877), pp. 157–158.

[574] W. Scheiber, *Uber Unendliche Reihen und deren Convergenz*, S. Hirzel, 1860.

[575] L. Schendel, *Zur theorie der kugelfunctionen*, J. reine angew. Math., 80 (1875), pp. 86–94.

[576] H. Schläfli, *Einige Bemerkungen zu Herrn Neumann's Untersuchungen über die Bessel'schen Functionen*, Math. Ann., 3 (1871), pp. 134–149.

[577] ———, *Ueber die allgemeine Möglichkeit der conformen Abbildung einer von Geraden begrenzten ebenen Figur in eine Halbebene*, J. reine angew. Math., 78 (1874), pp. 63–80.

[578] ———, *Sull'uso delle linee lungo le quali il valore assoluto di una funzione è costante*, Ann. di Mat., (2)VI (1875), pp. 1–20.

[579] ———, *Ueber die zwei Heine'schen Kugelfunctionen*, Bern, 1881.

[580] O. Schlömilch, *Notiz über ein bestimmtes Integral*, Zeitschrift für Math. und Phys., 2 (1857), pp. 67–103.

[581] ———, *Ueber die Bessel'sche Funktion*, Zeitschrift für Math. und Phys., 2 (1857), pp. 137–165.

[582] ———, *Über Fakultätenreihen*, Zeitschrift für Math. und Phys., 4 (1859), pp. 390–431.

[583] ———, *Compendium der Höheren Analysis*, Braunschweig: Vieweg, 1862.

[584] ———, *Ueber eine Kettenbruchenwickelung für unvollständige Gammafunctionen*, Zeitschrift für Math. und Phys., 16 (1871), pp. 261–262.

[585] ———, *Ueber eine Verwandte de Gammafunction*, Zeitschrift für Math. und Phys., 25 (1880), pp. 335–350.

[586] ———, *Ueber den Quotienten zweier Gammafunctionen*, 25, pp. 351–415.

[587] ———, *Handbuch der Algebraischen Analysis*, Frommen, Leipzig, 1889.

[588] J. J. Schonholzer, *Ueber die Auswerthungbestimmte Integrale mit Hülfe von Veranderungen des Integrationsweges*, Dissertation, Bern (1877).

[589] H. A. Schwarz, *Formeln und Lehrsätze zum Gebrauche der elliptische Functionen*, Springer-Verlag, Berlin–Heidelberg, 1893.

[590] P. L. Seidel, *Note über eine Eigenschaft der Reihen, welche Discontinuirliche Functionen Darstellen*, Abhandlunden der Mathematisch-Physikalischen Klasse der Kóniglich Bayerischen Akademie der Wissenschaften, 5 (1848), pp. 382–393.

[591] A. Seifert, *Ueber eine Neue Geometrische Einführung in die Theorie der Elliptischen Funktionen*, Charlottenburg, 1896.

[592] J. H. Silverman, *The Arithmetic of Elliptic Curves*, Springer Verlag, New York, first ed., 1986.

[593] ———, *Advanced Topics in the Arithmetic of Elliptic Curves*, Springer-Verlag, New York, first ed., 1994.

[594] B. A. Smith, *Tables of Bessel functions Y_0 and Y_1*, Mess. Math., 26 (1897), pp. 98–101.

[595] H. J. S. Smith, *On a formula for the multiplication of four theta functions*, Proc. London Math. Soc., 1 (1865), pp. 85–96.

[596] ———, *Note on the formula for the multiplication of four theta functions*, Proc. London Math. Soc., 10 (1879), pp. 91–100.

[597] J. Soldner, Extract of a letter. In Monatlicher Correspondenz zur beförderung der Erd- und Himmels-Kunde, Gotha, 23 (1811), pp. 182–188.

[598] N. Sonine, *Recherches sur les fonctions cylindriques et le développement des fonctions continues en séries*, Math. Ann., 16 (1–80), p. 1880.

[599] M. Spivak, *Calculus*, Publish or Perish Inc., 2nd ed., 1980.

[600] G. Springer, *Introduction to Riemann Surfaces*, American Mathematical Society, 1st ed., 2002.

[601] W. Stekloff, *Sur un problème de la théorie analytique de la chaleur*, C. R. Acad. Sci. Paris, 126 (1898), pp. 1022–1025.

[602] E. A. Stenberg, *Sur un cas spécial de l'équation de Lamé*, Acta Math., 10 (1887), pp. 339–348.

[603] T. J. Stieltjes, *Quelques remarques sur l'integration d'une équation différentielle*, Astr. Nach., 109 (1884), pp. 146–152; 261–266.

[604] ———, *Sur certains polynômes: Qui vérifient une équation différentielle linéaire du second ordre et sur la theorie des fonctions de Lamé*, Acta Math., 6 (1885), pp. 321–326.

[605] ———, *Recherches sur quelques séries semi-convergentes*, Ann. de l'École norm. Sup., 3 (1886), pp. 201–258.

[606] ———, *Sur le développement de $\log \Gamma(a)$*, Journal de Mathématiques Pures et Appliquées, Ser. 4, 5 (1889), pp. 425–466.

[607] J. Stirling, *Methodus Differentialis*, London, 1764.

[608] G. G. Stokes, *On the critical values of the sums of periodic series*, Trans. Camb. Phil. Soc., 8 (1849), pp. 533–583.

[609] ———, *On the dynamical theory of diffraction*, Trans. Camb. Phil. Soc., 9 (1856), pp. 1–62.

[610] ———, *On the numerical calculation of a class of definite integrals and infinite series*, Trans. Camb. Phil. Soc., 9 (1856), pp. 166–187.

[611] ———, *Mathematical and Physical Papers*, vol. I, Cambridge University Press, 1901.

[612] O. Stolz, *Ueber unendliche Doppelreihen*, Math. Ann., 24 (1884), pp. 157–171.

[613] O. Stolz and J. A. Gmeiner, *Theoretische Arithmetik*, vol. 4 of *Sammlung von Lehrbüchern auf dem Gebiete der Mathematischen Wissenschaften mit Einschuluss ihrer Anwendungen*, B. G. Teubner, Leipzig, 1902.

[614] C. Störmer, *Sur une généralisation de la formule $\frac{\varphi}{2} = \frac{\sin\varphi}{1} - \frac{\sin 2\varphi}{2} + \frac{\sin 3\varphi}{3} - \cdots$*, Acta Mathematica, 19 (1895), pp. 341–350.

[615] C. Sturm, *Mémoire sur la résolution des équations numériques*, Mém. présentés par divers Savans à l'Académie Royale des Science de l'Institute de France, 6 (1835), pp. 271–318.

[616] J. J. Sylvester, *Note on spherical harmonics*, Phil. Mag. Series 5, 2 (1876), pp. 291–307.

[617] ———, *The Collected Mathematical Papers*, edited by H. F. Baker, vol. I, Cambridge University Press, 1908.

[618] ———, *The Collected Mathematical Papers*, edited by H. F. Baker, vol. III, Cambridge University Press, 1909.

[619] J. Tannery, *Leçons d'Algèbre et d'Analyse*, Gauthier-Villars, 1906.

[620] J. Tannery and J. Molk, *Fonctions Elliptiques* (2 volumes), Reprinted by Chelsea, New York, 1972, 1893.

[621] B. Taylor, *Methodus Incrementorum Directa et Inversa*, London: William Innys, 1715.

[622] F. G. Teixeira, *Sur les séries ordonnées suivant les puissances d'une fonction donnée*, J. reine angew. Math., 122 (1900), pp. 97–123.

[623] L. W. Thomé, *Ueber die Reihen, welche nach Kugelfunctionen fortschreiten*, J. reine angew. Math., 66 (1866), pp. 337–343.

[624] ———, *Zur Theorie der linearen Differentialgleichungen*, J. reine angew. Math., 75 (1873), pp. 266–291.

[625] W. Thompson, *On the equations of the motion of heat referred to curvilinear coordinates*, Camb. Math. Journal, 4 (1845), pp. 33–42.

[626] ———, *On the rigidity of the earth*, Phil. Trans. R. Soc. Lond., 153 (1863), pp. 573–582.

[627] ———Lord Kelvin, *Mathematical and Physical papers*, vol. 3rd, Cambridge University Press, 1882.

[628] W. Thompson and P. T. Tait, *Treatise on Natural Philosophy*, Clarendon Press, 1867.

[629] E. C. Titchmarsh, *Hankel transforms*, Proc. Camb. Phil. Soc., 21 (1923), pp. 463–473.

[630] ———, *A contribution to the theory of Fourier transforms*, Proc. London Math. Soc., 23 (1925), pp. 279–289.

[631] I. Todhunter, *An Elementary Treatise on Laplace's Functions, Lamé's Functions and Bessel's Functions*, MacMillan and Co., London and Cambridge, 1875.

[632] C. A. Tracy and H. Widom, *Painlevé functions in statistical physics*, Publ. RIMS Kyoto Univ., 47 (2011), pp. 361–374.

[633] A. Transon, *Réflexions sur l'évenement scientifique d'une formule publiée par Wroński en 1812 et démontrée par Cayley en 1873*, Nouvelles Annales de Mathématiques, 13 (1874), pp. 161–174.

[634] I. Tweddle, *James Stirling's Methodus Differentialis. An Annotated Translation of Stirling's Text*, Sources and Studies in the History of Mathematics and Physical Sciences, Springer Verlag, 2003.

[635] C. Tweedie, *Nicole's contribution to the foundation of the Calculus of Finite Differences*, Proc. Edin. Math. Soc., 36 (1918), pp. 22–39.

[636] H. Umemura, *Painlevé equations and classical functions*, Sugaku Expositions, 11 (1998), pp. 77–100.

[637] C. J. d. l. Vallée Poussin, *Etudes sur les intégrales à limites infinies pour lesquelles la fonction sous le signe est continue*, Ann. Soc. Scient. Brux., 16 (1892), pp. 150–180.

[638] ———, *Démonstration simplifiée du théoréme de Dirichlet sur la progression arithmétique*, vol. 53, Mém. de l'Acad. de Belgique, 1896.

[639] ———, *Cours d'Analyse Infinitésimale*, Louvain, 1914.

[640] W. VAN ASSCHE, *Orthogonal Polynomials and Painlevé Equations*, vol. 27 of Australian Mathematical Society Lecture Series, Cambridge University Press, 2018.

[641] V. S. VARADARAJAN, *Linear meromorphic differential equations: a modern point of view*, Bull. Amer. Math. Soc., 33 (1996), pp. 1–42.

[642] V. VON DANTSCHER, *Vorlesungen über die Weierstrassche Theorie der Irrationalen Zahlen*, Gruck und Verlag von B. G. Teubener, Leipzig, 1908.

[643] H. VON KOCH, *Sur les déterminants infinis et les équations différentielles linéaires*, Acta Math., 16 (1892), p. 217.

[644] J. WALLIS, *Logarithmotechnia Nicolai Mercatoris. Concerning which we shall here deliver the account of the judicious Dr. I. Wallis, given in a Letter to the Lord Viscount Brouncker, as follows*, Phil. Trans., 3 (1668), pp. 754–764.

[645] J. WALLIS, *Opera Mathematica*, I. (1695).

[646] G. N. WATSON, *The general solution of Laplace's equation in n dimensions*, Mess. Math., 36 (1906), pp. 98–106.

[647] ———, *The cubic transformation of the hypergeometric function*, Quart. J. Math., 41 (1909), pp. 70–79.

[648] ———, *A theory of asymptotic series*, Phil. Trans., 211 (1911), pp. 279–313.

[649] ———, *The continuation of functions defined by the generalized hypergeometric series*, Camb. Phil. Trans., 21 (1912), pp. 281–299.

[650] ———, *Complex Integration and Cauchy's theorem*, Cambridge University Press, 1914.

[651] ———, *The limits of applicability of the principle of stationary phase*, Proc. Camb. Phil. Soc., 19 (1917), pp. 49–55.

[652] ———, *Asymptotic expansions of hypergeometric functions*, Trans. Camb. Phil. Soc., 22 (1918), pp. 277–308.

[653] ———, *A Treatise on the Theory of Bessel Functions*, Cambridge University Press, 1966.

[654] H. WEBER, *Ueber einige bestimmte Integrale*, J. reine angew. Math., 69 (1868), pp. 222–237.

[655] ———, *Ueber die Integration der partiellen Differentialgleichung* $\frac{\partial^2 u}{\partial x^2} + \frac{\partial^2 u}{\partial y^2} + k^2 u = 0$, Math. Ann., 1 (1869), pp. 1–36.

[656] ———, *Ueber die Besselschen Functionen und ihre Anwendung auf die Theorie der elektrischen Ströme*, J. reine angew. Math., 75 (1873), pp. 75–105.

[657] ———, *Ueber eine Darstellung willkürlicher Functionen durch Bessel'sche Functionen*, Math. Ann., 6 (1873), pp. 146–161.

[658] ———, *Ein Beitrag zu Poincaré's Theorie der Fuchs'schen Functionen*, Nachrichten von der Königl. Gesellschaft der Wissenschaften und dre Georg-Augusts-Universität zu Göttinger Nach., (1886), pp. 359–370.

[659] K. WEIERSTRASS, *Über die Theorie der analytischen Facultäten*, J. reine angew. Math., 51 (1856), pp. 1–60.

[660] ———, *Zur Functionenlehre*, Berliner Monatsberichte, (1880), pp. 719–743.

[661] ———, *Abhandlungen aus der Funktionenlehre*, Verlag von Julius Springer, 1886.

[662] ———, *Mathematische Werke*, vol. I, Berlin, Mayer and Muller, 1894.

[663] ———, *Mathematische Werke*, vol. II, Berlin, Mayer and Muller, 1895.

[664] C. WESSEL, *On the analytical representation of direction: an attempt applied to solving plane and spherical polygons*, 1797. Translated by Damhus, Flemming. Copenhagen: C. A. Reitzels, 1997. Editor Bodil Branner et al.

[665] H. WEYL, *The concept of a Riemann surface*, New York, Dover Publications, 3rd. ed., 2000.

[666] R. L. WHEEDEN AND A. ZYGMUND, *Measure and Integral. An introduction to Real Analysis*, Marcel Dekker, New York, 1977.

[667] A. N. WHITEHEAD, *Axioms of Projective Geometry*, no. 4 in Cambridge Math. Tracts, Cambridge University Press, 1906.

[668] A. N. WHITEHEAD AND B. RUSSELL, *Principia Mathematica*, Cambridge University Press, 1910–1913.

[669] E. T. WHITTAKER, *On the connection of algebraic functions with automorphic functions*, Phil. Trans. Roy. Soc., 192 (1899), pp. 1–32.

[670] ———, *On the functions associated with the parabolic cylinder in harmonic analysis*, Proc. London Math. Soc., 35 (1902), pp. 417–427.

[671] ———, *An expression of certain functions as generalized hypergeometric functions*, Bull. Amer. Math. Soc., 10 (1903), pp. 125–134.

[672] ———, *On the partial differential equations of mathematical physics*, Math. Ann., 57 (1903), pp. 333–355.

[673] ———, *On the functions associated with the elliptic cylinder in harmonic analysis*, in Proc. Inter. Congress Math., vol. 1, Cambridge, 1912, pp. 366–371.

[674] ———, *On the general solution of Mathieu's equation*, Proc. Edin. Math. Soc., 32 (1914), pp. 75–80.

[675] ———, *On Lamé's Differential Equation and Ellipsoidal Harmonics*, Proc. London Math. Soc., 14 (1915), pp. 260–268.

[676] ———, *Bessel functions and Kapteyn series*, Proc. London Math. Soc., 16 (1917), pp. 150–174.

[677] ———, *On the numerical solutions of integral-equations*, Proc. Roy. Soc. London, 94 (1918), pp. 367–383.

[678] ———, *Analytical Dynamics*, Cambridge University Press, 4th ed., 1959.

[679] E. T. WHITTAKER AND G. N. WATSON, *Modern Analysis*, Cambridge University Press, 1962.

[680] H. WILBRAHAM, *On a certain periodic function*, The Cambridge and Dublin Mathematical Journal, 3 (1848a), pp. 198–201.

[681] H. S. WILF, *generatingfunctionology*, Academic Press, 1st ed., 1990.

[682] R. H. WILSON AND B. O. PEIRCE, *Table of the first forty roots of the Bessel equation $J_0(x) = 0$ with the corresponding values of $J_1(x)$*, Bull. Amer. Math. Soc., 33 (1897), pp. 153–157.

[683] E. M. WRIGHT, *An enumerative proof of an identity of Jacobi*, Jour. London Math. Soc., 40 (1965), pp. 55–57.

[684] J. M. H. WROŃSKI, *Introduction à la Philosophie des Mathématiques et Technie de l'Algorithmique*, Courcier, Paris, 1811.

[685] A. W. YOUNG, *On the quasi-periodic solutions of Mathieu's differential equation*, Proc. Edinburgh Math. Soc., 32 (1913), pp. 81–90.

[686] W. H. YOUNG, *On series of Bessel's functions*, Proc. London Math. Soc., s2-(18) (1920), pp. 163–200.

[687] W. H. YOUNG AND G. C. YOUNG, *The theory of sets of points*, Cambridge University Press, 1906.

[688] W. ZUDILIN, *One of the numbers $\zeta(5)$, $\zeta(7)$, $\zeta(9)$, $\zeta(11)$ is irrational*, Russian Math. Surveys, 56 (2001), pp. 774–776.

[689] ———, *Apèry's theorem. Thirty years after. (An elementary proof of Apèry's theorem)*, Int. J. Math. Comput. Sci., 4 (2009), pp. 9–19.

[690] A. ZYGMUND, *Trigonometric series, Two volumes. Third edition. With a foreword by Robert A. Fefferman*, Cambridge University Press, 2002.

Author index

Subject index